CONCERTED EUROPEAN ACTION ON MAGNETS

(CEAM)

CONCERTED EUROPEAN ACTION ON MAGNETS (CEAM)

Edited by

I. V. MITCHELL

Commission of the European Communities, Brussels, Belgium

J. M. D. COEY

Trinity College, Dublin, Ireland

D. GIVORD

CNRS, Laboratoire Louis Néel, Grenoble, France

I. R. HARRIS

University of Birmingham, UK

R. HANITSCH

Technische Universität Berlin, Federal Republic of Germany

ELSEVIER APPLIED SCIENCE
LONDON and NEW YORK

ELSEVIER SCIENCE PUBLISHERS LTD
Crown House, Linton Road, Barking, Essex IG11 8JU, England

Sole Distributor in the USA and Canada
ELSEVIER SCIENCE PUBLISHING CO., INC.
655 Avenue of the Americas, New York, NY 10010, USA

WITH 113 TABLES AND 533 ILLUSTRATIONS

© 1989 ECSC, EEC, EAEC, BRUSSELS AND LUXEMBOURG
Softcover reprint of the hardcover 1st edition 1989

British Library Cataloguing in Publication Data

Concerted European Action on Magnets (CEAM).
1. Electrical engineering components: Magnets
I. Mitchell, I. V.
621.34

Library of Congress CIP data applied for

ISBN-13: 978-94-010-7003-4   e-ISBN-13: 978-94-009-1135-2
DOI: 10.1007/978-94-009-1135-2

Publication arrangements by Commission of the European Communities, Directorate-General Telecommunications, Information Industries and Innovation, Scientific and Technical Communication Service, Luxembourg

EUR 12047

LEGAL NOTICE
Neither the Commission of the European Communities nor any person acting on behalf of the Commission is responsible for the use which might be made of the following information.

DEDICATION

This book is dedicated to the memory of Professor René Pauthenet who was chief co-ordinator of the Concerted European Action on Magnets project from its inception until his untimely death on April 19th 1987.

FOREWORD

The beneficial impact of the European Communities involvement in scientific research and technology is wide-ranging and pervasive. There are high hopes of major advances in scientific knowledge and technological processes, while the emergence of a genuine tradition of collaborative research holds out great and continuing promise for the future. Close, frequent and long-term cooperation between universities, research centres and industry is already generating new synergies, forging a truly European scientific community. Many of tomorrows industrial developments, destined to be determinant for our economic success and prosperity, will spring from this research.

The **Concerted European Action on Magnets** - CEAM - project is a prime example of collaborative research and development. Financed from the Communities STIMULATION action and implemented with the help of EURAM, the advanced materials programme, CEAM will bestow great benefits on European industrial competitiveness, providing a channel for high quality basic research to find its way into commercial products.

This remarkable cooperative enterprise brought together 58 laboratories and more than 120 scientists and engineers in a sustained thirty month effort. It spanned every aspect of new iron-based high performance magnets - from theoretical modelling of their intrinsic magnetic properties to the design and construction of novel electrical devices and machines.

Besides adding a new European dimension to advanced magnetic technology, CEAM also ensured that a whole new generation of young researchers and technicians have been trained in applied magnetism.

CEAM's real successes, the way in which very diverse interdisciplinary skills and expertise were marshalled across frontiers and united in a joint endeavour, represent a fine example of what Europe can achieve with realistic goals and determination. The story of CEAM, of which CODEST, the European Committee on Science and Technology, is proud, will serve in the future as a benchmark by which other projects can be judged.

Professor Umberto Colombo
ENEA, Rome

ACKNOWLEDGMENTS

All those involved in making the Concerted European Action on Magnets project such a success deserve our thanks.
In particular, the Commission of the European Communities for providing the support and substantial funding for such a collaboration to come about and the Spanish Consejo Superior de Investigaciones Cientificas (CSIC) for their key role in organising the final meeting in Madrid. We owe a special debt of gratitude to José Serratosa, Jesús Gonzalez and their associates on the Spanish committees for all their efforts which produced such a successful final meeting. Finally, to all our wives and families for the fortitude with which they have borne so many lonely weekends because of CEAM.

TABLE OF CONTENTS

STRUCTURE OF THE BOOK

This final report of the Concerted European Action on Magnets
(1985-1988) consists of three main sections:
Section I describes the overall structure of the project,
its membership and aims and objectives, its organisation and
funding arrangements and details of publications and
collaborations.
Section II highlights some of the many scientific and
technical contributions CEAM has made in the field of
rare-earth iron permanent magnets and points out several areas
where further work is required.
Section III contains the scientific reports of all the
participating organisations as presented at the final
contractors meeting. The subdivision of Section III into
three parts (A - C) reflects the organisation of CEAM whereby
participants initially joined a research grouping that was
primarily interested in either;
A. New magnetic materials and their physical properties
    (The Materials Group);
B. Magnet fabrication and processing (The Processing Group)
C. Engineering applications of rare-earth iron magnets in
    electrical machines and static devices. (The Applications
    Group).
The contributions within these three parts, are arranged in
ten chapters, each centred on a specific theme, with a brief
summary written by CEAM members at the end of each chapter.
This neat arrangement inevitably involves some arbitary
choices since some contributions relate to the subject matter
of more than one chapter. However, since these types of
overlap were essential for the collaborations intended within
CEAM, it is hoped that the reader will use his best judgement
and where necessary refer to the extensive subject index which
is provided in Appendix .iv. at the end of the volume.

SECTION I

RESEARCH COLLABORATION IN EUROPE
—THE EXAMPLE OF CEAM

RESEARCH COLLABORATION IN EUROPE - THE EXAMPLE OF CEAM

I.V. Mitchell

1. INTRODUCTION

This book is the outcome of a comprehensive European research project on rare-earth iron-rich permanent magnets and their applications(*). It contains reports of the work carried out over a period of 30 months and which were presented at the final contractors meeting held in Madrid from April 13-16th 1988.

The legal basis to the programme was a decision of the Council of the European Communities on the 12th March 1985 (OJ No.L 83/13) giving approval for a multi-annual plan for stimulation of European scientific and technical cooperation and exchange over the period 1985-1988. The invitation to submit proposals was published in the Official Journal of the European Communities on 9th March 1985 No: C 73/2.

Major industrial nations such as Japan and the USA have long recognised the enormous potential for technological development offered by new materials. They are actively carrying out ambitious materials research and development with long-term aims. New and advanced materials developments and their associated processing technologies provide the powerful technological "push" which together with a corresponding market "pull" helps develop and maintain a healthy industrial and manufacturing base.
In this context, the European Commission is continually seeking ways to break down the political and linguistic barriers which exist in Europe and which hinder scientific innovation and the adoption of new technologies.

The Concerted European Action on Magnets (CEAM) reflected the desire of the European Commission to encourage collaborative research and development projects, involving universities, research institutes and industry, which could offer potential advantages to European industrial competitiveness. Such collaborations are intended to provide a channel for basic research to find its way into commercial products.

In the face of the challenge brought about by the discovery of a new material exhibiting remarkable magnetic properties, the CODEST committee of the Commission recommended to fund a Concerted Action on Magnets project as a STIMULATION operation, and the EURAM programme provided the overall project management.

(*) Contract No. ST2P-0064-1-F/2-IRL/3-UK/4-D

The STIMULATION action (1985-1988), open to all fields of the exact and natural sciences, and thus responsive to a wide range of research initiatives, aims to encourage European cooperation and scientific and technical interchange. It makes it possible for teams of researchers to undertake advanced and innovative projects which may evolve into more 'market oriented' fields. EURAM (1986-1989) is a shared-cost research and development sectorial programme on advanced materials, specifically aimed at raising the technological level of products of European manufacturing industries and thus helping them to compete better on world markets.

2. BACKGROUND

Remarkable advances in all fields of materials science and technologies have been made in the past half century. Major developments in hard magnetic materials over the years extended the scope of application of permanent magnets to include a diverse range of industrial and domestic devices such as electric motors, industrial machines, loudspeakers, sensors and actuators. Nevertheless, in most cases, the electro-magnet reigned supreme and continued to be used for most applications where a magnetic field was required.

The past twenty years have seen a magnetic mini-revolution with the development of new alloys, initially based on samarium and cobalt (Sm-Co). The opportunities provided by this new generation of permanent magnets provided the technolgical "push" which forced equipment manufacturers to take a fresh look at their products. Unfortunately, at just about the time that Sm-Co magnets were beginning to be applied, the world supply of cobalt proved to be unreliable and rapid escalation in its price left magnet users in a quandary. Manufacturers were understandably reluctant to commit themselves to incorporating the new high performance magnets into their products because they were wary of shortages and uncontrollable price fluctuations of the raw materials. Consequently, the element of market "pull" was absent as industry adopted a 'wait and see' attitude.
Partly in response to this situation, the Commission of the European Communities (CEC) included research into cobalt substitution for permanent magnets in the SUBSTITUTION programme (1982-85), aimed at finding suitable materials to replace scarce and strategic raw materials. In late 1983, Sumitomo Special Metals Co. of Japan and General Motors Corporation, USA announced almost simultaneously the breakthrough discovery of a new alloy based on neodymium, iron and boron (Nd-Fe-B) that exhibited astonishing magnetic properties and set new standards for permanent magnet performance.
This opened up the prospect of cheap, high-performance magnets which could find new applications in robotics, office automation, many types of motors and generators, scanners and specialised medical equipment and in transportation (e.g. magnetically-levitated transport systems).

At the request of several interested organisations and fully alert to the discovery of these 'supermagnets'and the fact that they offered the potential of a real breakthrough, the Commission organised the first European workshop on 'Nd-Fe Permanent Magnets: Their Present and Future Applications' in Brussels on October 25th 1984(*). As a direct consequence of the meeting, a plan of action was drawn up with a strong recommendation for a concerted European activity in this field. With the encouragement of the Commission, a group of researchers and engineers from universities, research institutes and industry throughout Europe prepared a detailed scientific proposal for a specific programme of research.
Their proposal was submitted to the STIMULATION action of the Commission during June 1985, where it received the unanimous approval of its advisory committee for science and technology (CODEST) at their meeting of September 23rd 1985 (A list of the current CODEST membership is given in **Appendix .i.**).
Thus, in less than a year from the date of the workshop, Community funding was forthcoming from the Stimulation programme for a major collaborative research and development project on permanent magnets. The scientific management of the project was taken in charge by the European Commission's research programme on Advanced Materials (EURAM), which had a priority area in the field of high performance permanent magnets.

Figure 1. Researchers from the 58 laboratories involved in the Concerted European Action on Magnets project

(*) 'Nd-Fe Permanent Magnets: Their Present and Future Applications', edited by I.V.Mitchell, Elsevier Applied Science Publishers, (1985) pp.i-xx, 270

3.   THE PROJECT:- CONCERTED EUROPEAN ACTION ON MAGNETS

CEAM associated 58 institutes throughout Europe - including, most of the laboratories with special expertise in rare-earth iron permanent magnets.   A full list of the participating groups is given in **Appendix .ii.**
The project received funding of 2.5 million ECU. for a programme of research covering a period of 30 months. Industrial companies made up about one third of the group with the others coming from universities and national laboratories. **Figure 2.** indicates the breakdown by country and type of organisation.

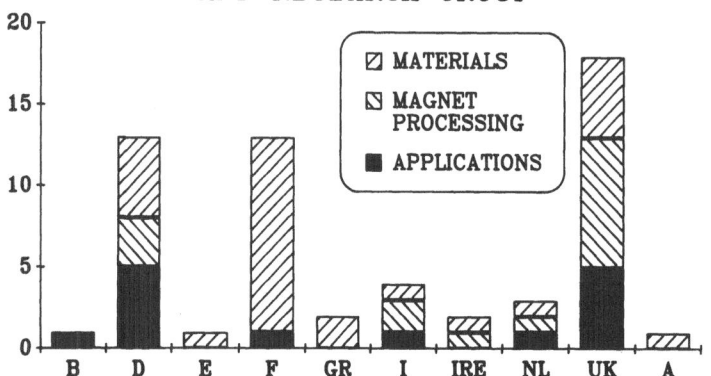

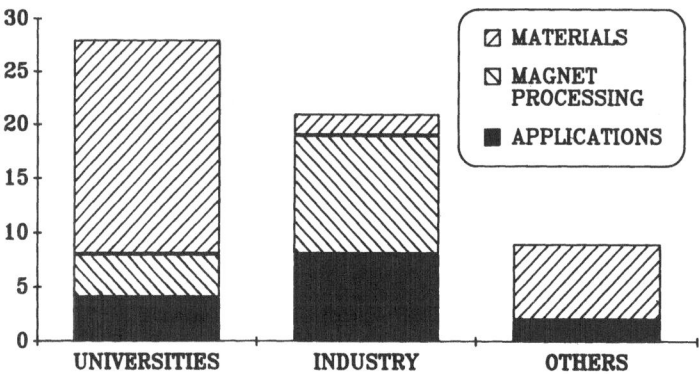

Figure 2.a) Distribution by Country
b) Distribution by Type of Organisation

4. AIMS AND OBJECTIVES

Participation in CEAM was open to laboratories which had an active interest in the field and which wished to join the project when it was first established.

The primary aims were:

- o to develop high performance iron-based rare earth permanent magnets and to design novel devices which exploit their exceptional properties;

- o to generate European collaboration by the exchange of scientists and stimulate a new generation of researchers to undertake projects in applied magnetism of industrial relevance;

- o to provide a skills and information base to permit European industry to exploit the advanced magnets effectively.

5. ORGANISATIONAL ASPECTS

The research programme was divided into the three broad areas of:

- **MATERIALS**
- **MAGNET PROCESSING**
- **APPLICATIONS**

The **MATERIALS** group was composed largely of physicists and chemists working on phase diagrams, searching for new alloys and examining the intrinsic and extrinsic magnetic properties of rare earth alloys with particular reference to those with the $Nd_2Fe_{14}B$ structure.

The **MAGNET PROCESSING** group mainly involved metallurgists and materials scientists and included significant industrial participation. They were primarily concerned with the microstructure of magnet alloys and the numerous problems of magnet processing and stability.

The third group on **APPLICATIONS** focussed on both electromagnetic and magnetostatic applications of the new magnets. Many of the participants in this group were electrical engineers and specialists in computer-aided design (CAD) working in industrial companies and universities.

Regular bi-annual meetings were held at different locations throughout the project life and consolidated technical progress reports were prepared on a six- monthly basis for the Commission. Close and continuous liaison with the scientific project manager in Brussels was maintained at all times.

6. PROGRAMME MANAGEMENT

Professor René Pauthenet of the CNRS, laboratoire Louis Neél, Grenoble acted as chief coordinator of CEAM up until his death in April 1987. Professor J.M.D. Coey of Trinity College, Dublin assumed the rôle thereafter. The Materials Group coordination was jointly managed by Dr. D.Givord of laboratoire Louis Neél, Grenoble and Professor J.M.D. Coey. Professor I.R. Harris of the University of Birmingham and Professor R. Hanitsch of the Technische Univeität Berlin coordinated the Magnet Processing Group and Application Group, respectively. From the part of the Commission, the overall scientific management was taken in charge by Dr. I.V.Mitchell of the Euram - advanced materials programme and the administration by Mr L. Bellemin of the Stimulation action.

At its inception, about 50 organisations chose to participate in CEAM. However during the project, several organisations decided to withdraw and other new groups asked to join. In all, more than 120 scientists and engineers from 58 laboratories directly participated in the project. Nine of the twelve EC Member States were represented, as well as Austria. Organisationally, CEAM rested on the four main coordinating laboratories. The structure of the project is illustrated schematically in **Figure 3.**

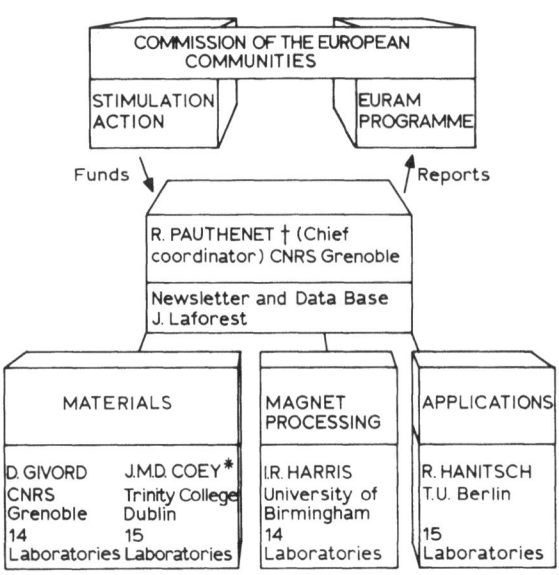

Figure 3. The Organisational Structure of CEAM

All other participants acted as subcontractors of one or more of these four co-ordinating institutes. **Table 1.** gives a summary of the organisations which actively participated in CEAM over a substantial period. A complete list of all subcontractor participants is given in **Appendix .ii.**

Table 1. Summary of Contractors participating in CEAM

CONTRACT No.	ORGANISATION	PLACE	NAME	PARTICIPATION

MATERIALS
=========

1.01	Centre Nat. de la Recherche Scientifique	Grenoble	D. Givord	Co-ordinator
1.02	Universite de Savoie	Annecy	J.M. Moreau	
1.03	Fulmer Research Institute	Slough	R.I. Saunderson	Partial
1.04	Vacuumschmelze GmbH	Hanau	W. Rodewald	
1.05	Centre d'Etudes Nucleaires de Grenoble	Grenoble	R. Chamberod	
1.06	Universite de Bordeaux	Talence	J. Etourneau	
1.07	Institut National Polytechnique	St. Martin d'Heres	C. Allibert	
1.08 a	Universite Grenoble I	St. Martin d'Heres	U. Berthier	
1.08 b	Universite Grenoble I	St. Martin d'Heres	F. Hartmann-Boutron	
1.09	Centre Nat. de la Recherche Scientifique	Grenoble	D. Fruchart	
1.10	Centre de Recherches Nucleaires	Strasbourg	J.P. Sanchez	
1.11	Centre d'Etudes Nucleaires de Grenoble	Grenoble	M. Boge	
1.12	Institut National Polytechnique	St. Martin d'Heres	R. Fruchart	
1.13	University of Zarragoza	Zaragoza	J. Bartolome	
1.14	Centre d'Etudes Nucleaires de Grenoble	Grenoble	R. Gillet	
1.21	Trinity College Dublin	Dublin	J.M.D.Coey	Co-ordinator
1.22	University of Amsterdam	Amsterdam	J.M.M Franse	
1.23	Kerforschungszentrum Karlsruhe	Karlsruhe	G. Czjzek	
1.24	University of Parma	Parma	G. Asti	
1.25	Ruhr Universitaet Bochum	Bochum	M. Rosenberg	
1.26	Philips Research Laboratories	Eindhoven	K.H.J Buschow	
1.27	University of Durham	Durham	W.D. Corner	
1.28	Nuclear Research Centre "Demokritos"	Attiki	A. Kostikas	
1.29 a	Max-Planck Inst. f. Metalforschung	Stuttgart	E.T. Henig	
1.29 b	Max-Planck Inst. f. Metalforschung	Stuttgart	H. Kronmueller	
1.30	University of Salford	Salford	P.J. Grundy	
1.31	University of Birmingham	Birmingham	S. Abell	
1.33	Imperial College London	London	D.G. Pettifor	
1.34	University of Crete	Crete	G.C. Hadjipanayis	
1.35	Technische Universitaet Wien	Vienna	H.R. Kirchmayr	
1.36	Universitaet Giessen	Giessen	C. Heiden	

MAGNET PROCESSING
==================

2.01	University of Birmingham	Birmingham	I.R. Harris	Co-ordinator
2.02	University of Sheffield	Sheffield	H.A. Davies	
2.03	SG Magnets Ltd.	Rainham	A.J. Ward	
2.04	Rare Earth Products Ltd	Widnes	D. Kennedy	
2.05	Gesellscaft f. Elektrometallurgie	Nuernberg	S. Sattleberger	
2.06	Industria Ossidi Sinterizzati SpA	Malgesso	A. Cartocetti	Partial
2.07	Mullard Southport Ltd	Southport	E. Rozendael	
2.08	Sunderland Polytechnic	Sunderland	A.G. Clegg	
2.09	Lucas Engineering and Systems Ltd.	Solihull	M. Ward	
2.10	General Electric Co.	Wembley	J.M. Vincent	
2.11	Trinity College Dublin	Dublin	J.M.D.Coey	
2.12	Krupp Widia GmbH	Essen	W. Ervens	Partial
2.13	Thyssen Edlestahlwerke AG	Dortmund	H. Nagel	
2.14	Ing. C. Olivetti SpA	Ivrea	S. Tori	

Table 1. continued

CONTRACT No.	ORGANISATION	PLACE	NAME	PARTICIPATION

APPLICATIONS
============

3.01	Technische Universitaet Berlin	Berlin	R. Hanitsch	Co-ordinator
3.02	University of Sheffield	Sheffield	D. Howe	
3.03	University of Manchester	Manchester	B.J. Chalmers	
3.04	K.U.Leuven	Heverlee	W. Geysen	
3.05	ERA Technology Ltd.	Leatherhead	M. Bradford	
3.06	General Electric Co.	Wembley	A.J. Walkden	
3.09	Kindervater & Sohn KG	Berlin	T. Kindervater	Partial
3.10	Dornier System GmbH	Friedrichshafen	I. Kitzmann	
3.11	Philips Research Aachen	Aachen	E.M.H.Kamerbeek	
3.12	Centre Nat. de la Recherche Scientifique	Grenoble	J. laForest	
3.13	Krupp Widia GmbH	Essen	W. Baran	
3.14	Motori ed Apparecchiature Elettriche	Offanengo	G. Ioppolo	Partial
3.16	University of Liverpool	Liverpool	K.J. Binns	
3.17	Philips Research Laboratories	Eindhoven	R.P. Van Stapele	

7. BREAKDOWN OF FUNDING

The total funding of CEAM was 2.5 million ECU, initially for a period of 24 months. The project was later extended for two further periods of six months.
The funding was divided between the three main research areas as follows:

Materials: 975,000 ECU (29 participating groups)
Magnet Processing: 900,000 ECU (14 participating groups)
Applications: 625,000 ECU (15 participating groups)

The funding was primarily directed at promoting the collaboration: defraying the cost of meetings, subsidising research and staff exchanges, providing information and easing the often difficult task of encouraging cross-fertilization of ideas and initiatives. Limited support was included for materials, equipment and running costs, where these were not covered in the normal way from internal sources.

8. INFORMATION EXCHANGE

Two guiding principles underlay the progress and efficient running of the collaborations:

   o   **RAPID INFORMATION EXCHANGE**

   o   **REGULAR MEETINGS**

8.1. CEAM INFORMATION CENTRE

The CEAM Information Centre was located at the CNRS, Grenoble and was charged with providing a regular newsletter service and maintaining a technical database and bibliography. The Information Centre produced the CEAM newsletter, under the editorial responsibility of Dr J. laforest, at regular 2 -3 monthly intervals. In general, contributions were made from within CEAM although on occasions specialist articles were solicited from outside. Circulation was restricted to CEAM members but single issues were made available to other interested parties, on request. The newsletter contained amongst other items, CEAM meetings and group reports, news of new products and relevant advances in permanent magnet technologies and applications; summaries of relevant sessions at other conferences; articles assessing progress in the various fields; a constantly enlarged and updated technical and bibliographical database; the latest available technical specifications of materials and products as well as off-the-record news items.
In total, 11 editions of the CEAM newsletter were published.

The database was principally a list of references to relevant papers and patents which was updated regularly, and made available on paper or floppy disc to the participants. It could be accessed on PC using widely available software. Manufacturers' technical data sheets were also kept, and copies made available on request.
The Information centre also acted to a certain extent as the hub of a research wheel with an exchange of information and a transfer of experimental results flowing along the spokes. This was achieved primarily by requiring all CEAM participants to send details, at the same time as their periodic reports, of scientific and technical publications submitted; theses awarded; conference presentations and patent applications, during the relevant period. In this way the Information centre added value to the extensive network of informal contacts that took place amongst participants at the regular meetings.
A presentation of the project was compiled by the CEC scientific manager during January 1988 in the form of a 24-page illustrated brochure entitled 'The Concerted European Action on Magnets'. A series of abridged versions in English, French, German and Spanish were made available in March 1988. A 15 minute video film entitled "Magnetism - A European Force", was commissioned by the EURAM programme and produced in Grenoble by Scop-Manivelle SA, a film production unit, supervised by CCST of Grenoble in close collaboration with the CEAM project coordinators. The film provided an insight into the international scientific and technical cooperation that has been at the heart of CEAM. Filming was taken on location in Grenoble, Dublin, Birmingham, Sheffield, Leuven, and Stuttgart at participating institutes.
Both the brochure and video served to present information

about CEAM to interested outsiders and was used to provide an impression of a Community project that succeeded in fostering European cooperation over a wide spectrum of scientific, technical and industrial interest, centered around a common theme. Several other films were prepared on the initiative of different participating groups.

8.2. REGULAR MEETINGS

Regular six monthly meetings of each of the three subject groups were an essential element in the project organisation. Their aims were to critically assess the goals attained and the progress made during the previous period; to encourage new collaborations by the exchange of personnel and information between laboratories and to strengthen existing co-operations.

These group meetings were supplemented by joint annual reunions which aimed to encourage scientific and technical cohesion between the disparate groups. They helped to breakdown the natural barriers that are prevalent between such different disciplines as materials science and engineering design and promoted the camaraderie which was so important to the ultimate success of the project. Meetings were always held in different Member States, each time hosted by a different participant. In all 12 meetings took place, as given in **Table 2.**

MEETING	PLACE	COUNTRY	DATE
Annual	Grenoble	France	October 1985
Group 1	Amsterdam	The Netherlands	March 1986
Group 2	Nürnberg	Germany	April 1986
Group 3	Manchester	United Kingdom	May 1986
Annual	Birmingham	United Kingdom	September 1986
Group·3	Berlin	Germany	March 1987
Group 1	Dublin	Ireland	March 1987
Group 2	Stresa	Italy	April 1987
Group 1	Crete	Greece	September 1987
Group 2	Eindhoven	The Netherlands	October 1987
Group 3	Leuven	Belgium	November 1987
Final	Madrid	Spain	April 1988

Table 2. **List of CEAM Group Meetings held within the EC**

8.3. MADRID MEETING

The culmination of the CEAM project was the final contractors' meeting and Open Day. Originally conceived as a

two-year programme, it was decided to prolong CEAM for two
further periods of 6 months, in order to make the most of the
promising developments in the early stages. The first
extension, up to the spring of 1988, was a period of full
activity. The existing budget was stretched to cover the
extension period and the Commission made a special contract
with the Spanish Consejo Superior de Investigaciones
Cientificas (CSIC) to help organise a final general meeting
in Madrid. Responsibility for the organisation was assumed
by Professor J. Serratosa, aided by Spanish local and
national committees.

8.4. THE OPEN DAY

The Open Day held at the Palacio de Congressos y Exposiciones
was aimed at presenting CEAM to a wider audience, of
students, industrialists, government officials and public
representatives, including members of the European Parliament
and CODEST. Professor U.Colombo who had been president of
CODEST when the proposal for CEAM was initially submitted,
presided at the morning session, which was attended by more
than 300 people. Speeches of a general nature were made by
representatives of the European Commission, the Spanish
ministry responsible for science, the CSIC and the government
of the Madrid region. Dr I.V.Mitchell, the Commissions'
Scientific Manager for CEAM, paid tribute to the late
Professor R.Pauthenet who was CEAM's Chief Coordinator up
until his untimely death. Professor J.M.D. Coey, who had
been acting as chief coordinator of CEAM following the death
of Professor Pauthenet, then gave an overview of recent
developments in permanent magnets, and outlined the scope,
organisation and achievements of the project. This was
followed by the première showing of the official CEAM video
film. The three other coordinators, Dr D.Givord, Professor
I.R.Harris and Professor R.Hanitsch then gave more detailed
presentations of the achievements of the Materials, Magnet
Processing and Applications groups, respectively. The
plenary session was concluded by Professor Colombo and was
followed by a press conference. Extensive coverage was given
to the events in the Spanish press and television.

The exhibition was a highpoint of CEAM. Organized in the
main foyer of the Palacio de Congressos y Exposiciones in the
format of a small trade fair, there were seventeen exhibits
broadly covering the subject areas of CEAM. The exhibition
symbolized the approach developed during the course of the
collaboration; each display was put together to a high
standard under the responsibility of one participant, but
presented the accomplishments of several laboratories working
in a particular area. These elements had a distinct unity,
reflecting the shape of the whole programme. Continuity was
provided by poster panels on each subject area which also
served as a guide to the exhibition. Besides text, pictures
and diagrams, the display material included models, PC-based
demonstrations, several video films which illustrated

particular technical points (hydrogen decrepitation, domain motion), or the activities of an entire group (Applications).

There were also a number of electrical machines on display, prototype motors generators and static magnetic devices built using Nd-Fe-B magnets. A rich and varied insight into CEAM activities was provided for the visitor. In addition, there were displays from the European Commission, several Spanish universities and research organizations, and a number of industrial companies with an interest in permanent magnets.
A selection of material was made from the Exhibition, including most of the Application Group exhibits, for subsequent display on the Commission of the European Communities stand at the Hanover Trade Fair from 20-27 April 1988. A specially prepared booklet with information on CEAM was prepared for this event.

8.5. THE TECHNICAL SESSIONS

The technical sessions took place in the CSIC establishment in Madrid from 13-16 April, following the Open day.
On this occasion, the format was different to that used at earlier meetings in that each subcontractor was allocated fifteen minutes in a plenary session in which to present the CEAM achievements of their laboratory. There were no parallel sessions. Related reports were grouped into sessions on a topic and each session concluded with a summary of achievements and prospects from a designated rapporteur. The final reports and summaries from the Madrid meeting are given in **Section III**. The final plenary session on the afternoon of Saturday 16th April gave an opportunity for everybody to review progress made since the general meeting held in Birmingham in September 1986. A list of goals defined there was constantly referred to as a touchstone of progress.

9. COLLABORATIONS, EXCHANGES AND PUBLICATIONS

A questionnaire was circulated at the end of the final meeting to obtain participant's evaluation of the different aspects of CEAM, and to find out how successfully the participants thought it had met its original aims.
There were more than forty responses; Comment on the achievements and coordination of the programme were overwhelmingly favourable. Exchange of personnel between different institutes has worked well, both at the level of graduate students, and of professionally qualified engineers and scientists. Overall, there were plenty of exchanges: 21 students were involved, for an average of 8 weeks each, and 35 staff, for an average of 4 1/2 weeks - more than 6 man-years in all. 38 students took part in CEAM related thesis work. The bulk of the time spent on exchanges (80%) was in the Materials group, but the distribution of people involved was more even (66:23:11% for groups Materials,

Magnet processing and Applications, respectively). It should be remembered that the number of subcontractors were split 50:24:26% among the three groups, whereas the budget split was 40:33:27%. In all, 181 researchers have been involved in CEAM since its inception. In general, the CEAM meetings were highly rated as was the CEAM Newsletter, with most readers reading most articles (51%) or cover-to-cover (32%).

Opinion on the question of competitiveness was divided; 38% of those having an opinion thought that CEAM had exerted an important effect in improving industrial competitiveness with respect to the US and Japan, 35% thought there had been some benefit and 8% thought not.

Turning to publications and patents, 476 papers are recorded in **Appendix .iii.**, as having appeared in the open literature or having been submitted for publication since CEAM's inception. However, since these have been arranged by subcontractor group, there is undoubtedly an element of double counting since some of them result from the joint effort of two (or more) CEAM laboratory groups. Excluding this factor results in a total of 356 unique scientific and technical papers recorded during the contract period.

The distribution among groups was (80%:5%:14%). Eight applications for patents were made during the life of CEAM and it is interesting to note that 7 of the 8 came from the Materials group. The replies to the questionaire confirmed that, for the most part, CEAM had played a significant rôle in the work which led up to these publications and patents. **Figure 4.** indicates the breakdown of publications submitted by an individual group or collaboratively by two or more subcontractors.

CEAM PUBLICATIONS

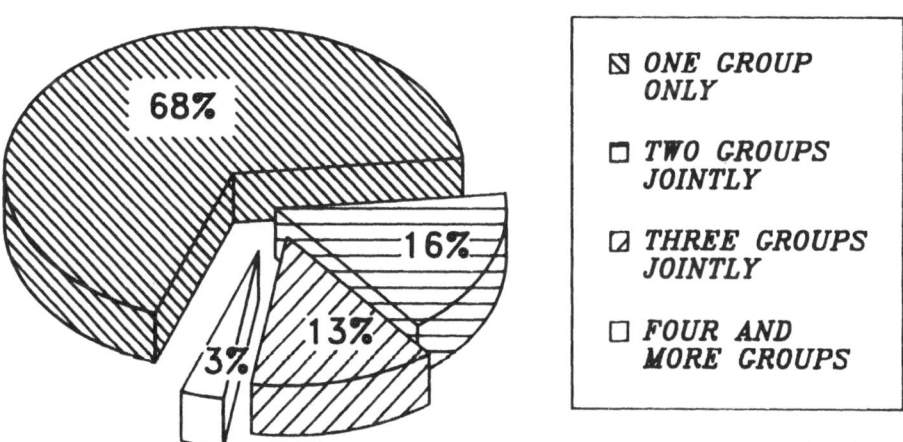

Figure 4. Breakdown of publications by No. of collaborations

10. FRUITS OF CO-OPERATION

Apart from the technical achievements already described, other goals of CEAM have been attained:

A new generation of young technicians and researchers have been trained in the field of applied magnetism - this has helped to revitalize a rather small and fragmented industry which is, nevertheless, critical to the economic and strategic wellbeing of Europe.

CEAM has helped to create the right conditions for a 'snowball' effect to be brought about - allowing the rapid exploitation of new results into industrial development and manufacture.

As far as advanced magnetic technology is concerned, CEAM has re-established a European dimension to the subject, which is indispensable in the face of intense competition from overseas.

The underlying principle within CEAM of free exchange of information and expertise, across a wide range of disciplines, has encouraged genuine pan- European cooperations. Once established, these links are not easily broken and lead to new collaborations - as evidenced by the number of joint project proposals received in recent Community R&D programmes.

CEAM's successes and the manner by which skills and expertise across so many frontiers, political as well as disciplinary, were brought - and kept - together, is a fine example of what can be achieved in Europe. This will serve in the future as a benchmark by which other projects may be judged.

SECTION .II.

RARE EARTH PERMANENT MAGNETS
– THE CONTRIBUTION OF CEAM

RARE EARTH IRON PERMANENT MAGNETS - THE CONTRIBUTION OF CEAM

J.M.D. Coey, D. Givord, I.R. Harris, R. Hanitsch

The impact of the discovery in 1983 of a high-performance iron-rich rare-earth permanent magnet has been rapid and far-reaching. This new material is probably the most significant advance in permanent magnetism since the development of ferrites in the 1950s. It is progressively supplanting samarium-cobalt (1:5 and 2:17) in many applications, and may eventually replace alnicos. Exciting new products and engineering prospects are opened up by the availability of magnets with an energy product in excess of 250 kJ.m^{-3} and a service temperature of up to 150°C. In CEAM, research has been conducted in parallel with a view to understanding and improving the magnetic properties of $Nd_2Fe_{14}B$, and related phases, searching for new iron-rich phases that may prove to be equally interesting, improving the processing and corrosion resistance of Nd-Fe-B magnets and using the magnets in prototype devices. This section is designed to highlight the contribution of CEAM and guide the reader to what is of interest to them in the rest of the book.

At present, all high-performance iron-based magnets are composed largely of the tetragonal $Nd_2Fe_{14}B$-structure phase whose existence was first detected by Kus'ma and co-workers in the USSR in the late 1970s. In 1983, Sagawa in Japan and Croat in the USA discovered that good permanent magnets could be made from this tetragonal compound, whose structure, shown in figure 1, was elucidated soon after.

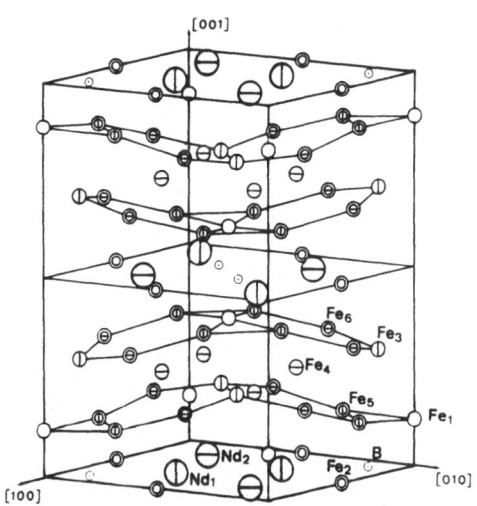

Figure 1.   The crystal structure of $Nd_2Fe_{14}B$

A. MATERIALS

The discovery of $Nd_2Fe_{14}B$ begs the question of whether other phases might exist with magnetic properties that are as good or better. One route in the search for new phases has been to look for pseudobinaries with structures related to known structure-types such as Th_2Ni_{17}, $BaCd_{11}$, $ThMn_{12}$ and $NaZn_{13}$. In this respect, structure maps based on existing databases and the concept of Mendeleev number (1.1)[*] have proved quite useful. The maps also emphasise the point that more ternaries of the rare-earths and late 3d transition elements form with boron than with any other third element, so that $Nd_2Fe_{14}B$ should not be seen exactly as a chance occurrence. The existence of the 2:14:1 phase is further documented in a literature survey of over 600 systems (1.2).

One promising place to look for a rival to $Nd_2Fe_{14}B$ seems to be among pseudobinaries with the $ThMn_{12}$ structure. About a hundred such compounds have been investigated in CEAM (1.3, 1.4, 1.6, 1.9); $Sm(Fe_{11}Ti)$ has been identified as one of the most promising. Most of its intrinsic magnetic properties are comparable to those of $Nd_2Fe_{14}B$, with the exception of magnetisation (and hence potential energy product) (1.4). A range of new $BaCd_{11}$-type compounds, some having their structure stabilized by interstitial carbon, likewise possess rather small iron moments (1.5, 1.9). $R(Fe_4B)$ materials are also of some interest (1.6). A particularly fruitful general method for seeking new magnetic materials has been melt-spinning followed by annealing and thermomagnetic analysis (1.7, 1.8). Several new metastable neodymium-iron borides and metastable pseudobinaries have been discovered in this way, but attempts to find a suitable ferromagnetic manganese-rich alloy have proved disappointing (1.10).

To summarize, novel compounds capable of bearing the record for energy product beyond the 500 kJ.m^{-3} mark may be awaiting discovery, but it has not proved easy to find them. About a fifth of the CEAM participants spent some of their time looking for new phases.

Intrinsic Magnetic Properties: The $Nd_2Fe_{14}B$ structure compounds provide a great opportunity for building a systematic understanding of the **intrinsic** magnetic properties of a series of rare-earth iron intermetallics because the whole series of fourteen compounds exists with iron, for rare earths from La to Lu (plus Y, but excluding Eu). There are also seven cobalt compounds. Despite the complexity of the structure, with its two rare earth sites and six iron sites, a good systematic understanding has been developed of the atomic moments, spin structures, exchange interactions, Curie

---

[*] The numbers in brackets refer to the contributions in the different chapters of Section III.

temperatures, spin reorientation transitions, crystal-field interactions and anisotropy fields for the series. These are the intrinsic magnetic properties of the compounds, specifically, those that are independent of microstructure. For these fundamental magnetic studies, the growth and dissemination of single crystals (2.1, 2.2) among the different laboratories has been invaluable.

The average iron moments in $R_2Fe_{14}B$ are discussed in terms of the concepts of strong ferromagnetism and magnetic valence (2.1), while the iron-iron exchange is determined by the iron coordination (1.1, 2.1) and spin fluctuations (1.1). Large invar anomalies in the thermal expansion (1.3, 2.7) are also a reflection of the magnetic state of the iron, and there is a correspondingly anomalous variation in the iron anisotropy (2.3).

Rare-earth iron exchange interactions decrease along the series in a manner now recognised as typical of 3d-4f intermetallics (2.1). One approach to determining crystal field parameters has been to fit the magnetisation curves measured as a function of temperature and applied field (1.4, 2.2). First-order magnetisation processes have been discovered in $Nd_2Fe_{14}B$ and $Pr_2Fe_{14}B$ (2.2, 2.3). The singular point detection technique in pulsed fields is particularly helpful for characterising these transitions, and for rapidly evaluating the anisotropy field of powders (2.3, 2.4). Spin reorientation transitions (1.4, 2.3, 2.7) and measurements of the ground-state magnetic structures such as the complex spin structure of $Ho_2Fe_{14}B$ (2.5) further constrain possible models of the magnetic interactions. Sets of crystal field parameters and exchange parameters that are capable of reproducing the observed behaviour have been deduced for all the members of the $R_2Fe_{14}B$ series (1.4).

One useful result of the effort that has been devoted to understanding the intrinsic magnetic properties is that it is now possible to predict with a fair degree of certainty what will be the effect of any particular substitutional impurity, whether on Nd or Fe sites. Cobalt is of particular interest because, in small quantities, it raises the Curie temperature of the iron compound without degrading the magnetisation or anisotropy field at room temperature. In order to extend the temperature-range of useful permanent magnet properties it is necessary also to increase the anisotropy field which can be done (at the cost of a small decrease in magnetisation) by replacing a few percent of neodymium by a heavy rare earth such as terbium or dysprosium which has a stronger second-order crystal field interaction.

Numerous solid solution series have been studied by CEAM investigators (e.g. 2.4). There is also one interstitial substitution - hydrogen - that has been studied extensively (1.9, 2.5, 2.6). Hydrogen has the effect of increasing the Curie temperature and iron magnetisation as it dilates the lattice. The effect on the Curie temperature is most pronounced in compounds like Nd_2Fe_{17} and $Ce_2Fe_{14}B$ where T_c is unusually low to begin with. Unfortunately the anisotropy field at room temperature is generally reduced, but there are interesting effects on the various terms in the crystal-field interaction which may lead to spin reorientations where none were present before.

Many studies of the atomic-scale magnetism of $R_2Fe_{14}B$, $RFe_{11}Ti$, RFe_4B_4 and related compounds have been carried out in CEAM by means of techniques that measure hyperfine interactions. ^{57}Fe Mössbauer spectroscopy is an informative technique which is available in many laboratories that have an interest with magnetic materials (1.4, 1.6, 2.5, 3.2, 3.3, 3.4). Different crystallographic sites are distinguished by their different hyperfine interactions. Rare earth resonances including ^{155}Gd, ^{161}Dy, ^{166}Er and ^{174}Er have also been exploited (2.5, 3.3, 3.4, 3.5) and some new ones, ^{157}Gd and ^{145}Nd, have been developed (3.6). Much of value has been learnt from these measurements about site preferences, atomic moments, electric field gradients (which are related to the leading terms in the crystal-field interactions), electronic structure of the rare earth and spin reorientation transitions. Nuclear magnetic resonance is an alternative technique for determining hyperfine interactions which can be applied for nuclei that possess a magnetic moment in their ground state. It has been used to investigate compounds with the $Nd_2Fe_{14}B$, $BaCd_{11}$ and other structure-types (3.1, 3.2) by means of the ^{89}Y, ^{143}Nd, ^{175}Lu, ^{57}Fe, ^{59}Co, ^{10}B and ^{13}C resonances. Some use has also been made of the muon spin rotation technique (2.6).

Extrinsic Magnetic Properties: It is well known that the coercivity in permanent magnets is much larger than in homogeneous bulk magnetic materials. This property arises from a specific microstructure. In particular, large coercivity may be achieved if the magnetic grains are sufficiently small and isolated one from another. To analyse these effects in Nd-Fe-B magnets, microstructural parameters (grain sizes, secondary phases, etc.,) have been studied. A good knowledge of the phase diagrams is a prerequisite for this. The ternary Nd-Fe-B diagram has been studied in great detail (4.1), and the changes associated with the substitution of some additional elements such as Al, Cu, Nb, Zr and O have been examined (4.1, 4.2, 4.3).

Another common observation is that the coercive field in permanent magnets is much lower than the anisotropy field, which is its maximum theoretical value. This reduction is believed to be due to the influence of defects, but the mechanisms involved in magnetisation reversal are not well understood. They have been studied in Nd-Fe-B magnets from both the theoretical and experimental viewpoints.

From the extensive investigation of the Nd-Fe-B phase diagram (4.1), see for example figure 2, it has been shown that, at the magnet sintering temperature (of about 1100°C) the three

phases (liquid, $Nd_2Fe_{14}B$ and $Nd_{1+\epsilon}Fe_4B_4$) coexist in
equilibrium when alloys of composition $Nd_{15}Fe_{77}B_8$ are used.

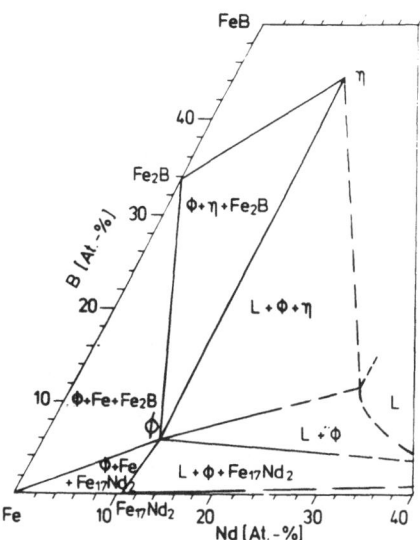

Figure 2. **Part of the Nd-Fe-B phase diagram**

If an alloy of composition $Nd_{18.5}Fe_{75}B_{6.5}$ is used, the n-phase
($Nd_{1+\epsilon}Fe_4B_4$) is no longer present and improved magnetic
properties may be obtained. In these magnets, the influence
on coercivity of sintering time and sintering temperature has
been correlated with grain size. It was concluded that the
mean grain diameter must be smaller than 12 µm to achieve
optimum properties (4.1).

In practice, the real microstructure of magnets involves more
than the three elements Nd, Fe and B. On the one hand, some
oxygen is always present and it has been shown that oxygen
favours the appearance of free Fe through rare-earth oxidation
(4.3). On the other hand, additional elements are often
introduced in order to improve the magnet properties. The
influence of many additive elements on the microstructure as
well as on the magnetic properties has been studied (2.4, 4.1,
4.2, 4.3, 5.3, 6.7). Additives are important on practical
grounds as they may permit the preparation of magnets with
improved properties, but they also allow to explore the
interplay between microstructure and coercivity. In sintered
magnets, the best elements for increasing coercivity are Dy
and Al. Dysprosium enters the rare earth sites in the
$Nd_2Fe_{14}B$ phase, leading to increased magnetocrystalline
anisotropy. However, the increase in coercivity observed in
magnets is larger than that expected from the increase of
anisotropy alone. Furthermore, it has been found that Dy has
a significant influence on magnet microstructure and some of
it enters secondary phases (5.4).

Additions of aluminium has a detrimental effect on the intrinsic magnetic properties of the $R_2Fe_{14}B$ phase. The increase in coercivity in Al-containing magnets is thus a microstructural property. Aluminium is found to improve the melting angle of the eutectic liquid phase, which leads to a better coverage and magnetic isolation of the $Nd_2Fe_{14}B$ grains (4.1).

The influence on extrinsic magnetic properties of several other elements has also been studied. Cobalt substitution would seem to be very attractive as it leads to a dramatic increase in the Curie temperature. Unfortunately, the coercivity of sintered magnets decreases, although that of Co-containing melt-spun ribbon can be very large, higher than 2 T (2.4). Copper does not dissolve in the $Nd_2Fe_{14}B$ phase, but in sintered magnets it leads to an increase in the size of the $Nd_2Fe_{14}B$ grains which may explain why there is a decrease in coercivity (4.2). In other magnets (ribbons, extruded magnets), a large coercivity may be obtained in Cu-containing alloys, showing that the precise influence of Cu on microstructure needs to be further examined. Other elements, such as Nb, Mo or Zr, which do not enter in the $Nd_2Fe_{14}B$ phase also have some influence on the microstructure (5.3). In particular, large Nb or Zr precipitates are observed in sintered magnets which do not greatly affect the magnet properties, but, in ribbons, very small particles containing Nb are formed in the matrix and some increase in coercivity may be obtained.

So far, it has not been possible to obtain extended coherent precipitates as required for coercivity to appear in the bulk (precipitation hardening). No phase with a structure closely related to that of $Nd_2Fe_{14}B$ has been discovered; TEM observations of $Nd_2Fe_{14}B$ grains in magnets indicate high crystalline quality. Both these facts suggest that precipitation hardening in the Nd-Fe-B system is rather unlikely to occur.

Analysis of the domain structure in Nd-Fe-B sintered magnets has shown that the domain walls always finish up at grain boundaries. This demonstrates that exchange interactions between grains are negligible. It should thus be possible to analyse coercivity mechanisms in isolated $Nd_2Fe_{14}B$ grains of the same size as those in the magnets. In fact, such grains turn out to have a much reduced coercivity, which probably results from oxidation at the surface. Consequently coercivity mechanisms have been examined through analysis of various magnetic measurements on magnets (5.1, 5.2). The theory of nucleation from the saturated state has been applied to the temperature and angular dependences of the coercivity. Large stray fields were shown to be present which strongly reduce the experimental coercive field, and a key role in magnetisation reversal has been attributed to misoriented grains (5.2). An alternative approach (5.1) has been to discuss the coercivity in terms of an activation volume where magnetisation reversal is initiated. This volume is

approximately proportional to the cube of the domain wall
width. The coercive field varies approximately as the
inverse cosine of the angle between the magnetic axis and
the magnetising field, suggesting that the anisotropy field
in the activation volume is much larger than the coercive
field. This is unlike usual nucleation theory and shows that
we are still far from a complete understanding of coercivity
mechanisms. However, it must be emphasized that the
discussions and collaborations which took place within CEAM
have significantly advanced our understanding of these
phenomena.

B. MAGNET PROCESSING

Nd-Fe-B-type alloys are remarkable in that, unlike their
predecessors based on $SmCo_5$ and $Sm_2(Co, Fe, Cu, Sr)_{17}$, they
can be processed into permanent magnets by a number of routes.
These are summarised as follows:

Route 1. Conventional powder/sinter.
Route 2. Calciothermic (Goldschmidt) powder/sinter.
Route 3. Hydrogen decrepitation (HD) powder/sinter.
Route 4. Atomisation powder/sinter.
Route 5. Melt spinning.
Route 6. Hot working.

In the research and development work carried out in the CEAM
programme emphasis was placed initially on route 3, the HD
powder/sinter process (6.3, 6.4, 6.5) and route 5, the melt
spinning process (6.7) whereas process 6, the hot working
route (6.7, 6.8, 6.9) emerged strongly towards the end of the
programme. In addition to the studies of the processing
aspects of the Nd-Fe-B type magnets, the corrosion behaviour
was also examined together with means of providing protection
against such corrosion (7.3, 7.4).

Powder Processing: In the conventional process (route 1), the
starting material is in the form of ingots which have been
produced by induction melting in argon. The detailed nature
of the ingot material was fully characterised to establish
if this had any bearing upon the properties of the resulting
sintered magnets (6.1). Thus, ingots were produced under a
wide range of conditions which resulted in a wide range of
grain sizes. No marked correlation between initial
microstructure and the final properties of the sintered
magnets was observed although some advantage might accrue from
using ingots with a fine grain size. Alloys from the two CEAM
materials suppliers gave excellent reproducibility as far as
final magnet properties were concerned (6.1, 6.2).
Microanalytical studies (6.3) confirmed the complex nature of
the grain boundary regions and a range of Nd/Fe ratios were
observed but limited light element analysis was carried out.
The extensive free iron observed in the "as cast" $Nd_2Fe_{14}B$
ingot could be removed by an appropriate homogenisation
treatment.

All the Nd-Fe-B-type ingots containing Nd-rich material readily absorbed hydrogen at room temperature and about 1 bar pressure. The heats of absorption and associated weight changes have been convincingly correlated with the amounts of Nd-rich material in the alloy (6.3). Oxidation substantially reduced the activity of the alloys and the single phase $Nd_2Fe_{14}B$ could not be activated at room temperature. It was concluded that hydrogen absorption and desorption could usefully be employed to provide information on the constitution of the alloys.

The presence of free iron leads to considerable difficulty in coarse crushing because of the increased toughness of the alloy. This problem was solved by the HD-process (6.3), which was successfully employed by three CEAM magnet producers (6.4, 6.5, 7.2), to produce sintered magnets. In essence the process has proved to be very simple to operate and large lumps of ingot material were reduced to relatively fine powder by exposure to hydrogen at room temperature and one bar (or/less) pressure. The hydrided material was very friable but easy to handle and could be milled readily to the final particle size of around 2µm, required for sintered magnets. The dramatic effect of hydrogen on a sintered Nd-Fe-B magnet is shown in figure 3.

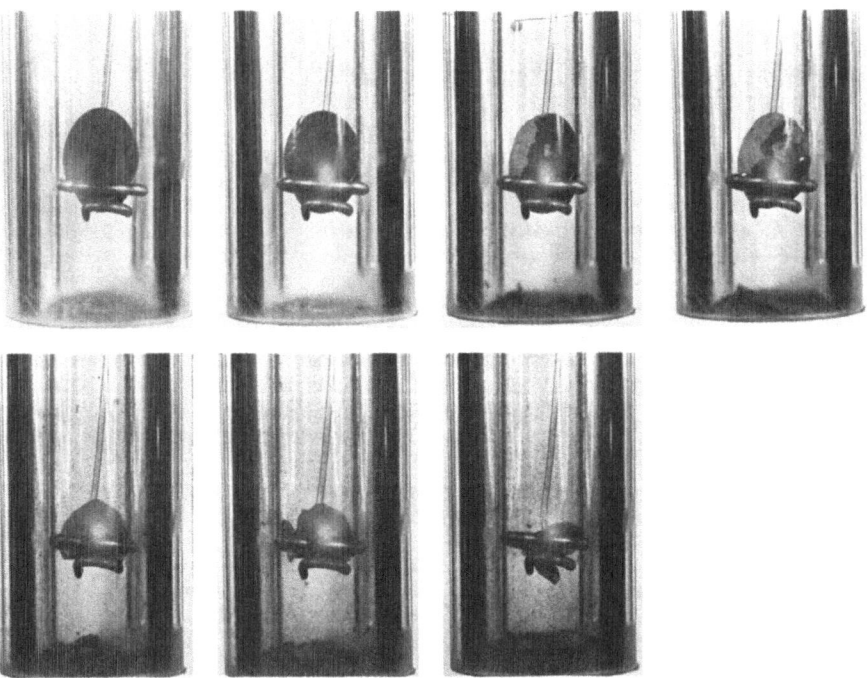

Figure 3.   Hydrogen decrepitation of Nd-Fe-B

The incorporation of the HD-step into the powder processing route is shown in figure 4. The HD-process is now being adopted by permanent magnet manufacturers worldwide.

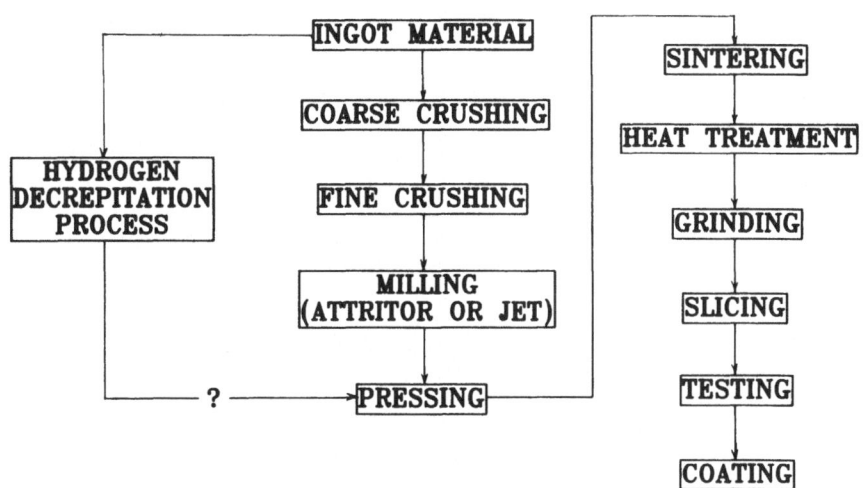

Figure 4. **Processing routes for Nd-Fe-B magnets**

One of the critical factors in magnet production by the powder route is the particle size distribution. The usual procedure is to obtain an average particle size by Fisher sub-sieve measurements. A novel technique for the characterisation of magnetic powder was developed (6.6) in the CEAM programme. This involved controlled oxidation of the powder and its associated gravimetric measurement. Analysis of the date gave information on the particle size distribution.

Melt Spinning: Commercial interest in bonded magnets based on melt spun material is growing and there is tremendous potential for further development. In particular, production of anisotropic ribbon direct from the wheel is a major priority. Research carried out in CEAM (6.8, 6.9) has shown that the preferred texture tends to be restricted to near surface regions of the ribbon. Other work (6.7) has shown that small additions of silicon can enhance the remanence.
These effects were difficult to reproduce and hence it was essential to define the precise process parameters. Various process parameters such as melt temperature, wheel speed, etc. were studied and one interesting observation (6.7) was that it was possible to achieve changes in magnetic properties

without corresponding modifications in the grain size. This may be due to microcompositional alterations in the material, possibly related to the microhardness changes observed on annealing the quenched ingot (6.3). Magnets could be made from flakes of the melt spun ribbon by bonding with plastic, with soft metals and by hot pressing techniques. All three methods were examined and, in particular, very promising magnets have been obtained by cold compaction of melt spun ribbon bonded with aluminium (8.3).

Commercial Magnets: In order to place the CEAM research in an appropriate context, a number of commercial magnets, both sintered and bonded, were examined (7.3) in terms of their magnetic properties, microstructure and mechanical strength. This work showed that the smallest grain sizes were associated with magnets fabricated using the atomisation powder/sinter route (route 4). A wide variety of microstructures were apparent in all the magnets investigated and in some cases, "clusters" of Nd-rich material were observed. These "clusters" are probably undesirable with regard to the corrosion resistance. The mechanical strength of the magnets correlated well with their density and wide variations were observed in the transverse rupture strength of magnets originating from different manufacturers.
A limitation on the wider use of the current generation of commercial Nd-Fe-B-type magnets is their high temperature magnetic characteristics. With this in mind, the reversible losses of a range of commercial magnets were investigated (7.5) and within the range of magnets studied, those produced by the hydrogen decrepitation powder/sinter process (route 3) exhibited the best high temperature properties.

Novel Processes: The use of route 6 (the hot working process) to induce radial magnetisation in a rod of Nd-Fe-B alloy by hot extrusion (8.2) was first reported at a CEAM meeting in September 1986 (8.1). Independent work has led to the production of aligned, hot-extruded, rectangular cross-section bar from cast Nd-Fe-B alloy, encapsulated in a steel sheath (8.2). This has the potential of becoming a very direct and economical means of manufacturing magnets.

Rotary forging was also employed (8.3) as a means of making cold compacts of melt spun ribbon with soft metal. One of the attractive features of this method was that high densities (up to 95% of theoretical) could be achieved without sintering, thus avoiding uncontrollable chemical interactions between the magnetic and bonding materials. There were also indications that these compacts possess corrosion resistant properties particularly when combined with such metals as aluminium and zinc. Moreover, it was also possible under appropriate conditions to induce anisotropy in the ribbon with very significant improvements in the energy products.

Corrosion: A major problem associated with the presently available Nd-Fe-B based magnets is their poor corrosion resistance. Possible solutions are (a) the

development of an intrinsically more stable material and (b) an effective coating procedure for the magnets.
In order to study their relative corrosion behaviour, the electrochemical corrosion potentials of the magnets (both sintered and bonded types) and alloys were measured in comparison with a wide variety of materials including zinc and iron. This work (7.4) clearly showed the much greater inherent stability of single phase $Nd_2Fe_{14}B$ compared with a multiphase $Nd_{15}Fe_{77}B_8$ alloy. Studies of various coating treatments highlighted the importance of the pretreatment procedures. Hydrogen, in particular, must be avoided or removed by an appropriate annealing treatment to eliminate the danger of hydrogen embrittlement. Electrolytic and non-electrolytic means of depositing a zinc layer were examined and both processes produced effective protective layers (7.4). However, the long-term stability of Nd-Fe-B-type magnets to corrosion remains as yet unsolved.

C. APPLICATIONS

Engineering research and development was centred on the application of Nd-Fe-B magnets in various types of rotating machines, linear machines and static devices. The technical achievements can be divided broadly into:

- New and improved design and engineering capabilities;
- Design and construction of a range of electrical machines;
- Design and construction of prototypes of static devices such as hexapoles, clamping devices and magnetic resonance image scanners.

There is a general consensus that Nd-Fe-B will become the material of choice for many new permanent magnet machines.

Design capabilities: Good design is the key to efficient use of Nd-Fe-B, exploiting the excellent magnetic properties while, at the same time, minimising the corrosion susceptibility and compensating for the temperature dependence of the properties of the present generation of magnets.
Sophisticated computer-aided-design (CAD) procedures have been developed to calculate complex magnetic fields, based on both analytical techniques and finite element numerical methods (9.3, 9.4, 9.11). A wide range of electrical machines (9.1, 9.2, 9.3, 9.5, 9.8, 9.9, 9.10, 9.11) and static devices (10.1, 10.2) have been designed which exploit the superior properties of Nd-Fe-B.

Details of improved magnet specifications and properties coming from CEAM and elsewhere (9.5) have been taken into account to refine the characteristics of the prototypes. It has become clear that the design engineer has to work on "coupled problems" as indicated in figure 5.

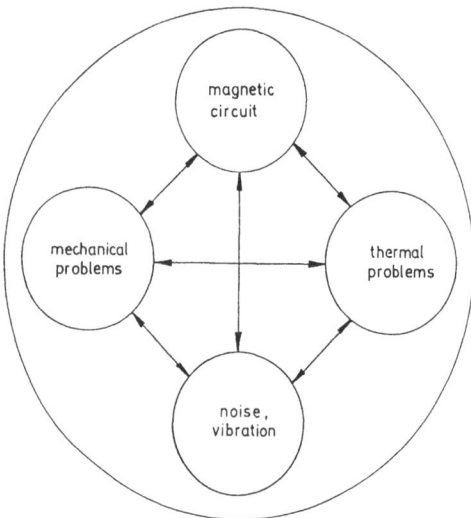

Figure 5.   Coupled problems in machine design

Engineering capabilities: The high remanence of Nd-Fe-B and
its large coercivity at room temperature makes permanent
magnet excitation particularly attractive in large electrical
machines.   However, handling and assembly of even small
magnetised parts is difficult and large magnetised components
present an added safety hazard as well as severe risk of
damage   during handling.   For   these reasons,   assembly
strategies and fixing procedures have been developed and
experience accumulated through the construction of prototypes
(9.1, 9.3, 9.7, 9.11).
The preferred strategy is to magnetise the material after
assembling the device but the high field needed to magnetise
Nd-Fe-B makes it difficult to produce the complex patterns of
magnetisation required   for some types of small motor.
Nevertheless, examples of miniature brushless-dc and stepper
motors have been constructed which demonstrate the benefit of
using the new material.   The sensitivity of the material to
high temperatures calls for increased attention to the thermal
aspects of design (9.4).   Special heat sinks may be needed
(9.2).

Small machines: In computer peripherals, such as printers and
storage devices, the trend is to small packages, low power
consumption, high efficiency and low cost.   Consequently,
drives must be designed to provide for the smallest possible
rotor inertia, mass and volume.   Moreover, there is an
increasing demand for low-noise electric motors.   A natural
solution is to use small Nd-Fe-B magnets with slotless stators
and either cup-type or disc-type air gap windings (9.1, 9.3,
9.10).   Due to the good demagnetisation characteristics of
Nd-Fe-B magnets, the flux density in the air gap remains high,
providing high torque with low noise.   However, the high
air-gap flux density can cause saturation and therefore a
finite-element approach is needed to optimise the design. An
alternative approach, is to bury the magnets in the rotor, as

in the design of a line-start synchronous motor (9.11). This novel configuration results in an air gap flux higher than in the magnet itself and virtually eliminates demagnetisation effects.

Industrial machines: There are extensive possibilities for employing Nd-Fe-B magnets in machines for industrial applications ranging from axis- and spindle-drives for machine tools to the electrodynamic braking of high-speed railway vehicles. Electronically-commutated brushless dc motors, which are likely to be a most important growth technology for European industry, featured strongly in the activities of the Applications group but detailed design studies have also been made on traditional brushed dc motors, line-start synchronous motors, stepper motors and permanent magnet generators (9.1, 9.2, 9.3, 9.5, 9.8, 9.9, 9.10, 9.11).

Nd-Fe-B is being considered for radial and axial designs of both slotted and slotless configurations. Prototype machines have been constructed, to validate, amongst other things, the thermal characteristics under both peak and continuous performance (9.1, 9.8), since currently available grades of Nd-Fe-B are restricted to operating at temperatures below 150°C. Amongst the advantages found for machines equipped with Nd-Fe-B are improved performance factors such as higher torque per frame size, improved efficiency and better dynamic response.
In some cases, motors have been evaluated in drive systems using pulse-width-modulation supply modules and incorporating torque, velocity and position control (9.6, 9.11). Figure 6 illustrates the system design approach.

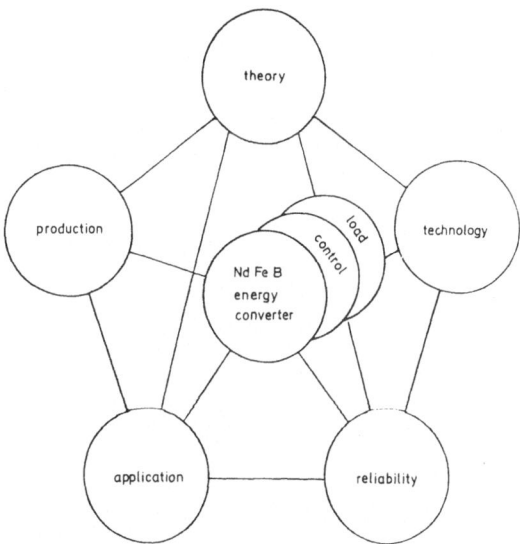

Figure 6. Systems approach to designing electrical machines

Static devices: The traditional use of permanent magnets has been in static applications. Common examples are clamping or holding devices for ferrous materials, loudspeaker and headphone elements, focussing and steering of charged particles and medical diagnosis equipment such as the magnetic resonance imaging scanner (MRI). In CEAM, several efficient permanent magnet devices have been designed to take advantage of the superior properties of Nd-Fe-B (10.1, 10.2).

A new range of workholding tools has been developed (9.2). The magnet-cost per unit-of-attractive-force for Nd-Fe-B is much better than for earlier designs based on traditional magnetic materials.

A new hexapole device has been constructed for use in an electron cyclotron resonance ion source. Here, ions are produced in a plasma which is contained in a 'magnetic bottle'. Previously, the magnetic bottle was created by a hexapolar field of permanent magnets and an axial field of electric coils. Using Nd-Fe-B magnets, stronger hexapole fields can be obtained - and denser plasmas contained - with the added advantage that the electrical coils and power supplies can be eliminated. The new powerful ion source can be easily mounted on the high voltage platforms used in nuclear physics facilities. Figure 7. shows a flux plot from the hexapole device.

Figure 7.   Flux plot for the hexapole ion source for a heavy-ion accelerator

Another noteworthy achievement of the group is the design of a permanent magnet configuration for whole-body magnetic resonance imaging (MRI) (10.2). This requires approximately 5 tonnes of Nd-Fe-B in place of 100 tonnes of ferrite. The device is designed to produce in its centre volume a very homogeneous magnetic field (uniform to 1 part in 10,000) with a strength of 0.2 Tesla. This permanent magnet-based MRI offers an attractive alternative to the more costly superconducting magnet systems presently used by hospitals to make proton resonance images for medical diagnostic examinations.

A special effect of European synergy has been brought into being by the transnational CEAM collaboration. It may serve as a model for European developments in others areas of basic research.

SECTION . III .

FINAL REPORTS OF PARTICIPANTS

SECTION III

– MATERIALS

CHAPTER 1

NEW PHASES,

STRUCTURES AND PROPERTIES

THE PREDICTION OF STRUCTURE-TYPE AND MAGNETIC PROPERTIES OF IRON-BASED ALLOYS.

D. G. Pettifor [*], P. Mohn [**], J. Staunton [***], P. Collins [****],

[*] Department of Mathematics, Imperial College of Science and Technology, London SW7 2BZ, England.
[**] Institute for Technical Electrochemistry, Technical University, Vienna, Austria.
[***] Department of Physics, University of Warwick, Coventry, CV4 7AL, England.
[****] Department of Pure and Applied Physics, Trinity College, Dublin, Ireland.

ABSTRACT

The search for new magnetic phases has centred on the rare-earth-iron pseudo-binaries with non-cubic crystal structures and the ternary borides. We show that the ordering of the known empirical structural data base within a limited number of two-dimensional structure maps helps focus the search for iron based alloys with a particular structure-type. In particular, the tetragonal binary structure-types $BaCd_{11}$ and $ThMn_{12}$ and the cubic or tetragonal $NaZn_{13}$ structure-type are located in fairly well-defined domains within a single structure map containing the A_2B_{17}, AB_{11}, AB_{12}, and AB_{13} stoichiometries, thereby suggesting alloying additions to stabilize iron-rich alloys with uniaxial structure-types. The domains of stability of ternary structure-types of transition metal-rare earth borides are displayed within a further two structure maps. The microscopic origin of the structural stability and intrinsic magnetic behaviour of transition metals and their monoborides is investigated theoretically. A simple expression for the Curie temperature is derived in terms of the magnetic moment and high field susceptibility. It accounts for the observed trends in Curie temperature of iron and cobalt alloys.

INTRODUCTION

The 1983 discovery of the new ternary phase $Nd_2Fe_{14}B$ sparked off a world-wide search for other compounds of iron which could be developed as good, cheap permanent magnets. The European effort, co-ordinated and supported through the Concerted European Action on Magnets program, has been searching for new magnetic phases amongst the rare earth-iron pseudo-binaries and the ternary borides which have uniaxial crystal structure, a prerequisite for large magnetic anisotropy and, hence, permanent magnetism. In this report we show that the ordering of the empirical structural data base of Villars and Calvert (1985) within a limited number of two-dimensional structure maps helps focus the search for iron-based alloys with a required struture-type. In particular, the tetragonal binary structure-types $BaCd_{11}$ and $ThMn_{12}$ and the cubic or tetragonal $NaZn_{13}$ structure-type are located in fairly well-defined domains within a single structure map containing the A_2B_{17}, AB_{11}, AB_{12}, and AB_{13} stoichiometries, thereby suggesting alloying additions to stabilize iron-based alloys with uniaxial structures. The domains of stability of ternary structure-types of transition metal-rare earth borides are displayed within a further two structure maps. The microscopic origin of the structural stability and intrinsic magnetic behaviour of certain transition metal systems is investigated theoretically.

STRUCTURE MAPS

Pseudo-binaries.

Numerous groups within the CEAM project have turned their attention to finding rare earth-iron-based alloys with the tetragonal structure-types $BaCd_{11}$ and $ThMn_{12}$ and the cubic or tetragonal $NaZn_{13}$ structure-type, since these structures offered the possibility of good magnetic anisotropy. However, since no binary rare-earth-iron alloys exist with these three structure types, the immediate problem facing the alloy designer was how to choose the ternary element X in order to stabilize $R(Fe,X)_{11}$, $R(Fe,X)_{12}$, $R(Fe,X)_{13}$ alloys with these particular structure types.

The choice of X is suggested by the plotting of a structure map. Recently it has been shown that the structures of all binary compounds with a given stoichiometry A_mB_n may be ordered within a two-dimensional structure map ($\mathcal{M}_A, \mathcal{M}_B$), where $\mathcal{M}$ is a purely phenomenological co-ordinate (Pettifor, 1986a,b). This so-called Mendeleev number $\mathcal{M}$ was obtained by running a string through the periodic table as shown in table 1. Pulling the ends of the string apart orders all the elements along a one-dimensional axis, their sequential order being given by $\mathcal{M}$. We must stress that the string in Table 1 runs from right to left through the rare earth series, thereby reflecting the increasing number of valence d-electrons which is known

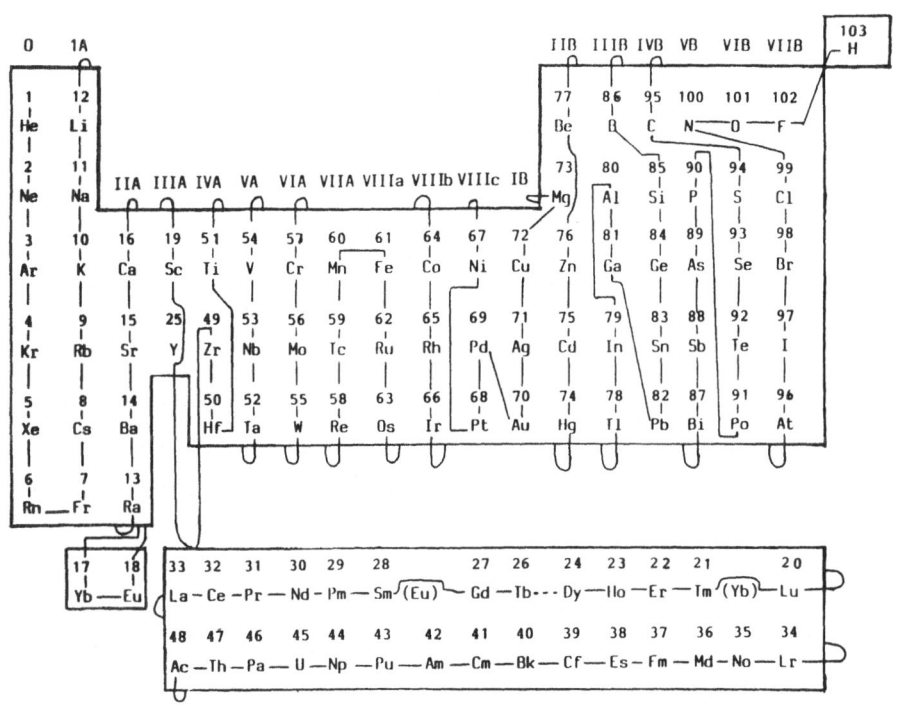

TABLE 1: The string running through this modified Periodic Table puts all the elements in sequential order, given by the Mendeleev number.

to directly affect structural stability (Duthie and Pettifor, 1977). Running the string from La to Lu in the direction of decreasing core size would lead to many examples of poor structural separation (Pettifor 1986a).

This simple procedure has been found to provide excellent structural separation for all binary compounds with the stoichiometries AB, AB_2, AB_3, AB_4, AB_5, AB_6, AB_{11}, AB_{12}, AB_{13}, A_2B_3, A_2B_5, A_2B_{17}, A_3B_4, A_3B_5, A_3B_7, A_4B_5 and A_6B_{23} respectively. Moreover, ternary or quaternary alloys $(A_xC_{1-x})_m$ $(B_yD_{1-y})_n$ with binary structure types A_mB_n may be treated as pseudo-binaries characterized by average Mendeleev numbers $\overline{\mathcal{M}}_A$ and $\overline{\mathcal{M}}_B$, where

$$\overline{\mathcal{M}}_A = x\,\mathcal{M}_A + (1-x)\,\mathcal{M}_C$$

and

$$\overline{\mathcal{M}}_B = y\,\mathcal{M}_B + (1-y)\,\mathcal{M}_D.$$

The pseudo-binaries are then found to fall into the same structural domains as the pure binaries (Pettifor 1988a). The structure maps can, therefore, be used to guide the search for new ternary and quaternary systems with a desired binary structure-type.

Fig. 1 shows the structure maps for the A_2B_{17}, AB_{11}, AB_{12} and AB_{13} stoichiometries respectively using the binary and ternary data base of Villars and Calvert (1985). The full and open symbols refer to binary and pseudo-binary systems respectively. We see that iron with a Mendeleev number of 61 takes the 2:17 stoichiometry with either the $Ni_{17}Th_2$ or the $Zn_{17}Th_2$ hexagonal structure-types. Unfortunately, these iron alloys have too low a Curie temperature for use as permanent magnets (e.g. $T_C = 324°K$ for Y_2Fe_{17}). The four structure maps in fig. 1 can be regarded as two-dimensional cross-sections through the three-dimensional space $(\overline{\mathcal{M}}_A, \overline{\mathcal{M}}_B, x)$ corresponding to the concentrations x = 0.105, 0.083, 0.077, and 0.071 of the different stoichiometries A_xB_{1-x}. These points in three-dimensional space may be projected down onto the $(\overline{\mathcal{M}}_A, \overline{\mathcal{M}}_B)$ plane over this limited range of concentration, thereby displaying all the data in fig. 1 on a <u>single</u> structure map. By assigning a

43

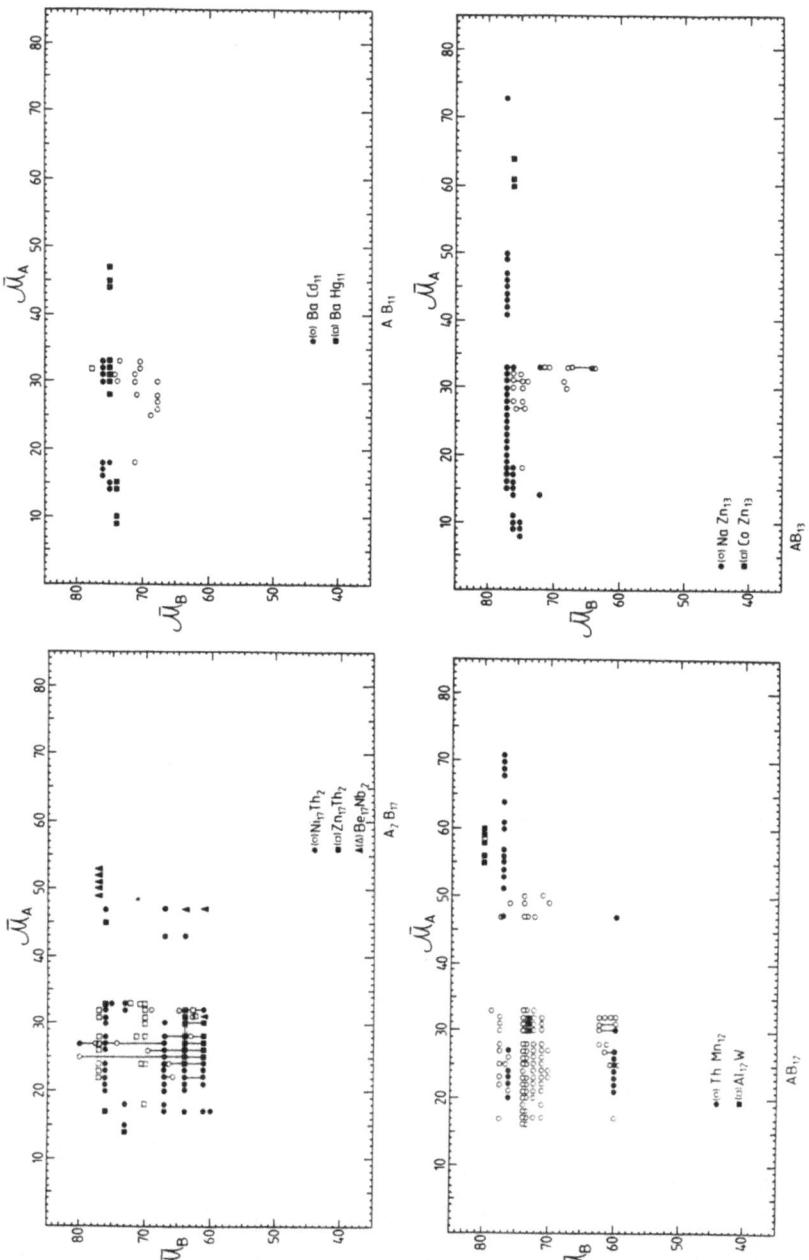

FIGURE 1: The structure maps for the A_2B_{17}, AB_{11}, AB_{12} and AB_{13} stoichiometries.

different colour to each stoichiometry, figure 5 of Pettifor (1988b) shows clearly that most (A,B) pairs choose one or other of the four stoichiometries A_2B_{17}, AB_{11}, AB_{12} or AB_{13}.

Different structure-types are located in well-defined domains, although there is some overlap between them. In particular, the large 2:17 domain is bounded from below by the $ThMn_{12}$ domain centred on $\bar{e}\bar{n}_B$ = 60(Mn) and from above by the $ThMn_{12}$ domain centred on $\bar{e}\bar{n}_B$ = 72(Cu). It is bounded on the right by the narrow vertical strip of $NaZn_{13}$ stability centred on $\bar{e}\bar{n}_A$ = 32(La). A small horizontal domain of $BaCd_{11}$ stability is centred on $\bar{e}\bar{n}_B$ = 67.5(Ni-Pt). Villars and Calvert's 1985 data-base which was used for constructing fig. 1 contains a few ternary <u>iron-based</u> alloys with 1:12 or 1:13 stoichiometry such as, for example, $Yb(Fe_{0.83}W_{0.13})_{12}$, $Y(Fe_{0.67}Mo_{0.25})_{12}$, $R(Fe_{0.33}Al_{0.67})_{12}$, $R(Fe_{0.5}Ga_{0.5})_{12}$, $La(Fe_{0.5}Al_{0.5})_{13}$, and $La(Fe_{0.88}Si_{0.12})_{13}$. The 1:11 stoichiometry can be stabilized with cobalt-rich compounds $R(Fe_xCo_{1-x})_9Si_2$ for $x \leqslant 0.2$ (Chevalier et al., 1988).

The single coloured structure map (Pettifor, 1988b) helps focus the search for new magnetic phases of iron with the AB_{11}, AB_{12}, or AB_{13} stoichiometries. If we are looking for iron-based alloys in which the <u>majority</u> of the atoms are iron, then it would be best to search in two domains: firstly, the lower $ThMn_{12}$ domain by replacing iron by transition elements to its left in the periodic table; secondly, the $NaZn_{13}$ domain by using early rare earths and replacing iron by metalloid elements to its right in the periodic table. As will be reported elsewhere at this meeting Buschow and his colleagues have done extensive work in just these areas. They have stabilized R $Fe_{12-x}M_x$ compounds with the tetgragonal $ThMn_{12}$ structure for the cases where R = Y or Gd and M ≡ Ti, V, Cr, Mo, or W, which would all lie in the lower $ThMn_{12}$ domain of fig. 1 (de Mooij and Buschow, 1987, 1988). However, they have also stabilized the $ThMn_{12}$ structure with M = Si which would fall in the 2:17 domain. Nevertheless, this appears to be exceptional since the $ThMn_{12}$ structure was not found for the other sp bonded elements Al, Ga, Ge, or Sn. These tetragonal $ThMn_{12}$-type pseudo-binaries look very promising for development as permanent magnets. In particular, $Sm(Fe_{12-x}M_x)$ compounds have Curie temperatures in the vicinity of $600^{O}K$ and an intrinsic magnetic anisotropy only somewhat less than $Nd_2Fe_{14}B$ (Buschow et al., 1988). On the other hand, the 1:13 pseudo-binaries $R(Fe_{1-x}Al_x)_{13}$ and $R(Fe_{1-x}Si_x)_{13}$ appear unsuitable for permanent magnet use because their $NaZn_{13}$ structure remains cubic rather than undergoing a tetragonal distortion (de Mooij and Buschow, 1987).

Ternaries.

Recently it has been shown that most of the non-H, C, N, or O containing phases with ternary structure-types listed by Villars and Calvert (1985) can be separated and displayed explicitly within only seven two-dimensional structure maps (Pettifor, 1988a). It quickly becomes apparent on looking at these maps that the ternary borides offer the majority of the structure-types containing iron and rare-earth constituents. This is illustrated in fig. 2 which shows the $A_{l<m}$ $B_m(Boron)_n$ structure map. Because the late transition metals with rare earths take so many different stoichiometries, a separate inset is required to display the information for compounds containing less than 33 at % boron. The discovery of the new ternary phase $Nd_2Fe_{14}B$ by General Motors Corp. and Sumitomo Special Metals Co. should, therefore, not be seen as a chance occurrence! (The known $R_2Co_{14}B$ compounds have not been included in fig. 2 because the $Nd_2Fe_{14}B$ symbol would mask that of the $Ce_2Co_7B_3$ phases listed by Villars and Calvert (1985)). The magnetic properties of alloys with the structure-types $SrNi_{12}B_6$, $Ce_2Co_7B_3$, $CeCo_4B$ and, of course, $Nd_2Fe_{14}B$ will be presented by other CEAM workers at this meeting.

MICROSCOPIC THEORY.

The search for new magnetic phases involves seeking alloys with specific crystal structures and good intrinsic magnetic properties such as Curie temperature, magnetic moment and anisotropy. In this section we discuss recent theoretical predictions of the structural stability and magnetic behaviour of certain transition metal systems.

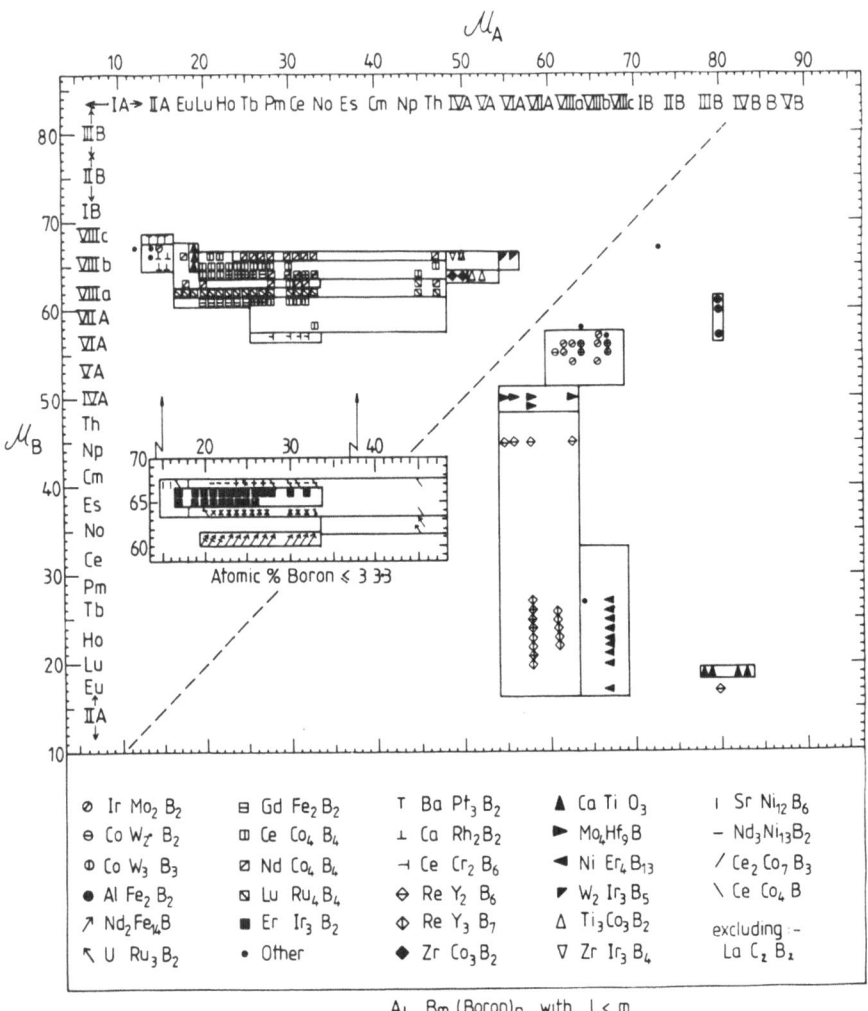

Figure 2: The $A_{l<m} \, B_m \, (Boron)_n$ structure map.

Transition Metal monoborides.

As a first step towards understanding the structural stability of $Nd_2Fe_{14}B$, the stability of the 3d-transition metal monoborides was studied by Mohn and Pettifor (1988) using Local Spin-Density Functional Theory and the Augmented Spherical Wave method. The total energy as a function of volume was calculated for the five different structure-types NaCl, CsCl, CuAu, FeB and CrB respectively and are compared in Figure 3. MnB and FeB were found to order magnetically with moments of $2.00\mu_B$ and $1.25\mu_B$ respectively, which is in good agreement with the experimental values of $1.83\mu_B$ and $1.12\mu_B$. Fig. 3 shows that the quantum mechanical calculations correctly predict the observed structural trend from CrB → FeB → CrB across the 3d series, the FeB structure-type being most stable for the binary compounds MnB, FeB, and CoB. The electronic density of states revealed the importance of the strong boron-boron bonding along the zig-zag chains in stabilizing the CrB and FeB lattices, which agrees with the simpler Tight-Binding calculations of Pettifor and Podloucky (1986). Similar bandstructure calculations have been performed for the 3d transition metal semi-borides T_2B in the Al_2Cu structure (Mohn 1988). The hyperfine field, calculated from the Fermi contact term alone, for FeB and Fe_2B were predicted to be -90 and -173 kGauss respectively in reasonable agreement with the experimental values of -110 and -248 kGauss.

Magnetic Anisotropy.

Recently progress has been made towards the first-principles calculation of the magnetic anisotropy of 3d-transition metals. Magnetic anisotropy arises from relativistic interactions, spin-orbit coupling linking the spin of an electron to the crystal field. It is thus necessary to solve the quantum mechanical problem of relativistic electrons moving in an array of metal atoms. Such a first principles approach for examining magnetic anisotropy effects has been developed which is based upon a fully relativistic spin-polarised multiple scattering framework.

48

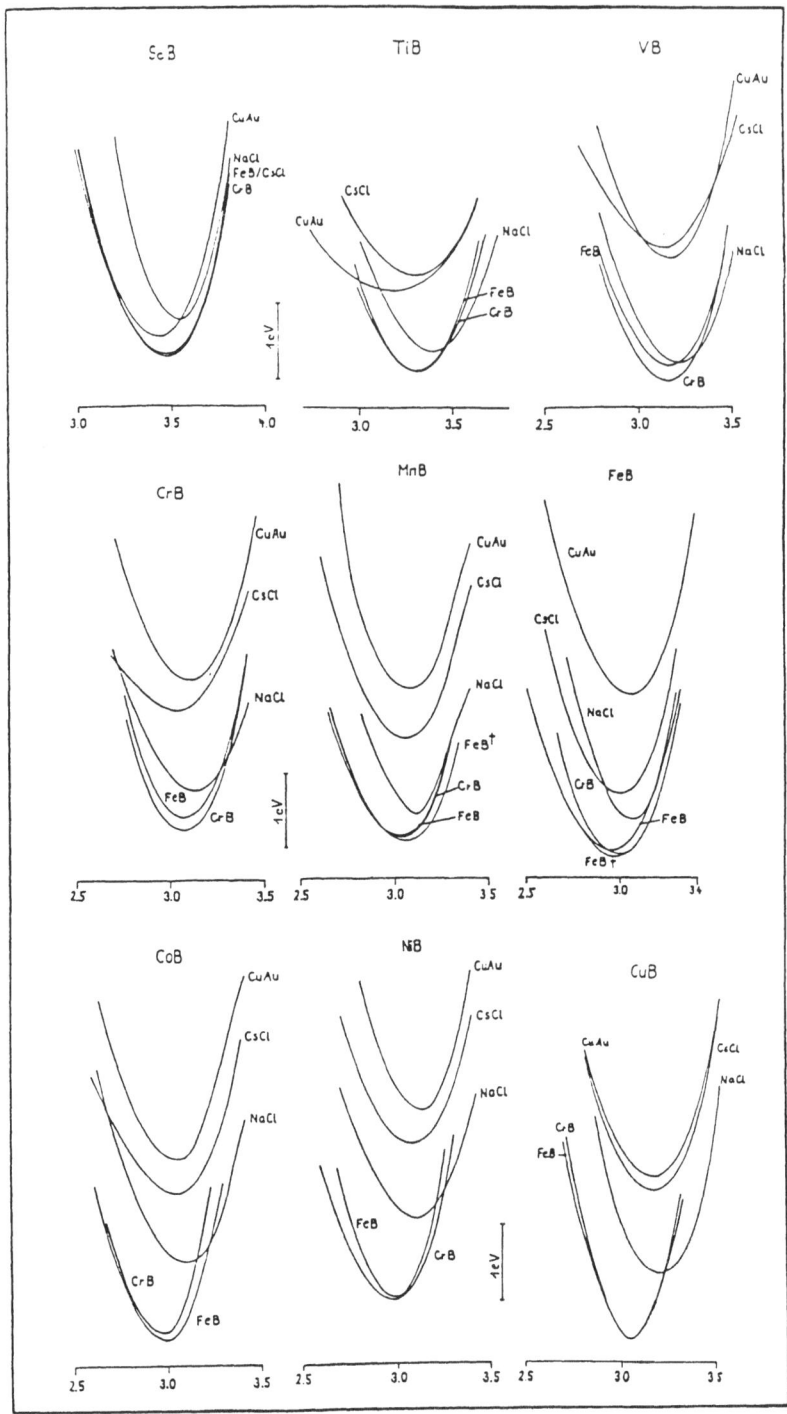

Figure 3: Comparison of the 3-d transition metal monoboride total energies for the five different structure-types NaCl, CsCl, CuAu, FeB and CrB respectively.

As a first step, the formalism has been applied to a study of the interaction between two magnetic impurities in a metallic host in which the mediating electrons are modelled by relativistic jellium, thereby providing a relativistic generalisation of the famous RKKY interaction. It not only depends depends on the usual spatially isotropic spin-spin term,but it is also a polynomial function of anisotropic pseudo-dipolar and squared Dzyaloshinski-Moriya type terms (Staunton et al., 1988).

The magnetic anisotropy energies of hcp cobalt and fcc nickel were then calculated. They were interpreted in terms of changes to the bandstructure as the orientation of the magnetization alters from one crystal axis to another (Strange et al., 1988a). Recently the magnetic anisotropy of iron atoms on a simple tetragonal latice has been computed as a function of the tetragonality c/a. A flip in the easy axis of magnetization occurs at c/a = 1.29 (Strange et al., 1988b). These calculations will be extended to more complicated tetragonal lattices in the near future.

Curie temperature.

A simple expression for the Curie temperature T_C has been derived by including the effect of spin fluctuations in the free energy (Mohn and Wohlfarth 1987). The Curie temperature satisfies

$$\frac{T_C^2}{T_{St}^2} + \frac{T_C}{T_{sf}} = 1 \tag{1}$$

where T_{St} is the Curie temperature predicted by Stoner theory and T_{sf} is the spin-fluctuation temperature

$$T_{sf} = m_0^2/(10k_B\chi_0) . \tag{2}$$

T_{sf} is given in terms of the ground state quantities, m_0 the magnetic moment and χ_0 the high-field susceptibility, which may be written

$$\chi_0^{-1} = (4\mu_B^2)^{-1} [N_\uparrow^{-1}(\epsilon_F) + N_\downarrow^{-1}(\epsilon_F)] - I/(2\mu_B^2) \tag{3}$$

where I is the Stoner exchange integral $N_\uparrow(\epsilon_F)$, $N_\downarrow(\epsilon_F)$, are the majority, minority spin density of states at the Fermi energy ϵ_F. Using the values of T_{st} and T_{sf} calculated from band theory, expression (1) leads to surprisingly accurate predictions for the Curie temperature of iron and cobalt alloys, as illustrated by table 2.

TABLE 2: Comparison between experimental and calculated values of the Curie temperature.

	T_{st}	T_{sf}	T_c(calc)	T_c(expt.)
Fe	2560	1293	1068	1043
FeB	1223	792	600	582
$Fe_{17}Y_2$	612	615	379	317
$Fe_{23}Y_6$	1312	523	458	481
Co	2240	2439	1436	1388
$Co_{17}Y_2$	2390	1706	1243	1185
Co_5Y	2081	1235	968	987
Co_7Y_2	1218	899	646	640

Rare earth-iron compounds will only have Curie temperatures approaching the $1000^\circ K$ of elemental iron once their spin fluctuation temperature T_{sf} is raised by decreasing χ_o and hence the density of states of the spin-split bands at the Fermi energy (c.f. eq. 3). This probably requires the local enivornment about the iron atoms to be closer to the relative orientation of the eight nearest and six second-nearest neighbours of the bcc lattice rather than the twelve nearest neighbours of the close-packed lattices.

CONCLUSION

We have shown that the ordering of the known experimental structural data within two-dimensional structure maps can help guide the search for new magnetic phases with particular structures. The theoretical prediction of structural stability and magnetic anisotropy of most ternary systems is still beyond the reach of first principles calculations. However, theory can provide insights into the origin of structural and magnetic behaviour, hopefully providing the experimentalist with reliable concepts for analysing the data to the best advantage.

REFERENCES

Buschow, K.H.J., de Mooij, D.B., Brouha, M., Smit, H.H.A., and Thiel, R.C. 1988. Magnetic properties of ternary Fe-rich rare earth intermetallic compounds. IEEE Transactions on Magnetics (to be published).

Chevalier, B., Gurov, G., Fournes, L., and Etourneau, J. 1988. J. Chem. Res. (to be published).

De Mooij, D.B. and Buschow, K.H.J. 1987. A new class of ferromagnetic materials: $RFe_{10}V_2$. Philips J. Res. $\underline{42}$, 246-251.

De Mooij, D.B., and Buschow, K.H.J. 1998. Some novel ternary $ThMn_{12}$-type compounds. J. Less-Common Metals $\underline{136}$, 207-215.

Duthie, J.C., and Pettifor, D.G. 1977. Correlation between d-band occupancy and crystal structure in the rare earths. Phys. Rev. Lett. $\underline{38}$, 564-567.

Mohn, P. and Pettifor D.G. 1988. The calculated electronic and structural properties of the transition-metal monoborides. J. Phys. C (in press).

Mohn, P. 1988. The calculated electronic and magnetic properties of the tetragonal transition metal semiborides. J. Phys. C (in press).

Mohn, P. and Wohlfarth, E.P. 1987. The Curie temperature of the ferromagnetic transition metals and their compounds. J. Phys. F$\underline{17}$, 2421-2430.

Pettifor, D.G. 1986a. The structures of binary compounds: I. Phenomenological structure maps. J. Phys. C19, 285–313.

Pettifor, D.G. 1986b. New alloys from the quantum engineer. New Scientist 110, No. 1510, 48–53.

Pettifor, D.G. 1988a. Structure maps for pseudo-binary and ternary phases. Mat. Sc. and Tech. (in press).

Pettifor, D.G. 1988b. Structure maps in magnetic alloy design. Physica B149 (in press).

Pettifor, D.G. and Podloucky, R. 1986. The structure of binary compounds: II. Theory of the pd-bonded AB compounds. J. Phys.C19, 315–33.

Staunton, J.B., Gyorffy, B.L., Poulter, J., and Strange, P. 1988. A relativistic RKKY interaction between two magnetic impurities – origin of a magnetic anisotropic effect. J. Phys. C (to be published).

Strange, P., Staunton, J. B., and Ebert, H. 1988b. A first principles theory of magneto-crystalline anisotropy in metals (in preparation).

Villars, P. and Calvert, L.D. 1985. Pearson's handbook of crytallographic data for intermetallic phases. Vols. 1–3. ASM Ohio.

A SURVEY OF THE EXPERIMENTAL LITERATURE ON THE B-Fe-Nd SYSTEM AND RELATED TOPICS

V.G. Rivlin* and A.P. Miodownik

Department of Materials Science and Engineering,
University of Surrey.

*Now at Fulmer Research Ltd.

ABSTRACT

A literature survey has been carried out in relation to the ferromagnetic ternary phase, $Nd_2Fe_{14}B$ (tetragonal symmetry). The question of the occurrence of similar tetragonal phases (2:14:1 phases) has been investigated for other ternary systems, insofar as the available information permits, for the series R-TM-X, where R = Rare Earth, TM = Transition Metal and X was taken from Groups IIIB to VB of the Periodic Table. For the well-documented boride and silicide ternary systems, R-Fe-B and R-Fe-Si, each series has a distinctive pattern of ternary phases. In the boride series the 2:14:1 phase occurs for most of the rare earths. Although there is some overlap between the silicides and the borides, this does not include the 2:14:1 phase which is absent from the silicide series. Consequently, Si does not appear to stabilize the symmetry associated with the 2:14:1 phase. A search of the R-Fe-Al series yielded little information but did indicate that Al would serve as a quaternary substitute for Fe and not for B in $Nd_2Fe_{14}B$. The very limited data for the R-Fe-C series did include one example of a 2:14:1 phase in the C-Fe-Gd system and what is known of the phase equilibria is briefly summarized for this system. It is predicted that the 2:14:1 phase will be found in the phase diagrams of other members of the R-Fe-C series.

A wider search of over 600 systems where TM = Mn, Fe, Co or Ni and X = Ga, In, Tl, Ge, Sn, Pb, P, As, Sb, Bi revealed no further examples of the 2:14:1 phase.

INTRODUCTION

This report summarizes the findings of a literature survey of experimental work carried out upon the B-Fe-Nd system and on allied systems. The rapid development of the subject (Mitchell (ed.), 1984, Strnat (ed.), 1985) means that most of the available information on the B-Fe-Nd system has been published in the last five years. The current position is given by a recent indexed bibliography (Capellan et al, 1986). The potential importance of magnetic phases analogous to the $BFe_{14}Nd_2$ phases has not escaped workers in this field and a whole series of rare earth analogues has now been established with the general formula $R_2Fe_{14}B$, where R = rare earth (Sinnema et al., 1984). The next question was whether a 2:14:1 phase with the tetragonal structure could be found in systems not containing boron, that is in the series R-M-X, where M is a transition metal like Mn, Fe, Co or Ni, and X is a non-metal or perhaps a metalloid.

The major problem, to which this work has been devoted, is the distribution of the crucial tetragonal symmetry in ternary alloy systems of the general formula R-M-X. As was subsequently discovered the experimental literature is too limited for firm conclusions to be drawn. However, the information available served to provide some interesting clues as to where experimental work might prove rewarding and this aspect is dealt with also.

The list of references gives standard sources consulted as part of the literature survey and also quotes references important to the development of the subject. References to individual systems would be to numerous to list here and for these the reader may wish to consult the summary report and bibliography compiled by the authors (Rivlin and Miodownik, 1986).

THE DISTRIBUTION OF THE 2:14:1 PHASE IN THE R-Fe-X SERIES

The main issue, as presented in discussions with CEAM and subsequent correspondence, was whether the element boron could be replaced by other non-metals or even by metalloids in the 2:14:1 phase. The question was to be answered, if possible, by means of a survey of the available experimental literature and without resorting to a priori generalizations. This meant that quite a wide selection of elements could be considered as candidates as a substitute for boron. In what follows the results are given for an agreed selection of candidate elements.

The initial survey was made for the element Al (following B in Group III) and for the non-metals C and Si which are also close to B in the periodic classification.

The survey was conducted by reference to the recent reviews by Chabot and Parthe, 1984 and by Rogl, 1984, (Chapters 49 and 51) and to the standard sources, (Villars and Calvert, 1985, Haughton and Prince, 1956, Prince, 1978, 1981). These were supplemented by computer searches of the literature via both the Surrey University and the NPL computer service.

For completeness searches were also made for the R-Fe-B series.

The main results of the survey are summarized in charts illustrating the type and distribution of different crystallographic phases occurring in the four series examined. These charts are shown in Figs. 1-3. Note that they refer only to the three series R-Fe-B, R-Fe-Si and R-Fe-C.

R-Fe-Al

The documentation on the series R-Fe-Al is largely confined to the study of pseudobinary solid solutions of symmetry such as the cubic and hexagonal Laves phases. There is one example of partial replacement of Fe by Al in $Nd_2Fe_{14}B$.

R-Fe-B

Fig. 1 shows that with the exception of the elements Eu and Yb a 2:14:1 phase occurs for all the rare earths in this series (Sinnema et al., 1984). In fact the 2:14:1 phase and its symmetry appear to be a characteristic of the R-Fe-B series. Similarly, certain other symmetries recur throughout most of this series. It is probable that the remaining gaps will be filled in by further experimental work but Fig. 1 gives a reasonably complete picture of the distribution of crystal symmetries in the R-Fe-B series.

Figure 1. Well – Established Ternary Phases in the Series R – Fe – B

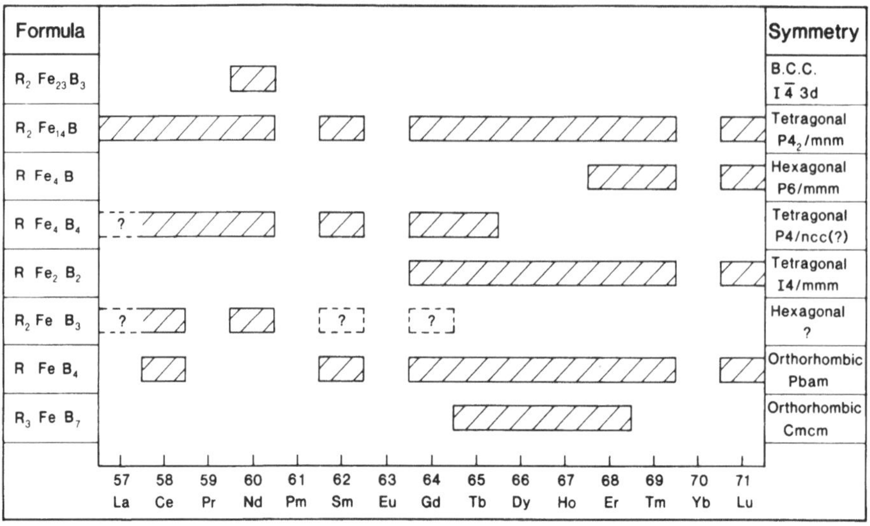

Formula	57 La	58 Ce	59 Pr	60 Nd	61 Pm	62 Sm	63 Eu	64 Gd	65 Tb	66 Dy	67 Ho	68 Er	69 Tm	70 Yb	71 Lu	Symmetry
$R_2 Fe_{23} B_3$			▨													B.C.C. $I\bar{4}3d$
$R_2 Fe_{14} B$	▨▨▨▨			▨		▨▨▨▨▨▨								▨		Tetragonal $P4_2/mnm$
$R Fe_4 B$									▨▨					▨		Hexagonal $P6/mmm$
$R Fe_4 B_4$	? ▨▨▨			▨		▨▨										Tetragonal $P4/ncc(?)$
$R Fe_2 B_2$						▨▨▨▨▨								▨		Tetragonal $I4/mmm$
$R_2 Fe B_3$	? ▨		▨		[?]		[?]									Hexagonal ?
$R Fe B_4$	▨			▨		▨▨▨▨▨▨								▨		Orthorhombic Pbam
$R_3 Fe B_7$						▨▨▨▨										Orthorhombic Cmcm

R-Fe-Si

The data are very extensive and Fig. 2 gives only the results for well-established phases. From Fig. 2 it can be seen that no representatives of the 2:14:1 phase have been found for the silicon series. It is a matter of opinion whether the tetragonal symmetry associated with this phase can be synthesised. However, it is clear that, although a great deal of information exists on the silicides, no example of the 2:14:1 symmetry has been reported. On the other hand, when one examines the other silicide phases listed in Fig. 2 a certain overlap with those listed in Fig. 1 may be seen. For instance the tetragonal symmetry, 14/mmm, occurs in Fig. 1 as RFe_2B_2 and in Fig. 2 as RFe_2Si_2. In other words, each series, R-Fe-B and R-Fe-Si, has a distinctive pattern of ternary phases with only a partial overlap between the two. The 2:14:1 phase is found in the former series but not in the latter and, to judge from the existing information, it is not stabilized by Si.

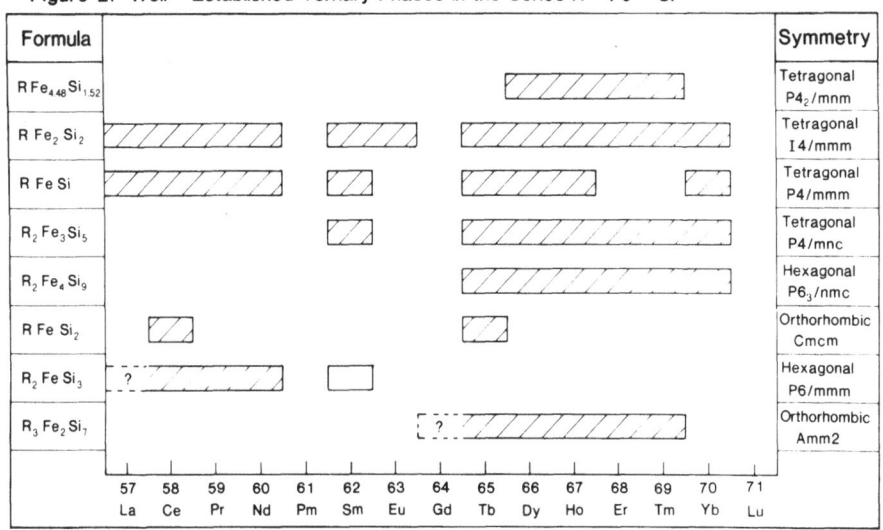

Figure 2. Well – Established Ternary Phases in the Series R – Fe – Si

R-Fe-C

Information available is very limited - only nine references were found. These do permit one to draw certain tentative conclusions. Thus, a 2:14:1 phase has indeed been reported in this series for the C-Fe-Gd system. The remaining references which deal with the crystal structure

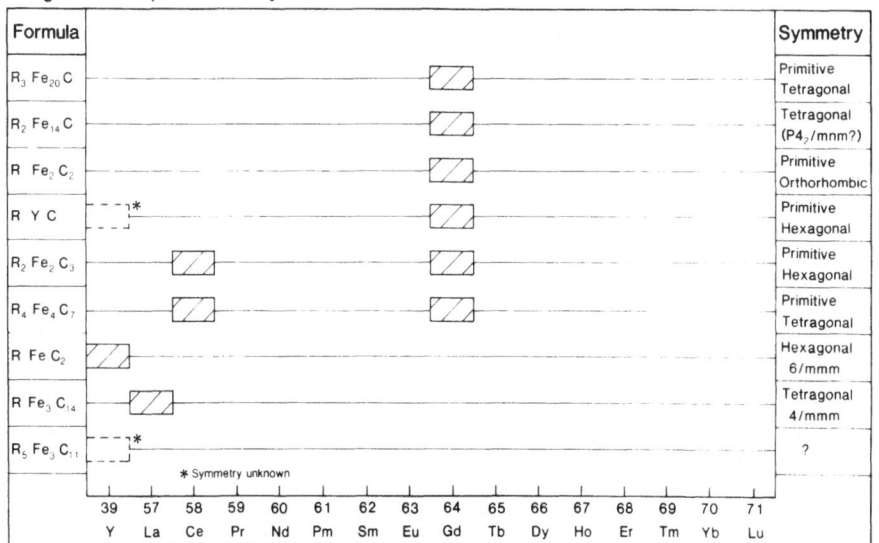

Figure 3. Reported Ternary Phases in the Series R – Fe – C

and stoichiometry of the ternary carbides are summarized in Fig. 3. It remains to be seen whether further examples of a 2:14:1 phase can be synthesized in the R-Fe-C series. The indications are that, when the gaps are filled, this will be the case.

A SURVEY OF THE SERIES R-TM-X

A wider enquiry was carried out into the state of the literature on alloys in the R-TM-X series, where R = Y, La-Lu, M = transition metal, Mn, Fe, Co or Ni and X is from Groups IIIB, IVB or VB (Ga, In, Tl, Ge, Sn, Pb, P, As, Sb or Bi). The sources searched were Haughton and Prince, 1956, Prince, 1978, 1981 and Ageev, 1959-1985. In a search of over 600 systems only 26 references were found. No further examples of a 2:14:1 phase were found. In view of the limited data the future discovery of new examples cannot be ruled out. However, individual papers cover a great deal of ground and permit one to discern tentatively some helpful trends. Thus a prominent symmetry is 14/mmm, already met with in the R-Fe-B and R-Fe-Si series. It occurs in the R-M-Ge series where M includes Fe, Co and Ni. It is also reported that in the series RMn_2Ge_2, where R = La, Ce, Pr and Nd the Curie temperature varies from 306 to 334 K. Instances of symmetries which have no parallel in the R-Fe-B and R-Fe-Si series include a Fe_2P-type structure, D m-P62m (R-M-Al, R-M-Ga and R-M-In where M = Co, Ni, Pt, Pd, Rh and Ir). Also in the series R-Ni-Sb, R-Pt-Sb and R-Ni-Bi there are 30 examples of a cubic symmetry, F43n.

DISCUSSION AND CONCLUSIONS

The literature search for information on the phase equilibria of ternary systems in the R-TM-X series has shown that information on their phase diagrams is, for the most part, very sparse. In certain cases the phase diagrams have received detailed attention because of their topical interest; the B-Fe-Nd system being the most prominent example but such instances are few. On the other hand the situation with respect to the crystal structures of ternary phases occurring in these systems is quite different. A considerable amount of information is available, particularly for the borides and the silicides. Recent reviews (Parthe and Chabot, 1984, Rogel, 1984, Sinnema et al, 1984, Villars and Calvert, 1985) have expedited the process of classifying this information for the purpose of extracting systematic trends which would be of predictive value in guiding future experimental work.

Principal conclusions are as follows:

* There is a distinctive pattern of ternary phases in each of the two series R-Fe-B and R-Fe-Si and some degree of overlap which, however, does not include 2:14:1 phase, the key phase of interest to this enquiry.

* The 2:14:1 phase is a typical phase in the R-Fe-B series.

* The 2:14:1 phase does not occur at all in the R-Fe-Si series.

* There is one known example of the 2:14:1 phase in the R-Fe-C series and it is believed that when the gaps in the information are filled more examples will be discovered in this series.

* A general survey of the R-TM-X ternary systems where X comes from periodic groups IIIB to VB has revealed no further examples of the 2:14:1 phase.

The restrictions on the formation of the 2:14:1 phase can only be accounted for in very general terms. It was at first thought quite likely that silicide examples would be found but the available evidence does not support this proposition. If one seeks reasons for the difference between silicon and boron the first parameter to assess would be the atomic size. A comparison of the candidate elements, X, reviewed in this report in terms of their atomic radii shows that most of them will differ substantially in size from boron. The actual differences will depend on the scale used for comparison (see, for example Pearson, 1972, Sutton, 1958). However, the elements closest to boron in terms of atomic radii include, as might be expected, carbon, nitrogen and oxygen followed, possibly, by phosphorus. It is suggested that a necessary condition for the formation of a 2:14:1 phase is an atomic diameter close to that of boron.

ACKNOWLEDGEMENTS

Thanks are due to Dr. T.G. Chart of the NPL for his assistance with the computer searches and to Dr. R.I. Saunderson of Fulmer Research for his interest and encouragement.

REFERENCES

Ageev, N.V.: "Diagrammy Sostoyniya Metallicheskikh Sistem", No's: 1-28, 1959-1985, VINITI, Moscow.

Capellen, J., Menzel, K.A. and Gschneider, K.A.: "Source Book on Neodynium-Iron-Boron Permanent Magnets", April1, 1986, Rep. 1s-RIC-9, Iowa, Rare Earth Information Centre.

Elliott, R.P. "Constitution of Binary Alloys, First suppl."; 1958, New York, McGraw-Hill.

Hansen, M. and Anderko, K.: "Constitution of Binary Alloys"; 1958, New York, McGraw-Hill.

Haughton, J.L. and Prince, A.: "The Constitutional Diagrams of Alloys", 2 ed; 1956, London, The Institute of Metals.

Kubaschewski, O.: "Iron-Binary Phase Diagrams", 1982, Dusseldorf, Springer-Verlag.

Mitchell, I.V. (Ed.): "Nd-Fe Permanent Magnets - Their Present and Future Applications". Report and proceedings of a Workshop meeting, Brussels 25.10.84. Commission of the European Communities. Pre-Print Version.

Parthe, E. and Chabot, B.: Ch.48 in "Handbook on the Physics and Chemistry of Rare Earths", Vol.6, Crystal structures and crystal chemistry of ternary rare earth-transition metal borides, silicides and homologues, pp.113-334, (K.A. Gschneider and K. Eyring, Eds.), 1984, Amsterdam, Elsevier.

Pearson, W.B.: "A Handbook of Lattice Spacings and Structures of Metals and Alloys", Vol.1, Vol.2, 1958, 1967, Oxford, Pergamon Press.

Pearson, W.B.: "The Crystal Chemistry and Physics of Metals and Alloys", 1972, New York, Wiley.

Prince, A., "Multicomponent Alloy Constitution Bibliography", 1955-1973; 1978, London, The Metals Society.

Prince, A., "Multicomponent Alloy Constitution Bibliography", 1974-1977; 1981, London, The Metals Society.

Rivlin, V.G. and Miodownik, A.P., 1986. Progress Report 11, Department of Materials Science and Engineering, University of Surrey (Sponsor's Reference NPL 82/0417).

Rogl, P.: Ch.49 in "Handbook on the Physics and Chemistry of Rare Earths", Vol.6, Phase equilibria in ternary and higher order systems with rare earth elements and boron, pp.335-523, (K.A. Gschneider and L. Eyring, Eds.) 1984, Amsterdam, Elsevier.

Rogl, P.: Ch.51 in "Handbook on the Physics and Chemistry of Rare Earths", Vol.7, Phase equilibria in ternary and higher order systems with rare earth elements and silicon, pp.1-264, (K.A. Gschneider and L. Eyring, Eds.), 1984, Amsterdam, Elsevier.

Shunk, F.A.: "Constitution of Binary Alloys, Second suppl."; 1969, New York, McGraw-Hill.

Sinnema, S. et al.: Magnetic properties of ternary rare-earth compounds of the type $R_2Fe_{14}B$, J.Magnetism Magn. Mater., 1984, $\underline{44}$, 333-341.

Smithells Metals Reference Book, 6th Ed., E.A. Brandes (Ed.); 1983, London, Butterworths.

Strnat, K.J. (Ed.): Proc. Eighth Int. Workshop on Rare earth Magnets and their Applications and the Fourth International Symposium on Magnetic Anisotropy and Coercivity in Rare Earth-Transition Metal Alloy, May, 1985, Dayton, Ohio, University of Dayton.

Sutton, L.E. (Ed.): "Tables of Interatomic Distances and Configurations in Molecules and Ions", 1958, Special Publication No.11, London. The Chemical Society.

Villars, P. and Calvert, L.D.: "Pearson's Handbook of Crystallographic Data for Intermetallic Phases", 1985, Vols 1.2.3; American Society for Metals, Metals Park, Ohio.

NOTE IN PROOF

Since the above report was written the author's attention has been drawn to a communication, published in 1985, by Marusin et al (1) announcing the discovery of a carbide phase, $La_2Fe_{14}C$, isotypic with the tetragonal 2:14:1 phases found in the borides. The space group is $P4_2/mnm$, $\underline{a}$ 0.8819 (2), $\underline{c}$ 1.2142 (6) nm. In the last two years K.H.J. Buschow and his group have established that a complete series of $R_2Fe_{14}C$ phases can be synthesised where R is almost any rare earth. The carbon-based phases are isotypic with their boron analogues and have very similar magnetic properties. See Buschow et al (2) for a recent communication on this topic in which it is noted that Fe can be replaced by Mn in the formula $Nd_2Fe_{14-x}Mn_xC$.

1. E.P. Marusin, O.I. Bodak, A.O. Tsokol and V.S. Fundamenskii : Kristallografiya, 1985, 30, (3) 581-583 (Chemical Abstracts, 103, 62912c).

2. K.H.J. Buschow, D.B. de Mooij and C.J.M. Denissen : J. Less-Common Metals, 1988, 142, L13-L17.

NOVEL TERNARY FE-RICH RARE EARTH INTERMETALLICS

K.H.J. Buschow and D.B. de Mooij

Philips Research Laboratories
P.O.Box 80.000, 5600 JA Eindhoven, The Netherlands

ABSTRACT

The structures and the magnetic properties of some novel ternary compounds were studied. A large number of compounds have the approximate composition $RFe_{10}T_2$, where T represents Cr, V, Ti, Mo, W or Si. The crystal structure of these latter compounds can be derived from the tetragonal $ThMn_{12}$ structure type. The crystallographic positions occupied by the different T atoms show marked differences. For 80 compounds investigated the lattice constants and Curie temperatures were determined. The invar properties of some representative compounds were investigated. Special attention was paid to the analysis of the magnetocrystalline anisotropy in these materials.

1. INTRODUCTORY SURVEY

A straightforward way of finding new ternary phases is based on binary intermetallic compounds in which at least one of the constituent elements has two or more crystallographically nonequivalent sites. Preferential occupation of one of these sites by a third element X may then lead to a ternary compound. From the viewpoint of permanent magnet applications the interest is primarily in compounds of relatively high Fe concentration which have crystal structures of sufficiently low symmetry to make uniaxial anisotropy possible. The choice then becomes rather limited and remains confined to compounds of the following structure types: Th_2Ni_{17} (or Th_2Zn_{17}) $SrNi_{12}B_6$, $CeMn_6Ni_5$, $BaCd_{11}$ and $ThMn_{12}$.

Compounds of the Th_2Ni_{17} (or Th_2Zn_{17}) type suffer from a too low Curie temperature and a too low magnetocrystalline anisotropy. Both quantities can be raised by suitable substitution of third elements into the Fe sites of R_2Fe_{17}. A prospect for using substituted R_2Fe_{17} compounds for permanent magnet materials was presented elsewhere (Buschow 1987).

Fe-base ternary compounds of the type $SrNi_{12}B_6$ do not exist. Compounds of the type $RCo_{12}B_6$ do exist but their Curie temperatures are fairly low (Rosenberg et al. 1988, Jurczyk et al. 1987). Some relevant data reported have been listed in Table 1. It may be derived from these data that substitution of Fe into the Co sites of $RCo_{12}B_6$ does not improve their magnetic properties.

There is only a limited number of Fe-base compounds of the $CeMn_6Ni_5$ and $BaCd_{11}$ types. A limited amount of data are shown in Table 1, the data of the $BaCd_{11}$ compounds being taken from the paper by Le Roy et al 1988. It can be seen that Co substitution in RFe_9Ti_2 raises T_c considerably. The value of the $CeMn_5Ni_5$ type compounds for permanent magnet application may be limited, however, owing to the need to use fairly high concentrations of the rather expensive Co.

Compounds of the tetragonal $ThMn_{12}$ structure form quite an extensive class of materials. The main purpose of the present report is to review their properties and evaluate their relevance for permanent magnet applications.

2. TERNARY COMPOUNDS WITH $ThMn_{12}$ STRUCTURE

2.1. Composition and crystal structure

Examination of numerous ternary Fe-rich alloys by means of X-ray diffraction has shown that Fe-rich compounds of the tetragonal $ThMn_{12}$ structure are formed quite generally when rare earth and Fe are combined with Ti, V, Cr, Mo, W or Si. The lattice constants of the compounds investigated are given in Table 2. The X-ray intensities observed for several representative examples were used in a structure determination employing the $ThMn_{12}$ structure type as trial structure. Some of the results are shown in the top part of Table 3. It follows from these results that the lowest value of the reliability factor R is obtained ($R = \Sigma_i |I_o - I_c| \Sigma_i I_o$ where I_o and I_c represent the observed and calculated intensities respectively) when the Si atoms share the 8j and 8f positions with the Fe atoms in $RFe_{10}Si_2$. However, Mo shows a preference for occupying the 8i position. The site occupancies in $RFe_{10}V_2$ compounds were studied by Helmholdt et al. (1988) using neutron diffraction data of $YFe_{10}V_2$, $TbFe_{10}V_2$ and $DyFe_{10}V_2$ and by Moze et al. (1988) using neutron diffraction data of $YFe_{11}Ti$. It follows from these results that Ti, V and Mo have a strong preference for occupying the 8i site.

The structure observed for the various $RFe_{10}T_2$ compounds may be analysed in terms of metallic radii and enthalpy effects. For this purpose we show in Fig. 1 schematic respresentations of the unit cell of the underlying structure. Arguments were presented earlier (Buschow 1988) to show that size considerations are not sufficient to explain the site preferences observed. Enthalpy effects have to be included. It then follows that the difference in structure between $RFe_{10}Si_2$ on the one hand and $RFe_{10}T_2$ (T = Ti, V, Mo) on the other is related to the fact that the heat of mixing between R and Si is negative and the heat of mixing between R and Ti, V or Mo is positive (Niesen et al., 1983). Since all R atoms in $RFe_{10}T_2$ are equivalent, one may easily visualize their coordination from that of the central atom in Fig. 1. It may be derived from this figure that the central R atom is in the centre of two interpenetrating tetragonal prisms. One of these prisms is due to the 8(f) sites occurring at $z = \frac{1}{4}c$ and $z = \frac{3}{4}c$, the other prism is due to the 8(j) sites occurring at $z = 0$ and $z = 1$. Each of these sites therefore gives rise to eight nearest neighbour T atoms for the R sites in RT_{12}. The nearest 8(i) site atoms are located at the same z level as the central atom and there are only four nearest 8(i) site neighbours to an R atom. This means that the 8(i) sites have by far the smallest area of contact with the R sites. In view of the positive enthalpy contribution associated with R and Ti(V,Mo) contacts one may expect therefore that the 8(i) site will be preferred by these three elements. This agrees with the corresponding structure determinations (see Table 3). The comparatively large negative heat of mixing between R and Si leads to an equally large stabilizing effect if the contact between the R and Si atoms is as large as possible. From enthalpy considerations one would therefore expect the Si atoms to prefer the 8j and 8f sites. This too agrees with our structure determination (see Table 3).

To conclude this section we note that the occurrence of $ThMn_{12}$ type compounds is not limited to the composition $RFe_{10}T_2$. Fairly extensive solid solution ranges are observed in the systems $RFe_{12-x}V_x$, where the range extends from $x \approx 1.4$ to $x \approx 3.5$. For the Si and Mo compounds the solid solution ranges are much smaller and for T = Ti and W the $ThMn_{12}$ structure is not observed for $x = 2$ but for slightly smaller x values. Fe-rich compounds of

Table 1: Lattice constants and Curie temperatures of ternary compounds with CeMn$_6$Ni$_5$, BaCd$_{11}$ or SrNi$_{12}$B$_6$ structure.

Compound	a($\overset{\circ}{A}$)	c($\overset{\circ}{A}$)	T$_c$(K)
CeMn$_6$Ni$_5$-type compounds:			
Gd Fe$_9$Ti$_2$	8.238	4.821	280
Gd Fe$_8$CoTi$_2$	8.245	4.855	330
Gd Fe$_6$Co$_3$Ti$_2$	8.226	4.857	450
BaCd$_{11}$-type compounds:			
CeFe$_{10}$SiC$_{0.5}$	10.049	6.528	390
PrFe$_{10}$SiC$_{0.5}$	10.107	6.534	430
NdFe$_{10}$SiC$_{0.5}$	10.083	6.528	410
SmFe$_{10}$SiC$_{0.5}$	10.029	6.538	460
SrNi$_{12}$B$_6$-type compounds:			
CeCo$_{12}$B$_6$	9.479	7.426	154
PrCo$_{12}$B$_6$	9.485	7.468	177
NdCo$_{12}$B$_6$	9.489	7.475	174
NdCo$_{10}$Fe$_2$B$_6$	9.506	7.484	105
NdFe$_{12}$B$_6$	9.605	7.545	230
SmCo$_{12}$B$_6$	9.470	7.458	172
GdCo$_{12}$B$_6$	9.454	7.453	166
Y Co$_{12}$B$_6$	9.443	7.435	163

Table 2: Lattice constants and Curie temperatures of several ternary compounds with ThMn$_{12}$ structure

Compound	a($\overset{\circ}{A}$)	c($\overset{\circ}{A}$)	T$_c$(K)	Compound	a($\overset{\circ}{A}$)	c($\overset{\circ}{A}$)	T$_c$(K)
Sm Fe$_{10}$Si$_2$	8.467	4.755	606	Dy Fe$_{10}$Si$_2$	8.404	4.748	566
Sm Fe$_8$Co$_2$Si$_2$	8.447	4.742	714	Ho Fe$_{10}$Si$_2$	8.390	4.749	558
Sm Fe$_7$Co$_3$Si$_2$	8.431	4.729	765	Er Fe$_{10}$Si$_2$	8.386	4.743	550
Sm Fe$_5$Co$_5$Si$_2$	8.420	4.708	850	Tm Fe$_{10}$Si$_2$	8.369	4.733	546
Gd Fe$_{10}$Si$_2$	8.437	4.757	610	Lu Fe$_{10.5}$Si$_{1.5}$	8.371	4.751	528
Tb Fe$_{10}$Si$_2$	4.415	4.747	585	Lu Fe$_{10}$Si$_2$	8.370	4.740	540

Table 2 (continued)

Compound	a($\overset{\circ}{A}$)	c($\overset{\circ}{A}$)	T_c(K)	Compound	a($\overset{\circ}{A}$)	c($\overset{\circ}{A}$)	T_c(K)
Y $Fe_{10}Si_2$	8.373	4.737	540	Nd $Fe_{10}Cr_2$	8.532	4.769	530
Y $Fe_8Co_2Si_2$	8.424	4.743	670	Sm $Fe_{10}Cr_2$	8.496	4.760	565
Y $Fe_6Co_4Si_2$	8.421	4.730	790	Gd $Fe_{10}Cr_2$	8.515	4.766	580
Y $Fe_4Co_6Si_2$	8.359	4.684	820	Tb $Fe_{10}Cr_2$	8.444	4.745	525
Y $Fe_2Co_8Si_2$	8.328	4.665	788	Dy $Fe_{10}Cr_2$	8.419	4.733	495
Ce $Fe_{10.8}Ti_{1.2}$	8.543	4.787	495	Ho $Fe_{10}Cr_2$	8.392	4.730	485
Nd $Fe_{10.8}Ti_{1.2}$	8.594	4.793	545	Er $Fe_{10}Cr_2$	8.432	4.749	475
Sm $Fe_{10.8}Ti_{1.2}$	8.561	4.792	585	Tm $Fe_{10}Cr_2$	8.416	4.732	465
Gd $Fe_{10.8}Ti_{1.2}$	8.523	4.783	600	Lu $Fe_{10}Cr_2$	8.412	4.736	450
Er $Fe_{10.8}Ti_{1.2}$	8.494	4.794	500	Y $Fe_{10.5}Cr_{1.5}$	8.442	4.745	540
Y $Fe_{10.8}Ti_{1.2}$	8.509	4.789	520	Y $Fe_{10}Cr_2$	8.463	4.756	510
Ce $Fe_{10}V_2$	8.502	4.754	440	Y $Fe_{9.3}Cr_{2.7}$	8.463	4.739	420
Nd $Fe_{10.5}V_{1.5}$	8.565	4.774	600	Y $Fe_{9.0}Cr_{3.0}$	8.445	4.751	385
Nd $Fe_{10}V_2$	8.569	4.778	570	Y $Fe_{8.5}Cr_{3.5}$	8.442	4.748	360
Sm $Fe_{10.5}V_{1.5}$	8.533	4.774	620	Y $Fe_{10}Cr$ V	8.473	4.758	515
Sm $Fe_{10}V_2$	8.537	4.772	610	Ce $Fe_{10}Mo_2$	8.567	4.786	260
Gd $Fe_{10.5}V_{1.5}$	8.524	4.778	635	Pr $Fe_{10}Mo_2$	8.634	4.808	385
Gd $Fe_{10}V_2$	8.518	4.778	61	Nd $Fe_{10.5}Mo_{1.5}$	8.590	4.791	440
Gd $Fe_{9.5}V_{2.5}$	8.518	4.776	545	Nd $Fe_{10}Mo_2$	8.606	4.798	400
Tb $Fe_{10}V_2$	8.502	4.774	570	Nd $Fe_{9.5}Mo_{2.5}$	8.612	4.809	350
Dy $Fe_{10}V_2$	8.493	4.768	540	Sm $Fe_{11}Mo$	8.566	4.778	510
Ho $Fe_{10}V_2$	8.479	4.767	525	Sm $Fe_{10.5}Mo_{1.5}$	8.572	4.788	460
Er $Fe_{10}V_2$	8.475	4.766	505	Gd $Fe_{10.5}Mo_{1.5}$	8.557	4.791	520
Tm $Fe_{10}V_2$	8.455	4.760	496	Gd $Fe_{10}Mo_2$	8.581	4.806	430
Lu $Fe_{10}V_2$	8.447	4.761	483	Y $Fe_{10.5}Mo_{1.5}$	8.513	4.783	460
Y $Fe_{10.6}V_{1.4}$	8.487	4.769	577	Y $Fe_{10}Mo_2$	8.541	4.792	360
Y $Fe_{10.4}V_{1.6}$	8.488	4.769	575	Y $Fe_{9.5}Mo_{2.5}$	8.576	4.815	300
Y $Fe_{10.2}V_{1.8}$	8.488	4.770	556	Sm $Fe_{10.5}W_{1.5}$	8.557	4.791	520
Y $Fe_{10}V_2$	8.496	4.773	540	Gd $Fe_{10.8}W_{1.2}$	8.540	4.773	570
Y $Fe_{9.8}V_{2.2}$	8.497	4.774	516	Gd $Fe_{10.5}W_{1.5}$	8.565	4.777	550
Y $Fe_{9.4}V_{2.6}$	8.500	4.775	478	Gd $Fe_{10.4}W_{0.8}Mo_{0.8}$	8.561	4.780	550
Y $Fe_{9.2}V_{2.8}$	8.506	4.778	455	Gd $Fe_{10.5}W_{0.5}Mo$	8.569	4.786	500
Y Fe_8V_4	8.529	4.796	350	Y $Fe_{10.5}W_{1.5}$	8.516	4.763	500
Th $Fe_{10}V_2$	8.583	4.775	515	Y $Fe_{10}W_2$	8.548	4.779	500

the type $GdFe_{12-x}Al_x$ can be obtained by meltspinning followed by annealing (Wang Xiang-Zhong et al. 1988).

2.2. Results of magnetic measurements

The compounds of the series $RFe_{10}V_2$ were studied in much detail by de Boer et al. (1987), using high magnetic fields and aligned powders. They found that the Fe sublattice anisotropy is about 4 T and hence of the same magnitude as in $R_2Fe_{14}B$. Similar conclusions were reached for $YFe_{11}Ti$ by Moze et al. (1988). The rare earth sublattice anisotropy is much lower in $RFe_{10}V_2$ than in $R_2Fe_{14}B$. The Curie temperatures fall into the range 483 to 616K. It may be derived from the results listed in Table 3 that similarly high magnetic ordering temperatures are encountered in the series $RFe_{10}Si_2$ and $RFe_{10.8}Ti_{1.2}$. Slightly lower values are observed in $RFe_{10}Cr_2$ while in $RFe_{10}Mo_2$ and $RFe_{10.8}W_{1.2}$ the ordering temperatures have decreased considerably. Examples of temperature dependences of the magnetization are shown in Fig. 2. When measured in low magnetic fields (full lines; H = 40 kA/m) the magnetic phase transitions are fairly sharp but they become much more gradual when measured in higher fields (broken lines; H = 1000 kA/m).

The effect of composition on the ordering temperatures of some $ThMn_{12}$ compounds is shown in Figs. 3 and 4. It is seen in Fig. 3 that the effect of substitution of nonmagnetic elements for Fe in RFe_{12} leads to a drastic reduction in T_c for Mo whereas the effect is less strong for V.

2.3. Neutron diffraction

Neutron diffraction performed on $YFe_{10}V_2$ showed that this compound orders ferromagnetically with an easy magnetization direction parallel to the c-axis. The magnetic moments observed at 4.2 K at the three Fe sites are 1.70 μ_B for the 8(i) site, 1.53 μ_B for the 8(f) site and 1.16 μ_B for the 8(j) site. Neutron diffraction measurements were also made on the compounds $TbFe_{10}V_2$ and $ErFe_{10}V_2$ (Helmholdt and Buschow 1988). At room temperature both compounds have an easy magnetization direction parallel to the c-axis. At 4.2 K the two compounds behave differently. In $TbFe_{10}V_2$ the easy magnetization direction deviates from the c-axis by 50 degrees. In $ErFe_{10}V_2$ the easy magnetization direction is perpendicular to the c-axis. These data were analysed in terms of the single ion contributions due to the magnetocrystalline anisotropy associated with the rare earth atoms. Using crystal field theory the following expressions can be derived (Lindgard and Danielsen 1975):

$$K_1 = - \frac{3}{2} \alpha_J <r^2> A_2^0 <O_2^0> + 5\beta_J <r^4> A_4^0 <O_4^0> \tag{1}$$

$$K_2 = \frac{35}{8} \beta_J <r^4> A_4^0 <O_4^0> \tag{2}$$

where O_n^m are operator equivalents tabulated by Hutchings (1964). Terms of higher order than A_4^0 are left out of account. The quantities α_J and β_J are the well-known Stevens coefficients, and $<r^2>$ and $<r^4>$ are the expectation values for the 4f radii. The crystal field

Table 3: Reliability factors (in %) obtained after refinement of the X-ray data of $LuFe_{10}Si_2$ and $GdFe_{10}Mo_2$. Mo or Si occupy the various positions specified in the first column.

Si (Mo) position	$R(LuFe_{10}Si_2)$	$R(NdFe_{10}Mo_2)$
8i	26.8	**10.0**
8j	21.9	34.2
8f	16.0	31.5
8j + 8f	**11.9**	28.7
8i + 8j + 8f	13.5	20.2

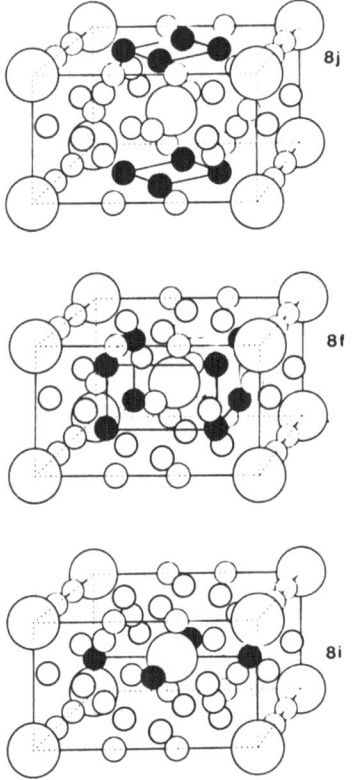

Fig. 1. Schematic representation of the $ThMn_{12}$-type unit cell, showing the three different Fe positions. Large circles: R atoms; small circles: Fe atoms.

69

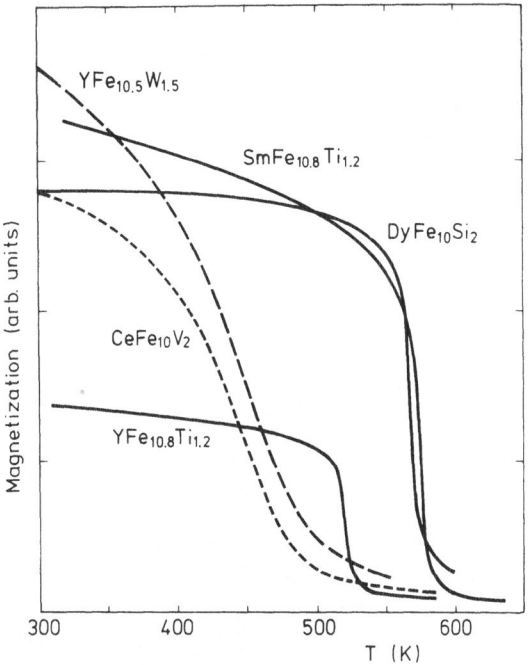

Fig. 2. Temperature dependence of the magnetization in several $RFe_{10}T_2$ compounds. Full lines H = 40kA/m; broken lines, H = 1000kA/m.

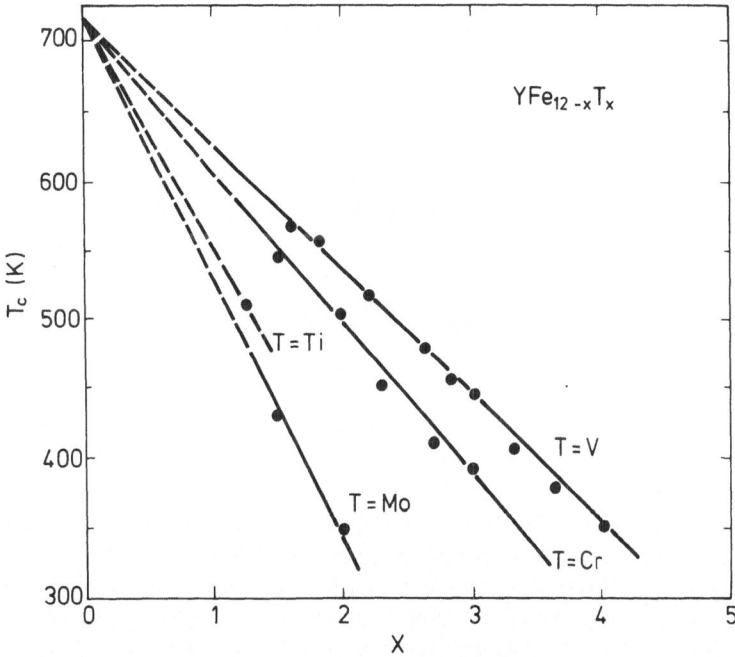

Fig. 3. Composition dependence of the Curie temperature in compounds forming part of the homogeneity region of $YFe_{12-x}T_x$.

parameters A_2^0 and A_4^0 can be regarded as being approximately constant within a given series of isostructural compounds. From the fact that $ErFe_{10}V_2$ and $TbFe_{10}V_2$ adopt magnetic structures at 4.2 K with easy magnetization directions completely different from that found in $YFe_{10}V_2$ at 4.2 K it may be concluded that at 4.2 K the rare earth sublattice anisotropy prevails. If one uses a negative value for A_2^0 (as suggested earlier, Buschow et al. 1988) in the expressions for K_1 and K_2 (eqs. 1 and 2) one finds that there exists no common value of A_4^0 for $ErFe_{10}V_2$ and $TbFe_{10}V_2$ suitable for describing simultaneously the preferred magnetization directions observed in both of these compounds at 4.2 K. Provided that the assumption $A_2^0 < 0$ is correct, this means that the description of the magnetocrystalline anisotropy in $RFe_{10}V_2$ compounds requires terms of even higher order than considered here.

2.4 Mössbauer spectroscopy

[155]Gd Mössbauer spectroscopy was performed on several ternary Gd compounds of the type $GdFe_{10}T_2$ with T = V, Mo an Si (Buschow et al. 1988). In view of the easy magnetization direction being parallel to the c-axis, and in view of the fact that there is only one single R site in the $ThMn_{12}$ type compounds, it was relatively easy to analyse the [155]Gd Mössbauer spectra made on several of these compounds. Only small deviations between fit and experiment were found and attributed to the occurrence of distributions of T atoms over the available 3d sites. The parameters derived from the corresponding fits were used to obtain an estimate of the crystal field parameters A_2^0 (derived from the V_{zz}). These values are included in Table 4. Comparison with the values listed for the $Nd_2Fe_{14}B$ type compounds shows that the A_2^0 values in $RFe_{10}T_2$ are substantially smaller, and have a negative sign. The smaller magnitude of A_2^0 indicates that the R sublattice anisotropy is expected to be smaller in $RFe_{10}T_2$ than in the corresponding $R_2Fe_{14}B$. The negative sign of A_2^0 derives from the fact that the R sublattice anisotropy was found to favour an easy magnetization direction parallel to the c-axis only in those $RFe_{10}T_2$ in which the R component has a positive second-order Stevens factor α_J. For the compounds in which R is a light rare-earth element this means that compounds with R = Sm are suitable, whereas Nd and Pr based materials are not suitable for permanent magnet purposes. Several samples of the $RFe_{10}V_2$ family were studied by Gubbens et al. (1987) using [161]Dy, [166]Er and [169]Tm Mössbauer spectroscopy. The temperature dependences of the hyperfine parameters were determined together with those derived from [57]Fe Mössbauer spectroscopy. The results were used in a mean field analysis to obtain the magnetic coupling constants between the Fe moments (J_{FeFe}) and between R and Fe moments (J_{RFe}). These values are compared with similar data pertaining to $R_2Fe_{14}B$ and R_2Fe_{17} in Table 5.

Experimental evidence was found for a spin reorientation in $DyFe_{10}V_2$, occurring near 190 K.

2.5 Invar properties

Experimental results obtained on several representatives of the series $R_2Fe_{14}C$ and $RFe_{10}T_2$ can be compared in Fig. 5. Compounds of the types $R_2Fe_{14}C$ and $R_2Fe_{14}B$ give rise to a strong anomalous thermal expansion at temperatures close to the corresponding Curie temperatures. With regard to technological applications, it may noted that a low thermal expansion is desirable in modern magnetomotive constructions. Regarding fundamental aspects of magnetism, we recal that Invar type anomalies, such as shown by $R_2Fe_{14}B$, $R_2Fe_{14}C$ and $RFe_{10}T_2$ compounds, may be interpreted in terms of local moment magnetostriction or in

Table 4: Structural and magnetic data of various ternary Fe-rich rare earth compounds and related compounds. $T_c(Y)$ is the Curie temperature for the Y compound. If no data are available for the Y compound, T_c for the Lu compound has been listed, indicated by an asterisk. $T_c(Gd)$ is the Curie temperature of the corresponding Gd compound. Two values of the second-order crystal field parameters are given for structures with two R sites

Compound	Structure	$T_c(Y)$ [K]	$T_c(Gd)$ [K]	A_2^0 [K a_o^{-2}]	3d sublattice anisotropy
$R_2Fe_{14}B$	$Nd_2Fe_{14}B$	571	660	+ 680; + 661	easy axis
$R_2Fe_{14}C$	$Nd_2Te_{14}B$	495*	620	+ 714; + 586	easy axis
$R_2Co_{14}B$	$Nd_2Fe_{14}B$	1010	1053	+ 580; + 406	easy plane
$RFe_{10}V_2$	$ThMn_{12}$	532	616	− 140	easy axis
$RFe_{10}Cr_2$	$ThMn_{12}$	515	580	−	easy axis
$RFe_{10}Mo_2$	$ThMn_{12}$	350	400	− 120	easy axis
$RFe_{10.8}W_{1.2}$	$ThMn_{12}$	500	570	−	easy axis
$RFe_{10.8}Ti_{1.2}$	$ThMn_{12}$	520	600	−	easy axis
$RFe_{10}Si_2$	$ThMn_{12}$	530	610	− 120	easy axis
RFe_9Ti_2	$CeMn_6Ni_5$	−	280	−	−
R_2Fe_{17}	Th_2Ni_{17}	324	476	− 70; + 70	easy plane
R_2Co_{17}	Th_2Ni_{17}	1186	1218	− 120; − 50	easy plane
RCo_5	$CaCu_5$	987	1014	− 698	easy axis
RFe_4B	$CeCo_4B$	573*	−	− 775; − 1221	easy axis
$RCo_{12}B_6$	$SrNi_{12}B_6$	156	163	−	−

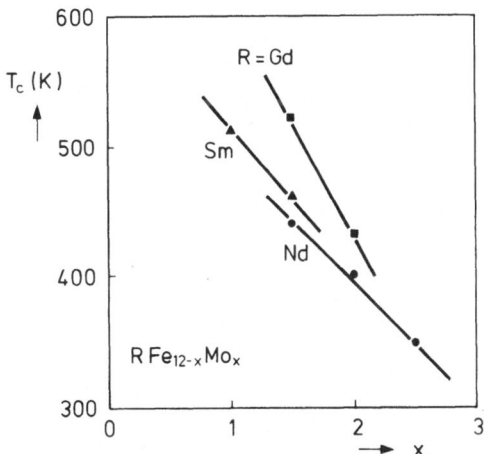

Fig. 4. Composition dependence of the Curie temperature of compounds of the type $RFe_{12-x}Mo_x$.

terms of magnetovolume effect associated with itinerant electrons. In a more detailed investigation of the thermal expansion behaviour of $R_2Fe_{14}B$ compounds made by Grössinger and Buschow (1987) in quite an extended temperature range (4.2 - 1000 K), it was found that the spontaneous volume magnetrostriction derived from our data extends to temperatures considerably above the corresponding Curie temperatures. The spontaneous volume magnetostriction at 4.2 K was obtained after subtraction of the lattice contribution. The latter was estimated from the Grüneisen function defined by a linear high-temperature slope and the Debye temperature. Values of the spontaneous volume magnetostriction range from 2.5 to 3.5% in the various compounds investigated, and consist of contributions due to the Fe sublattice and the rare earth sublattice (Grössinger and Buschow 1987). The results shown for $RFe_{10}T_2$ in Fig. 5 seem to indicate that the spontaneous volume magnetostriction is considerably lower in $RFe_{10}T_2$ than in $R_2Fe_{14}B$. This is the case in particular for compounds in which T represents Mo and W.

3. DISCUSSION

In the mean-field approximation the Curie temperature can be expressed as

$$3kT_c = a_{FeFe} + (a_{FeFe}^2 + 4a_{FeR}a_{RFe})^{1/2} \tag{3}$$

where

$$a_{FeFe} = ZJ_{FeFe}S_{Fe}(S_{Fe} + 1) \tag{4}$$

and

$$a_{RFe}a_{FeR} = Z_1Z_2S_{Fe}(S_{Fe} + 1)(g_J - 1)^2 J(J + 1) J_{RFe}^2. \tag{5}$$

From the crystal structure of the $RFe_{10}V_2$ compound it follows that each R atom has an average number of $Z_1 \approx 17$ Fe nearest neighbour atoms while each Fe atom has an average number of $Z_2 \approx 2$ R nearest neighbour atoms and an average number of $Z \approx 8$ Fe neighbours.

It follows from these equations that T_c depends in a complicated way on the de Gennes factor $G_J = (g_J - 1)^2 J(J + 1)$. By contrast, experimental results on ternary rare earth compounds are currently interpreted as behaving in accordance with a simple linear relationship between T_c and G_J (Buschow 1988; Pedziwiatr and Wallace 1986). This apparent discrepancy is based on the following approximation. Further elaboration of eqs. 3-5 shows that

$$T_c^J \cdot \Delta T_c = \text{constant} \cdot G_J \tag{6}$$

where $\Delta T_c = T_c^J - T_c^0$ is the difference in Curie temperature between a compound in which the R component is magnetic ($J \neq 0$) and a compound in which R is nonmagnetic ($J = 0$).

Equation (6) may be written in the form

$$T_c^J = \text{constant} \cdot \frac{G_J}{T_c^J} + T_c^0 . \tag{7}$$

The apparent linear behaviour between T_c^J and G_J then follows from the fact that G_J varies roughly two orders of magnitude when passing through the lanthanide series whereas T_c^J varies only little. This is true in particular for compounds in which the main contribution to T_c^J derives from the 3d sublattice. For instance, in $RFe_{10}T_2$ compounds T_c^J varies only about 10%.

When analysed by means of the full expressions (3)-(5) the Curie temperatures of the various series behave in much the same way as shown for the series $RFe_{10}Cr_2$ in Fig. 6. In this figure the full line represents the calculated data normalized to the experimental value of $GdFe_{10}Cr_2$. It is seen that experimental data are higher than the calculated values when R represents a light rare-earth element while they are lower when R represents a heavy rare-earth element. This behaviour can be interpreted as a dependence of J_{RFe} on R which tends to decrease J_{RFe} as one passes through the lanthanide series. A theoretical explanation for this was recently offered by Belorizky et al. (1987). Restricting ourselves to the J_{RFe} values for R = Gd we find by means of the mean field analysis that J_{RFe} values are slightly higher in $RFe_{10}V_2$ than in $R_2Fe_{14}B$. The same holds for the coupling constant J_{FeFe} between the Fe moments (see Table 5).

For a description of the hard magnetic properties at room temperature only the first term of eq. (1) needs be taken into consideration owing to the fact that the expectation values $<O_2^0>$ and $<O_4^0>$ decrease with temperature as m_R^3 and m_R^{10}, respectively (m_R is the reduced R magnetization). Application of the experimental values of A_2^0 derived from Mössbauer spectroscopy (Table 4) requires the introduction of a screening constant which accounts for the screening of the 4f electrons by means of the rare earth valence electrons. This leads to the expression

$$K_1 \approx -\frac{3}{2} \alpha_J <r^2> N_R A_2^0 <O_2^0> (1 - \sigma_2) . \tag{8}$$

When the A_2^0 values of the series $R_2Fe_{14}B$ are compared with those of $RFe_{10}V_2$ (Table 4) it would appear that the former materials are better by a factor of about 4. However one has to take into account that for permanent magnet purposes $Nd_2Fe_{14}B$ qualifies in the former series whereas the compounds $SmFe_{10}T_2$ qualify in the latter series. For Sm the factor $\alpha_J <r^2> <O_2^0> (1 - \sigma_2)$ is better than for Nd by a factor of about 2, so that the overall expectation for the hard magnetic properties of $SmFe_{10}T_2$ is still reasonable. Large values for the room temperature anisotropy field in $SmFe_{10}Ti$ were reported by Ohashi et al. (1988). Although these authors reported a different composition and crystal structure, it can be derived from their data that they have actually investigated the properties of $ThMn_{12}$ type ternaries. The observation of coercive forces in $ThMn_{12}$ type ternaries was made only incidentally on $SmFe_{10}Mo_2$ (Schultz 1988).

Table 5: Results of a mean field analysis of the magnetic data of various types of compounds. For the meaning of the different parameters, see text. (After Gubbens et al. 1987.)

Compound	Z	Z_1	Z_2	S_{Fe}	J_{FeFe} $(10^{-22}J)$	J_{RFe} $(10^{-22}J)$
$RFe_{10}V_2$	8	17	2	0.75	6.5	− 2.0
$R_2Fe_{14}B$	10	18	2.5	1.1	4.8	− 1.8
R_2Fe_{17}	10	19	2	2	3.2	− 1.2

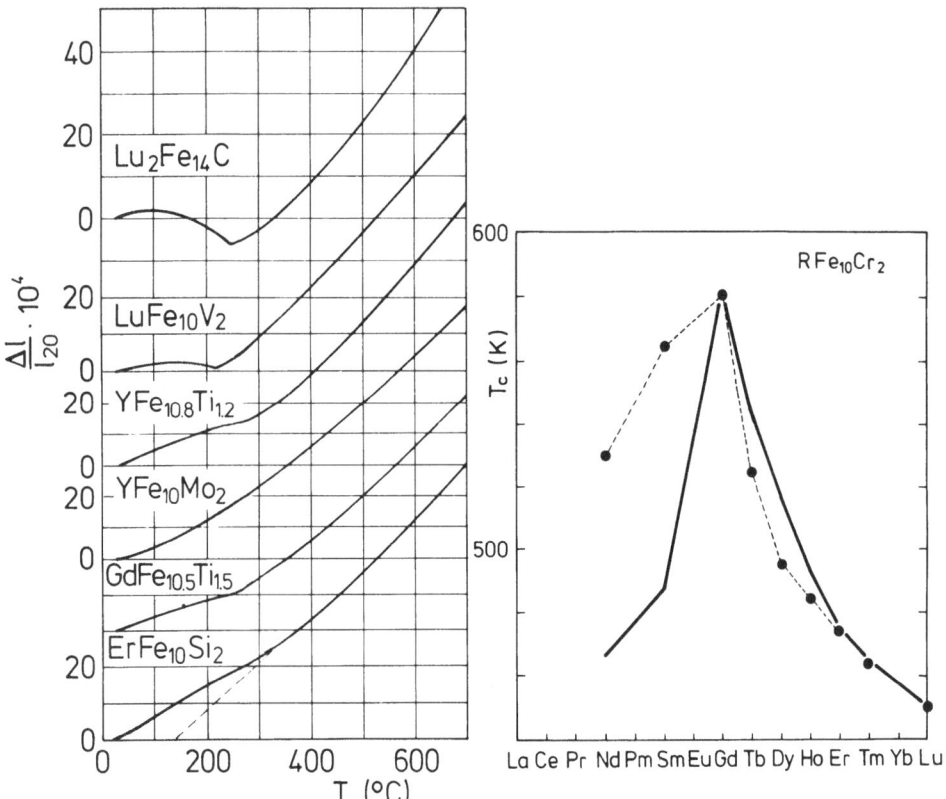

Fig. 5. Thermal expansion behaviour of several $RFe_{10}T_2$ compounds and $Lu_2Fe_{14}C$ (left).

Fig. 6. Curie tempatures of $RFe_{10}Cr_2$ compounds as a function of the R component. The full line represents the function $G_J = (g_J − 1)^2 J(J + 1)$ normalized to the value of T_c for $GdFe_{10}Cr_2$ (right).

REFERENCES

Belorizky, E., Fremy, M.A., Gavigan, J.P., Givord, D. and Li, H.S. 1987. J. Appl. Phys. **61**, 3971.

Buschow, K.H.J., de Mooij, D.B., Brouha, M., Smit, H.H.A. and Thiel, R.C. 1988 March. IEEE Trans. Mag. MAG.

Buschow, K.H.J., Proceedings of the 9th International Workshop on Rare Earth Magnets and their Application. Bad Soden, FRG, August 31 - September 2, 1987. Deutsche Physikalische Gesellschaft, Bad Honnef, FRG.

Buschow, K.H.J., 1988 April. J. Appl. Phys. **63**, 3130.

Chevalier, B., Gurov, G., Fournes, L. and Etourneau, J., J. Chem. Res. (to be published).

de Boer, F.R., Ying-kai Huang, de Mooij, D.B. and Buschow, K.H.J., 1987. J. Less-Common Met. **135**, 199.

de Mooij, D.B. and Buschow, K.H.J., 1988. J. Less-Common Met. **136**, 207.

Grössinger, R. and Buschow, K.H.J., 1987. J. Less-Common Met **135**, 39.

Gubbens, P.C.M., van der Kraan, A.M. and Buschow, K.H.J., 1987. Proc. 5th Intl. Symposium on Magnetic Anisotropy and Coercivity in Rare Earth Transition Metal Alloys, Bad Soden. Deutsche Physikalische Gesellschaft, Bad Honnef, FRG.

Helmholdt, R.B., Vleggaar, J.J.M. and Buschow, K.H.J., 1988. J. Less-Common Met., **138**, L11.

Helmholdt, R.B. and Buschow, K.H.J. J. Less-Common Met. (to be published).

Hutchings, M.T. 1964. Solid State Phys. **16**, 227.

Jurczyk, M., Pedziwiatr, A.T., Sankar, S.G. and Wallace, W.E. (1987). J. Magn. Magn. Mater **68**, 257.

Le Roy, J., Moreau, J.M., Bertrand, C. and Fremy, M.A., (1988). J. Less Comm. Met **136**, 19.

Lindgard, P.A. and Danielsen, O'. 1975. Phys. Rev. **B11**, 351.

Moze, O., Pareti, L., Solzi, M. and David, W.I.F. (1988). Solid State Comm. (to be published).

Niessen, A.K., de Boer, F.R., Boom, R., de Châtel, P.F., Mattens, W.C.M. and Miedema, A.R. 1983. Calphad **7**, 51.

Ohashi, K., Yokoyama, T., Osugi, R. and Tawara, Y., IEEE Trans. Magn. (to be published).

Pedziwiatr, A.T. and Wallace, W.E. 1986. J. Less-Comm. Met. **126**, 41.

Rosenberg, M., Mittag, M. and Buschow, K.H.J., April 1988. J. Appl., Phys.

Schultz, L., (private communication).

Wang, Xiang-Zhong, Chevalier, B., Berlureau, T., Etourneau, J., Coey, J.M.D., and Cadogan, J.M. (1988). J. Less Common Met. **138**, 235.

INTRINSIC MAGNETIC PROPERTIES of IRON-RICH COMPOUNDS

WITH THE $Nd_2Fe_{14}B$ or $ThMn_{12}$ STRUCTURE

J. M. D. Coey, H. S. Li,[+] J. P. Gavigan,[+] J. M. Cadogan and B. P. Hu,

Department of Pure and Applied Physics, Trinity College,
Dublin 2, Ireland

Intrinsic magnetic properties including magnetic ordering temperatures, spin reorientation transitions, transition metal moments, ^{57}Fe hyperfine fields, anisotropy fields and magnetization curves in high applied fields are compared for the $R_2Fe_{14}B$ and $R(Fe_{11}Ti)$ series of compounds. Exchange interactions, transition-metal anisotropy and rare-earth crystal field parameters are deduced for the two series. Single-crystal magnetization curves of eight $R_2Fe_{14}B$ compounds and $Nd_2Co_{14}B$ are included in the analysis, as well as data on some solid solution series. At room temperature, K_1 is similar in $Sm(Fe_{11}Ti)$ and $Nd_2Fe_{14}B$ but it may be increased by rare earth substitution only in the latter case. The theoretical upper limit on energy product $\mu_0 M_s^2/4$ is 259 kJ m^{-3} for $Sm(Fe_{11}Ti)$, half that of $Nd_2Fe_{14}B$, mainly on account of the lower iron moment in the $ThMn_{12}$ structure.

INTRODUCTION

The intrinsic magnetic properties of a rare-earth transition-metal intermetallic compound set limits to the performance that can be achieved by any permanent magnet made from the material. The intrinsic coercivity $_MH_c$ will be less than the anisotropy field B_a/μ_0, the coercivity on the B:H loop, $_BH_c$, cannot exceed the spontaneous magnetization M_s, and the energy product $(BH)_{max}$ will be less than $\mu_0 M_s^2/4$.

Much of our work in CEAM has been aimed at achieving a consistent understanding of the intrinsic magnetic properties of compounds with the $Nd_2Fe_{14}B$ or $ThMn_{12}$ structures. This systematic investigation underlines the potential of rare-earth iron intermetallic compounds [1,2] and guides the choice of chemical substitutions to modify their properties. Our investigations have included measurements in high magnetic fields, made in Grenoble in close collaboration with D. Givord and other members of group 1.01. Single crystals of several members of the $R_2Fe_{14}B$ series (R = Pr, Nd, Gd and Dy) were grown there by the Czochralski method. Other CEAM connections included those with L. Pareti (1.24, Parma), D. Fruchart (1.09, Grenoble), A. Yaouanc (1.11, Grenoble) and J. Etourneau (1.06, Bordeaux).

The crystal structures of $Nd_2Fe_{14}B$ [3] (or $La_2Fe_{14}C$ [4]) and $ThMn_{12}$ are illustrated in the first two figures, together with the magnetic structures as a function of temperature, showing reorientations of the magnetization

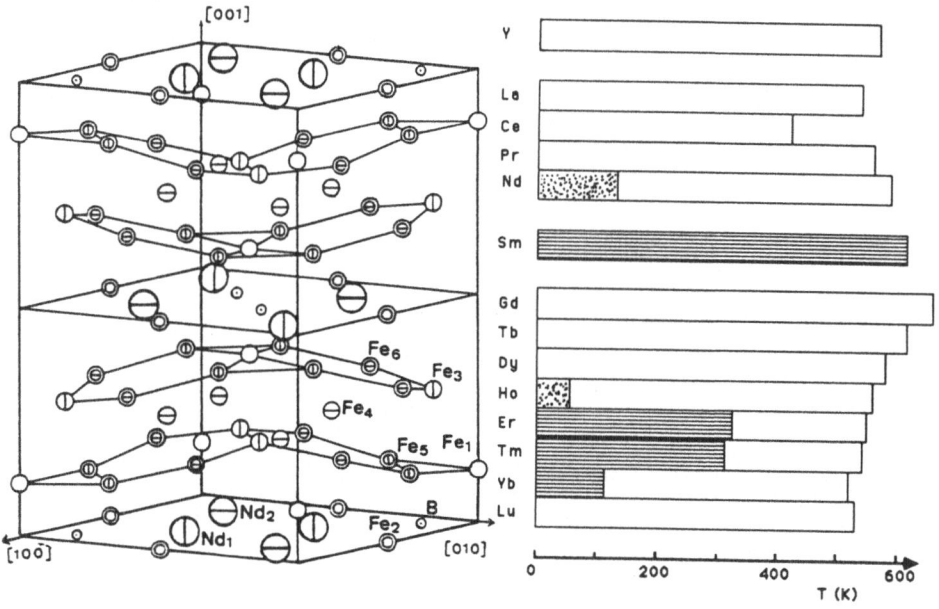

Figure 1 Crystal structure (a) and magnetism (b) of the $R_2Fe_{14}B$ series.

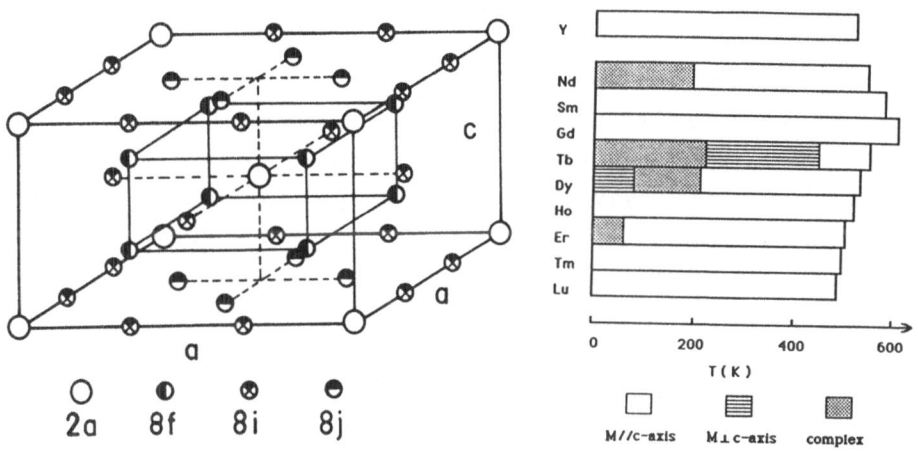

Figure 2 Crystal structure (a) and magnetism (b) of the $R(Fe_{11}Ti)$ series.

direction from c-axis to the plane (T_{sr}) or to some intermediate orientation (T_{st}). We will explain these transitions, and those occurring as a function of applied field. Note that in $Nd_2Fe_{14}B$ there are two rare earth (R) sites (4f and 4g), each with *mm* symmetry, and six transition metal (T) sites, 16 k_1, 16

k_2, $8j_1$, $8j_2$, 4c and 4e, whereas in $ThMn_{12}$ there is a single 2a R-site with symmetry $4/mmm$, and three T-sites 8f, 8i and 8j. Titanium substitutes preferentially on 8i sites [5]. The point symmetry of the rare-earth site fixes the terms present in the expression for the rare-earth crystal field, the most important source of anisotropy in these compounds.

2. MODELS

2.1 <u>Exchange interactions</u>

A two sublattice mean field model is used to treat the exchange interactions between unpaired spins of the magnetic constituents. The exchange interactions are represented by effective fields acting on the T or R sublattices.

$$B_{ex}(T) = n_{TT} M_T + n_{RT}\gamma M_R \qquad (1a)$$
$$B_{ex}(R) = n_{RT}\gamma M_T + n_{RR}\gamma^2 M_R \qquad (1b)$$

where M_i is the sublattice magnetization (in $J.T^{-1}m^{-3}$ or $A.m^{-1}$), n_{ij} are the mean field coefficients (with the same dimensions as μ_0) and $\gamma = 2(g-1)/g$ is the ratio of spin moment to total rare earth moment where g is the Landé g-factor. It turns out that $n_{TT} > n_{RT} \gg n_{RR}$, so the R-R interaction can be neglected.

When the rare-earth is nonmagnetic (Y, La, Lu, Th), the Curie point T_c is given in terms of n_{TT} as

$$T_c = n_{TT}C_T, \text{ with } C_T = N_T 4S^*(S^*+1)\mu_B^2/3k \qquad (2)$$

where N_T is the number of atoms per m^3, S^* is the effective spin of the transition metal, which is assumed to have no orbital moment, and μ_B is the Bohr magneton. S^* may be deduced either from the Curie-Weiss susceptibility above T_c, or by setting $<\mu_T>$, the average atomic moment at zero temperature, equal to $2S^*\mu_B$. Here we adopt the latter convention.

When R is magnetic, the theory gives

$$T_c = (1/2)\{[T_T + T_R + [(T_T - T_R)^2 + 4T_{RT}^2]^{1/2}\} \qquad (3)$$

where $T_T = n_{TT}C_T$; $T_R = n_{RR}\gamma^2 C_R$; $T_{RT} = n_{RT}|\gamma|(C_R C_T)^{1/2}$. Here C_R is defined in terms of the rare-earth quantum number J as $N_R g^2 J(J+1)\mu_B^2/3k$. When T_R is negligible, the expression for n_{RT} reduces to

$$n_{RT} = (1/|\gamma|) [T_c(T_c-T_T)/C_R C_T]^{1/2} \qquad (4)$$

Alternatively, n_{RT} can be deduced from the temperature variation of the total magnetization. On the basis of collinear ferrimagnetic configurations of R and T spins for the heavy rare earths and ferromagnetic configurations for the light rare earths, $M_R(T)$ is deduced from the temperature variation of the total magnetization, taking $M_T(T/T_c)$ to be that of the Y, La or Lu compound. The rare-earth sublattice magnetization may be calculated in terms of the Brillouin function $B_J(x)$ and n_{RT} (Eq 1b).

$$M_R(T) = M_R(0) \, B_J(gJ\mu_B B_{ex}(R)/kT) \qquad (5)$$

n_{RT} is chosen to give the best agreement between experiment and theory. No account was taken of crystal field interactions at this stage.

2.2   Crystal field interactions

In order to calculate changes in magnetic structure as a function of temperature or applied field B_0, we use two coupled equations for the R and T sublattices which include terms to represent the magnetocrystalline anisotropy [6]. A phenomenological term $K_{1T}\sin^2\theta$ is used for the T-sublattice, where θ is the angle between $<M_T>$ and c; this has been shown to be adequate to describe the anisotropy in $Y_2Fe_{14}B$ [7]. For the rare-earth, the crystal field Hamiltonian for a single J multiplet is $\Sigma B_{nm}O_{nm}$ where B_{nm} are the crystal field parameters for a given rare earth and O_{nm} are the Stevens operators.

$$E_T = K_{1T}\sin^2\theta + (B_{ex}(T) - B_0).<M_T> \qquad (6a)$$
$$H_R = \Sigma B_{nm}O_{nm} + (B_{ex}(R) - B_0).\mu_R \qquad (6b)$$

The latter equation is for a single site i where the rare earth moment is μ_R. It yields the free energy from the partition function, $F_i = -kT\ln Z_i$, and the total energy for the system

$$E_{tot} = E_T + \Sigma F_i - B_{ex}(T).<M_T> \qquad (7).$$

For a given set of parameters $\{B_{nm}, n_{RT}\}$, the energy can be minimized as a function of (θ,ϕ), the orientation of $<M_T>$, to yield the magnetization curve for fixed values of B_0 (fig 3a) or the temperature variation of the magnetic structure.

In practice it proved more efficient to fit magnetization curves by first parameterizing them in terms of the effective anisotropy constants $\{K_i\}$ which appear in the expression appropriate for tetragonal symmetry

$$E_a = K_1\sin^2\theta + (K_2 + K_2{}'\cos4\phi)\sin^4\theta + (K_3 + K_3{}'\cos4\phi)\sin^6\theta \quad (8)$$

and then calculating the constants $\{K_i{}^{calc}\}$ associated with a trial set of crystal field parameters $\{B_{nm}\}$. Minimizing the difference between the sets of K_i leads to an acceptable set of B_{nm}. The experimental anisotropy constants $\{K_i\}$ are deduced from magnetization curves measured in hard directions. The total energy is now

$$E_{tot} = E_a - M_s.B_0 \qquad (9)$$

where M_s is the spontaneous magnetization assumed to be independent of θ, ϕ and B_0. Minimizing this gives

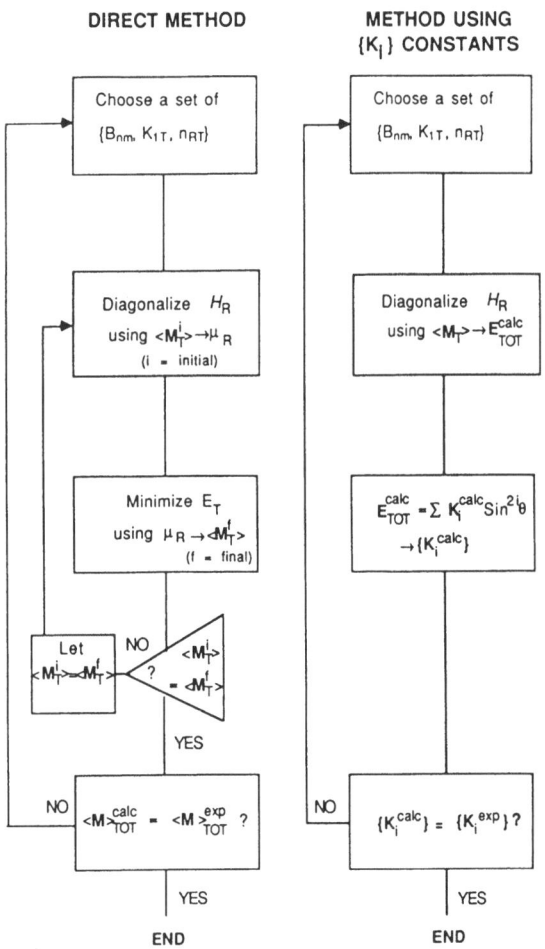

DIRECT METHOD

METHOD USING
$\{K_I\}$ CONSTANTS

Figure 3   Procedures for calculating the magnetization curves in terms of the model: a) directly and b) in terms of effective anisotropy constants $\{Ki\}$.

$$M_S B_0/2\sigma = K_1 + 2(K_2 \pm K_2')\sigma^2 + 3(K_3 \pm K_3')\sigma^4 \qquad (10)$$

where $\sigma = M/M_S$ is the reduced magnetization. The calculated anisotropy constants are obtained by working out E_{tot} from Eq. 7 for five appropriate orientations of $<M_T>$, and then obtaining five simultaneous equations involving $\{K_i^{calc}\}$ from Eq. 9. Since the 'constants' $\{K_i\}$ are purely fitting parameters which have no physical significance because the magnetic structure changes as a function of applied field [8] it is important to choose configurations for calculating $\{K_i^{calc}\}$ that closely resemble the real situation. As a check, the magnetization curve is recalculated from $\{B_{nm}\}$.

3. NONMAGNETIC RARE EARTHS

a) $R_2Fe_{14}B$ Table 1 lists Curie temperatures, magnetization, anisotropy and hyperfine field data for $T \approx 0$, and the derived exchange paramaters. Data are taken from our own work [9-11] and the papers cited in references 1. The average iron moment $<\mu_{Fe}>$ derived from magnetization curves at 4.2 K and the average ^{57}Fe hyperfine field $<B_{hf}>$ derived from Mössbauer spectra at 4.2 K are approximately proportional, with a conversion factor of 15.6 T/μ_B for the 4f series. The iron moment decreases with the lanthanide contraction (La $\rightarrow$ Lu), although T_c is practically unchanged. Following the sequence Y, La, Th, the average iron moment remains practically the same, but T_c decreases with increasing 3d-nd mixing. (An attempt to make $Sc_2Fe_{14}B$ was unsucessful). The R element does however have an influence on the moment of the 4c iron with which it shares the basal plane. That this moment decreases in the sequence Y, La, Th is best seen from the transfered hyperfine field on ^{11}B [12]. Iron on 4c sites provides the exchange bridge between the hexagonal double layers in the structure.

The proportionality between average iron moments and hyperfine fields leads us to deduce the atomic moments on different iron sites from fits to the

Table 1. Magnetic Properties of Compounds with Nonmagnetic Rare Earths

	T_c (K)	m (μ_B/fu)	$<\mu>$ (μ_B)	$<B_{hf}>$ (T)	n_{TT} (μ_o)	$B_{ex}(T)$ (T)	B_a (T)
$Y_2Fe_{14}B$	568	30.5	2.18	34.1	394	606	(1.30)
$La_2Fe_{14}B$	543	31.0	2.21	33.6	386	577	(1.1)
$Lu_2Fe_{14}B$	538	28.2	2.01	31.2	411	600	(1.9)
$Ce_2Fe_{14}B$	427	30.0	2.14	31.9	306	461	(2.7)
$Th_2Fe_{14}B$	480	30.3	2.16	(31.0)	338	526	(2.6)
$Y_2Co_{14}B$	1015	19.4	1.39		1219	1273	
$La_2Co_{14}B$	955	19.7	1.41		1267	1289	
$Y(Fe_{11}Ti)$	524	19.0	1.73	(27.0)	486	622	(4.0)
$Lu(Fe_{11}Ti)$	488	(18.2)	(1.65)	(25.8)	484	595	(4.5)

Measurements at 4.2K, except values in brackets that are extrapolated from higher temperatures

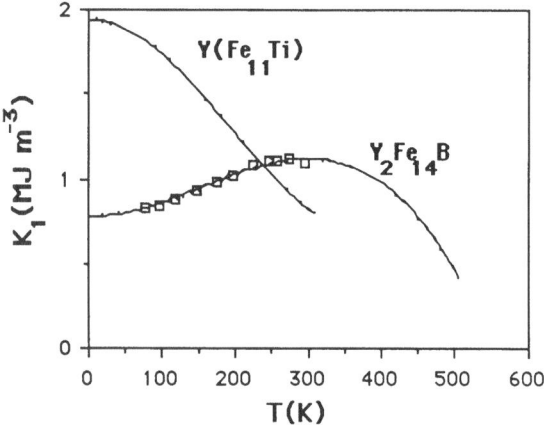

Figure 4 Temperature variation of the iron sublattice anisotropy constant K_1 in the $Y_2Fe_{14}B$ (points from Ref. 11, solid line from Ref. 23) and $Y(Fe_{11}Ti)$ (solid line from Ref. 10).

Mössbauer spectra. The weak components from 4c and 4e sites are difficult to resolve, but iron moments on the $16k_1$, $16k_2$, $8j_1$, $8j_2$, 4c and 4e sites in $Y_2Fe_{14}B$ at 4.2 K deduced in this way are 2.1, 2.2, 2.3, 2.4, 1.9 and 2.3 μ_B, respectively [9]. The hyperfine fields are slightly anisotropic particularly at 16k and $8j_2$ sites, reflecting a small orbital contribution to the iron moments ($\leq 0.1\mu_B$) that is responsible for the 3d anisotropy.

The anisotropy constant K_{1T} determined by Givord et al. from measurements as a function of temperature on an $Y_2Fe_{14}B$ crystal [7] are shown in Fig. 4. Values of the anisotropy field at $T\approx0$ included in Table 1, are related to K_{1T} by $B_a = 2K_{1T}/M_S$. The unusual temperature dependence of K_{1T} may be related to the giant volume magnetostriction ω_S = 2.9%[13]. Analysis of the pressure derivative of M_S, $(1/M_S)(dM_S/dP)$ = -0.02 GPa^{-1}, leads to the conclusion that much of the pressure dependence of T_c, $(1/T_c)(dT_c/dP)$ = -0.05 GPa^{-1} [14]. is due to the reduction in iron moment, although part of it may be attributable to a positive volume derivative of the exchange interactions,(dn_{TT}/dV). A similar analysis has been made of the effect of hydrogen, which dilates the lattice and increases the iron moment [15]. The pressure dependences of iron moments and exchange interactions are undoubtedly anisotropic, varying not just with volume but with the c/a ratio.

b) $R_2Co_{14}B$ The cobalt compounds have very high Curie temperatures and an average transition-metal moment of 1.4 μ_B that is rather insensitive to volume; there is no anomalous volume magnetostriction [16]. The main differences between the iron and cobalt magnetism may be understood if the

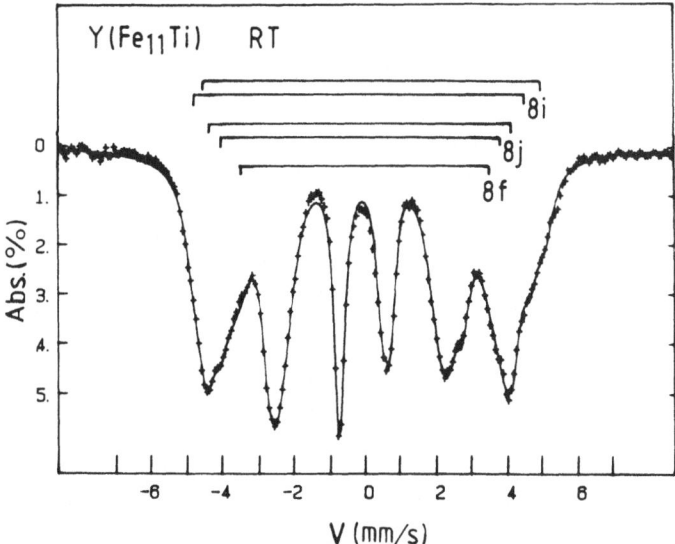

Figure 5   Mössbauer spectrum of $Y(Fe_{11}Ti)$ at room temperature , showing the components of the fit.

iron compounds are weak ferromagnets (both ↑ and ↓ 3d electrons at the Fermi level) whereas the cobalt compounds are strong ferromagnets, with a full ↑ 3d band. The sign of K_{1T}, opposite to that of iron, has been discussed in terms of a localized model for the 3d shell [17]. The cobalt has hard c-axis anisotropy.

c) R(Fe₁₁Ti)   Curie temperatures of the iron-rich $ThMn_{12}$ - structure compounds are comparable to those of $R_2Fe_{14}B$ [10,18], but the iron moments are significantly smaller (table 1). Average iron moments deduced from magnetization are 1.7 μ_B, and the conversion factor from hyperfine field is again 15.6 T/μ_B. Fitting the Mössbauer spectrum is even more complicated than it was for $Y_2Fe_{14}B$ because random substitution of Ti on 8i sites [5] leads to environments for iron on each of the three transition-metal sites with different numbers of Fe and Ti nearest-neighbours.   A fit with two components for each of the three sites and a 4:3:4 intensity constraint (figure 5) leads to hyperfine structures with $<B_{hf}>$ of 23.4 T, 30.8 T and 27.3 T at 80 K which are attributed to 8f, 8i and 8j sites respectively.   The iron has easy c-axis anisotropy in $R(Fe_{11}Ti)$, and the anisotropy constant K_1 deduced from measurements of the magnetization curves of oriented polycrystalline material with the field applied perpendicuular to the orientation direction [19] are shown in figure 4. Anisotropy fields at T≈0 are included in table 1.

4. MAGNETIC RARE EARTHS

4.1 Exchange interactions

The R-T exchange interactions may be deduced from T_C or from a fit to the temperature variation of the total magnetization. Values obtained by the two methods are usually in quite good agreement. Data on the Gd compounds are presented in table 2. Note that the exchange field acting on the R-sublattice is much the same in $Gd_2Fe_{14}B$ and $Gd_2Co_{14}B$. While cobalt substitution in the iron compounds sharply increases T_C it will do nothing to increase the R-sublattice magnetization at ambient temperature (and hence the intrinsic rare-earth anisotropy).

The variation of n_{RT} across the $R_2Fe_{14}B$ and $R(Fe_{11}Ti)$ series is plotted in figure 6. The systematic decrease by a factor two or more from the beginning to the end of the rare-earth series has been observed for other rare-earth intermetallic compounds and it is discussed by Belorizky *et al.* in terms of 4f-5d overlap [20].

4.2 Magnetization and crystal field.

It is useful to distinguish three regions of temperature where different

Table 2. Magnetic Properties of Gadolinium Compounds at 4.2K

	T_C (K)	m (μ_B/fu)	n_{RT} (μ_0)	$B_{ex}(R)$ (T)
$Gd_2Fe_{14}B$	665	17.8	215	348
$Gd_2Co_{14}B$	1050	6.9	326	352
$Gd(Fe_{11}Ti)$	607	(12.9)	250	341

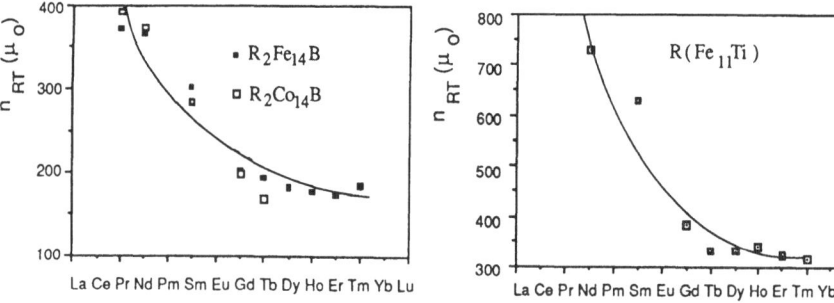

Figure 6   Variation of n_{RT} deduced from T_C of a) $R_2T_{14}B$ and b) $R(Fe_{11}Ti)$.

terms may dominate the magnetocrystalline anisotropy. From figure 4, it is clear that the iron sublattice anisotropy falls off only in the vicinity of the Curie point. The terms in the rare-earth anisotropy depend on the thermal averages $<O_{nm}>$; for example

$$K_{1R} = -\{(3/2)\ B_{20}<O_{20}> + 5B_{40}<O_{40}> + (21/2)\ B_{60}<O_{60}>\}$$
$$K_{2R} = (7/8)\ \{5B_{40}<O_{40}> + 27\ B_{60}<O_{60}>\} \qquad (11)$$
$$K_{3R} = -(231/16)B_{60}<O_{60}>$$

Expressions relating the other terms in Eq. 8 to $<O_{nm}>$ are given in reference 10. At the lowest temperatures, second, fourth and sixth-order terms in the rare-earth anisotropy may be important. The decline in rare-earth sublattice magnetization on increasing temperature is more rapid than that of the transition-metal sublattice because $B_{ex}(R) < B_{ex}(T)$, hence averages of the fourth and sixth order Stevens operators fall off rapidly with increasing temperature. Eventually, the more rapid decline in R sublattice magnetization compared with that of the iron or cobalt sublattices may lead to a situation where the iron or cobalt anisotropy overrides that due to the second-order crystal field. Regions of temperature where different contributions may dominate the anisotropy of the intermetallic compounds are indicated below;

2,4,6 order	2 order	
R anisotropy	R anisotropy	T anisotropy
0		T_c

a) $R_2Fe_{14}B$. The spin reorientation transitions from axis to plane for ions with positive second order Stevens coefficient α_J, R = Er, Tm and Yb (figure 1b), are due to a change in relative importance of the second order rare earth anisotropy which favours the plane at both 4f and 4g sites [1] and the iron anisotropy, which favours the c-axis. The parameters B_{nm} in Eq. 6b) depending on the rare earth are related to the crystal field parameters A_{nm} depending on the lattice by relations such as

$$B_{20} = \alpha_J<r^2> A_{20};\ \ B_{40} = \beta_J<r^4> A_{40};\ \ B_{60} = \gamma_J<r^6> A_{60} \qquad (12)$$

where $<r^n>$ is an average over the 4f shell and α_J, β_J and γ_J are the second, fourth and sixth-order Stevens factors for the rare-earth ion. Table 3 illustrates the relative strengths of some crystal-field terms for the different ions. Values of A_{20} may be deduced from the reorientation transiton temperatures T_{sr}. The magnetic ground state for Er, Tm and Yb is a ferrimagnetic fan structure with different angles between the magnetization of the 4f and 4g sublattices and the iron magnetization [21].

86

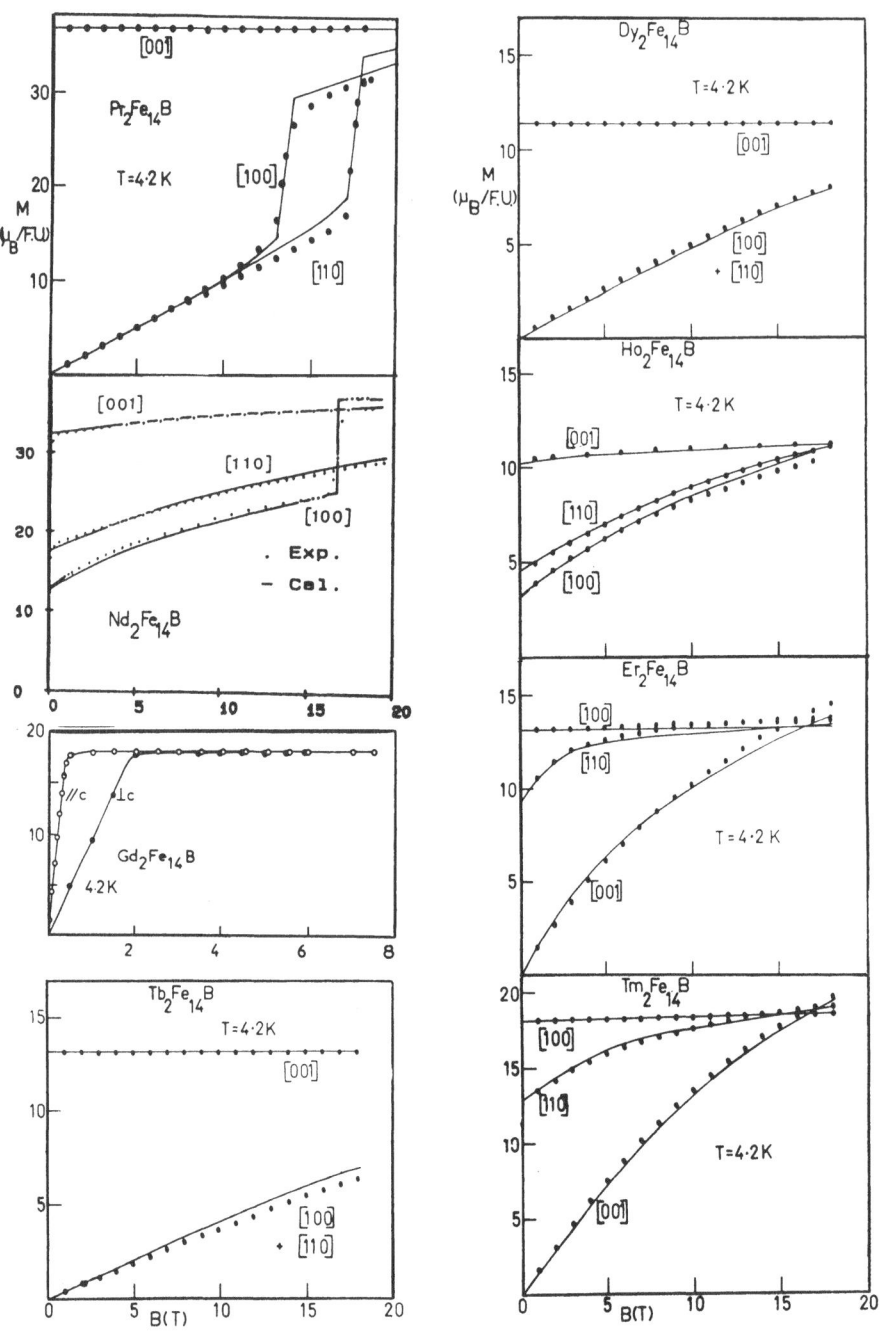

Figure 7 Magnetization curves measured on single crystals of $R_2Fe_{14}B$ at 4.2 K. Fits obtained from the crystal field model are shown by solid lines.

Table 3: Properties of Magnetic Trivalent Rare Earth Ions (Zero Temperature)

	J	S	γ	$\langle r^2 \rangle \alpha_J O_{20}$	$\langle r^4 \rangle \beta_J O_{40}$	$\langle r^6 \rangle \gamma_J O_{60}$
Ce	5/2	1/2	-1/2	-0.74	1.51	-
Pr	4	1	-1/2	-0.71	-2.12	5.9
Nd	9/2	3/2	-3/4	-0.26	-1.28	-8.6
Sm	5/2	5/2	-5	0.40	0.34	–
Sm*	7/2	5/2	-9/26	0.33	-0.19	2.0
Gd	7/2	7/2	1	–	–	–
Tb	6	3	2/3	-0.55	1.20	-1.3
Dy	15/2	5/2	1/2	-0.52	-1.46	5.6
Ho	8	2	2/5	-0.20	-1.00	-10.0
Er	15/2	3/2	1/3	0.19	0.92	9.0
Tm	6	1	2/7	0.45	1.14	-4.0
Yb	7/1	1/2	1/4	0.43	-0.79	0.7

*Excited state. $\langle r^n \rangle$ values from A J Freeman and J P Desclaux J Magn Magn Mat **12** 11 (1979)

The spin tilt transition to a noncollinear ground state at low temperatures for R = Nd and Ho, ions with a negative α_J, is due to the higher order terms in the crystal field [1]. A determination of the crystal field parameters A_{nm} may be made from the single-crystal magnetization curves shown in figure 7. Note the first-order magnetization processes observed for Pr and Nd when the field is applied in a hard direction.

These data have all been fitted in the way described in section 2. The complete expression for the crystal field at a site with mm symmetry is:

$$H_{cf} = B_{20}O_{20} + B_{22}O_{22} + B_{40}O_{40} + B_{42}O_{42} + B_{44}O_{44} + B_{60}O_{60}$$
$$+B_{62}O_{62}+B_{64}O_{64}+B_{66}O_{66} \qquad (13)$$

There are two such sites, and therefore eighteen crystal field parameters in $R_2Fe_{14}B$! To reduce these to a managable number, we first make use of the single-crystal ^{155}Gd Mössbauer determination of the nuclear electric field gradients by Bogé et al. [22] to fix $|A_{22}/A_{20}| = \eta$, the asymmetry parameter of the field gradient at either site. (Note that η is greater than 1 if **c** is taken as the principal axis at the 4g site [1]) .The ratio of V_{zz} for 4f and 4g sites is used to determine the ratio of A_{20}. This reduces the free second-order parameters to one. Similarly, for the fourth-order terms point charge calculations are used to fix the ratios A_{44}/A_{40}, and the ratio of values of A_{40}

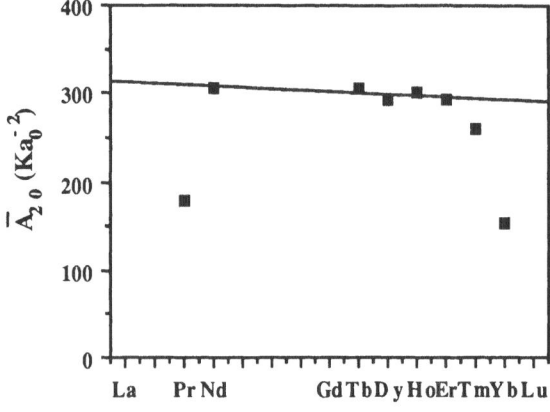

Figure 8   Variation of A_{20} across the $R_2Fe_{14}B$ series.

Table 4: Crystal Field Parameters for $R_2Fe_{14}B$

	A_{20}	A_{22}^s	A_{40}	A_{44}^c	A_{60}	A_{64}^c	A_{20}	A_{22}^s	A_{40}	A_{44}^c	A_{60}	A_{64}^c
Pr	176	-116	3	-9	-7	-42	179	351	3	8	-7	-4
Nd	304	-200	-15	43	-2	-33	308	605	-13	-41	-2	-13
Tb	304	-200	-15	43	-2	-33	308	605	-13	-41	-2	-13
Dy	292	-192	-14	42	-1	-20	296	581	-12	-39	-1	-8
Ho	298	-196	-9	26	-1	-19	302	593	-8	-24	-1	-7
Er	292	-222	-16	37	-1	-6	296	689	-14	-39	-1	-3
Tm	258	-205	-12	35	-2	-27	262	617	-11	-33	-2	-10
Yb	151	-100	-7	22	-1	-17	154	303	-6	-20	-1	-6

Units of A_{nm} are Ka_o^{-n}: The first six values are for the 4f site, the second six are for the 4g site.

for the two sites. A_{42} is neglected. Among the sixth-order terms, identical values of A_{60} are assumed for the two sites, A_{62} and A_{66} are neglected, but A_{64} is allowed to take different values. The justification for ignoring terms with m = 2 or 6 and n>2 is that there is no term in cos 2ϕ or cos 6ϕ in the *macroscopic* anisotropy of a tetragonal crystal. Values of A_{nm} for the different compounds are shown in table 4.

Figure 8 shows the variation of A_{20} across the $R_2Fe_{14}B$ series; There is a slight decrease with atomic number, but Pr and Yb stand out as exceptions. Some 4f mixing is possible for these rare earths so that the ground state configuration does not involve a precisely integral number of 4f electrons.

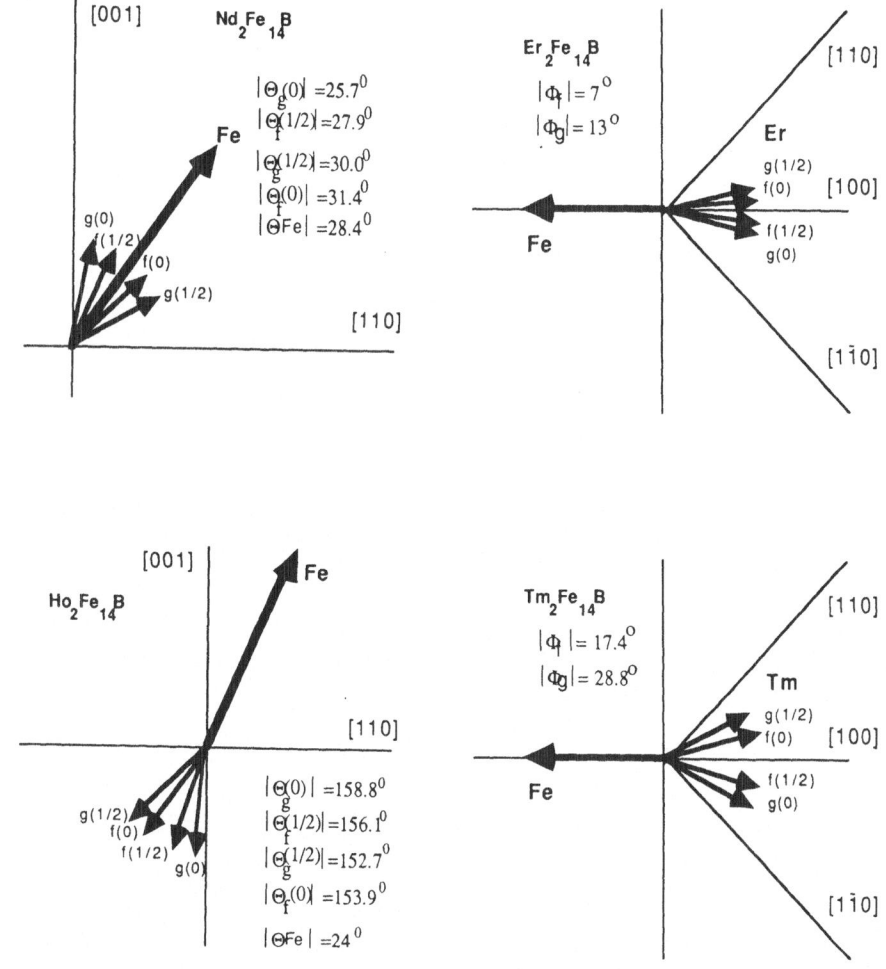

Figure 9   Ground state magnetic structures for four members of the $R_2Fe_{14}B$ series derived from the crystal field parameters of table 3.

Contributions to the macroscopic in-plane anisotropy from B_{44} at the two sites cancel. The easy direction in the plane for those rare earths with α_J negative is decided by the B_{64} terms. It is different for Pr on the one hand (γ_J positve) and Nd and Ho on the other (γ_J negative).

Finally, we summarize in figure 9 the ground state magnetic structures predicted for the series from the crystal field parameters of table 4.

b) $R_2Co_{14}B$. Our fit to the magnetization curves of $Nd_2Co_{14}B$ at 4.2 K, measured by Hirosawa *et al.* [23], is shown in figure 10. As in the iron compound, the magnetization is not collinear with the c-axis at the lowest temperatures. Fitting these curves, and the spin tilt and spin reorientation

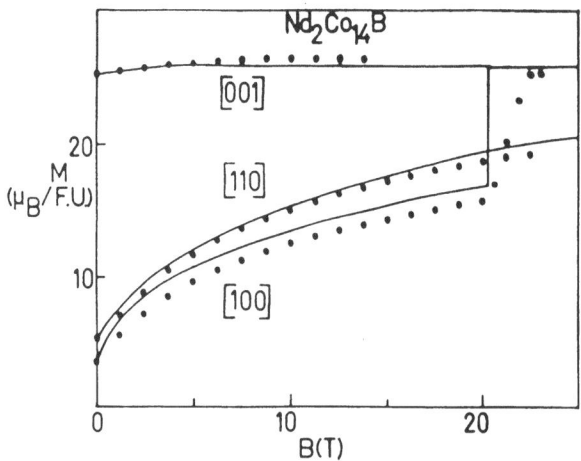

Figure 10 Magnetization curves measured on a single crystal of $Nd_2Co_{14}B$ at 4.2 K (from Ref.23). The solid lines are the fit obtained from the crystal field model.

transitions at 50 K and 545 K, leads to a set of crystal field parameters related to those of $Nd_2Fe_{14}B$ by the multiplicative factors $a_2=1.14$, $a_4=-0.25$, $a_6=0.80$ [24].

c) $R(Fe_{1-x}Co_x)_{14}B$. The behaviour of these solid solution series was of particular interest in view of the high Curie temperatures of $R_2Co_{14}B$ and the large anisotropy field for $Pr_2Co_{14}B$, originally reported by Sinnema et al [25].

The variation of K_{1T} and T_c across the $Y(Fe_{1-x}Co_x)_{14}B$ series has been reported by Bolzoni et al [26]. The change of sign of K_{1T} occurs at $x \sim 0.7$. The variation of T_c with x can be accounted for in the mean field model provided $n_{FeCo} = 938 \mu_0$, i.e. it is practically the same as n_{CoCo}.

The magnetic phase diagram for the praseodymium solid solutions is shown in figure 11a), while the variations in magnetization at 4.2 K and room temperature anisotropy field across the series are shown in the other parts of the figure. Taking the crystal field parameters determined for $Nd_2Co_{14}B$, the room temperature anisotropy field calculated for $Pr_2Co_{14}B$ is 13.7 T, in good agreement with observation, (13.8 T). We have seen in figure 8 that the uniaxial anisotropy of $Pr_2Fe_{14}B$ is anomalously low, which we ascribe to incipient valence instability. This valence instability appears to persist up to x = 0.8.

d) $(Er_xR'_{1-x})_2Fe_{14}B$. The effect of substitutions of rare-earth elements with positive α_J (Sm) or negative α_J (Tb) on the spin reorientation in $Er_2Fe_{14}B$ is quite straightforward. The former raises T_{sr} and stabilizes the state with $M \perp c$ whereas the latter depresses T_{sr} in favour of the state with $M \| c$, as shown in figure 12.

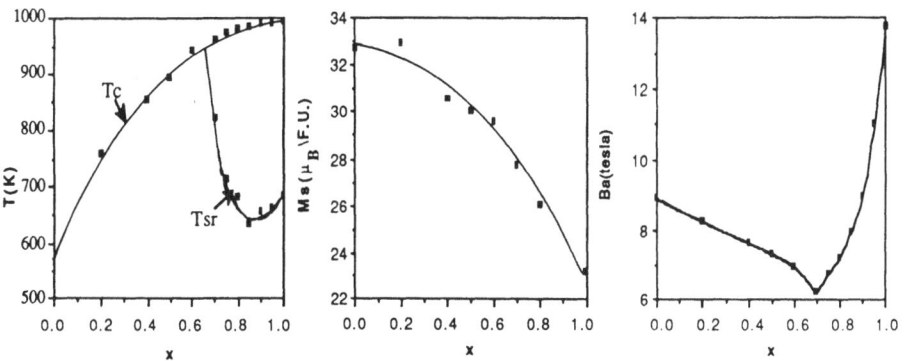

Figure 11 Magnetic properties of $Pr_2(Fe_{1-x}Co_x)_{14}B$ solid solutions. a) Magnetic phase diagram, b) Magnetization at 4.2 K, c) anisotropy field at room temperature.

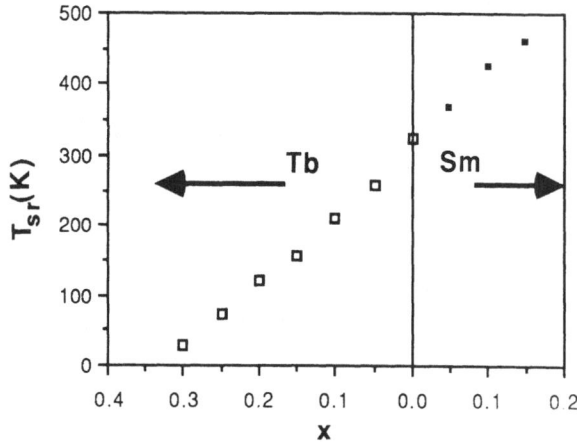

Figure 12 Spin reorientation transition in $(Er_{1-x}R'_x)_2Fe_{14}B$ with $R' = Sm$ or Tb.

e) $R(Fe_{11}Ti)$. Two main points of difference between the $ThMn_{12}$ structure and the $Nd_2Fe_{14}B$ structure which are evident from an examination of figures of 1b and 2b are (i) the sign of A_{20} is opposite in the two structures and (ii) the magnitude of the rare-earth crystal field interaction at the 2a site is rather small. All compounds except $Tb(Fe_{11}Ti)$ appear to have easy c-axis anisotropy at room temperature, which means that the iron anisotropy overides that of the rare earths for which α_J is negative. (Contrast this with $R_2Co_{14}B$ where the rare earth anisotropy overrides that of the cobalt at room temperature for R = Pr, Nd and Tb; T_{sr} for these compounds is 660, 545, and 760 K, respectively).

The complete expression for the crystal field at the 2a site with *4/mmm*

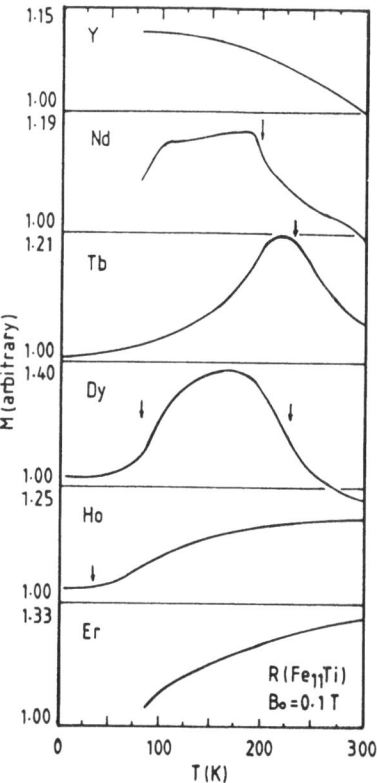

Figure 13 Thermomagnetic scans of polycrystalline $R(Fe_{11}Ti)$ materials showing changes due to spin reorientations.

point symmetry is

$$H_{cf} = B_{20}O_{20} + B_{40}O_{40} + B_{44}O_{44} + B_{60}O_{60} + B_{64}O_{64} \quad (14)$$

Point charge calculations are used to fix the ratios of terms of the same order, i.e. $A_{44}/A_{40} = -2.4$ and $A_{64}/A_{60} = 1.6$, so there are just three crystal field parameters to be deduced from experiment. Experimental data that show the spin reorientations are thermomagnetic scans and Mössbauer spectra on oriented absorbers [10,27] (fig. 2b). We also have magnetization curves on oriented powders [10]. Some results are shown in figures 13 and 14.

The spin reorientation away from the c-axis towards the plane starts at the point where K_1 (the sum of R and T contributions) changes sign. However, the transition may be extended in temperature if K_2 is opposite in sign to K_1. The angle between the moment direction and the c-axis is given by

$$\sin \theta = \{(-K_2 \pm [K_2^2 - 3K_1K_3]^{1/2}/3K_3\}^{1/2} \quad (15)$$

or by $\sin \theta = (-K_1/2K_2)^{1/2}$ if K_3 is negligible.

The reorientation for Dy begins at about 220 K, and it is complete at 80

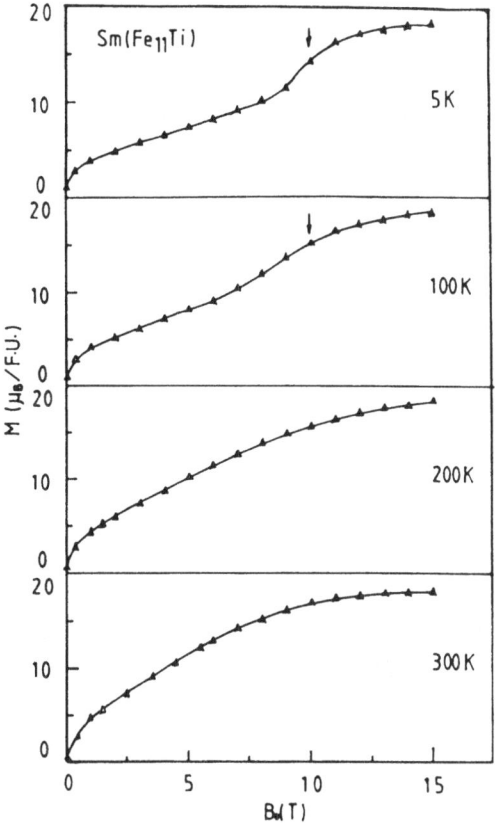

Figure 14 Magnetization curves measured on oriented samples of R(Fe$_{11}$Ti).

K [27]; this can be reproduced with A$_{20}$= -61 Ka$_0$$^{-2}$ and A$_{40}$ = -5.9 Ka$_0$$^{-4}$. If A$_{60}$ is greater than 0.24 Ka$_0$$^{-6}$, the limiting value of θ is 90°, otherwise it is 69°.

Holmium has a particularly small value of α_J, and the reorientation there has its onset only below 40 K. The case of terbium is a little more complicated because β_J is positive (whereas it is negative for Dy and Ho) and K$_2$ is therefore negative. The spin reorientation is abrupt as it was for R$_2$Fe$_{14}$B with R = Tm or Er, and it occurs just above room temperature. The second transition at about 220 K, is caused by a second sign change of K$_1$ due to the fourth order terms.

Values of A$_{nm}$ deduced from the behaviour of these compounds are summarized in table 5. The temperature variation of the phenomenological anisotropy constants of Eq. 8 are shown for several of the compounds in figure 15.

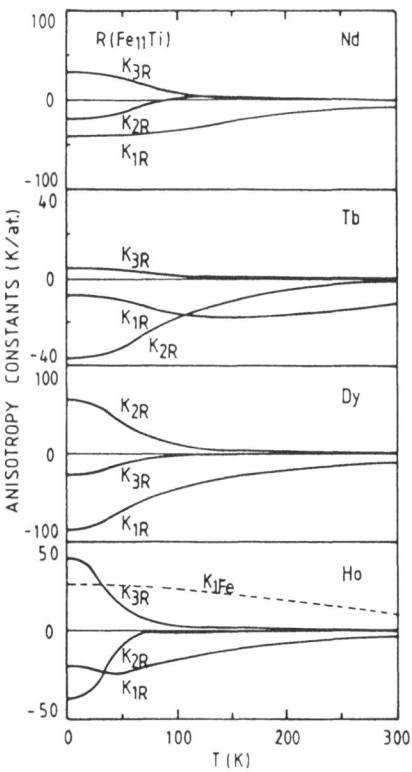

Figure 15 Temperature dependence of the rare earth anisotropy constants for R(Fe$_{11}$Ti). K$_1$(Fe) is shown by the dashed line.

Table 5. Crystal Field Parameters for R(Fe$_{11}$Ti)

A$_{20}$ (Ka$_0^{-2}$)	A$_{40}$ (Ka$_0^{-4}$)	A$_{44}$ (Ka$_0^{-4}$)
-61	-5.9	14.2

Turning, finally, to the light rare earths, a spin reorientation for the Nd compound is predicted at 190 K with the parameters of table 4, and it actually occurs at 200 K. A first order magnetization process of the second type is shown for Sm in figure 14 in a field of 10 T. An explanation of this behaviour requires that the J = 7/2 excited multiplet of samarium be considered in the calculations (J mixing). It may be seen from table 3 that this state has a significant sixth-order interaction and a *positive* value of γ.

5. OTHER ACHIEVEMENTS

Besides the results on the intrinsic magnetic properties of compounds with the $Nd_2Fe_{14}B$ or $ThMn_{12}$ structure, reported above, CEAM-related work includes:

i)   Determination of the hydrogen absorption and desorption characteristics of $Nd_2Fe_{14}B$ [28] by thermopiezic analysis [29],

ii)   Determination of the influence of hydrogen [30], or hydrogen and nonmagnetic substitution [31], on the magnetic properties of Nd_2Fe_{17},

iii)   Identification of an hexagonal ferromagnetic phase of approximate composition $NdFe_3O_8$ with a=20.21Å, c=12.35Å and T_c=504 K, formed by crystallization from the Nd=Fe eutectic[32],

iv)   Preparation of amorphous and coercive microcrystalline Nd-Fe-B by melt spinning [33],

v)   Preparation of $Gd(Fe_nAl_{12-n})$ with the $ThMn_{12}$ structure and 6<n<10, by melt spinning [34].

6. CONCLUSIONS

We now consider the intrinsic magnetic properties of the $R_2Fe_{14}B$ and $R(Fe_{11}Ti)$ series from the point of view of permanent magnet applications. In table 6 the magnetic properties at room temperature of the yttrium compound and the best magnetic rare earth compound with uniaxial anisotropy are compared.

i)   The Curie temperatures of the two families are rather similiar. They are quite adequate for magnets operating up to about 100°C, but pose a problem at higher temperatures. T_c is greatly enhanced by cobalt substitution in both cases, albeit with some loss of anisotropy. There may be a chance of increasing T_c further by incorporating less titanium (down to 0.8 Ti per formula), by heat treatments or by using another element to stabilize the $ThMn_{12}$ structure.

ii)   The anisotropy fields of $Nd_2Fe_{14}B$ and $Sm(Fe_{11}Ti)$ at room temperature are also quite similar. However, there is plenty of scope for increasing B_a in $Nd_2Fe_{14}B$ by substituting another rare earth with a strong second-order crystal field interaction and negative α_J. It can be seen from Table 3 that Tb and Dy are best. The cost is a reduction in magnetization since alloys with heavy rare earths are ferrimagnetic. No such flexibility exists for $Sm(Fe_{11}Ti)$ because Sm is already the ion with positive α_J that exhibits the strongest anisotropy. A_{20} is about five times greater in the $Nd_2Fe_{14}B$ structure than it is in the $ThMn_{12}$ structure.

Table 6. Intrinsic Magnetic Properties at Room Temperature

	T_c (K)	K_1 (MJ/m^3)	B_a (T)	$\mu_0 M_s$ (T)	$\mu_0 M_s^2/4$ (KJ/m^3)
$Y_2Fe_{14}B$	568	1.10	2.0	1.36	368
$Y(Fe_{11}Ti)$	524	0.89	2.0	1.12	250
$Nd_2Fe_{14}B$	592	5.05	9.1	1.60	509
$Sm(Fe_{11}Ti)$	584	4.78	10.5	1.14	259

iii)   The main advantage of $Nd_2Fe_{14}B$ lies in the large value of its magnetization (table 5). The room-temperature average iron moment of $2.0\mu_B$, as opposed to 1.6 μ_B for $Sm(Fe_{11}Ti)$, and the larger rare-earth contribution to the magnetization lead to a value of $\mu_0 M_s$ that is 40% higher in $Nd_2Fe_{14}B$ than in $Sm(Fe_{11}Ti)$. The quantity $\mu_0 M_s^2/4$ in table 6 represents the theoretical uper limit on the energy product of a magnet made from the alloy. This figure for $Sm(Fe_{11}Ti)$ is less than values that have already been achieved for Nd-Fe-B magnets.

iv) $Sm(Fe_{11}Ti)$ is likely to become a serious candidate as a material for making high-performance permanent magnets if useful coercivity can be developed, particularly in oriented melt-spun ribbons or in precipitation-hardened alloys, both of which have proved difficult to achieve with Nd-Fe-B.

ACKNOWLEDGEMENTS.

We are grateful to Dr M. Sagawa for making available crystals of $R_2Fe_{14}B$ (R = Tb, Ho, Er and Tm) for measurements at the S N C I, Grenoble. We are also grateful to Dr J.A.A.J Perenboom for facilitating our measurements on some of the $R(Fe_{11}Ti)$ compounds at the High Field Magnet Laboratory of the University of Nijmegen.

+ : The Laboratoire Louis Néel, BP166x, 38042 Grenoble Cedex, France.

REFERENCES

[1]   J M D Coey, J Less-Common Metals **126** 21 (1986); Physica Scripta **19** 426 (1987)
[2]   K H J Buschow, Mater. Sci. Repts **1** 1 (1986)
[3]   D Givord, H S Li and J M Moreau, Solid State Commun **50** 497 (1984)
[4]   E P Marusin, O I Boidak, A O Tsokol and V F Fundamenskii, Soviet Physics — Crystallography **30** 338 (1985)

[5]   O Moze, L Pareti, M Solzi and W I F David, Solid State Commun **66** 465 (1988)

[6]   J M Cadogan, J P Gavigan, D Givord and H S Li, J Phys F **18** 779 (1988)

[7]   D Givord, H S Li and R Perrier de la Bathie, Solid State Commun **51** 857 (1984)

[8]   S Rinaldi and L Pareti, J Appl Phys **50** 7719 (1979)

[9]   R Fruchart, P l'Heritier, P Dalmas de Réotier, D Fruchart, P Wolfers, J M D Coey, L P Ferreira, R Guillen, P Vulliet and A Yaouanc, J Phys F **17** 484 (1987)

[10]   B P Hu, H S Li, J P Gavigan and J M D Coey J Phys C (1989)

[11]   F Bolzoni, J P Gavigan, D Givord, H S Li, O Moze and L Pareti, J Magn Magn Mater **66** 158 (1987)

[12]   K Erdman, P Deppe, M Rosenberg and K H J Buschow, J Appl Phys **61** 4340 (1987)

[13]   D Givord, H S Li, J M Moreau and P Tenaud, J Magn Magn Mater **54-57** 445 (1986); K H J Buschow and R Grössinger, J Less-Common Metals **135** 39 (1987)

[14]   J Kamarad, Z Arnold and J Schneider J Magn Magn Mater **67** 29 (1987)

[15]   J M D Coey, A Yaouanc and D Fruchart, Solid State Commun **58** 413 (1986)

[16]   D LeRoux, Thèse d'Etat, Grenoble 1986

[17]   J J M Franse, N P Thuy and N M Hong (to be published)

[18]   K H J Buschow, D B de Mooij, M Brouha, H H A Smit and R C Thiel, IEEE Trans Magn 1988)

[19]   H S Li and B P Hu, J Physique (1989)

[20]   E Belorizky, M A Frémy, J P Gavigan, D Givord and H S Li, J Appl Phys **61** 3971 (1987)

[21]   J M Cadogan, J Less-Common Metals, **135** 269 (1987)

[22]   M Bogé, G Czjzek, D Givord, C Jeandey, H S Li and J L Oddou, J Phys F **16** L67(1986)

[23]   S Hirosawa, K Tokuhara, H Yamamoto, S Fujimura, M Sagawa and H Yamauchi, J Appl Phys **61** 3571 (1987)

[24]   H S Li, J P Gavigan, J M Cadogan, D Givord and J M D Coey, J Magn Magn Mater **72** L241 (1988)

[25]   S Sinnema, J J M Franse, R J Radwanski, K H J Buschow and D B de Mooij, J Physique **46** C6 301 (1985)

[26]   F. Bolzoni, F Leccabue, O Moze, L Pareti and M Solzi, J Appl Phys **61** 5367 (1987)

[27]   H S Li, B P Hu and J M D Coey, Solid State Commun **66** 133 (1988)

[28]   J M Cadogan and J M D Coey, Appl Phys Lett **48** 442 (1986)

[29]   D H Ryan and J M D Coey, J Phys E **19** 693 (1986)

[30]   X Z Wang, K Donnelly, J M D Coey, B Chevalier, J Etourneau and T Berlureau, J Mater Sci **23** 329(1988)

[31]   B P Hu and J M D Coey, J Less Common Metals **142** 295 (1988)

[32]   B P Hu, J M D Coey, C J Cardin, E J Devlin and I R Harris, J Less Common Metals **144** L29 (1988)

[33]   J M Cadogan, D H Ryan and J M D Coey, Mater Sci Eng **99** 143 (1988)

[34]   X Z Wang, B Chevalier, J Etourneau, J M D Coey and J M Cadogan, J Less Common Metals **138** 235 (1987)

A SEARCH FOR NEW PERMANENT MAGNET MATERIALS :

CRYSTALLOGRAPHIC AND MAGNETIC PROPERTIES OF NEW COMPOUNDS

J. Allemand, C. Bertrand, J. Le Roy, J.M. Moreau, D. Paccard, L. Paccard

Laboratoire "Structure de la Matière", Université de Savoie, 9, rue de l'Arc-en-Ciel, 74019 ANNECY-le-Vieux (France)

M. A. Frémy, D. Givord

Laboratoire "Louis Neel", CNRS, 166 X, Grenoble Cedex (France)

ABSTRACT

Permanent magnet materials are characterized by high Curie temperature, large spontaneous magnetization and large magnetic anisotropy. The best natural elements to produce these properties are rare earth and transition elements (Fe and Co). New materials could be obtained by addition of others elements such as B, Si or C which stabilized the corresponding atomic structures. New compounds have been synthetized and their crystallographic and magnetic properties have been studied : $Nd(Co_xMn_{1-x})_{12}$, $Nd(Ni_xMn_{1-x})_{12}$, $RFe_{10}SiC_{0.5}$ (R = Ce, Pr, Nd, Sm), $DyFe_2SiC$, $DyFe_2Si_2C$, $R(Fe_xCo_{1-x})_{10}SiC_{0.5}$. Curie temperatures of 950 K for $NdCo_{10}SiC_{0.5}$ and 670 K for $SmCo_{10}SiC_{0.5}$ were measured.

INTRODUCTION

Permanent magnet materials are characterized by large exchange interactions, high spontaneous magnetization and large magnetic anisotropy. The best materials contain rare earth and transition elements such as $SmCo_5$ and Nd_2Fe_{17}. Moreover uniaxial anisotropy occurs only with atomic structures having a uniaxial symetry. Thus obtaining new permanent magnet materials means synthetizing new phases containing rare earth, iron or cobalt and other elements such as B, Si or C to stabilize an atomic structure with a uniaxial symmetry. Our first success was the synthesis and structure determination of $Nd_2Fe_{14}B$ [Givord et al., 1984] in cooperation with "Laboratoire Louis Neel, CNRS, Grenoble". Then in order to obtain new compounds we substituted Co and Ni for Mn in $NdMn_{12}$ with the tetragonal $ThMn_{12}$ type structure. We also obtained quaternary compounds with the tetragonal $BaCd_{11}$ type structure.

Chevalier et al. [1986] studied the crystallographic and magnetic properties of $NdCo_9Si_2$ and $NdCo_{9-x}Fe_xSi_2$ with $BaCd_{11}$ type structures as reported [Bodak and Gladyshevskii, 1969].The maximum amount of iron that could be substituted for cobalt corresponded to x = 5. Attempts to prepare $NdFe_9Si_2$ were unsuccessful. Then we tried to obtain quaternary compounds by adding small amounts of Si and C and succeeded with $RFe_{10}SiC_{0.5}$ [Le Roy et al., 1987],

Dy_2Fe_2SiC [L. Paccard et al., 1987], and $Dy_2Fe_2Si_2C$ [L. Paccard et al., 1988]. Substitution of cobalt for iron was made in $RFe_{10}SiC_{0.5}$ giving Curie temperature of 350 K for $NdCo_{10}SiC_{0.5}$ and 670 K for $SmCo_{10}SiC_{0.5}$. A neutron diffraction study is in progress to determine precisely the ferromagnetic or ferrimagnetic like behaviour of these compounds. Magnetic measurements were made at the "Laboratoire Louis Neel, CNRS, Grenoble".

METHODS AND TECHNIQUES

Samples were prepared by conventional arc melting and induction melting techniques. The phase determination was made using X - ray diffraction techniques on powders and single crystals obtained by mechanical fragmentation from the crushed melt. The Curie temperatures were measured using a translation balance. The saturation magnetization and the magneto crystalline anisotropy were measured in a field up to 8 T using a superconducting coil. For the anisotropy measurements the samples were prepared by aligning powders (< 25 µm) in a magnetic field (H = 2T) and fixing them in epoxy resin. X - ray diffraction was used to determine the easy direction of the oriented samples.

ACHIEVEMENTS

A - Structural and magnetic properties of $Nd(Co_x Mn_{1-x})_{12}$ and $Nd(Ni_x Mn_{1-x})_{12}$

The compound $NdMn_{12}$ crystallizes in the tetragonal $ThMn_{12}$ -type structure. Mn and Zn are the only transition metals in forming this type structure. $NdMn_{12}$ is reported to be antiferromagnetic with $T_N = 135$ K. However, it would be interesting to substitute different 3d transition metals for Mn in order to determine their extent of solubility in the $NdMn_{12}$ system and to stabilize a ferromagnetic phase. For this reason the pseudo-binary compounds $Nd(Fe_xMn_{1-x})_{12}$, $Nd(Co_xMn_{1-x})_{12}$ and $Nd(Ni_xMn_{1-x})_{12}$ were prepared over the whole range of concentration by induction melting in a cold crucible under an argon atmosphere. They were annealed at 600°C for two days under vacuum. We investigated, at first, the range of solubility In the $Nd(Fe_xMn_{1-x})_{12}$ system, compounds of the $ThMn_{12}$ -type were not found.

For the systems $Nd(Co_xMn_{1-x})_{12}$ and $Nd(Ni_xMn_{1-x})_{12}$, the phase $ThMn_{12}$ exists in the ranges $0.3 \leq \times \leq 0.7$ and $0.3 \leq \times \leq 0.6$, respectively. Within these limits, the replacement of manganese by cobalt or nickel leads to a decrease of the lattice parameters for a constant c/a value. The decrease of parameters is in accordance with the metallic radii of the individual atoms (Mn has a larger atomic radius than Co or Ni). The magnetic behaviours of $Nd(Co_xMn_{1-x})_{12}$ were only investigated with x varying from 0.3 to 0.7. For x = 0.5 to 0.7, the compounds are ferromagnetic (fig. 1). At 4.2 K, the spontaneous magnetization attains a value of 0.37 µB/F.U, 5.14 µB/F.U and 11.16 µB/F.U. respectively. The corresponding Curie temperatures are ≈5 K, 128 K and 370 K.

The compounds with x = 0.3 and 0.4 are not magnetic and do not exhibit any spontaneous magnetization. The concentration dependence of saturation magnetization and Curie temperatures are shown in figure 2. They rise nearly linearly with increasing x.

A possible interpretation of these results could be that the magnetic moment of cobalt is affected by the surrounding of Mn atoms which are either not magnetic or are coupled antiferromagnetically with the Co moments.

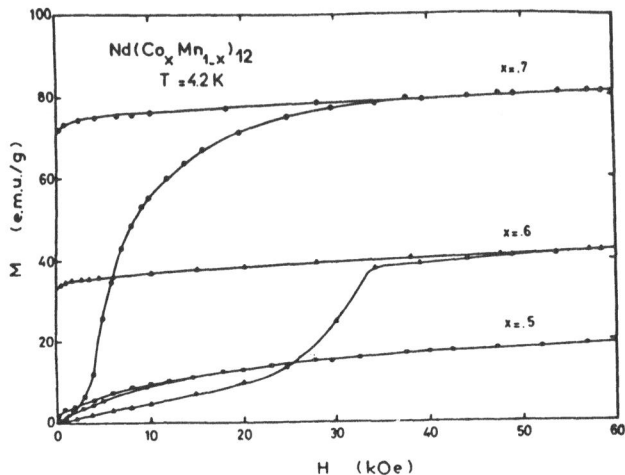

Fig. 1 : Variation of the magnetization of $Nd(Co_xMn_{1-x})_{12}$ with field, T = 4.2 K.

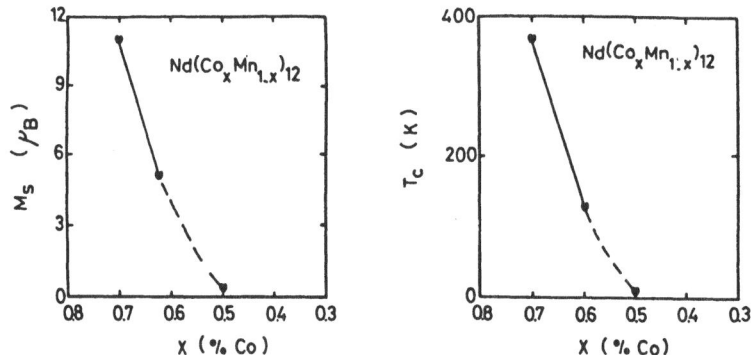

Fig.2: Variation of saturation magnetization(a)and T_c(b)as a function of x in $Nd(Co_xMn_{1-x})_{12}$.

B - **A new quaternary iron-rich phase stabilized by carbon : DyFe$_2$SiC**

X-ray diffraction studies of single crystal DyFe$_2$SiC were used to establish the crystal structure, with space group *Cmcm* and lattice constants $a = 3.712$ (0) Å, $b = 10.531$ (3) Å, $c = 6.863$ (1) Å. The crystal structure is derived from the YNiAl$_2$ type (Re$_3$B-type derivative) with an ordered replacement of aluminium atoms by iron atoms and of nickel atoms by silicon atoms and with an addition of carbon atoms between the infinite columns of silicon centred DyFe$_2$ prisms.

TABLE 1

Positional and thermal parameters for DyFe$_2$SiC, space group *Cmcm*

atom	x	y	z	B(A^2)
Dy(4c)	0.0	0.0463(1)	1/4	1.09(2)
Fe(8f)	0.0	0.6667(3)	0.5645(5)	1.28(4)
Si(4c)	0.0	0.764(1)	1/4	1.6(1)
C(4b)	0.0	0.5	0.0	1.0(2)

B is the isotropic equivalent thermal parameter of the anisotropically refined atoms. The estimated standard deviations are given in parenthesis.

TABLE 2

Comparison of lattice parameters for Re$_3$B and DyFe$_2$SiC

Re$_3$B (*Cmcm*)	DyFe$_2$SiC (*Cmcm*)
$a = 2.890$ Å	$a = 3.712$ Å
$b = 9.313$ Å	$b = 10.531$ Å
$c = 7.258$ Å	$c = 6.863$ Å

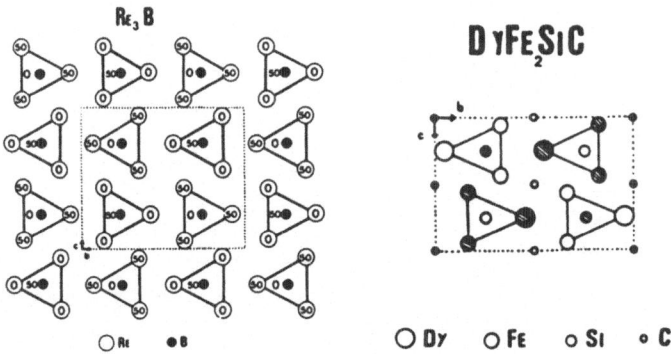

Fig. 3 : Comparison of the structures of DyFe$_2$SiC and Re$_3$B (space group *Cmcm*).

C - $Dy_2Fe_2Si_2C$: A new structure derived from Ge_2Os and stabilized by carbon

[L. Paccard et al., 1988]

X-ray diffraction studies of $Dy_2Fe_2Si_2C$ single crystals were used to establish the crystal structure, with space group $C2/m$ and lattice constants $a = 10.581$ (3) Å, $b = 3.912$ (2) Å, $c = 6.734(1)$ Å, $\beta = 129.15$ (3)° and $V = 216$ (2) Å^3. The crystal structure is derived from the Ge_2Os type with an ordered replacement of germanium atoms by dysprosium and iron atoms and of osmium atoms by silicon atoms and with an addition of carbon atoms between the isolated infinite columns of silicon-centred double prisms. One carbon atom is at the centre of an octahedron where the six vertices are occupied by four dysprosium and two iron atoms.

$Pr_2Fe_2Si_2C$, $Nd_2Fe_2Si_2C$, $Gd_2Fe_2Si_2C$, $Tb_2Fe_2Si_2C$, $Ho_2Fe_2Si_2C$, $Er_2Fe_2Si_2C$, $Y_2Fe_2Si_2C$ are isostructural with $Dy_2Fe_2Si_2C$.

TABLE 3

Positional and thermal parameters for $Dy_2Fe_2Si_2C$ ($C2/m$)

Atom	x	y	z	B (Å^2)
Dy(4i)	0.06106(5)	0.0	0.79354(9)	0.344(7)
Fe(4i)	2.2970(2)	0.0	0.4031(3)	0.34(2)
Si(4i)	0.6562(3)	0.0	0.2074(5)	0.49(5)
C(2d)	0.0	0.5	0.5	0.6(2)

The estimated standard deviations are given in parentheses.

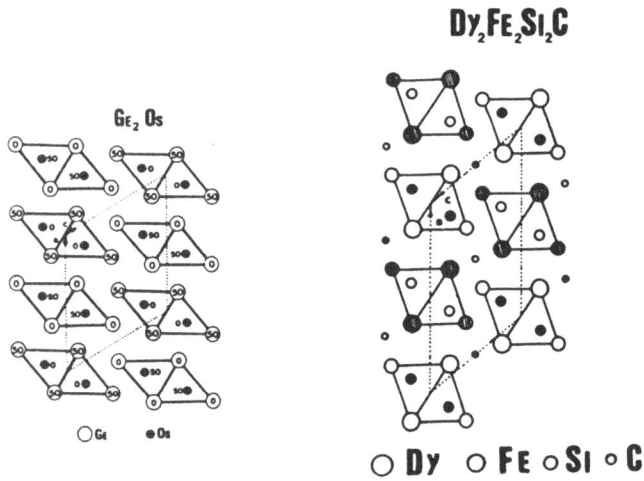

Fig. 4 : Comparison of the structures of $Dy_2Fe_2Si_2C$ and Ge_2Os (space group $C2/m$).

TABLE 4

Comparaison of lattice parameters for Ge_2Os and $Dy_2Fe_2Si_2C$

Ge2Os (C2/m)	Dy2Fe2Si2C (C2/m)
$a = 8.995$ Å	$a = 10.581$ Å
$b = 3.094$ Å	$b = 3.912$ Å
$c = 7.685$ Å	$c = 6.734$ Å
$\beta = 119.16°$	$\beta = 129.15°$

D. Crystallographic and magnetic properties of a new series $RFe_{10}SiC_{0.5}$ (R = Ce, Pr, Nd, Sm) [Le Roy et al., 1987]

X-ray diffraction studies on single crystals were used to establish the crystal structure of the new compound $NdFe_{10}SiC_{0.5}$, which has space group $I4_1/amd$, with $Z = 4$ formula units per cells. The lattice constants are $a = 10.037$ (7) Å and $c = 6.495$ (4) Å. The final value of R was 0.05 for 283 independent intensities. The positions of the neodymium, iron and silicon atoms correspond to the $BaCd_{11}$ structure. The carbon atoms fill octahedral vacancies formed by four iron and two neodymium atoms, with an occupancy of 25%. The series $RFe_{10}SiC_{0.5}$ (R ≡ Ce, Pr, Nd, Sm) are iso-typic with $NdFe_{10}SiC_{0.5}$ and all these compounds are isotypic with $La\ Mn_{11}C_{2-x}$.

TABLE 5

Positional and thermal parameters for $NdFe_{10}SiC_{0.5}$ with $BaCd_{11}$ filled type (space group $I4_1/amd$)

Atom	Occupation factor	x	y	z	B (Å^2)
Nd in 4 (a)	1	0	0.75	0.125	0.26(2)
Fe(1) in 4 (b)	1	0	0.25	0.375	0.46(2)
Fe(2) in 32(i)	1	0.1254(2)	0.0449(2)	0.1886(3)	0.64(2)
Fe(3) in 8 (d)	0.5	0	0	0.5	0.5(1)
Si in 8(d)	0.5	0	0	0.5	0.3(1)
C in 8 (c)	0.25	0	0	0	0 3(2)

TABLE 6

Lattice parameters of $RFe_{10}SiC_{0.5}$ compounds

R	$a(\text{Å})$	$c(\text{Å})$	$(V/N)^{1/3}\ (\text{Å})^a$	c/a
Ce	10.049(4)	6.528(3)	2.387	0.650
Pr	10.107(3)	6.534(2)	2.396	0.647
Nd	10.083(3)	6.529(2)	2.392	0.648
Sm	10.092(7)	6.538(4)	2.395	0.648

[a] V is the unit cell and N is the number of atoms in unit cell ($N = 42$).

Magnetic measurements

Magnetic susceptibilities were measured on polycrystalline samples sealed in quartz tubes using a translation balance. For all the compounds, the temperature dependence of the susceptibility shows, at a critical temperature T_c, an anomaly characteristic of a transition from a ferromagnetic or ferrimagnetic state to a paramagnetic state. The Curie temperatures, listed in Table 3, lie between 390K ($CeFe_{10}SiC_{0.5}$) and 460K ($SmFe_{10}SiC_{0.5}$).

Magnetic measurements were performed between 1.5 K and 300 K with a superconducting coil providing a maximum field of 7.5 T. The field dependances of magnetization obtained at 4.2 K and 300 K in the different compounds on free powders are shown in Fig. 5 and 6. At room temperature ferromagnetic or ferrimagnetic behaviour is observed for all the $RFe_{10}SiC_{0.5}$ compounds; in low fields, a large susceptibility is found, corresponding to the expected slope of the demagnetizing fields. In higher fields, the magnetization tends to saturate. The spontaneous magnetization values deduced from these data in the different compounds are given in Table 7.

Similar behaviour is observed in $CeFe_{10}SiC_{0.5}$ at all temperatures down to 4.2 K. The deduced value of the spontaneous magnetization M_s is 13.2 μ_B (formula unit)$^{-1}$ at 4.2 K which corresponds to 1.32 μ_B (iron atom)$^{-1}$, assuming ferromagnetic alignment of all the iron moments

TABLE 7

Curie temperature T_c, spontaneous magnetization M_s at 4.2 K and 300 K, theoretical moment for the rare earth ion R^{3+} and value of iron moment of the compound $R Fe_{10}Si C_{0.5}$

Compound	T_c (K)	M_s(4.2 K) μ_B(formula unit)$^{-1}$	M_s(300 K) μ_B(formula unit)$^{-1}$	gj^J	M_{Fe} μ_B(Fe atom)$^{-1}$
$Ce Fe_{10} Si C_{0.5}$	390	13.2	6.5	2.14	1.32
$Pr Fe_{10} Si C_{0.5}$	430	17.1	9.2	3.2	1.39
$Nd Fe_{10} Si C_{0.5}$	410	17.5	8.4	3.27	1.42
$Sm Fe_{10} Si C_{0.5}$	460	14.8	11.1	0.71	1.41

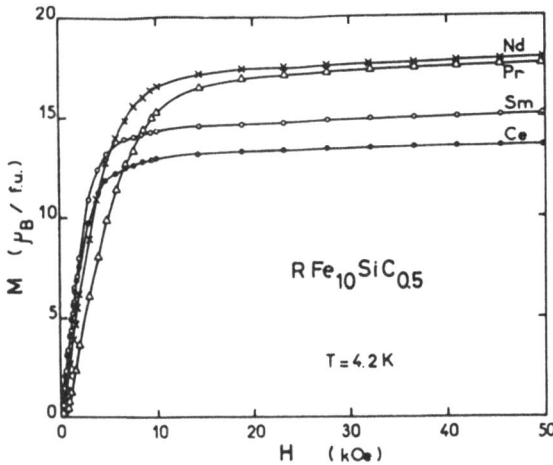

Fig. 5 : Field dependence of the magnetization at 4.2 K for $RFe_{10}SiC_{0.5}$ compounds.

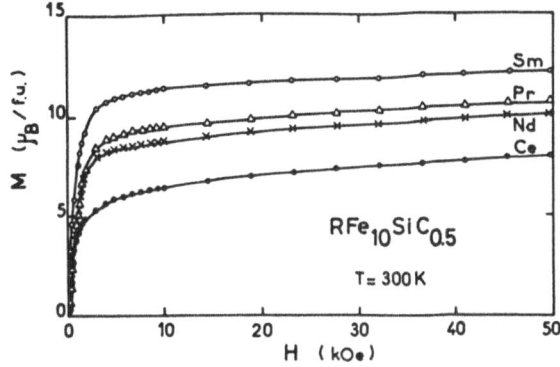

Fig. 6 : Field dependence of the magnetization at 300 K for $RFe_{10}SiC_{0.5}$ compounds.

For the compounds with a magnetic R atom (R ≡ Pr, Nd, Sm), at low temperature, the susceptibility in low field is much weaker than that for $CeFe_{10}SiC_{0.5}$; above a critical field of the order of 1 kG, the magnetization increases rapidly and tends to saturate above 5 kG. Measurements of hysteresis loops have confirmed that this phenomenon results from an intrinsic coercivity which occurs at low temperatures in highly anisotropic compounds. The compounds are ferromagnetic down to very low temperatures. The spontaneous magnetizations obtained at 4.2 K are listed in Table 7. Assuming the theoretical moment for the R^{3+} ion and ferromagnetic coupling between the rare earth and the ion moments, the iron moment in these compounds is deduced to be $1.4\mu_B$ (iron atom)$^{-1}$, as in $CeFe_{10}SiC_{0.5}$.

E   Crystallographic and magnetic properties of $Nd(Fe_{1-x}Co_x)_{10}SiC_{0.5}$ and $Sm(Fe_{1-x}Co_x)_{10}SiC_{0.5}$

$Nd(Fe_{1-x}Co_x)SiC_{0.5}$ and $Sm(Fe_{1-x}Co_x)_{10}SiC_{0.5}$ compounds exist in the tetragonal $BaCd_{11}$-type structure through the whole composition range ($0 \leq \times \leq 1$).

TABLE 8 : Lattice parameters and cell volume for $Nd(Fe_{1-x}Co_x)_{10}SiC_{0.5}$ alloys

$$Nd(Fe_{1-x}Co_x)_{10}SiC_{0.5}$$

x	a (Å)	c (Å)	v (Å^3)
0	10.060	6.521	659.95
0.143	10.050	6.507	657.22
0.286	10.040	6.498	655.00
0.428	10.001	6.487	648.83
0.571	9.964	6.440	639.37
0.714	9.923	6.407	630.87
0.857	9.900	6.404	627.65
1	9.874	6.379	621.92

TABLE 9 : Lattice parameters and cell volume for $Sm(Fe_{1-x}Co_x)_{10}SiC_{0.5}$

$$Sm(Fe_{1-x}Co_x)_{10}SiC_{0.5}$$

x	a (Å)	c (Å)	v (Å^3)
0	10.070	6.524	661.56
0.2	10.021	6.486	651.32
0.4	9.992	6.477	646.66
0.5	9.978	6.458	642.96
0.6	9.946	6.434	636.47
0.8	9.887	6.397	625.32
1	9.874	6.396	623.58

TABLE 10 : Curie temperature and saturation magnetization at 4.2 K for $R(Fe_{1-x}Co_x)_{10}SiC_{0.5}$ compounds with R = Nd and Sm

$$Sm(Fe_{1-x}Co_x)_{10}SiC_{0.5}$$

x	T_c (K)	Ms (4.2 K) μ B/F.U
0	460	14.80
0.2	490	13.16
0.4	560	11.40
0.5	590	11.65
0.6	624	10.55
0.8	660	9.32
1	670	7.50

$$Nd(Fe_{1-x}Co_x)_{10}SiC_{0.5}$$

x	T_c (K)	Ms (4.2 K) μ B/F.U
0	410	17.50
0.143	468	17.52
0.286	540	16.75
0.428	626	16.31
0.571	718	14.71
0.714	738	11.34
0.857	760	10.08
1	950	9.02

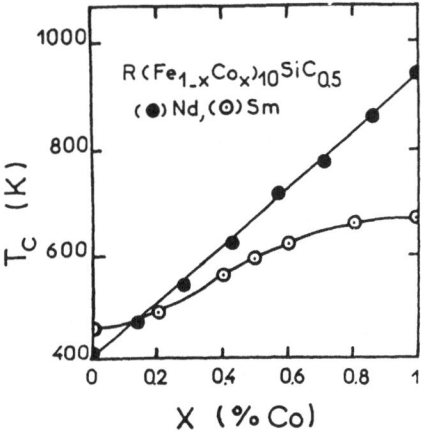

Fig. 7 : Composition dependance of Curie temperature of $R(Fe_{1-x}Co_x)_{10}SiC_{0.5}$

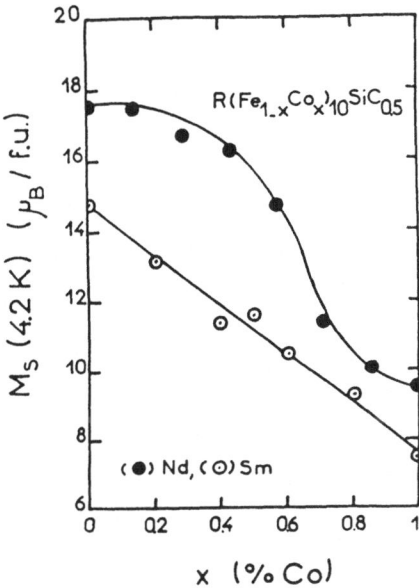

Fig. 8 : Composition dependance of saturation magnetization at 4.2 K for $(Fe_{1-x}Co_x)_{10}SiC_{0.5}$

CONCLUSION

Synthesis of new iron-rich and cobalt-rich quaternary compounds has been made by conventional arc-melting techniques and crystallographic properties have been studied by X - ray diffraction. The $R(Fe_{1-x}Co_x)_{10}SiC_{0.5}$ compounds can be used as permanent magnet materials. A large amount of work is now necessary to study the unpredictable existence of others iron-rich or cobalt-rich phases stabilized by B, Si and C.

REFERENCES

Bodak, O.I. and Gladyshevskii, F.I. Dopov. Akad. Nauk. Ukr. RSR, Ser A, 5 (1969) 452-455.

Chevalier, B. Gurov, G.Fournes, L. and. Etourneau, J. Communication au congrès de la société française de chimie, Paris, September 8-12, (1986).

Givord, D. Li, H.S. Moreau, J.M. Solid State Commun 50, 497, (1984).

Le Roy, J. Moreau, J.M. Bertrand, C. and Frémy, M.A. Journal of the Less Common Metals, 136, (1987), 19-24.$C_{0.5}$

Paccard, L. Paccard, D. and Bertrand, C. Journal of the Less Common Metals, 135, (1987), L5-L8.

Paccard, L. and Paccard, D. Journal of the Less Common Metals, 136, (1988), 297-301.

INTRINSIC AND EXTRINSIC MAGNETIC PROPERTIES OF RARE EARTH-TRANSITION METAL-METALLOID ALLOYS

C. Christides, A. Kostikas, D. Niarchos, A. Simopoulos, G. Zouganelis
"Demokritos" National Research Center, Institute of Materials Science,
153 10 Ag. Paraskevi, Athens, Greece

ABSTRACT

Several groups of rare earth-iron-boron or transition metal alloys have been examined by various experimental techniques with the aim to understand the magnetic properties at the atomic level and relate them to the macroscopic magnetic properties. Investigations of substituted $R_2Fe_{14-x}Co_xB$ alloys (R=Y,Dy; x=0,1,2) have shown that Co enters preferentially in the c or k_2 site of the $R_2Fe_{14}B$ structure. The change of the easy direction of magnetization from the plane to the c axis has been studied in a series of pseudoternary alloys of the type $Er_{2-x}Dy_xFe_{14}B$ with $0 < x < 1.5$. Within the R-Fe-B system the compounds of the type $R_{1+e}Fe_4B_4$ and RFe_4B were also investigated. The first group forms across the whole lanthanide series. Magnetic ordering occurs at temperatures between 4.2 and 25 K for R=Pr, Nd, Sm, Gd, Tb, Dy, Ho. The RFe_4B compounds (R=Er,Tm) have axial anisotropy. The distribution of Fe and B in the i and (c,d) sites of the hexagonal P6/mmm structure was investigated by Moessbauer and X-ray measurements. A series of new alloys of the pseudobinary type $RFe_{12-x}T_x$ (R=Gd, Nd; T=Ti, V, Cr, Mo) crystallizing in the $ThMn_{12}$ structure has been prepared. They show axial anisotropy and anisotropy fields $H_A \simeq 3$ kA/m. A two phase system prepared by annealing melt spun samples with composition $R_2Fe_7B_3$ was found to possess high coercivities at room temperature.

INTRODUCTION

The full understanding of the outstanding magnetic properties of R-Fe-B alloys and more particularly $Nd_2Fe_{14}B$ requires the knowledge of magnetic parameters at the atomic level. This knowledge is also essential in the search of new alloys with similar or improved properties. With this aim in mind, the work accomplished by our group within the CEAM project was focused in the following areas:

(a) Investigations of modified $R_2Fe_{14}B$ compounds by substitution of Fe or the R element in order to establish the site preferences, if any, and the effect on magnetic anisotropy.

(b) Studies of other compounds in the R-Fe-B system as $R_{1+e}Fe_4B_4$ and RFe_4B.

(c) Search for new magnetic phases with the $ThMn_{12}$ type structure.

(d) Studies of the high coercivities developed in two phase systems prepared by proper thermal treatment of alloys with composition $Nd_2Fe_7B_3$.

A variety of preparative and analytical techniques were used for these investigations including arc melting and melt spinning, X-ray diffraction (XRD), Moessbauer and magnetization measurements and differential thermal analysis (DTA). A summary of the results obtained is given in the following sections. Part of this investigation has been given in greater detail in journal articles. (Kostikas et al., 1985, Niarchos et al., 1986a, Niarchos and Simopoulos, 1986b, Simopoulos and Niarchos, 1988).

EXPERIMENTAL PROCEDURES

Ingots of the studied alloys were prepared by arc melting stoichiometric amounts of the elements (purity 3N for the rare earths, 4N for Fe and B) in Ar atmosphere. Bulk samples were studied as prepared or after annealing at appropriate temperatures for several days. Melt spun samples were prepared in a locally constructed melt spinning system with the following features: Cu wheel diameter 17 cm, width 1.5 cm, speed variable up to a maximum of 80 m/s.

Moessbauer spectra were obtained with polycrystalline or oriented absorbers using a conventional constant acceleration spectrometer with a ^{57}Co in Rh source. Magnetization measurements were carried out with a PAR 55 vibrating sample magnetometer in fields up to 2.0 T. The phase constitution and crystal structure was checked by X-ray powder patterns using Co-Kα radiation. DTA data were obtained with a Perkin Elmer Thermal Analyzer.

Oriented samples were prepared by mixing fine powder with epoxy and letting the mixture to harden in an applied field of 1.8 T.

RESULTS

<u>Studies of substituted $R_2Fe_{14-x}Co_xB$ alloys (x=1,2,4; R=Y,Dy).</u>

Substitution of Fe by Co in the $Nd_2Fe_{14}B$ alloy leads to an increase of the Curie temperature while the spontaneous magnetization decreases. (F. Bolzoni et al. 1987). In connection with this effect it is of interest to determine the site preference, if any of the Co atoms. We have studied a series of Co substituted alloys of the type $R_2Fe_{14-x}Co_xB$ near the low Co concentration side (x=1,2,4). As it has been shown by several studies (e.g. Kostikas et al.,1985), the Moessbauer spectra of the unsubstituted compounds have sufficient resolution to distinguish different Fe sites in the $Nd_2Fe_{14}B$ structure by their hyperfine parameters. The sites where Co enters can therefore be inferred from the changes in the ^{57}Fe spectrum. Since the reliability of this assignment depends on the

knowledge of the hyperfine parameters for the unsubstituted compounds, we
have reexamined the fitting of Moessbauer spectra especially with regard to
the parameters of the c and e sites for which there is no general
agreement. We have shown that in $Dy_2Fe_{14}B$ and $Nd_2Fe_{14}B$ a better fit can be
obtained by assigning a value of 31.0 T for the c-site which is higher than
that used in other simulations (Fruchart et al., 1987). This value,
however, gives better agreement with neutron diffraction data (Givord et
al., 1985).

Another point that must be taken into account in simulating spectra
of the substituted compounds is the line broadening resulting from a
possible distribution of Co. We have accounted for this by allowing for a
spread ΔH in hyperfine fields. A comparison of the spectra of the Dy
compound for x=0 (unsubstituted) and x=2 shows clearly that there is a
reduction in intensity in the line near -5 mm/s. Satisfactory fits can be
obtained by assuming that the Fe atoms are substituted either at the k_2 or
c site. It is interesting to note that these are the sites with the
shortest average Fe-Fe distances (0.2535 and 0.2532 nm respectively).
Similar conclusions were reached with the analysis of the Y alloy.

Complementary information on site substitution was obtained by
Moessbauer measurements on an ^{57}Fe doped sample of $Nd_2Co_{14}B$ with
stoichiometry $Nd_2{}^{57}Fe_{0.75}Co_{13.25}B$ (Simopoulos and Niarchos,1988). We
conclude that Fe enters preferentially into the j_2 site, in agreement with
a similar previous investigation (Van Noort, 1985a). As noted by Bolzoni
et al. (1987), this is the site with the highest Fe-Fe distance (0.2698 nm)
and fewest Nd neighbors.

Spin Reorientation in $Er_xDy_{2-x}Fe_{14}B$ alloys

Crystalline electric field (CEF) interactions induce uniaxial
anisotropy in $R_2Fe_{14}B$ compounds for the rare earths with negative Stevens
factor ($a_j<0$) while the elements with $a_j>0$ display basal anisotropy. In
the latter case basal anisotropy prevails at low temperatures, but as the
temperature increases the uniaxial anisotropy of the Fe sublattice
dominates. The change of direction of the magnetization from the plane to
the c axis, known as spin reorientation, occurs for $Er_2Fe_{14}B$ ($a_j<0$) near
325 K and is absent in $Dy_2Fe_{14}B$ ($a_j>0$). The effect on the spin
reorientation of the coexistence of two rare earth elements with competing
anisotropy was studied in the pseudoternary system $Er_{2-x}Dy_xFe_{14}B$ with
Moessbauer spectroscopy of oriented samples (Niarchos and Simopoulos,

1986b). The spin reorientation temperature was determined by the change in the intensity of the $\Delta m=0$ Moessbauer absorption lines. Since the absorbers are polarized at 320 K, well above the spin reorientation temperature (Tr), in a direction parallel to the gamma rays, this direction coincides with the c axis of the crystallites. Above T_r therefore, the magnetization is parallel to the gamma rays and the $\Delta m=0$ lines must vanish. This effect is shown in Fig.1 for the compound $Er_{1.95}Dy_{0.05}Fe_{14}B$. The disappearance of the $\Delta m=0$ lines proceeds gradually within an interval of 12 K. The width of the transition may be attributed to inhomogeneities depending on sample preparation and the distribution of Dy. In fact water quenched samples display a narrower width of the transition than slowly cooled samples. Alternatively this effect may be due to a gradual spin reorientation which increases with Dy concentration.

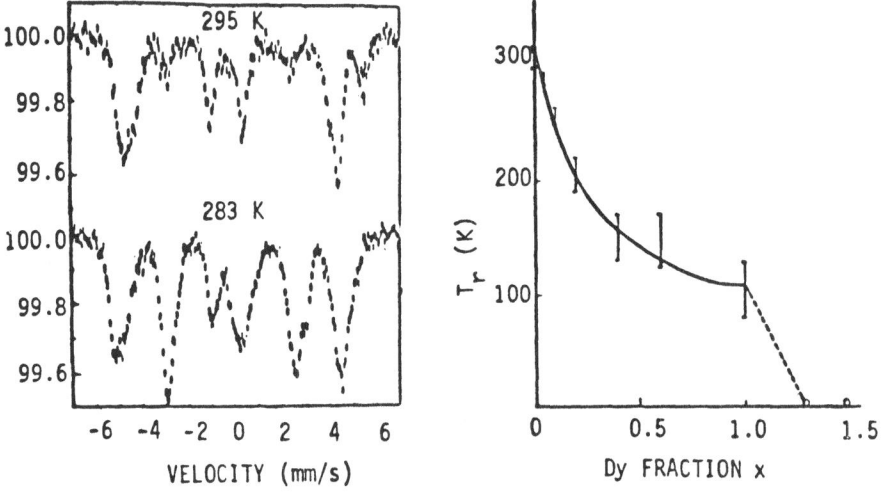

Fig. 1 (left) Moessbauer spectra above (T=295 K) and below (T=283 K) the spin reorientation temperature for an oriented sample of $Er_{1.95}Dy_{0.05}Fe_{14}B$.

Fig. 2 (right) The variation of the spin reorientation temperature in the series of $Er_{2-x}Dy_xFe_{14}B$ as a function of x.

The dependence of T_r on the concentration of Dy is shown in Fig. 2. The error bars indicate the temperature range where intermediate intensities were observed for the $\Delta m=0$ Moessbauer lines. The two samples with values of x equal to 1.3 and 1.5 did not show any spin reorientation down to temperatures of 6 K. This behavior is different than that observed for the $Er_{2-x}Gd_xFe_{14}B$ system (Vasquez and Sanchez, 1987) where T_r decreases smoothly over the whole concentration range of Gd. The reason is presumably that the CEF of Dy dominates that of Er at concentrations larger than 50% while Gd as an S state ion does not contribute to the CEF anisotropy energy and acts only in "diluting" the CEF anisotropy of the Er sublattice.

Another point of interest in Fig 1 is the disappearance of the line at the most positive velocity during the transition from 295 to 283 K. This corresponds to the j_2 iron site in the notation of Herbst et al., 1984 which displays the largest quadrupole interaction in the system. Its disappearance as the temperature is lowered is consistent with a spin reorientation from the c-axis to the basal plane which results effectively in a change of the quadrupole coupling by a factor of -1/2 and therefore shifts the line to lower velocities. This effect can be used to determine the spin reorientation with non oriented absorbers.

Magnetic Properties of $R_{1+e}Fe_4B_4$.

The phase diagram of Nd-Fe-B contains other ternary phases than the $Nd_2Fe_{14}B$ (Φ phase), among them most notable the η phase with approximate composition $NdFe_4B_4$. This phase is present in most commercially produced magnets (Kostikas et al.,1985) and plays an important role, as a precipitate near grain boundaries, in magnetization reversal processes (Wallace, 1984).

We have performed a systematic study of magnetic properties of the series of compounds with general formula $R_{1+e}Fe_4B_4$ which were prepared for all rare earths from La to Tm, with the exception of Eu. According to previous crystallographic studies (Givord et al., 1985) the structure of this phase consists of two closely interpenetrating sublattices of tetragonal symmetry with common basal plane parameters containing the Fe-B and rare-earth atoms respectively. On this basis the X-ray powder diagrams were indexed as two sets of reflections with common l=0 components corresponding to the space groups I4/mmm and $P4_2$/ncm for the R and Fe-B sublattice respectively. Using six reflections for the first sublattice and five for the second we have determined the common basal lattice parameter a and the c axis parameters c_R and c_{Fe}. The parameter e, is then calculated as $e=c_{Fe}/c_R-1$. We find that c_{Fe} is essentially constant for the whole series while c_R decreases consistently as R varies from La to Tm. This result can be attributed to the contraction of atomic radius along the rare earth series.

Magnetization measurements have shown that with the exception of La, Ce, Er and Tm all other compounds exhibit magnetic ordering transitions in the range between 4.2 and 25 K. Transition temperatures, determined by Arrot plots are given in Table 1. In the same table are also listed the magnetic moments per formula unit as calculated from the value of the magnetization at 4.2 K in a field of 1.8 T.

TABLE 1. Magnetic data and hyperfine parameters of $R_{1+e}Fe_4B_4$

R	$T_c(K)$		M(BM)	I.S.[a]	ΔE_Q
	Exp	Cal.		(mm/s)	(mm/s)
				at 300 K	
La	-	-	-	0.034(5)	0.632(5)
Ce	(b)	0.3	-	0.010(5)	0.530(5)
Pr	7.5(5)	1.2	1.50	0.032(5)	0.584(5)
Nd	16.0(5)	2.9	2.40	0.026(5)	0.582(5)
Sm		7.1		0.023(5)	0.566(5)
Gd	24.0(5)	24.0	6.02	0.022(5)	0.556(5)
Tb	16.0(5)	16.0	3.50	0.018(5)	0.528(5)
Dy	11.5(5)	11.2	5.70	0.014(5)	0.532(5)
Ho	6.5(5)	7.11	5.50	0.037(5)	0.544(5)
Er	(b)	4.03	-	0.024(5)	0.548(5)
Tm	(b)	1.84	-	0.024(5)	0.546(5)

a. With respect to iron at room temperature.

b. No transition observed down to 4.2 K.

In a simple molecular field modeling of the magnetic interactions, the transition temperatures are expected to follow the de Gennes values, i.e. $T_c \propto (g-1)^2 J(J+1)$. Values calculated from this relation and normalized to the experimental result for Gd are compared with experimental values in Fig. 3. It is seen that the de Gennes rule is followed very well by the heavy rare earth compounds but significant discrepancy is observed for the light rare earths. Deviations of this type may be attributed to CEF effects (Noakes et al., 1983). The presence of CEF effects is also indicated by the reduced values of the high field magnetic moments listed in Table 1.

The magnetization results imply that the Fe atoms do not carry a magnetic moment in agreement with reported measurements of Givord et al., (1986). A clear demonstration of the absence of magnetic moment for Fe has been given by Moessbauer measurements. The Moessbauer spectra of all compounds consist of a simple quadrupole doublet down to 4.2 K, with

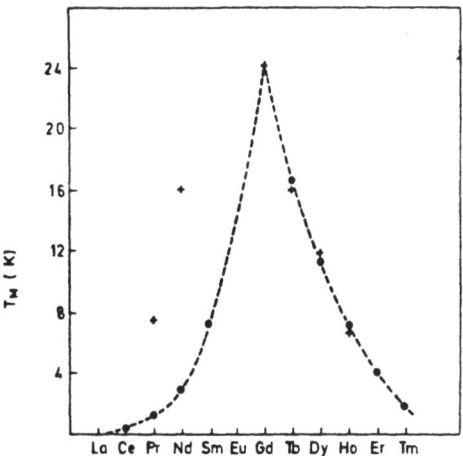

Fig. 3 Magnetic ordering temperatures of the $R_{1+e}Fe_4B_4$ compounds; (+) experimental results; (o) values calculated from T $(g-1)^2J(J+1)$ and normalized to the value for Gd.

parameters listed in Table 1. The possibility of a small magnetic moment at the Fe atom was further probed by the application of an external field of 0.65 T perpendicular to the gamma rays. For the Gd compound at 4.2 only a broadening of the quadrupole doublet was observed. A simulation of this spectrum assuming an effective field at various directions to the EFG and taking a powder average yields an estimate of the effective field of 0.7±0.2 T which, within the experimental accuracy, is equal to the applied field.

Structure and magnetic properties of RFe₄B alloys (R=Er,Tm)

Previous work on RFe_4B compounds (R=Lu,Er,Tm) has established the uniaxial ferrimagnetic structure of these alloys and their basic magnetic parameters (Van Noort et al. 1985b, Vaishnava et al. 1985). Hyperfine magnetic fields have been obtained from Moessbauer spectra but there are discrepancies in site assignments, mainly due to the possibility of Fe and B disorder in the i and (c,d) sites of the $CeCo_4B$ structure assumed for these compounds.

We have obtained magnetization data vs T in the temperature range of

4.2 to 700 K and in applied fields of 40 and 1600 kA/m. The ErFe$_4$B data (Fig. 4) show a clearly defined compensation point arising from antiferromagnetic coupling of the Er and Fe sublattices. The compensation point in the Tm alloy is apparent only in the high field data. There is no anomaly, indicative of a spin reorientation, as suggested in previous investigations (Van Noort et al., 1985b). The absence of spin reorientation was verified by Moessbauer spectra of an oriented sample at 77 K and room temperature.

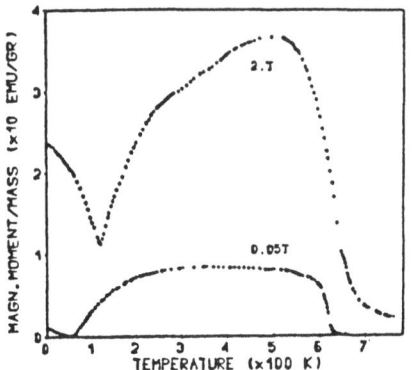 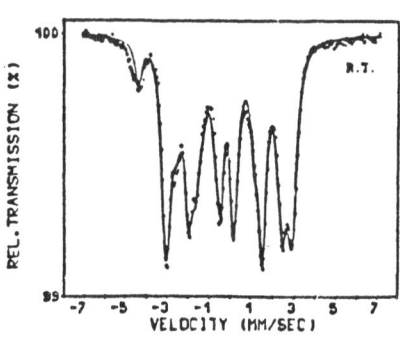

Fig. 4 (left) Temperature dependence of the magnetization of a ErFe$_4$B alloy in fields of 0.05 and 2.0 T.

Fig. 5 (right) Moessbauer spectrum of a TmFe$_4$B alloy at room temperature. The solid line is a computer fit with the distribution model described in the text.

Powder X-ray diffraction diagrams were indexed with the reflections of the space group P6/mmm corresponding to the CeCo$_4$B-type structure. A notable feature of the data is the absence of (h,k,2l+1) lines. This indicates that instead of a fully ordered CeCo$_4$B structure, a disordered configuration of Fe and B in the (c,d) and i sites may occur. To test this assumption we have calculated intensities with occupation probabilities as determined by the Moessbauer data (see below). This resulted in an improvement of the reliability factor from 15% for a fully ordered structure to 7% for the disordered structure.

Moessbauer spectra were obtained with polycrystalline absorbers at room temperature and 4.2 K. The spectra for the two alloys at both temperatures are similar. We have analyzed these spectra with a model which assumes different probabilities p_i and p_c for occupation by B of the sites i and (c,d). The 2c and 2d sites were taken equivalent since they correspond to the same site of the parent CaCu$_5$ structure. Assuming binomial distribution of the B atoms in the i and (c,d) sites we can readily calculate the probabilities of nearest neighbor configuration for an Fe atom in the i and (c,d) sites. A more detailed account of this calculation will be given in a forthcoming publication.

The Moessbauer lineshape calculated with this model is compared with the experimental data in Fig. 5. The main result of this analysis is that the values of p_i and p_c are significantly different (p_i=0.033, p_c=0.339), i.e. there is a preference of B to occupy the (c,d) site. For a statistical distribution p_i=p_c=0.2. The observed values of p_i and p_c imply an average site occupation in the unit cell as follows: i-site (0.2B, 5.8Fe), (c,d)-site (1.4B, 2.6Fe). These values do not agree exactly with stoichiometry but the discrepancies are probably within the uncertainty of the determination of p_i and p_c.

Magnetic properties of RFe$_{12-x}$T$_x$ alloys

Within the scope of the search for new permanent magnet materials we have started an extensive study of pseudobinary alloys with general formula RFe$_{12-x}$T$_x$ which may crystallize in the tetragonal ThMn$_{12}$ type structure.

The value of x was chosen near the composition predicted by the phenomenological method of Pettifor (1986), applied to pseudobinary alloys with the ThMn$_{12}$ structure. Samples were prepared for the following combinations of R and T: (R=Gd; T=Ti, Cr, Mo), (R=Nd; T=V), (R=Y; T=Mo,Ti), (R=Sm, T=Mo,Ti). In the case of the Gd-Mo alloys a series of samples with x=1.8, 2.0, 2.2 was prepared in order to test the homogeneity, patterns can be indexed with the reflections of the ThMn$_{12}$ type structure as the dominant phase.

TABLE 2. Lattice constants and magnetic data for RFe$_{12-x}$T$_x$ alloys

Alloy	a (nm)	b (nm)	T_c (K)	M_s (μ_B)
Nd$_{1.2}$Fe$_{9.8}$V$_{2.2}$	0.855	0.477	525	15.05
Gd$_{1.1}$Fe$_{9.5}$Cr$_{2.5}$	0.845	0.475	608	10.62
Gd$_{1.1}$Fe$_{10.2}$Mo$_{1.6}$	0.854	0.479	446	7.00
Gd$_{1.1}$Fe$_{10}$Mo$_2$	0.856	0.480	410	5.69
Gd$_{1.1}$Fe$_{9.8}$Mo$_{2.2}$	0.858	0.480	345	4.52

We report here results for the Nd and Gd alloys. The lattice
constants, Curie temperatures and magnetization in Bohr magnetons per
formula unit at room temperature are listed in Table 2. Magnetization
measurements with oriented samples in applied fields up to 2T give an
extrapolated estimate of $H_c=250$ kA/m. The temperature dependence of the
magnetization for the Mo alloys is shown in Fig. 6.

Moessbauer spectra show considerable broadening of the magnetic
transition metal T in the sites i,j and f of the $ThMn_{12}$ structure. Fig. 8
shows the simulation of the spectrum of $GdFe_{11}Ti$ with the assumption that
Ti is statistically distributed over the lattice sites. A spread of

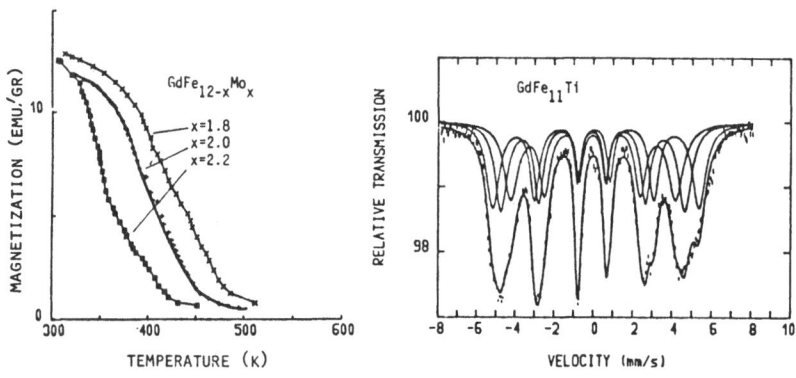

Fig. 6 (left) Temperature dependence of the magnetization of
$GdFe_{12-x}Mo_x$ alloys in a field of 50 mT.

Fig. 7 (right) Moessbauer spectrum of a $GdFe_{11}Ti$ alloy at room
temperture. The solid line is a simulation assuming that Ti is
statistically distributed in the i,j and f sites.

hyperfine fields has been taken into account for each site in this
simulation.

Similar results on this type of alloys have been reported by
de Boer et al. (1987). It appears that these alloys have sufficiently high
anisotropy but their saturation magnetization is low compared to the
$R_2Fe_{14}B$ phases.

We have made also preliminary measurements on a series of Sm alloys
with T=V, Si, Ti and x=2 provided within the CEAM project by RARE EARTH
PRODUCTS. XRD and Moessbauer measurements show significant iron content in
the V and Si alloys.

High Coercivity R-Fe-B two phase alloys (R=Pr,Nd,Sm).

The possibility of preparation of high coercivity alloys with more than one crystalline phases and the appropriate microstructure was explored in R-Fe-B alloys with nominal stoichiometry $Nd_2Fe_7B_3$. Samples were prepared from ingots of this composition by melt spinning and subsequent annealing at various temperatures.

Differential thermal analysis (DTA) measurements on the melt spun samples show a sharp peak near 870 K indicating the precipitation of a new phase. The nature of this transformation was examined by magnetization and Moessbauer measurements. The magnetization of the melt spun samples of the three alloys studied drops to a value close to zero in the range of 310 to 330 K. Upon further heating to the temperature of the transformation and cooling back, the presence of a new phase with a Curie temperature near 580 K is ascertained. Moessbauer spectra were obtained at room temperature for a $Nd_2Fe_7B_3$ sample before and after the thermal cycling. The spectrum of the melt spun sample is typical of an amorphous phase. The spectrum of the annealed sample on the other hand shows clearly a magnetic and a para-magnetic phase. The magnetic pattern is typical of $Nd_2Fe_{14}B$ while the parameters of the paramagnetic doublet fit those of $Nd_{1+e}Fe_4B_4$ (Kostikas et al., 1985, Niarchos et al., 1986a). A computer fit of the spectrum with two components verifies this assignment and yields an approximate phase

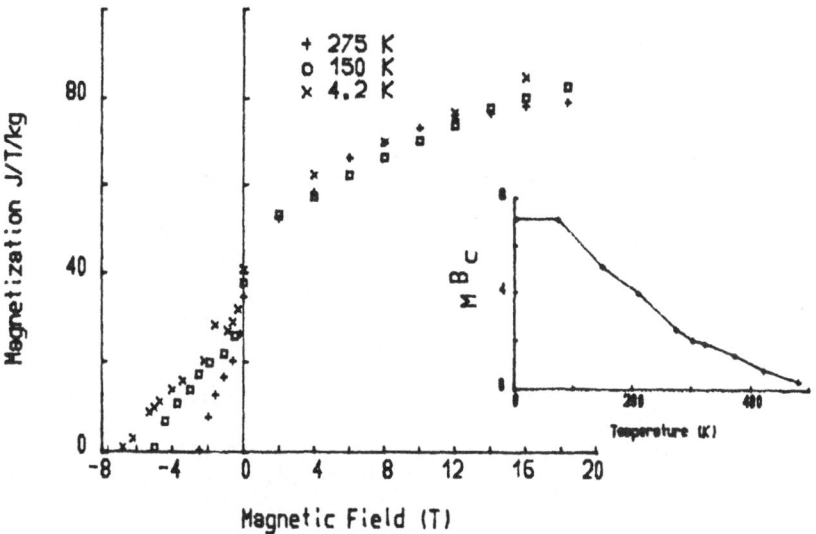

Fig. 8 Coercivity of two phase $Pr_2Fe_7B_3$ alloys. The insert shows the variation of $_MB_c$ with temperature.

composition for the iron containing phases of 30% $Nd_2Fe_{14}B$ and 70% $Nd_{1+e}Fe_4B_4$.

The coercivity of the melt spun and annealed samples was studied in the temperature range of 4.2 to 500 K. Results obtained at three different temperatures at the high field facility of SNCI are shown in Fig. 8 for a sample of $Pr_2Fe_7B_3$. The coercive field $_MB_c$ rises from about 2T at room temperature to 7T at 4.2K. The insert shows the temperature variation of $_MB_c$ up to 500 K. Preliminary measurements on samples annealed at different temperatures show that the coercive field is strongly dependent on the annealing temperature. The remanent magnetization, however, is low, 30 J/T/kg presumably due to the presence of the paramagnetic phase. Further study is required to optimize the properties of these alloys with respect to composition and thermal treatment. The data of Fig. 8 also show that the sample must contain up to 30% of an additional magnetic phase if we assume similar M_s. The same result could also be explained with about 15% Fe content. The magnetization and Moessbauer data, however, do not show evidence for an additional magnetic iron containing phase. The reason for this discrepancy is not clear at present.

ACKNOWLDEGEMENT

We thank Dr. J. Gavigan for the hysteresis data of $Pr_2Fe_7B_3$ obtained at the SNCI facility.

REFERENCES

Bolzoni F., Coey, J.M.D., Gavigan, J., Givord, D., Moze, O., Pareti, L. and Viadieu, T. 1987. J. Magn. Magn. Mater. 65, 123-127.

De Boer, F.R., Huang Ying-Kai, de Mooij, D.B. and Buschow, K.H.J. Phillips J. Res. 42, 246-251.

Fruchart, R., L'Heritier, P., Dalmas de Reotier, P., Fruchart, D., Wolfers, P., Coey, J.M.D., Ferreira, L.P., Guillen, R., Vulliet, P. and Yaouonc, A. 1987. J. Phys. F: Met. Phys. 17, 483-501.

Givord, D., Moreau, J.M. and Tenaud, P. 1985. Solid State Commun. 55, 303.

Givord, D., Li, H.S., Moreau, J.M. and Tenaud, 1986. J. Magn.Magn. Mater. 54-57, 131.

Herbst, J.F., Croat, J.J., Pinkerton, F.E. and Yelon, W.B. 1984. Phys. Rev. B29, 4176.

Kostikas, A., Papaefthymiou, V., Simopoulos, A. and Hadjipanayis, G.C. 1985. J. Phys. F: Met. Phys., 15, L129-L133.

Niarchos, D., Zouganelis, G., Kostikas, A. and Simopoulos, A. 1986a. Sol. State Commun. 59, 389-391.

Niarchos, D. and Simopoulos, A. 1986b. Sol. State Commun. 59, 669-672.

Noakes, D.R., Shenoy, G.K., Niarchos, D., Umarji, A.M. and Aldred, A.T. 1983. Phys. Rev. B27, 4317.

Pettifor, D.G., 1986. J. Phys. C: Solid State Phys. 19, 285-313.

Simopoulos, A. and Niarchos, D. 1988. Hyperfine Interactions (in press).

Van Noort, H.M. and Buschow, K.H.J. 1985a. J. less-Common Met. <u>113</u>, L9-L13.

Van Noort, H.M., De Mooij, D.B. and Buschow, H.J. 1985b. J. Less Common Met., <u>111</u>, 87-95.

Vaischnava, P.P., Kimball, C.W., Umarji, A.M., Malik, S.K. and Shenoy, G.K. 1985. J. Magn. Magn. Mater. <u>49</u>, 286-290.

Vasquez, A. and Sanchez, J.P. 1987. J. Less Common Met. <u>127</u>, 71.

Wallace, W.E. 1984. J. Less-Comm. Met. <u>100</u>, 85.

A SEARCH FOR NEW Fe-RICH PHASES FOR THE DEVELOPMENT OF PERMANENT MAGNETS

G.C. Hadjipanayis[*]

Research Center of Crete
Heraklion, Crete, GREECE

ABSTRACT

Melt-spinning has been used as a quick tool to search for new Fe-rich anisotropic phases in R-Fe-B(Si), R-Fe-C and R-Fe-Ti systems. In $PrFe_{9.8}B_{3.2}$ a metastable phase with $T_c \sim 380°C$ has been observed which transforms to the stable Fe_2B upon a longer heat-treatment. In R-Fe-Ti the two ternary phases observed are associated with the $ThMn_{12}$ type structure. The high coercivities obtained in R-Fe-C(B) alloys are possibly due to the small size of grains and the presence of an unknown phase with composition R-Fe-C which is uniformly distributed throughout the sample.

[*]Currently at the Department of Physics, Kansas State University.

INTRODUCTION

Since the discovery of R-Fe-B magnets (Hadjipanayis et al. 1983; Sagawa et al. 1984) there has been an increased effort in the search for new magnetic phases with high magnetization, large anisotropy and high Curie temperature. Most of the studies have been focused on Fe-rich alloys containing some rare-earth (for the high anisotropy) and some other element to stabilize the "new" ternary phase.

In this article we summarize our search efforts in the systems R-Fe-B, R-Fe-Si, R-Fe-C and R-Fe-Ti.

EXPERIMENTAL

Rapid solidification processes such as melt-spinning and splat-cooling are widely used in the preparation of amorphous and metastable phases. Rapid solidification is a quick and powerful technique in the search for new phases. We have used melt-spinning in Fe-R alloys to prepare metastable phases because no stable compounds of Fe exist with light rare earths. This procedure has led to the discovery of the 2:14:1 structure (Croat et al. 1986, Givord et al. 1984).

The crystallization of amorphous samples was studied by differential scanning calorimetry (DSC) using a Dupont 1090 thermal analyzer. The crystallization temperatures obtained from the DSC data were used to heat-treat the melt-spun samples for optimum magnetic hardening. A thermomagnetic experiment was used first as a simple heat treatment. In this experiment, the amorphous sample was heated up to a temperature greater than the crystallization temperature, while the magnetization in a constant magnetic field, $M_H(T)$ data showed any magnetic phases or magnetic transitions that existed in the system over the temperature range studied. These results were compared with DSC data and X-ray diffraction measurements for the complete identification of the phases resulting after crystallization. The hysteresis loops of melt-spun and heat-treated samples were compared to determine the effect of crystallization on the hard magnetic properties of the sample.

RESULTS AND DISCUSSION

$\underline{RFe_{9.8}M_{3.2}}$ (R=Pr,Nd M=B,Si)

This composition was chosen to stabilize the ternary phases with the cubic $NaZn_{13}$ and tetragonal $BaCd_{11}$ structures (Bodak and Gladyshevskij, 1969).

The melt-spun samples were found to be mostly amorphous with a Curie temperature 180 and 280°C for the $NdFe_{9.8}Si_{3.2}$ and $PrFe_{9.8}B_{3.2}$ samples, respectively (Fig. 1 and Fig. 2). Upon a heat-treatment at around 700°C the $PrFe_{9.8}B_{3.2}$ sample showed an interesting behavior. After the M vs T experiment up to 700°C, two new phases appeared with the majority phase having a Curie temperature T_c around 385°C and the minority phase a T_c above 700°C. With increasing heat-treatment, the amount of higher-T_c phase increased at the expense of the phase with lower-T_c (Fig. 2). After a heat treatment to 780°C the lower T_c phase disappeared and the only phase present is the one with $T_c \sim 740$°C which is believed to be Fe_2B. The metastable phase with $T_c \sim 385$°C has not yet been identified. For the $NdFe_{9.8}Si_{3.2}$ alloy, the phase obtained after the M vs T experiment (to 740°C) have a $T_c \sim 710$°C. This phase is believed to be an Fe-Si phase. Similar results were reported by Niarchos et al. (1986).

We have tried different heat-treatments to produce a single phase microstructure or a fine microstructure that is a mixture of the two phases in an attempt to increase H_c. The best results are shown in Fig. 3 where the coercivity is only 600 Oe.

$\underline{R_2(Co,Fe)_7B_3}$

The as-spun samples are mostly amorphous and magnetically soft. The amorphousness of the ribbons was checked by differential scanning calorimetry, which showed an irreversible exothermic peak at 562°C. This is a much higher temperature than the magnetic ordering temperature of the corresponding amorphous phase. X-ray diffraction of the as-spun ribbons

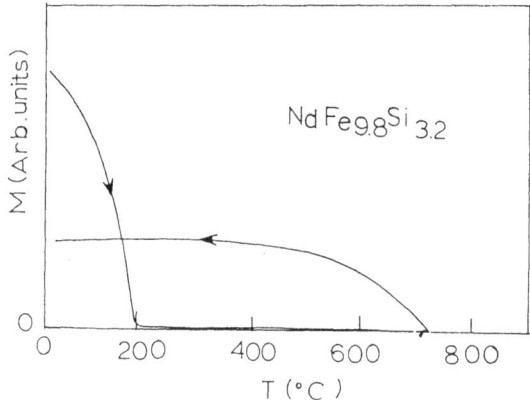

Fig. 1   Thermomagnetic data (M vs T).

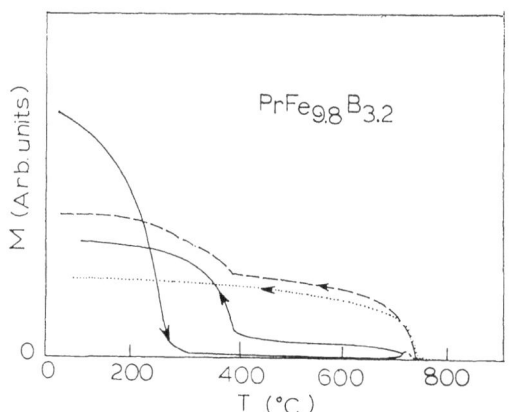

Fig. 2   Thermomagnetic data (M vs T).

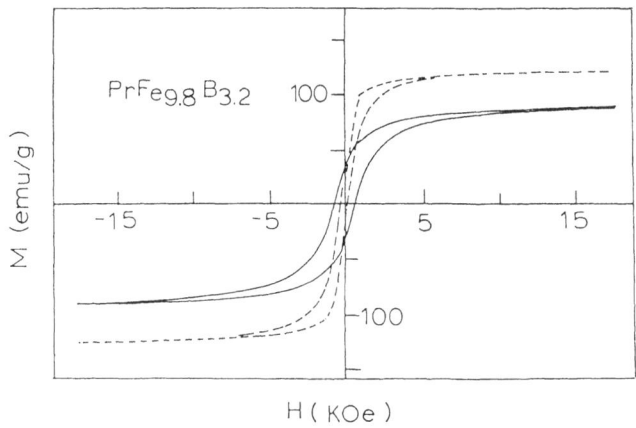

Fig. 3   Hysteresis loops in a heat-treated $PrFe_{9.8}B_{3.2}$ sample.

showed diffuse rings, indicative of an amorphous material. Similar
studies on crystallized ribbons showed the existence of a major phase,
which is identified as hexagonal   (Masry, 1980) $Sm_2Co_7B_3$, with the
exception of $Pr_2Fe_7B_3$ and $Nd_2Fe_7B_3$ which show a pattern slightly different
from the one above. In addition, some Co-α phase, in the cobalt
containing samples, was detected. This is consistent with the thermomagnetic data (Fig. 4) which showed a finite magnetization at temperature
as high as ~800°C. A similar behavior is also observed for the
$Nd_2Fe_5Co_2B_3$ sample. It is interesting to note that this compound has an
ordering temperature similar to that of $Sm_2Fe_2Co_5B_3$ sample, around 430°C.
In contrast the $(PrNd)_2Fe_7B_3$ compounds have a Curie temperature around
310°C, lower than the Co containing samples.

The hysteresis loops of the samples after the M vs T experiment are
shown in Fig. 5. The maximum available field of 17 kOe was not enough to
fully saturate the magnetization of samples. From the minor hysteresis
loop of a $Nd_2Fe_5Co_2B_3$ sample, a value of 13.8 kOe is obtained for H_c . The
coercivity could not be measured for the $Sm_2Fe_2Co_5B_3$ sample. Hysteresis
loops at lower temperatures and higher fields (70 kOe) were measured for
some of these samples to further study their hysteresis behavior, but even
this high magnetic field was not enough to saturate the magnetization.
The Fe-rich compounds have a much higher coercivity indicating a larger
magnetic anisotropy.

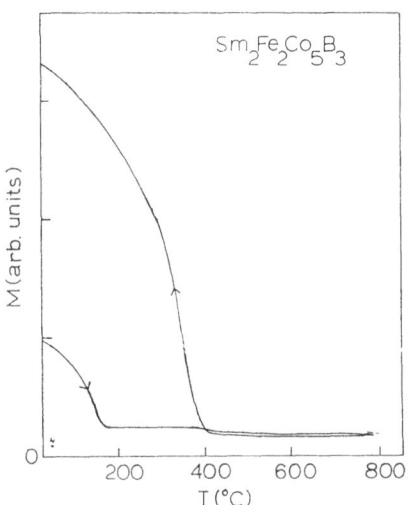

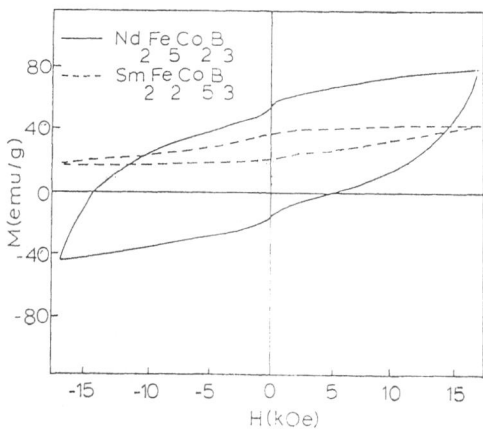

Fig. 4 Thermomagnetic data in
$Sm_2Fe_2Co_5B_3$.

Fig. 5 Hysteresis loops in heat-
treated samples.

R–Fe–C

Large coercivities have been reported (Liu and Stadelmaier, 1986) in as-cast Dy(Nd)–Fe–B(C) alloys. We have investigated the origin of these coercivities by correlating the magnetic properties with the micro-structure as observed with transmission electron microscopy.

The as-cast samples did not have a high coercivity. However, after a heat–treatment at around 900°C for 72 hours the samples became magnet-ically hard. Figure 6 shows the hysteresis loops of a $Dy_{15}Fe_{77}C_8$ and a $Nd_9Dy_6Fe_{77}C_{7.2}B_{0.8}$ sample. The maximum coercivities obtained for these samples in a field of 55 kOe are 23 and 8.3 kOe, respectively.

Fig. 6a

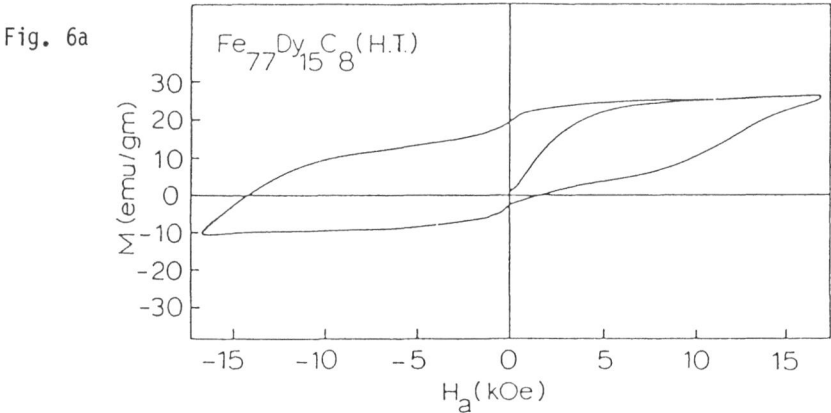

Fig. 6b

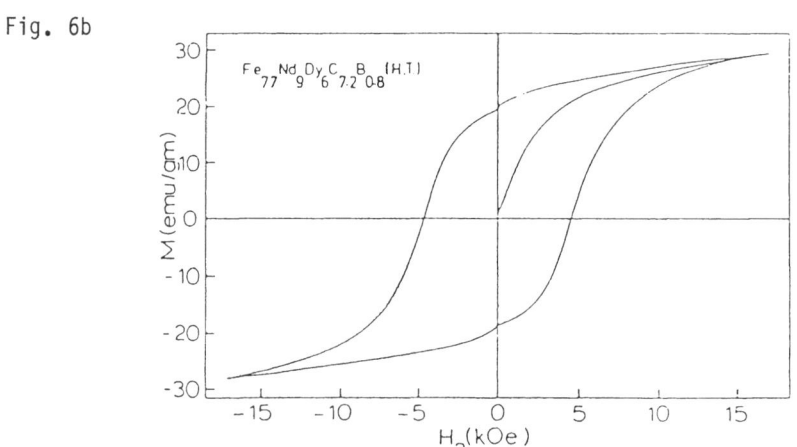

Fig. 6 Hysteresis loops in heat-treated as-cast samples (a) $Dy_{15}Fe_{77}C_8$ and (b) $Fe_{77}Nd_9Dy_6C_{7.2}B_{0.8}$.

Microstructure studies (Fig. 7) on as-cast samples showed the presence of several phases with the majority phase having the $R_2Fe_{14}C$ structure. Other phases include an "unknown" phase with possible composition RFeC and R_2Fe_{17}. It is interesting to note that the grain size of the carbon substituted samples is much lower than that of R-Fe-B samples. This fact together with the uniform distribution of R-Fe-C phase might be responsible for the high H_c.

R-Fe-Ti

As-cast alloys and melt-spun ribbons were prepared with compositions $R_yFe_xTi_z$ where $78 \leq x \leq 87$, $3 \leq y \leq 15$, $3 \leq z \leq 13$ and R=Sm, Nd. Thermomagnetic data on Fe-Sm-Ti as-cast samples showed the presence of two major magnetic phases with Curie temperatures T_c around 200 and 320°C, a minor phase with $T_c \sim 420$°C and some a-Fe (Fig. 8). In Nd-Fe-Ti as-cast samples two major magnetic phases with T_c about 150 and 280°C, and some α Fe have been observed. X-ray diffraction studies together with electron microprobe analysis indicate that the phase with $T_c \sim 420$°C is $SmFe_2$ while the other magnetic phases are ternary R-Fe-Ti compounds. In Nd-Fe-Ti a non-magnetic phase with composition Fe_2Ti and a Nd-rich phase is also observed. The two ternary phases have the compositions of Fe 83-85 at.% and the balance of Sm (or Nd) plus Ti with the ratio of Sm (or Nd) or Ti as 1:0.9 and 3:1, respectively. The Curie temperatures of these two ternary phases are 320°C and 200°C for the Sm system, and 280°C and 150°C for the Nd system with the higher temperature corresponding to the 1:0.9 rare earth to titanium ratio. In the past it was thought (Cadieu et al., 1984) that these phases have a stabilized $Sm(Fe,Ti)_5$ structure. Today it is believed (de Mooij and Buschow, 1987) that the phases have the $ThMn_{12}$ type phase with a=8.56 Å and c=4.79 Å. The differences in T_c may be related to the different amount of Ti present in the phases. A similar observation has been made by Buschow (1987). Compared to $Nd_2Fe_{14}B$, the $Fe_{83-85}(Sm+Ti_{0.9})_{17-15}$ phase has similar Curie temperature (320°C), but lower saturation moment (7000 G), much lower anisotropy field (20-30 kOe) and coercivity (maximum of 2.5 kOe in melt-spun ribbons); thus it is not as interesting a permanent magnetic material as $Nd_2Fe_{14}B$. Attempts to make a sintered magnet did not have much success. A post-sintering heat treatment of $Fe_{72.5}Sm_{14.5}Ti_{13}$ magnets increased the amount of α-Fe which explains the reduction of coercivity and the increase of magnetic moment after heat treatment. This problem might be solved by adding an excess amount of Sm.

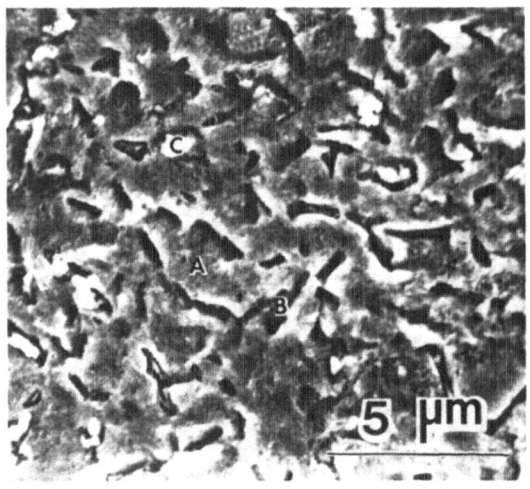

Fig. 7 Microstructure of as-cast $Dy_{15}Fe_{77}B_8$ A≡Matrix $Dy_2Fe_{14}C$
B≡Carbon rich Dy-Fe-C and C≡Dy rich region.

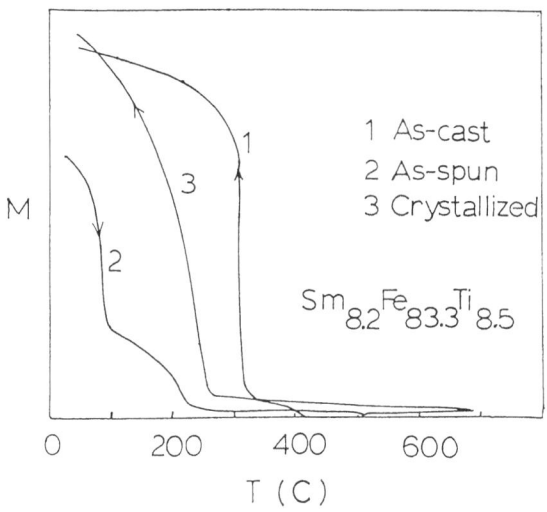

Fig. 8 Thermomagnetic data in R-Fe-Ti samples.

REFERENCES

Bodak, O.I. and Gladyshevskij. E.I. 1969. Dopov. Akad. Nauk, Ukr RSR, Ser. A12, 1125.

Bodak, O.I. and Gladyshevskij, E.I. 1969. Dopov. Akad. Urk. RSR, Ser. A5, 452.

Buschow, K.H.J. 1987. 3M Conference.

Cadieu, F.J., Cheung, T.D., Wickvamasekara, L. and Aly, S.H. 1984. J. Appl. Phys. 55, 2611.

Croat, J.J., Herbst, J.F., Lee, R.W. and Pinkerton, F.E. 1984. J. Appl. Phys. 55, 2079.

de Mooij, D.B., Buschow, K.H.J. and Philips, J. 1987. Res. 42, 246.

Givord, D., Li, H.S. and Perrier de La Bathie, R. 1984. Solid State Commun. 51, 857.

Hadjipanayis, G.C., Hazelton, R.C. and Lawless, K.R. 1983. Appl. Phys. Lett. 43, 797.

Liu, N.C. and Stadelmaier, H.H. 1986. Mater. Lett. 4, 377.

Masry, N.A.E. 1986. Ph.D. Thesis, North Carolina State University.

Niarchos, D., Roig, A., Rao, K.V. and Hadjipanayis, G. 1986. Proc. First Int. Workshop on Non Cryst. Solids, Felia de Guixols, Spain.

Sagawa, M., Fujimura, S., Yamamoto, H., Matsuura, Y. and Hiraga, K. 1984. IEEE Trans. Magnn. MAG–20, 1584.

NOVEL TERNARY Gd-Fe-Ti ALLOYS WITH ThMn$_{12}$ STRUCTURE

D. Cochet-Muchy and S. Païdassi

Centre d'Etudes Nucléaires de Grenoble
Département de Métallurgie
85X, 38041 Grenoble Cedex, France

ABSTRACT

Ingots of the nominal compositions GdFe10.5Ti1.5 and GdFe11Ti have been studied in their as-cast and annealed forms. They are primarily constituted by a ternary ThMn12-type compound. The samples show different secondary phases, among which another ternary compound. The ThMn12 structure has also been observed in a melt-spun GdFe10.5Ti1.5 ribbon studied by TEM. Preliminary magnetic measurements have been performed.

INTRODUCTION

In the search for new hard magnetic materials, it has been pointed out recently that ferromagnetic compounds of rare-earth and iron, with interesting Curie temperatures and uniaxial structure (tetragonal ThMn12 structure), are obtained by adding a third element (De Mooij *et al.*, 1988; Xiang-Zhong *et al.*, 1988). The choice of this third element (among Si, Ti, V, Cr, Mo, W, Al) is a critical problem due to its effects on Tc and on the Fe moment. We have chosen to perform a preliminary study on cast and rapidly quenched Gd-Fe-Ti alloys, with compositions surrounding the stoichiometry of the ThMn12-type compound.

CAST ALLOYS

Induction-melted ingots of the nominal atomic composition GdFe10.5Ti1.5 and GdFe11Ti are obtained from commercially available products (Gd,Fe 99.9%,Ti 99%). Some of them are vacuum annealed at 850°C for 250 hours. For all the powdered samples, the X-ray diffraction patterns show a set of reflections which are satisfactorily indexed on the basis of the tetragonal ThMn12 structure. The corresponding lattice constants are listed in Table 1. Weak lines corresponding to the Fe2Ti phase are also identified. A complementary phase determination has been performed by SEM and X-EDS semi-quantitative analysis, allowing the identification of other secondary phases. The results are listed in Table 1. T1 refers to the ternary ThMn12-type compound, for which the estimated Fe:Ti ratio is between 8.4 and 9.4, *i.e.* the stoichiometry is comprised between GdFe10.7Ti1.3 and GdFe10.8Ti1.2. T2 refers to another ternary compound in the Gd-Fe-Ti system, depleted in iron and richer in Ti (estimated composition: 6.5 at.% Gd, 81 at.% Fe, 12.5 at.% Ti). This phase could be related to the CeMn6Ni5-type compound reported by De Mooij *et al.* (1988) in the same system. These authors have also found a stoichiometry near GdFe10.8Ti1.2 for the T1 phase. Fig. 1 shows typical microstructures of the annealed samples (backscattered electron images). From the previous results, the isothermal section at 850°C of the ternary Gd-Fe-Ti phase diagram can be roughly drawn (Fig. 2): GdFe10.5Ti1.5 belongs to the three-phase region T1 + T2 + Fe2Ti. Due to its Gd:(Fe+Ti)=1:12 ratio, GdFe11Ti should be in the T1 + Gd2Fe17 + Fe region. The observation of Fe2Ti can be explained by a shift between the real and nominal compositions (Gd loss = x) which has led to overcome the very near binary limit Fe + T1.

TABLE 1   Phases present in the cast alloys and lattice constants of the ThMn12-type compound (T1).

	phases	a (nm)	c (nm)
$GdFe_{10.5}Ti_{1.5}$			
as-cast	T_1, Fe_2Ti	0.856	0.479
annealed	T_1, T_2, Fe_2Ti	0.855	0.480
$Gd_{1-x}Fe_{11+x}Ti$			
as-cast	T_1, Fe_2Ti, Fe	0.856	0.479
annealed	T_1, Fe_2Ti, Fe	0.854	0.479

TABLE 2   Saturation magnetizations at 4.2 K and Curie temperatures of the annealed cast alloys.

	M_S (μ_B/F.U.)	T_C (K)
$GdFe_{10.5}Ti_{1.5}$	11.8	600
$Gd_{1-x}Fe_{11+x}Ti$	13.7	600

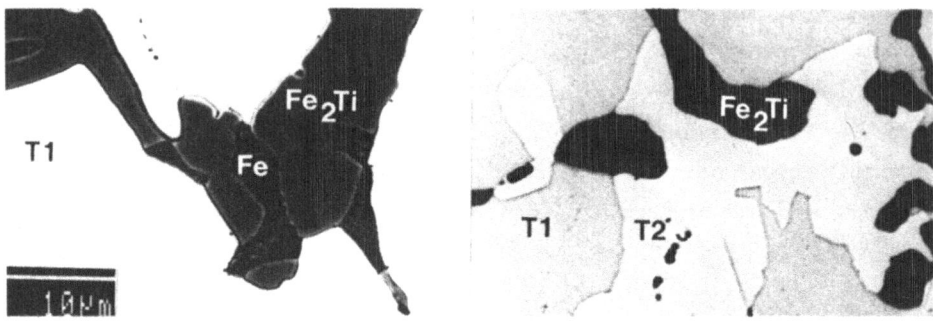

Fig. 1   Typical microstructures of the annealed $Gd_{1-x}Fe_{11+x}Ti$ (left) and $GdFe_{10.5}Ti_{1.5}$ (right). Backscattered electron images.

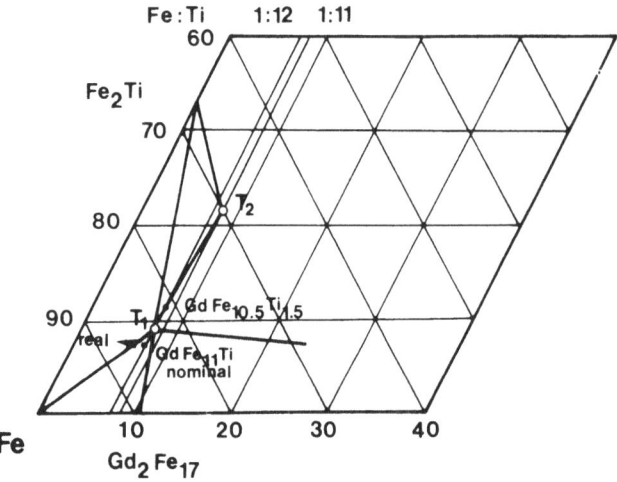

Fig. 2   Gd-Fe-Ti ternary phase diagram: Fe-rich part of an hypothetical isothermal section at 850°C.

MELT-SPUN RIBBON

A 15 μm thick ribbon of the GdFe10.5Ti1.5 nominal composition has been obtained by using the melt-spinning process under vacuum (wheel speed = 42 m/s). Pieces of the as-quenched ribbon have been thinned by one-beam or two-beams ion-milling, allowing TEM observations of regions at different depths. The electron micrographs show an homogeneous microstructure, with a nearly constant grain size around 20 nm (Fig. 3). All the rings appearing on the corresponding electron diffraction pattern are well fitted assuming a tetragonal ThMn12-type structure with a=0.85 nm and c=0.48 nm.

The material seems to be single phase. Its behaviour upon heating has been determined by DTA: two exothermal events are detected, at 720°C and 870°C (heating rate = 20 K/mn). Using X-ray diffraction, the first event is associated with the crystallization of a tetragonal CeMn6Ni5-type compound with a=0.826 nm and c=0.485 nm. The second one corresponds to the crystallization of Fe2Ti. This evolution to the equilibrium state is consistent with the phase constitution of the corresponding cast alloy reported in the previous section. Moreover, it shows that ThMn12-type single phase materials can be produced with a metastable composition by rapid quenching, which is very promising to investigate new compositions. After heating to 950°C, the average grain size has raised to 150 nm (Fig. 3) and the lattice constants of the major ThMn12-type phase are found to be a=0.854 nm and c=0.480 nm.

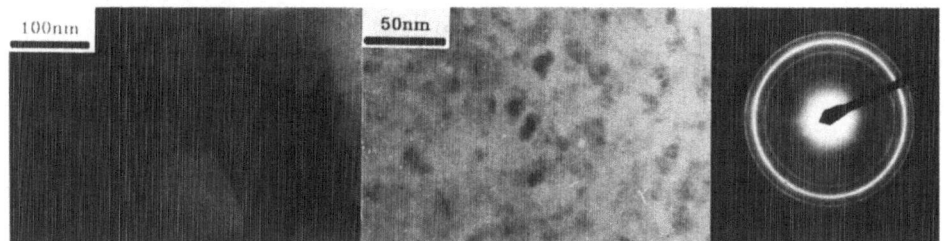

Fig. 3   Electron micrographs and diffraction pattern of the as-quenched ribbon (right) and heated to 950°C (left).

MAGNETIC PROPERTIES

Curie temperatures of 600 K have been observed by DTA for the cast alloys of both compositions and the ribbon. From magnetometric measurements, no coercivity has been found in the as-quenched ribbon. The saturation magnetizations measured at 4.2 K on the annealed cast-alloys are listed in Table 2.

REFERENCES

De Mooij D.B., Buschow K.H.J., 1988, J. Less-Common Met., 136, 207-215.
Xian-Zhong W., Chevalier B., Berlureau T., Etourneau J., Coey J.M.D., Cadogan J.M., 1988, J. Less-Common Met., 138, 235-240.

STRUCTURAL AND MAGNETIC PROPERTIES OF RE$_2$Fe$_{17}$H$_x$ (RE = Nd,Sm)

HYDRIDES AND IRON-RICH COMPOUNDS Nd(Co$_{1-x}$Fe$_x$)$_9$Si$_2$ and Gd(Fe$_x$Al$_{1-x}$)$_{12}$

B. CHEVALIER[*], J. ETOURNEAU[*] and J.M.D. COEY[**]

[*] Laboratoire de Chimie du Solide du C.N.R.S.
Université de Bordeaux I, 351 cours de la Libération
33405 Talence Cedex, France.

[**] Physics Department, Trinity College, Dublin 2, Ireland.

ABSTRACT

The influence of the hydrogen absorption on the magnetic properties of RE$_2$Fe$_{17}$ (RE = Nd,Sm) compounds has been studied. After hydrogenation, the Curie temperature increases by well over 100°C for both intermetallic materials. On the other hand the new compounds Gd(Fe$_x$Al$_{1-x}$)$_{12}$ and Nd(Co$_{1-x}$Fe$_x$)$_9$Si$_2$ have been obtained by melt-spinning or arc-melting. For x < 0.83, the Gd(Fe$_x$Al$_{1-x}$)$_{12}$ materials crystallize in tetragonal ThMn$_{12}$ structure type and are ferrimagnetic. The tetragonal BaCd$_{11}$ structure type is found for Nd(Co$_{1-x}$Fe$_x$)$_9$Si$_2$ silicides. In this system, the Curie temperature goes through a maximum as x increases (T$_c$ ≃ 560 K for x = 0.22).

I - HYDROGEN ABSORPTION AND DESORPTION IN RE$_2$Fe$_{17}$ (RE = Nd,Sm) INTERMETALLIC COMPOUNDS.

I-1- INTRODUCTION

Ion-rich intermetallic compounds with rare-earth elements are usually ferromagnetic, with a relatively low Curie temperature. For example, T$_c$ for Y$_2$Fe$_{17}$ is 51°C, whereas T$_c$ is 913°C for the isomorphic cobalt compound Y$_2$Co$_{17}$. The low Curie temperatures of the RE$_2$Fe$_{17}$ series preclude their use as permanent magnets. Furthermore, the pressure coefficient (dT$_c$/dP) for Y$_2$Fe$_{17}$ is large and negative, - 5.0 K/kbar (K.H.J. Buschow et al., 1977). This suggests that the shorter iron-iron distances are associated with weaker ferromagnetic (or even antiferromagnetic) coupling, in agreement with the Slater-Néel curve.

Surprisingly, there seem to be no reports in the literature of the effect of hydrogen on the magnetic properties of RE$_2$Fe$_{17}$ compounds, although the hydrogenation behaviour of Sm_2Co_{17} and Pr_2Co_{17} has been studied (J. Evans et al., 1985). RE_2Fe_{17} is the most iron-rich intermetallic phase with the rare-earths, and in view of the large negative pressure coefficient of T_c, one might expect a substantial increase on hydrogenation if $RE_2Fe_{17}H_x$ phases can be prepared. In this paper we report just such an effect for compounds with the light rare-earths RE = Nd,Sm. We monitor the hydrogen uptake of the RE_2Fe_{17} compounds and compare results with those of $R_2Fe_{14}B$ (Wang et al., 1988).

I-2- SAMPLE PREPARATION AND HYDROGENATION

Nd_2Fe_{17} and Sm_2Fe_{17} are prepared by arc-melting the 99.9 % pure elements under argon, with a 5 % excess of the rare-earth and then annealing for a week at 900°C in quartz tubes sealed under vacuum. The X-ray diagrams were indexed on an hexagonal cell (Th_2Zn_{17}-type structure) containing three formula units. Lattice parameters are given in table 1.

TABLE 1 Crystallographic and magnetic data of RE_2Fe_{17} and $RE_2Fe_{17}H_x$ compounds (RE = Nd,Sm).

Compound	a(Å)	c(Å)	$\Delta V/V$ (%)	T_c ($^\circ$C)	M_s (μ_B/formula) at 20°C	$<B_{hf}>$ (T) at 20°C
Nd_2Fe_{17}	8.578	12.462		57	22.93	17.5
$Nd_2Fe_{17}H_{2.4}$	8.681	12.510	2.8	175	30.90	22.5
Sm_2Fe_{17}	8.553	12.442		115	23.20	25.0
$Sm_2Fe_{17}H_{2.0}$	8.653	12.506	2.9	253	31.33	26.5

The samples of the alloys ($\simeq$ 50 mg) were hydrogenated in the Thermopiezic Analyser TPA (D.H. Ryan et al., 1986). A typical TPA trace for Nd_2Fe_{17} showing the temperature dependence of the hydrogen pressure and its derivative, is shown in Fig.1. There are two distinct steps in the curve that closely resemble those found earlier in $Nd_2Fe_{14}B$ (J.M. Cadogan et al., 1986). Stage I hydrogen absorption occurs at about

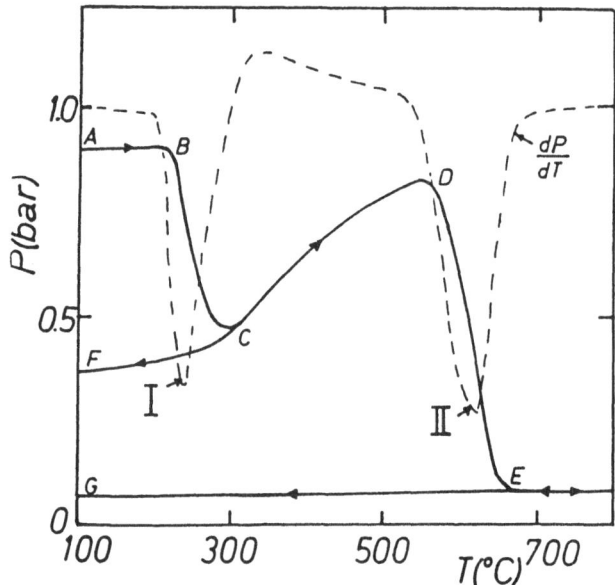

Fig.1 - TPA trace for Nd_2Fe_{17} showing the temperature dependence of the hydrogen pressure (solid line) and its derivative (broken line).

250°C. On cooling (segment CF in Fig.1) the sample is found to retain the Th_2Zn_{17} structure with increased lattice parameters, listed in table 1. Under these conditions the hydrogen content in the formula $RE_2Fe_{17}H_y$ is typically x = 2.2. The increase in volume $\Delta V/V$ = 2.8 - 2.9 % corresponds to an extra volume per absorbed hydrogen of 3.4 $\overset{\circ}{A}^3$. Near 600°C the sample reabsorbs a large amount of hydrogen (segment DE in Fig. 1). This stage II absorption corresponds to decomposition of the material into $RH_{2-\epsilon}$ + α-Fe.

I-3- MAGNETIC PROPERTIES

Magnetic thermal scans were therefore carried out on samples of hydrogenated material sealed in very small double-walled quartz tubes with almost no free volume, so that hydrogen desorption is minimal. Typical results are shown in Fig.2 where it can be seen that T_c increases by well over 100°C for both rare-earths (Table 1). The large increase in magnetization at 20°C ($\simeq$ 50 %) essentially reflects the Curie temperature change.

I-4- DISCUSSION

The hydrogenation behaviour of the R_2Fe_{17} phase is very similar to that of $RE_2Fe_{14}B$. The same two-stage hydrogen absorption occurs, although the hydrogenation and decomposition temperatures are somewhat different. The Curie temperature enhancement of Nd_2Fe_{17} is greater than

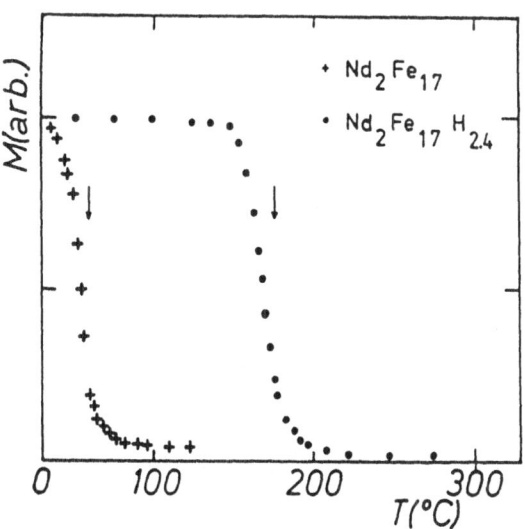

Fig.2 - Temperature dependence at the magnetization of Nd_2Fe_{17} before and after hydrogenation.

that of $Nd_2Fe_{14}B$, although the absolute value remains lower. Thus, it appears that despite the increase of the rather weak ferromagnetic coupling in Nd_2Fe_{17}, it still remains less than that in $Nd_2Fe_{14}B$. The same mechanism of exchange enhancement, related to an increase in average Fe-Fe distances, seems to operate here as it does in $RE_2Fe_{14}B$.

II - IRON-RICH PSEUDOBINARY ALLOYS WITH THE ThMn$_{12}$ STRUCTURE OBTAINED BY MELT-SPINNING : Gd(Fe$_x$Al$_{1-x}$)$_{12}$, x = 0.50, 0.67 and 0.83.

II-1- INTRODUCTION

The search for iron-rich rare earth alloys for magnetic applications recently received added impetus with the discovery of the tetragonal $Nd_2Fe_{14}B$ phase (M. Sagawa et al., 1984). Since then, several other new ternary neodymium iron borides have been discovered and there has been renewed interest in the $BaCd_{11}$, $ThMn_{12}$ and $CdZn_{13}$ structure types. For example, a $BaCd_{11}$-structure compound $Nd(Fe_{10}Si)C_{0.5}$ has recently been stabilized with interstitial carbon and $Pr(Fe_{11-x}B_x)$ has been prepared for $0.5 < x < 1.5$.

Pettifor has recently developed two-dimensional structure maps based on the "Mendeleev number" as a means for aiding the search for new pseudobinary compounds (D. Pettifor , 1986). In these maps, the $ThMn_{12}$ structure is stabilized over quite a wide area in two separate regions. It is therefore of interest to try to make iron-rich $RE(Fe_xAl_{1-x})_{12}$ alloys with RE = rare earth and M an element with a Mendeleev number slightly smaller (eg Cr , V) or much larger (eg Al) than that of iron. Previously it has been reported by Felner, Nowik et al (1983) that $RE(Fe_xAl_{1-x})_{12}$ alloys can be prepared by arc melting for $0.33 < x < 0.50$ with RE = many of the rare earth series. In this work we show that the stability range of the $ThMn_{12}$ structure in these compounds can be extended to $x < 0.83$ for RE = Gd and Dy using a rapid-quench technique (Wang et al., 1988).

II-2- EXPERIMENTAL METHODS

The elements were first melted in an arc furnace in an atmosphere of Ti-gettered argon to give alloys of composition $RE(Fe_xAl_{1-x})_{12}$ with $0.50 < x < 0.83$ and RE = Gd or Dy.

Pieces of ingot of order 1g with $x = 0.67$ and 0.83 were then melt-spun in helium through a quartz orifice 300 μm in diameter onto a copper wheel moving with a surface speed of 50 m/s. Fragments of ribbon 5-10 mm in length were thus obtained. These were subsequently annealed at 450°C for 100 hours, since differential thermal analysis showed that the first thermal event occurs at about 400°C. Samples were studied before and after annealing by X-ray diffraction, magnetization measurements and Mössbauer spectroscopy.

II-3- RESULTS

Before spinning, a mixture of a $ThMn_{12}$-structure phase, a 2-17-structure phase and Fe were present when $x = 0.67$ or 0.83 although a pure $ThMn_{12}$-structure phase was found for $x = 0.50$. After spinning there was no longer any free iron; only the reflections of the tetragonal structure were present and the a and c parameters has decreased. In spun $Gd(Fe_{0.83}Al_{0.17})_{12}$ the X-ray reflections were broadened and only the most intense $ThMn_{12}$ structure reflections were visible. Lattice parameters of the $Gd(Fe_xAl_{1-x})_{12}$ series are listed in Table 2. On annealing at 450°C, the X-ray patterns become sharper, but some free Fe appears for x = 0.83. Similar results are obtained for RE = Dy.

Table 2 - Lattice parameters of $Gd(Fe_xAl_{1-x})_{12}$.

x	0.33	0.42	0.50	0.67	0.83
a (Å)	8.74	8.71	8.64	8.57	8.49
c (Å)	5.05	5.03	5.00	4.95	4.89

The Curie temperatures of the alloys were determined from thermal scans of the magnetization. They are plotted in Figure 3 (for RE = Gd) as a function of x, where it can be seen that the Curie temperature of the ideal " $GdFe_{12}$ " compound would be 600 K. The saturation magnetization at 300K is 2.11, 6.07 and 10.6 μ_B/formula for x = 0.50, 0.67 and 0.83 respectively.

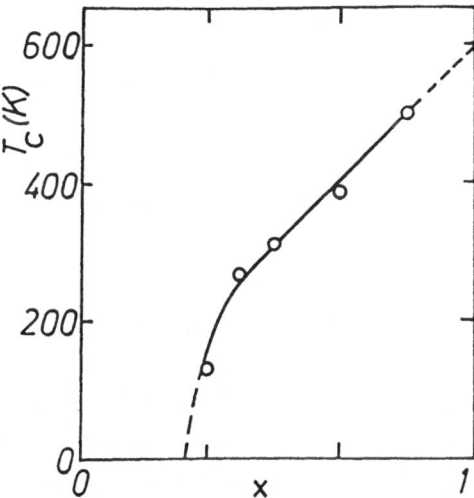

Fig.3 - Curie temperature of the $Gd(Fe_xAl_{1-x})_{12}$ compounds as a function of x.

Room temperature Mössbauer spectra of the $Gd(Fe_{0.83}Al_{0.17})_{12}$ and after annealing are shown in Figure 4. The average hyperfine field of the spun alloy is 20.9 T, and this changes on annealing at 450°C to 23.0 T. A fit to Mössbauer spectrum of annealed $Gd(Fe_{0.83}Al_{0.17})_{12}$ with three hyperfine components is summarized in Table 3. The average Fe moment, at room temperature, deduced from the average hyperfine field is 1.5 μ_B.

Table 3 - Hyperfine parameters deduced from fit to room temperature Mössbauer spectrum of the annealed $Gd(Fe_{0.83}Al_{0.17})_{12}$ sample. B_{hf} = hyperfine field, Δ = quadrupole splitting and δ = isomer shift (relative to αFe)

Subspectrum	B_{hf} (T)	$\|\Delta$ (mm.s^{-1})	$\|\delta$ (mm.s^{-1})	%
I	26.8(4)	0.03(4)	0.05(2)	44(5)
II	18.8(4)	0.07(4)	-0.07(2)	36(5)
III	22.2(4)	-0.06(4)	-0.09(2)	20(5)

The observed room temperature magnetization of $Gd(Fe_{0.83}Al_{0.17})_{12}$ implies a Gd moment of 4.5 μ_B, assuming a ferrimagnetic structure in which the Gd sublattice is antiparallel to the Fe sublattice. The Gd moment at room temperature is substantially reduced from its zero temperature value (~ 7 μ_B) because of relatively weak Gd-Fe exchange (T/T_c 0.6).

Our fit to Mössbauer spectrum of annealed $Gd(Fe_{0.83}Al_{0.17})_{12}$ (Table 3) indicates that two of the three non rare-earth sites in this compound (subspectra I and II) are fully occupied by Fe whilst the third site (subspectrum III) is equally occupied by Fe and Al. From an inspection of the nearest neighbour environments of the 8i, 8j and 8f sites in the $ThMn_{12}$ structure, we assign subspectrum I, having the largest hyperfine field, to Fe in the 8i site. It is more difficult to make an unambiguous assignment for the other two sites, but coordination numbers of magnetic and nonmagnetic neighbours are more clearly differentiated if Al substitutes preferentially in the 8j site (ie subspectrum II is 8f and

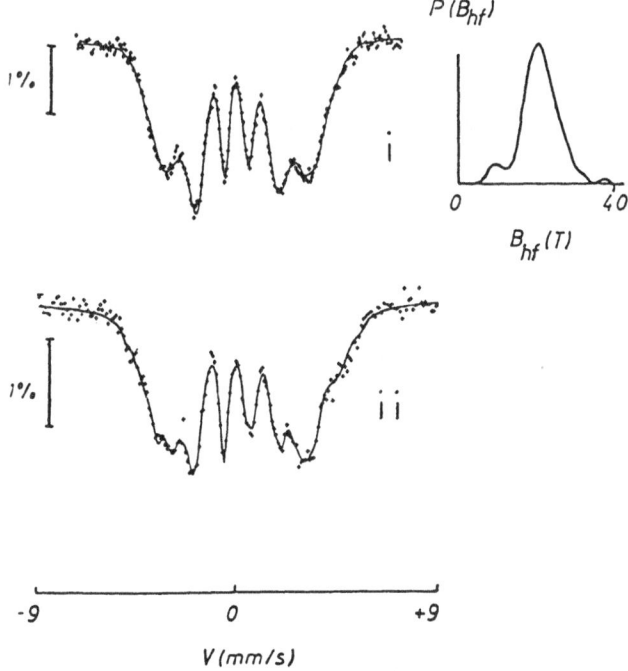

Fig.4 - Room temperature ^{57}Fe Mössbauer spectra of melt-spun $Gd(Fe_{0.83}Al_{0.17})_{12}$: (i) before annealing and (ii) after annealing at 450°C.

subspectrum III is 8j). None of the sites has a near-zero iron moment, as was found for $RE(Fe_xAl_{1-x})_{12}$ alloys containing less iron (I.Felner et al., 1983). Site preferences of Fe and Al atoms in $RE(Fe_xAl_{1-x})_{12}$ were recently discussed by Melamud et al.,(1987). These authors have carried out Wigner-Seitz constructions for this system and argue that a "substantial" moment is associated with Fe atoms occupying the 8i sites, in accord with our assignment. These authors also conclude that (i) the 8i site is the preferred site for Fe and (ii) Fe in the 8j or 8f sites does not carry a large moment. However, this latter conclusion is in conflict with the results of a neutron diffraction study of $Y(Fe_xAl_{1-x})_{12}$ by Wang et al. (1981) who find significant moments on Fe atoms occupying the 8i and the 8j sites with a near-zero moment on the 8f site Fe atoms. Controversy obver the Fe and Al positions in RFe_nAl_{12-n} is still not fully resolved.

II-4- CONCLUSION

We have demonstrated that it is possible to extend the range of stability of the $ThMn_{12}$ structure in $RE(Fe_xAl_{1-x})_{12}$ compounds by melt spinning. The gadolinium compounds order ferrimagnetically. The Curie temperature of the $x = 1$ end member would be approximately 600 K, which is sufficient to make it worthwhile to evaluate similar materials with a non S-state rare earth for potential permanent magnet applications.

III - **THE NEW TERNARY SILICIDES** : $Nd(Co_{1-x}Fe_x)_9Si_2$ ($0 < x < 0.55$).

III-1- INTRODUCTION

Refering to the famous high performance permanent magnet $Nd_2Fe_{14}B$, it is worthwhile mentioning that 3d-4f interactions play an important role in the magnetic properties of the compounds (D. Givord et al., 1984). The large 3d magnetic interactions determine the ordering temperature and the magnetic coupling between transition metal and rare earth atoms, allows the large anisotropy, characteristic of 4f ions.

In view of **searching** for new permanent magnet materials, we have investigated a Nd-Co(Fe)-Si system in the region of this pseudo-ternary phase diagram which is rich in transition metal. In this region, the ternary compound $NdCo_9Si_2$ has been reported previously by Bodak et al. (1969).

We have studied the crystallographic and magnetic properties of the $Nd(Co_{1-x}Fe_x)_9Si_2$ silicides (B. Chevalier et al., 1987).

III-2- EXPERIMENTAL PROCEDURES

$Nd(Co_{1-x}Fe_x)_9Si_2$ samples were prepared from powder elements by cold-crucible levitation melting and quenching. After melting, the samples were annealed at 850-900°C under vacuum for five days, in evacuated quartz tubes.

The magnetic properties of the compounds were studied in the range 4.2-700 K by using a Faraday type balance and a vibrating sample magnetometer.

III-3- RESULTS AND DISCUSSION

We have prepared the $Nd(Co_{1-x}Fe_x)_9Si_2$ alloys for $0 < x < 1$. After melting and quenching, the structural and microprobe analyses reveal that these ternary silicides are single phase only for $0 < x < 0.55$. For samples with x 0.55, X-ray patterns indicate the presence of α-Fe, $Nd_2(Fe,Co,Si)_{17}$ and $Nd(Fe,Co)_2Si_2$. This study confirms that it is not possible to prepare $NdFe_9Si_2$ using the experimental conditions mentioned above. After annealing at 900°C for 5 days, the samples decompose for $x > 0.22$ giving α-Fe and $Nd(Co,Fe)_2Si_2$ as the main products.

For $0 < x < 0.55$, the ternary silicides obtained by melting and quenching are isostructural with $NdCo_9Si_2$. The structure of $NdCo_9Si_2$ which is tetragonal with the $I4_1/amd$ space group derives from the $BaCd_{11}$-type. Cobalt atoms occupy two different sites : 32(i) and 4(b), while neodymium and silicon atoms are located in 4(a) and 8(d) positions respectively. Each cobalt atom is surrounded by 8 cobalt atoms but the Co-Co distance for these two sites are sensibly different.

The variation with x of the lattice parameters for $Nd(Co_{1-x}Fe_x)_9Si_2$ is shown in Table 4. The a and c parameters increase with iron concentration. The increase of the lattice parameters probably occurs because iron atoms are bigger in size than cobalt atoms.

Table 4 - Crystallographic and magnetic data of $Nd(Co_{1-x}Fe_x)_9Si_2$: a,c lattice constants; T_c Curie temperature and $M_{sat.}$ saturation magnetization per mole at 300 K and 4.2 K and per transition metal atom at 4.2 K.

x	Lattice constants		T_c (K)	$M_{sat.}$ ($\mu_B mole^{-1}$)		$M_{sat.}$ (μ_B at.trans^{-1}) at 4.2 K
	a (Å)	c (Å)		300 K	4.2 K	
0	9.794	6.328	*			
0.11	9.819	6.360	521(3)	8.98	11.97	0.97
0.22	9.842	6.365	556(3)	10.34	13.32	1.12
0.33	9.885	6.400	551(3)	11.39	14.69	1.27
0.44	9.900	6.408	521(3)	12.12	15.31	1.34
0.55	9.913	6.423	492(3)	12.91	16.24	1.44

* see text.

All these compounds have been found to be ferromagnetic. The thermomagnetic curve obtained for $NdCo_9Si_2$ reveals the presence of a parasitic phase in this ternary silicide. In our experimental conditions, it is not possible to prepare $NdCo_9Si_2$ as a single phase.

The composition dependence of the Curie temperatures for the system investigated are shown in Figure 5(a) and summarized in Table 4. T_c is maximum for $x = 0.22$ and decreases with a further increase of the iron concentration. In many binary compounds as $RE_2(Co_{1-x}Fe_x)_{17}$ (RE = rare earth) decrease of the Curie temperature is generally found when Co is

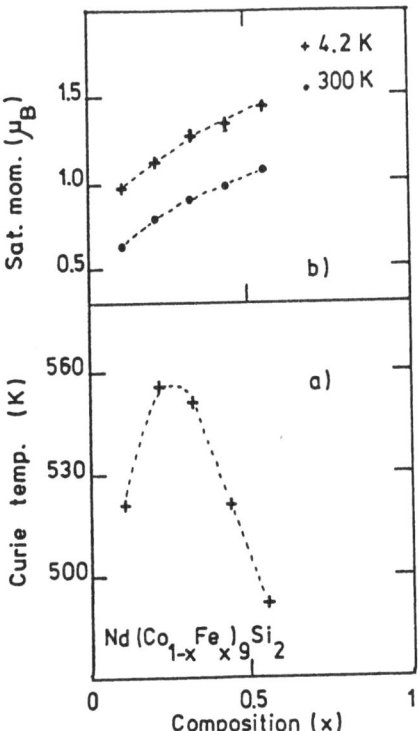

Fig.5 - Composition dependence of the Curie temperature (a) and saturation moment per transition metal atom at 300K and 4.2K(b).

replaced by Fe (H.R. Kirchmayr et al., 1979). The iron-iron exchange interactions are very sensitive to the Fe-Fe distances. When these distances are less than 2.45 Å, the Fe-Fe exchange interactions may be antiferromagnetic according to the Bethe-Slater curve for localized moments. In these conditions, the T_c dependence versus composition for $Nd(Co_{1-x}Fe_x)_9Si_2$ silicides can be explained as follows :

- for small x values (x < 0.22), the iron atoms tend to occupy preferentially the 4(b) sites having the biggest volume,

- for x > 0.22 the iron atoms are distributed statistically in the two sites and because the interatomic distances for 32(i) sites are small the number of antiferromagnetic coupling increases with x.

The spontaneous magnetization per transition metal atom is plotted in Figure 5(b) as a function of x. For determination of this magnetization, we have considered a ferromagnetic coupling between neodymium and transition metal moments. A $g_J J = 3.27$ μ_B value for Nd^{3+} atom has been taken. The transition metal moment increases with x. The magnetic behaviour of these ternary silicides is similar to that observed in $RE(Co_{1-x}Fe_x)_{17}$ and $RE_2(Co_{1-x}Fe_x)_{14}B$ compounds. These results are in qualitative agreement with the Slater-Pauling curve for Fe-Co alloys. For x = 0.55 and T = 4.2 K, the transition metal moment (1.44 μ_B) is very close to that observed for the iron moment (1.42 $_B$) in another isotypic ternary silicide $NdFe_{10}SiC_{0.5}$ (J. Le Roy et al., 1988).

REFERENCES

Bodak, O.J. and Gladyshevskij, F.J. 1969. Dopov. Akad. Nauk. Ukr. RSR., 5, 452.

Buschow, K.H.J. 1977. Rep. on Prog. in Phys., 40, 1179.

Chevalier, B., Gurov, G., Fournès, L. and Etourneau, J., 1987. J. Chem. Research, 5, 138.

Cadogan, J.M. and Coey, J.M.D. 1986. Appl. Phys. Lett., 48, 442.

Evans, J., King, C.E. and Harris, J.R. 1985. J. Mat. Sci., 20, 817.

Felner, I., Nowick, I. and Seh, M. 1983. J. Magn. Magn. Mater., 38, 172.

Kirchmayr, H.R. and Poldy, C.A. 1979. Handbook on the Physics and Chemistry of Rare Earths. eds. Gschneidner, K.A. Jr. (North-Holland, Amsterdam), 2.

Le Roy, J., Moreau J.M., Bertrand, C. and Fremy, M.A. 1988, J. Less Comm. Met. (to be published).

Melamud, M., Bennett, L.H. and Watson, R.E. 1987, J. Appl. Phys., 61, 4246.

Pettifor, D. 1986. New Scientist, 29 May, pp 48-53

Ryan, D.H. and Coey, J.M.D. 1986, J. Phys. E, 19, 693.

Sagawa, M., Fujimura, S., Togawa, N., Yamamoto, H. and Matsuura, Y. 1984, J. Appl. Phys., 55, 2083.

Yang, Y.C., Kebe, B., James, W.J., Deportes, J. and Yelon, W. 1981, J. Appl. Phys., 52, 2077.

Wang, X.Z., Donnely, K., Coey, J.M.D., Chevalier, B., Etourneau, J. and Berlureau, T. 1988, J. Mat. Science (to be published).

Wang, X.Z., Chevalier, B., Etourneau, J., Berlureau, T., Coey, J.M.D. and Cadogan, J.M. 1988, J. Less-Comm. Met. (to be published).

STRUCTURAL AND MAGNETIC PROPERTIES OF SOME RE-(Fe,Mn)-(B,X) ALLOYS

W. Rodewald

Vacuumschmelze GmbH, P.O. 2253, 6450 Hanau, FRG

ABSTRACT

The effect of a partial substitution of Fe by Mn on the magnetic properties of sintered Nd-Fe-B magnets has been examined. The intrinsic magnetic properties of RE modified Heusler alloys, RE: Nd, Sm, Dy and of Mn-rich compounds of different RE-Mn-X systems, X: B, C, were investigated. Additions Nb or Mo to $Nd_2(Fe,Co)_{14}B$ result in precipitation of hexagonal NbFeB or tetragonal Mo_2FeB_2 compounds. The coercivity of such magnets is still determined by nucleation of reversed domains.

INTRODUCTION

The discovery of the magnetic properties of the intermetallic $Nd_2Fe_{14}B$ compound by Sagawa (1984) and Croat (1984) demostrates, that intermetallic compounds may have favourable intrinsic properties for the production of permanent magnets. Hence it is a challenge to investigate other RE-TM-X systems; RE: rare earth metal, TM: transition metal, X: metalloid, whether there exist TM-rich compounds with appropriate hardmagnetic properties.

The temperature stability of sintered Nd-Fe-B magnets may be improved by pinning the magnetic domain walls by precipitates. In alloys containing additions of Nd or Mo precipitates of Nb-Fe- or Mo-Fe-compounds have been detected by Parker (1987) and Allibert (1987), respectively. The influence of Nb or Mo additions on the magnetic properties of sintered Nd-Fe-B or Nd-Fe-Co-B magnets have been investigated. The results and the composition of the precipitates are presented in this report.

EXPERIMENTAL

The $Nd_{15}(Fe,Mn)_{77}B_8$ alloys were melted in a vacuum induction furnace. The RE-Mn-X alloys, X: B, C or Cu, were melted in an arc furnace in small quantities. Anisotropic magnets have been prepared by powder metallurgy, the processing is demonstrated in Fig. 1. Melting points and phase transitions were determined by DTA. The microstructure of RE-Mn-X alloys have been examined by metallographical analysis.

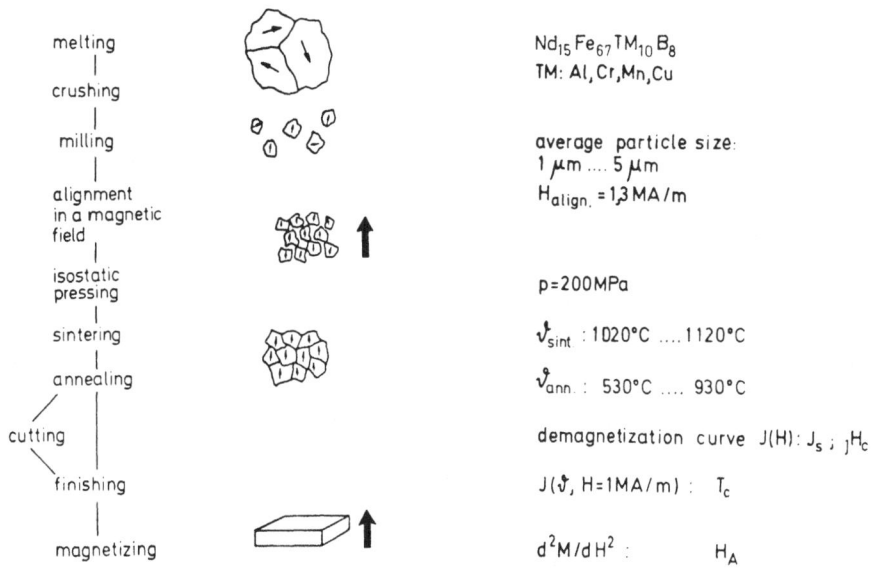

melting

crushing

milling

alignment in a magnetic field

isostatic pressing

sintering

annealing

cutting

finishing

magnetizing

$Nd_{15}Fe_{67}TM_{10}B_8$
TM: Al,Cr,Mn,Cu

average particle size:
$1\,\mu m \ldots 5\,\mu m$
$H_{align.} = 1{,}3\,MA/m$

$p = 200\,MPa$

ϑ_{sint} : 1020°C ....1120°C

$\vartheta_{ann.}$: 530°C .... 930°C

demagnetization curve $J(H)$: J_s ; $_jH_c$

$J(\vartheta, H=1MA/m)$: T_c

d^2M/dH^2 : H_A

Fig. 1   Processing and characterization of sintered
RE-(Fe,Mn)-B magnets.

The composition of the different compounds were analysed by an electron microprobe. Small precipiates have been investigated by TEM.

The magnetic properties of sintered magnets, of molten or of annealed alloys were measured in a Foner vibrating sample magnetometer. Curie temperatures were determined from the temperature dependence of the polarization J in a magnetic field by plotting J^2 versus temperature. The anisotropy field strength H_A was measured by the singular point detection (SPD-) technique at room temperature (Asti et al. 1974).

RESULTS

At first the effect of a partial substitution of Fe by Mn in sintered magnets of $Nd_{15,2}(Fe_{1-x} Mn_x)_{77,5}B_{7,3}$ alloys has been studied. Since there is no improvement of the intrinsic magnetic properties, a research for RE-Mn-X compounds with X: B, C or Cu has been carried out. Finally the composition of some ternary precipitates in $Nd_2Fe_{14}B$ grains will be reported.

Magnetic properties of $Nd_{15,2}(Fe_{1-x} Mn_x)_{77,5}B_{7,3}$ magnets.

The dependence of the saturation polarization, the Curie temperature, the anisotropy field strength and the intrinsic coercivity of sintered Nd-Fe-Mn-B magnets on the Mn-concentration has been examined. The relation of the intrinsic coercivity to the anisotropy field strength and to localized stray fields will be discussed. Magnets with different Mn-concentrations have been prepared by powder metallurgy, see Fig. 1.

In order to determine the saturation polarization of the $Nd_2(Fe_{1-x} Mn_x)_{14}B$ compounds the measured saturation polarization of sintered magnets was related to the fraction of the matrix phase. A partial substitution of Fe by Mn decreases the saturation polarization by about - 5,2% per at% Mn in the compound. The reduction of the saturation polarization is larger than expected by a simple dilution model. Assuming a replacement of Fe-atoms on 16 k-sites and an antiferromagnetic coupling between the magnetic moments of the Mn-atoms and the Fe-atoms, the Mn-atoms must carry a magnetic moment of about 4 μ_B in order to get such a strong reduction of the saturation polarization. Since in Heusler alloys (Uhl, 1982) and in RE_6Mn_{23} compounds (Wallace, 1973) Mn-atoms carry magnetic moments of 4 μ_B and since the distance between Fe- and Mn-atoms in the $Nd_2(Fe,Mn)_{14}B$ compounds is smaller than 0,28 nm such an antiferromagnetic coupling is to be expected (Huang et al. 1986 and Slater, 1930).

The effect on the Curie temperature is of the same magnetude: there is a decrease by - 4,8% per at% Mn. This indicates a reduction of the exchange interaction. The anisotropy field strength is also reduced by - 3,6% per at% Mn. The intrinsic coercivity of sintered magnets, annealed at 700°C, does not decrease proportional to the Mn-concentration; but magnets containing about 8 at% Mn have an intrinsic coercivity of about 2 kA/cm. Some demagnetization curves of $Nd_{15,2}(Fe_{1-x},Mn_x)_{77,5}B_{7,3}$ magnets are compiled in Fig. 2.

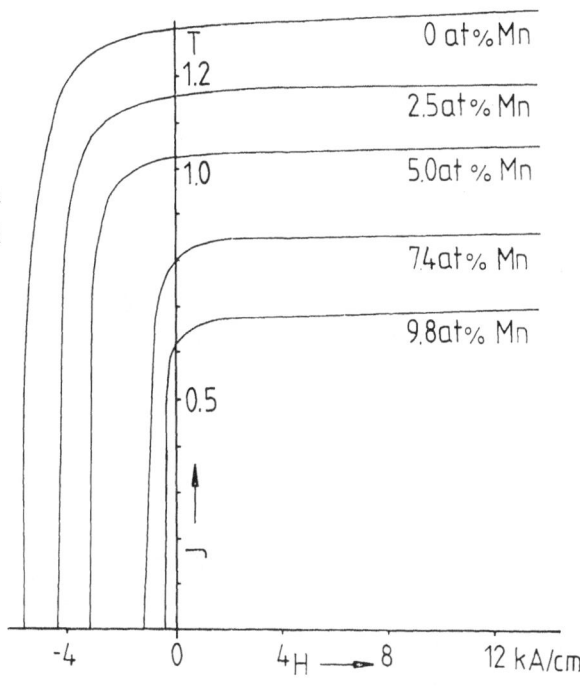

Fig. 2 Demagnetization curves J(H) of sintered $Nd_{15,2}(Fe,Mn)_{77,5}B_{7,3}$ magnets, annealed at 700°C.

How does the coercivity depend on the anisotropy field strength? According to Herzer (1986) the coercivity is given by:

$$\mu_o \cdot {}_JH_c = c \cdot \mu_o \cdot H_N - N \cdot J_s \qquad (1)$$

$_J H_c$ denotes the coercivity of the polarization, H_N the nucleation field, an N an effective demagnetization coefficient.
The nucleation field depends on the anisotropy coefficients
K_1 and K_2. If K_2 can be neglected in respect to K_1, the nucleation field strength corresponds to the anisotropy field
strength and equation (1) can be rewritten:

$$\mu_o \cdot {}_J H_c = c \cdot \mu_o \cdot H_A - N \cdot J_s \quad (2)$$

The coefficient c is a measure for the magnetic isolation of individual grains what is caused by Nd-rich constituents between the grains. The second term represents localized stray fields which may facilitate nucleation of reversed
domains. An estimate of the maximum stray fields at the cor-
ners of crystals gives fields of about $- 2 \cdot J_s$ (Adler,
1985). Under these assumptions the dependence of the coefficient c on the composition of sintered magnets can be calculated what gives some information about the ratio of the coercivity to the anisotropy field strength. In Mn-containing magnets the anisotropy field strength as well as the coefficient
c decrease strongly with increasing Mn-concentration, see
Fig. 3, whereas in Al-containing magnets the coefficient c
increases with Al-concentration (Rodewald et al. 1987).

Fig. 3 Dependence of
the anisotropy field
strength H_A and of the
coefficient c on the
Mn-concentration in
sintered Nd-Fe-Mn-B
magnets.

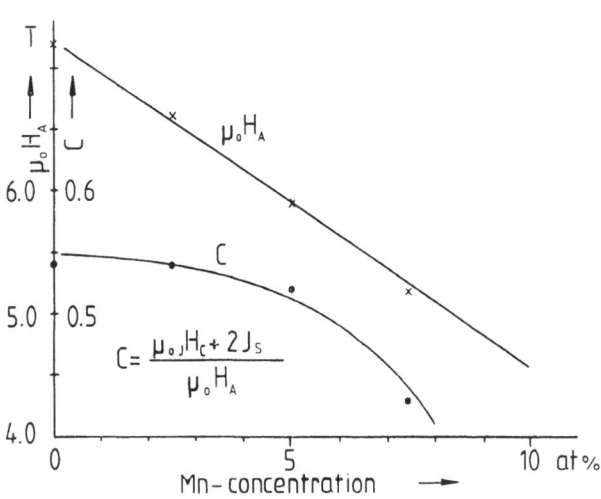

Search for RE-Mn-X compounds

In the literature there are only a few informations on
RE-Mn-X compounds with X: B, C or a non-ferromagnetic metal.
Therefore different RE-Mn-B alloys were examined, if there
exist Mn-rich compounds. The various alloys are marked by
circles in Fig. 4.

Fig. 4 Review of examined
RE-Mn-X alloys (O) with
X: B, C or Cu, and known
RE-Mn-borides (*).

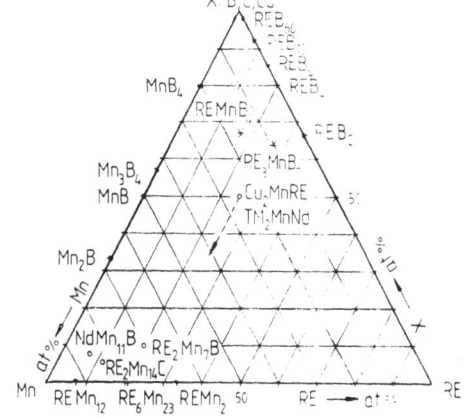

The Cu_2MnAl type Heusler alloys have a rather large
solubility range for Cu as well as for Al, so that RE modified Heusler alloys were examined. The existence of RE_2Mn_7B
and $RE_2Mn_{14}C$ compounds for different RE metals was checked
and a replacement of Rhenium in $YRe_{11}B$ compounds by Mn, which
both belong to group VII B of the Periodic Table, was tested.
The results are compiled in this section.

In Heusler alloys Mn-atoms carry a magnetic moment of
about 4 μ_B. The unit cell contains four formula units of
Cu_2MnAl. The Cu-atoms form a cubic face-centered superstructure, which surrounds a body centered cubic structure of Mn-
and Al-atoms, see Fig. 5.

Fig. 5 Unit cell of the
Cu_2MnAl Heusler alloy.

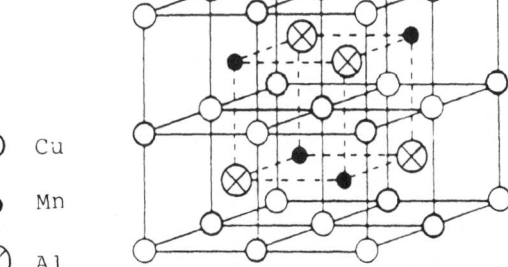

○   Cu

●   Mn

⊗   Al

The homogeneity range of Cu and Al enables a substitution of Cu by other ferromagnetic metals, such as Fe, Co, and Ni; Al can be replaced by other non-ferromagnetic metals, for instance Ga, In, Si, Ge, Sn, Pb, Sb (Uhl, 1982).
A substitution of Al by RE-metals has been investigated. Alloys of Cu_2MnAl, Cu_2MnRE with RE: Pr, Nd, Sm, Dy and TM_2MnNd with TM: Fe, Co, Ni, were melted in an arc furnace in a water-cooled Cu crucible. The ingots were crushed and remelted twice in order to improve the homogeneity.

Metallographic analysis proved that the alloys contained up to four different phases, besides the Cu_2MnAl compound which was single-phase. The saturation polarization and the Curie temperature of the Cu_2MnAl compound amount to 0,68 T and 380°C respectively. No signifant anisotropy field strength was detected by the SPD technique. All RE modified Heusler alloys contained many dendrites which could not be dissolved by annealing at temperatures between 680°C and 850°C. Electronprobe microanalysis proved the existence Cu_2MnRE, RE: Nd, Sm, Dy, and TM_2MnNd, TM: Fe, Co, Ni, compounds, but none of them had favourable ferromagnetic properties at room temperature.

According to a literature review by Rogl (1986) in the ternary Y-Mn-B system there exist three intermetallic compounds: Y_2Mn_7B, Y_3MnB_7 and $YMnB_4$. The magnetic properties of these compounds are not known. In order to examine the existence and the magnetic properties of the Mn-rich RE_2Mn_7B compounds, it was melted with the following RE metals: Ce, Pr, Nd, Sm or Dy. Metallographic analysis and DTA however revealed, that the alloys contain up to five different phases.

After annealing at 650°C for 60h the demagnetization curves J(H) have been measured. The Ce_2Mn_7B and the Pr_2Mn_7B alloy contain no ferromagnetic compound. However in a magnetic field of 13,5 kA/cm the polarization of the Nd-, Sm- or Dy-containing alloys amounts to about 0,04 T, 0,14 T and 0,06 T respectively. The Curie temperatures, determined in an external field of 10 kA/cm, vary between 160°C and 170°C and coincide with the Curie temperatures of the corresponding

RE_6Mn_{23} compounds. Since Ce and Pr do not form RE_6Mn_{23} compounds, those alloys have not been ferromagnetic at room temperature.

From a comprehensive literature review Rivlin (1987) concluded, that in RE-TM-C systems, TM: Fe, Mn, Co, Ni, the tetragonal compound $RE_2Mn_{14}C$ may exist. $RE_2Mn_{14}C$ compounds have been melted from a Mn-C master-alloy and RE metals, RE: Ce, Pr, Nd, Sm or Dy, in a water-cooled Cu crucible in an arc furnace. The ingots were remelted twice in order to improve the homogeneity. The $RE_2Mn_{14}C$ ingots were not stable in air and disintegrated within a few days. The Ce-, Pr-, and Nd-containing alloys do not contain any ferromagnetic compounds at room temperature. Only in the $Sm_2Mn_{14}C$ and $Dy_2Mn_{14}C$ ingots some ferromagnetic constituents were detected, but these could not be identified.

In a review paper Rogl (1984) reported the existence of a $YRE_{11}B$ compound. Rhenium and manganese belong both to the same group of the Periodic Table. Therefore the existence of $REMn_{11}B$ compounds has been examined for RE: Ce, Pr, Nd, Sm. Metallographic analysis and DTA revealed that the molten alloys contain various compounds. In the $NdMn_{11}B$ ingot binary borides Mn_2B, MnB, ternary borides $NdMn_4B$ and $NdMn_2B$ or $NdMn_2B_2$, the Nd_6Mn_{23} compound and Nd-rich as well as Mn-rich solid solutions have been detected. A $NdMn_{11}B$ compound could not be identified.

Summing up: the existence of RE modified Heusler alloys has been proved, but they are not ferromagnetic at room temperature. Up to now no Mn-rich ferromagnetic RE-Mn-X compounds have been detected.

PRECIPITATES IN $Nd_2Fe_{14}B$

The temperature dependence of the coercive field strength of nucleation-type magnets is mainly determined by the temperature dependence of the anisotropy energy K, whereas for pinning-type magnets the temperature dependence of the coercivity is proportional to $\sqrt{K}$. Hence the temperature stability in sintered Nd-Fe-B magnets may be improved by pinning the domain walls at precipitates. In Nd-Fe-B alloys with additions of Nb, Mo or Zr, precipitates in $Nd_2Fe_{14}B$ grains have been reported by Parker and Grundy (1987) and Allibert (1987). The effect of Mo-additions on the properties of sintered Nd-Fe-Co-B magnets and the composition of Nb-Fe-B or Mo-Fe-B precipitates has been examined.

For the investigations following alloys were melted in a vacuum induction furnace: 33 wt% Nd, 10 wt% Co, 1.1 wt% B, Fe bal. and 33 wt% Nd, 10 wt% Co, 2 wt% Mo, 1.1 wt% B, Fe bal. Both alloys were milled to fine alloy powders. In order to vary the Mo concentration the alloy powders were blended in different ratios. The mixed alloy powders were aligned in a magnetic field, sintered at temperatures between 1000°C and 1100°C and annealed at 630°C. The magnetic properties of sintered and of annealed magnets have been measured in a Foner vibrating sample magnetometer.

The saturation polarization of the Nd-Fe-Co-Mo-B magnets decreases by about - 1,2% per wt% Mo in the alloy what roughly corresponds to the dilution of the magnetic moments. The Curie temperature amounts to about 415°C as is to be expected for Nd-Fe-Co-B alloys with a Co concentration of 10 wt% Co. The anisotropy field has not been changed by Mo: $H_A \cong$ 55 kA/cm. With increasing Mo-concentration there is only a small increase of coercivity, see Fig. 6. The coercivity of these Nd-Fe-Co-Mo-B magnets is still determined by nucleation.

Fig. 6 Dependence of the intrinsic coercivity on the Mo-concentration of sintered and of annealed Nd-Fe-Co-Mo-B magnets.

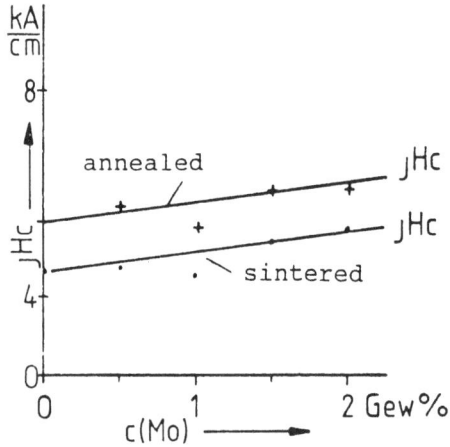

The microstructure of sintered magnets has been investigated by transmission electron microscopy. The $Nd_2(Fe,Co)_{14}B$ compound dissolves at sintering temperature up to 1 wt% Mo. The $Nd_{1,1}(Fe,Co)_4B_4$ compound does not dissolve Mo. There are some precipitates in the interior of $Nd_2(Fe,Co)_{14}B$ grains. Microanalysis of the precipitates yielded an atomic ratio of Mo: Fe = 2:1 and some B. By electron diffraction the precipitates could be identified as the tetragonal compound Mo_2FeB_2; a = 0,5782 nm, c = 0,3148 nm, (Rieger, 1964). Fig. 7 gives a comparison of precipitates of the NbFeB compound with a hexagonal structure, a = 0,6 nm, c = 0,32 nm (Schrey, 1987) in Nb-containing Nd-Fe-B magnets and the tetragonal Mo_2FeB_2 compound in Mo-containing Nd-Fe-Co-B magnets.

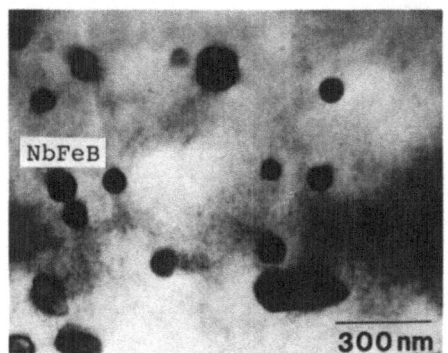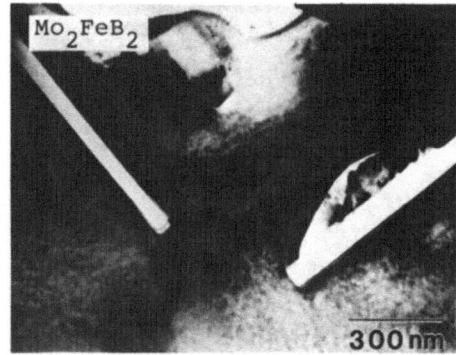

Fig. 7   TEM micrographs of NbFeB and Mo_2FeB_2 precipitates in sintered Nd-Fe-B and Nd-Fe-Co-B magnets.

The dimensions of the NbFeB precipitates range from 100 nm up to 200 nm in diameter, whereas the Mo_2FeB_2 plate-like precipitates are about 150 nm thick with a diameter up to 650 nm. Such precipitates are much too large for an efficient pinning of domain walls, δ_w = 5,2 nm (Sagawa, 1985), in Nd-Fe-B magnets.

Annealing of the Nd-Fe-Co-Mo-B magnets at 630°C reduces the solubility of Mo in $Nd_2(Fe,Co)_{14}B$ grains to less than 0,5 wt%. Hence the size of the Mo_2FeB_2 precipitates increases or additional precipitates are formed. Fig. 8 gives a TEM-micrograph of a Mo_2FeB_2 particle after an anneal at 630°C for 60h. The particle is enclosed by an additional phase whose composition is $Nd(Fe,Co)_2$ according to electron microprobe analysis. The growth of the Mo_2FeB_2 precipitate probably consumes some B so that a local B-deficiency occurs and pseudo-binary Nd-(Fe,Co) compounds result.

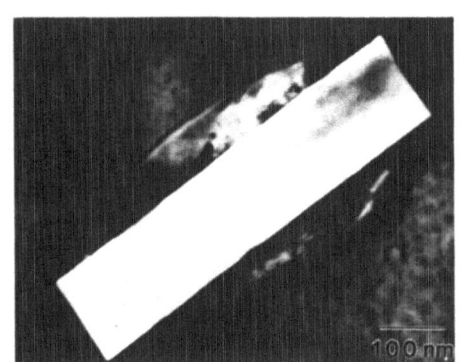

Fig. 8   TEM micrograph of a Mo_2FeB_2 precipitate in a $Nd_2(Fe,Co)_{14}B$ grain after a 60h anneal at 630°C. The particle is enclosed by the $Nd(Fe,Co)_2$ compound.

CONLUSIONS

A partial substitution of Fe by Mn in sintered Nd-Fe-B magnets result in a strong decrease of the saturation polarization, the Curie temperature and the anisotropy field strength. The decrease of the polarization may be due to an antiferromagnetic coupling between the magnetic moments of the Mn- and of the Fe-atoms. Since in the $Nd_2Fe_{14}B$ compound the distance between Fe-and Mn-atoms is shorter than 0,28 nm, there is some evidence for an antiferromagnetic coupling (Huang, 1986).

In RE-Mn-X systems, X: B, C or Cu, no Mn-rich compounds with favourable intrinsic properties for permanent magnets could be detected.

Additions of Nb and Mo to $Nd_2Fe_{14}B$ or $Nd_2(Fe,Co)_{14}B$ compounds result in the precipitation of hexagonal NbFeB or tetragonal Mo_2FeB_2 respectively. These ternary compounds are incoherent to the $Nd_2Fe_{14}B$ compound. The dimensions of the precipitates are much too large compared to the thickness of domain walls in $Nd_2Fe_{14}B$, so that there is no effective pinning.

ACKNOWLEDGMENTS

The author thanks W. Kirchner for melting the RE-Mn-alloys, P. Schrey and J. Nahm for the investigations of the microstructure by SEM and TEM, and J.P. Jacquet and B. Wall for the preparation of sintered magnets and for the measurements of the magnetic properties.

REFERENCES

Adler, E. et al. 1985, Proc. 8th Int. Workshop on REPM, Dayton, p. 757.
Allibert, C.H., 1987, CEAM Report, March-Sept. 1987
Asti, G. et al., 1974, J. Appl. Phys. 45, 3600
Croat, J.J. et al., 1984, Appl. Phys. Lett. 44 148
Herzer, G. et al. 1986, J. Magnet. Magnet. Mat. 58, 48.
Huang, M.Q. et al., 1986, J. Less Common Metals 124, 55.
Parker, S.F.H. et al., 1987, J. Magnet. Magnet. Mat 66, 74.
Rieger, W. et al., Monatshefte Chemie 95, 1502.
Rodewald, W. et al., 1987, IEEE Trans. Magnet. (submitted).
Rogl, P. 1984, Handbook on the Physics and Chemistry of Rare Earths, North-Holland Phys. Publ. Vol. 6 p. 335.
Rogl, P. 1986, Report of the 1st Ann. CEAM Meeting, Birmingham.
Sagawa, M. et al., 1984, J. Appl. Phys. 55, 2083.
Sagawa, M. et al., 1985, Proc. 8th Int. Workshop on REPM, Dayton, p. 587.
Slater, J.C., 1930: Phys. Rev. 36, 57.
Uhl, E., 1982, J. Solid State Chem. 43, 354.
Wallace, W.E., 1973, Rare Earth Intermetallics, Academic Press New York and London, p. 191.

SUMMARY OF DISCUSSION :

NEW PHASES, STRUCTURES AND PROPERTIES

K.H.J. Buschow, D.G. Pettifor

Three important perspectives on the possibility of obtaining novel intermetallic phases emerged from the different contributions to this session.

1 - From the theoretical point of view, the two-dimensional structure maps of Pettifor in conjunction with the data set of Villars and Calvert may be a great help in finding novel iron-rich phases,

2 - From the practical point of view, it was reported that a number of ternary phases can not be formed by following standard procedures such as casting and annealing, but these phases do form when a path is taken that leads via the metastable amorphous state. Examples are Fe-rich compounds of the series $R(Fe_{1-x}Al_x)_{12}$ with $ThMn_{12}$ structure, Fe-rich compounds of the series $R(Fe_{1-x}Si_x)_{11}$ with $BaCd_{11}$ structure and Fe-rich compounds of the series $R(Fe_{1-x}Co_x)_{12}B_6$ with the $SrNi_{12}B_6$ structure. In all these cases the initial step in the preparation consisted of melt spinning.

3 - A second point with regard to synthesis of novel materials was made by showing that a given structure-type can be stabilized by applying a fourth component small enough to fill up empty sites in a given crystal structure. Examples are Fe-rich ternaries of the type $R(Fe_{1-x}Si_x)_{11}Co_{0.5}$ having the (partially filled) $BaCd_{11}$ structure.

Magnetic properties were reported for several series of ternaries with $ThMn_{12}$ structure. From the data presented it became clear that these materials have anisotropy energies and saturation magnetizations at room temperature that are inferior to those of $R_2Fe_{14}B$. It has not yet been generally possible to develop coercivity in the $ThMn_{12}$-type materials. Based on the intrinsic properties it is not expected that permanent magnets made from $ThMn_{12}$-type compounds (with R = Sm) will be as good as $Nd_2Fe_{14}B$-type magnets. Nevertheless the former may become an acceptable alternative for $Nd_2Fe_{14}B$-type magnets in applications where a high energy product is not all that mattters. The same holds for permanent magnets based on rare-earth-modified Fe_3B.

SECTION III

— MATERIALS

CHAPTER 2

MAGNETIC PROPERTIES AND

CHEMICAL SUBSTITUTION

MAGNETIC PROPERTIES OF THE $R_2Fe_{14}B$ COMPOUNDS (R = RARE EARTH), A SERIES OF R-FE INTERMETALLICS

J.P. Gavigan[+Δ], D. Givord[+], H.S. Li[+]

[+] Laboratoire Louis Néel
C.N.R.S.
166 X, 38042 Grenoble Cedex, France
[Δ]Physics Department
Trinity College
Dublin, Ireland

ABSTRACT

The magnetic properties of the $R_2Fe_{14}B$ compounds are discussed together with those observed in binary R-Fe alloys. The decrease of the Fe moment as a function of the amount of R atoms alloyed is discussed in terms of the Friedel interpretation of the Slater-Pauling curve for transition metal alloys. The Fe-Fe exchange interactions are essentially determined by local environment effects. In Y-Fe and Lu-Fe compounds, it is shown that the molecular field coefficient n_{FeFe} between Fe atoms decreases as the Fe coordination number increases. In compounds with magnetic R elements, the coefficient n_{RFe} is obtained for a number of series. n_{RM} is not a constant going across a given series but decreases by a factor of 2. This variation is shown to be related to the variation of the exchange interactions between 4f and 5d electrons.

INTRODUCTION

Four different series of compounds (RFe_2, RFe_3, R_6Fe_{23}, R_2Fe_{17}) exist among binary R-Fe alloys. In these different compounds, the Fe 3d moment ranges from about 1.5 μ_B (RFe_2) to 2 μ_B (R_2Fe_{17}). The Fe-Fe exchange interactions dominate over the R-Fe and R-R ones. The Curie temperature is always significantly lower than that of pure Fe, it decreases as the Fe percentage increases. In the R_2Fe_{17} compounds, a large magnetovolume anomaly is observed in the ordered magnetic state. The above properties show that the 3d-3d exchange interactions are strongly dependent on environment effects and interatomic distances. In these compounds, the R-Fe exchange interactions are the most important of those involving the rare-earths ; combined with CEF interactions, they determine the magnetic behavior of the rare-earth sublattice.

The $R_2Fe_{14}B$ compounds constitute a new series of compounds. They crystallize in a structure of tetragonal symmetry (Herbst et al., 1984; Givord et al., 1984). Their properties are discussed in this paper in the light of those observed in binary R-Fe alloys. The $R_2Fe_{14}B$ series constitutes the only complete series of R-Fe compounds. This facilitates the discussion of the observed behaviours and permits some general conclusions, valid for all R-Fe alloys, to be drawn.

3d MAGNETISM IN R-M AND $R_2M_{14}B$ COMPOUNDS

Magnetic moments

In R-Fe intermetallic compounds, when R is a non-magnetic element, the value of the magnetization is characterized by a decrease of the magnetic moment as the percentage of R element is increased. This is illustrated in figure 1 for Y_yFe_{1-y} alloys.

Williams et al. (1983a) have discussed the variation of the magnetization in several series of alloys using Friedel's interpretation (Friedel, 1958) of the Slater-Pauling curve for transition metal alloys. In this approach, the bulk magnetization of the alloy is not considered in terms of magnetic and non-magnetic atoms, but rather in terms of the average magnetic moment per each atom present. Let $N\uparrow$ and $N\downarrow$ be the number of valence electrons of each spin state. The chemical valence is

$$Z = N\uparrow + N\downarrow \tag{1}$$

and the magnetic moment per atom is

$$M = N\uparrow - N\downarrow \tag{2}$$

$$= 2N\uparrow - Z \tag{2'}$$

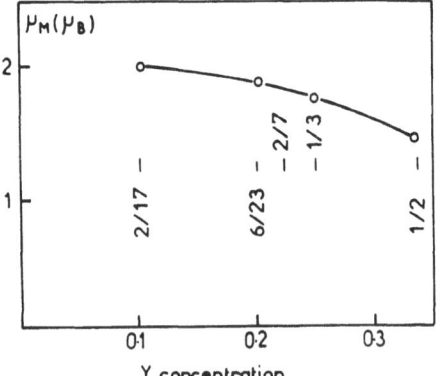

Fig. 1   Variation of the 3d magnetic in Y-Fe alloys as a function of the percentage of Y atoms alloyed. Meaning of the different abbreviations : 1/2 = RM_2 ; 1/3 = RM_3 ; 6/23 = R_6M_{23} ; 2/14/1 = $R_2M_{14}B$; 2/17 = R_2M_{17}.

In transition metals, the $N\uparrow$ electrons can be separated into the itinerant electrons and the more localized d electrons. The number $N_d^\uparrow$ of d electrons in the $\uparrow$ spin state band is precisely equal to 5 per atom in strong ferromagnets (Co, Ni). Following Williams et al. (1983a), the concept of magnetic valence, Z_m, may be introduced.

$$Z_m = 2N_d^\uparrow - Z \tag{3}$$

is an integer, and M is deduced to be

$$M = Z_m + 2N_{sp}^\uparrow \tag{4}$$

where $N_{sp}^\uparrow$ is the number of sp electrons in the $\uparrow$ spin state band.

The basic idea of Friedel's theory (Friedel, 1958) for interpreting magnetism in transition metal alloys is that by replacing an atom of Fe, Co or Ni by an atom of an early transition metal (Cr, V, Ti,...) in an Fe, Co or Ni matrix, one removes precisely five d states from below the Fermi level E_F and introduces five impurity d states above E_F. In strong ferromagnets, the magnetic moment can thus be evaluated using equation (4) provided Z_m is replaced by :

$$\bar{Z}_m = 2N_d^\uparrow(1 - y) - Z_M(1 - y) - yZ_A \tag{5}$$

where Z_M and Z_A refer to the chemical valence of the element M and of the alloyed element A, respectively.

Provided $N_{sp}^\uparrow$ is a constant, M varies linearly with $\bar{Z}_m$. It also varies linearly with the electron per atom ratio as used in the familiar Slater-Pauling curve. An important result, illustrated by the above approach, is that, as long as strong ferromagnetism is maintained, the magnetization can be evaluated without a detailed knowledge of electron transfer and hybridization effects.

The variation of the magnetization in R-Fe compounds may be interpreted assuming that rare-earth elements, whose external electronic configuration is $5d^2 6s^1$ ($4d^2 5s^1$ for yttrium) are early transition elements. Thus, in the case of Y-Fe, $Z_{Fe} = 8$, $Z_Y = 3$. Assuming, from band structure calculation, that $2N_{sp}^{\uparrow} = 0.9$ (Williams et al., 1983a), the calculated variation of the Fe magnetic moment as a function of $\bar{Z}_m$ corresponds to the straight line in figure 2. Experimental data for various Y-Fe compounds are also shown in figure 2. In the different compounds, the magnetic moments lie below the calculated line. This corresponds to the occurrence of weak ferromagnetism. However, as the Y concentration is increased, the magnetic moments approach the calculated values which reveals that Fe magnetism tends towards strong ferromagnetism. This result, previously noted in Y-Fe amorphous alloys (Malozemoff et al., 1983), is confirmed by band structure calculations (Williams et al., 1983b).

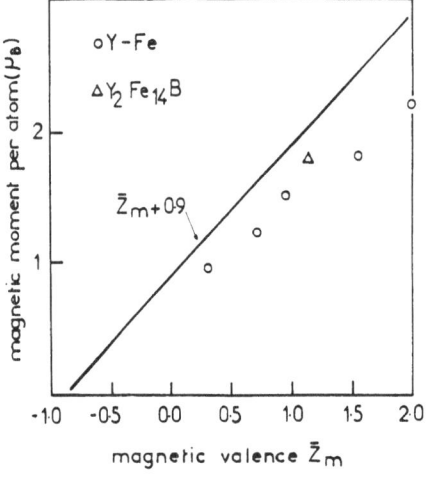

Fig. 2   Magnetic moment per atom in Y-Fe alloys as a function of the magnetic valence Z_m.

The above analysis can be extended to the $Y_2Fe_{14}B$ compound assuming that the chemical valence of the boron atoms, Z_B, is equal to 2. It is deduced that weak ferromagnetism characterises this compound. However, the magnetic moment per atom is closer to the line calculated for strong ferro-magnetism than in binary compounds. This effect may tentatively be attributed to the lower density of compounds crystallizing in the $Nd_2Fe_{14}B$-type structure compared to that of the R-Fe binary compounds. Indeed, it is an experimental fact confirmed by band structure calculations that Fe atoms arranged in a close-packed configuration are not strongly ferromagnetic (Malozemof et al., 1983).

Dependence of the Fe-Fe interactions on the Fe coordination number

The Curie temperatures in Y-Fe alloys are shown in figure 3 as a function of the amount of Y atoms alloyed. T_c is always significantly lower than that of pure Fe, it increases with increasing percentage of Y atoms. This variation in T_c can be due either to variations in the values of the local moments, μ_M or to variations in the strength of interactions between moments, represented by the molecular field coefficient n_{FeFe}.

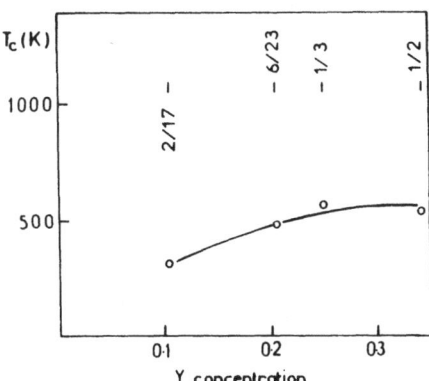

Fig. 3   Curie temperature in Y-Fe alloys as a function of the percentage of Y atoms alloyed. For meaning of abbreviations see caption of Fig. 1.

TABLE 1 : T_c/μ_{Fe}^2 (K/μ_B^2) in R-Fe compounds.

	RFe$_2$		RFe$_3$		R$_6$Fe$_{23}$		R$_2$Fe$_{14}$B		R$_2$Fe$_{17}$	
	Y	Lu	Y	Lu	Y	Lu	Y	Lu	Y	Lu
T_c/μ_{Fe}^2 (K/μ_B^2)	258	277	186		137	140	125	129	81	71

In order to separate out the respective contributions of μ_{Fe}^2 and n_{FeFe} in the value of the magnetic interactions, it is meaningful to determine the value of the parameter T_c/μ_{Fe}^2 which is proportional to n_{FeFe}. These values are listed in table 1 for Y and Lu compounds, they decrease with the percentage of rare-earth atoms. Although the Fe magnetic moment is larger in Y_2Fe_{17} than in YFe_2, this strong reduction in the molecular field coefficient

n_{FeFe} for Fe-rich compounds is responsible for the value of T_c being lower in Y_2Fe_{17} (310 K) than in YFe_2 (570 K). These large variations in n_{FeFe} from one compound to another can be attributed to a dependence of n_{FeFe} either on interatomic distances or on local environment effects. It is unlikely that the variations in Fe-Fe interatomic distances from R_2Fe_{17} to RFe_2 which are of about 2 % may be responsible for the observed increase in n_{FeFe} by about a factor 3. Thus, it must be concluded that the values of n_{FeFe} are strongly influenced by local environment effects. Such a behaviour has already been observed in pure Fe, bcc α-Fe being a ferromagnet which orders at 1043 K while fcc γ-Fe is not magnetically ordered. These phenomena may be described in terms of local coordination of Fe atoms, $c_n \cdot T_c / \mu_{Fe}^2$ is plotted in figure 4 as a function of $\bar{c}_n$ in different Y-Fe and Lu-Fe inter-metallic compounds. The interactions between Fe moments decrease regularly as $\bar{c}_n$ increases from $6(RFe_2)$ to 10.3 (R_2Fe_{17}). it is very striking indeed that the ferromagnetism of pure α-Fe ($\bar{c}_n = 8$) and the absence of ferroma-gnetism in γ-Fe ($\bar{c}_n = 12$) is predicted by this plot. Band structure calcu-lations (Kübler, 1981) permit the understanding of the above behaviour. Strong ferromagnetic interactions are associated with open structures in which the local coordination of Fe atoms is low. The interactions are strongly reduced and antiferromagnetism may even occur in the case of close packed structure, with high local coordination for the Fe atoms (γ-Fe, R_2Fe_{17} compounds).

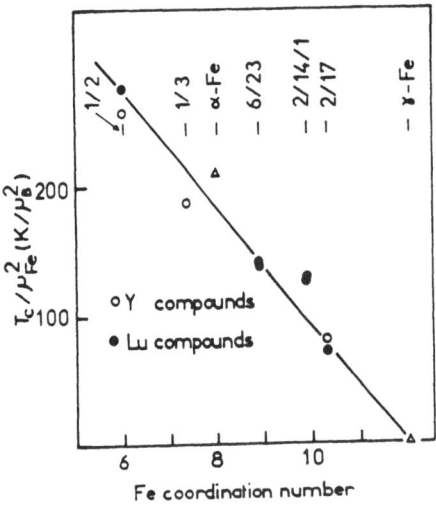

Fig. 4 T_c / μ_{Fe}^2 in Y-Fe and Lu-Fe compounds as a function of the mean local coordination of Fe atoms. For meaning of the abbreviations see caption of Fig. 1.

EVIDENCE FOR A SYSTEMATIC DEPENDENCE OF R-Fe EXCHANGE INTERACTION

In R-Fe intermetallics the highest value of T_C always occurs in the Gd compound where the spin magnetic moment is a maximum for the rare earths. This means, as is generally expected, that the interactions occur between spins, $E_{RFe} = -n_{RFe} M_R^S M_{Fe}^S$, where M_R^S and M_{Fe}^S are the spin moments of the rare earth and Fe, respectively, and n_{RFe} is the relevant molecular field coefficient.

In a given series, due to the expected similarity of the rare earth 5d band structure, it is usually assumed that n_{RFe} is a constant. An appropriate ferrimagnetic molecular field treatment leads to an expression for T_C given by :

$$T_C = \frac{1}{2} [T_{Fe} + T_R + (T_{Fe} - T_R)^2 + 4 T_{RFe}^2] \qquad (6)$$

where T_{Fe}, T_R and T_{RFe} represent the contributions to T_C due to Fe-Fe, R-R and R-Fe interactions respectively. T_{RFe}^2 is proportional to the de Gennes factor of the rare earth $(G(J) = (g_J - 1)^2 J(J + 1))$. Thus, in the instance where $T_{RFe} \ll T_{Fe}$, $T_C - T_{Fe}$ should be approximately proportional to $G(J)$. However, for compounds with the light rare earths the experimental T_C values are always found to be larger than the theoretically predicted ones (Sinnema et al., 1984). It was therefore decided to look more closely at this phenomenon on a number of R-Fe series (RFe_2, RFe_3, R_6Fe_{23}, R_2Fe_{17}, $R_2Fe_{14}B$).

Experimental evidence for a systematic dependence of n_{RFe}

The values of T_C for compounds in a number of RFe series are listed in table 2. In such systems where we consider for simplicity two magnetic sublattices of different nature, the magnetization M_{Fe} and M_R in an applied field H at $T > T_C$ may be written as :

$$M_{Fe} = \chi'_{Fe} [H + (2(g_J - 1)/g_J)n_{RFe} M_R] \qquad (7)$$

$$M_R = \chi'_R [H + (2(g_J - 1)/g_J)n_{RFe} M_{Fe}] \qquad (8)$$

TABLE 2   Experimental T_C(K) and deduced n_{RFe} (Oe cm^3/emu) for compounds in a number of R-Fe series. Small numbers for T_C correspond to extrapolated values.

	R₂Fe₁₄B		R₂Fe₁₇		R₆Fe₂₃		RFe₃		RFe₂	
	T_C	n_{RM}	T_C	n_{RM}	T_C	n_{RM}	T_C	n_{RM}	T_C	n_{RM}
La	542ᵃ		235		400		528		500	
Ce	427ᵃ									
Pr	565ᵃ	3490	283	3334					543	3828
Nd	592ᵃ	3435	327	3288					578	3586
Sm	621	2812	385	2927			650	3159	680	3955
Gd	664ᵃ	1874	460	2044	655	2116	728	2228	790	2791
Tb	620	1794	408	2054	574	1863	648	2016	702	2594
Dy	589	1686	363	1997	529	1663	600	1849	638	2298
Ho	569	1639	325	1923	509	1621	565	1640	602	2109
Er	554ᵃ	1595	293ᵃ	1802	489	1317	550	1692	579ᵃ	1877
Tm	545ᵃ	1703	265ᵃ	1330	483	967	535	1506	563ᵃ	972
Yb	527ᵃ								543	
Lu	534ᵃ		257ᵃ		492		526		570ᵃ	

ᵃ This study.

where $\chi'_{Fe} = \chi_{Fe}/(1 - n_{FeFe}\chi_{Fe})$ and $\chi'_R = \chi_R/[1 - 4((g_J - 1)^2/g_J^2)n_{RR}\chi_R]$ are the exchange susceptibilies and n_{FeFe}, n_{RFe}, n_{RFe} are molecular field coefficients representing the different exchange interactions between spin moments. χ_{Fe} and χ_R are the intrinsic susceptibilities without interactions. From equations (7) and (8) the condition which determines the value of T_C is :

$$1 - \chi'_{Fe}\chi'_R 4[(g_J - 1)^2/g_J^2]n_{RFe}^2 = 0 \;. \tag{9}$$

It is noted that Crystalline Electric Field interactions may be neglected, as calculations showed that their effect on the value of T_C is less than 1 K. The expression for T_C in Eq. 6 is deduced from

$$T_{Fe} = n_{FeFe}C_{Fe} \tag{10}$$

$$T_R = 4[(g_J - 1)^2/g_J^2]n_{RR}C_R \tag{11}$$

and

$$T_{RFe} = [2(g_J - 1)/g_J]n_{RFe}\sqrt{C_R C_{Fe}}$$

$$= \sqrt{(T_C - T_{Fe})(T_C - T_R)} \;. \tag{12}$$

where C_R and C_{Fe} are the rare earth and Fe Curie constants respectively (Burzo, 1973; Burzo, 1985). In order to determine n_{RFe} in a given compound using equation (12) values of T_C, T_{Fe} and T_R are needed. T_C is the experimental value of the ordering temperature. In the R₂Fe₁₄B series values of T_{Fe} for each compound were obtained by a linear interpolation between the T_C values for compounds with non-magnetic La and Lu. For the other series the compound with La does not exist. T_C values for hypothetical La compound deduced from the values of T_C for the other members of the series are listed in table 2. The values of T_R for the different compounds were determined using the value of n_{RR} deduced from the ordering temperature in R-Ni compounds (n_{RR} = 226 Oe cm^3/emu) and equation (11).

The values of n_{RFe} thus obtained are listed in table 2 and those for the $R_2Fe_{14}B$ series are plotted in figure 5a.

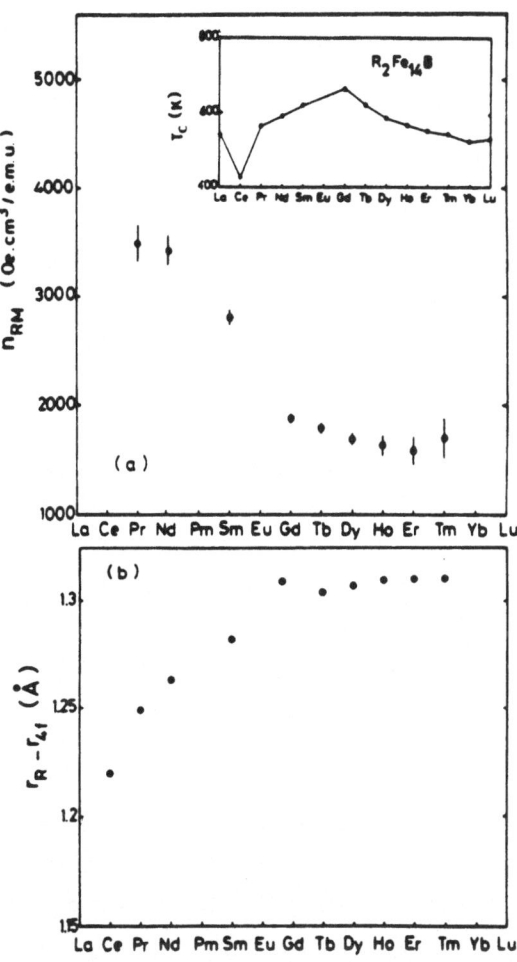

Fig. 5 (a) Calculated n_{RFe} in $R_2Fe_{14}B$
(b) r_{4f}-r_{5d} across the 4f series.

Origin of the variation of n_{RFe} across a given series of compounds

Striking features are evident from table 2 and figure 5 : (i) n_{RFe} is not constant going across a given series but decreases by about a factor of 2. (ii) For a given rare earth similar values of n_{RFe} are obtained in the different series. Given the indirect nature of the R-Fe interactions, two possible explanations may be proposed to give account of (i) above : either the 3d-5d or the 5d-4f interactions vary from compound to compound. As a general rule exchange interactions are distance dependent. Thus the 3d-5d interactions are expected to vary with $d_1 = d_{RFe} - r_{5d} - r_{3d}$ and the 5d-4f exchange interactions with $d_2 = r_{5d} - r_{4f}$ where d_{RFe} is the distance between R and Fe atoms and r_{3d}, r_{5d}, r_{4f} are the radii of the 3d, 5d and 4f shells respectively. Since 5d and 3d electrons are in the outermost shells, we assumed, as a first approximation, that $r_{5d} \simeq r_R$ and $r_{3d} \simeq r_{Fe}$, where r_R and r_{Fe} are the metallic radii of the R and Fe atoms. Finally, r_{4f} was approximated by $\sqrt{\langle r^2_{4f} \rangle}$, where $\langle r^2_{4f} \rangle$ has been calculated by Desclaux and Freeman (1979).

In a given series r_{Fe} is a constant. The variations of d_1 and d_2 are thus determined by the separate variations of d_{RFe} and r_R for d_1, r_{4f} and r_R for d_2. There is a 40 % decrease in r_{4f} across the series and a corresponding decrease in r_R of only 4 %. Thus the lanthanide contraction is ten times smaller than the corresponding contraction of the 4f shell. Because of the more localised character of the 4f electrons with respect to that of the 5d electrons, the former are more sensitive to the increasing nuclear charge. The result is that d_2 increases continuously across the lanthanide series. Values of d_2 are plotted in figure 5b. It is clear that the variation of n_{RFe} remarkably mimicks that of d_2. It can also be shown that, to a first approximation, the variation of d_1 across a series is negligible.

We are thus led to conclude that the deduced variation in the strength of R-Fe interactions is linked intrinsically to the rare earth. It is attributed to the variation of the exchange interactions between 4f and 5d electrons which are larger for light rare earth elements since the difference between the spatial extent of the 4f and 5d electrons is reduced.

REFERENCES

Burzo, E. 1973. Stud. Cercet. Fiz., $\underline{25}$, 425.

Burzo, E., Oswasd, E., Huang, M.Q., Boltich, E. and Wallace, W.E. 1985.
      J. Appl. Phys., $\underline{57}$, 4109.

Desclaux, J.P. and Freeman, A.J. 1979. J. Magn. Magn. Mat., $\underline{12}$, 11.

Friedel, J. 1958. Nuovo Cim. Suppl. n° 2, 287.

Givord, D., Li, H.S., Moreau, J.M. 1984. Solid Stat. Commun., $\underline{50}$, 497.

Herbst, J.F., Croat, J.J., Pinkerton, F.E., Yelon, W.B. 1984. Phys. Rev.,
      $\underline{B29}$, 4176.

Kübler, J. 1981. Phys. Lett., $\underline{A81}$, 81.

Legvold, S. 1980. in "Ferromagnetic Materials" (North-Holland, Amsterdam),
      chap 3.

Malozemoff, A.P., Williams, A.R., Terakura, K., Moruzzi, V.L. and
      Fukamichi, K. 1983. J. Magn. Magn. Mat., $\underline{35}$, 192.

Sinnema, S., Radwanski, R.J., Franse, J.J.M., de Mooij, D.B. and
      Buschow, K.H.J. 1984. J. Magn. Magn. Mat., $\underline{45}$, 335.

Williams, A.R., Moruzzi, V.L., Malozemoff, A.P. and Terakura, K. 1983a.
      IEEE Trans. Magn., $\underline{MAG-19}$, 1983.

Williams, A.R., Moruzzi, V.L., Gellatt, D.C. and Kübler, J. 1983b,
      J. Magn. Magn. Mat., $\underline{31-34}$, 88.

MAGNETIC INTERACTIONS IN RARE-EARTH - 3d INTERMETALLICS

J.J.M. Franse, S. Sinnema, R. Verhoef, R.J. Radwański,

F.R. de Boer, A. Menovsky

Natuurkundig Laboratorium der Universiteit van Amsterdam,
Valckenierstraat 65, 1018 XE Amsterdam,
The Netherlands

ABSTRACT
    In the scope of the CEAM project progress has been made in the follow-
ing subjects:
- preparation, especially single crystal growth,
- study of magnetization processes in magnetic fields up to 40 T,
- evaluation of the experimental results by developing a theoretical framework.
At the start of the CEAM project, the production of single crystals was
mainly focussed on the R_2T_{17} compounds (T = Fe, Co). A series of thirteen
sizable single crystals has been prepared with volumes of the order of 1
cm^3. High-magnetic-field studies revealed a new type of exchange driven
magnetic transition in fields above 20 T. These transitions can succesfully
be described by a model involving the exchange and crystal field interactions. The values for the interaction parameters as derived from magnetisation measurements have been checked by inelastic neutron diffraction studies at the Institut Laue Langevin in Grenoble on a large single-
crystalline sample of Dy_2Co_{17}. The potency of the model is shown by the
applicability to other series of RE-3d intermetallics. At the same time,
field-oriented powders of polycrystalline $RE_2Fe_{14}B$ and $RE_2Co_{14}B$ compounds
were studied in order to establish the magnetic parameters of both series.
The production of high-quality single crystals of the 2:14:B series turned
out to be less trivial than in case of the 2:17 compounds and the pulling
rate in the Czochralski technique had to be reduced by a factor of ten
(3 mm/hour). Sizable single crystals of the 2:14:B series have been pro-
duced   for $Y_2Fe_{14}B$, $Nd_2Fe_{14}B$, $Gd_2Fe_{14}B$, $Tb_2Fe_{14}B$, $Dy_2Fe_{14}B$ and
$Nd_2(Fe_{0.8}Co_{0.2})_{14}B$. These samples have been investigated in high magnetic
fields and, on the basis of these results, values for the exchange and
crystal-field parameters have been deduced for both series. A new technique
has been developed for measuring the exchange stiffness between the rare-
earth and 3d sublattices by performing long-pulse high-field measurements
on free powders. Finally, an extensive study has been started in order to
determine the individual site contributions to the anisotropy of the 3d
sublattice in several series of Y-(Fe,Co) compounds, in particular in the
$Y_2(Fe,Co)_{14}B$ pseudo-ternaries.

I. INTRODUCTION.

    At the time the CEAM programme started, research was carried out at

the Natuurkundig Laboratorium of the University of Amsterdam on two dif-

ferent series of rare earth - transition metal compounds: R_2T_{17} and $R_2T_{14}B$.

For the R_2T_{17} series, the production of single crystals was planned for

most of the cobalt and iron compounds, whereas for the $R_2T_{14}B$ series the attention was mainly devoted to polycrystalline and powdered samples. The Amsterdam High Field Facility turned out to be extremely well suited for studying the crystal-field and exchange interactions in these materials. A new type of exchange-driven transition has been observed, for instance, in the magnetisation curves of some of the R_2T_{17} compounds and most recently in $Er_2Fe_{14}B$. Transitions of this type turn out only to occur in the field range between 20 T and 100 T so high-magnetic-field facilities are essential in carrying out these investigations. Supported by the CEAM programme, The Natuurkundig Laboratorium also included the production and study of single-crystalline $R_2T_{14}B$ compounds in its research programme. Attempts in cooperation with the Laboratoire Louis Néel in Grenoble to produce single crystalline material of the compound $Er_2Fe_{14}B$ were not succesful. This compound was chosen because our model calculations predicted an exchange-driven transition in this compound, just below or above 40 T, depending on the precise values of the exchange and crystal-field parameters. Other compounds were better suited for the single-crystal production and good-quality single crystals exist at present for a number of 2:14:B compounds. In the meantime a new technique was developed for observing the exchange-driven transitions in a lower-field range by using free monocrystalline powder particles. The transition that is calculated to occur in $Er_2Fe_{14}B$ for a field of 40 T along the easy direction of magnetisation, shifts in the free-powder sample to a field of 32.1 T that can easily be reached in our installation. Experiments on several other single crystals of the $R_2T_{14}B$ series have now been performed in magnetic fields up to 40 T along the main crystallographic axes, among them $Nd_2Fe_{14}B$ and $Pr_2Fe_{14}B$. These magnetisation data turn out to be a good starting point for the evaluation of the crystal-field and exchange parameters that dominate the magnetic behaviour of these compounds. Finally, a study has been started on the 3d anisotropy in different series of R_xT_y compounds (R = Y, Lu, Th), including the pseudo-ternaries $Y_2(Fe,Co)_{14}B$. In this study, the contributions to the anisotropy arising from the different crystallographic sites are examined.

II. SAMPLES AND SAMPLE PREPARATION.

Single crystalline samples.

Within the project period the research programme on the R_2T_{17} compounds has been completed. This programme included the growth of single crystals of the cobalt and iron compounds. In total thirteen sizable single crystals of the R_2T_{17} compounds have been produced: eight single crystals of the R_2Co_{17} series with $R = Y$, Pr, Nd, Gd, Tb, Dy, Ho and Er, four single crystals of the R_2Fe_{17} series with $R = Y$, Dy, Ho and Er and finally, a pseudo-binary crystal with the composition $Ho_2Co_{14}Fe_3$. These single crystals have a typical volume of $1\,cm^3$.

At the same time attempts were made to grow single crystals of the $R_2T_{14}B$ series. The growth of single crystals with this composition turned out to be less trivial than in the case of the R_2T_{17} compounds. Nevertheless, up till now six sizable single crystals, with volumes of about $0.2\,cm^3$, have been grown. The produced single crystals are: $Y_2Fe_{14}B$, $Nd_2Fe_{14}B$, $Gd_2Fe_{14}B$, $Tb_2Fe_{14}B$, $Dy_2Fe_{14}B$ and $Nd_2(Fe_{0.8}Co_{0.2})_{14}B^\dagger$.

All these single crystals have been produced in an adapted tri-arc Czochralski apparatus (Menovsky and Franse, 1983). Buttons of 10-20 gr, melted together in a separate furnace, are used for the Czochralski method. The purity of the starting materials is 99.9 % or better.

The constituents of the R_2T_{17} compounds were melted together in an almost stoichiometric ratio; most of the R_2T_{17} compounds melt congruently. For these compounds a pulling rate of 30 mm/hour could be realised.

The $R_2T_{14}B$ compounds do not melt congruently and an off-stoichiometric melt has to be used. The composition of the melt was the same or close to the composition suggested by Givord et al. (1984): $R_{15}T_{77}B_8$. Due to the off-stiochiometry of the melt a low pulling rate has to be used: 3 mm/hour, which is ten times smaller than in case of the R_2T_{17} compounds.

In spite of all precautions, rare-earth oxides are always present in the melt. These oxides have a tendency to float on the surface, and can disturb the growth of the single crystal in cases when they come too close to the liquid-solid interface or stick on it where they act as parasitic nuclei for the crystal growth. In the single crystals produced, no oxides have ever been found inside the material.

To confirm the quality of the single crystals, Laue photographs were taken from different positions on the as-grown crystals. These Laue pictures reveal mosaic structures to be absent or to be less than $1°$. Out of the single crystals, spheres (3 mm diameter) and slices were machined by spark erosion. The spheres are used for the magnetic measurements. The slices serve to determine lattice parameters and (if necessary) to perform metallographic analyses. Lattice parameters were determined by a standard powder diffractogram technique, by which also the absence of second phases could be confirmed. These lattice-parameter values were compared with values calculated from the experimentally determined densities of the samples. No significant deviations were found to occur.

More details about the growth of these single crystals can be found in Sinnema et. al (1987a).

Polycrystalline samples.

The polycrystalline samples of the $R_2T_{14}B$ and $R_2T_{14}C$ series[††] were made by melting together the appropriate amounts of 99.9 % pure starting materials, in an arc furnace under a purified argon gas atmosphere. After arc melting, the samples were wrapped in Ta-foil, sealed into an evacuated quartz tube and annealed for several weeks. The $R_2T_{14}B$ compounds were annealed for two weeks at 900 °C and the $R_2T_{14}C$ compounds were annealed for several weeks at temperatures between 850 and 900 °C.

In order to confirm the proper phase and to check whether the samples were single-phase, X-ray diffraction measurements were performed with a standard powder diffractometer. The $R_2Fe_{14}B$ compounds were found to have the tetragonal $Nd_2Fe_{14}B$ structure for all the rare-earths and turned out to be single-phase (Sinnema et al., 1984). In the $R_2Co_{14}B$ series, the tetragonal $Nd_2Fe_{14}B$ structure was only found for the compounds with R = Y, La, Pr, Nd, Sm, Gd and Tb; with R = Ce, Dy, Ho, Er, Tm and Lu the tetragonal $Nd_2Fe_{14}B$ structure is not formed (Buschow et al. 1985). Most the compounds of the $R_2Fe_{14}C$ series (R = Sm, Gd, Tb, Dy, Ho, Er, Tm and Lu) crystallize in the $Nd_2Fe_{14}B$ structure (De Boer et al., 1988). Recently, also $Nd_2Fe_{14}C$ has been stabilised (De Boer et al., to be published). The $R_2Fe_{14}C$ samples were approximately single phase. The amount of impurity phases was about 5 %.

The alignment of the powders was carried out, at room temperature, in a field of 1 T by fixing their direction with epoxy resin. The powders used for the magnetisation measurements on free powders of the $Er_2Fe_{14-x}Mn_xB$ compounds were sieved after powdering in order to get particles which are smaller than 40 μm.

III. EXPERIMENTAL METHODS.
The Amsterdam High-Field Installation.

High-field magnetisation measurements were performed in the High-Field Installation of the University of Amsterdam (Gersdorf et al., 1983). With this facility it is possible to reach magnetic fields up to 40 T, constant within one part in a thousand during 0.1 second.

The field is generated by a current through a copper coil, from which the Joule heat is removed after each pulse. Liquid neon, which is supplied by a neon liquifier in a closed circuit, is used as the cooling medium. The duration of the pulse is limited, because the resistance of the coil increases almost exponentially as a function of time during the pulse and can become so high, that it is no longer possible to maintain constant current within one part in a thousand. In the extreme case of a magnetic field of 40 T the field can be kept constant during 0.1 s. The duration of the pulse can be longer if a field lower than 40 T is chosen. Typical pulse time is 2 seconds in total. The measurements are carried out by means of step-wise field pulses with field intervals that can be changed (7 steps of 1,2,3,4 or 5 T intervals were used). During the steps the field is constant for about 40 ms, a time long enough to let decay eddy currents

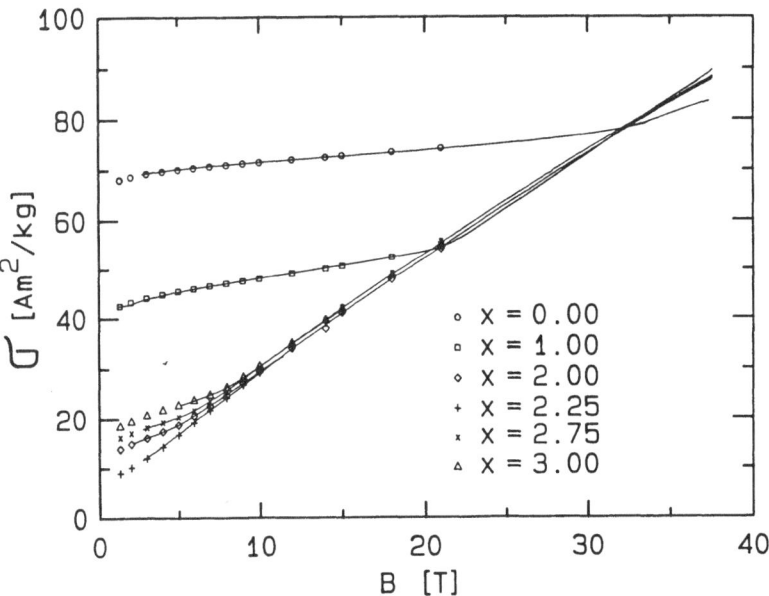

Fig. 1. *High-field magnetisation curves of free powders of* $Er_2Fe_{14-x}Mn_xB$ *at 4.2 K. "Step-wise" pulses (represented by the different markers) are combined with "continuously decreasing" pulses (solid lines).*
These curves are analysed within the two-sublattice model for **free** *powders. Within this model, the slope after the transition is inverse proportional to the molecular field coefficient* n_{RT} *(see eq.1). From the experimental curves a value of 0.45 kgT/Am^2 has been derived. These high-field magnetisation measurements on free powders constitute a new experimental technique to determine, in a direct way, the crucial coefficient* n_{RT}.

in metallic specimens. For a more precise determination of transition
fields other pulses can be programmed. In these pulses the field decreases
continuously from a given value (even 38 T) with a slope that can be programmed within the limits of the installation. The influence of eddy
currents in the compounds investigated turns out to be negligibly small.
As an example, magnetisation measurements on $Er_2Fe_{14-x}Mn_xB$ free powders are
presented in fig. 1.

The magnetisation is measured by an induction method. The voltage
induced in a pick-up coil by the time variation of the magnetisation is
electronically integrated to yield the magnetisation itself.
A limited range of temperatures can be reached by immersing the sample in a
cryogenic liquid (He, H_2, N_2).

The Two-Coil Extraction Magnetometer.

The magnetisation measurements as a function of temperature in fields
up to 7 T are performed in an apparatus that is based on a set of oppo-
sitely wound pick-up coils in a superconducting solenoid. The sample
holder is moved between the centres of the pick-up coils and, by integrating the induced voltage, the magnetisation of the specimen is determined.

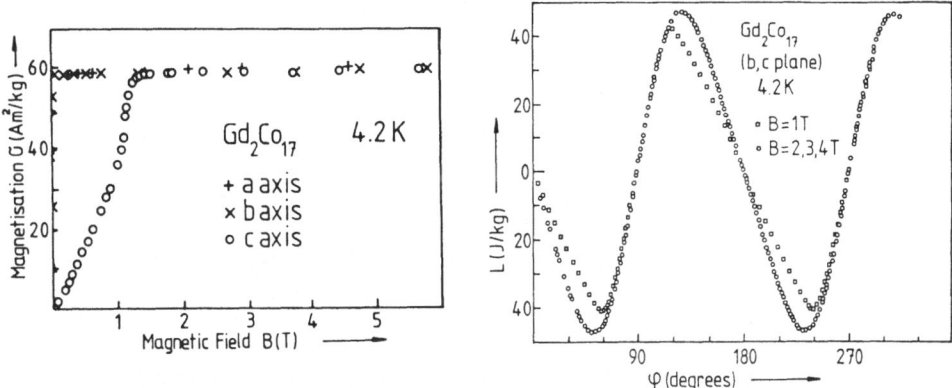

Fig. 2. a. Magnetisation curves of Gd_2Co_{17} at 4.2 K.
The discontinuity in the curve with the applied field along the
c-axis at 1.10 T is identified as a FOMP transition. The transition field and the magnetisation at which the transition takes
place can perfectly be reproduced by applying the theory
developed by Asti and Bolzoni (1980), using the anisotropy constants derived from the torque measurements on this sample (see
fig 2. b.).
Fig. 2. b. The torque measured in the bc plane of Gd_2Co_{17} at 4.2 K
at different values of the applied magnetic field.

The pick-up coils and sample holder are placed in an insert dewar, in which the temperature is controlled between 4.2 K and 300 K. An example of a magnetisation measurement on Gd_2Co_{17} measured with this apparatus is shown in figure 2 a. Experiments under He pressures up to 10 kbar can be performed as well in the same temperature range.

The Torque Magnetometer.

Torque measurements are performed in a torque magnetometer with a rotating superconducting solenoid. The field (5 T at most) of this solenoid is perpendicular to the length direction of the torque wire. The rotation of the magnet is controlled by a stepper-motor and its position is optically detected using a gray-code disk together with a set of light sensors. A mirror that is stiffly connected to the sample reflects a light beam. The position of the reflected light beam is observed by autocollimation. A typical torque measurement on Gd_2Co_{17} is shown in figure 2 b. With this equipment torque measurements can be performed in the temperature range between 4.2 K and 300 K and under He gas pressure up to 8 kbar.

IV. THEORETICAL OUTLINE AND EXPERIMENTAL RESULTS.

There is growing evidence that the magnetic behaviour of $Nd_2Fe_{14}B$ as well as other compounds of the $R_2Fe_{14}B$ family, is to a large extent the result of an interplay of crystal-field and exchange interactions. In a macroscopic picture the crystal-field effect is related to the magnetocrystalline anisotropy. To investigate the magnetocrystalline anisotropy, compounds with light rare earths are best suited, because the rare-earth moment can be rotated in that case by the external fields over a wide range of orientations. The transformation of the macroscopic anisotropy constants $\{K_i\}$ into the microscopic crystal-field parameters, B_n^m, is not a trivial problem due to the large scatter in the data reported for the anisotropy constants. The origin of this large scatter is the inappropriate description of magnetisation curves in terms of a constant value of the resultant magnetisation. Compounds with the heavy rare earths are better suited for the study of the strength of the 3d-4f exchange interaction due to their ferrimagnetic structure. The field-induced non-collinearity of the ferrimagnetic system has a large impact on the magnetisation curve along the hard axis.

The magnetisation curves are analysed within a phenomenological model by considering the free energy expression of the 3d-4f magnetic system composed of the rare-earth(s) and transition-metal sublattice magnetisations in an external field **H**:

$$E_{R-T} = E_a^{R(i)} + E_a^T - \mu_o \, n_{RT} \, \mathbf{M}_T \, \mathbf{M}_R^{(i)} - \mathbf{M}_T \, \mathbf{B}_o - \mathbf{M}_R^{(i)} \, \mathbf{B}_o \qquad (1)$$

$\mathbf{M}_T$ and $\mathbf{M}_R$ denote the magnetisation of the 3d and the rare-earth sublattices (in a number of i), respectively, and $\mathbf{B}_o = \mu_o \mathbf{H}$. The 3d-4f interaction is expressed in a molecular-field approximation with the parameter n_{RT} as the intersublattice molecular-field coefficient.

The angle dependence of the anisotropy energy E_a is written in the form:

$$E_a(\theta,\alpha) = \sum_{n=0}^{\infty} \sum_{m=0}^{n} \kappa_n^m \, P_n^m(\cos\theta) \, \cos m\alpha \qquad (2)$$

where P_n^m are Legendre functions and κ_n^m the corresponding anisotropy coefficients. θ and α are the polar and azimuthal angles of the sublattice magnetisation with respect to the crystallographic directions <001> and <100>, respectively.

For the tetragonal symmetry, expression (2) for the magnetocrystalline anisotropy energy leads to the form:

$$E_a(\theta,\alpha) = \kappa_2^0 \, P_2^0 + \kappa_4^0 \, P_4^0 + \kappa_6^0 \, P_6^0 + (\, \kappa_4^4 \, P_4^4 + \kappa_6^4 \, P_6^4 \,) \, \cos 4\alpha \qquad (3)$$

The parameters κ_4^4 and κ_6^4 are responsible for the anisotropy within the tetragonal plane. Different signs of κ_2^0 and κ_4^0 lead to a competition which determines the easy-axis direction of the rare-earth moment and, due to the 3d-4f coupling, of the resultant magnetisation. For a positive value of κ_4^0, the term $\kappa_4^0 \, P_4^0$ has a minimum at 49 degrees with respect to the c-axis, while a negative value of κ_2^0 favours the easy c-axis.

The equilibrium state of the magnetic system is described by the angles θ and α for each of the i+1 sublattice magnetisations. Values for θ and α provide the components of the sublattice magnetisations along the applied field. The equilibrium state is found by minimizing eq.(1) with respect to the angles of the magnetisations. The set of anisotropy coefficients $\{\kappa_n^m\}$ and the intersublattice molecular-field coefficient n_{RT} are determined simultaneously by a minimizing procedure. Values for the anisotropy energy E_a^T and the magnetic moment M_T are, in general, deduced from the isostructural compounds with a non-magnetic rare-earth element (Y, La or Lu). The exact evaluation of E_a^T is not crucial as it is usually one order of magnitude smaller than the rare-earth anisotropy. Finally, a magnetic rare-earth partner causes an enhancement of the magnetism of the 3d sublattice but this effect in 3d-rich compounds turns out to be of minor importance.

The above-presented equations (1) and (2) form a good base for a realistic phenomenological approach of the magnetisation curves. The model can be used for all 3d-4f compounds with well-defined values of the magnetic moments. This situation is realized in all 3d-based rare-earth compounds ($R_2Fe_{14}B$, $R_2Co_{14}B$, R_2Fe_{17}, R_2Co_{17}, RCo_5 ...). The model takes into account the quite obvious fact that an external field disturbs the internal ferrimagnetic structure if present. It turns out that even a relatively small deviation from the ferrimagnetic structure has a large impact on the initial slope of the magnetisation curve along the hard magnetic axis.

In the $Nd_2Fe_{14}B$-type of structure there are two crystallographic sites for the rare-earth atoms, denoted by f and g (Givord et al., 1984). Mössbauer studies of $Gd_2Fe_{14}B$ performed by Bogé et al. (1986) reveal that the leading parameter of the crystal-field interactions A_2^0 is almost equal for both sites. Due to this fact, a two-sublattice system is able to provide a good insight into the magnetic behaviour of this class of compounds and, especially, into the high-field magnetisation process.

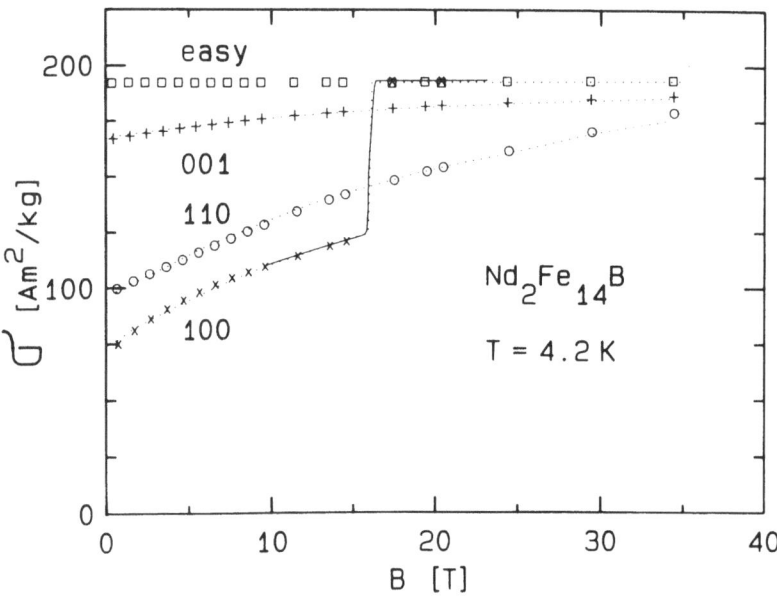

Fig. 3. High-field magnetisation curves of $Nd_2Fe_{14}B$ at 4.2 K, with the applied magnetic field along the three main crystallographic directions and along the easy direction of magnetisation (different markers and solid lines). Dashed lines represent theoretical fits to the experimental data.

The anisotropy coefficients of the neodymium sublattice, derived from an analysis of magnetisation curves of $Nd_2Fe_{14}B$ measured along four directions in fields up to 35 T, (see fig.3) enable us to construct the angle dependence of the anisotropy energy of the neodymium sublattice as the sum of contributions originating from the different coefficients κ_n^m. There is a minimum for a direction of the internal (molecular) field at $30.5°$ from the c axis in the (110) plane in agreement with the experimental result. The role of κ_4^0 in the formation of the minimum is decisive although the contribution from the coefficient κ_6^4 is so substantial that the κ_4^0 term itself is not able to produce the cone structure. The coefficient κ_6^4 plays a major role in the formation of the minimum at $90°$ for the <100> direction. The existence of this minimum is required for the abrupt field-induced spin-reorientation (FOMP) transition at 16.3 T. Values (in units of MJm^{-3}) of -9.7, $+2.2$, $+2.1$, $-0.6\ 10^{-3}$ and $+3.6\ 10^{-3}$ have been deduced for the anisotropy coefficients: κ_2^0, κ_4^0, κ_6^0, κ_4^4 and κ_6^4, respectively. The anisotropy coefficients lead to estimates for the CEF parameters, using relations like $A_2^0 = \kappa_2^0/2J_2\alpha_J<r_{4f}^2>$ and $A_2^2 = 3\kappa_2^2/J_2\alpha_J<r_{4f}^2>$; where $J_2 = J(J-\frac{1}{2})$. According to our analysis, the best set of crystal-field coefficients (in unit of $K a_o^{-n}$) is given by: $A_2^0 = +350$, $A_4^0 = -21$, $A_6^0 = -2$, $A_4^4 = +3.7$ and $A_6^4 = -26$. These results are in reasonable agreement with the values obtained by Cadogan et al. (1988). Diagonalization of the crystal-field and molecular field Hamiltonian leads to a cone angle of $30.8°$ as well as to a vanishing cone structure at 135 K in good agreement with experimental data. At a field of 35 T along the <110> direction, the neodymium moments are rotated to $\theta = 54°$ and the canting angle between the neodymium and iron sublattices amounts to $11°$ for this field value. A cone structure for the resultant magnetisation also exists in $Ho_2Fe_{14}B$ with a cone angle of $24°$. A set of crystal-field parameters close to that derived for $Nd_2Fe_{14}B$ shows a) a minimum in the ground-state energy of the Ho^{3+} ion at $24.7°$ from the c axis in the plane (110) and b) a vanishing cone structure at 57 K. The magnetic structure of $Ho_2Fe_{14}B$ is found to be non-planar as an effect of the large A_2^2 parameter and the finite value of the intersublattice interactions. The four holmium sublattice moments are spread around a direction antiparallel to the iron magnetisation that is tilted out by $24.3°$ from the c axis. In the absence of an external field the spread in orientation of the holmium moments is estimated to be a few degrees (Radwański and Franse, to be published). A similar spatial fan structure of a ferromagnetic type should exist in $Nd_2Fe_{14}B$. In general it is true that the parameter A_2^2 is more effective in the formation of the fan structure for easy-plane systems with relatively weak intersublattice interactions. Preliminary magnetisation curves of $Pr_2Fe_{14}B$ (fig. 4) indicate a complex magnetic structure within

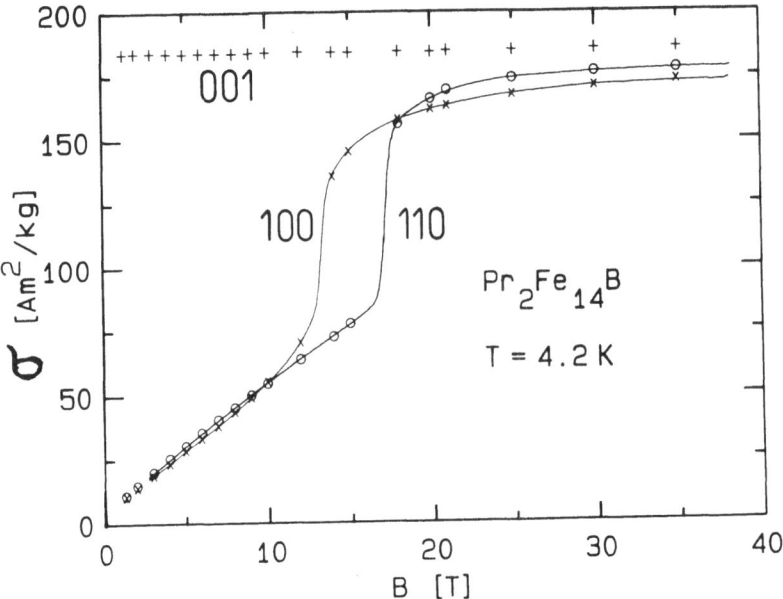

Fig. 4. *High-field magnetisation curves of $Pr_2Fe_{14}B$ at 4.2 K. The single crystal that has been used for the experiments was grown by H.S. Li and kindly provided by dr. D. Givord (Laboratiore Louis Néel, Grenoble).*

the tetragonal plane; a field of 38 T is apparently not able to align the praseodymium moments.

The same model has been applied to examine the magnetisation curves of R_2Co_{17} compounds with the hexagonal structure. The high-field magnetisation curves along the main crystallographic directions are shown in fig. 5 for Ho_2Co_{17} as an example (Franse et al., 1985). The magnetisation curves as well as the field-induced spin-reorientation transitions at 22.0 T and 29.5 T are well reproduced by the model introduced above. The spin-reorientation transition is governed by the molecular-field coefficient n_{RT} and by the in-plane anisotropy (the parameter B_6^6). A cone structure also exists in this holmium compound. A cone angle $\theta_c = 85°$ results from the CEF interactions (Radwański et al., to be published). The magnetisation measured along the c direction, the hard axis, exceeds the value of the spontaneous magnetisation already above 12 T. This explicitly proves that the ferrimagnetic structure is disturbed. The canting angle at a field of 35 T amounts to $34°$. The intersection of the hard- and easy-axis magnetisation

(12 T in case of Ho_2Co_{17}) is often incorrectly called an anisotropy field. A similar intersection has been observed for $Dy_2Fe_{14}B$. The present study shows that this intersection field is hardly related to the anisotropy energy. It rather reflects the strength of the intersublattice interactions (Radwański et al., 1987).

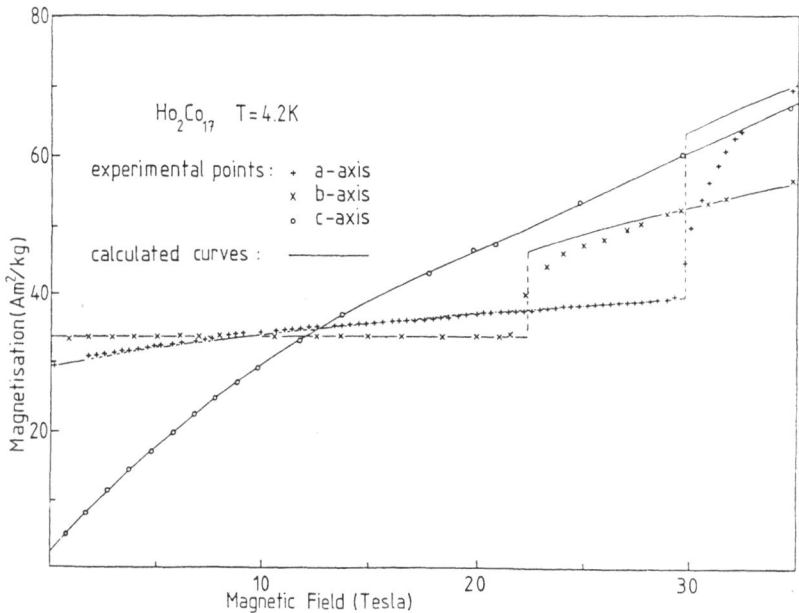

Fig. 5. High-field magnetisation curves of Ho_2Co_{17} at 4.2 K. The solid lines represent theoretical fits to the experimental data.

The analysis of the high-field behaviour of Ho_2Co_{17} results in a value of 159 T for the exchange field experienced by the holmium spin-moment (Franse et al., 1985). A value of 215 T has been derived for the exchange field experienced by Ho in $Ho_2Fe_{14}B$. The exchange field is nicely proportional to the value of the 3d moment ($1.65 \mu_B$ and $2.24 \mu_B$, respectively). The present study confirms that the strength of the 3d-4f interaction does not depend on the nature of the 3d spin and that it is hardly sensitive to the crystallographic structure.

The exchange interactions in 3d-4f intermetallic compounds are postulated to be described by a Heisenberg-like Hamiltonian:

$$\mathcal{H}_{ex}^{RT} = -2 \; \Sigma \; J_{RT} \; \mathbf{S}_T \cdot \mathbf{S}_R \qquad (5)$$

where $\mathbf{S}_T$ and $\mathbf{S}_R$ represent the 3d and rare-earth spins, respectively. This interaction can be re-written as the interaction of the R spin-moment with an effective field (= exchange field) defined as:

$$H_{ex}^{RT} = - \frac{S_T}{\mu_o \mu_B} \sum_T J_{RT} \qquad (6)$$

Assuming the exchange interaction J_{RT} to be isotropic with non-zero value for the nearest neighbours it can be written as:

$$H_{ex}^{RT} = - \frac{z \, J_{RT} \, S_T}{\mu_o \mu_B} \qquad (7)$$

The value of 215 T for the exchange field in $Ho_2Fe_{14}B$ results in a value of -129 K for $\sum J_{RT}$. Taking into account a value of 18 for the number of nearest neighbours in the $Nd_2Fe_{14}B$ structure, one obtains a value of -7.2 K for the exchange interaction parameter J_{RT} between the iron and holmium spins in $Ho_2Fe_{14}B$. In Ho_2Co_{17} J_{RT} amounts to -6.8 K (Sinnema et al., 1987b). The values for the interaction parameters as derived from magnetisation measurements, have been checked by performing inelastic neutron diffraction measurements on a large single-crystalline sample of Dy_2Co_{17} at the Institut Laue Langevin in Grenoble (Colpa et al., to be published).

Investigations of the pseudobinary compounds $Y_2(Fe,Co)_{14}B$ have been undertaken in order to understand the 3d anisotropy. The dependence of the anisotropy energy as a function of the cobalt concentration for three different temperatures is shown in fig. 6. This quite complex composition and temperature dependences are analyzed in terms of individual-(3d)-site contributions to the anisotropy energy and are thought to be due to competing effects arising from the different 3d crystallographic sites. Preliminary results show that the composition dependence can be quantitatively understood by taking into account a) the preferential occupation of the different crystallographic sites by iron and cobalt atoms, b) the different sign of the second-order Stevens factor attributed to the iron and cobalt ions and c) for each site a different anisotropy energy. See also Thuy et al. (1986).

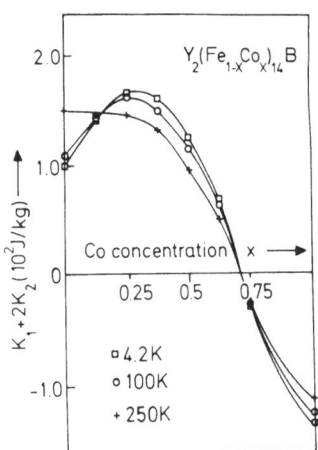

Fig. 6. Composition dependence of the anisotropy constants of $Y_2(Fe_{1-x}Co_x)_{14}B$ compounds at three different temperatures. Solid lines represent fits to the experimental data.

Recently, field-oriented powders of the $R_2Fe_{14}C$ series have been studied in cooperation with the Philips Research Laboratories.

† Cooperation with Ji Song-quan is highly acknowledged.

†† Samples have been prepared at the Philips Research Laboratories, Eindhoven.

V. REFERENCES.

Asti, G. and Bolzoni, F. 1980. Theory of First Order Magnetization Processes: Uniaxial anisotropy. J. Magn. Magn. Mater., 20, 29-43.

Bogé, M., Czjzek, G., Givord, D., Jeandey, C., Li, H.S. and Oddou, J.L. 1986. A ^{155}Gd Mössbauer study of a $Gd_2Fe_{14}B$ single crystal. J. Phys. F: Met. Phys., 16, L67-L72.

Buschow, K.H.J., Mooij, D.B. de, Sinnema, S., Radwański, R.J. and Franse, J.J.M. 1985. Magnetic and crystallographic properties of ternary rare earth compounds of the type $R_2Co_{14}B$. J. Magn. Magn. Mater., 51, 211-217.

Cadogan, J.M., Gavigan, J.P., Givord, D. and Li, H.S. 1988. A new approach to the analysis of magnetization measurements in rare earth transition metal compounds. Application to $Nd_2Fe_{14}B$. in press.

Boer, F.R. de, Huang, Ying-kai, Zhang, Zhi-dong, Mooij, D.B. de and Buschow, K.H.J. 1988. Magnetic and crystallographic properties of ternary rare-earth compounds of the type $R_2Fe_{14}C$. J. Magn. Magn. Mater., 72, 167-173.

Franse, J.J.M., Boer, F.R. de, Frings, P.H., Gersdorf, R., Menovsky, A., Muller, F.A., Radwański, R.J. and Sinnema, S. 1985. Magnetic transitions in single-crystal Ho_2Co_{17} studied in high magnetic fields. Phys. Rev., B31, 4347-4349.

Gersdorf, R., Boer, F.R. de, Wolfrat, J.C., Muller, F.A. and Roeland, L.W. 1983. The high magnetic field facility of the University of Amsterdam. In "High Field Magnetism" (Ed. M.Date). (North Holland, Amsterdam). pp. 277-287.

Givord, D., Li, H.S. and Perrier de la Bathie, R. 1984. Magnetic properties of $Y_2Fe_{14}B$ and $Nd_2Fe_{14}B$ single crystals. Solid State Commun., 51, 857-860.

Menovsky, A. and Franse, J.J.M. 1983. Crystal growth of some rare earth and uranium intermetallics from the melt. J. Crystal Growth, 65, 286-292.

Radwański, R.J., Franse, J.J.M. and Sinnema, S. 1987. Effective anisotropy constants in rare earth - 3d intermetallics. J. Magn. Magn. Mater., 70, 313-315. also Radwański, R.J. and Franse, J.J.M. 1987. Rare-earth contribution to the magnetocrystalline anisotropy energy in $R_2Fe_{14}B$., Phys. Rev., B36, 8616-8621.

Sinnema, S., Radwański, R.J., Franse, J.J.M., Mooij, D.B. de and Buschow, K.H.J. 1984. Magnetic properties of ternary rare-earth compounds of the type $R_2Fe_{14}B$. J. Magn. Magn. Mater., 44, 333-341.

Sinnema, S., Verhoef, R., Menovsky, A.A. and Franse, J.J.M. 1987a. Crystal Growth of R_2T_{17} and $R_2Fe_{14}B$ intermetallics (R = Rare Earth, T = Co, Fe). J. Crystal Growth, 85, 248-251.

Sinnema, S., Franse, J.J.M., Radwański, R.J., Menovsky, A. and Boer, F.R. de 1987b. High-field magnetisation studies on the $Ho_2(Co,Fe)_{17}$ intermetallics. J. Phys. F: Met. Phys. 17, 233-242.

Thuy, N.P. and Franse J.J.M., 1986. The magneto crystalline anisotropy of $Y_2(Co_xFe_{1-x})_{17}$. J. Magn. Magn. Mater., 54-57, 915-916.

MACROSCOPIC STUDIES OF MAGNETIC ANISOTROPY IN RARE-EARTH INTERMETALLIC COMPOUNDS

L. Pareti, O. Moze, M. Solzi, F. Bolzoni, G. Asti, G. Marusi

Istituto MASPEC del CNR

Via Chiavari 18/A, 43100, Parma, Italy

ABSTRACT

The Singular Point Technique and high pulsed field magnetometry have been extensively utilized in order to study the composition and temperature dependence of both the anisotropy field and the critical field of first order field induced magnetic processes (FOMP) in rare-earth Fe based intermetallics. In particular, type I FOMP in $Nd_2Fe_{14}B$ and type II FOMP in $Pr_2Fe_{14}B$ were investigated . High order anisotropy energy constants are needed to account for the FOMP. Planar anisotropy energy constants are also necessary to describe FOMP in $Nd_2Fe_{14}B$. In conjunction with determinations of preferential site occupations, the contribution and behaviour of 3d and 4f magneto-crystalline anisotropy to the overall anisotropy have been systematically investigated.

INTRODUCTION

The announcement in 1984 of the discovery of an Fe based Rare-Earth intermetallic compound having excellent permanent magnet properties had the effect of revitalizing Rare-Earth intermetallic research. Renewed attention has been focussed on this field with a correspondingly large number of both academic institutes and industries being involved in a very active research and development effort. In addition to a substantive effort in the U.S.A and Japan, Europe launched an intensive research effort with the creation and support of the Concerted European Action On Magnets (CEAM). Among the results produced by such an effort is the realization that many binary intermetallic compounds which are not interesting from the point of view of magneto-crystalline anisotropy could be rendered most useful by forming isomorphous compounds of suitable Rare-Earths with a stabilizing element.

The participation of MASPEC in the CEAM program was directed to a systematic study of 3d and 4f magneto-crystalline anisotropy, magnetization processes and preferential site occupations in numerous compounds with the tetragonal $Nd_2Fe_{14}B$ structure and the new family of ternary Rare-Earth intermetallic compounds based on the tetragonal $ThMn_{12}$ structure.

MATERIALS PREPARATION AND EXPERIMENTAL TECHNIQUE

Materials preparation has consisted of a combination of in-house preparative techniques and various materials obtained in active collaborations with other CEAM partners. Materials at MASPEC were obtained by argon arc melting of appropriate elements followed by standard furnace annealing. Collaborations were performed with Laboratoire Louis Néel, Grenoble, Laboratoire de Crystallographie du CNRS, Grenoble, Philips Research, Eindhoven, Trinity College, Dublin and ICMA, CSIC-Univ. de Zaragoza. Sample monophasicity and homogeneity was controlled by thermomagnetic analysis and X-ray diffraction. Thermomagnetic analysis was also used to measure the Curie temperature in all compounds.

Magnetization curves of polycrystals and single crystals have been measured in a low field (up to 20kOe) vibrating sample magnetometer and in a high pulsed field magnetometer (up to 300kOe). The unique singular point detection technique (SPD, Asti et al, 1974) enabled precise measurements of the anisotropy field to be obtained from polycrystalline materials. The typical error in the determination of the anisotropy field H_a is not more than 2%. Futhermore, exotic behaviour of the magneto-crystalline anisotropy such as First Order Magnetization Processes (FOMP, Asti et al, 1980) can be accurately quantified. The shape of the SPD peak is also often quite useful in that it allows an estimate to be made of the signs of high order anisotropy constants.

High resolution neutron diffraction has been used to investigate magnetic and crystal structures and to correlate this with the observed behaviour of the magnetic anisotropy. Particular emphasis has been placed

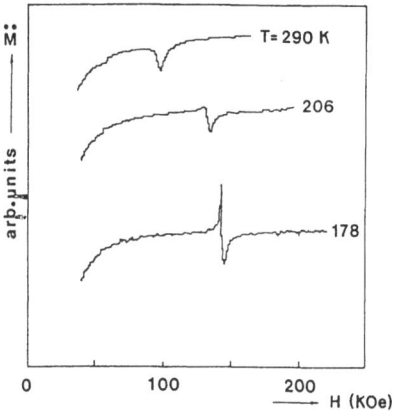

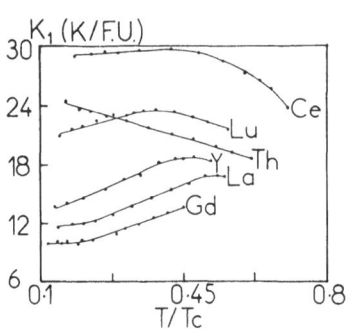

Fig.1. Examples of SPD signals: 2nd derivative of M with respect to time versus applied field for various T in $Nd_2Fe_{14}B_{0.5}C_{0.5}$. (Bolzoni et al.1985)

Fig.2. Variation of K_1 in Kelvin/f.u. with T/T_c for $RE_2Fe_{14}B$. (Bolzoni et al. 1987a).

on studies of preferential site occupancies of both Rare-Earth and Transition Metal sublattices. The neutron facilities of the Spallation Neutron Source, ISIS at the Rutherford Appleton Laboratory, U.K and of the Institut-Laue-Langevin, Grenoble have been extensively utilized.

RESULTS AND DISCUSSION

Early investigations were concentrated on $Nd_2Fe_{14}B$ (Asti et al, 1985), and C (Bolzoni et al, 1985) and Y (Bolzoni et al, 1986) substituted compounds. The use of the SPD technique enabled a measurement of the temperature dependence of H_a to be carried out on $Nd_2Fe_{14}B$. It also provided the first evidence that at T<200K a FOMP takes place (Fig. 1). From a comparison between the Y and Nd compounds it became evident that Nd is the main source of the anisotropy. However, the Fe sublattice has an unusually high anisotropy. Substitution of Nd by Y results in a decrease, as expected , of all the important magnetic characteristics, H_a, σ_s and T_c . An anomalous temperature dependence of H_a in $Y_2Fe_{14}B$ was observed and was attributed to a competition of the anisotropies at the six different Fe sites. Carbon was found to substitute B up to at least 75% in the tetragonal $Nd_2Fe_{14}B$ structure. Carbon doping also gave rise to

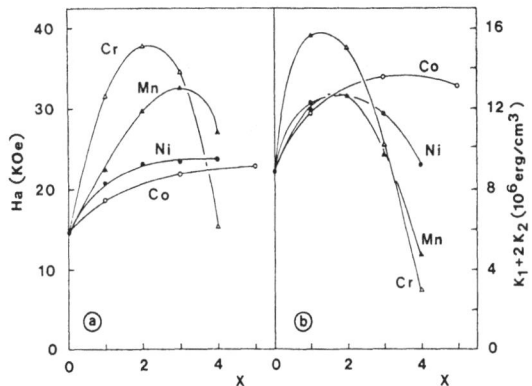

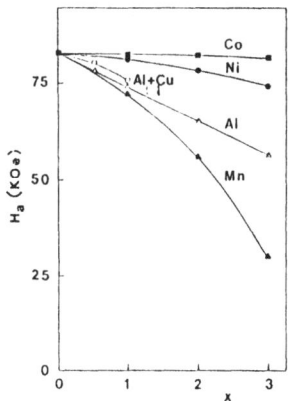

Fig.3. Composition dependence of (a) H_a and (b) K_1 for $Y_2Fe_{14-x}Me_xB$ where Me=Co, Ni, Mn, Cr. (Pareti et al. 1987a).

Fig.4. Composition dependence of H_a at 293K for $Nd_2Fe_{14-x}Me_xB$.

a slight decrease in T_c and σ_s whilst it increased H_a at room temperature from 83kOe to 96.5kOe for 75% B substitution.

In order to study the transition metal sublattice anisotropy, a systematic investigation of $RE_2Fe_{14}B$ (with non-magnetic RE) and $Y_2Fe_{14-x}Me_xB$ (Me=Co, Ni, Mn, Cr, Cu, Al and Zr) was carried out. In the compounds $RE_2Fe_{14}B$ with non-magnetic RE the variation of T_c was interpreted in terms of the dependence of the magnetic interactions with interatomic distances and on the type of RE (Bolzoni et al, 1987a). In all compounds except for $Th_2Fe_{14}B$ the temperature dependence of the anisotropy exhibits a maximum at $T/T_c=0.4$ (Fig.2). This effect is ascribed to competitions between different Fe sites and/or to a change in CEF interactions associated with a magneto-volume anomaly which occurs below T_c.

By a partial substitution of Fe by Co, Ni, Mn and Cr in $Y_2Fe_{14}B$, a common behaviour in the variation of the anisotropy was surprisingly found (Pareti et al 1987a). In all cases a maximum in the composition dependence of the anisotropy was observed (Fig.3). Another common feature in all the substitutions was that in all cases the anomalous temperature dependence of H_a disappeared. These common features are difficult to account for considering that the various ions have different effects on

σ_s and T_c and different site occupancies in the transition metal sublattice. Neutron diffraction and Mossbauer studies (Moze et al, 1988a; Bolzoni et al, 1987b) clearly indicate a different preferential occupation between Mn, Cr on one hand and Co on the other hand. In the former case, Mn and Cr have a strong preference for the $8j_2$ sites in contrast to Co. Enhanced local environment effects of the distribution of magnetization in Mn and Cr substituted $Y_2Fe_{14}B$ compounds were observed.

The same types of transition metal substitutions were performed on $Nd_2Fe_{14}B$ compounds in order to clarify the different roles and contributions to the anisotropy from the Rare-Earth and Transition metal sublattices (Bolzoni et al, 1987b,c,d,e). Complete isomorphous compounds of the above substitutions were found to exist only for Co substitution. Substitutions with x>3 in $Nd_2Fe_{14-x}M_xB$ resulted in the formation of the 2:17 phase for M= Ni, Mn. The Curie temperature was enhanced strongly by Co and only slightly by Ni substitutions whilst it decreased with Mn, Cr and Al. Cobalt was found to give a planar contribution to the anisotropy (40kOe at 293K) larger than that of the axial Fe contribution (22.5kOe at 293K). With the exception of low Co content the saturation magnetization is reduced in all cases, particularly for Mn; the replacement of 25% of Fe by Mn results in a non-ferromagnetic compound at 293K. The overall axial anisotropy appears also to be reduced in all cases (Fig.4). It should also be noted at this point that the increase in coercive force observed in permanent magnets when some Al is substituted for Fe, can not be attributed to an increase in H_a (which is indeed decreased) but rather to a particular microstructure of $Nd_{15}Fe_{77-x}Al_xB$.

Two different types of contributions to the anisotropy from transition metal ions have been proposed. Apart from a direct contribution it has been found that TM provide an indirect influence on the Nd anisotropy. In particular, taking into account the effect of the variation of T_c with the given substitutions, Ni induces a decrease of the Nd anisotropy whilst Mn has no effect. The situation with Co is such that there is a complex dependence of the Nd anisotropy with Co substitution. A minimum in the Nd anisotropy is observed when 80% of the Fe is replaced by Co. For Co substitutions larger than 80% a first order spin reorientation transition (easy axis to easy plane) takes place that is a temperature exists above which the Nd axial anisotropy is overcomed by the Co planar anisotropy. For lower Co content the Curie temperature is reached before the SRT takes place. This interesting behaviour of the $Nd_2Fe_{14-x}Co_xB$ magnetic phase diagram is displayed in Fig 5, where a ɔicritical point seems to be present.

Concerning the low temperature axis-to-cone SRT observed at 135 K in $Nd_2Fe_{14}B$, it has been found that in all the considered cases of iron substitutions (Co, Ni, Mn, Cr, Al, Cu, Si) there is a decrease of both the SRT temperature and of the cone angle value (fig.6).

In contrast to the behaviour of the Nd anisotropy in Co substituted compounds, the opposite behaviour has been observed in $Pr_2Fe_{14-x}Co_xB$ compounds (L. Pareti et al, 1985a; F. Bolzoni et al, 1987d; J. P. Gavigan et al, 1988). In spite of the strong Co planar anisotropy, the overall axial anisotropy of $Pr_2Co_{14}B$ is found to be much larger than that of the $Pr_2Fe_{14}B$. This effect, which is further evidence that two contributions to the anisotropy have to be taken into account for the transition metal, is due to an increase in the Pr ion anisotropy for x>0.7. The direct contribution was easily quantified by taking into consideration the anisotropy of Co substituted $Y_2Fe_{14}B$ compounds. The indirect contribution was evaluated by taking into consideration this anisotropy in the overall behaviour of the Co substituted $Pr_2Fe_{14}B$ compounds. In the range 0.7<x<1, there is a corresponding variation in c/a (Fig.8), suggesting that the change of the Pr anisotropy can be attributed to the change in 2nd order CEF terms which could be associated with preferential Fe and Co occupancies. An analysis of the SPD peak shape allowed an estimate to be made of the signs of the high order anisotropy constants. The analysis indicated that in $Pr_2Fe_{14}B$ at T<100 K a FOMP takes place. Differently from $Nd_2Fe_{14}B$, the FOMP is of type II i.e. after the discontinuity in the magnetization, the system is not in the saturated state. Type II FOMP

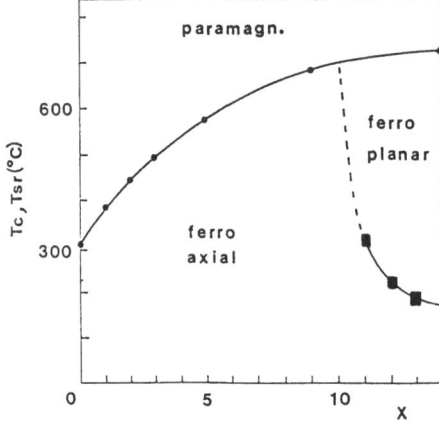 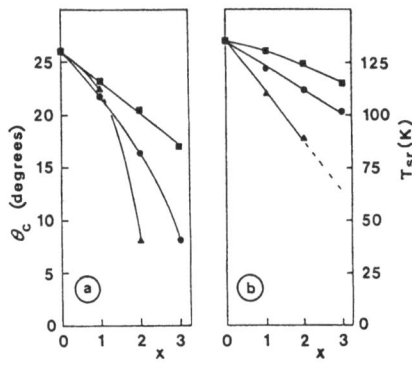

Fig.5. Composition dependence of the Curie temperature T_c (●) and the easy-axis to easy-plane spin re-orientation transition temperature T_{sr} (■) in $Nd_2Fe_{14-x}Co_xB$.

Fig.6. Variation (a) of the cone angle θ_c at 77K and (b) of T_{sr}, the spin re-orientation temperature, in the system $Nd_2Fe_{14-x}Me_xB$ for Me=Ni(●),Mn(▲) and Co(■). (Bolzoni et al. 1987c).

were also directly observed in $Pr_2Fe_{14-x}Co_xB$ for $x \leq 0.6$ compounds by SPD measurements. However , FOMP at very high fields (20 Tesla) cannot be excluded for Co richer compounds. A computer simulation established that contrary to Nd compounds, a negative 4th order and positive 6th order anisotropy energy constants are required to describe the observed SPD peak shape. (Fig.7). From this analysis the presence of type II FOMP were predicted for single crystal $Pr_2Fe_{14}B$.

High field magnetization measurements on a single crystal gave the first direct evidence of the presence of FOMP in $Nd_2Fe_{14}B$ (Pareti et al, 1985b). An analysis of the magnetization curves along the three crystallographic directions allowed a determination to be made of the axial and planar anisotropy constants in the phenomenological anisotropy energy expansion:

$$E_a = K_1\sin^2\theta + (K_2 + K_2'\cos4\varphi)\sin^4\theta + (K_3 + K_3'\cos4\varphi)\sin^6\theta$$

Positive 4th order (K_2) and negative 6th order (K_3) uniaxial together with planar terms are required to describe the observed discontinuities

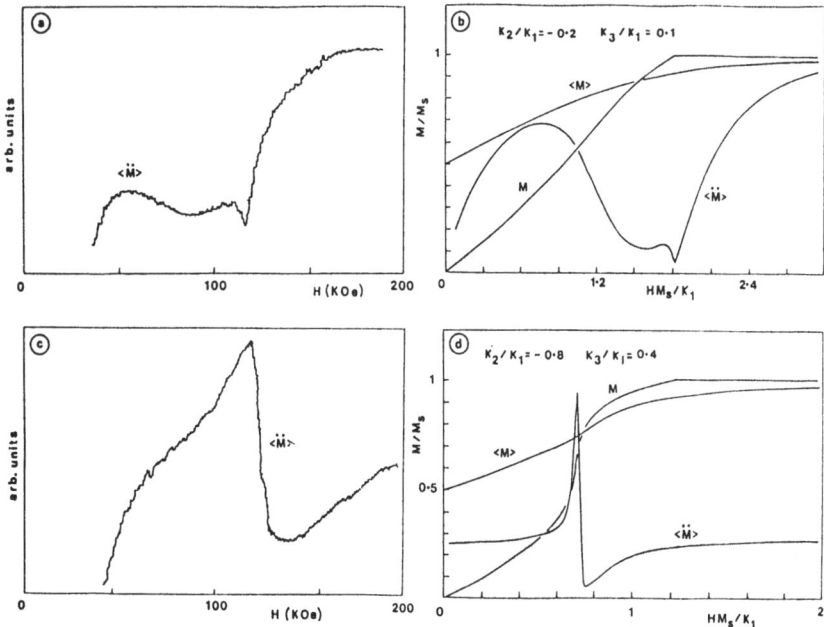

Fig.7. Measured SPD peakshapes at 220K (a) and 78K (c) respectively for $Pr_2(Fe_{0.6}Co_{0.4})_{14}B$. A computer simulation with selected values of the ratios of the anisotropy energy constants, K_2/K_1 and K_3/K_1 is displayed in figures (b) and (d). The computer simulation consists of the expected shape of the magnetization curve for a polycrystal and the second derivative $\ddot{M}$ with respect to the applied field. It can be seen that the given parameters reproduce very well the measured SPD peakshape. Additionally, the expected magnetization curve for a single crystal is also displayed. The FOMP is of type II.

in the magnetization curves (Bolzoni et al 1987f). In table 1 are displayed the results of the analysis of the magnetization curves whilst a typical example of the fit of the model to the observed data is displayed in Figure 9.

An SPD study of Tb substituted $Nd_2Fe_{14}B$ compounds revealed the huge contribution of Tb to the overall axial anisotropy. A 50% increase in H_a was observed with a substitution of only 20% of Tb (Pareti et al, 1987b).

The competition between planar (Er) and axial (Dy) anisotropies in the 2:14:B structure has been investigated with a study of the system $(Er_{1-x}Dy_x)Fe_{14}B$. (Algarabel et al, 1988). Substitution of Er by Dy very rapidly decreases the temperature of the EP-EA transition observed in $Er_2Fe_{14}B$ at 328 K. Additionally the transition is no longer one from EP to EA but rather from EC to EA. The value of the cone angle is also found

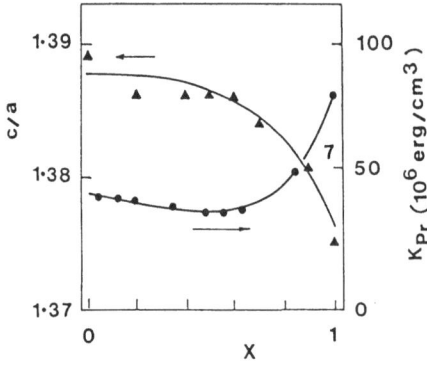

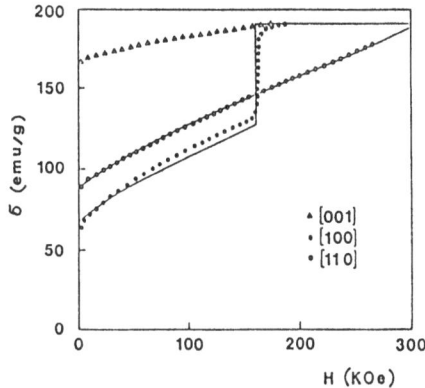

Fig.8. Variation of the Pr anisotropy (a) and of the c/a ratio (b) in $Pr_2(Fe_{1-x}Co_x)_{14}B$ compounds.

Fig.9. Magnetization curves at 77K for $Nd_2Fe_{14}B$ along the 3 principal crystallographic directions. The solid lines are the computed best fits of the experimental data. (Bolzoni et al. 1987f)

to correspondingly decrease rapidly with Dy substitution. At the same time, Dy markedly increases the axial anisotropy field (at 293K, H_a=17.4kOe for x=0.1 and for x=0.5, H_a=142.4kOe). Calculations based on a CEF-exchange model agree very well with the observed behaviour of the spin transitions.

The effects of H absorption on the Fe and Re sublattice anisotropies in $RE_2Fe_{14}B$ (RE=Y, Nd, Ho, Tm) were investigated by the SPD technique. Hydrogen is found to substantially depress both the Fe and RE magneto-crystalline anisotropy (Pareti et al, 1988). However such a reduction appears to be very large for Ho and particularly for Nd which both give a resultant axial anisotropy whilst it is less important in the case of Tm which gives a resultant planar anisotropy. The observed H composition dependence of the Ho anisotropy implies the existence of a particular H filling sequence (Table 2).

An attempt to distinguish between 4f and 4g site anisotropies in $RE_2Fe_{14}B$ compounds was made by making dilute substitutions of non-magnetic Y and La of corresponding small and large ionic sizes in Pr and Nd 2:14:B compounds (Moze et al, 1988b).

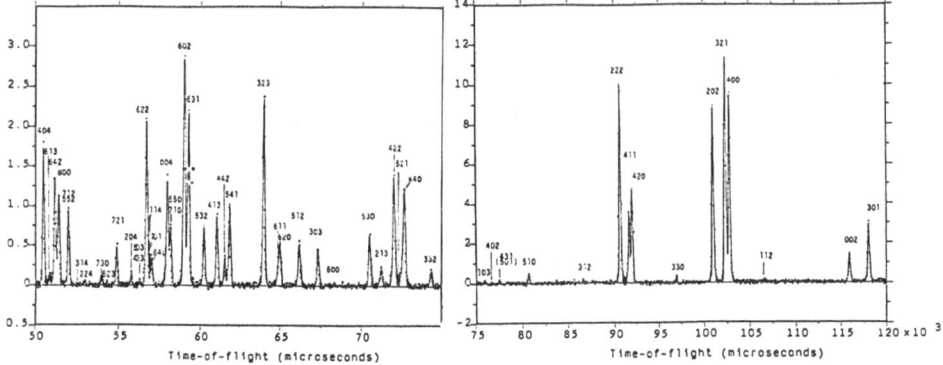

Fig.10. Observed and refined neutron diffraction patterns for $YFe_{11}Ti$ at 573K. The range in d-spacing is from 1 to 2.4Å. A neutron time-of-flight of 50000 microseconds corresponds to a d-spacing of approx.1Å. The peaks are indexed on the basis of the tetragonal $ThMn_{12}$ structure. (Moze et al. 1988c).

Any preferential substitution, in the presence of different contributions from 4f and 4g sites would result in a different modification of the overall anisotropy due to the small or large non-magnetic ion substitution. High resolution neutron powder diffraction was used to quantify accurately the preferential occupation of 4f and 4g sites in Y and La substituted Pr compounds. This in conjunction with accurate SPD measurements of the behaviour of H_a with Y and La substitution allowed a reliable estimate to be made of the 4f and 4g site anisotropies. The 4f site anisotropy was found to be very much larger than the 4g site anisotropy in $Nd_2Fe_{14}B$ whilst both sites have equal site anisotropy contributions in $Pr_2Fe_{14}B$.

Structural and magnetic investigations of compounds of the new $ThMn_{12}$ structure have been recently initiated. High resolution neutron powder diffraction has been used to determine precisely the crystal structure and preferential site occupation in $YFe_{10}Ti$, $YFe_{11}Ti$ and $YFe_{10}V_2$ compounds (Moze et al, 1988c,d). In all cases the 8j and 8f sites are occupied by Fe whilst the Ti and V reside with Fe atoms at the remaining 8i site. Compounds of $RE(TM_{1-x}M_x)_{12}$, RE=Y, Er and Sm, TM=Fe, Co and M=Ti, V and Si have been investigated by the SPD technique in order to evaluate the RE and TM sublattice anisotropies and to compare Fe and Co

Table I. Temperature dependence of some magnetic characteristics of single crystal $Nd_2Fe_{14}B$. (Bolzoni et al. 1987f).

T (K)	δ (emu/g)	H_{cr} 100 (kOe)	H_a 110 (kOe)	K_1 (10^6 erg/cm³)	K_2 (10^6 erg/cm³)	K_3 (10^6 erg/cm³)	K'_2 (10^6 erg/cm³)	K'_3 (10^6 erg/cm³)	θ_c (°)
293	168.0	82.5*	82.5	52.3	0	0	0	0	0
271	171.5	91.1*	91.1	50.0	9.0	−3.0	0	0	0
247	175.0	100.8*	107.0	49.6	19.0	−5.8	2.0	1.84	0
220	178.0	109.3	122.4	47.6	38.8	−15.9	−4.52	4.47	0
176	182.9	120.3	151.2	30.0	100.7	−47.0	−15.71	15.00	0
163	184.0	128.0	162.3	15.2	126.4	−56.7	−18.43	17.18	0
148	185.1	133.1	178.6	6.0	161.7	−76.7	−26.70	26.20	0
124	187.5	143.0	203.6	−25.1	244.0	−125.0	−50.00	48.75	14.8
104	189.0	150.0	252.0	−63.0	326.0	−162.0	−85.00	81.90	22.0
95	190.0	153.0	270.0	−98.0	378.0	−184.0	−93.00	91.00	25.8
78	190.5	161.5	318[b]	−123.7	421.3	−201.1	−111.00	109.00	28.0

* H_a measured along the 100 direction.
[b] Extrapolated value.

Table II. Composition dependence of some magnetic properties at 293 K in $RE_2Fe_{14}BH_x$. K(total) and K(RE) represent $K_1+2K_2+\ldots$. For Tm compound K(total) is K_1. (Pareti et al. 1988).

RE	x	H_a (kOe)	σ_s (emu/g)	K (total) (10^6erg/cm³)	K (RE) (10^6erg/cm³)	K_1 (Fe) (10^6erg/cm³)	K_1 (RE) (10^6erg/cm³)
Y	0	21.0	157.0	11.5	0	11.5	0
	1.68	13.3	169.0	7.8	0	7.8	0
	2.56	8.7	172.3	4.5	0	4.5	0
	3.0	4.6	173.8	2.8	0	2.8	0
Ho	0	103.5	79.3	33.3	21.8	11.5	---
	1.0	63.9	94.2	24.4	14.9	9.5	---
	2.0	45.3	98.4	18.1	11.1	7.0	---
	3.0	27.0	111.8	12.6	9.8	2.8	---
	4.1	21.8	117.0	10.4	10.4	0	---
Nd	0	82.5	168.0	52.3	40.8	11.5	---
	2.8	17.6	•185.5	12.4	8.4	4.0	---
Tm	0	80.1	110.5	−36.4	---	11.5	−47.9
	2.96	74.6	135.0	−41.5	---	3.3	−44.8

anisotropies in the $ThMn_{12}$ structure. (Solzi et al, 1988). As in the 2:14:B structure, the Fe anisotropy is axial whilst the Co is planar in $YFe_{11}Ti$ and $YCo_{11}Ti$. However, there is no anomalous temperature dependence of H_a in $YFe_{11}Ti$. Furthermore the Co anisotropy is smaller than the Fe anisotropy (at 293K, H_a=14kOe for $YCo_{11}Ti$ and H_a=21kOe for $YFe_{11}Ti$). This is in contrast to the situation in the tetragonal 2:14:B structure. Samarium is found to give a strong axial anisotropy in

SmFe$_{11}$Ti (Li et al, 1988), with H$_a$=90kOe at 293K. Furthermore, a type II
FOMP is observed at T<150K in SmFe$_{11}$Ti which is associated , as in the
case of Pr$_2$Fe$_{14}$B, with large negative 4th order and large positive 6th
order anisotropy constants. Contrary to the expectation Er ion is found
to give a surprisingly low anisotropy in ErFe$_{10}$V$_2$. However a type I FOMP
has been detected in this system by SPD technique, below 120K (Solzi et
al. 1988).

SOME GENERAL REMARKS

The magnetic anisotropy of Rare-Earth intermetallic compounds is in most
cases the result of concomitant different mechanisms that contribute and
interfere with each other in a variety of ways. This situation is evident
if one considers the rich phenomenology displayed by these class of
materials, with the presence of magnetic transitions of first and second
order, both spontaneous and field induced, with a frequency that finds no
equivalent in other classes of magnetic materials. There is at present a
better understanding of some basic mechanisms on a microscopic scale, but
the test of the interpretation is strongly affected by the way that the
experimental measurements are performed and the way one carries out the
analysis of the experimental data.

Apart from considerations on the reliability and accuracy of the
obtained data and the inherent limitations of the experimental method, an
important source of errors is often in the connection between the
measured experimental parameters and the true physical constants
pertaining to specific mechanisms. This connection is not trivial and the
use of the classical phenomenological treatments for multisublattice
systems of this kind can be the source of numerous artificial
discrepancies, such as in the case of ferrimagnetic order (G.Asti
and A.Deriu,1981).

The presence in these compounds of exchange interactions of the same
order of magnitude as the anisotropy energy implies deviations from
collinear magnetic order. Even for small canting angles the free energy

of the system is strongly modified and the macroscopic anisotropy properties are influenced in various ways. Non-monotonic dependences of the apparent anisotropy constants in pseudo-binaries have been erroneously considered as evidence of the non single-ion origin of the anisotropy energy. A simple first order pertubation theory, assuming small angle canting (SAC) between sublattice moments, provides analytical solutions and allows an easy interpretation of various non-linear effects. The SAC model explains in a straightforward way the easy cone in $PrCo_5$ (Pareti et al., 1988a) and the aforementioned clamorous discrepancy between anisotropy constants in Ho_2Co_{17} and Ho_2Fe_{17} obtained by spin wave spectra and magnetic measurements. Exact expressions can be obtained for some magnetic phase transitions of second order, namely the saturation at the anisotropy field H_a along a hard direction and spin re-orientation transitions. In addition field induced transitions (FOMP) can be adequately described within such a phenomenological framework (G.Asti et al,1987;G.Asti,1987).

REFERENCES

Algarabel, P. A., Ibarra, M. R., Marquina, C., Marusi, G., Moze, O., Pareti, L., Solzi, M., Arnaudas, J. I and del Moral, A., 1988. High Pulsed Magnetic Field Measurements Of The Magnetic Anisotropy In $(Er_{1-x}Dy_x)Fe_{14}B$ Compounds. To be presented at the 2nd International Symposium On High Field Magnetism, 1988, Leuven.

Asti, G. and Rinaldi, S., 1974. Singular points in the magnetization curve of a polycrystalline ferromagnet. J. Appl. Phys., 45, 3600-3610.

Asti, G. and Bolzoni, F., 1980. Theory of first order magnetization processes:uniaxial anisotropy. J. Magn. Magn. Mater., 20, 29-43.

Asti, G., Bolzoni, F., Leccabue, F., Pareti L. and Panizzieri R., 1985. Anisotropy measurements of Nd-Fe-B by the Singular Point Detection technique, Proc. Workshop on Nd-Fe permanent magnets- present and future applications. Brussels, 1984, Ed. I.V. Mitchell. p.161.

Asti, G. and Deriu, A., 1983. Magnetic anisotropy in RE transition metals, Proc. III Intern.Symp. on magnetic anisotropy and coercivity in RE-TM alloys. Baden/Vienna. 1982.

Asti, G.,Bolzoni, F. and Pareti, L.,1987. Magnetic anisotropy of RE magnets. IEEE.Trans.Mag-23, no5, 2521-2526.

Asti, G. 1987. Magnetic Anisotropy and Magnetization Processes in Rare-Earth Metal Compounds. V Int.Symp. on magnetic anisotropy and coercivity in RE-TM alloys. Bad Soden, FRG, Sept.3,1987.

Bolzoni, F., Leccabue, L., Pareti, L. and Sanchez, J. L. 1985. Magnetic Anisotropy Of Carbon Doped $Nd_2Fe_{14}B$. Journal de Physique, C6, 305-307.

Bolzoni, F., Deriu, A., Leccabue, F., Pareti, L. and Sanchez, J. L. 1986. Magnetic and Mossbauer Study Of $Y_{2-x}Nd_xFe_{14}B$. Journal Of Mag. Magnetic Matls., 54-57, 595-596.

Bolzoni, F., Gavigan, J. P., Givord, D.,Li, H. S., Moze, O. and Pareti.,L. 1987a. 3d Magnetism In $R_2Fe_{14}B$ Compounds. Journal Of Mag. Magnetic Matls., 66, 158-162.

Bolzoni, F., Leccabue, F., Moze, O., Pareti, L., Solzi, M. and Deriu, A. 1987b. 3d and 4f Magnetism In $Nd_2Fe_{14-x}Co_xB$ and $Y_2Fe_{14-x}Co_xB$ Compounds. J. Appl. Phys., 61(12), 5369-5373.

Bolzoni, F., Leccabue, F., Moze, O., Pareti, L. and Solzi, M., 1987c. Magnetocrystalline Anisotropy Of Ni and Mn Substituted $Nd_2Fe_{14}B$ Compounds. Journal Of Mag. Magn. Matls., 67, 373-377.

Bolzoni, F., Coey, J. M. D., Gavigan, J.,Givord, D.,Moze, O., Pareti, L. and Viadieu, T. 1987d. Magnetic Properties Of $Pr_2(Fe_{1-x}Co_x)_{14}B$ Compounds. Journal Mag. Magn. Matls., 65, 123-127.

Bolzoni, F., Moze, O., Pareti, L., Solzi, M., 1987e, CEAM-Report, Dublin, Ireland, March 1987.

Bolzoni, F., Moze, O. and Pareti, L. 1987f. First-Order Field Induced Magnetization Transitions In Single-crystal $Nd_2Fe_{14}B$. J. Appl. Phys., 62(2), 615-620.

Gavigan, J. P., Li, Hong-Shuo., Coey, J. M. D., Viadieu, T., Pareti, L., Bolzoni, F. and Moze, O. 1988. Magnetic Transitions and Anomalous Behaviour Of Pr in $Pr_2(Fe_{1-x}Co_x)Fe_{14}B$. To be presented at ICM'88, Paris.

Li Hong-Shuo, Hu Bo-Ping, Gavigan, J., Coey, J.M.D., Pareti, Moze, O. 1988. First Order Magnetization Process in $Sm(Fe_{11}Ti)$. To be presented at ICM '88.

Moze, O., Pareti, L., Solzi, M., Bolzoni, F., David, W. I F., Harrison, W. T. A. and Hewat, A. 1988a. Magnetic Structure and Preferential Site Occupation In Mn and Cr Substituted $Y_2Fe_{14}B$ Compounds. J. Less-Common Met., 136, 375-383.

Moze, O., Marusi, G., Pareti, L., Solzi, M. and David, W. I. F. 1988b. 4f and 4g Site Anisotropy In $Nd_2Fe_{14}B$ and $Pr_2Fe_{14}B$ Compounds. To be presented at The International Conference On Neutron Scattering, 1988, Grenoble.

Moze, O., Pareti, L., Solzi, M. and David, W. I. F. 1988c. Neutron Diffraction and Magnetic Anisotropy Study Of Y-Fe-Ti Intermetallic Compounds. Sol. Stat. Comm. 66,5 , 465-469.

Moze, O., Pareti, L., Solzi, M. and David, W. I. F. 1988d. Neutron Diffraction Determination Of The Crystal Structure Of The Intermetallic Compound $YFe_{10}V_2$. To be presented at the International Conference On Neutron Scattering, 1988, Grenoble.

Pareti, L., Solzi, M., Bolzoni, F., Moze, O. and Panizzieri, R. 1987a. 3d Magnetism In $Y_2Fe_{14-x}Me_x$ with Me=Co, Ni, Mn, Cr. Sol. Stat.

Comm., $\underline{61}$, 761-766.

Pareti, L., Szymczak, H. and Lachowicz, H. K. 1985a. High Field Magnetization Processes In $Pr_2Fe_{14}B$. Phys. Stat. Sol. (a),$\underline{92}$, K65-67.

Pareti, L., Bolzoni, F. and Moze, O. 1985b. Direct Observations Of First-Order Magnetization Processes In Single Crystal $Nd_2Fe_{14}B$. Phys. Rev. B., $\underline{32}$, no 11, 7604-7607.

Pareti, L., Bolzoni, F., Solzi, M. and Buschow, K. H. J. 1987b. Magnetocrystalline Anisotropy In $Nd_{2-x}Tb_xFe_{14}B$. J. Less-Common Met., $\underline{132}$, L5-L8.

Pareti, L., Moze, O., Fruchart, D., L'Heritier, Ph. and Yaouanc, A. 1988. Effects Of Hydrogen Absorption On The 3d and 4f Anisotropy in $RE_2Fe_{14}B$ (RE=Y, Nd, Ho, Tm). Accepted for publication in The Journal Of Less-Common Met..

Pareti, L., Moze, O., Solzi, M. and Bolzoni, F. 1988a. Magneto-crystalline Anisotropy in $Pr_{1-x}Y_xCo_5$. J. Appl. Phys., $\underline{63}$, 172-175.

Solzi, M., Pareti, L. and Moze, O. Rare-Earth and Transition Metal Magnetic Anisotropy In Some Intermetallic Compounds of The Tetragonal $ThMn_{12}$ Structure. To be presented at ICM '88. Paris.

INTRINSIC AND EXTRINSIC MAGNETIC PROPERTIES OF SUBSTITUTED $Nd_2Fe_{14}B$ ALLOYS AND MAGNETS

K.H.J. Buschow*.S. Heisz, G. Hilscher, P. Hundegger, R. Grössinger, H. Kirchmayr, R. Krewenka, R. Leeb, O. Mayerhofer, H. Sassik, G. Wiesinger, A. Wimmer

Institut für Experimentalphysik, Technische Universität Wien
Wiedner Hauptstrasse 8-10, A-1040 Vienna, Austria
*Philips Research Laboratory, Eindhoven, Netherlands

ABSTRACT

Nd-Fe-B alloys have been prepared in which Nd has been substituted mainly by Dy or Fe by Co, Al, Ga etc. The intrinsic property, anisotropy field H_A has been determined in pulsed magnetic fields up to 30 Tesla (T) between 4 K and the Curie-temperatures by means of the Singular Point Detection technique (SPD). The full hysteresis loop has been measured in static fields up to 5 T at temperatures between 100 K and 400 K. The extrinsic properties, coercive field, initial magnetization, energy product, etc. have been evaluated. The measurements are discussed with respect to the influence of substitutions on H_A and on the hysteresis loop and possible mechanisms responsible for anisotropy and coercivity are proposed.

INTRODUCTION

At the Institute of Experimental Physics, Technical University Vienna we tried to link our scientific interests and competence in rare earth intermetallics with our long standing experience with $SmCo_5$ and Sm_2Co_{17} magnets in order to help to overcome the shortcomings of the newly developed Nd-Fe-B magnets. One possibility for improvements are systematic substitutions of the different positions in the $Nd_2Fe_{14}B$ lattice. Any substitution modifies, however, intrinsic as well as extrinsic properties in a complex manner.

With the advent of the Nd-Fe-B-magnets we therefore adapted well proofed and developed new methods for characterizing these different materials. An intrinsic basic property of all modern permanent magnetic materials is the anisotropy field H as a function of temperature and composition.

The most important extrinsic properties i.e. properties not only depending on the composition, intermetallic phases and crystal structures, but also depending on production parameters as e.g. heat treatment, grain size etc. can be evaluated from the full hysteresis loop as a function of temperature. Typical data deduced from a hysteresis loop are the saturation magnetization, the remanence, the coercivities

$_B H_c$ and $_M H_c$, but also the initial magnetization curve after thermal demagnetization or counter field demagnetization as well as minor loops or recoil curves.

From all the intrinsic as well as extrinsic parameters models for the origin of anisotropy as well as coercivity can be deduced and therefore a better understanding for possible improvements can be achieved.

EXPERIMENTAL

The anisotropy field H_A was measured on arc melted and vacuum annealed samples using the SPD (Singular Point Detection) technique. The theory of this method predicts a singularity in the second derivative of the magnetization $(d^2 M/dH^2)$ at the anisotropy field H_A if the external field is applied parallel to the hard plane (anisotropy parameter $K_1 > 0$). From the fact that this method samples only the properties of the correctly oriented grains (H_{ext} parallel to the hard axis), it follows that H_A can be measured on aligned powder samples but also on polycrystalline samples. This method is well suited for uniaxial materials as e.g. Nd-Fe-B alloys.

On sintered magnets as well as melt spun material, compacted to cubes with the size 10x10x10mm in static fields from -5T to +5T the full hysteresis loop as well as the initial magnetization curve and minor loops have been measured between 100K and 400K.

EXPERIMENTAL RESULTS AND DISCUSSION

The influence of Co on the anisotropy field of $La_2Fe_{14-x}Co_xB$ as a function of temperature is given in fig.1. In this case the nonmagnetic La does not contribute to the anisotropy. The anisotropy field of the corresponding Pr-compounds has also been measured (Grossinger et al., 1987b).

Dy is especially effective for improving the anisotropy field (fig.2, Grossinger et al. 1987d).

Co is effective in improving the Curie temperature, (fig. 3, Grossinger et al. 1986a), but decreases the saturation magnetization (fig.4, Grossinger et al. 1988). The simultaneous addition of Dy and Co improves H_A and T_c, however, the actual coercivity at elevated temperatures may not be improved, depending on the processing parameters of the magnet.

Some replacement of Nd by Ce or mixtures of Rare Earths containing Ce is possible, as can be deduced from the temperature dependence of the anisotropy field H_A in $(Nd_xCe_{1-x})_{15}Fe_{77}B_8$ alloys (Fig. 5, Grossinger 1986b).

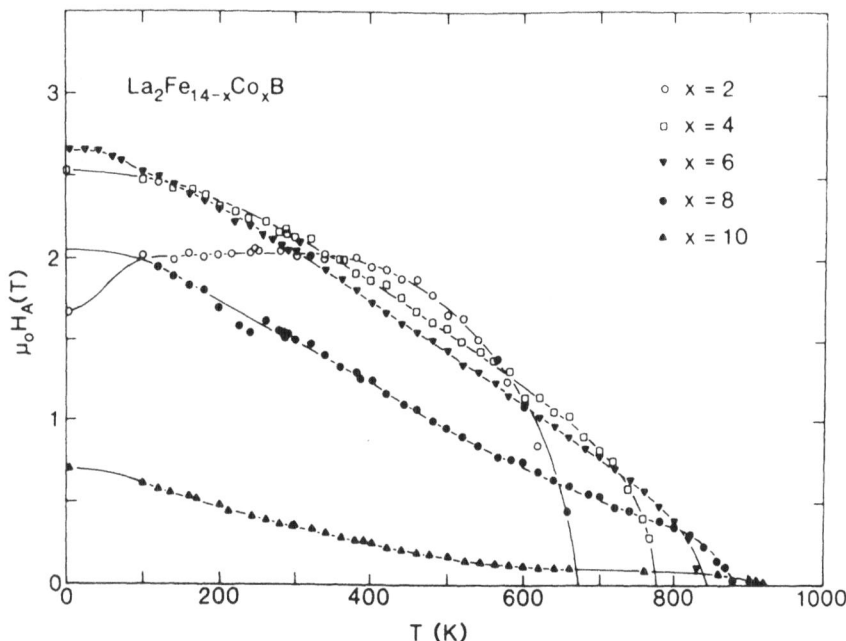

Fig.1 Temperature dependence of the anisotropy field in various compounds of the series $La_2Fe_{14-x}Co_xB$ (Grössinger et al. 1988)

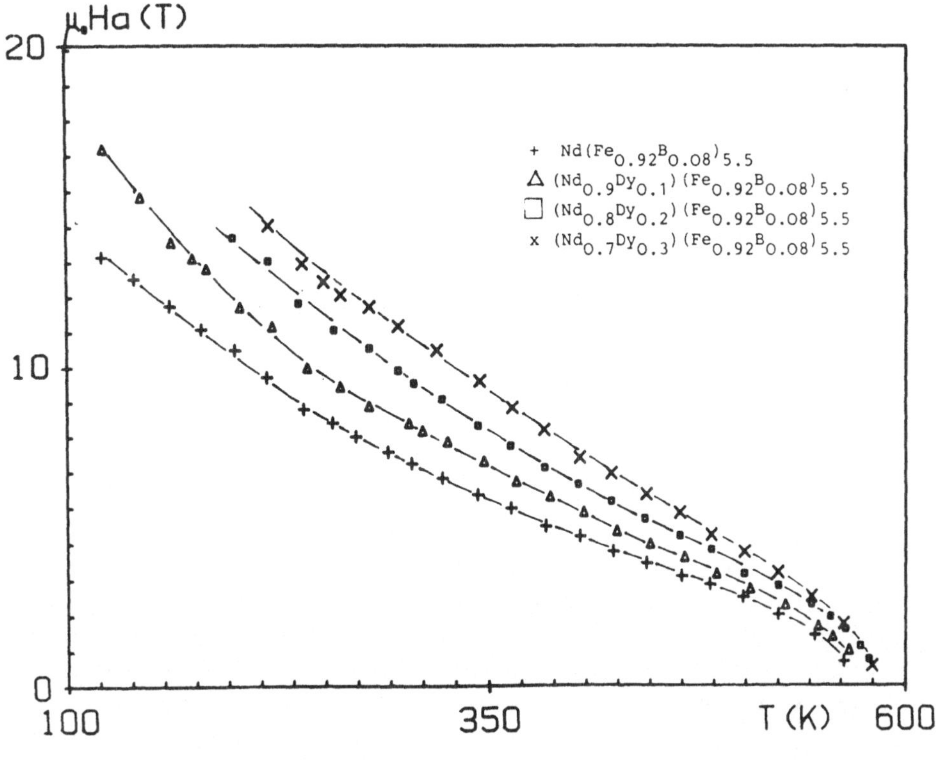

Fig.2 Temperature dependence of the anisotropy field of Dy-substituted Nd-Fe-B alloys

The mechanism of magnetic hardening can be studied in melt spun Nd-Fe-B and Nd-(Fe,Co)B-material.Fig.6 shows the influence of wheel velocity, which is indicative of the cooling rate on the coercivity (Heisz et al. 1987c).

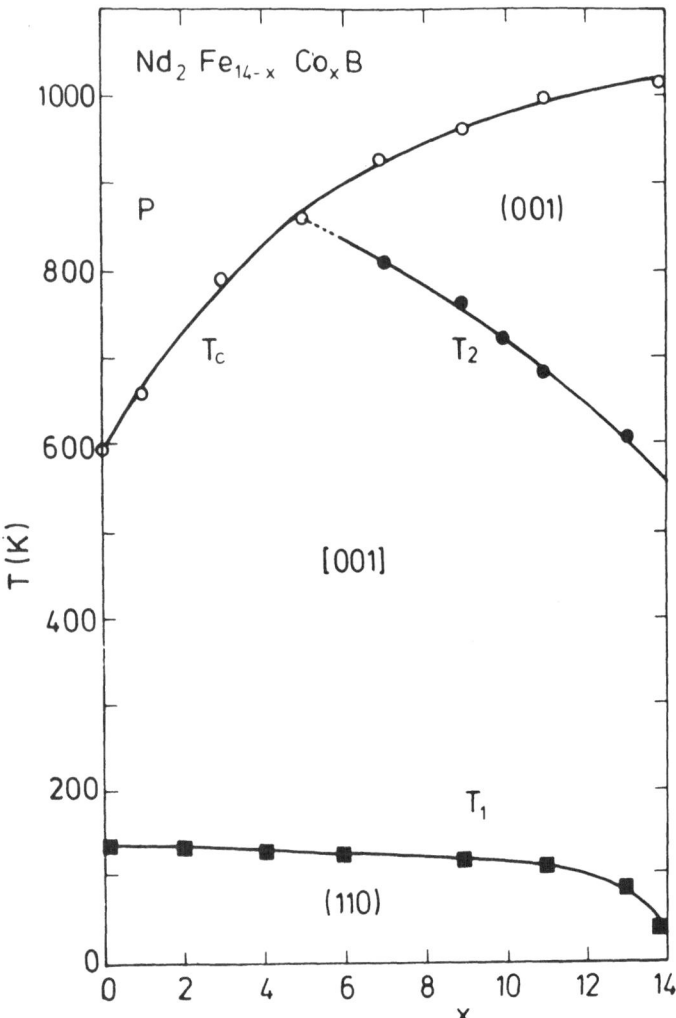

Fig.3 Magnetic phase diagram of the $Nd_2Fe_{14-x}Co_xB$ system (P paramagnetic, T_1 und T_2 spin reorientation temperatures)

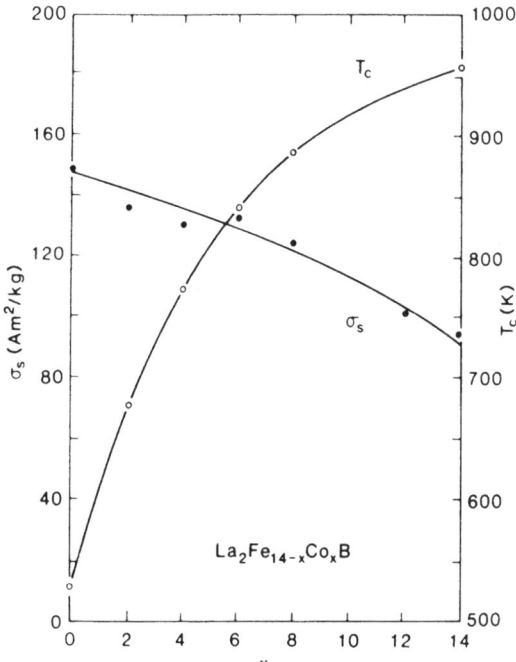

Fig.4 Concentration dependence of the Curie temperature T_C and room temperature saturation magnetization σ_s in $La_2Fe_{14-x}Co_xB$.

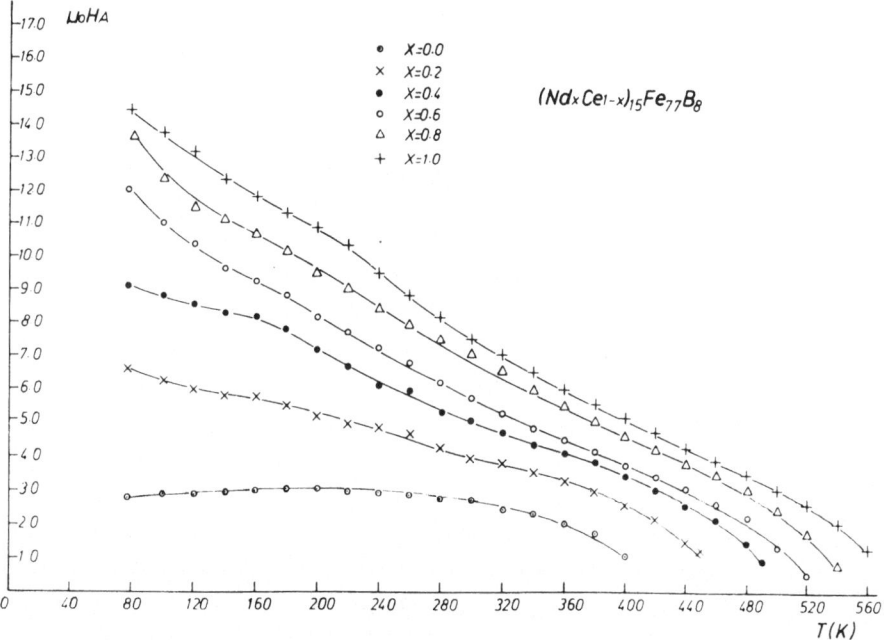

Fig.5 Temperature dependence of the anisotropy field $\mu_oH_A(T)$ of $(Nd_xCe_{1-x})_{15}Fe_{77}B_8$

Minor loops (fig.7) as well as Mössbauer spectra (fig.8) of overquenched, optimal quenched or annealed ribbons show that the magnetic properties are extremely sensitive to the microstructure of the material (Heisz 1987c). A magnified high energy part of the Mössbauer spectrum of the overquenched ribbon with a coercivity of 0.1T is shown before and after a heat treatment at 700°C in fig.8a and fig.8b respectively. The heat treatment results in an increase of the coercivity from 0.1T up to 1.4T. An even larger jump of the coercivity is observed for overquenched ribbons with the 15:77:8 stoichiometry (from 0.1T to 2.2T). These spectra demonstrate that the α-Fe spectrum is drastically reduced after the heat treatment at 700°C.

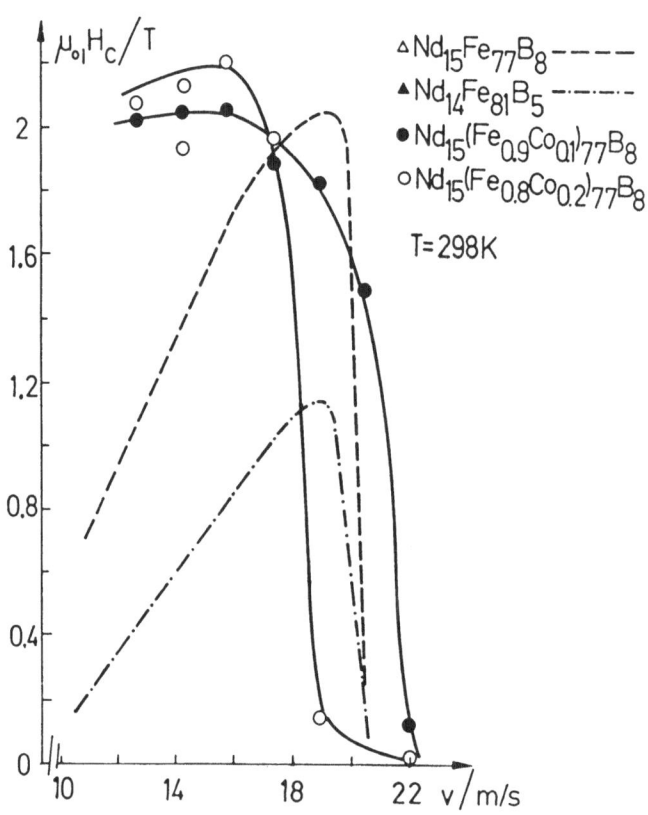

Fig.6 Coercivity of melt spun $Nd_{15}Fe_{77}B_8 \cdot Nd_{14}Fe_{81}B_5 \cdot$ $Nd_{15}(Fe_{0.9}Co_{0.1})_{77}B_8$ and $Nd_{15}(Fe_{0.8}Co_{0.2})_{77}B_8$ as a function of the wheel velocity.

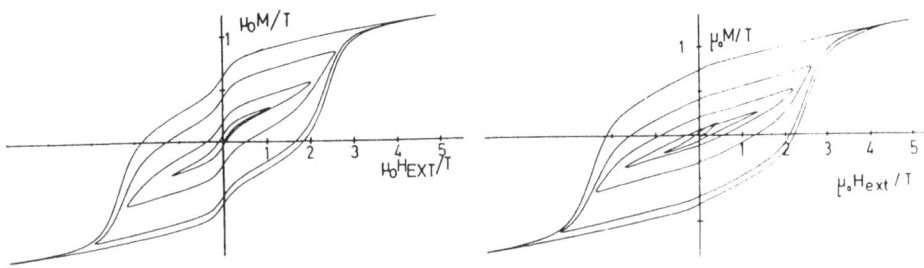

Fig.7 Successive minor loops of a) an overquenched $Nd_{15}(Fe_{0.9}Co_{0.1})_{77}B_8$ vs - 17.3m/s and b) minor loops of an optimal quenched $Nd_{15}Fe_{77}B_8$.

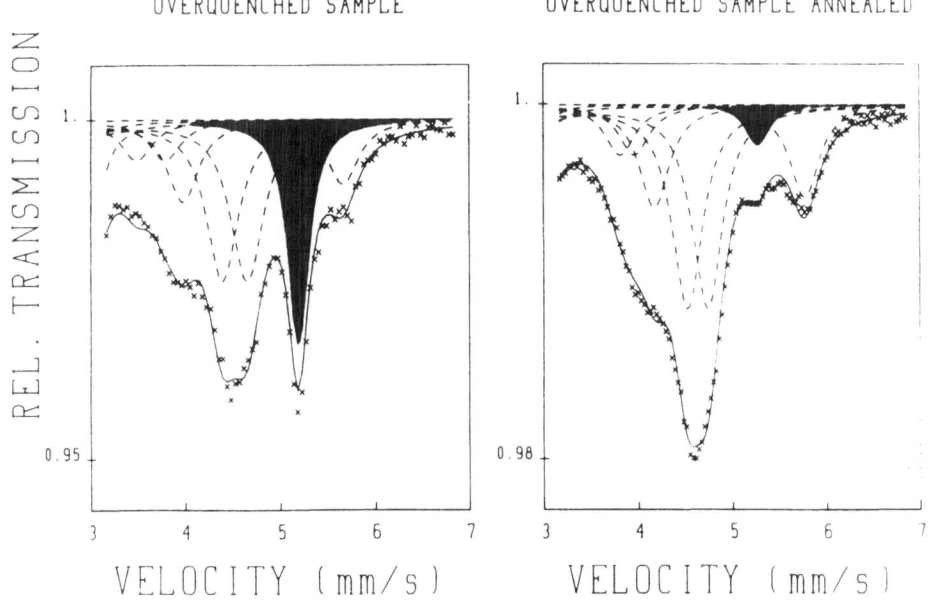

Fig. 8 Magnified part of a Mössbauer spectrum of an overquenched $Nd_{14}Fe_{81}B_5$ ribbon at room temperature a) before and b) after annealing. The black area indicates the outer line of the α-Fe spectrum in both figures.

By assuming a model of randomly distributed non-interacting particles the hysteresis loop for homogeneous rotation can be calculated. Fig. 9a (Heisz et al. 1987c) shows such a loop calculated with K_1, K_2 and M_s corresponding to $Nd_2Fe_{14}B$ at room temperature, together with both virgin curves and the recoil curve which leads to the counter field demagnetized state. The thermal demagnetized state represents the case of randomly distributed demagnetization vectors with vanishing stray field. After a thermal demagnetization the irreversible rotation processes are completed, when the magnetizing field reaches the anisotropy field. The kinks in the virgin- and in the demagnetization curve originate from the occurance of irreversible processes at H_{i1}. This field H_{i1} rises with increasing K_1 and K_2. For $K_2=0$ irreversible processes start at a lower angle of $\alpha=45°$ and therefore at a lower field (α....angle between c-axis and external field) while for $K_2=0.7MJ/m^3$ λ attains $50°$. Up to H_{i1} both virgin

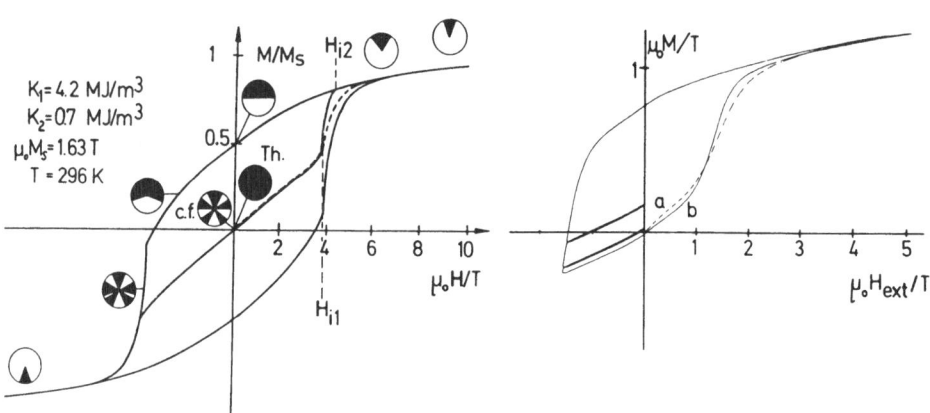

Fig.9 Calculated hysteresis loop (a) for randomly distributed noninteracting particles with K_1, K_2, μ_oM_s, corresponding to Nd-Fe-B at 296K. The dashed curve indicates the thermal virgin curve, the recoil curve leads to the counter field demagnetized state (c.f.). The dark area of the circles indicate schematically the angular distribution of the magnetization vectors. (Th.) indicates the thermal demagnetized state. (b): Hysteresis loop of a nearly optimal quenched $Nd_{14}Fe_{81}B_5$ sample with both thermal ($----$) and counter field demagnetized virgin curves.

curves coincide; only reversible processes take place. Above H_{i1} the counter field virgin curve increases faster than the thermal virgin curve because of the different distributions of the magnetization vector of the thermal and field demagnetized states as shown schematically in fig.9a. This is also the reason that irreversible processes in the field demagnetized state are completed below H_A at H_{i1} which is the counter field necessary to obtain a vanishing stray field at H=0.

In real melt spun material (see fig.9b) where grain-grain interactions, a soft magnetic phase and large grains are present, the irreversible magnetization processes caused by domain wall motions occur already at low fields. All three phenomena enhance the initial susceptibility of both virgin curves. Large grains increase primarily the slope of the thermal virgin curve because of their multi domain structure, where walls can be moved easily. Therefore the thermal virgin curve exhibits always the same or a steeper slope than the counter field virgin curve. This is indeed observed in all our experiments.

As a final example the low temperature hysteresis loops of sintered (Nd,Dy)Fe-B-magnets are presented (fig.10, Heisz et al. 1987b).

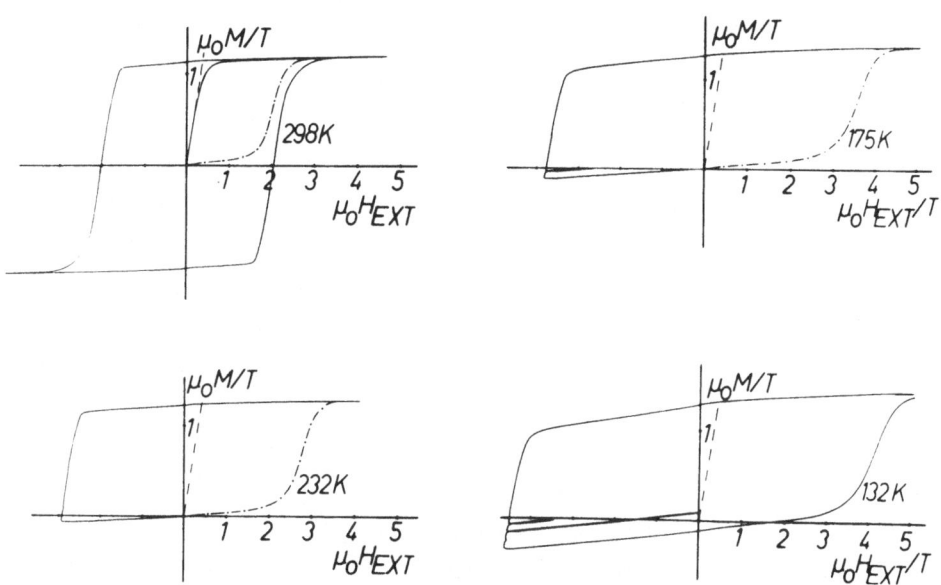

Fig.10 Hysteresis loops of a sintered $(Nd_{0.8}Dy_{0.2})(Fe_{0.92}B_{0.08})_{5.5}$ magnet at various temperatures. The dashed dotted curve indicates the virgin curve after counter field demagnetization.

As a result of these and similar investigations one can conclude, that by systematic studies of the above mentioned intrinsic and extrinsic properties in recent years a much deeper understanding of the origin and mechanism of anisotropy and coercivity has been obtained. Therefore also systematic improvements of the practical properties of Nd–Fe–B based magnets have been possible.

Further detailed results are found in the references listed below.

REFERENCES

Burzo, E. and Kirchmayr, H.R. 1988. Physical properties of $R_2Fe_{14}B$–based alloys. vol.12 of the "Handbook on the Physics and Chemistry of Rare Earths", ed. by K.A. Gschneidner and LeRoy Eyring, North–Holland Physics Publ. in print

Coey, J.M.D. 1987. New permanent magnet materials. Physica Scripta vol.T19, 426–434

Grössinger, R., Sun, X.K., Eibler, R., Buschow, K.H.J. and Kirchmayr, H.R. 1985. The temperature dependence of the anisotropy field in $R_2Fe_{14}B$ compounds (R = Y, La, Ce, Pr, Nd, Gd, Ho, Lu). Journal de Physique, 46, C6, 221–224

Grössinger, R., Krewenka, R., Sun, X.K., Eibler, R. Kirchmayr, H.R.and Buschow, K.H.J. 1986a. Magnetic phase transitions and magnetic anisotropy in $Nd_2Fe_{14-x}Co_xB$ compounds.J. Less Common Metals 124, 165–172

Grössinger, R., Hilscher, G. and Kirchmayr, H. 1986b. Rare Earth Permanent Magnets. Radex Rundschau 1986, 120–129

Grössinger, R., Krewenka, R., Kirchmayr, H.R., Naastepad, P. and Buschow K.H.J. 1987a. Note on the coercivity in Nd–Fe–B–magnets. J. Less Common Metals 134, L17–L21

Grossinger, R., Krewenka, R., Kirchmayr, H.R., Sinnema, S, Yang Fu–Ming, Huang Ying–Kai, DeBoer, F.R. and Buschow K.H.J. 1987b. Magnetic anisotropy in $Pr_2(Fe_{1-x}Co_x)_{14}B$ compounds. J. Less Common Metals 132, 265–272

Grössinger, R., Krewenka, R., Haslinger, F., Sagawa, M., Kirchmayr, H.R. 1987c. Temperature dependence of the coercivity and the anisotropy of Nd–Fe–B based magnets. IEEE Transactions on Magnetics, MAG–23, 2114–2116

Grössinger, R., Harada, H., Keresztes, A., Kirchmayr, H.R., Tokunaga M. 1987d. Anisotropy and hysteresis studies of highly substituted Nd–Fe–B based permanent magnets. IEEE Transactions on Magnetics, MAG–23, 2117–2119

Grössinger, R., Kirchmayr, H.R. and Buschow, K.H.J. 1988. Magnetic anisotropy in the System $La_2Fe_{14-x}Co_xB$ and its relation to the system $Nd_2Fe_{14-x}Co_xB$. J. Less Common Metals 136, 367–373

Heisz, S., Hilscher, G. 1987a. The origin of graduated demagnetization curves of Nd–Fe–B magnets. Journal of

Magnetism and Magnetic Materials <u>67</u>, 20-28

Heisz, S., Hilscher, G., Kirchmayr, H.R. 1987b. The effect of the spin reorientation upon the demagnetization curves of Nd—Fe—B magnets. IEEE Transactions on Magnetics <u>MAG—23</u>, 3110—3112

Heisz, S., Hilscher, G., Wiesinger, G. and Sassik, H., 1987c.The initial magnetization process of melt spun Nd—Fe—B material. Proceedings of the "Ninth International Workshop on Rare—Earth Magnets and Fifth International Symposium on Magnetic Anisotropy and Coercivity in Rare Earth—Transition Metal Alloys", Bad Soden, Federal Republic of Germany, 31. August — 3. September 1987, <u>Part 2</u>, 267—274

Hilscher, G., Grössinger, R., Heisz, S., Sassik, H., Wiesinger, G. 1986. Magnetic and anisotropy studies of Nd—Fe—B based permanent magnets. J. Mag. Mag. Materials <u>54—57</u>, 577—578

Leonowicz, M., Heisz, S., Hilscher, G., 1988. Effect of Al addition on magnetic properties and demagnetization behaviour of sintered Nd—Fe—B magnets, Proceedings of the International Conference on Magnetism 88 Paris, August 1988, submitted

CRYSTAL CHEMISTRY, PHYSICAL PROPERTIES, HYDRIDATION PROPERTIES OF Nd$_2$Fe$_{14}$B-TYPE OF COMPOUNDS

D. Fruchart*, P. Wolfers*, S. Miraglia*, L. Pontonnier*, F. Vaillant*, H. Vincent*,

D. Le Roux*, A. Yaouanc[+], P. Dalmas de Reotier[+], P. l'Héritier[x], R. Fruchart[x].

* Laboratoire de Cristallographie du C.N.R.S., associé à l'Université J. Fourier, 166 X, 38042 Grenoble Cedex, France.
[+] M.D.I.H., Département de Recherches Fondamentales, C.E.N.G., 85 X, 38041 Grenoble Cedex, France.
[x] U.A. 1109, E.N.S.P.G., Institut National Polytechnique de Grenoble, BP 46, 38402 Saint-Martin d'Hères, France.

ABSTRACT

Structure and magnetic characteristics of RE$_2$Fe$_{14}$B have been determined on powder and single crystal samples, both by X-ray and neutron diffraction. All the magnetic and structural parameters have been measured versus the absorbed hydrogen. Hydrogen acts indirectly on the magnetic structure via the magnetoelastic couplings, but also directly via the "screening" effect of hydrogen filling interstices between RE and Fe. Results from various techniques as Mössbauer spectrocopy, μ+SR, magnetization measurements... obtained on pure and hydrided samples are reported.

INTRODUCTION

In order to better understand the magnetic properties of materials, structural investigations using both X-ray and neutron scattering experiments are highly of importance, since it exists large magnetoelastic forces as in alloys or intermetallics. By the way, hydrogen absorption in 3d-RE alloys and the subsequent changes induced in their magnetic properties is also a good probe permitting to act on the crystal and magnetic parameters. From this point of view, we have investigated the structural properties, the magnetic behaviour versus temperature and magnetic field, and the impact of hydrogen on the properties of RE$_2$M$_{14}$B compounds (M = Fe, Co).

SYNTHESIS OF THE MATERIALS :

All our compounds were synthesized using the HF-furnace cold crucible technique. The compounds were annealed for several hours to several days depending on the sample. The composition of the ingots was verified by using $\lambda_{Cr}(k_\alpha)$ - X-ray diffraction. In some cases, electron diffraction and electron microscopy have permitted to analyse the microstructure at the scale of intergranular phases.

STRUCTURE DETERMINATION

Simultaneously to the structure determination of Herbst, Givord and Shoemaker (84), the same type of work has been performed using $Nd_2Fe_{14}B$ and $Nd_2Co_{14}B$ sphere-shaped crystals.

As this work was performed very accuratly, significative improvements in the structure characteristics and the interatomic distances have been obtained ($\Delta d/d \approx 5~10^{-5}$) (Le Roux, 86) (Table I). Some doubts about the reality of the tetragonal symmetry (SG $P4_2/mnm$) have been expressed, more particularly with the cobalt compound. Nevertheless, in spite of the goodness of the statistics, the present data left inconclusive the question of the symmetry lowering. The final result was proposed to the BEEVERS Co (Edimburgh) in order to built the representative model of the series.

TABLE 1

Atom parameters measured in $Nd_2Co_{14}B$

(anisotropic thermal factors were used in the fits,

but equivalent Debye Waller factors are given here).

Atom	position	x	y	z	$B_{eq.}$
Nd_1	4f	0.14414(04)	x	0	0.49
Nd_2	4g	0.72471(04)	-x	0	0.46
Co_1	4c	0	1/2	0	0.60
Co_2	$16k_1$	0.72375(04)	0.06925(04)	0.37357(03)	0.50
Co3	$16k_2$	0.46300(04)	0.13985(04)	0.32192(03)	0.43
Co_4	$8j_1$	0.18165(08)	x	0.25336(05)	0.54
Co_5	$8j_2$	0.40123(08)	x	0.29495(05)	0.48
Co_6	4e	0	0	0.38420(06)	0.47
B	4f	0.37665(96)	x	0	0.70

More recently, studying the magnetic structure of $Ho_2Fe_{14}B$ (Wolfers, 88) at 300, 100 and 20 K by neutron diffraction, we simultaneously determine the atomic arrangement. A very high quality single crystal, 1.6 mm in diameter, sphere shaped was used and a very large lot of Bragg peaks (≈ 2200 per temperature) was recorded in the whole reciprocal space. Accurate structure refinements were performed. It appeared that at low temperature, when the Ho magnetic moments deviate from the c easy axis, the tetragonal symmetry is lowered. This deviation can be fully interpreted in the Cm monoclinic space group. Relative shifts of some 10^{-3} the previously determined high temperature atom parameters are measured. This distortion results in the high magnetostrictive forces already evidenced by the anomalous thermal behaviour of the cell parameters, the anisotropy constants expressed in terms of c/a (T) (Li, 87 ; Andreev, 85). An analysis of the a versus c, c/a, V cell parameters versus Z_{RE} permits us to evidence strong correlations with the nature of the RE atom (l'Héritier, 88). It is clear that RE – magnetism via a de Gennes-like law influences the anisotropic part of the elastic forces.

HYDROGENATION PROPERTIES

We discover, for the first time, that these new RE-M-B materials were able to absorb significant amounts of hydrogen, at room temperature and reasonnable pressure (fraction of a bar) (l'Heritier, 84). The interest reveals to be double :

- first, hydrogen modifies the metal lattice (via elastic constants, electric charges, electron transfers...) and it is interesting to use it as a perturbating probe.
- second, the relatively large expansion in lattice parameters induces a decrepitation process and consequently offers the technical interests of easily produce very fine magnetic particles (≤ 0.1 µm).

We used neutron diffraction in order to localize the hydrogen (deuterium) atoms and thus test the model of filling interstitial tetrahedra (disposed around the RE) that we developped (Fruchart, 84). The maximum filling is 5.5 H/f.u., and it could be reached via a two steps process (two levels of the binding energy). In the $RE_2Fe_{14}B$ system, we systematically observed that before a critical concentration $X_c \approx 2H/f.u.$, the cell parameter c increases slowly, then fastly and after saturates progressively. Simultaneously **a** increases almost linearly versus x the amount of absorbed hydrogen. The consequently modified elastic constants, interatomic distances..., yield to large changes in the magnetic characteristics. As for the pure compounds, the $\Delta V/V$ and $\Delta(c/a)/(c/a)$ parameters measured in the hydrided compounds, depend on the nature of the RE metal (l'Héritier, 88)... Moreover, the RE-magnetism via the magnetoelastic forces (Buschow, 87) seems to play a weak but non negligible role on the total amount of absorbed hydrogen (Fruchart, 88). We have also analysed the decrepitation process in order to obtain extra fine particles. Electron diffraction and microscopy reveal that this milling process is essentially intergranular for stoichiometric materials (2–14-B) and of onion skin type for the magnet compositions ($\approx$ 2-10-B). Laves-type interstitial phases are revealed in Nb-doped compounds. The so-called coercitive interstitial phases with composition close to RE Fe_4B_4 remains insensitive to hydrogen activity. High temperature cycles performed under hydrogen atmosphere show that a demixion process occurs for $T \geq 550$ C . During this treatment, very oxidable iron particles are formed and rather stable RE-hydrides (Fruchart, 87).

MAGNETIC PROPERTIES

Here, we distinguish the microscopic properties (neutron diffraction, Mössbauer spectroscopy, μ^+SR spectroscopy, XPS and X-AES...) from the macroscopic properties (Curie point determination, bulk magnetization, anisotropy, reorientation effects...).

Microscopic properties

1. Neutron diffraction.

Both powder and single crystal techniques were used. With the first one, only $Y_2Fe_{14}{}^{11}B$, $Ce_2Fe_{14}{}^{11}B$, $Nd_2Fe_{14}{}^{11}B$, $Er_2Fe_{14}{}^{11}B$ and corresponding hydrides were studied. On the other hand $Y_2Co_{14}{}^{11}B$ and $Nd_2Co_{14}{}^{11}B$ were also analysed. The low temperature magnetic structure of the Nd-compounds were not elucidated since the deviation from collinearity of the magnetic structure greatly increased the number of the fitted variables. For the other compounds, no great difference appeared between the uncharged and the H–charged compounds. Our measurements are in fair agreement with ^{57}Fe Mössbauer spectroscopy (Dalmas, 85 ; Fruchart, 87), and they do not exhibit the too large distribution in magnetic moments as elsewhere claimed. The Fe_1 (4c), Fe_2 (16 k), Fe_5 (4e) bond with each other by "classical" Fe-Fe distances : they share magnetic Fe moments equal or slightly lower than those observed in Fe-metal. On the contrary, to the shortest and longest bonds correspond higher moments (Fe_3(16 k), Fe_4(8 j), Fe_5(8 j)). In the light of the fits, it seems that the Er compound with an easy-plane anisotropy (Stevens factor $\alpha_j > 0$ for Er^{3+}) is not fully collinear in the basal (001) plane (Dalmas, 85, 87). For the $Y_2Co_{14}{}^{11}B$, here again the values of the 3d magnetic moments are distributed apart from the mean value $< \mu_{Co} > \approx 1.4 \ \mu_B$, both Co_3 and Co_5 (shortest bonds) exhibit high values (Le Roux, 86). (Table I).

Using the neutron 4-circles diffractometry technique we first analysed a $Nd_2Fe_{14}B$ single crystal unsuccessfully. If the magnetic structure determination remains inconclusive (quality and shape of the crystal, weakness of the RE moments...) we have therefore observed some weak forbidden lines (in terms of the SG $P4_2/mnm$). Then, a very carefully prepared experiment on $Ho_2Fe_{14}B$ (supplied by the Sagawa's group) and already mentionned, was analyzed (Wolfers, 88). We concluded that the spin reorientation (SR) process starting in $Ho_2Fe_{14}B$ at $T = 58$ K, operates via a continuous process on the almost collinear 6 Fe-sublattices. But there exists large deviations to collinearity on the different types of Ho sites (as deduced through the lowered symmetry). (Table II). (Figure 1). However the resulting $n_{RE-Fe}M_{RE}$ molecular field acting on the Fe sites is quasi parallel to the Fe moments. A temperature-constant deviation between the resulting magnetization as deduced from this experiment and the bulk-magnetization experiments, well evidences a net conduction electron polarisation of ≈ -0.25 μ_B/Fe. The accurracy of the magnetic moment determination is particularly high and cannot be obtained using other technique (Table III). We have to remark that the a priori precise polarized neutron technique is unable to fit correctly such a complicated structure. First, it cannot attain the depolarising components (non collinear components), secondly the nuclear factor structures attached to the lowered symmetry prevent from determining the flipping ratio (F_M/F_N) correctly.

TABLE II
Magnetic structure of $Ho_2Fe_{14}B$ at T = 20 K

Atom	Moment (μ_B)	Theta (°)	Phi (°)
Ho1__1	-10.0 (0.5)	46 (1)	45 (fixed)
Ho1__2	"	13 (1)	"
Ho1__3	"	16 (1)	61 (4)
Ho1__4	"	16 (1)	29 (4)
Ho2__1	-10.0 (0.5)	25 (1)	34 (3)
Ho2__2	"	25 (1)	56 (3)
Ho2__3	"	25 (1)	45 (fixed)
Ho2__4	"	25 (1)	"
Fe1 (4c)	2.44 (0.04)		
Fe2 (16k_1)	2.45 (0.04)		
Fe3 (16k_2)	2.51 (0.04) →	22 (fixed)	45 (fixed)
Fe4 (8j_1)	3.07 (0.05)	(from magnetization measurment)	
Fe5 (8j_2)	2.28 (0.04)		
Fe6 (4e)	2.56 (0.05)		

RESULTING MAGNETIZATION /FORMULA :

M	=	16.0 (0.5)	20.1 (0.5)	45 (fixed).

FIGURE 1

Magnetic structure of Ho$_2$Fe$_{14}$B at 20 K.

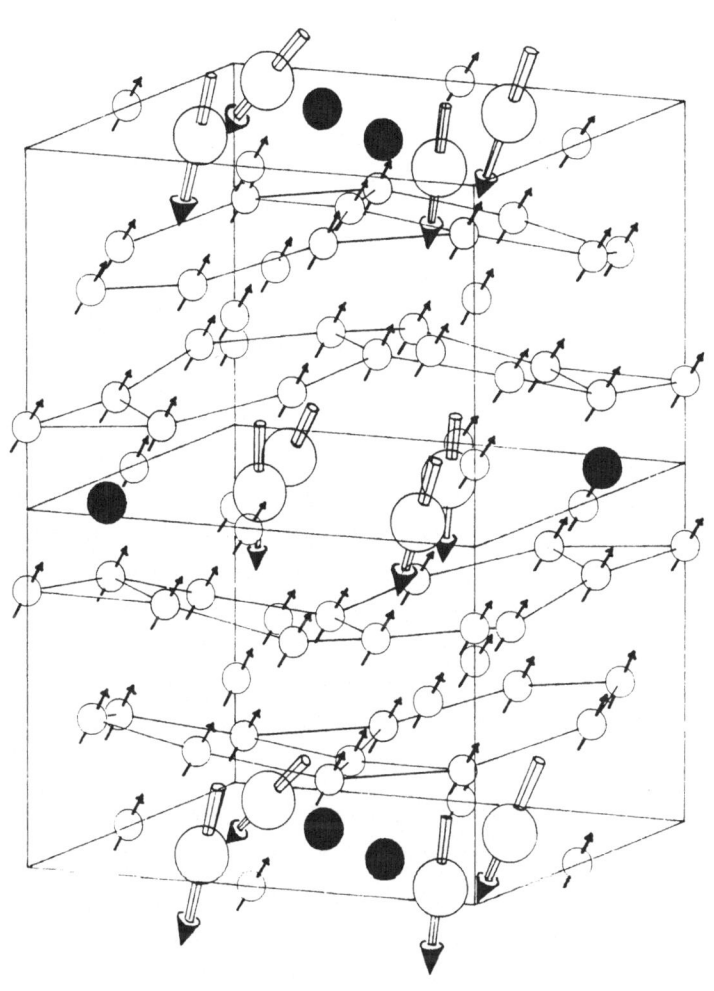

Deviations from the tetragonal symmetry (crystal structure) and non-collinear, non-axial magnetic structure are fully associated via the magnetoelastic couplings. The stability of such a structure could partly depends on negative Ho-Ho exchange interactions (Coey, 84).

2. Mössbauer spectroscopy : ^{57}Fe, ^{161}Dy, ^{166}Er spectroscopies (Dalmas, 85 ; Ferreira, 85 ; Fruchart, 87).

The spectra of the whole $RE_2Fe_{14}B$ series (excepted with Pm, Tm, Yb) have been recorded both at R.T. and 4.2 K. Results are presented on Table III.

The hyperfine parameters exhibit a strong RE dependence, to connect to a de Gennes-type law. The hyperfine fields are strongly anisotropic. This is particularly true for the Fe_5 (8 j_2) site that is the shortest bound, by two Fe_3(16 k_2) sites. This anisotropy probably comes from the hyperfine coupling tensor in place of dipolar field effects (Dalmas, 85).

Additional ^{57}Fe measurements have been performed on oriented powder samples with RE = Y, Pr and Ho. For the case $Y_2Fe_{14}B$, the measurements at 4.2 K were performed under applied magnetic fields up to 4 T. From all these measurements we deduce that the only compounds with an easy c axis at room temperature, which show a SR of the iron sublattice at lower temperature, are the Nd ($\theta_{4.2\ K} = 30° \pm 3°$) and Ho ($\theta_{4.2\ K} = 22° \pm 2°$). Iron hyperfine fields are anisotropic and the anisotropy has different signs on the different sites (Fruchart, 87).

An analysis of the non-4f contribution to the electric field gradient at the ^{161}Dy or ^{166}Er nuclei in $Dy_2Fe_{14}B$ and $Er_2Fe_{14}B$ respectively, is consistent with published results on ^{155}Gd in $Gd_2Fe_{14}B$ (Bogé, 86). In order to explain that a spin rotation is found only for RE = Nd and Ho, we invoke the relative small quadrupole moment of the 4f shell for these two ions and crystal field terms of higher than second order. (Fruchart, 87 ; Li, 87).

Second order crystal field terms B^0_2 have been deduced to be opposite in sign on each RE sites from ^{161}Dy Mössbauer spectroscopy. The fourth and may be the sixth order terms have been recognized to be the key of the RE crystal field Hamiltonian developped to elucidate the origin of the SR. But, we observed that this B^0_2 parameter decreases with hydrogen, much faster for the 4f site than the 4g (Ferreira, 85). Nevertheless, comparison between the impact of hydrogen on the SR process in $Ho_2Fe_{14}BH_x$ and $Dy_2Fe_{14}BH_x$ (induced by H) shows that the high order CEF terms relative to the B^0_2 increase with hydrogen uptake. Since induction of a SR phenomena have been also observed in the Gd compound, it turns out that the 3d anisotropy terms must be simultaneously accounted for. (Yaouanc, 87 ; Fruchart, 87 ; Zhang, 88).

TABLE III

Mössbauer spectroscopy of $RE_2Fe_{14}B$

Comparison between the iron magnetic moments deduced from our [57] Mössbauer measurements and data from neutron scattering experiments and theoritical calculations. The moments are given in Bohr magnetons. Additional references are from Fruchart et al. (1987).

	4c	$16k_1$	$16k_2$	$8j_1$	$8j_2$	4e	T (K) references
Y	1.90 (10)	2.07 (10)	2.23 (3)	2.31 (10)	2.31 (10)	2.28 (10)	4.2 Fruchart et al (1987)
	1.95 (5)	2.25 (3)	2.25 (3)	2.80 (4)	2.40 (4)	2.15 (5)	4.2 Givord et al (1985)
	3.18	1.63	2.16	3.20	2.64	0.90	0 Gu and Ching (1986)
Ce	1.72 (10)	1.96 (3)	2.10 (3)	2.36 (3)	1.95 (10)	2.19 (10)	4.2 Fruchart et al (1987)
	2.4 (1)	2.7 (1)	2.2 (1)	3.4 (1)	2.7 (2)	2.1 (2)	77 Herbst and Yelon (1986)
	1.9 (1)	1.9 (1)	2.5 (1)	2.5 (1)	2.3 (1)	2.1 (1)	6 Dalmas de R. et al (1986)
Pr	1.77 (10)	2.07 (3)	2.15 (3)	2.37 (3)	2.01 (10)	2.18 (10)	4 2 Fruchart et al (1987)
	1.7 (2)	2.6 (1)	2.5 (1)	3.3 (1)	2.4 (1)	1.7 (2)	77 Herbst and Yelon (1985)
Nd	1.97 (10)	2.08 (3)	2.16 (3)	2.43(3)	2.06 (10)	2.28 (6)	4.2 Fruchart et al (1987)
	2.75 (6)	2.60 (4)	2.60 (4)	2.85 (5)	2.30 (5)	2.10 (6)	4.2 Givord et al (1985 a)
	2.2 (3)	2.4 (1)	2.4(1)	3.5 (1)	2.7 (1)	1.1 (2)	77 Herbst et al (1984)
	2.49	2.40	2.4	2.61	2.33	2.42	0 Szpunar and S. (1985)
Dy	1.69 (10)	2.06 (3)	2.17 (3)	2.49 (3)	2.10 (10)	2.28 (10)	4.2 Fruchart et al (1987)
	2.5 (1)	2.6 (1)	2.5 (1)	3.0 (1)	2.5 (1)	2.4 (1)	77 Herbst and Yelon (1985)
Er	1.96 (16)	2.08 (6)	2.21 (5)	2.25 (5)	2.15 (16)	2.25 (16)	4.2 Fruchart et al (1987)
	1.95 (10)	2.25 (1)	2.9 (1)	2.7 (1)	2.05 (10)	1.8 (1)	6 Dalmas de R. et al (1986)
Lu	1.70 (10)	1.95 (3)	1.99 (3)	2.32 (3)	2.00 (10)	2.06 (10)	4.2 Fruchart et al (1987)
	2.2 (2)	2.8 (2)	2.4(2)	3.6 (2)	2.9 (2)	1.7 (2)	77 Herbst and Yelon (1986)

We measure that charge transfers are not responsible for the increase in the Fe-magnetic moments as observed when hydriding the sample. But it could result in the narrowing of the d-bands with the expansion of the cell.

3. μ^+ SR spectroscopy

This technique reveals to be very sensitive to the rotation phenomena. The muons clearly measure the c-component of the magnetization, and the change is fairly observed for $Nd_2Fe_{14}B$ between 120 and 140 K, but nothing has appeared in the case of $Pr_2Fe_{14}B$ (Yaouanc, 87). For $Ho_2Fe_{14}B$, the SR phenomenon is also evidenced lower than 58 K, but the dispersion of the frequencies give evidence of the non collinearity of the low temperature magnetic structure mentionned discussed in the part devoted to neutron diffraction results. At 4.2 K the frequency of the muon precession measured versus the RE element clearly evidences a linear relation with the RE-moment. It confirms that the Er and Tm compounds are fully planar structure. It indicates that the mean Fe magnetization is constant in the series.

4. XPS and X ray-AES measurements.

The XPS technique failed to evidence the change in the 4f filling state of (one) magnetic cerium atom in $Ce_2Fe_{14}BH_x$ (Fruchart, 87). The too high vaccum needed around the sample leads to destabilize the hydride. The neutron experiment was finally confirmed by X-ray AES technique and reveals a 4f-localized state onto Ce site (Yaouanc, 87).

Macroscopic properties

1. Bulk magnetization

Systematically performed on powder samples, pure and hydrided compounds, the magnetization measurements evidence :

- a net increase of the resulting magnetization (≈ 10 %) on hydriding

- a large softening of the materials. This results via the modifications of the intrinsic anisotropy properties and the H-reacted integranular phases (potential sources of coercivity). In most of the cases, the measurements performed on oriented and glued samples show that the anisotropy fields (3-4 T for non magnetic RE to some tens of T for magnetic RE) are strongly reduced to values close to 1 T for hydrided materials (Dalmas, 87 ; Coey, 86).

High magnetic field measurements (up to 18 T) performed on a single crystal sphere of $Ho_2Fe_{14}B$ has permitted to evidence the (110) easy plane with the hierarchy [001] > [110] > [100]. A single to multi-domains process occurs at higher and higher temperature ; it is characterized by a critical field H_c, measured when decreasing the applied field. Its thermal variation could be roughly related to the anisotropy constant K_1 (T).

2. Curie point (T_c) measurements.

Analyzed as a function of x, the hydrogen content, the Curie point T_c of the $RE_2Fe_{14}BH_x$ increases systematically. First, the increase is rather fast for $0 < x \leq 2$, and for $x \geq 2$ a relative saturation effect is observed (Figure 2). This must be correlated to the stepping effect of hydrogen when occupying different tetrahedral sites. But the total ΔT_c increase is fairly in proportion with the square of the saturation magnetization increase. This proves that in spite of enlargements of the interatomic distances upon hydridation, the Fe-Fe exchange interactions are kept rather constant. The measured increases are $\Delta T_c \approx 50\text{-}70$ K, they are up to 100 K with RE = Ce, in connection with the rise of a magnetic moment on one Ce atoms (Dalmas, 87).

3. Anisotropy measurements.

We have already indicated that the relative increase in saturation magnetization Ms was close to 10 % . In fact this value applies well only if $M_{RE} = J = 0$ (non magnetic RE) and $M_{RE} \propto |L - S|$ (first row RE). On the contrary with $M_{RE} \propto |L + S|$ (second row RE) a weakening of the negative exchange interactions $J_{RE\text{-}Fe}$ is induced by hydrogen absorption. The external magnetic field is able to flip the opposite two-sublattices magnetic system, all the more so the local anisotropies are weakened. (Coey, 86).

Actually, the anisotropy field Ha, is systematically lowered by hydrogen insertion, for all the series. It is particularly clear, when the anisotropy is relatively high, in connection with the nature of the RE metal.

Changes in anisotropy characteristics versus hydrogen content are more appreciable for rather low hydrogen absorption rate. The decrease tends to a saturation level (as for T_c) at the highest H-contents. (Pareti, 88).

An extensive study performed on the $Dy_2Fe_{14}BH_x$ system, shows that at 300 K, K_1 and K_2 slightly increase from x = 0 to x = 4.7 (Coey, 86). But at low temperature, K_1 changes in sign ($+ \rightarrow -$: a spin reorientation has been induced) and K_2 is more than doubled, comparison with the H–uncharged material. An estimation of the $K_1{}^{Fe}(x)$ and $K_1{}^{RE}(x)$ terms was made, assuming that the Fe sublattice term is fully determined from the Y-compound. At 293 K, it is shown that $K_1{}^{Fe}(x)$ continuously decreases with x, but there is a large decrease in $K_1{}^{Nd}(x)$, and a relatively small one for $K_1{}^{Ho}(x)$. For $K_1{}^{Tm}(x)$, the change remains small, in fact this term is negative in agreement with the positive sign of the α_J Stevens' second order term. For RE = Dy, it slightly increases. In summary, we consider that the anisotropy constants $K_1{}^{Fe}(x)$, $K_1{}^{RE}(x)$, $K_2{}^{RE}(x)$...are perceptibly dependent on the content in hydrogen atoms sitting

FIGURE 2

Curie temperature of the hydrides $RE_2Fe_{14}BH_x$ versus x, the amount of hydrogen. All the compounds refer to the right scale, excepted for RE = Ce (left scale). The magnetic RE are represented by black symbols, contrary to that of the non-magnetic RE (empty symbols).

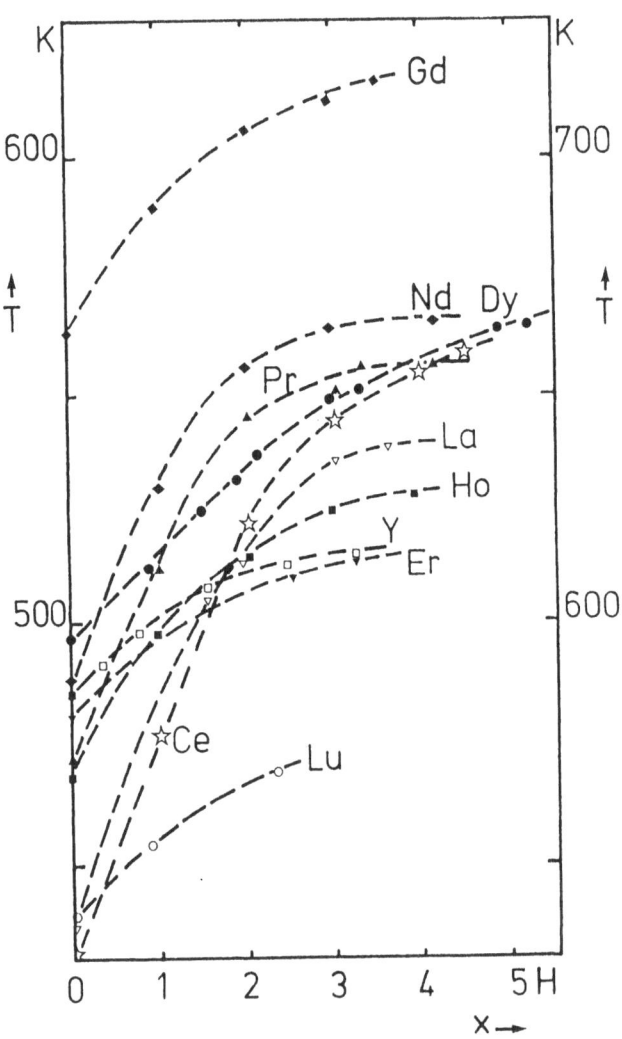

around the RE-atoms. This has to be related to the thermal and hydrogen-induced changes in the CEF terms B^0_2, B^0_4....(Pareti, 88).

The existence or the induction by hydrogen of spin reorientation (SR) phenomena well enlight the competition between the different terms of anisotropy.

4. Spin reorientation temperature (SRT) (Fruchart, 87).

In figure 3, all available information to date is given. The results come from different techniques as Mössbauer spectroscopy, magnetization measurements both on oriented and non-oriented samples. But the more sensitive technique is the χ_{ac} initial susceptibility (Rillo, 88). In a first case there is no SRT in the pure compound and then it is induced for $x \geq 1$ (Pr, Gd, Dy) ; the increase in molecular field term n_{Fe-RE} is not able of producing the SRT by itself. On the other hand, the appearance of SRT can be explained as caused by an increase of the B^0_4 term, by a factor six (Pr) and three (Dy) respectively. This later result well agrees with the observed increase of the K_2 anisotropy constant, which is proportional to B^0_4, and decrease of K_1 in $Dy_2Fe_{14}BH_x$ at low temperature, as determined by magnetization measurements (Fruchart, 87).

The increase of SRT observed in Ho is again caused by an increase of B^0_4, which in this case cannot be countered by the reduction of n_{RE-Fe} as for the Nd compound. Indeed a smaller effect due to the exchange field could be expected for the heavy RE (Li, 87). The case of $Ho_2Fe_{14}BH_x$ is peculiar since it is the only studied case with $\alpha_J > 0$. Contrary to the other coupounds studied its SRT is observed at much higher temperature $T_{SR} = 320$ K, where the Fe sublattice anisotropy K_{1Fe} becomes dominant and strongly temperature dependent. However this simple model (Rillo, 88) is not able to explain the "anomalous" behaviour of $Gd_2Fe_{14}BH_x$ versus x. In principle, there is no CEF interactions on Gd. The observed SRT (Figure 3) could come from competing anisotropy terms relative to the 6-different Fe sublattices and different in strength J_{Gd-Fe} exchange interactions. Contrary to Zhang, 88, the reorientation process is not total at high H-content, since our Mössbauer data indicate that the resulting hyperfine fields only deviate of $\theta \approx 40°$ from the c-axis.

Conclusion

The magnetoelastic forces play a very important role on the strength of the magnetic characteristics of the high performance magnet materials $RE_2Fe_{14}B$. This is confirmed by our structural and magnetic structure determination performed on $Ho_2Fe_{14}B$ single crystal by neutron diffraction. Moreover, hydrogen absorption reveals as a pertinent tool used for

FIGURE 3

The spin reorientation temperature (SRT) behaviour of some $RE_2Fe_{14}BH_x$ versus x, the amount of absorbed hydrogen. The data come from ^{57}Fe Mössbauer spectroscopy, $\chi_{a.c.}$ susceptibility and magnetization measurements. The arrows disposed close to the RE symbols refer to the left or right temperature scale. Both scales are used for Gd compound.

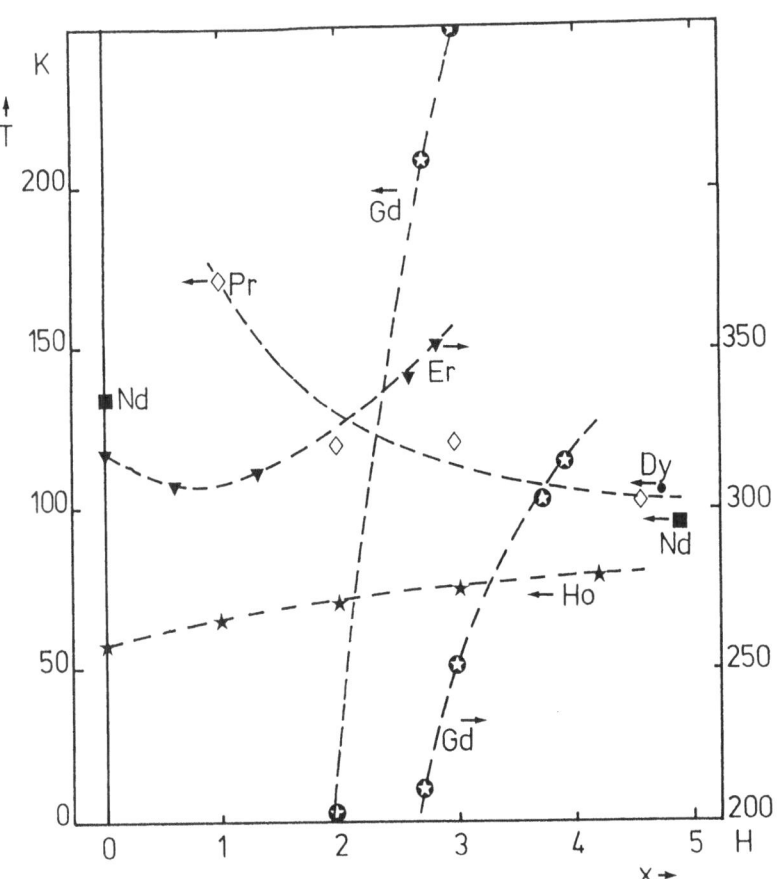

modifying the structural and elastic force parameters and consequently the competing terms involved in the magnetic structure stability.

Ackowledgments :

Parts of this work are the result of strong collaborations with the CEAM-groups of ZARAGOZA, DUBLIN and PARMA. We thank the other members of CEAM for hepfull discussions. We warmly thank SAGAWA and his coworkers for kindly preparing us the single crystal used for the neutron diffraction experiment.

References

- Andreev A. V. , Deryagin A. V. , Zadvorkin S.R , and Terent'ev S. V.,
1985, Sov. Phys. Solid State **27,** 6, 987.

-Boge M., Czjzek G. , Givord D. , Jeandey C. , Li H. S. , Oddou J. L.,
1986, J. Phys. F : Met. Phys, **16** L, 67.

-Buschow K. H. J. , Grössinger R. ,
1987, J. of Less Com. Met., **135**, 39-46.

- Coey J. M. D. , Cadogan J. M. and Ryan D. H.,
Rep. Proc. Workshop Meet. Brussels, 1984 in I. V. Mitchell (ed.)
Nd-Fe Permanent Magnets. Their present an future applications,
Elsevier, Amsterdam, 1986, 143.

- Coey J. M. D. , Yaouanc A. , Fruchart D.,
1986, Solid State Comm., **58**, 7, 413-416.

- Dalmas de Réotier P. , stage Res. Ing., Institut National Polytechnique de Grenoble,
June 21, 1985.

- Dalmas de Réotier P. , Fruchart D. , Wolfers P. , Guillen R. , Vuillet P., Yaouanc A. ,
Fruchart R. and l'Héritier P.
1985, Colloque C6, suppl. au N° 9, **46**, C6-323.

- Dalmas de Réotier P. , Fruchart D. , Pontonnier L. , Vaillant F. , Wolfers P. , Yaouanc A. ,
Coey J. M. D. , Fruchart R. and Ferreira C. P. , Guillen R. , Vuillet P. ,
1985, J. of Magn & Magn. Mat., **53**, 145.

- Fruchart D. , Wolfers P. , Vuillet P. , Yaouanc A. , Fruchart R. and l'Héritier P.,
Rep. Proc. Workshop Meet. Brussels, 1984 in I. V. Mitchell (ed.),
Nd-Fe Permanent Magnets. Their present an future applications,
Elsevier, Amsterdam, 1986, 173.

- Fruchart D. , Vaillant F. , Yaouanc A. , Coey J. M. D. , Fruchart R. , l'Héritier P. ,
Riesterer T., Osterwalder J. , Schlapbach L. J.,
1987, J. of Less Com. Met., **87**, 130.

- Fruchart D. , Pontonnier L. , Vaillant F. , Bartolome J. , Fernandez J. M. , Puertolas J. A. ,
Rillo C. , Regnard J. R. , Yaouanc A. , Fruchart R. , l'Héritier P. , (EMMA Manchester 87),
IEEE Trans. on Magn., march 1988, **24**, 2, 1641-1643.

- Fruchart R. , l'Héritier P. , Dalmas de Réotier P. , Fruchart D. , Wolfers P. , Coey J. M. D. ,
Ferreira L. P. , Guillen R. , Vuillet P. and Yaouanc A.,
1987, J. Phys. F : Met. Phys., **17**, 483-501.

- Givord D. , Li H. S. and Moreau J. M.,
1984, Solid State Comm., **51,** 857.

- Herbst J. F. , Croat J. J. , Pinkerton F. E. and Yelon W. B. ,
1984, Phys. Rev., B **29**, 4176.

- Le Roux D., Doctor Thesis U.S.T.M.G. (Grenoble 1986).

- Li H. S., Doctor Thesis U.S.T.M.G. (Grenoble 1987).

- l'Héritier P., Chaudoüet P. , Madar P. , Rouault A., Senateur J. P. and Fruchart R.,
1984, C. R. Acad. Sci. Paris, **299** (II), 849.

- l'Héritier P. , Fruchart R. , Fruchart D. , Wolfers P., Yaouanc A. ,
1988 (CEAM report, this book).

- Pareti L. , Moze O. , Fruchart D. , l'Héritier P. , Yaouanc A.,
J. of Less Com. Met. (to appear).

- Rillo C. , Bartolomé J. , Chaboy J. , Fernandez J. M. , Navarro R.,
1988 (CEAM report, this book).

- Shoemaker C. B. , Shoemaker D. and Fruchart R. ,
1984, Acta Crystallogr., Sect. C **40**, 1665.

- Wolfers P., Fruchart D., Miraglia S., Bartolomé J., Pannetier J., Hirosawa S., Sagawa M., (
to be published).

- Yaouanc A. , Budnick J. , Albert E. , Hama M. , Weidinger A. , Fruchart R. , l'Héritier P. ,
Fruchart D. , Wolfers P.

1987, J. of Magn & Magn. Mat., **67**, 286-290.

- Zhang S. G, Pourarian F. and Wallace W. E. ,
1988, J. of Magn & Magn. Mat., **71**, 203-211.

ELABORATION AND CHARACTERIZATION OF $Nd_2Fe_{14}B$-TYPE PHASES

AND THEIR HYDRIDES

Robert Fruchart, Philippe l'Héritier

U.A. 1109 CNRS, Ecole Nationale Supérieure de Physique de Grenoble
BP 46, 38402 Saint Martin d'Hères (France)

Daniel Fruchart, Pierre Wolfers

Laboratoire de Cristallographie du CNRS
166 X, 38042 Grenoble Cedex (France)

Alain Yaouanc

Département de Recherches Fondamentales, MDIH, CENG
85 X, 38041 Grenoble Cedex (France)

ABSTRACT

The $RE_2Fe_{14}B$ compounds can absorb hydrogen without structure change.
The basic characteristics are very sensitive to hydrogen absorption ; the
effects are more or less strong with respect to a content x of about 2.5
H/f.u., depending on the nature of the effect ; itself and the cell volume
depend on the RE radius and the nature of the RE have an effect upon the
lattice variation in the series, separating light from heavy RE. On occupying different interstitial sites (depending on the RE and the H
content), hydrogen expands the lattice, modifying the exchanges Fe-Fe and
Fe-RE. The Curie temperature, saturation magnetization are enhanced, the
anisotropy fields and coercivity decrease non linearly with x. The spin
reorientation phenomena are affected : they are even induced in Pr, Dy and
Gd hydrided compounds where a planar anisotropy occurs up to the spin reorientation temperature.

INTRODUCTION

We are interested in tetragonal $Nd_2Fe_{14}B$- type phases and their hydrides since 1983 : we contributed to the determination of the crystal structure of $Nd_2Fe_{14}B$ (Shoemaker, 1984) and we have studied the hydridation and
the series of fully hydrogenated compounds $RE_2Fe_{14}BH_x$, the isostructural
series $RE_2Co_{14}B$ and their hydrides $RE_2Co_{14}BH_x$ for the first time
(l'Héritier, 1984). We usually work with the group 1-09 (we form a separated group for administrative reasons). Nevertheless, we present here some
of our own results, mainly experimental.

AIMS AND INTEREST

First of all, the controlled introduction of hydrogen expands the lattice constants in the $RE_2Fe_{14}B$ compounds, resulting in an increase of the magnetization (+ 8 %) and Curie temperature, and a decrease in the anisotropy : spin reorientation phenomena are modified or induced, depending the case ; so the hydrogen absorption allows to test the magnetostructural and magnetoelastic terms. Secondly, a practical aspect is the interest of hydrogen in the metallurgical process of these sorts of alloys, such as reducing atmosphere in sintering, ingot decrepitation by reversible formation of hydrides before the milling (Fruchart, 1984 ; Harris, 1985) or obtaining fine particles for recording media. We have therefore undertaken a systematical study of hydrided $RE_2Fe_{14}B$ phases. By another way, some attempts have been made to examine two systems of the same structure, with Gd, where Co replaces Fe and where Mn replaces Co. Finally, several $RE_2Fe_{14}B$ phases have been tested by positive muon spin rotation spectroscopy.

The results of a lot of experiments (Mössbauer, neutrons, XPS, AES, μ^+ SR...) are presented and discussed in papers and reports with mainly our colleagues of groups 1-09 (D. Fruchart, A. Yaouanc et al.), and 1-10 (J.P. Sanchez et al.).

EXPERIMENTAL

The ingots $RE_2Fe_{14}B$ were prepared from 3N elements or boron-iron alloys with the use of the rf induction heating in a water-cooled copper crucible (levitation technique) under argon atmosphere. As-cast samples were annealed in evacuated quartz ampoules at 900°C for 1 or 2 weeks. X-ray diffraction analysis were achieved using a Guinier-Hägg camera (Cr-Kα radiation). Thermomagnetic analysis (TMA) were carried out using a Faraday balance under an about 500 Oe field. Hydridation of the samples at a maximum hydrogen content was performed under several bar of H_2 gas pressure, after degasing under vacuum and sometimes activation heating at 150°C. The hydridation rate was measured by gravimetry. Intermediate hydrogen contents were obtained by heating calculated amounts of pure and hydrided compounds at about 350°C.

Some samples were magnetically oriented and glued under fields up to 2 T, in order to measure the anisotropy or for obtaining Mössbauer spectra. The laboratory has an electromagnet applying fields up to 2.6 T from 4.2 K to 300 K, using the axial extraction method.

ACHIEVEMENTS

1- Solid solutions $RE_2Fe_{14}BH_x$

Lattice properties behaviour

The increase in the a parameter is linear with the hydrogen content x, whereas the x dependence of c exhibits a smaller increase for x ⟩ 2H/f.u. Moreover, for a given $RE_2Fe_{14}BH_x$ system, the variation of c as a function of a seems to be the same, whatever the sample may be : a increases more first than c (Fig. 1) ; the expansion of the lattice is anisotropic, favouring the basal plane first. So hydrogen must fill different interstitial sites, competing with each other in the progressive hydrogen charge ; these sites consist of tetrahedra close to the RE atoms and have been identified by neutron diffraction technique (Fruchart 1986, Dalmas 1987) ; The theoritical filling in the tetragonal cell cannot exceed 5.5 H per formula unit, but the maximum content depends on the R.E. The volume expansion ΔV_H per hydrogen, $1,5 < \Delta V_H < 2.9$ $Å^3/H$, is lower in these phases than in simple metals (about 2.9 $Å^3/H$).

The RE dependence of the unit cell volume follows the well-known "lanthanide contraction" (Sagawa 1984, Sinnema 1984, l'Héritier 1984). However the variation of c as a function of a makes two straight lines to appear in $RE_2Fe_{14}B$ series, for light and heavy RE respectively (Fig. 2) ; they meet with the point of $Gd_2Fe_{14}B$ for the virgin alloys ; therefore the RE magnetism acts on the anisotropic part of the cell elastic forces via a sort of De Gennes' law. For the same compounds, fully hydrogenated, the straight lines are shifted upper with larger slopes, and meet with a point close to $Sm_2Fe_{14}BH_x$. The $RE_2Co_{14}B$ series only exists with light RE and gives a single straight line for virgin compounds and another one for fully hydrogenated compounds, although the maximum hydrogen content differs from one RE compound to the next, in the same preparation conditions.

The Curie temperature

Throughout all the series $RE_2Fe_{14}BH_x$, the hydrogen rate dependence of the increase in T_c is distributed along 3 sets of curves (Fig. 3). The upper set includes the La and Ce compounds, the medium the Pr and Nd com-

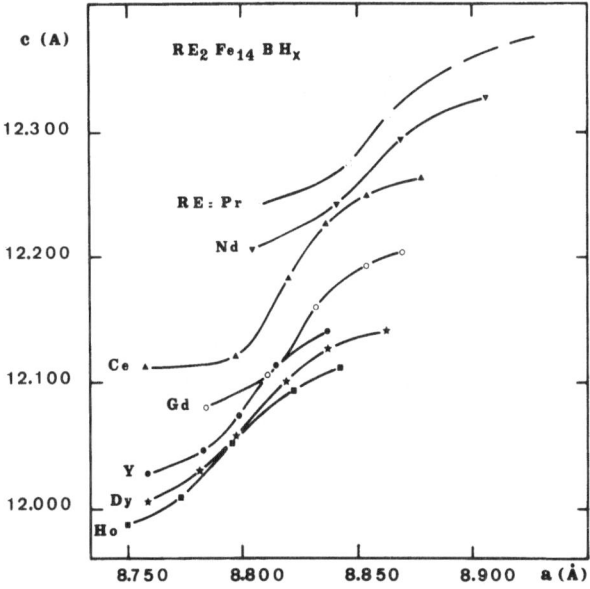

Fig. 1   The c-parameter variation as a function of the a-parameter
in the tetragonal cell of some $RE_2Fe_{14}BH_x$ series (0 ≪ x ≪ 5).

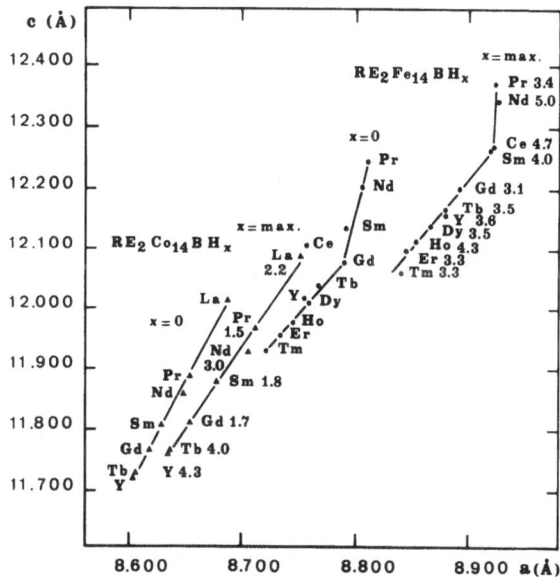

Fig. 2   The c-parameter variation as a function of the a-parameter
in $RE_2Co_{14}B$ (triangles) and $RE_2Fe_{14}B$ (dark circles) series before
and after hydriding at maximum content (the values are indicated).

pounds, the lower includes the others with RE = Y, Gd, Ho, Dy.

For a same series $RE_2Fe_{14}BH_x$, the main increase in Tc corresponds to x
< 2.5 ; the increase Δ Tc, related to the expansion of the lattice, is
attributed to a reduction of the negative exchange interaction between Fe
atoms in such a structure where some Fe-Fe distances are lower than 2.5 Å.
However the fact that Δ Tc is proportional to the increase in the saturation magnetization is to be noticed and suggests that the exchange interactions are not strongly modified.

Effect of hydrogen on the spin reorientation phenomena

The existence or the absence of SRT in the non-hydrided compounds can
be explained by some competing anisotropies between the iron and the RE
sublattices ; now four cases of SRT occur with the hydrogen insertion,
allowing to discuss the origins of the SRT (Fruchart 1987) :

   a) induction of SRT by hydrogen (RE = Pr, Dy, Gd)

   b) decrease in the existing SRT (Nd) or after its appearance (Pr),
      when raising the hydrogen content.

   c) increase in SRT with x (Ho : from T = 60 K (x=0) to T = 80 K
      (x=4.1).

   d) first decrease (x<1) and then increase with x (Er)

The rotation of the magnetization evidenced by TMA and ^{57}Fe Mössbauer
spectroscopy upon oriented powders is not only a crystal field effect, at
least this cannot explain the rotation seen in $Gd_2Fe_{14}BH_x$ for x > 2
(Yaouanc 1987, Zhang 1988) ; the 3d anisotropy or the electronic structure
of the conduction electrons (Fermi surface) could be important factors.

2- Positive muon spin rotation spectroscopy

The μ^+ SR measurements on polycristalline samples in zero applied
field have been performed on $Nd_2Fe_{14}B$ and $Pr_2Fe_{14}B$ first (Yaouanc 1987)
and other $RE_2Fe_{14}B$ (Niedermayer 1987). A single sharp μ^+SR line is observed ; the temperature dependence of the muon frequency reflects the spin
reorientation (Nd, Ho) and can be explained by assuming that only the c-
axis component of a magnetization is sampled by the muon. Let B_μ be the
local field sampled by the muon and B_L the Lorentz field, in $RE_2Fe_{14}B$
(Fig. 4), the R.E. magnetic moment dependence of $(B_\mu - B_L)$ is linear when

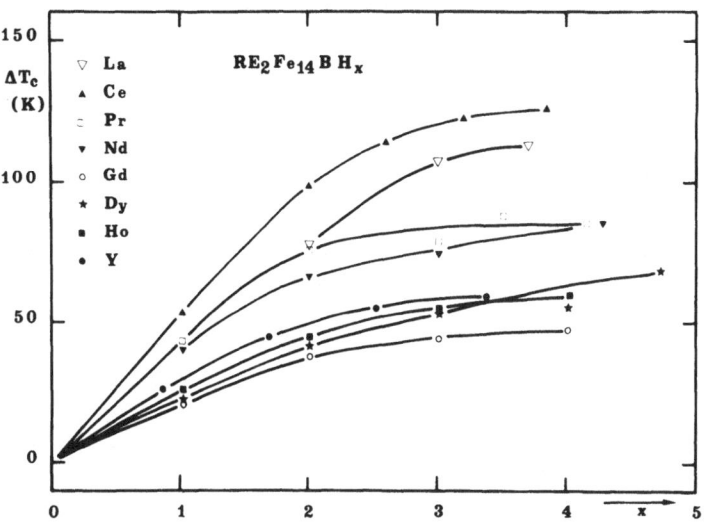

Fig. 3   Hydrogen concentration dependence of the increase in the Curie temperature in several $RE_2Fe_{14}BH_x$ systems.

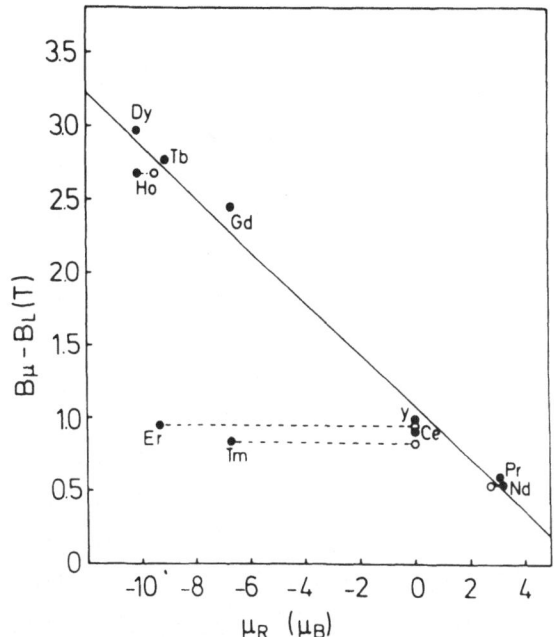

Fig. 4   Rare-earth magnetic moment dependence of $(B\mu - B_L)$ at T = 15 K in $RE_2Fe_{14}B$. The open circles correspond to the c-axis components of the RE moments (Niedermayer 1987)

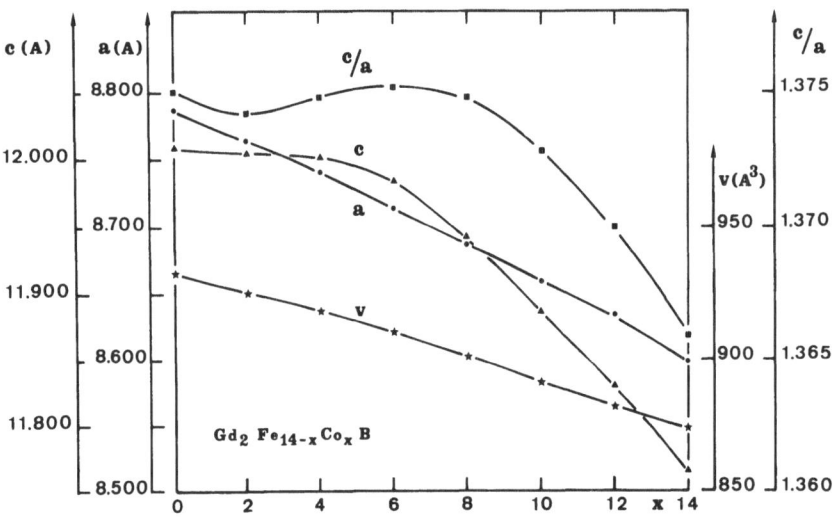

Fig. 5   Composition dependence of the lattice dimensions a, c, c/a,
v in the $Gd_2(Fe_{14-x}Co_x)B$ system

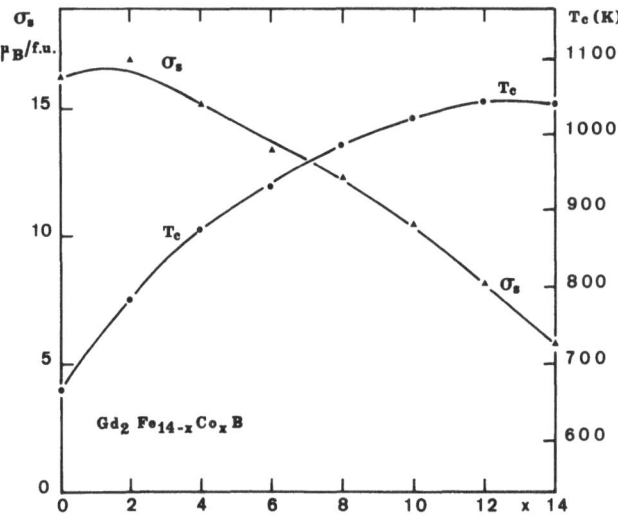

Fig. 6   Composition dependence of the Curie temperature and
saturation magnetization at T = 4.2 K in the $Gd_2(Fe_{14-x}Co_x)B$ system

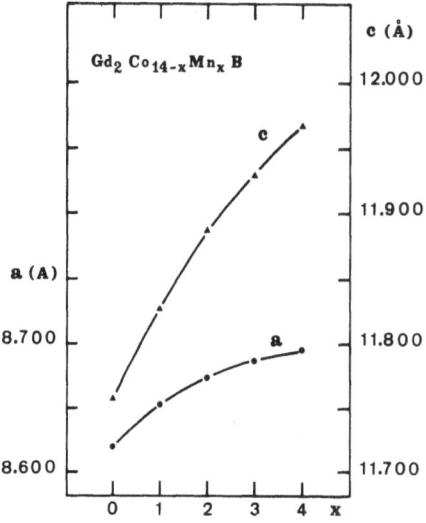

Fig. 7　Composition dependence of a and c-parameters in the
$Gd_2(Co_{14-x}Mn_x)B$ system .

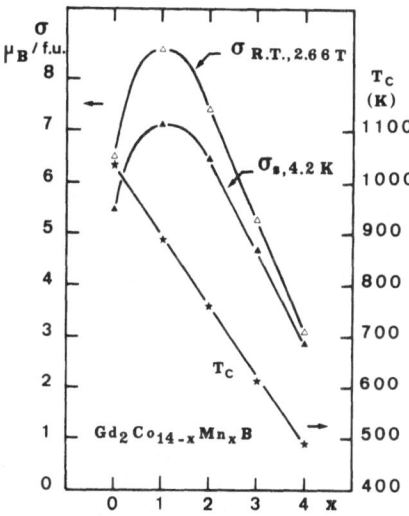

Fig. 8　Composition dependence of the Curie temperature and
magnetization (at R.T. under an 2.6 T external field and saturation
magnetization at 4.2 K)in the $Gd_2(Co_{14-x}Mn_x)B$ system.

considering the c-axis magnetic component of the RE (Niedermayer 1987); so the average iron sublattice magnetization remains identical in the series.

3- $Gd_2(Fe_{14-x}Co_x)B$ solid solutions

Cobalt is continuously substituted for iron. The unit cell volume follows the Vegard's law, but c/a decreases for 8 < x < 14 (Fig. 5). It has been checked by X-ray diffraction ($\theta - 2\theta$ diffractometer) upon magnetically aligned samples that the easy magnetization direction is parallel to the c-axis for 0 < x < 8 and perpendicular for 10 < x < 14, at room temperature ; therefore the change of slope of the c/a ratio (and also of c as a function of a) is related to a different easy magnetization direction.

T_c strongly increases for 0 < x < 4 (in +50 K per Co atom substituted for Fe). The magnetization decreases from 16.2 to 5.8 μ_B between x = 0 and x = 14 at T = 4.2 K (Fig. 6). A slight increase in σ_s is noticed at about x = 2, which could originate in an electron transfer from Fe to Co, raising the average Fe moment, as in $Nd_2 (Fe_{14-x}Co_x)B$ (Abache 1986). Any ordering of 3d elements has not been evidenced yet by ^{57}Fe Mössbauer effect.

4- $Gd_2(Co_{14-x} Mn_x)B$ solid solutions

The manganese substitution for cobalt in $Gd_2Co_{14}B$ is maximum for x = 4, leading to an increase in the unit cell volume (Fig. 7). T_c decreases by 135 K per substituted Co atom and coupling changes are noticed. The magnetization under 150 KOe at T = 4.2 K decreases from 5.4 μB (x = 0) to 2.85 μB (x = 4). The easy direction turns out to be parallel to the c-axis for x > 3. The temperature dependence of the magnetization increases except for x = 4 (maximum at T = 200 K).

CONCLUSION

The introduction of hydrogen in $RE_2Fe_{14}B$ contributes to clear some fundamental properties of these phases, mainly in spin rotation phenomena. The μ^+ SR proves to be an attractive means of investigating the SRP by being sensitive to the c axis component of the magnetization.

ACKNOWLEDGMENTS. The authors are grateful to A. Weidinger and C. Niedermayer for their collaboration and for providing us with μ^+ SR facilities at the S.I.N. (Zürich).

REFERENCES

Abache, C. and Oesterreicher, H., 1986. J. Appl. Phys. 60, 1114-1117.

Dalmas de Réotier, P., Fruchart, D., Pontonnier, L., Vaillant, F., Wolfers, P., Yaouanc, A., Coey, J.M.D., Fruchart, R. and L'Héritier Ph., 1987. J. Less-Common Met., 129, 133-144.

Fruchart, R., Madar, R., Rouault, A., L'Héritier, Ph., Taunier, P., Boursier, D., Fruchart, D. and Chaudouët, P., 1984, Institut National de la Propriété Industrielle (Paris) déposé le 9 juin 1984. n° 84 10387. European Patent Office (The Hague, The Netherlands), 20 June, 1985, No 85 401230.9 (Applicant : CNRS). U.S. Patent and Trademark Office, Washington D.C., Filed June 19, 1985, serial No 746, 360.

Fruchart, D., Wolfers, P., Vuillet, P., Yaouanc, A., Fruchart, R. and l'Héritier, Ph., 1986. Rep. Proc. Workshop Meeting, Brussels, October 1984, "Nd-Fe Permanent Magnets, Their Present and Future Applications" Mitchell, I.V. ed., Elsevier, Amsterdam, 143.

Fruchart, D., Pontonnier, L., Vaillant, F., Bartolomé, J., Fernandez, J.M., Puertolas, J.A., Rillo, C., Regnard, J.R., Yaouanc, A., Fruchart, R. and L'Héritier, Ph., E.M.M.A.'87 Conference, University of Salford (G.B.), 1988. IEEE Trans-Mag., 24, 1641-1643.

Harris, I.R., Noble, C. and Bailey, T., 1985, J. Less-Common Met., 106 L1

L'Héritier, Ph., Chaudouët, P., Madar, R., Rouault, A., Sénateur, J.P. and Fruchart, R., 1984. C.R. Acad. Sc. Paris, 299 II 849-851.

Niedermayer, C., 1987, μSR Messungen an R_2 Fe_{14} B - Verbindungen, Diplomarbeit, Universität Konstanz (FRG).

Sagawa, M., Fujimura, S., Yamamoto, H., Matsuura, Y. and Hiraga, K., 1984. IEEE Trans-Mag. MAG-20, 1584.

Shoemaker, C.B., Shoemaker, D.P. and Fruchart, R., 1984. Acta Cryst., C40, 1665-1668

Sinnema, S., Radwanski, R.J., Franse, J.J.M., de Mooij, D.B. and Buschow, K.H.J., 1984. J.M.M.M., 44, 333-341

Yaouanc, A.J., 1987. Report on the scientific activity of the CEAM, Meeting in Dublin, March 1987, p. 14.

Yaouanc, A.J., Budnick, J., Albert, E., Hamma, M., Weidinger, A., Fruchart, R., L'Héritier, Ph., Fruchart, D. and Wolfers, P., 1987. J.M.M.M. 67, L286-290.

Zhang, L.Y., Pourarian, F. and Wallace, W.E., 1988. J.M.M.M. 71, 203-211.

MAGNETOSTRICTION AND LOW FIELD MAGNETIZATION PROCESSES IN

$RE_2Fe_{14}B$ COMPOUNDS.

P.A. Algarabel, J.I. Arnaudas, J. Bartolomé, J. Chaboy, A. Del Moral

J.M. Fernández, M.R. Ibarra, C. Marquina, R. Navarro, C. Rillo,

Instituto de Ciencia de Materiales de Aragón.
C.S.I.C. - Universidad de Zaragoza.
50009 Zaragoza, Spain.

ABSTRACT

In the magnetic ordered phase some of the $RE_2Fe_{14}B$, $(RE_xRE'_{1-x})_2Fe_{14}B$ and $RE_2Fe_{14}BH_x$ compounds show spin reorientation transitions (SRT). An extensive experimental study of the SRT's using thermal expansion, magnetostriction and a.c. initial magnetic susceptibility measurements has been performed. The thermal expansion shows strong Invar anomalies for the whole $RE_2Fe_{14}B$ series. From the magnetostriction parallel and perpendicular data the anisotropic and volume striction have been obtained. Finally the SRT temperature has been accuratelly determined from the a.c. susceptibility experiments. Moreover the experimental results has been analized and compared with available models.

INTRODUCTION

The recent development of $Nd_2Fe_{14}B$ based permanent magnets has brought

a huge amount of research in the fundamental properties of $RE_2Fe_{14}B$ (Sagawa

et al. ,1984; Sinnema et al., 1984), $(RE_xRE'_{1-x})_2Fe_{14}B$ (Rechenberg et al.,

1987; Niarchos and Simopoulos, 1986; Ibarra et al., 1987) and $RE_2Fe_{14}BH_x$

(Pourarian et al., 1986; Fruchart et al., 1988) compounds (RE=rare earth).

These systems belong to the space group $4P_2/mnm$, where the RE^{3+} ions occupy

two different crystallographic sites with point symmetry mm, and the Fe

atoms six different ones.

They order ferro or ferrimagnetically with critical temperatures, T_c,

varying between 527 and 664 K for the Yb and Gd respectively. Above room

temperature they present axial anisotropy essentialy produced by the Fe sub-

lattices. At lower temperatures the anisotropy produced by the interaction

of the RE^{3+} with the crystal field gradient (CEF) begins to be important and

originates at some temperature, T_s, a spin reorientation phase transition

(SRT).

Below T_c, the easy axis is the c direction except for the Sm compound

which lays on the perpendicular plane. Besides, at lower temperatures some members undergo a SRT, changing the easy axis from the **c**-direction to the perpendicular plane (Er,Tm,Yb) or to an intermediate orientation (Nd,Ho).

Our contribution under the CEAM scheme has been devoted to the study of $RE_2Fe_{14}B$, $(RE_xRE'_{1-x})_2Fe_{14}B$ and $RE_2Fe_{14}BH_x$ compounds using thermal expansion, magnetostriction and a.c. initial magnetic susceptibility measurements. Polycristalline samples of all members of the series and $Ho_2Fe_{14}B$ and $Nd_2Fe_{14}B$ single crystals has been used. Moreover the results have been compared with available theoretical models.

THERMAL EXPANSION

Linear thermal expansion measurements, $\Delta l/l$, on polycrystalline $RE_2Fe_{14}B$ samples, between 3.5 and 350 K, have been performed using the strain-gauge technique and a d.c. sensitive bridge. In fig. 1a the results of $Er_2Fe_{14}B$ and $Pr_2Fe_{14}B$ are shown. The data for other compounds in the series can be found elsewhere (Givord and Li, 1985; Buschow et al., 1988; Ibarra et al., 1988). A large Invar effect has been observed for all the series and this behaviour has likely its origin in the strong spacial dependence of the exchange interactions. In the case of $Pr_2Fe_{14}B$, for instance, at room temperature $\alpha(T)=1/l(\partial l/\partial T)$, is very small and becomes negative for $Er_2Fe_{14}B$.

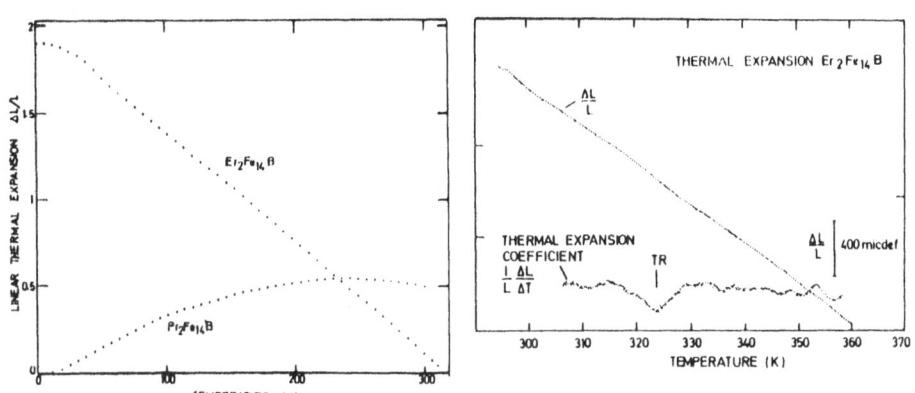

Fig. 1 (a) Linear thermal expansion, $\Delta l/l$, for $Er_2Fe_{14}B$ and $Pr_2Fe_{14}B$ (b) $\alpha(T)$ anomaly at T_s for $Er_2Fe_{14}B$.

On the other hand $\alpha(T)$ shows slight anomalies associated with T_s as could be seen in fig. 1b for $Er_2Fe_{14}B$. At $T_s=(322\pm1)K$ the magnetization spontaneously rotates from the easy direction (**c**-axis) to the basal plane below T_s.

MAGNETOSTRICTION

Magnetostriction measurements have been carried out on $RE_2Fe_{14}B$ (RE= Pr, Nd, Dy, Er and Y) polycrystalline samples and on $Nd_2Fe_{14}B$ single crystal. Temperatures between 4.2 and 300 K, and pulsed magnetic fields up to 15 Tesla were used. The magnetostrictive strains were measured using the strain-gauge technique and a sensitive a.c. bridge.

<u>Polycrystalline materials</u>

In the polycrystalline sample the magnetostriction parallel ($\lambda_{\parallel}$) and perpendicular ($\lambda_{\perp}$) to the applied field have been measured in order to separate the anisotropic magnetostriction $\lambda_t = \lambda_{\parallel} - \lambda_{\perp}$ and the isotropic volume one, $\omega = \lambda_{\parallel} + 2\lambda_{\perp}$. In fig. 2a,b the isothermes for the $Er_2Fe_{14}B$ compound are shown.

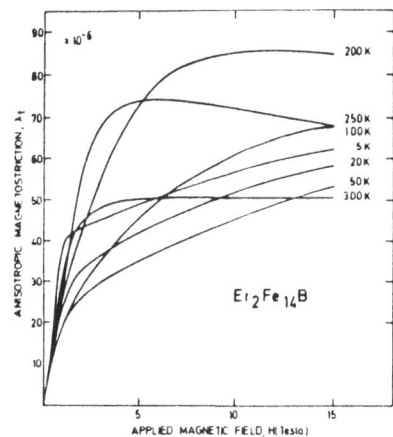 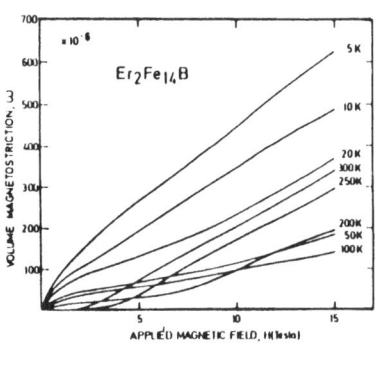

Fig. 2 (a) Anisotropic λ_t and (b) volume ω magnetostriction isotherms for $Er_2Fe_{14}B$.

For a single crystal with axial symmetry the magnetostrictive strain is given by the expression (Callen and Callen, 1965),

$$\lambda(\alpha,\beta) = 1/3\lambda^{\alpha}_{11} + 1/2\sqrt{3}\ \lambda^{\alpha}_{12}(\alpha_z^2 - 1/3) + 2\ \lambda^{\alpha}_{21}(\beta_z^2 - 1/3) + \sqrt{3}\ \lambda^{\alpha}_{22}(\alpha_z^2 - 1/3)(\beta_z^2 - 1/3)$$
$$+ 2\lambda^{\gamma}\{(\alpha_x^2 - \alpha_y^2)(\beta_x^2 - \beta_y^2)/4 + \alpha_x\alpha_y\beta_x\beta_y\} + 2\lambda^{\varepsilon}\{\alpha_x\alpha_z\beta_x\beta_z + \alpha_y\alpha_z\beta_y\beta_z\} \quad [1]$$

where $\alpha(\alpha_x, \alpha_y, \alpha_z)$ gives the direction cosines of the spontaneuous magnetization, $\beta(\beta_x, \beta_y, \beta_z)$ the strain measurement direction and $\lambda^\Gamma_{ij}(H,T)$ represent the different deformation modes compatibles with the point symmetry. An average of λ over all the possible orientations of the grains in the polycrystal gives,

$$\lambda_t = 6\sqrt{3}/45\ \lambda^\alpha_{22} + 6/15\lambda^\gamma + 4/15\lambda^\varepsilon \qquad [2a]$$

$$\omega = \lambda^\alpha_{11} \qquad [2b]$$

It can be see that from λ_t and ω polycrystalline measurements it is possible to deduce the exchange mode λ^α_{11} whereas the anisotropy CEF modes λ^α_{22} (tetragonal), λ^γ (breaking of cylindrical symmetry) and λ^ε (shear), always appear mixed in λ_t. In order to separate such modes single crystal $\lambda(\alpha, \beta)$ measurements are needed.

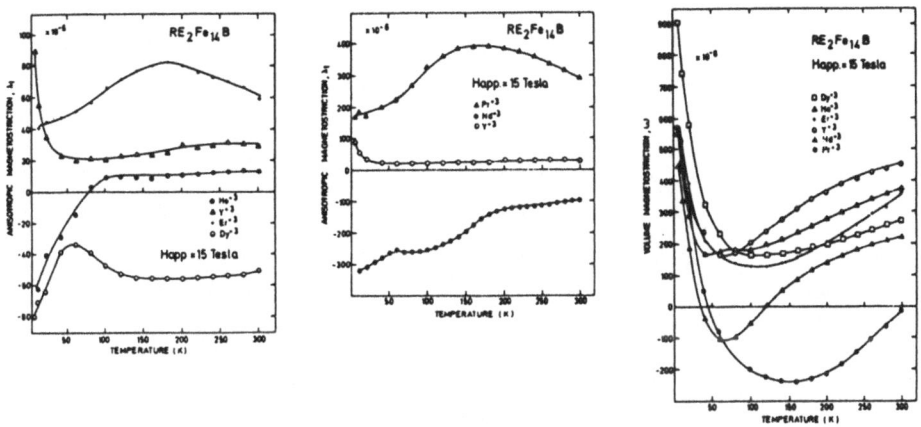

Fig. 3 Thermal dependence of anisotropic λ_t (3a and 3b) and volume ω (3c) magnetostrictions at the maximum field applied 15 T.

In fig. 2a some selected isotherms of λ_t are shown for $Er_2Fe_{14}B$. Saturation at temperatures near T_s can be observed as a consequence of the anisotropy K_1 constant softening, which results from the competition between the axial Fe sublattices anisotropy and the planar one of the Er^{3+} ions. In fig. 2b a strong volume striction at low temperatures ($\partial\omega/\partial H \approx 30x$ 10^{-10} Oe^{-1} at 5 K) can be observed. According to equation [2b] the origin of such behaviour of ω should be related with the strain dependence for the exchange interactions. For $Pr_2Fe_{14}B$ the onset (around 15 T) of a second class first order magnetization processes (FOMP) transition is observed (Hirogoshi et al., 1987), although smeared out due to the polycrystalline nature of our samples.

In fig. 3a,b the temperature variation of λ_t at the maximun field of 15 T for the whole series is depicted. Comparison with the $Y_2Fe_{14}B$ compound clearly shows that λ_t is essentially contributed by the RE^{3+} ion. Also the peculiar decreasing of λ_t at low tempeatures for some of the compounds can be a consequence of the mixture of different deformation modes (equation [2a]). However (see fig. 3c) the thermal variation of ω for the series is quite similar to the Y one, indicating a dominance of the Fe sublattices. Also the anomalous variation at high temperatures indicates a complex competition between the six Fe sublattices as well.

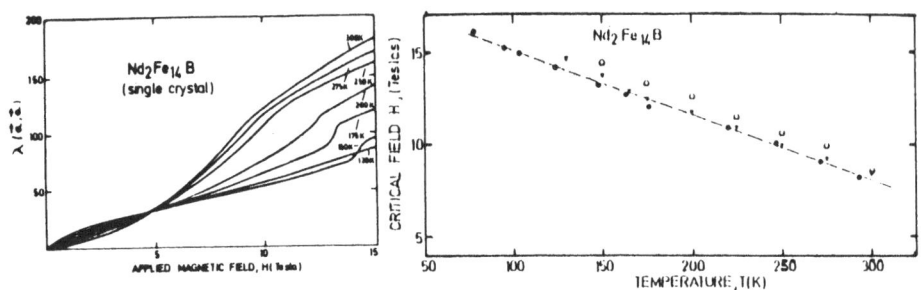

Fig. (4a) Magnetostriction isotherms for a $Nd_2Fe_{14}B$ single crystal. (4b) Thermal dependence of the critical field for the FOMP transitions. Full symbols obtained from magnetostrictions, open ones from Pareti et al., 1985.

$Nd_2Fe_{14}B$ Single crystal

Magnetostriction measurements, $\lambda(\alpha,\beta)$ have been performed on $Nd_2Fe_{14}B$ along the main symmetry directions. In fig. 4a selected isotherms for the different modes are depicted. The strong low temperature increase below

T_s=126 K is clearly related with the SRT, which produces an intermediate direction of easy magnetization. It is noteworthy the observation of a FOMP transition in the $\lambda(a,a)$ mode. In fig. 4b the temperature dependence of the critical field for SRT has been plotted being in good agreement with those values of the singular point detection (SPD) technique (Pareti et al., 1985).

MAGNETIC PHASE DIAGRAMS OF $(Er_xRE_{1-x})_2Fe_{14}B$ SERIES (RE= Nd,Dy)

As it has been well established for Er^{3+}, the Stevens factor, α_J is positive and has planar anisotropy. Therefore for Nd^{3+} and Dy^{3+}, $\alpha_J<0$, give rise to axial anisotropy. On the other hand the Fe sublattices give also rise to axial anisotropy. Due to the existence of competitive anisotropies, SRT processes take place in almost the whole series. In order to detect T_s, a.c. low field magnetic susceptibility has been measured at f=15 Hz (Alga-rabel et al., 1988). From figs. 5 and 6 we can observe that the substitu-tion of Er by Nd and Dy produces a decrease of T_s, as a consequence of the reinforcement of the axial anisotropy at low temperatures. In the $(Er_xNd_{1-x})_2Fe_{14}B$ series two peak anomalies have been observed, for the range $0.2 \le x \le 0.6$ being both associated with SRT's, at respectively T_{s1} (high) and T_{s2} (low) temperatures. At T_{s1} a SRT from axial (A) to cone-1 (C1) magnetic

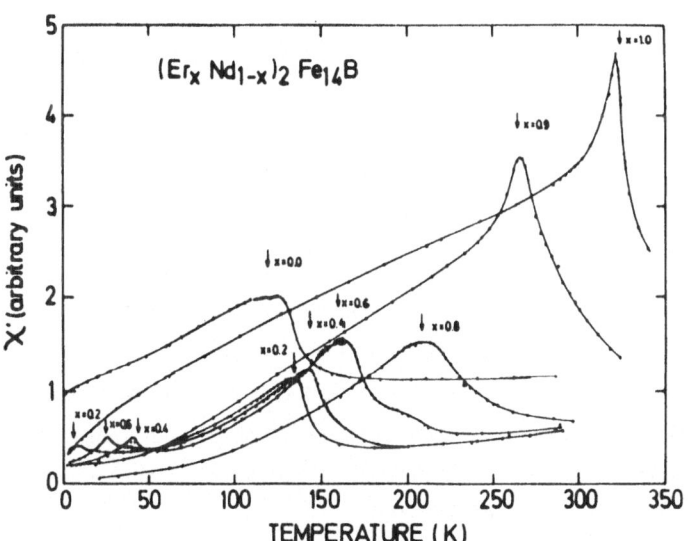

Fig. 5 A.c. magnetic susceptibility of $(Er_xNd_{1-x})_2Fe_{14}B$.

structures took place, and at T_{s2} from C1 to cone-2 (C2). Perpendicular and parallel magnetization measurements and also from the SPD technique, per-for-med on magnetically aligned powder samples, has been carried out in order to obtain the cone angle (see fig. 7). They have confirmed the exis-tence of such SRT transitions. In the case of the $(Er_xDy_{1-x})_2Fe_{14}B$ series, for $x>0.8$ a A→P SRT is observed, which becomes incomplete for $x\leq0.8$, where a cone phase is stabilized. The transition disappears for $x\leq0.5$ in good agreement with Mossbauer results (Rechenberg et al., 1987; Niarchos and Simopoulos, 1986). A full account of the magnetic phase diagram so deduced for both series is given in fig.8 a,b.

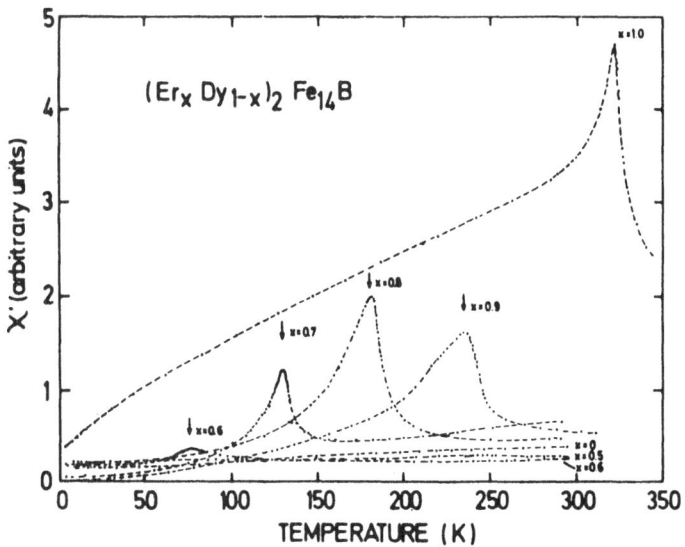

Fig. 6 A.c. magnetic susceptibility of $(Er_xDy_{1-x})_2Fe_{14}B$.

Outline of the CEF theory of SRT transitions

In order to explain the behaviour of SRT a single ion CEF model have been used to describe the RE contribution to the anisotropy energy and a molecular field confined into the x-z plane for the exchange interaction. Any position of the magnetic moment in the plane, has been characterized by the angle with the **c** axis, θ. Moreover it has been assumed that the exchan-ge coupling between effect ve moments for Fe, RE and RE' sublattices is stabilized, being collinear and either ferromagnetic, in the case of light rare, or ferrimagnetic for heavy ones.

The CEF Hamiltonian for the different RE ions, compatible with the mm symmetry, has been considered up to fourth order

$$H_{CEF} = B^0_2 O^0_2 + B^2_2 O^2_2 + B^0_4 O^0_4 \qquad [3]$$

where B^m_n are CEF parameters and O^m_n Stevens operators. The exchange Hamiltonian is given by

$$H_{ex} = -g_J \mu_B |H_{mol}| (J_z \cos\theta + J_x \sin\theta) \qquad [4]$$

where H_{mol} is the molecular field originated by the Fe which is related with the

exchange field, H_e, by $H_{mol} = [2(g_J-1)/g_J] H_e$. The f-d exchange is considered

constant for the series of isostructural compounds. For H_{mol} the values

given by Rechenberg et al., 1987, have been used with a temperature

dependence, $H_{mol}(T) = H_{mol}(0)(1-0.5(T/T_c)^2)$. J_z and J_x are operators

associated with the components of the angular momentum. Diagonalization of

the total Hamiltonian $H = H_{CEF} + H_{ex}$ within the ground manifold J and the use

of the partition fuction $Z(\theta, T)$ produce a free energy per RE^{3+} ion, $F(\theta, T) =$

$-kT \ln Z(\theta, T)$ which is written as

$$F(\theta, T) = x F_{Er}(\theta, T) + (1-x) F_{RE}(\theta, T) \qquad [5]$$

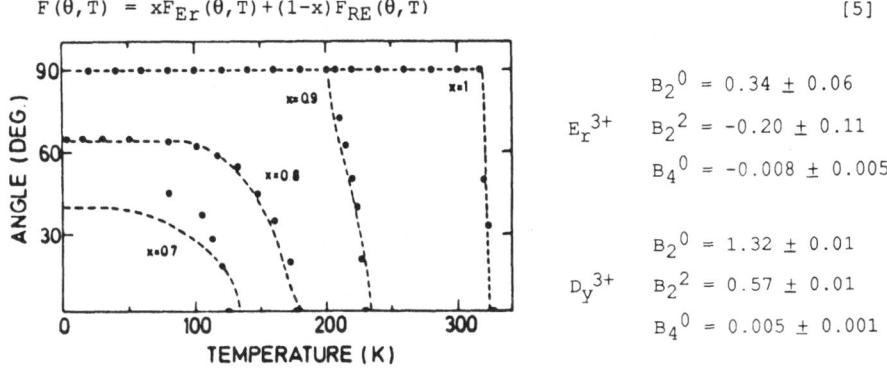

$$
Er^{3+} \quad
\begin{aligned}
B_2^0 &= 0.34 \pm 0.06 \\
B_2^2 &= -0.20 \pm 0.11 \\
B_4^0 &= -0.008 \pm 0.005
\end{aligned}
$$

$$
Dy^{3+} \quad
\begin{aligned}
B_2^0 &= 1.32 \pm 0.01 \\
B_2^2 &= 0.57 \pm 0.01 \\
B_4^0 &= 0.005 \pm 0.001
\end{aligned}
$$

Fig. 7 Temperature dependence of the SRT cone for $(Er_x RE_{1-x})_2 Fe_{14}B$ RE=Nd,Dy

In this way we are assuming that the Er and RE ions are randomly dis-

tributed over the two RE sites and also that the CEF parameters are effec-

tive ones for both sites. Later assumptions can be made in order to obtain

the values for the CEF in different 4f and 4g sites. The total free energy

for the system $F(\theta, T) = F_{RE}(\theta, T) + F_{Fe}(\theta, T)$ has been obtained considering the

contribution from the iron sublattice per RE ion,

$$F_{Fe}(\theta, T) = K_1(T) \sin^2\theta \qquad [6]$$

We have calculated for different temperatures the angular dependence of the total free energy and the minimum for that, gives the orientation for the Er, RE and Fe sublattice magnetic moments. This model has been applied, in order to explain the phase diagram of the serie Er-Dy explaining the thermal dependence, $\theta(T)$, as is showed in fig. 8a,b.

SRT CRITICAL BEHAVIOUR OF $(Er_xRE_{1-x})_2F_{14}B$

For $(Er_xDy_{1-x})_2Fe_{14}B$, $0.6 \leq x \leq 1.0$, a divergence of χ' at T_s (see figs. 5 and

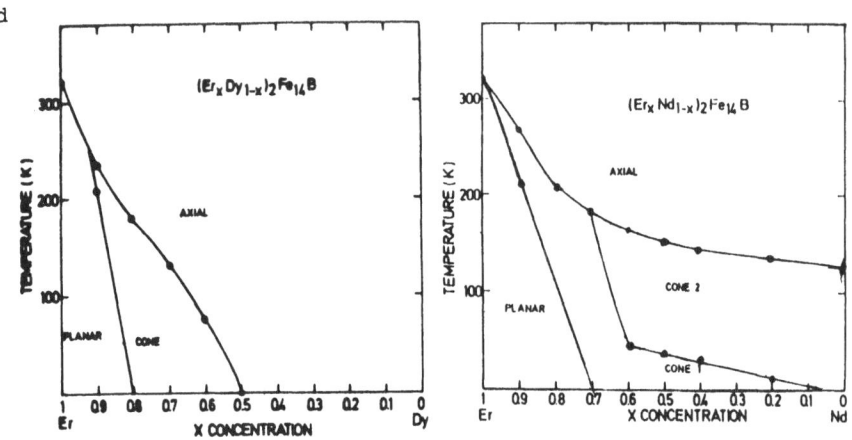

Fig. 8 Magnetic phase diagrams for $(Er_xRE_{1-x})_2Fe_{14}B$ RE= Dy(8a) and Nd(8b).

6) marks the SRT. For $0.9 \leq x \leq 1.0$ there are two SRT: $A \rightarrow C$, $C \rightarrow P$, and only one for the remainder. On the other hand (see figs. 8a and 8b) for $(Er_xNd_{1-x})_2Fe_{14}B$, $0.2 \leq x \leq 0.60$, two SRT: $A \rightarrow C1$, $C1 \rightarrow C2$, at T_{s1} and T_{s2}, respectively took place. For $x=0.6$ an angular point in $\theta(T)$ is observed at T_{s1} and T_{s2}.

Considering that the SRT order parameter is θ, a set of relations between critical exponents may be derived. For the field initial suscepti-bility $\chi_H = [\partial^2 F/\partial H^2]$ and considering a field parallel to the c-axis, $M = M_s\cos\theta$, and small θ

$$\chi_H = M_s \theta \chi_\theta \qquad [7]$$

where $\chi_\theta = (\partial\theta/\partial H)_T$ is the angular susceptibility, which should diverge at

T_{s1}. Now θ will be scale for H=0 calling as usual $t=(T-T_{s1})/T_{s1}$, and then

for $t\to0$

$$\theta \sim |t|^{\beta} f(H/|t|^{\beta\delta})\qquad\qquad[8]$$

β and δ being the critical exponents for $T>T_{s1}$, and $f(x)$ a scaling function.

The critical behaviour of χ_H will be

$$\chi_H \sim |t|^{-\omega'}\qquad\qquad(T<T_{s1})\qquad\qquad[9]$$

with the new exponent $\omega'=\gamma'-\beta$. For $T<T_{s1}$ $\beta=0$ and indeed $\omega=\gamma$.

For the SRT C1→C2, and for $T>T_{s2}$ the order parameter is $\varphi=\theta(T_{s2})-\theta$

which will scale as,

$$\varphi= |t|^{\beta} f(H/|t|^{\beta\delta})\qquad\qquad[10]$$

and for $t\to0$ it holds the more complex scaling,

$$\chi_H \sim |t|^{-\gamma} (k_1 + k_2|t|^{\beta} + k_3 |t|^{2\beta})\qquad\qquad[11]$$

where the second and third terms in the parenthesis are not important at

$t\to0$. For $T<T_{s2}$, θ, can be either constant for $(Er_xDy_{1-x})_2Fe_{14}B$ or $\theta=$

$\theta(T_{s2})+\varphi'$. Now the same scaling as [11] is obtained, but with an exponent

γ'.

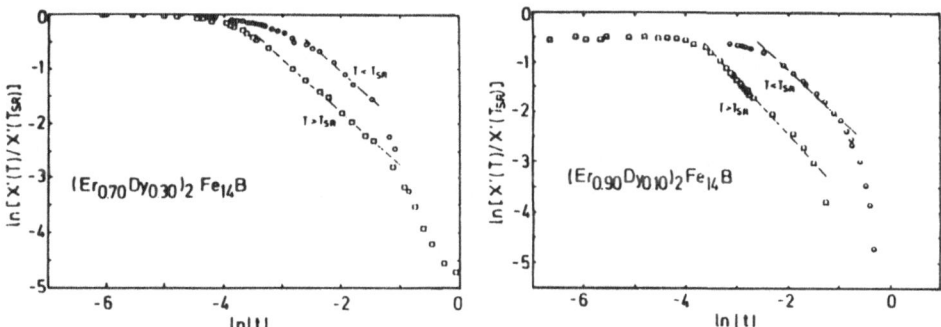

Fig. 9 Double logarithmic scaling of χ' for $(Er_{0.7}Dy_{0.3})_2Fe_{14}B$
$(T_{s1}=129.8$ K) (a) and $(Er_{0.9}Dy_{0.1})_2Fe_{14}B$ $(T_{s1}=235.0$ K) (b).

In order to show the critical behaviour of θ at T_{s1}, $\ln\chi_H$ vs $\ln|t|$ have

been ploted in figs. 9a,b. The part of χ_H showing anomalous behaviour is

superposed to a background due to the well known Hopkinson effect which has

been subtracted. As it can be observed, except for a region where χ_H is constant, limited by the demagnetizing field effects, shows good scaling and critical behaviour. The deduced critical exponents are shown in Table I.

TABLE I SRT transition temperatures and critical exponents.

	$(Er_xDy_{1-x})_2Fe_{14}B$			$(Er_xNd_{1-x})_2Fe_{14}B$					
x	T_{s1}	γ	ω'	T_{s1}	γ	ω'	T_{s2}	γ	ω'
0.				125.9	1.52	1.59	--	--	--
0.2				134.5	1.15	1.27	10.7	0.69	0.79
0.4				142.5	0.80	0.84	42.0	0.88	0.80
0.6	76.6	1.21	0.88	162.7	1.01	0.98	26.8	0.74	0.65
0.7	129.7	0.96	0.96						
0.8	181.1	1.02	1.04	206.7	0.89	1.16	--	--	--
0.9	235.0	1.11	0.94	267.1	1.31	0.90	--	--	--

In fig. 9a, we show the scaling of χ_H for x=0.40 in the   series $(Er_xNd_{1-x})_2Fe_{14}B$, at the A→C1 transition. The deduced critical exponents γ and ω' are displayed on Table I. On the other hand we show, in fig. 9b, the scaling at $T \sim T_{s2}$ (C1→C2) for some compounds of the series, where critical scaling is observed. γ and γ' values are displayed on Table I, where we should notice that $\gamma \approx \gamma'$.

A.C. SUSCEPTIBILITY OF POLYCRYSTALLINE $RE_2Fe_{14}B$ COMPOUNDS.

Compounds with RE= La, Ce, Pr, Nd, Gd, Dy, Ho, Er, Tm, Yb and Lu, as well as the Y one for comparative purposes, were measured. An exciting field amplitude h_0=1 Oe with frequency f=120 Hz and the temperature ranging between 77 to 300 K was used in all measurements.

Y,La,Ce,Gd,Lu

The orbital magnetic moment of these ions is L=0 and for Y, La, Ce and Lu the magnetic properties are due to the Fe sublattice exclusively. As expected, in the four compounds a continuous decay of χ' with the tempera- ture is observed. The χ' values at 100 K are; 40, 29, 28 and 30 emu/mol, for Y, La, Gd, and Lu respectivelly, which correspond very closely to the estimated value for a polycristalline sample of spherical particles when only demagnetizing effects are considered, $\chi'_{demag}=35$ emu/mol.

Two contributions are observed, $\chi'=\chi'_{dw}+\chi'_{rot}$, where χ'_{dw} is due to domain wall motions and χ'_{rot} is ascribed to reversible rotations of the moments excited by the small a.c. external field. For small internal stress there is a negligible pinning of the domain walls and the measured χ' value is mainly the inverse demagnetization factor. Since it depends exclusively on the geometry, one needs to consider χ'_{rot} to account for the temperature dependence. Indeed, χ'_{rot} depends as the inverse of the internal anisotropy field, H_A. This field (Hirosawa et al.,1986) show a gradual increase with the temperature, thus explaining the observed decreasing of χ'.

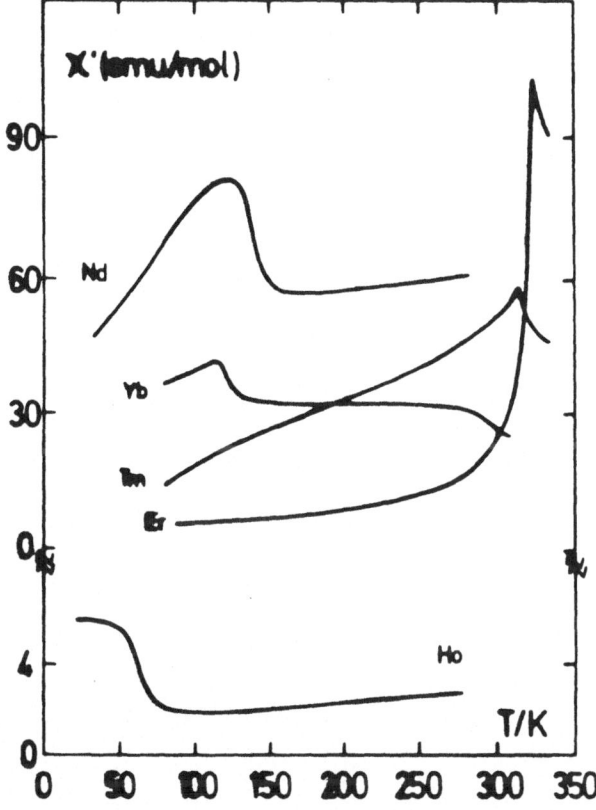

Fig.10 A.c. magnetic susceptibility of $RE_2Fe_{14}B$ for RE= Nd(135 K), Er(320 K), Ho(58 K), Tm(312 K), Yb(112 K). Numbers in brackets are the SRT temperatures.

For the Ce compound a small anomaly (2% of χ') was observed at 230 K which is probably due to some impurity in the sample, since no peculiar behavior is known in H_A. Our observations are consistent with the published results (Grossinger et al., 1986).

Pr, Dy

Below T_C, the strong CEF anisotropy of the RE atoms induce the magnetization allignment in the [001] direction. For both compounds the χ' measurements do not show any anomaly.

As above, the analysis of χ' in two terms is possible, but some differences are noticeable. The stronger increase of χ' with temperature and its concavity may be explained as due to the stronger decrease of H_A. The rather low χ' values below 100 K, (Pr: $\chi'=10$, Dy: $\chi'=3$ emu/mol) would indicate a lower mobility of the domain walls due to the rather large value of H_A (Hirosawa et al., 1986).

Er, Tm, Yb.

Cusp like SRT anomalies are detected in these compounds, at $T_s=(320\pm1)K$, $(312\pm1)K$ and $(112\pm1)K$, respectively (see fig. 10). Though the anomalies of the Er and Tm were rather sharp no hysteresis effect were observed. In contrast to all other members of the series, $\alpha_J>0$, consequently the SRT from [001] to the basal plane is caused by the competing RE and Fe anisotropies. The molecular field theory previously outlined gives values in agreement (within 10%) with the experimental T_s.

The magnetization SRT anomaly as a function of temperature for poly-crystalline samples has been computer simulated (Boltich et al., 1987). One may compare satisfactorily the present results with their model of a grain containing two domains with different easy axis of magnetization. The observed cases for Er, Tm and Yb would correspond to a second order aniso-tropy $K_2\leq0$, with the transition taking place for the temperature at which $K_1=0$.

Nd, Ho

Both compounds undergo a SRT to a conical orientation. They have similar Curie temperatures T_C=586 K and 573 K for the Nd and Ho respectively

but have different SRT temperatures (inflexion points in χ' at T_S=(135±1)K

and (58±1)K for the Nd and Ho respectively).

The measurements of polycrystalline bar shaped samples are depicted in

fig. 10, both show a rounded step-like anomaly in χ', with an inflexion

point almost at T_S and onset temperatures T_O=(144±1)K and (75±1)K for the Nd

and Ho compounds respectively. Following the simulations (Boltich et al.,

1987), the measured curves mimic the predictions for a disordered powder

with $K_2>0$ in both cases. Indeed, in the Nd compound K_2 has been derived to

be positive in the temperature range scanned. The conclusion of the

mentioned simulations is to ascribed the SRT transitions to the inflexion

point T_S rather than the onset T_O, For Nd, χ'= 80 emu/mol at 100K, is higher

than the estimated inverse demagnetization factor for grains considered as

spheres, thus the bar shape is probably causing an increased response in

this compound. For Ho, the low value, χ'=2 emu/mol at 100 K, is of the order

of the χ'_{rot} contribution and may be caused by a prefered orientation of the

grains in the bar shaped sample or a low mobility of magnetic domains as has

been observed in the single crystal.

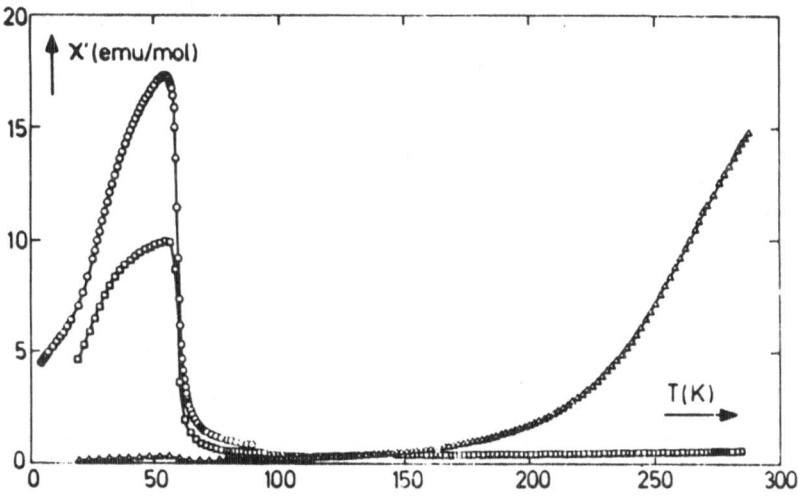

Fig.11 Magnetic susceptibility of $Ho_2Fe_{14}B$ single crystal with
the field applied in the [110](circles), [100](squares), [001]
(triangles) directions.

$Ho_2Fe_{14}B$ Single crystal.

A single crystal polished to a sphere of 2 mm of diameter, provided by Sumitomo S.M.C., was measured along three directions, [100], [001] and [110], between 4.2 and 300 K (see fig. 11). In the basal plane directions, the experimental results resemble the polycrystalline data; a step-like anomaly with inflexion point at $T_s = (57.6 \pm 0.5)$K. The anomaly at T_s is much more abrupt, as expected for a single crystal. Below T_s a continuous decay is present, at which inflexion point the imaginary component shows a maximum. The height of the maximum detected in the [110] direction is nearly a factor of two higher than along the [100] one. In contrast, along the [001] no anomalous feature is observed at T_s, while a strong continuous increase starts at 150 K up to room temperature. An increase in the complex component, χ'', appears simultaneously to χ'.

Evidently, the step at T_s is due to the disalignment of the magnetic moments from the c direction, giving a non-zero proyection in the basal plane, while in the easy axis the perturbation is too small to be detected. Considering the reversible displacement of magnetic domains and the reversible rotation of magnetic moments inside the domains the data are nicely fitted (Rillo et al., 1988).

HYDRIDED POLYCRISTALLINE COMPOUNDS $RE_2Fe_{14}BH_x$.

The uptake of hydrogen provokes important modifications in the magnetic properties of the $RE_2Fe_{14}B$ compounds (Fruchart et al 1988). T_C increases with x at rates depending on the particular RE atom. Three series have been studied RE=Pr, Er and Gd. The phase diagram of Pr and Er together with available data on the Ho (Regnard et al., 1987) and Nd (Oesterreicher and Oesterreicher, 1984; Pourarian et al., 1986) are depicted in fig. 12.

For the Pr non-hydrided compound no anomalous behaviour was oberved. However, for $x \geq 1$ a clear maximum arises in χ' which is ascribed to a SRT induced by hydridation as proposed by Pourarian et al., 1986. At higher temperatures a steady increase in χ' is detected, accompanied by a maximum in the magnetic energy absorption, χ''. This second effect is probably due to the onset of domain wall motions permitted by the decreased magnetic hardness caused by hydridation. For all the Er hydrides, a clear maximum similar to the SRT one for the pure compound was observed. Only the concentration x =0.6 shows a second maximum probably caused by small amount (10%) of unreacted pure compound in the sample.

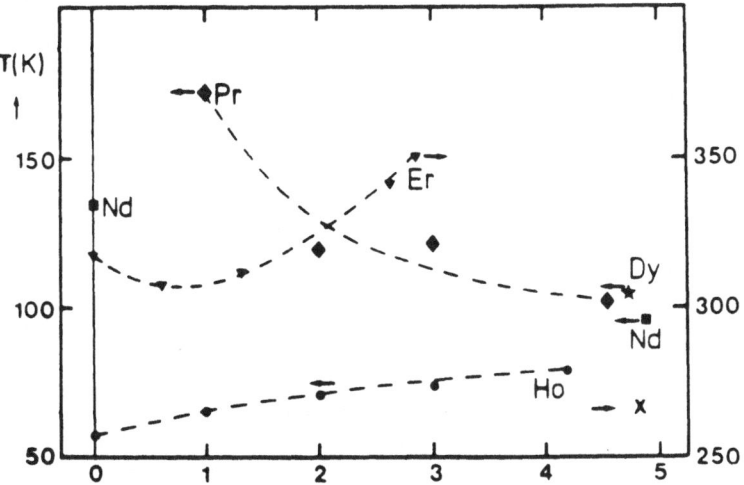

Fig.12 The spin reorientation temperature of the Pr, Nd, Dy, Ho and Er compounds versus the number of absorved hydrogen atoms.

As for the Pr, no anomalous behaviour is observed in the Gd non-hydrided compound. For $x \geq 1$ several anomalies in χ' with acompaning maxima in χ'' are detected for each compound of the series (see fig. 13). The broad maxima observed in χ' for x=3 and x=3.5, with inflexion points at T_S=(296±1)K and (258±1)K respectively, may be ascribed (Zhang et al.,1988) to SRT induced by hydrogen insertion. Those authors reported an SRT anomaly for x=3.8 which is above the highest temperature (350 K) of our χ data. For $T<T_S$ other anomalies are detected in χ' and χ'' for x =3.8, 3.5 and 3 indicative of a more complicated phase diagram than that given by Zhang et al., 1988. The same holds for x=2 and 1 for which no SRT has been reported.

The results given in fig. 12 have been analysed in terms of the same model succesfully applied to explain the SRT temperatures in the pure compounds (Cadogan, 1987). The following conclusions could be drawn:

The induction of SRT in the Pr and Dy compounds for $x \geq 1$ cannot be caused by the estimated increase of the molecular field constant between RE and Fe, H_{ex}. On the other hand, an increase of the B_4^0 by a factor of six (Pr) and three (Dy) suffice to give rise to the SRT's observed for $x \geq 1$. This result is in agreement with the observed increase in the K_2 constant, which is proportional to B_4^0, and the related decrease of K_1.

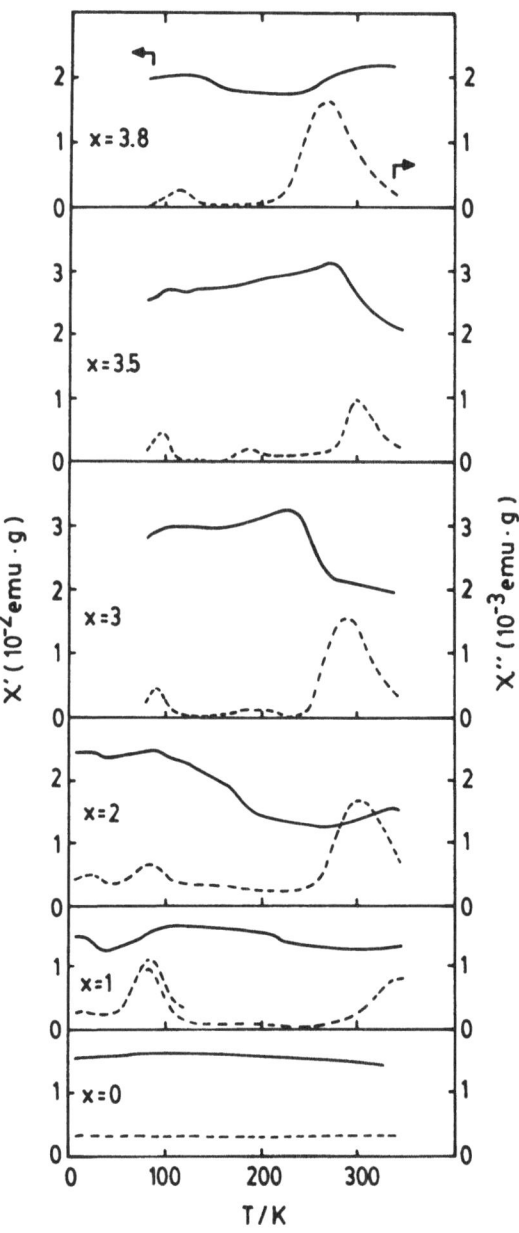

Fig.13 A.c. magnetic susceptibility of $Gd_2Fe_{14}BH_x$ for x=0, 1, 2, 3, 3.5 and 3.8.

The increase in SRT temperature observed for the Ho compound is again caused by an increase in B_4^0 which cannot be countered by the reduction of H_{ex} for x≥1, while in the Nd case it is capable of reducing T_s. Once the SRT is induced by hydridation on the Pr compound, the same behaviour as for the Nd compound is encountered. One may say that for the compounds with $\alpha_J<0$ hydridation provokes an increase in the K_2 anisotropy term (increase in $B_4{}^0$) and a related decrease in K_1 which tend to increase the SRT temperature. Besides, the variations in H_{ex} produce a proportional effect on T_s. The interplay of both effects would yield the observed SRT variation.

The Er series is the only case with $\alpha_J>0$ in which the effect of H content on T_s has been studied so far. For the pure compound $T_s=320$ K which is much higher than in the other studied cases. At this temperature the variations of K_2 affect much less the value of T_s. On the contrary, the anisotropy of the Fe sublattice $K_1(Fe)$ can be expected to be very temperature dependent similarly to the behaviour of $Y_2Fe_{14}B$. Another effect to be considered is the decrease in the absolute value of $K_1(Fe)$ with hydridation (Coey et al., 1986). The calculations show an increase in the SRT in spite of the decreasing RE-Fe interaction deduced from the T_c variation for $x>1$. However, this model has not been able to explain the T_s dip observed for $x=1$, a fact which remains contradictory.

Due to the S state of Gd in this series there is no CEF interaction and the SRT is ascribed to different competitive anysotropy contributions of the Fe sublattices with different temperature dependences affected by hydrogenation (Zhang et al., 1988). Our χ' data show typical second order SRT anomalies ($K_2>0$) from the easy axis to an easy cone as in the case of Pr, in contrast with a first order easy axis to easy plane suggested by Zhang et al., 1988. The rich phenomenology observed upon hydrogenation makes evident the necessity of systematic studies in Gd and other RE hydrides.

REFERENCES

Algarabel P.A., del Moral A., Ibarra M.R., Arnaudas J.I. 1988. J. Phys. Chem. Solids, 49, 213.

Boltich E.B., Pedziwiatr A.T., Wallace W.E. 1987. J. Magn. & Magn. Mat. 66, 317.

Buschow K.H.J., Grossinger R., Kirchmayr H.R. 1988. J. Magn.& Magn. Mat. in press.

Cadogan J.M., Coey J.M.D., Gavigan J.P., Givord D., Li H.S. 1987. J. Appl. Phys., 61, 3974.

Callen E.R. and Callen H.B. 1965. Phys. Rev. 139, A455.

Coey J.M.D., Yaouanc A., Fruchart D. 1986. Solid. St. Comm. 58, 413.

Fruchart D., Pontonnier L., Vaillant F., Bartolomé J., Fernández J.M., Regnard J.R., Yaouanc A., Fruchart R., L'Heritier Ph. 1988. I.E.E.E. Transactions on Magnetism in press.

Givord D., Li H.S. 1985. in "Nd-Fe-B permanent magnets", I.V. Mitchell editor (Elsevier-London).

Grossinger R., Sun X.K., Eibler R., Buschow K.H.J., Kirchmayr H.R. 1986. J. Magn. & Magn. Mat. 58, 55.

Hirosawa S., Matsuura Y., Yamamoto H., Fujimura S., Sagawa M., Yamauchi H. 1986. J. Appl. Phys., 59, 873.

Hirogoshi H., Kato H., Yamada M., Saito N., Nakawa Y., Hirosawa S., Sagawa M. 1987. Sol. St. Comm., 62, 7.

Ibarra M.R., Algarabel P.A., Alberdi A., Bartolomé J., del Moral A. 1987. J. Appl. Phys. 61, 3451.

Ibarra M.R., Marquina C., Algarabel P.A., Arnaudas J.I., del Moral A. 1988. submited to Sol. St. Comm.

Niarchos D. and Simopoulos 1986. Sol. St. Comm. 59, 669.

Oesterreicher K. and Oesterreicher H. 1984. Phys. Stat. Sol. (a) 85, K61.

Pareti L., Bolzoni F. and Moze O. 1985. Phys. Rev. B6, 3515.

Pourarian F., Huang M.Q., Wallace W.E. 1986. J. Less-Comm. Met. 120, 63.

Rechenberg H.R., Sanchez J.P., L'Heritier Ph., Fruchart R. 1987. Phys. Rev. 36B, 1865.

Regnard J.R., Yaouanc A., Fruchart D., Le Rouux D., L'Heritier Ph., Coey J.M.D., Gavigan J.P. 1987. J. Appl. Phys. 61, 3565.

Rillo C., Chaboy J., Navarro R., Bartolomé J., Fruchart D., Yaouanc A., Chenevier B., Sagawa M., Hirosawa S. (1988) submitted to J. Appl. Phys.

Sagawa M., Fujimura S., Togawa N., Yamamoto H., Matsuura Y. 1984. J. Appl. Phys. 55, 2083.

Sinnema S., Radwanski R.J., Franse J.J.M., de Moij D.B., Buschow K.A.J. 1984. J. Magn. & Magn. Mat, 44, 333.

Zhang L.Y., Pourarian F., Wallace W.E. 1988. J. Magn. & Magn. Mat. 71, 203.

SUMMARY OF DISCUSSIONS :

MAGNETIC PROPERTIES AND CHEMICAL SUBSTITUTION

J.J.M. Franse, H.R. Kirchmayr

The contributions to this session report progress in understanding the basic properties of $Nd_2Fe_{14}B$ and related compounds as well as the effects of chemical substitutions on magnetic moment, Curie temperature and anisotropy energy, as well as the effects of hydrogenation. Progress concerning the basic properties of the $Nd_2Fe_{14}B$ compound relates to the magnetism of the 3d and the 4f sublattices as well as the magnetic interactions between the two sublattices.

It is worth remarking that interest in 3d magnetism received added momentum with the discovery of new classes of 3d-4f intermetallic compounds that are of practical use. This revival of interest in 3d magnetism followed a period when experimentalists and theoreticians could not agree about some basic concepts concerning magnetic ordering and the persistence of itinerant moments at temperatures well above the magnetic ordering temperature. The systematic study of the iron moment and the Curie temperature in a series of yttrium-iron compounds with different compositions reveals some interesting phenomena. The atomic moment in this series can be interpreted rather well in terms of a band model with a transition from weak ferromagnetism for α-iron towards strong ferromagnetism for several of the Y-Fe compounds. The strength of the 3d-3d interactions, represented by the value for the Curie temperature, decreases with increasing Y content. The large reduction of Curie temperature with Y concentration cannot be explained by the magneto- volume effect which is known to be extremely large in the 2:14:B compounds. It was suggested by Givord et al that the reduction in the T_c values is related to the increase in the iron coordination number. In a plot of T_c/μ^2_{Fe} vs the iron coordination number, all Y-Fe compounds are situated on a single straight line, including α-Fe and γ-Fe.

Progress in the understanding of the anisotropy of the 3d sublattice is still rather limited. Some efforts have been reported in determining the anisotropy contributions of the six individual 3d sites by a study of the anisotropy in pseudoternary compounds.

The magnetism of the rare-earth elements in the 2:14:B compounds is determined by the crystalline electric field and the exchange field produced by the 3d species. A unique set of crystal-field parameters has been deduced which represents the magnetization process in almost all 2:14:B compounds studied so far. This successful approach was for a large part realised by performing high field studies on single-crystal samples. The European facilities for high magnetic fields in Amsterdam, Grenoble and Parma, as well as the single-crystal growth equipments in Amsterdam and Grenoble, played an essential role in these studies. These results must be considered as one of the achievements of the C.E.A.M. programme. The problem encountered interpretating effective anisotropy constants has also been resolved.

Values for the exchange field obtained by different groups now agree well, after some initial problems related to the definitions of the molecular field coefficient n_{RT} and the exchange parameter J_{RT}. The 2:14:B compounds with the heavy rare-earth elements can be described quite well with a value of approximately -7 K for this latter parameter a value about a factor of two larger for the high-rare-earth elements is found. Although some refinement in the description of the magnetization process is certainly needed, it can safely be said that the main parameters that govern the magnetization process are well established.

By the study of chemical substitutions, the variations of such magnetic parameters as the spontaneous magnetization, the magnetic anisotropy and the Curie temperature are now well documented. Substitutions offer the possibility of investigating the magnetic moment and the anisotropy contributions of the different crystallographic sites, provided that the site preferences are known from neutron, x-ray or Mössbauer measurements. Substitutions on the rare- earth sites in the $Nd_2Fe_{14}B$ compound have been exploited in order to study the critical behaviour of magnetic structure transitions.

In conclusion we mention the hydrogenation studies of the 2:14:B compounds. On inserting hydrogen atoms in interstitial sites the lattice parameters increase considerably, augmenting in this way the Curie temperature and the 3d magnetic moment, but decreasing the anisotropy energy. Hydrogen absorption is of special importance for the preparation of fine particles by a decrepitation process.

SECTION III

— MATERIALS

CHAPTER 3

ATOMIC SCALE MAGNETISM

N.M.R. STUDY OF INTERMETALLIC COMPOUNDS FOR PERMANENT MAGNETS

Y. Berthier

Laboratoire de Spectrométrie Physique (associé au C.N.R.S.)
Université Joseph Fourier Grenoble I
B.P. 87, 38402 Saint-Martin d'Hères Cédex, France

ABSTRACT

We report several investigations by N.M.R. of the hyperfine interactions in intermetallic compounds, which are or may be suitable for permanent magnet applications. i) Study of hyperfine interactions in $R_2Fe_{14}B$ compounds with R = Nd, Lu and Y. ii) Study of the cobalt substitution by iron in the $Nd(Co_{1-x}Fe_x)_9Si_2$ compounds. iii) Study of the 3d sublattice anisotropy in $Y_2(Fe, Co)_{14}B$ compounds.

HYPERFINE INTERACTIONS IN $R_2Fe_{14}B$ COMPOUNDS (R = Nd, Lu, Y)

(Collaboration with D. Givord, Group 1.01)

The aim of this work was the determination of the different components of the hyperfine fields on most of the atomic sites (magnetic and non magnetic) in these materials, in order to deduce the values of the magnetic moments of the rare earths and of the iron atoms and also to obtain a better understanding of their respective interactions.

In the compound $Nd_2Fe_{14}B$, the spin echo N.M.R. spectrum of ^{143}Nd (I = 7/2) was measured at 1.4 K (Berthier et al, 1986). Two set of seven broadened lines corresponding to the magnetic and the quadrupolar hyperfine interactions on the two rare earth sites were obtained as shown in Fig. 2. For each site the total hyperfine field measured on ^{143}Nd contained three components : the 4f electrons field H_{4f} which is proportional to the magnetic moment of Nd, the transferred field from iron atoms H_{nn}^{Fe} and the transferred field from rare earth neighbours H_{nn}^R. These two latter contributions were obtained from N.M.R. of ^{175}Lu (I = 7/2) in $(Nd_{.9}Lu_{.1})_2Fe_{14}B$ and $Lu_2Fe_{14}B$ compounds (see fig. 3) in which lutetium is non magnetic. In $Y_2Fe_{14}B$, two intense lines corresponding to ^{89}Y in sites 4g and 4f (Fig. 1) give also a direct measurement of the transferred hyperfine field from iron neighbours. The hyperfine parameters and the effective hyperfine field are summarized in table 1. Accounting for the ratio of the s electrons hyperfine constants, tabulated by Campbell (1969), the values of $H_{nn}^{Fe}(Nd)$ deduced from both lutetium and yttrium compounds are in perfect agreement. The different contributions to the hyperfine field on ^{143}Nd at the two rare earth

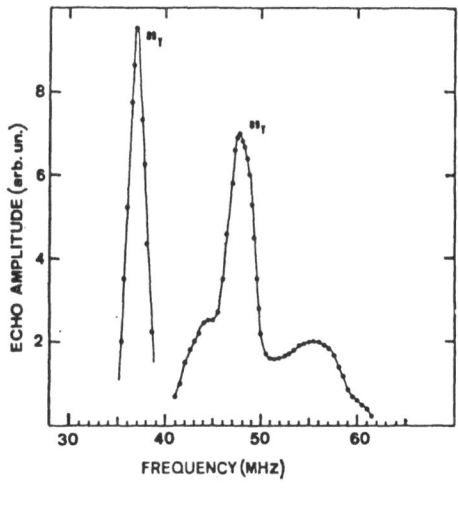

Fig. 1   Zero field N.M.R. of
^{89}Y in Y$_2$Fe$_{14}$B at 4.2 K.

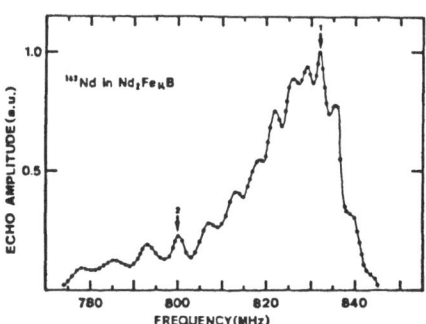

Fig. 2   Zero field N.M.R. of
^{143}Nd in Nd$_2$Fe$_{14}$B at 1.4 K.
(Arrows indicate the $-1/2 \rightarrow 1/2$
transition corresponding to the
sites 4f and 4g).

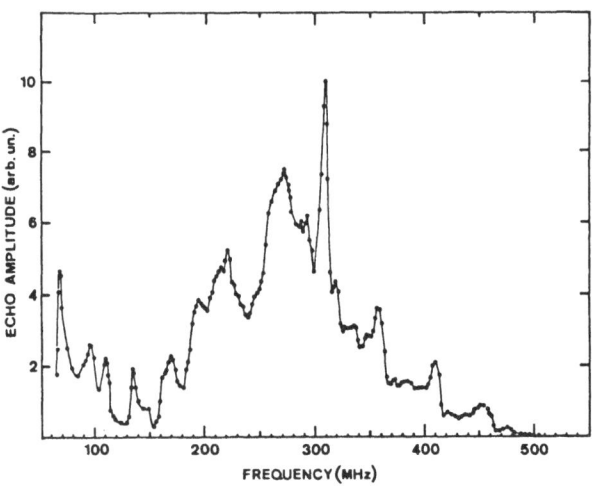

Fig. 3   Zero field N.M.R. of ^{175}Lu in Lu$_2$Fe$_{14}$B at 4.2 K.

sites, and the corresponding magnetic moment of Nd ions are reported in table 2.

TABLE 1   Hyperfine parameters : $\mathcal{H} = h\left\{a_t I_z + P_t\left[I_z^2 - \frac{1}{3} I(I + 1)\right]\right\}$

compounds	site	nucleus	a_t (MHz)	P_t (MHz)	H_{eff} (kG)
$Nd_2Fe_{14}B$	4f	^{143}Nd	832	1.5	3570
" "	4g	^{143}Nd	800	4.0	3434
$Lu_2Fe_{14}B$	4f	^{175}Lu	220	25	458
" "	4g	^{175}Lu	310	25	646
$(Nd_{.9}Lu_{.1})_2Fe_{14}B$	4f	^{175}Lu	216	20	450
" " "	4g	^{175}Lu	304	20	626
$Y_2Fe_{14}B$	4f	^{89}Y	37	–	177.4
" "	4g	^{89}Y	48	–	230

TABLE 2

Site	Symmetry	H_{eff} (kG)	H_{nn}^{Fe} (kG)	H_{nn}^{Nd} (kG)	H_{4f} (kG)	$\mu(\mu_B)$ RMN	$\mu(\mu_B)$ neutron
Nd_1	4f	3570	– 360	+ 12	3930	3	2.30
Nd_2	4g	3434	– 465	+ 12	3899	2.97	2.25

The component H_{nn}^R is rather small (12 kG) and reflects the weak interaction between rare earth atoms. The calculated moments for the two rare earth sites are derived from the μ/H_{4f} proportionality and the free-ion hyperfine field value of 4280 kG corresponding to $\mu = 3.27$ μ_B (see Bleaney, 1972). Although these moment values are somewhat larger than those obtained from polarized neutron diffraction (Givord et al, 1985), the ratio of their values for the two sites is the same. Furthermore, one must keep in mind that below 150 K, the spin canted structure of $Nd_2Fe_{14}B$ (E.B. Boltich et al, 1985) implies that polarized neutrons diffraction is measuring only the projection of the Nd moment along the direction of magnetization. Finally the very short distances between the rare earth site 4g and the iron atoms at site 4c and 16 k can account for the increase of H_{nn}^{Fe} on this site.

REFERENCES

Berthier, Y., Boge, M., Czjzek, G., Givord, D., Jeandey, C., Li, H.S. and
    Oddou, J.L. 1986. J. Magn. Magn. Mat., $\underline{54-57}$, 589.
Bleaney, B. 1972. "Magnetic Properties of Rare-Earth Metals (Plenum, New
    York). Chapt. 8.
Boltich, E.B., Oswald, E., Huang, M.Q., Hirosawa, S., Wallace, W.E. and
    Burzo, E. 1985. J. Appl. Phys., $\underline{57}$, 4106.
Campbell, I.A. 1969. J. Phys. C $\underline{2}$, 1338.
Givord, D. and Li, H.S. 1985. J. Appl. Phys., $\underline{57}$, 4100.

N.M.R. STUDY OF $Nd(Co_{1-x}Fe_x)_9Si_2$ COMPOUNDS

(Collaboration with B. Chevalier, Group 1.06 and J.P. Sanchez, Group
1.10).

This work was undertaken in the frame of a systematic investigation of
the Nd-Co(Fe)-Si system in the 3d metal rich side of the pseudo-ternary
phase diagram. All these compounds are isostructural with the tetragonal
$NdCo_9Si_2$ phase in which the transition metal atoms occupy two different sites : 32(i) and 4(b) while neodynium and silicon atoms are located in 4(a)
and 8(d) positions, respectively. Each transition metal M has 8M as nearest
neighbours but the M-M distances for the two sites are sensibly different.
All these compounds order "ferromagnetically" with Curie temperature in the
range 500-560 K for $0.11 < x < 0.5$. We have combined a detailed N.M.R.
(^{59}Co, ^{143}Nd) and Mössbauer (^{57}Fe) studies with the macroscopic data in order to determine the local 3d and 4f moment properties and to provide better understanding of their coupling mechanism (Berthier et al, 1988).

^{59}Co hyperfine interactions

Zero field spin echo N.M.R. spectra of ^{59}Co in $Nd(Co_{1-x}Fe_x)_9Si_2$ with
$x = 0.11$, 0.22, 0.33 and 0.44, measured at 4.2 K are shown in Fig. 4. The
resolved satellite lines observed as a function of x, correspond to the hyperfine field distribution (HFD) at cobalt nuclei, which depends on their
local environment. Two criteria were used in the site assignment : i) the
relation of the spin echo intensity to the site population ; ii) the consis-
tence of the frequency shift for each site, with the local environment. The
low frequency range (150-180 MHz) of the spectra corresponds to the small
number of Co atoms at the 4b sites $(Co(1))$. For frequencies larger than
180 MHz, a set of lines with larger intensities were attributed to the HFD
arising from the increase of the number of iron next neighbours as seen by
cobalt atoms at the 32i sites $(Co(2))$. A comparison of the relative

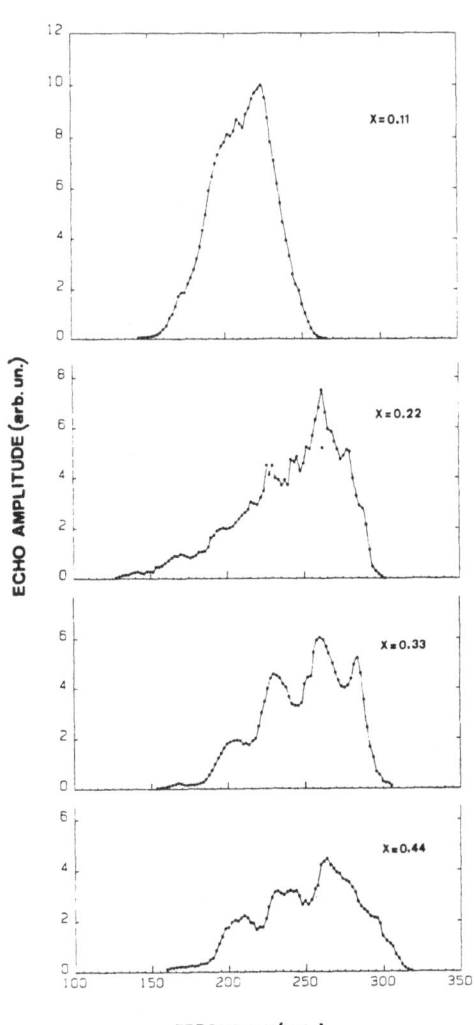

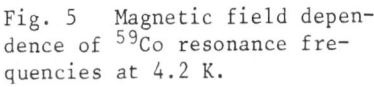

Fig. 4   Zero field N.M.R. of ^{59}Co in $Nd(Co_{1-x}Fe_x)_9Si_2$ at 4.2 K.

Fig. 5   Magnetic field dependence of ^{59}Co resonance frequencies at 4.2 K.

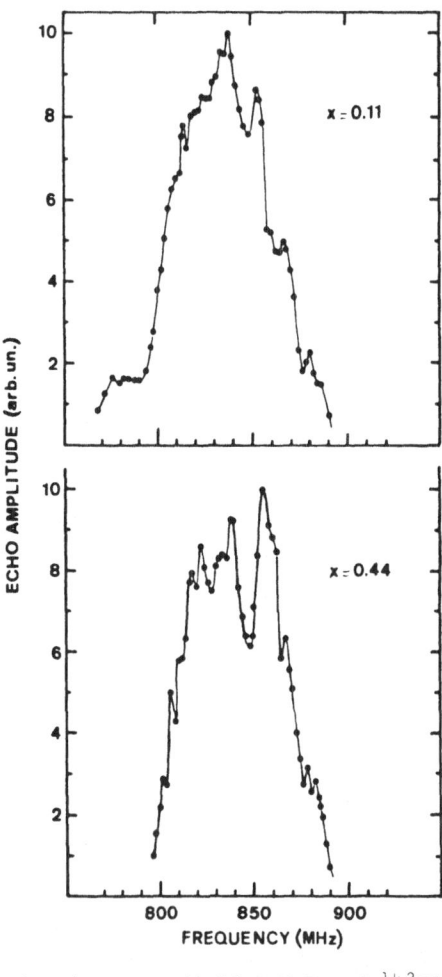

Fig. 6   Zero field N.M.R. of ^{143}Nd in $Nd(Co_{1-x}Fe_x)_9Si_2$ for x = 0.11 and x = 0.44 at 4.2 K.

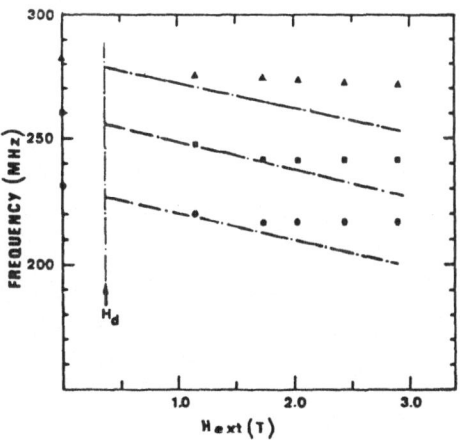

intensities of these lines to a binomial distribution of the eight iron next neighbours around Co(2) atoms illustrates a deviation from a random distribution of iron substituting the cobalt. This underlines the repulsive interaction between iron atoms when their concentration increases. The large dispersion (from 2.47 Å to 2.73 Å) of the distances between Co(2) atoms and their eight first neighbours is probably the cause of such an order in the iron substitution.

A preferential order in the Co(1) sites substitution by iron is also suggested by : i) the ratio between the respective intensities of the Co(2) and Co(1) lines which is lower than 32/4 expected from the crystallographic structure, and ii) by the extra line with low intensity which appears at 248 MHz for all concentration. This line corresponds to the Co(1) neighbours of Co(2) atoms which are substituted by iron as soon as $x = 0.11$.

A quantitative analysis of the cobalt moment was made using the proportionality between H_{eff} and the local Co spin moments (Hirosawa et al, 1982) given by :

$$H_{eff} = \alpha \, \mu_{Co}^{spin} \quad \text{with} \quad \alpha = - 130 \text{ kOe}/\mu_B \quad (1)$$

The local moments corresponding to the different lines were then analyzed as the sum of two contributions :

$$\mu_i \, (Co) = \mu_0 + \Delta\mu(i) \quad (2)$$

where μ_0 represents a constant value (1.54 μ_B) corresponding to Co atoms with eight first neighbours, and $\Delta\mu(i)$ represents the increase of the moment for Co with i first neighbours irons. $\Delta\mu(i)$ is not linear in i, as predicted by the simple model of Jaccarino and Walker (1963). For each sample we have calculated the mean value of the cobalt sublattice effective moment using the sum :

$$\bar{\mu} \, (Co) = \sum_i{}' a_i \, \mu_i \quad (3)$$

where the a_i represent the intensities of the different lines assigned to sites 1 and sites 2, and where μ_i are the local moments of Co deduced from the frequencies of the lines. These values are reported in table 3 with those obtained from the difference between the total moment measured from magnetization and the averaged moment of iron derived from Mössbauer experiment : $\overset{\sim}{\mu} \, (Co) = \mu_t - \bar{\mu} \, (Fe)$.

x	$\tilde{\mu}$ (Co) (μ_B)	$\bar{\mu}$ (Co) (μ_B)
0.11	0.90	1.62
0.22	0.96	1.89
0.33	1.04	1.915
0.44	1.14	1.915

TABLE 3

The magnetic field dependence of the resonance frequencies of ^{59}Co spectrum was studied in a $NdCo_6Fe_3Si_2$ sample. The nearly constant experimental values measured for the three most intense lines (see fig. 5) in an external field up to 3 T indicate that the hyperfine fields corresponding to these sites are not colinear to the applied field. Consequently, Co moments are not colinear with the total magnetization.

^{143}Nd hyperfine interactions

Zero field spin echo N.M.R. spectra of ^{143}Nd were measured in samples with x = 0.11 and x = 0.44 at 4.2 K. Figure 6 shows for each spectrum a pattern of seven broadened lines corresponding to the magnetic and the quadrupolar hyperfine interactions on the rare earth site 4(a). The hyperfine parameters derived from these spectra, show that for the two extreme iron concentration, the Nd magnetic moment is fully saturated (μ_S = 3.27 μ_B). As concerns the quadrupole hyperfine parameters, we note that for the two samples $|P_t|$ is larger than the free iron value. This extra ionic contribution to P_t can be due either to a lattice distorsion or more likely to an unusual anisotropy of the conduction electron polarization around the rare earth.

Discussion and conclusion

Our N.M.R. data confirm a preferential occupancy of sites (4b) by iron as soon as x reaches 0.11. They also indicate a not completely random distribution of iron at site (32i). These two preferential orders have probably the same origin ; iron atoms prefer the shorter 3d M-M sites. This means that for the (32i) 3d sites, whose first neighbours distances d_{nn} are distributed over a large range, iron atoms replace cobalt atoms at sites with the shortest d_{nn}.

The cobalt moment measured by N.M.R. increases monotically with x. This behaviour is the result of a cooperative effect which involves the number of Co and Fe atoms in their local environment in addition to the rare earth magnetic contribution. The contribution of the latter is constant as the neodynium ions remain in a fully saturated state for any concentration. Finally the data reported in table 3 are coherent with the anomalous behaviour of ^{59}Co hyperfine field under an external applied field if we suppose that the Co moments lie along a direction with an angle $\theta = 55°$ with respect to the principal magnetization axis. This canted magnetic structure is not surprising if the exchange interaction between iron and cobalt atoms is lower than the local anisotropy.

REFERENCES

Berthier, Y., Chevalier, B., Etourneau, J., Rechenberg, R. and Sanchez, J.P. 1988. To be published in J. Magn. Magn. Mat.
Hirosawa, S. and Nakamura, Y. 1982. J. Magn. Magn. Mat., 25, 284.
Jaccarino, V. and Walker, L.R. 1963. Phys. Rev. Lett., 15, 258.

N.M.R. SPECTROSCOPY OF $Y_2(Fe_xCo_{1-x})_{14}B$

(Collaboration with D. Givord and T. Viadieu, Group 1.01)

These compounds which have no magnetic moment on the rare earth site are interesting for studying the anisotropy of the 3d sublattice when iron atoms are progressively substituted by cobalt atoms. The substitution of cobalt for iron in these materials significaly enhances T_c (C. Abache et al, 1986) and affects also the magnetic anisotropy with a spin reorientation from c axis to a,b planes for x = 0.5 (M.Q. Huang et al, 1986). Extensive studies have been reported on the preferential sites substitution, using neutron diffraction (J.F. Herbst et al, 1986) and Mössbauer spectroscopy on ^{57}Fe (P. Deppe et al, 1987) but there is only few data concerning the cobalt moments. The aim of our study was to investigate the local moment variation of Co atoms versus x and to relate the orbital character of these moment to the 3d anisotropy in these compounds (Berthier et al, 1988 a & b).

The zero field N.M.R. of $Y_2(Fe_xCo_{1-x})_{14}B$ with x = 0, 0.2, 0.4, 0.6 and 0.8 were investigated in the frequency range 35-350 MHz at 4.2 K. In the spectra shown on figure 7, the lines in the low frequency range (below 50 MHz) correspond to the N.M.R. of ^{89}Y and those in the upper range corresponds to ^{59}Co. A complete assignment of the resonance lines to the appropriate cobalt sites was achieved, taking into account their respective abundances and the first neighbours of Co or Fe as a function of x.

For the pure cobalt compound (x = 0) we have calculated the local

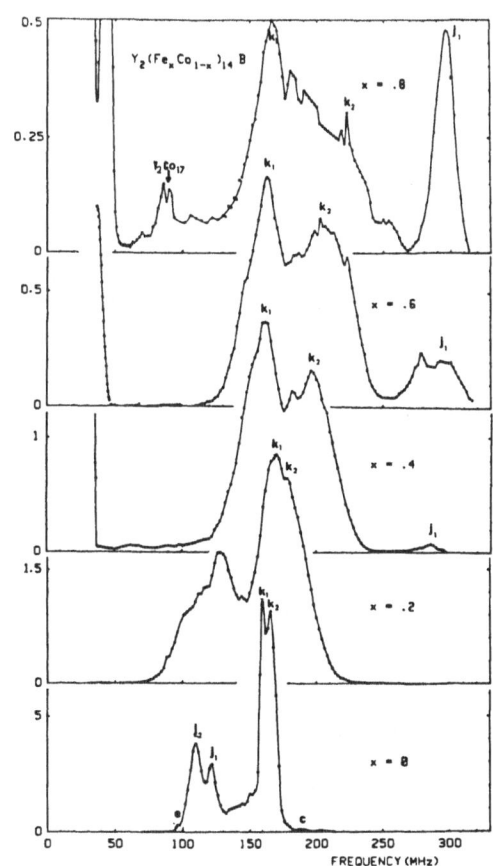

Fig. 7 Zero field N.M.R. spectra of $Y_2(Fe_xCo_{1-x})_{14}B$ at 4.2 K.

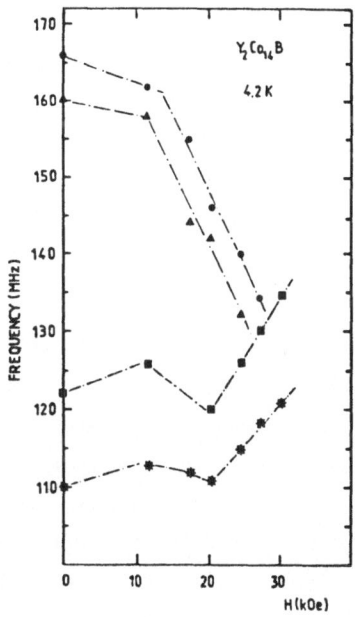

Fig. 8 Magnetic field dependence of ^{59}Co resonance frequency in $Y_2Co_{14}B$

●site k_1 ⎫
▲site k_2 ⎭ $H_{hf} < 0$

■site j_1 ⎫
✱site j_2 ⎭ $H_{hf} > 0$

moment of cobalt at the different sites, assuming that H_{hf} was proportional to μ_{Co}. In the case of a pure spin moment, equation (1) can be used. These values of μ_{Co} are compared in table 4 with those obtained by D. Le Roux (1986) using neutron diffraction. The agreement is good, except in the case

TABLE 4

Site	First neighbours	Freq (MHz)	H_{hf} (kOe)	μ_{Co} (μ_B) Neutron	R.M.N.
4 c	8Co, 2Y	188	186	1.50 (14)	1.43
16 k_1	10Co, 2Y	160	158.4	1.17 (11)	1.22
16 k_2	10Co, 2Y, 1B	166	164.3	1.36 (6)	1.26
8 j_1	9Co, 3Y	122	120.7	1.53 (8)	0.93
8 j_2	12Co, 2Y	110	109	1.23 (10)	0.84
4 c	9Co, 2Y, 2B	95	94	0.66 (16)	0.72

of $8j_2$ and $8j_1$ sites which correspond to the lower frequency lines. N.M.R. measurements under applied magnetic field (Fig. 8) have shown that the hyperfine field corresponding to these sites were positive, contrary to the case of the other sites where it were negative. This result reveals the partly orbital character of the cobalt moment which induce a positive contribution to the hyperfine field (Streever, R.L., 1979) given by :

$$H_{orb} (Co) = \beta \, \mu_{orb} (Co) \quad \text{with } \beta = + 650 \text{ kOe}/\mu_B \qquad (4)$$

When the iron concentration increases, all the lines are broadened by the dispersion of the moments beared by iron and cobalt neighbours surrounding cobalt in each 3d sites. The large shift of the lines assigned to sites $8j_1$ and $8j_2$, towards the high frequencies, is a consequence of the preferential occupancy of the site $8j_2$ by iron as soon as $x = 0.2$. The site j_1 has the larger number of first neighbours located on site j_2 , and this contribute to a large increase of the hyperfine field at this site.

The evolution with x of the lines assigned to sites k_1 and k_2 shows a larger shift for the k_2 site. This is also due to the predominancy of neighbours at sites j_2 around the site k_2. For $x > 0.4$, the two extra lines which appear at 250 MHz and 290 MHz correspond respectively to the Co on sites e and on sites j_1. For these two Co sites the larger amplitude of the lines is due to the preferential occupancy of these sites by cobalt and the large value of the hyperfine field is a consequence of the predominant j_2 sites as first and second neighbours.

As a conclusion, this study inlight the important role played by the 3d site $8j_1$ and $8j_2$ in the substitution of Fe by Co and shows the interest of ^{59}Co N.M.R. in order to study the orbital character of Co atoms on the different 3d sites of a compound, and to understand the microscopic mechanism of the anisotropy of the 3d sublattice.

REFERENCES

Abache, C. and Oesterreicher, H. 1986. J. Appl. Phys., 60, 1114.
Berthier, Y., Nassar, N. and Viadieu, T. 1988a. To be published in the proceedings of I.C.M. 88.
Berthier, Y., Nassar, N. and Viadieu, T. 1988b. To be published in J. Magn. Magn. Mat.
Deppe, P. Rosenberg, M., Hirosawa, S. and Sagawa M. 1987. J. Appl. Phys., 61, 4337.
Herbst, J.F. and Yelon, W.B. 1986. J. Appl. Phys., 60, 4224.
Huang, M.Q., Boltich, E.B. and Wallace, W.E. 1986. J. Magn. Magn. Mat., 60, 270.
Le Roux, D. 1986. Thesis, University Joseph Fourier-Grenoble 1.
Streever, R.L. 1979. Phys. Rev., B 19, 2704.

A MÖSSBAUER SPECTROSCOPY AND NUCLEAR MAGNETIC

RESONANCE STUDY OF $R_2(Fe,Co)_{14}B$ AND RELATED COMPOUNDS

M. Rosenberg, P. Deppe, K. Erdmann, Th. Sinnemann

Ruhr-Universität Bochum
Experimentalphysik VI NB O3/34
PB 10 21 48, 4630 Bochum
BR Deutschland

ABSTRACT

From the Mössbauer and NMR spectroscopy study of the hyperfine inter-action parameters in $R_2(Fe,Co)_{14}B$ and $RFe_{12-x}M_x$ the local magnetic moments at the 6 different Fe crystallographic sites in the former and at the 3 types of sites in the latter ones were derived and found in good agreement with the moments of the unit cells measured magnetically.
In the case of $RFe_{12-x}M_x$ the statistical distribution of M on one or more types of Fe sites gives rise to a splitting and broadening of the Mössbauer lines which makes necessary a substantial increase of the number of fitting lorentzian sextets. The position and the width of the B NMR signals in $R_2Fe_{14}B$ and $R_2Co_{14}B$ are mainly determined by the transferred hyperfine fields from the Fe or Co neighbours and in a large extent by the distribu-tion of the dipolar fields in the domain walls of the compounds with uni-axial or planar anisotropy.

INTRODUCTION

Our main interest was to find correlations between structure and local magnetic moments at the Fe-sites of $R_2Fe_{14}B$ and related compounds via hyper-fine interactions studied with both Mössbauer and pulsed spin-echo NMR spectroscopies.

Another matter of interest was to use the distributions of the transferred hyperfine field at the B-site in order to get more information about the domain walls in these materials for both the cases of axial and planar magnetic anisotropies occuring in the $R_2Fe_{14}B$ and $R_2Co_{14}B$ systems.

For these purposes we studied several compositional series as follows:

1. $R_2Fe_{14}B$ where R = Y, La,Ce,Pr,Nd,Sm,Gd,Tb,Dy,Ho,Er,Tm,Lu and Th.

2. $R_2Co_{14}B$ where R = Y, La,Ce,Pr,Nd,Sm,Gd,Er,Lu,Th.

3. $Nd_2Fe_{14-x}Co_xB$

4. $R_2Fe_{14}C$ where R = Gd,Dy

5. $Nd_5Fe_2B_6$

6. $RFe_{12-x}M_x$ where M = Si,Mo,V,Ti

The measurements were carried out on polycrystalline samples prepared by K.H.J. Buschow (Philips, Eindhoven), H. Stadelmeier (North Carolina State

University), S. Hirosawa and M. Sagawa (Sumitomo Special Metals Co.,Osaka)
and Johnson Matthey Company (REacton products).
The Mössbauer spectra were usually taken at different temperatures in the
range 4.2 - 300 K and the NMR spectra at 4.2 K.

HYPERFINE FIELDS AND MAGNETIC MOMENTS AT THE Fe SITES IN $R_2Fe_{14}B$ AND
$R_2Fe_{14}C$ (Rosenberg et al, 1985; Rosenberg et al, 1986; Erdmann et al,
1987, 1988; Deppe, 1987 b)

The Mössbauer spectra were fitted in reliable way to six lorentzian
sextets corresponding to the six different Fe-sites in the tetragonal
lattice of $R_2Fe_{14}B$ and $R_2Fe_{14}C$ compounds. The hyperfine fields HF (see
Fig. 1 for the sites j_2, k_1 and k_2) were derived after subtracting
dipolar and Lorentz field contributions. The highest values of HF of
about 38,5 T were found at the $Fe-j_2$ sites in the case of uniaxial anisotropy. The presence of a magnetic rare earth element has little influence
on the values of HF at the Fe-sites. The same can be said about the lanthanide contraction as long as the volume of the unit cell doesn't reach
values smaller than a critical one lying very close to the volume of the
unit cell of $Tm_2Fe_{14}B$.

An anisotropic contribution of the HF at the $Fe-j_2$ site could be
found after a careful comparison of the uniaxial anisotropic state with
the planar anisotropic one. In the former case the anisotropic hyperfine
interaction gives rise to a pseudo dipolar field of -1.3 T which leads
to a corresponding enhancement of HF at the j_2-sites in the state with
uniaxial anisotropy as compared to the one with planar anisotropy.

Taking for the hyperfine interaction constant the same value as in
alpha-Fe,values of the Fe magnetic moments in the range 2.1 - 2.5 μ_B were
derived, giving values of the average Fe moment of the unit cell in excellent agreement with those derived from measurements of the saturation magnetization.

In order to verify the reliability of the fitting procedure of the
Mössbauer spectra a pulsed spin-echo NMR study of the $R_2Fe_{14}B$ compounds
has been undertaken. The main complication was in this case the quite
strong overlap of the [57]Fe NMR signals with the very intense [11]B ones,
making rather difficult the separation and identification of the [57]Fe
contributions from the six Fe-sites. [10]B NMR spectra taken in the lower
frequency range provided through scaling evidence that in the majority

of cases for the compounds with uniaxial anisotropy the [11]B signals should be rather large, therefore overlaping the whole range of [57]Fe resonant frequencies. Fortunately we found that because of the quite large difference in relaxation times between [11]B and [57]Fe nuclei, by increasing the intervals between the rf-pulses up to 300 - 1000 μs a substantial decrease of the [11]B-background could be reached, allowing us to clearly identify the signals arising from the different Fe sites.

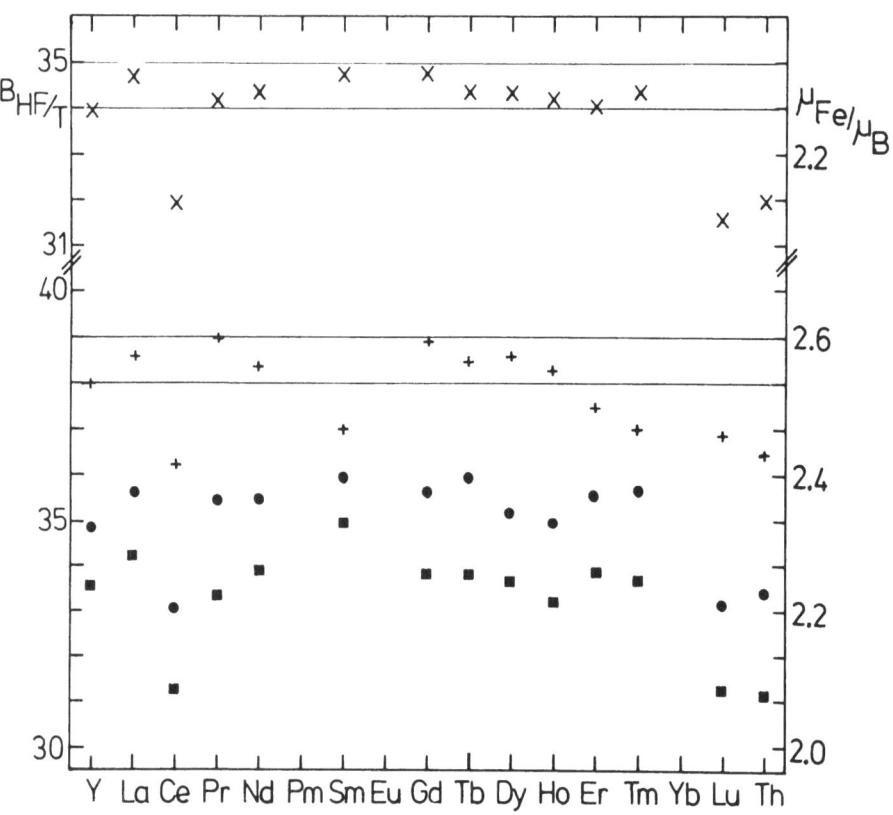

Fig. 1   Hyperfine fields B_{HF} and derived magnetic moments μ at selected Fe sites in $R_2Fe_{14}B$: (+) j_2 site; (■) k_1 site; (●) k_2 site; (x) average values.

Typical NMR spectra taken under such conditions are shown in Fig. 2. The values of HF obtained from the [57]Fe NMR frequencies were in good agreement with the ones derived from the Mössbauer spectra. A very favourable case was the one of $Gd_2Fe_{14}C$, where the NMR signal of ^{12}C occurs in a frequency range lower than that of [57]Fe and all the 6 Fe peaks were easily identified

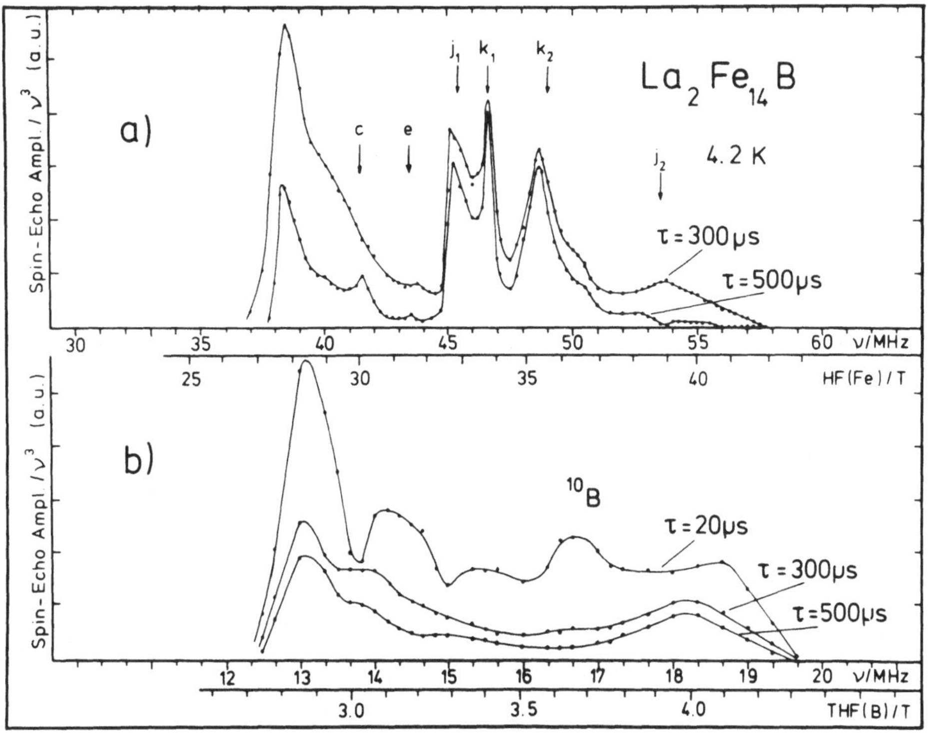

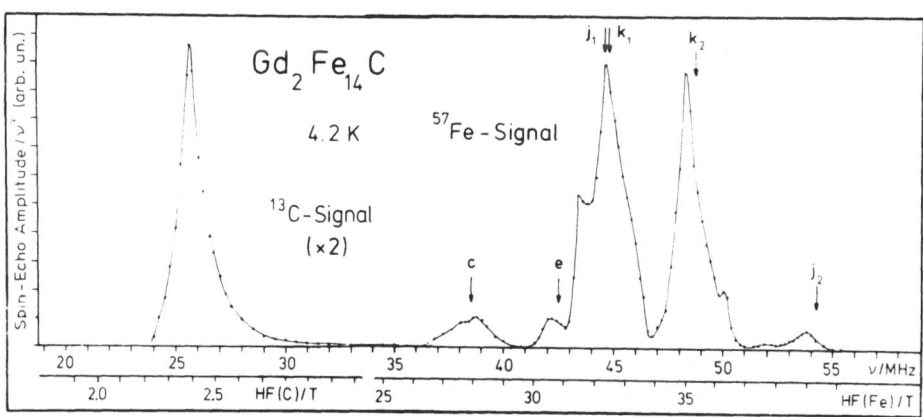

Fig. 2   NMR spectrum of $La_2Fe_{14}B$:
a) ^{11}B and ^{57}Fe. The arrows indicate the Mössbauer values of the HFs at Fe sites.
b) ^{10}B spectrum scaled with ^{11}B on the common THF scale.

Fig. 3   NMR spectrum of $Gd_2Fe_{14}C$. The arrows indicate the Mössbauer values of the HFs at Fe sites. The ^{13}C signal occurs at lower frequencies in contrast to the ^{11}B signal in $R_2Fe_{14}B$ which partially overlaps the ^{57}Fe spectrum.

(Fig. 3). The values of HF were in very good agreement with the ones obtained from a Mössbauer study of this compound.

TRANSFERRED HYPERFINE FIELDS AT B AND R SITES IN $R_2Fe_{14}B$ AND $R_2Co_{14}B$
(Rosenberg et al, 1986; Erdmann et al, 1987 a, 1988)

The B NMR spectra have peaks of maximal intensity at the low frequency side and widths which depend on the type of anisotropy, of the magnetic or nonmagnetic nature of R and on the nature of the 3d-component Fe or Co.

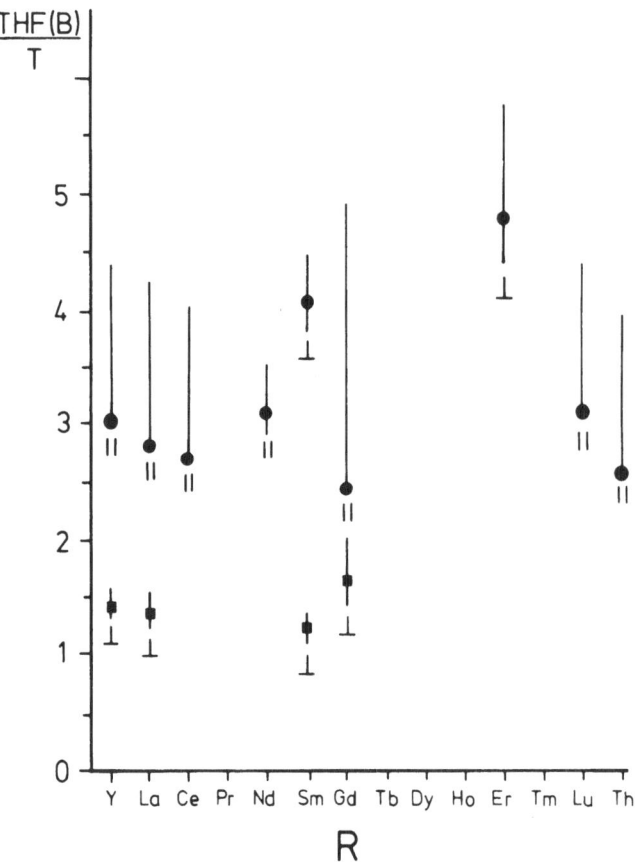

Fig. 4   Transferred hyperfine fields at B site in $RFe_{14}B$ (●)
and $RCo_{14}B$ (■). The bars indicate the widths of the hyperfine
field distributions. Note the drastic increase in THF accompanied
by a strong decrease in widths for $R_2Fe_{14}B$ with planar anisotropy.
For $R_2Co_{14}B$ all the THF values and spectral widths are   significantly
reduced as compared with $R_2Fe_{14}B$.

The value of the transferred hyperfine field (THF) at the B-site taken at the maximal intensity of the NMR signal and the width of the B spectra are given in Fig. 4. The main trends occuring in the systematics of THF(B) can be explained in terms of a contribution due to the polarization of the conduction electrons which is practically the same for compounds containing a given 3d-component (larger for the $R_2Fe_{14}B$ than for $R_2Co_{14}B$) and variable dipolar and Lorentz fields occuring at the B nuclei in the domain walls which are the main source of the NMR signals. In contrast to the Fe NMR peaks which are only slightly broadenend by the position dependent dipolar fields in the domain walls, in the case of B such effects are strongly amplified because of the larger gyromagnetic ratios of the ^{10}B and ^{11}B nuclei. In the case of $R_2Fe_{14}B$ compounds with uniaxial anisotropy we found besides the large width of the B signals a structure with several peaks which cannot be explained by the crystalline structure alone, because only one site is available for the B atoms. In order to understand the drastic variation of the THF values and spectral widths of the B signals of $R_2Fe_{14}B$ compounds with uniaxial anisotropy when a magnetic R with a large moment is substituted for a nonmagnetic R, we treated the cases of $Gd_2Fe_{14}B$ and $Lu_2Fe_{14}B$ quantitatively. The B spectra have maxima at their low frequency side corresponding to values of the transferred hyperfine field THF(B) of 2.40T and 3.16T for R = Gd and Lu respectively. Although B atoms occupy

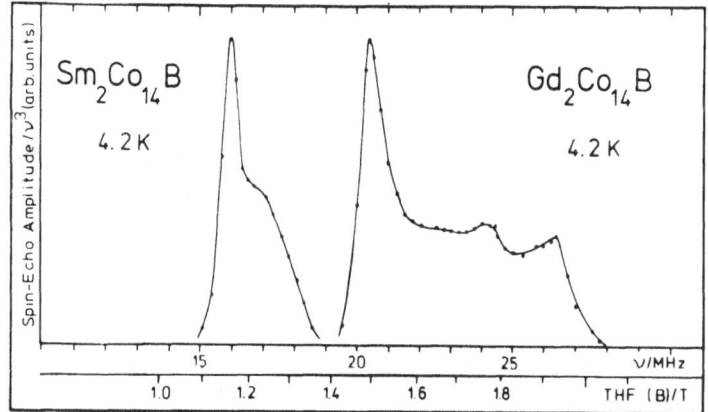

Fig. 5 ^{11}B spectrum in $Sm_2Co_{14}B$ and $Gd_2Co_{14}B$. ^{59}Co signals occur above 80MHz. The shift of THF forwards higher frequency values on going from Sm to Gd is mainly due to larger negative dipolar field contribution in the domain walls in the latter case.

only one site in the $R_2TM_{14}B$ structure, the whole range of B signals extends over ~ 2.6 and $\sim 1.3T$ for R = Gd and Lu respectively. These values can be compared with those determined in $La_2Fe_{14}B$ as a reference, namely THF(B) = 2.81T for the THF at maximum intensity and a range of 1.5T. In order to estimate domain wall effects in the NMR spectra we calculated the local dependence of the dipolar field in $180°$ walls choosing a reasonable width of 3nm and the classical continuum approximation for the magnetization rotation from (001) to (00$\bar{1}$) in the (110) plane because the (110) direction is the easy direction for the spins when turning out of the tetragonal c axis, at least for R = Gd above 110K. We took spheres of 10nm radius, an average value for μ_{Fe} of $2.2\mu_B$ (except $\mu_{Fe}(j_2) = 2.5\mu_B$) and a value for μ_{Gd} of $7\mu_B$. In the center of the walls the Fe sublattice gives rise to dipolar fields between $-0.1T$ and $+0.1T$, dependent on the B site in the unit cell where the negative sign denotes a direction opposite to the local magnetization. At the edge of walls and in domains, the dipolar fields reached values of respectively 1.1T and 0.85T, where the splitting is negligible. Therefore a difference of about 1.2T arises between the hyperfine fields at B nuclei located in the center and at the edge of the walls, corresponding to a ^{11}B frequency range of 16MHz if only Fe magnetic moments contribute to the dipolar field. The range of variation of the dipolar field drastically increased when the Gd sublattice contribution was taken into account. In the vicinity and in the center of the walls H_{Dip} was shifted to $-0.3T$ .. $- 0.75T$, whereas at the edge and in the domains it reached 1.65T and 1.5T respectively. The additional dipolar contributions of the Gd sublattice increases the range of values of the dipolar field up to 2.5T, corresponding to a ^{11}B frequency range of 34MHz. Both the values obtained for Fe contributions alone to the dipolar field and the one including the Gd contributions are in good agreement with the experimental data for the NMR spectral widths in $R_2Fe_{14}B$ with R = La, Lu and in $Gd_2Fe_{14}B$ respectively.

In the Co-containing samples pure ^{11}B spectra were obtained in the frequency range from 15 to 28MHz. The values of THF(B) at maximum intensity and the range of B signals are 1.16T and 0.2T for $Sm_2Co_{14}B$ and 1.49T and 0.5T for $Gd_2Co_{14}B$ (Fig. 5). The rather large shift of the B lines to lower frequencies mainly results from the smaller hyperfine coupling constant a_{Co-B} and the smaller average Co magnetic moment of about $1.4\mu_B$. Taking for the average Fe moment in $R_2Fe_{14}B$ a value of $2.2\mu_B$ and a ratio

a_{Fe-B}/a_{Co-B} =1.8, which was found in $Co_{3-x}Fe_{x}B$ alloys with similar environment of B nuclei, we obtain a ratio of 2.8 between THF(B) values in
$R_2Fe_{14}B$ and $R_2Co_{14}B$. Measurements of $Sm_2Fe_{14}B$ with planar anisotropy (as
$Sm_2Co_{14}B$) give a value of THF(B) = 4.03T which is about 3.5 times the value
we find in the latter compound, supporting the expectation of our estimation.

A difficulty arises when one considers the strong reduction of the
frequency width of the B spectrum to about 1/6 of its value in the $R_2Fe_{14}B$
compounds with axial anisotropy when Fe is replaced by Co. From the above
considerations one would expect reductions of only about 2/3 of the value
found in the Fe-containing compounds.

The reason for this discrepancy resides in the circumstance that the
investigated Co-containing compounds exhibit planar anisotropy. This gives
rise to a different magnitude and dependence of dipolar fields on the location of the resonant nuclei in the domain walls. In the case of planar anisotropy, with the magnetic moments in the walls rotating in the (001) plane,
we calculated the variation of the dipolar field taking $\mu_{Co} = 1.4\mu_B$, disre-
garding the rotation of magnetic moments inside the sphere because of the
small anisotropy constant K_3 in the plane. For $\mu_R = 0.3\mu_B$ ($Sm_2Co_{14}B$) the
dipolar field reached -0.24T with a splitting of $\pm$ 0.08T around this value
when the magnetization is parallel to (110). In the case of $\mu_R = 7\mu_R$
($Gd_2Co_{14}B$), the dipolar field reached -0.58T with a splitting of +0.23T.
Both the positive shift of THF(B) (0.34T) and the increase of the B hyperfine field range (0.3T) when comparing $Gd_2Co_{14}B$ with $Sm_2Co_{14}B$ are in quite
good agreement with the measured hyperfine field distribution at the B site.

The drastic increase of THF(B) in $Sm_2Fe_{14}B$ and $Er_2Fe_{14}B$ with planar
anisotropy up to -4.03 and -4.75T is also a result of the reduction of the
positive dipolar fields at the B nuclei in the regions of the broad 90
degree walls separating the domains of a sample with an easy plane but
with not too large anisotropy in the plane.

NMR signals from R nuclei were found in the $Y_2Fe_{14}B$ and $Y_2Co_{14}B$ compounds. Whereas in the latter one [89]Y signals were present at 14.6MHz (7T)
and around 21.2MHz (10.2T), they were found at 36.7MHz (17.6T) and 48.4MHz
(23.2T) in $Y_2Fe_{14}B$. Obviously the two Y signals have to be correlated with
the two different crystallographic sites (4g and 4f) of the Y atoms in the
$Y_2Fe_{14}B$ lattice. In other Y-Co compounds the THF at the Y site with the
closest Co neighbors reaches the highest value. For this reason we believe
that in $Y_2Co_{14}B$ the line at 14.6MHz belongs to Y nuclei located at 4g sites.

SITE PREFERENCE OF Fe IN $Nd_2(Fe_{1-x}Co_x)_{14}B$

(Deppe et al, 1987 a)

Recently, Van Noort and Buschow showed that in $Nd_2Co_{14}B$ only slightly substituted with Fe, the Fe atoms exhibit a strong preference for the j_2 sites. In order to determine the site preference of Fe and Co in the $Nd_2(Fe_{1-x}Co_x)_{14}B$ compounds and the influence of the composition on the hyperfine field at the Fe nuclei we have studied with Mössbauer spectroscopy the whole composition range of the $Nd_2(Fe_{1-x}Co_x)_{14}B$ series but with more emphasis put on the compositions with x>0.5 where a marked increase of the intensity of the Fe-j_2 sextet with x could be found (Fig. 6) thus offering direct evidence for a preferential distribution of the Fe atoms at j_2 sites in the presence of Co. A closer examination of Fig. 6 shows that the intensity of the Fe(k_2) line decreases drastically with increasing Co content. Rather unaffected is the intensity ratio $I(k_1)/I(j_1)$ which remains close to the ideal value of 2 and only slightly affected are the ratios $I(k_1)/I(e,c)$. The shaded area in Fig. 6 has to be proportional to the relative amount of Fe atoms distributed on j_2 sites. The value of 37% for x = 0.854 is in excellent agreement with that of 38% given previously by Van Noort and Buschow for a sample with x = 0.98.

HYPERFINE FIELDS AND MOMENTS AT THE Fe SITES IN $RFe_{12-x}M_x$

In order to better understand the influence of the local environment and of the substitution of M(V, Ti or Mo) for Fe on the Fe magnetic moments at the three different crystallographic sites 8i, 8f and 8j of $RFe_{12-x}M_x$ intermetallics with R=Y, Sm, Gd, Er, a Mössbauer and NMR study was undertaken. Because of the statistical distribution of the M atoms over the different crystallographic sites the Mössbauer spectra were rather complex. Previous X-ray studies provided evidence for preferential occupation of i-sites by V and Mo and of f-sites by Ti atoms. Therefore, in order to fit the Mössbauer spectra we started from a binomial distribution of the M component over the i or f sites and evaluated the influence of the different Fe environments on the hyperfine field (HF) at the Fe sites (Fig. 7). Fits with up to 10 values of HF were necessary for $RFe_{10}V_2$ and $GdFe_{10}Mo_2$. In the last case the fit with more than three HF values was also confirmed by the structure of the Fe-NMR signal. Using the averaged HF at Fe nuclei on 8i, 8f and 8j sites and a hyperfine interaction constant of $15T/\mu_B$ values of the Fe moments in $RFe_{10}V_2$ and $RFe_{10}Mo_2$ of 1.8-1.89, 1.48-1.57 and 1.16-1.33μ_B were obtained. The values of the average Fe moment in $RFe_{10}V_2$ are

Fig. 6   Part of the Mössbauer spectra of
$Nd_2(Fe_{1-x}Co_x)_{14}B$ with
$x \geq 0.61$. The shaded area
indicates the intensity of
the Fe(j_2) line proportional to the relative
amount of Fe distributed
on j_2 sites, which reaches
37% for $x = 0.85$.

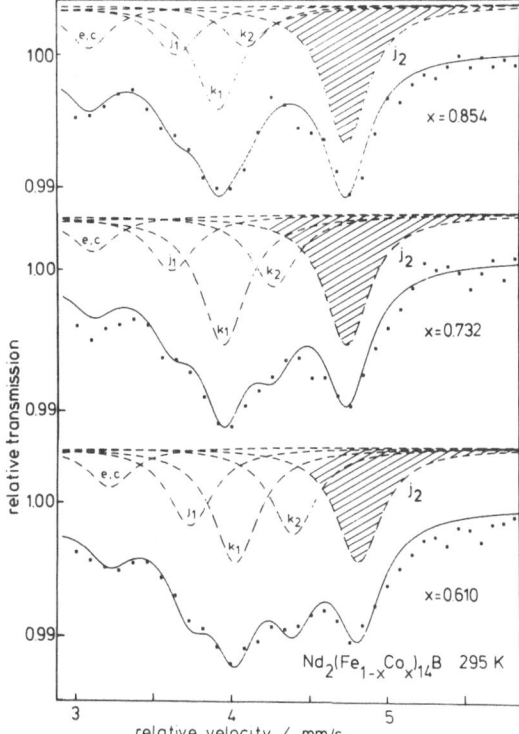

Fig. 7   Mössbauer spec-
trum of $Gd_2Fe_{10}V_2$ fitted
to 10 lorentzian sextets.

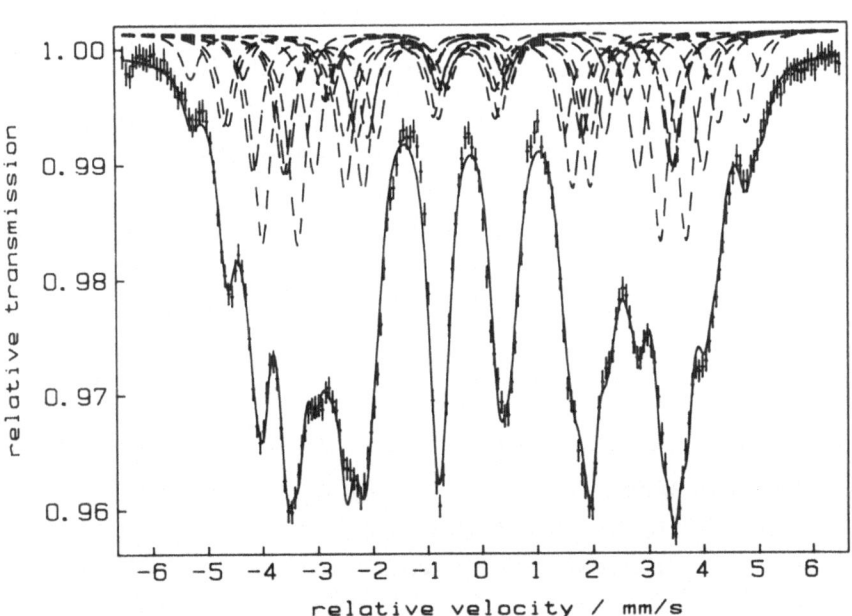

in good agreement with the results of magnetic measurement.

MÖSSBAUER SPECTROSCOPY, NMR AND MAGNETIC STUDIES OF RELATED COMPOUNDS
(Rechenberg et al, 1986; Deppe et al, 1987c; Erdmann et al, 1986; Rosenberg et al, 1988)

RFe_4B_4

The hyperfine interaction parameters of the RFe_4B_4 compounds were measured down to 4.2K. No sizeable magnetic splitting could be detected, in agreement with the magnetic measurements which showed that only the magnetic R elements order magnetically at low temperatures. A broadening of the Mössbauer [57]Fe quadrupolar doublets can be taken as evidence for the incommensurability of the Fe and R sublattices.

$Nd_5Fe_2B_6$

A [57]Fe Mössbauer and X-ray study of compounds around the composition $Nd_5Fe_2B_6$ confirmed the rhombohedral structure proposed for $Nd_5Fe_2B_6$ in the literature but showed that a homogeneity range of any significance is absent. The hyperfine field 2.4T at 5K that develops at the Fe located at 6c sites in $Nd_5Fe_2B_6$ is due to a conduction electron spin polarization induced by the RKKY exchange interaction between the Nd moments.

$RCo_{12}B_6$

The magnetic properties of polycrystalline intermetallics $YCo_{12}B_6$ and $GdCo_{12}B_6$ which crystallize in the hexagonal $SrNi_{12}B_6$ structure have been studied in dependence on temperature and magnetic field. $YCo_{12}B_6$ is ferromagnetic with $T_c = 151.5K$ and an average Co moment of $0.44\mu_B$, whereas $GdCo_{12}B_6$ is a ferrimagnet with $T_c = 161K$ and a compensation point at 47.8K, with an average Co moment of $0.43\mu_B$. The spin-echo nuclear magnetic resonance was used to study [59]Co hyperfine fields in these compounds. Two groups of signals were obtained and interpreted in terms of anisotropic orbital contributions to the hyperfine field in the domain walls and domains.

CONVERSION ELECTRON MÖSSBAUER SPECTROSCOPY AS A METHOD TO DETERMINE THE DEGREE OF ORIENTATION IN THE $Nd_2Fe_{14}B$ COMPOUND

We have applied the CEMS to the study of the magnetic orientation in bulk permanent magnets made from $Nd_2Fe_{14}B$. Using the anisotropy in the emission of gamma-rays and electrons for the Mössbauer transition of [57]Fe one can evaluate the degree of alignment of the magentic moments in mate-

rials with preferential orientations. In the case of our oriented anisotropic $Nd_2Fe_{14}B$ permanent magnets we found a disalignment of less than 25
degrees.

The applicability of this method depends obviously on the domain
structure at the surface of the magnet. In the case of the strongly anisotropic $Nd_2Fe_{14}B$ magnets no important deviations of the magnetization
from the easy axis are expected and consequently an evaluation of the
degree of orientation from CEMS spectra seems to be quite reliable.

REFERENCES

Deppe, P., Rosenberg, M., Hirosawa, S. and Sagawa, M. 1987a, J.Appl. Phys.
    61, 4337.
Deppe, P., 1987b, Ph.D. Dissertation Ruhr-Universität Bochum
Deppe, P., Rosenberg, M. and Buschow, K.H.J., 1987, Sol. Stat. Commun.
    64, 1247.
Erdmann, K., Deppe, P., Rosenberg, M. and Buschow, K.H.J.1 1987,
    J. Appl. Phys. 61, 4340.
Erdmann, K., Rosenberg, M. and Buschow, K.H.J., 1988a, paper GD-06 presen-
    ted at the 32nd M.M.M. Chicago 1987, to appear in J. Appl. Phys.
Erdmann, K., Rosenberg, M. and Buschow, K.H.J., 1988b, paper GD-05 presen-
    ted at the 32nd M.M.M. Chicago 1987, to appear in J. Appl. Phys.
Rosenberg, M., Deppe, P., Wojcik, M. and Stadelmaier, H., 1985, J. Appl.
    Phys. 57, 4124.
Rosenberg, M., Deppe, P. and Stadelmaier, H., 1986, Hyperfine Interactions
    28, 503.
Rosenberg, M., Mittag, M. and Buschow, K.H.J., 1988, paper DQ-04 presented
    at the 32nd M.M.M. Chicago 1987, to appear in J. Appl. Phys.

MÖSSBAUER STUDY BY ^{57}Fe AND ^{174}Yb MÖSSBAUER SPECTROSCOPIES

OF $Yb_2Fe_{14}B$, $RCo_{4-x}Fe_xB$ (R = Pr, Nd, Sm, Er, Tm, Lu)

AND $SmFe_{10}T_2$ (T = Ti, V, Si)

Y. Gros, F. Hartmann-Boutron, C. Meyer

Laboratoire de Spectrométrie Physique (associé au C.N.R.S.)
Université Joseph Fourier Grenoble I
B.P. 87, 38402 Saint-Martin d'Hères Cédex, France

ABSTRACT

i) Mössbauer investigation of $Yb_2Fe_{14}B$ with ^{57}Fe and ^{174}Yb shows the existence of a magnetization reorientation at 115 K and provides additional information on the exchange and crystal fields acting on the R.E. ions in the $R_2Fe_{14}B$ series. ii) In compounds of the form $RCo_{4-x}Fe_xB$, which are ferromagnetic for R = Pr, Nd, Sm (x $\lesssim$ 2) and ferrimagnetic for R = Er, Tm, Lu, Mössbauer spectroscopy yields the magnetization direction, which depends both on the R.E. anisotropy and on the transition metal anisotropy ; this last anisotropy is apparently influenced by the strong affinity of the iron for the 2c site, evidenced by ^{57}Fe Mössbauer spectra. The easy axis ferromagnet $SmCo_2Fe_2B$ has interesting magnetic characteristics. iii) Samples of the new family $SmFe_{10}T_2$ (T = Ti, V, Si) have been investigated ; those containing Ti are pure phases, with an easy axis both at R.T. and at 150°C ; their Curie temperature is $T_c \sim 590$ K.

MÖSSBAUER STUDY OF $Yb_2Fe_{14}B$

The compound $Yb_2Fe_{14}B$ belonging to the family of the supermagnet $Nd_2Fe_{14}B$ was successfully prepared by Givord [1.01] and Gavigan and Coey [1.21] in a sealed tantalum crucible under reduced argon atmosphere. Because of the great difficulty of the synthesis of ytterbium alloys, various impurities are present in the sample.

^{57}Fe Mössbauer study

^{57}Fe Mössbauer spectra have been obtained between 300 K and 5,4 K (Fig. 1). In the room temperature spectrum all the phases identified with X rays can be observed rather easily. Some unidentified ones perhaps also exist.

The spectrum of $Yb_2Fe_{14}B$ exhibits the classical qualitative features observed for the $R_2Fe_{14}B$ magnetized along the c axis, with a subspectrum, usually assigned to the $8j_2$ site, which is characterized by a hyperfine field (305 kOe) which is large compared with the average one observed for the other sites (250 kOe). 2% (in Fe percentage) of Fe metal is present, as well as 20% of Yb_2Fe_{17} as a paramagnetic doublet (T_c = 255 K). A good fit of the whole spectrum is therefore difficult to obtain.

At low temperature we find an evidence of a spin reorientation near T_{SR} = 115 K. As a matter of fact a clear change in the spectrum is occurring in the range 120 K-110 K in a relatively progressive way : the extra line characteristic of the $8j_2$ site is present down to 120 K. At 117 K it starts to move towards the major spectrum and at 115 K it has completely reached it. This behaviour is usually assigned to a spin reorientation from the c axis to the basal plane. Indeed it is admitted that the h.f. field is highly anisotropic for the $8j_2$ site, which leads to a jump in its modulus at the reorientation, (Fig. 2), accompanied by a change of the quadrupole parameter ε. The same features have been observed by Sanchez et al [1.10] in $Er_2Fe_{14}B$.

Such a reorientation is predicted for the $R_2Fe_{14}B$ which are magnetized in the basal plane at low temperature when the rare earth anisotropy dominates (the easy plane Rare Earths being those for which $\alpha_J > 0$: Sm, Er, Tm, Yb). When temperature increases, the easy plane Rare Earth anisotropy decreases much more rapidly than the 3d easy axis iron one and this results in a reorientation towards the c axis.

This has been experimentally observed in $Er_2Fe_{14}B$, $Tm_2Fe_{14}B$ and now in $Yb_2Fe_{14}B$ by Gavigan et al using magnetization and neutron diffraction measurements : i) there is an anomaly in the temperature dependence of the magnetization near 115 K ii) neutron diffraction without and with an external field clearly shows that the spins are along [001] above 115 K and along [100] at 4 K. Our Mössbauer measurements are an independent confirmation of this.

Another check of this spin rotation can be performed with Mössbauer spectroscopy on a magnetically aligned Mössbauer sample. Alignment was achieved by Gavigan et al at 300 K, in a field of 2 Teslas by freezing in an epoxy resin. At room temperature the magnetic moments are aligned perpendicular to the surface of the sample i.e. parallel to the γ ray direction. The Mössbauer spectrum should then be characterized by the vanishing of the lines corresponding to the $\Delta m = 0$ transitions (lines n° 2 and 5). On the other hand, if below 120 K the moments rotate and become perpendicular to the γ ray, lines 2 and 5 should contribute again to the resonance with $I_{2-5} = \frac{4}{3} I_{1-6}$. This is qualitatively observed in Fig. 3 (not quantitatively, because of impurities).

Yb$_2$Fe$_{14}$B   (Fe^{57})

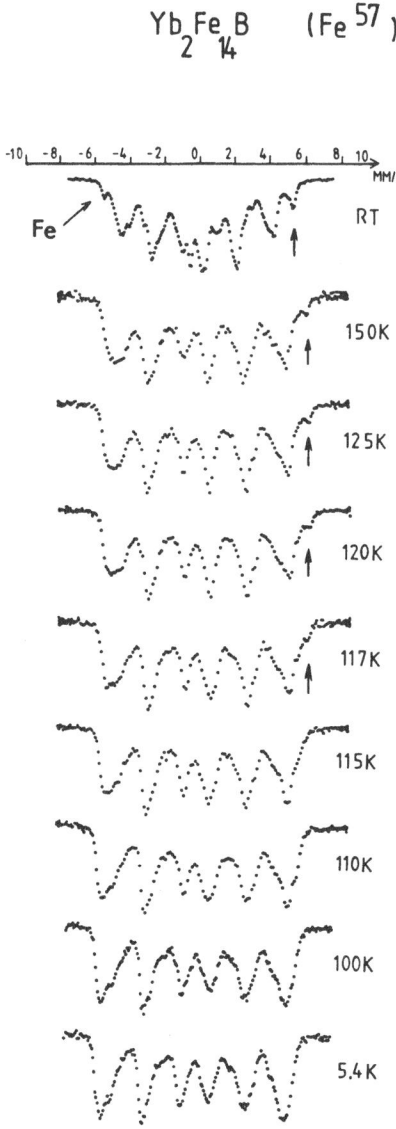

Fig. 1 ^{57}Fe Mössbauer spectra of Yb$_2$Fe$_{14}$B : thermal evolution, emergence of the 8j$_2$ line above 115 K.

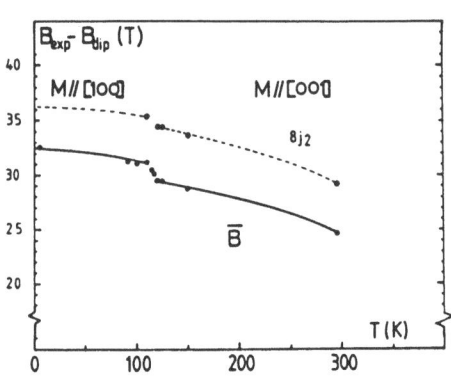

Fig. 2 ^{57}Fe hyperfine fields in Yb$_2$Fe$_{14}$B, above and below the reorientation temperature (individual field for site 8j$_2$ and average field for the other sites). The discontinuity is due to orbital effects.

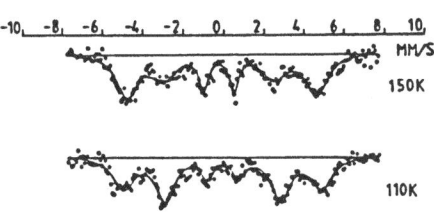

Fig. 3 ^{57}Fe Mössbauer spectra of a sample of Yb$_2$Fe$_{14}$B oriented in a magnetic field, above and below the reorientation temperature 115 K, showing the variation of the intensities of the intermediate lines.

Fig. 2 represents the variation of the hyperfine field at the $8j_2$ site and of the average hyperfine field at the other sites. At the reorientation temperature there is a discontinuity which is due only in part to dipolar effects. Indeed the hyperfine field is essentially the sum of two contributions : a spin contribution ($- 15$ T/μ_B) and an orbital contribution ($+ 65$ T/μ_B), this last one appreciably quenched by crystal field effects, so that the dominant contribution is the spin one. When, at $T = T_{SR}$, the Fe magnetization rotates from an easy direction to a difficult one, the orbital contribution to the hyperfine field decreases and the total h.f. field increases.

^{174}Yb Mössbauer study

It was performed in collaboration with G. Czjzek and coworkers [1.23]. Its aim was to get information on the Fe-Yb exchange and on the crystalline field acting on the Yb^{3+} ion, in order to complete the systematics of the series $R_2Fe_{14}B$ investigated by Givord and coworkers [1-01].

The Mössbauer transition of ^{174}Yb involves a γ ray with relatively high energy $E = 76.5$ keV ; it is therefore necessary to use a cooled source and it is not possible to get spectra at $T > 150$ K because the Lamb-Mössbauer factor becomes too small. Fig. 4 displays a set of spectra obtained jointly by our group and group [1.23]. Their analysis is not simple, because appreciable amounts of impurity phases (Yb_2Fe_{17}, Yb_2O_3) are present and also because ytterbium occupies two different sites with rather low local symmetry. Fig. 5 represents the thermal variations of the quadrupole coupling and of the hyperfine field for the two ytterbium sites deduced from preliminary fits of the spectra :

$$\Delta_Q(T) = e\ V_{zz}(T)Q$$

$$B(T) = A_J <J>/\hbar\ \gamma_J$$

The values obtained at 0 K are close to the free ion values, as expected when the Fe-R.E. exchange field is large compared with the crystalline electric field (C.E.F.) and when the lattice electric field gradient (E.F.G.) is small compared with the 4f E.F.G.

In order to get more quantitative information on exchange and C.E.F. effects, a much more elaborate analysis is required. Such an analysis, taking impurity phases into account, was developed by group [1.23]. Its metho-

dology and its results can be found in the report of this group, in which
the data deduced from ^{174}Yb Mössbauer Effect are compared with those obtained by magnetization measurements [1.01] and also with data relative to
other $R_2Fe_{14}B$ [1.01].

Publications related with the present work :

Burlet, P., Coey, J.M.D., Gavigan, J.P., Givord, D., Meyer, C. 1986.
    Solid State Comm. 60, 723.
Meyer, C., Gavigan, J.P., Czjzek, G., Bornemann, H.J. 1988, July.
    Communication submitted to the International Conference on Magnetism
    Paris.

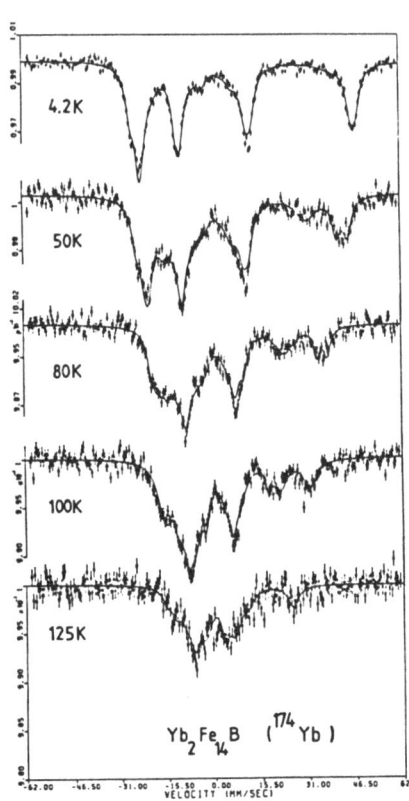

Fig. 4 ^{174}Yb Mössbauer spectra
of $Yb_2Fe_{14}B$ at various temperatures.

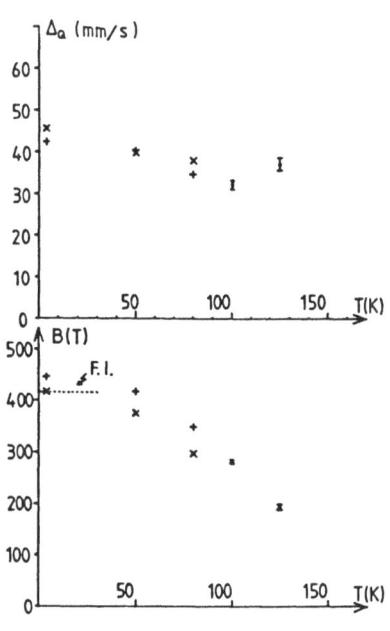

Fig. 5 ^{174}Yb hyperfine fields
and quadrupole effects in
$Yb_2Fe_{14}B$ (two ytterbium sites),
deduced from the Mössbauer spectra, versus temperature.

STUDY OF COMPOUNDS OF THE FORM $RCo_{4-x}Fe_xB$

This investigation was performed in collaboration with Fremy and Tenaud (Group 1.01).

Study of compounds with light rare earths ($PrCo_3FeB$, $NdCo_3FeB$, $SmCo_3FeB$, $SmCo_2Fe_2B$

Rare earth compounds RCo_4B with the hexagonal crystalline structure of $CeCo_4B$ were first prepared by Kuzma et al (1979). Mixed compounds $RCo_{4-x}Fe_xB$ with R = Pr and Sm were subsequently obtained by El Masry et al (1983), Spada et al (1984). This family looks attractive for permanent magnets since the magnetic moments of R.E. and transition metal atoms are ferromagnetically coupled and the Curie temperatures of the compounds are fairly high ($\sim$ 600-700 K). RFe_4B compounds do not exist for light R.E. and $x \lesssim 1.5$ for Pr, Nd, $x \lesssim 2$ for Sm. According to the literature, $PrCo_{4-x}Fe_xB$ compounds have an easy plane anisotropy while $SmCo_3FeB$, $SmCo_2Fe_2B$ have easy axes. This suggests that easy axes are associated with positive Stevens coefficient α_J (Sm, Er, Tm, Yb) and easy planes with negative α_J (other R.E.). However it has been reported by Pedziwiatr et al (1987) that $NdCo_4B$ ($\alpha_J < 0$) has an easy axis at 77 K and 295 K ; this could occur if the cobalt anisotropy favours an easy axis and if it dominates over that of the neodymium, either at all temperatures or above a certain temperature T_{SR} (< 77 K), in which case a reorientation should occur below T_{SR} ; the different behaviours of $NdCo_4B$ and $PrCo_4B$ could then be due to the fact that $|\alpha_J(Pr)| \sim 3|\alpha_J(Nd)|$. As a matter of fact, according to Smit, Thiel and Buschow (1988), $GdCo_4B$ (in which Gd has negligible anisotropy), has an easy axis, i.e. cobalt atoms favour an easy axis (as in the RCo_5 compounds).

The aim of our ^{57}Fe Mössbauer study was to identify the Mössbauer spectra corresponding to the two different sites of the crystalline structure and to look for site preference, hyperfine field orientation and hyperfine structure anisotropy. Three types of spectra were used : unoriented powder, powder oriented in 12 kOe and difference spectra. Assuming the magnetization to be parallel to the hexagonal axis in $SmCo_3FeB$ and $SmCo_2Fe_2B$, and perpendicular to it in $PrCo_3FeB$ and $NdCo_3FeB$ as indicated by magnetic measurements, the spectra could be satisfactorily interpreted in terms of subspectra associated with axial site 2c and orthorhombic site 6i (Figs 6,7). The electric field gradients at 2c and 6i sites are found to be quite similar to those at 2c and 3g sites in $NdCo_5$:Fe, except for an even larger

asymmetry at 6i site ($\eta \sim 0.9$). The magnetic hyperfine structure is anisotropic : in the fits of the present study the full orthorhombic character
of the 6i site was taken into account. The Mössbauer study indicates a
strong preference of iron for 2c sites in these compounds. This result is
interesting since, as noticed by El Masry and Stadelmaier (1983), the substitution of iron into $SmCo_4B$ is beneficial for certain magnetic properties,
leading to an increase of the Curie temperature, the saturation magnetization and the coercivity.

On the other hand, the fact that $NdCo_4B$ has an easy axis and $NdCo_3FeB$
has an easy plane could indicate that substitution of one Fe atom, which
mainly replaces cobalt in the 2c sites, has a detrimental effect on the
easy axis anisotropy of the transition metal (which in $SmCo_5$ is attributed
essentially to 2c cobalt atoms). Note that an anomaly is also observed in
$ErCo_3FeB$ (see below).

T_c' s, magnetizations M, and anisotropy fields H_A at 4.2 K for the easy
axis compounds $SmCo_2Fe_2B$, $SmCo_5$ and $Nd_2Fe_{14}B$, compiled from the literature,
are compared in the table. Apart from the fact that the present coercitive
field of $SmCo_2Fe_2B$ is rather low, $SmCo_2Fe_2B$ and $SmCo_5$ look comparable, with
the advantage of lower cobalt content in $SmCo_2Fe_2B$. With respect to the
supermagnet $Nd_2Fe_{14}B$, $SmCo_2Fe_2B$ has a higher T_c but a twice smaller M.

		$SmCo_5$	$SmCo_2Fe_2B$	$Nd_2Fe_{14}B$
	T_c (K)	1020	768	588
TABLE	M (4 K) Gauss	885	709	1480
	H_A (4 K) kOe	300	602	540

Study of compounds $R(Co_{4-x}Fe_x)B$ with heavy rare earths (Er, Tm, Lu)

Compounds of the form RFe_4B with R = Er, Tm, Lu were first prepared by
Kuzma et al (1983) who reported that they have the hexagonal $CeCo_4B$ structure. Their magnetic properties were investigated by various authors who
found that $LuFe_4B$ is ferromagnetic and $ErFe_4B$, $TmFe_4B$ are ferrimagnetic,
with Curie temperatures of the order 600 K. All three compounds have easy
magnetization axes along the hexagonal c axis, with anisotropy fields at
4 K which vary from 20-30 kOe for $LuFe_4B$ (in which only iron contributes)
to 80 kOe for $ErFe_4B$ ($\alpha_J > 0$ for Er, Tm). There is some ambiguity concerning the exact crystallographic structure, because the superstructure lines

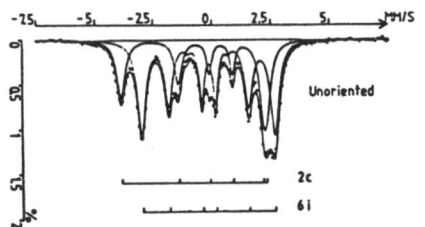

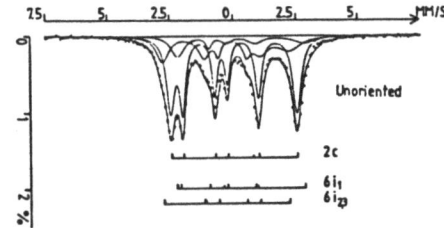

Fig. 6 ^{57}Fe Mössbauer spectrum of SmCo$_2$Fe$_2$B at 295 K
(easy axis).

Fig. 7 ^{57}Fe Mössbauer spectrum of
NdCo$_3$FeB at 295 K (easy plane).

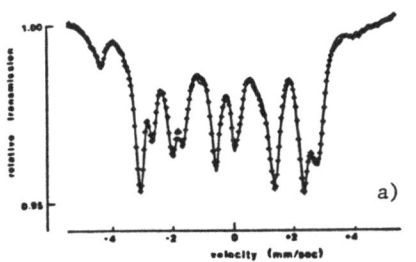

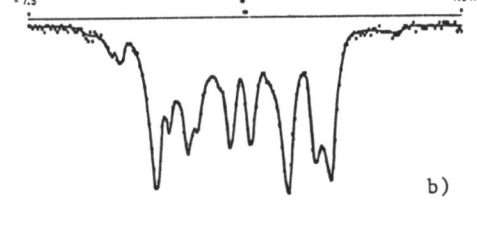

Fig. 8 ^{57}Fe Mössbauer spectra of TmFe$_4$B at 295 K. a) After Van Noort
et al (1985). b) This work.

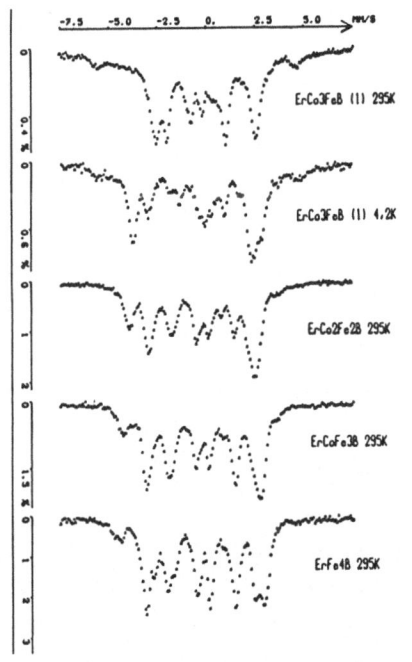

Fig. 9 ^{57}Fe Mössbauer spectra
of the series ErCo$_{4-x}$Fe$_x$B. At room
temperature there is an easy plane
for x = 3 and an easy axis for
x = 2, 1, 0. A reorientation
(plane → axis) occurs for the sample ErCo$_3$FeB (1) presented here
between 295 K and 4.2 K.

which differentiate SmCo$_4$B from SmCo$_5$ are too weak or even missing : some authors have attributed this feature to a boron disorder. Note that some RFe$_2$ impurity is often visible in the X ray spectra of our samples.

A ^{57}Fe Mössbauer study of ErFe$_4$B by Vaishnava et al (1985) led to a room temperature spectrum consisting of a main, essentially Zeeman, spectrum with broad structureless lines attributed to the 6i sites and of a small exotic external line attributed to the 2c site ; however the ratio of the two subspectra seems to be of the order 1/10 instead of the expected 1/3. Van Noort et al (1985) published a R.T. spectrum of TmFe$_4$B in which several lines of the main spectrum are double peaked. They interpret the whole spectrum (main spectrum + external line) in terms of boron disorder.

We have undertaken an investigation of RFe$_4$B. At room temperature, the spectra obtained for LuFe$_4$B, TmFe$_4$B, ErFe$_4$B look quite similar to that of Van Noort et al (1985) for TmFe$_4$B (Fig. 8), except for slight differences in the relative intensities of the two peaks in the structure of the main lines (these relative intensities also vary somewhat from sample to sample). The spectra change only slightly between R.T. and 4 K. The two inner lines seem to be too large (they increase upon annealing) ; this might be due to the presence of impurity phases with very low T_c s such as RFe$_2$B$_2$ and RFe$_4$B$_4$. The small exotic line clearly belongs to a spectrum (10% of the total intensity) characterized by both a large quadrupole effect and a large hyperfine field ($\sim$ 250 kOe at 4 K), while the main spectrum can be analyzed as a superposition of two subspectra (50% and 40% of the total intensity) with different hyperfine fields (196 kOe and 178 kOe for LuFe$_4$B at 4 K) and small apparent quadrupole effects, as already observed for 6i sites in compounds such as SmCo$_2$Fe$_2$B.

Additional information is provided by the study of the series ErCo$_{4-x}$Fe$_x$B. Magnetic measurements have been performed on the compounds with x = 0, 2, 3 by Drzazga et al (1987) who found that they have an easy axis as expected for Er ($\alpha_J > 0$) ; as concerns holmium compounds ($\alpha_J < 0$) at R.T., Drzazga et al find an easy axis for HoCo$_4$B (as in NdCo$_4$B at 77 K and 295 K) but an easy plane for HoCo$_3$FeB (as in NdCo$_3$FeB) and HoCo$_2$Fe$_2$B (homologous compound with Nd does not exist).

Mössbauer spectra have been obtained on ErCo$_{4-x}$Fe$_x$B with x = 1, 2, 3, 4 (Fig. 9). The spectrum of ErCo$_2$Fe$_2$B is quite similar to that of SmCo$_2$Fe$_2$B, confirming the existence of an easy axis. At room temperature the spectrum of ErCo$_3$FeB is quite similar to those of PrCo$_3$FeB and NdCo$_3$FeB indicating the presence of an easy plane. However at 4 K two absorbers prepared with different parts of the sample gave different spectra : absorber (1) has an easy axis (as $SmCo_3FeB$) and absorber (2) has an easy plane (as $NdCo_3FeB$). This probably arises from different iron concentrations in the two absorbers (note that absorber (1) contains a magnetic impurity with a large hyperfine field $\sim$ 313 kOe).

The evolution of the spectra at 4.2 K in the series $ErCo_3FeB$ (1), $ErCo_2Fe_2B$, $ErCoFe_3B$, $ErFe_4B$ is regular and coherent with an easy axis. By continuity it appears that the small external line of $ErFe_4B$ is a 2c line, while the main spectrum, or at least the subspectrum of it with highest hyperfine field, is due to the 6i site. However several things concerning the RFe_4B spectra remain unexplained. First : the decomposition of the main spectrum into double peaked lines when reaching x = 4 (observed by us and by Van Noort et al (1985) but not by Vaishnava et al (1985)). Second : the intensity ratio of the 2c subspectrum and of the main spectrum (if it is entirely due to 6i sites) which is too low. Third : the fact that while complete extinction of the intermediate b lines in oriented samples is well achieved for x = 2 and 3, there is a small remnant of the R.H.S. intermediate line for x = 4 (observed also for Lu and Tm). We have not be able to interpret these anomalies : is an impurity phase present in the RFe_4B, is the crystalline structure (boron positions for example) not the same in all parts of the sample, is the magnetic structure not colinear, etc..?

In conclusion the present study emphasizes the importance of the transition metal anisotropy in $RCo_{4-x}Fe_xB$ compounds. In RCo_4B, cobalt favours an easy axis and in RFe_4B (R = Er, Tm, Lu) iron also favours an easy axis. However there seems to exist an intermediate Fe concentration zone around x = 1, where the iron preference for the 2c site has a detrimental effect on the 3d anisotropy, which has important consequences on the magnetization direction at room temperature ; similar effects have been observed in $Y_2(Co_{1-x}Fe_x)_{17}$ (see fig.19 of Buschow (1977)).

BIBLIOGRAPHY

Buschow, K.H.J. 1977. Repts Prog. Phys. 40, 1179.
El-Masry, N.A. et al 1983. Zeit Metallkde 74, 86.
Drzazga, Z. et al 1987. Proceedings Internat. Symposium Bad-Soden, p. 297.
Kuzma, Yu B et al 1979. J. Less. Comm. Met. 67, 51.
Kuzma, Yu B et al 1983. J. Less. Comm. Met. 90, 217.
Pedziwiatr, A.T. et al 1987. J. Magn. Magn. Mat. 66, 69.
Smit, H.H. et al 1988. J. Phys. F Met. Phys. 18, 295.

Spada, F. et al 1984. J. Less. Comm. Met. 99, L21.

Vaishnava, P.P. et al 1985. J. Magn. Magn. Mat. 49, 286.

Van Noort, H.M. et al 1985. J. Less. Comm. Met. 111, 87.

Publications related with the present work :

Gros, Y., Frémy, M.A., Hartmann-Boutron, F., Meyer, C., Tenaud, P.
 - Paper submitted for publication to J. of Magn. Magn. Mat.
 - Communication accepted for presentation at the International Confe-
   rence on Magnetism, I.C.M., 1988, Paris.

STUDY OF COMPOUNDS $SmFe_{10}T_2$ (T = Ti, V, Si)

Several samples belonging to the new family of ferromagnetic compounds
with $ThMn_{12}$ structure were supplied by group 2.04. Their batch number and
weight composition (%) are as follows : 1780 : $Sm_{17.6}Fe_{76.7}Ti_{5.8}$; 1806 :
$Sm_{18.1}Fe_{74.6}Ti_{7.3}$; 1807 : $Sm_{20.7}Fe_{72.2}Ti_{7.1}$; 1803 : $Sm_{18.3}Fe_{68.9}V_{12.8}$;
1804 : $Sm_{14.1}Fe_{72.2}V_{13.7}$; 1809 : $Sm_{18.6}Fe_{74.7}Si_{6.7}$;
1810 : $Sm_{22.7}Fe_{70.3}Si_{7.0}$.

X ray spectra were obtained by J. Laforest (group 1.01) and M. Pernet
(Laboratoire de Cristallographie, Grenoble). They show that all three
Sm-Fe-Ti compounds are good quality samples with the $ThMn_{12}$ structure ;
sample 1807 apparently contains a small amount of $SmFe_2$. Both samples of
Sm-Fe-V are mainly amorphous, with very little crystallized phases
($SmFe_{10}V_2$, FeV, $SmFe_2$) ; ^{57}Fe Mössbauer spectroscopy indicates that under
long duration heating between 500 K and 700 K, sample n°1803 decomposes into
at least three phases : Fe metal, a phase which is paramagnetic at room tem-
perature and a phase (yet to be identified) which is still magnetic at 700
K. As concerns Sm-Fe-Si compounds, sample 1809 contains about 25%-30% of
Fe and FeSi alloys, while sample 1810 is rather good.

^{57}Fe Mössbauer spectroscopy of Sm-Fe-Ti (1780),Fig.10a,b,indicates that
it has an easy axis both at room temperature and at 150°C ; its Curie temperature is 590 K ; the lines of the Mössbauer spectrum are very broad, probably because of atomic disorder on the 8i, 8j, 8f sites : as a matter of
fact, at R.T. a fairly good fit is obtained with an amorphous type distribution of hyperfine fields, characterized by an average value $\overline{H}_n \sim 250$ kOe and
a FWHM of the order 100 kOe. $\overline{H}_n$ is about 15% lower than in the supermagnet
$Nd_2Fe_{14}B$ where $\overline{H}_n = 297$ kOe (Onodera et al, 1987). The same result must
hold for the iron magnetic moments in the two compounds.

^{57}Fe Mössbauer spectroscopy of an absorber of Sm-Fe-Si (sample 1810),
oriented in epoxy resin with a 12 kOe field, shows only partial extinction

of the median lines : the existence of an easy axis in this compound is therefore not proved. The hyperfine field at R.T. is of the order 220 kOe and the Curie temperature is 550 K. These characteristics are less favourable for permanent magnet application than those of Sm-Fe-Ti.

BIBLIOGRAPHY

Onodera, H. et al. 1987. J. Magn. Magn. Mat., 68, 6.

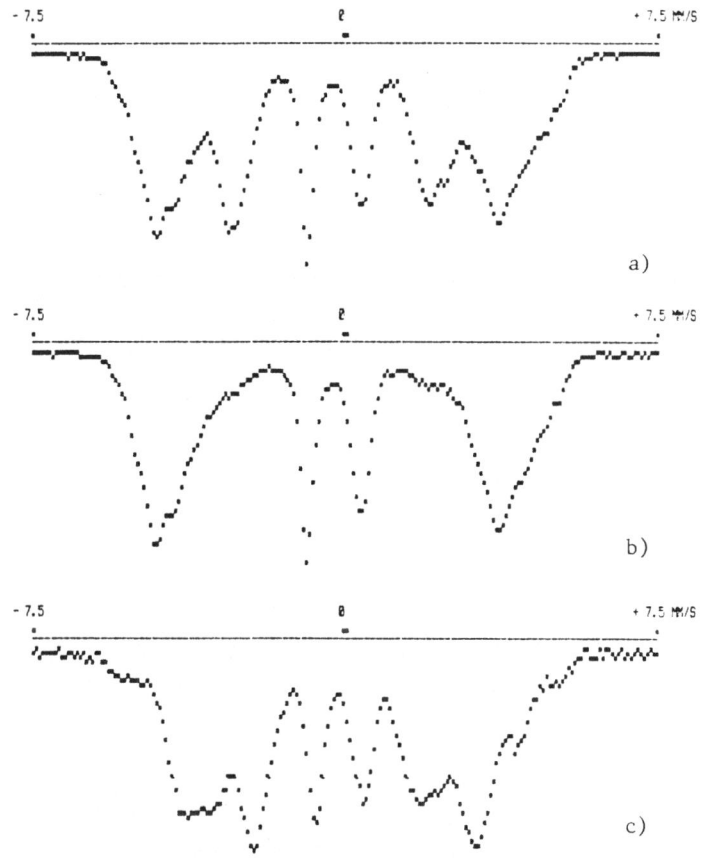

Fig. 10 ^{57}Fe Mössbauer spectra at room temperature. a) Sm-Fe-Ti (1780) : standard powder ; b) Sm-Fe-Ti (1780) : oriented powder (γ ∥ to the orienting field) showing the extinction of the intermediate lines (easy axis) ; c) Sm-Fe-Si (1810) : standard powder.

INTRINSIC PROPERTIES OF TERNARY RARE-EARTH COMPOUNDS
FROM ^{57}Fe, ^{161}Dy, ^{166}Er MÖSSBAUER SPECTROSCOPIES

J.P. Sanchez

Centre de Recherches Nucléaires (I N2 P3)
67037 Strasbourg Cedex, France

ABSTRACT

^{57}Fe, ^{161}Dy and ^{166}Er Mössbauer results are reported in the $R_2Fe_{14}B$ alloys and some of the corresponding hydrides. The local moments at the six inequivalent iron crystallographic sites are inferred from the ^{57}Fe hyperfine field data. The occurrence of fairly pure J_z = 15/2 state at the two Dy and Er sites is demonstrated from the near free ion values of the rare-earth hyperfine fields. Empirical estimates of the second order crystal field parameters are provided by the ^{161}Dy quadrupole interaction data. Informations on the magnetic structures and spin reorientation phenomena are obtained from the study of magnetically oriented samples. A detailed analysis of the spin reorientation in $(Er_{1-x}Dy_x)_2Fe_{14}B$ alloys allowed to estimate the importance of higher order crystal field terms (B_4^0 , B_6^0). The study of the $R_{1+\varepsilon}Fe_4B_4$ phases indicated that the iron atoms are essentially non-magnetic and revealed the occurrence of a giant value for the crystal field parameter A_2^0. The magnetic properties of the $Nd(Co_{1-x}Fe_x)_9Si_2$ alloys, inferred by combining bulk measurements with Mössbauer and NMR data, are presented.

INTRODUCTION

The elucidation of the basic properties of rare-earth intermetallic compounds, which is often complicated by the coexistence of large number of low symmetry atomic sites, requires a combination of experimental methods. Among these, Mössbauer spectroscopy is attractive for the determination of the local magnetic and structural properties from hyperfine interaction measurements at both the iron and rare-earth elements. This microscopic tool provides a variety of useful informations concerning the local magnetic moments, the magnetic structure and spin reorientation effects. The analysis of the quadrupole interaction data at the rare-earth sites allows an evaluation of the second order crystal field parameters, a key to understand the magnetocrystalline anisotropy of these materials.

In this paper, we will summarize our contribution to the knowledge of the intrinsic properties of the $R_2Fe_{14}B$ phases and related compounds and present the results we obtained from the study of the $R_{1+\varepsilon}Fe_4B_4$ phases. Finally, we will report on the magnetic properties of the $Nd(Co_{1-x}Fe_x)_9Si_2$ compounds.

MÖSSBAUER MEASUREMENTS IN THE $R_2Fe_{14}B$ ALLOYS AND RELATED PHASES

Iron sublattice properties : local moments

[57]Fe Mössbauer investigations of the $R_2Fe_{14}B$ alloys were undertaken with the emphasis to obtain informations on the magnetic properties of the Fe sublattice (Friedt et al.,1984 ; Friedt et al.,1986 a,b ; Sanchez, 1987). Complex Mössbauer spectra (Fig.1) were observed as expected owing to the occurrence of six iron sites of low symmetry in addition to the presence of impurity phases ($\alpha - Fe$, $R_{1+\epsilon}Fe_4B_4$ ...). Due to the complexities of the

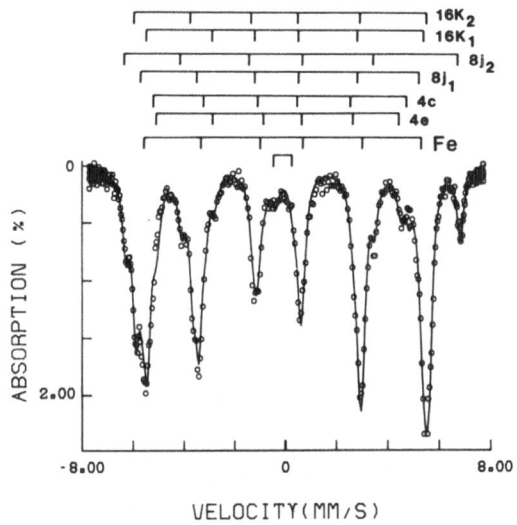

Fig.1 [57]Fe Mössbauer spectrum of $Dy_2Fe_{14}B$ at 4.2 K. The bar diagrams refer to the six iron sites and to the impurity phases, α-Fe and $Dy_{1+\epsilon}Fe_4B_4$. The notation of the sites follows the notation of Givord et al., 1984.

spectra and the limited resolution, there was no unique way to fit the data; but some general agreement exists for the hyperfine values of the three major sites $16k_1$, $16k_2$ and $8j_2$. Nevertheless, the average hyperfine field appears to be insensitive to the fitting model. The assignment of the different Fe sites was made by reference to the systematics of the hyperfine field parameters established in the R-Fe intermetallics. Friedt et al. (1986 a,b) showed from point charge calculation that the $8j_2$ site has the largest quadrupole interaction (nearly axial). This favourable circumstance was used to detect spin reorientation transition. The values of the iron moment estimated from the hyperfine field data (Table 1) are only in crude agreement with those determined from neutron measurements (Friedt et al.,1986 a,b ; Sanchez, 1987 ; Givord et al.,1985).

TABLE 1   Values of the hyperfine fields at the six iron sites in $Y_2Fe_{14}B$ at 4.2 K. The iron moment was estimated from the hyperfine field corrected for the dipolar field (1 $\mu_B \equiv$ 15 T). The last line corresponds to the moments deduced from neutron experiments (Givord et al.,1985).

	$16k_2$	$16k_1$	$8j_2$	$8j_1$	4c	4e	Average
H_{hf} (T)	34.4	32.5	38.1	32.1	32.4	32.0	33.7
$H_{hf}-H_{dip}$ (T)	34.6	32.5	38.8	32.7	31.5	32.0	33.9
μ (μ_B)	2.31	2.17	2.59	2.18	2.10	2.13	2.26
μ_n (μ_B)	2.25	2.25	2.80	2.40	1.95	2.15	2.32

Rare-earth sublattice properties : local moments and crystal field

^{161}Dy and ^{166}Er Mössbauer measurements (Fig.2) of the compounds $R_2Fe_{14}B$ have been reported by Friedt et al.(1986 a,b) and by Sanchez et al. (1986). The intraionic fields ($H_i = H_{4f} + H_{cp}$, where H_{4f} is the field produced by the 4f electrons and H_{cp} the core polarization contribution) at the ^{166}Er (757 T, average) and ^{161}Dy (568 and 566.5 T) have been shown to be very close to the free ion values (765 and 563 T for ^{166}Er and ^{161}Dy, respectively). This implies that J_z is nearly fully saturated in $R_2Fe_{14}B$ alloys as expected when the exchange dominates the crystal field interaction.

The ^{161}Dy quadrupole interaction data together with point charge calculations (the latter provides the orientation of the principal system of axes of the crystal field tensor as well as the ratio $B_2^2/B_2^0 = \eta$) were used by Friedt et al.(1986 a) to estimate the second - order crystal field parameter B_2^0 at the two Dy sites. The values of B_2^0 found for both Dy sites are of equal sign (negative) and magnitude and the corresponding A_2^0 values compare well with those obtained from ^{155}Gd quadrupole interaction data (Table 2). The deduced second order crystal field parameters predict that the c-axis is an easy axis for R ions with the second order Stevens coefficient $\alpha_J < 0$ while an easy plane is favoured when $\alpha_J > 0$ (e.g. Er).

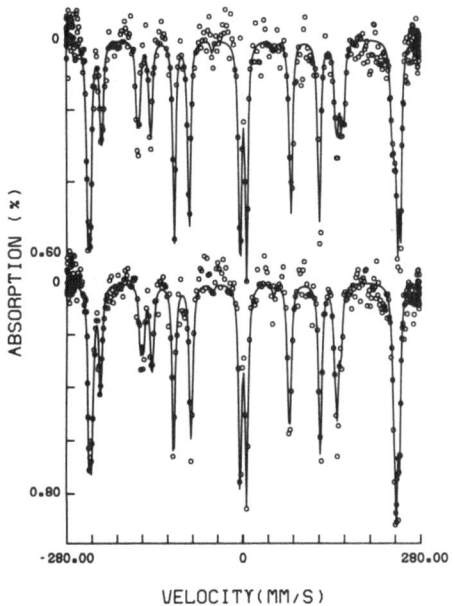

Fig.2 ^{161}Dy Mössbauer spectra at 4.2 K of $Dy_2Fe_{14}B$ (top) and of $Dy_2Fe_{14}B$ $H_{3.3}$ (bottom).

TABLE 2   Second order crystal field parameters (referred to c-axis) estimated from ^{161}Dy quadrupole interaction data. Comparison with ^{155}Gd data (Bogé et al.,1986) and point charge calculations.

		A_2^0 (K / a_0^2)	$|A_2^2|$ (K / a_0^2)
Site 4f	^{155}Gd	680	414
	^{161}Dy	650	171
	p.c.	1900	500
Site 4g	^{155}Gd	661	1298
	^{161}Dy	650	743
	p.c.	1400	1600

Hydrogen dependence of the hyperfine interactions

^{57}Fe Mössbauer measurements on the $R_2Fe_{14}B$ H_x (R = Dy, Er) alloys were reported by Friedt et al. (1984 ; 1986 a,b). The data showed that the average hyperfine field increases (2 - 3 %) on hydrogenation. The increase of the average Fe moment was however insufficient to account for the observed enhancement of the Curie temperature ; volume expansion and electronic factors must be considered too.

The influence of hydrogen on the hyperfine parameters at the rare earth site was investigated by ^{161}Dy and ^{166}Er Mössbauer spectroscopies (Friedt et al.,1986a; Sanchez et al.,1986). In the hydrides, the hyperfine fields were significantly reduced (~ 3 % for Er and ~ 1 % for Dy) in comparison to the virgin samples (Fig.2). This was ascribed to changes of the conduction electron polarization contribution and /or/ (Er hydride) to crystal field admixture. The decrease on hydrogenation of the ^{161}Dy isomer shift was interpreted as due to a charge transfer from the Dy 6s orbital onto the H atoms.

The large increase of the quadrupole coupling constant in the Dy hydride was related to the decrease of the K_1 anisotropy constant with hydrogen charging.

Magnetic structures and spin reorientation phenomena

At high temperature, the iron anisotropy overrules that due to the rare-earth. The iron moment is collinear with the c-axis (the iron easy axis). The low temperature magnetic structure is determined by the rare-earth anisotropy e.g. an easy plane perpendicular to the c-axis for Er. Spin reorientations may be observed when the rare-earth and iron anisotropies compete. The iron moment reorientation phenomena encountered in $R_2Fe_{14}B$ alloys have been investigated thoroughly by Mössbauer effect measurements in magnetically oriented samples. The reorientation temperature (T_r) was detected by following the intensity change of the intermediate resonance lines (Fig.3).

For $Er_2Fe_{14}B$ (Vasquez et al.,1985) the ^{57}Fe Mössbauer spectra were shown to change dramatically around 328 K. These data clearly evidenced that the spins rotate from the c-axis to basal plane on cooling the alloy. It was demonstrated, on the other hand, (Vasquez et al.,1985) that the Fe moments remained aligned parallel to the c-axis when cooling, down to 4.2 K,

$Ce_2Fe_{14}B$ and $Dy_2Fe_{14}B$ alloys.

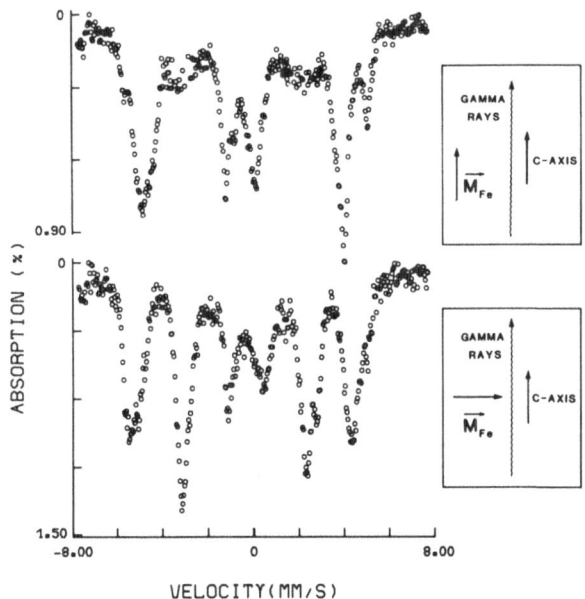

Fig.3 [57]Fe Mössbauer spectra of magnetically aligned $Er_2Fe_{14}B$ powder sample taken above (top) and below (bottom) T_r = 328 K.

The change of T_r by partial substitution of Er by another rare-earth in $(Er - R)_2Fe_{14}B$ alloys was investigated by [57]Fe Mössbauer spectroscopy in $(Er_{1-x}Gd_x)_2Fe_{14}B$ and $(Er_{1-x}Dy_x)_2Fe_{14}B$ magnetically oriented samples (Vasquez and Sanchez, 1987 ; Rechenberg et al.,1987a). The study of T_r as a function of x was used for testing microscopic models for the crystal field and exchange interactions in these alloys.

The data for $(Er_{1-x}Gd_x)_2Fe_{14}B$ were quantitatively explained in terms of a Hamiltonian which included second order crystal field terms, R-Fe interaction, plus an uniaxial anisotropy due to the Fe sublattice (Vasquez and Sanchez, 1987). The consideration of fourth and sixth-order crystal field terms in addition to the second order ones was found to be necessary for a quantitative account of the observed concentration dependence of T_r in $(Er_{1-x}Dy_x)_2Fe_{14}B$ (Rechenberg et al.,1987a ; Table 3).

TABLE 3   Crystal field parameters deduced from the analysis of the concentration dependence of T_r in $(Er_{1-x}Dy_x)_2Fe_{14}B$.

Ion (site)	B_2^0 (K)	B_2^2 (K)	B_4^0 (K)	B_6^0 (K)
Er(4f)	0.36	±0.22	-3.0×10^{-3}	5.0×10^{-5}
Er(4g)	0.34	±0.66	-3.0×10^{-3}	5.0×10^{-5}
Dy(4f)	-0.98	±0.60	4.7×10^{-3}	3.1×10^{-5}
Dy(4g)	-0.91	±1.80	4.7×10^{-3}	3.1×10^{-5}

MAGNETIC PROPERTIES OF THE $R_{1+\epsilon}Fe_4B_4$ ALLOYS

The $R_{1+\epsilon}Fe_4B_4$ alloys which appear as an impurity phase in the $R_2Fe_{14}B$ compounds have been extensively studied for their structure and magnetic properties (Givord et al.,1986 ; Niarchos et al.,1986 ; Rechenberg and Sanchez, 1987 ; Rechenberg et al.,1987b).

All compounds of the $R_{1+\epsilon}Fe_4B_4$ series were shown to be paramagnetic at 300 K but most of them order ferromagnetically at low temperatures (The highest ordering temperature was determined in $Sm_{1+\epsilon}Fe_4B_4$, $T_c \cong 37$ K). Such low temperatures are related to the absence of magnetic moment on Fe which was inferred from Mössbauer measurements (Rechenberg and Sanchez, 1987). ^{57}Fe Mössbauer spectra for all compounds are doublets. Line broadenings below T_c, due to the occurrence of a small magnetic splitting, were observed only in two cases, namely R = Dy and Sm. Analysis of the spectra revealed the magnetization to be parallel to the c-axis for Sm and perpendicular to the c-axis for Dy (Rechenberg and Sanchez, 1987). These results are consistent with a second order crystal field mechanism for anisotropy ; the negative sign of A_2^0 was deduced from ^{155}Gd and ^{161}Dy Mössbauer measurements (Rechenberg et al.,1987b). It is of interest to notice the giant value of the A_2^0 term (-2450 K / a_0^2) due to the peculiar structure of the $R_{1+\epsilon}Fe_4B_4$ alloys. The quasi-incommensurate nature of the R and Fe + B sublattices is reflected by a small spread of the R hyperfine interaction parameters. The importance of crystal field effects versus exchange was demonstrated from the reduction (with respect to the free ion values) of the 4f contribution to the ^{161}Dy hyperfine parameters.

The ground state Dy magnetic moment was estimated to amount to 9.2 μ_B.

MAGNETIC PROPERTIES OF $Nd(Co_{1-x}Fe_x)_9Si_2$ ALLOYS

The discovery of $R_2Fe_{14}B$ phases has initiated efforts to produce new iron-rich pseudo binary compounds related to phases with well-known structural types (e.g. $BaCd_{11}$). It was recently shown that $Nd(Co_{1-x}Fe_x)_9Si_2$ alloys $(0 < x \leqslant 0.55)$ with the $BaCd_{11}$ structure type can be stabilized ; preliminary account of their magnetic properties showed T_c to be around 500 - 560 K (Chevalier et al.,1987). This earlier report was supplemented by detailed NMR (^{59}Co, ^{143}Nd) and Mössbauer (^{57}Fe) studies (Berthier et al., to be published). Only the results of the Mössbauer investigations will be presented.

^{57}Fe Mössbauer measurements have been carried out on four samples of $Nd(Co_{1-x}Fe_x)_9Si_2$ with x = 0.11, 0.22, 0.33 and 0.44. It has been shown that : a) the Fe atoms are not randomly distributed for small x ($\leqslant 0.11$) ; b) the average hyperfine field (or magnetic moment) passes through a maximum with increasing x ; it decreases significantly for $x \geqslant 0.44$ (Table 4). The latter behaviour very similar to what occurs in $R(Co_{1-x}Fe_x)_n$ alloys (n = 2, 3) can be explained in terms of a rigid band model. Mössbauer measurements in applied external field demonstrated that the iron sublattice magnetization is collinear with the applied field when $H_{ext} \geqslant 1.5$ T (Fig.4).

The average Co magnetic moments have been estimated by combining the $\bar{\mu}_{Fe}$ data with the average transition metal moment deduced from the bulk magnetization measurements under the assumption of ferromagnetic alignment of the Nd, Fe and Co sublattices. The in-field ^{59}Co NMR data showed, however, that the cobalt moments are canted. These data combined with those of Table 4 allowed to estimate the Co canting angle (Berthier et al.,1988).

TABLE 4   Average ^{57}Fe hyperfine field and transition metal moments in $Nd(Co_{1-x}Fe_x)_9Si_2$ compounds.

x	$\bar{H}_{hf}$ (T)	$\bar{\mu}_{Fe}$ (μ_B)	$\bar{\mu}_{Co}$ (μ_B)
0.11	21.9	1.51	0.90
0.22	24.7	1.70	0.96
0.33	25.2	1.74	1.04
0.44	23.1	1.59	1.14

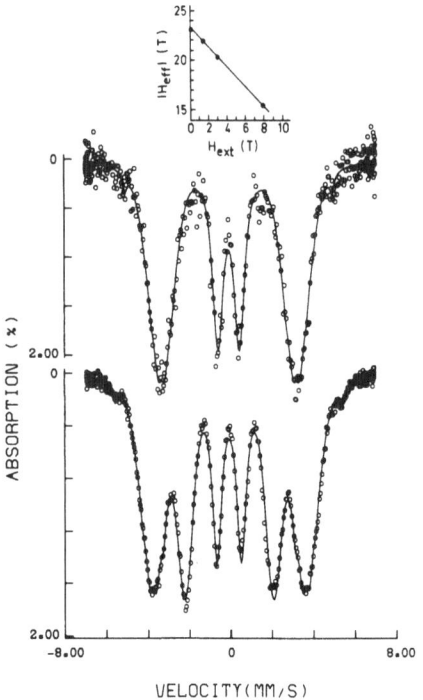

Fig. 4 ^{57}Fe Mössbauer spectra of $Nd(Co_{0.56}Fe_{0.44})_9Si_2$ at 4.2 K without (bottom) and within (top) a longitudinal external field of 3 T. The inset shows the external field dependence of the effective field.

ACKNOWLEDGMENTS

The collaboration of Y. Berthier, A. Bonnenfant, B. Chevalier, J. Etourneau, J.M. Friedt, R. Fruchart, P. L'Héritier, R. Poinsot, H. Rechenberg and A. Vasquez, at different stages of the CEAM project, is gratefully acknowledged.

REFERENCES

Berthier, Y., Chevalier, B., Etourneau, J., Rechenberg, H.R. and Sanchez, J.P. 1988. Magnetic properties of $Nd(Co_{1-x}Fe_x)_9Si_2$ alloys from magnetization, NMR and Mössbauer studies (to be published).

Bogé, M., Coey, J.M.D., Czjzek, G., Givord, D., Jeandey, C., Li, H.S. and Oddou, J.L. 1986. A ^{155}Gd Mössbauer study of a $Gd_2Fe_{14}B$ single crystal. J. Phys. F 16, L67-72.

Chevalier, B., Gurov, G., Fournes, L. and Etourneau, J. 1987. New ferromagnetic compounds : the ternary silicides $NdCo_{9-x}Fe_xSi_2$ ($0 \leqslant x \leqslant 5$). J. Chem. Research 5, 138.

Givord, D., Li, H.S. and Moreau, J.M. 1984. Magnetic properties and crystal structure of $Nd_2Fe_{14}B$. Solid State Comm. 50, 497 - 499.

Givord, D., Li, H.S. and Tasset, F. 1985. Polarized neutron study of the compounds $Y_2Fe_{14}B$ and $Nd_2Fe_{14}B$. J. Appl. Phys. <u>57</u>, 4100-4102.

Givord, D., Moreau, J.M. and Tenaud, P. 1986. Refinement of the crystal structure of $R_{1+\epsilon}Fe_4B_4$ compounds (R = Nd, Gd). J. Less Common Met. <u>123</u>, 109-116.

Friedt, J.M., Sanchez, J.P., L'Héritier, P. and Fruchart, R. 1984. Mössbauer spectroscopy in the $RE_2Fe_{14}BH_x$ materials. In "Proceedings of the Workshop on Nd-Fe Permanent Magnets" (Ed. I.V. Mitchell). (Commission of the European Communities, Brussels). pp.179-182.

Friedt, J.M., Vasquez, A., Sanchez, J.P., L'Héritier, P. and Fruchart, R. 1986a. Magnetism and crystal field properties of the $RE_2Fe_{14}BH_x$ alloys (RE = Y, Ce, Dy, Er) from Mössbauer spectroscopy. J. Phys. F <u>16</u>, 651-667.

Friedt, J.M., Vasquez, A., Sanchez, J.P., L'Héritier, P. and Fruchart, R. 1986b. Mössbauer spectroscopy on $RE_2Fe_{14}BH_x$ alloys. Hyperfine Interactions <u>28</u>, 611-614.

Niarchos, D., Zouganelis, G., Kostikas, A. and Simopoulos, A. 1986. Magnetic properties of the $R_{1+\epsilon}Fe_4B_4$ compounds (R = rare earth) from magnetization and Mössbauer measurements. Solid State Comm. <u>59</u>, 389-391.

Rechenberg, H.R. and Sanchez, J.P. 1987. Effective magnetic fields on Fe in $R_{1+\epsilon}Fe_4B_4$ alloys. Solid State Comm. <u>62</u>, 461-464.

Rechenberg, H.R., Sanchez, J.P., L'héritier, P. and Fruchart, R. 1987a. Crystal-field interactions and spin reorientation in $(Er_{1-x}Dy_x)_2Fe_{14}B$. Phys. Rev. B <u>36</u>, 1865-1871.

Rechenberg, H.R., Bogé, M., Jeandey, C., Oddou, J.L., Sanchez, J.P. and Tenaud, P. 1987. [155]Gd and [161]Dy Mössbauer study of $R_{1+\epsilon}Fe_4B_4$ alloys (R = Gd, Dy). Solid State Comm. <u>64</u>, 277-281.

Sanchez, J.P., Friedt, J.M., Vasquez, A., L'Héritier, P. and Fruchart, R. 1986. [166]Er Mössbauer spectroscopy in the $Er_2Fe_{14}BH_x$ alloys. Solid State Comm. <u>57</u>, 309-313.

Sanchez, J.P. 1987. New permanent magnets investigated by Mössbauer spectroscopy. In "Proceedings of XXII Zakopane School on Physics, Part 2 : condensed matter studies by nuclear methods" (Eds. K. Krolas and K. Tomala) (Institute of Nuclear Physics and Jagiellonian University, Krakow). pp.156-193.

Vasquez, A., Friedt, J.M., Sanchez, J.P., L'Héritier, P. and Fruchart, R. 1985. Spin reorientation phenomena in $RE_2Fe_{14}B$ (RE = Ce, Dy, Er) alloys from [57]Fe and [161]Dy Mössbauer spectroscopies. Solid State Comm. <u>55</u>, 783-786.

Vasquez, A. and Sanchez, J.P. 1987. Spin reorientation phenomena in $(Er_{1-x}Gd_x)_2Fe_{14}B$ alloys. J. Less. Common. Met. <u>127</u>, 71-78.

HYPERFINE INTERACTIONS ON R-Fe-B COMPOUNDS

STUDIED BY MOSSBAUER SPECTROSCOPY

M. Bogé, C. Jeandey, J.L. Oddou

Centre d'Etudes Nucléaires de Grenoble
D.R.F./Service de Physique/M.D.I.H.
85X F-38041 GRENOBLE (FRANCE)

ABSTRACT

Hyperfine parameters in the $R_2Fe_{14}B$ compounds (R=Gd, Dy), have been determined by Mössbauer spectroscopy (^{155}Gd and ^{161}Dy nuclei). As Gd is an S ion, hyperfine fields measured yield information on transferred fields and quadrupolar effects on the distribution of charges in the environment of the ion. On the other hand, for Dy, the orbital terms are dominant in the hyperfine parameters. In the $R_2Fe_{14}B$ compounds, two rare earth sites have to be considered, in accordance with the known crystal structure.

Gd resonance study shows the main axis of the electric field gradient to have different directions at the two rare earth sites. Moreover, crystal electric field (C.E.F.) parameters A_2^0 and A_2^2 may be deduced very accurately from Mössbauer spectra taken on a $Gd_2Fe_{14}B$ single crystal absorber. A comparison between spectra taken in zero field and in a field of 8 Tesla shows the magnetic hyperfine field at Gd nuclei to be positive at both lattice sites and primarily caused by a strong polarisation due to Fe moments. Experimental temperature dependence of the Dy hyperfine field B_{hf} is in an excellent agreement with the calculated curve of temperature dependence deduced for the Dy moments. C.E.F. effects can be neglected.

From ^{155}Gd Mössbauer experiments in $Gd_{1+\epsilon}Fe_4B_4$, a crystal field term A_2^0 equal to- 2450 K/a_0^2 is deduced. This value is to our knowledge the largest found in Gd intermetallic compound. No unique direction between the hyperfine field and the c axis may be established. The hyperfine field becomes perpendicular to the c axis when Nd is substituted to Gd and parallel to the c axis when Sm is substituted to Gd. This is inferred to the different sign of the second order Stevens factor α_J in Nd and Sm (whereas $\alpha_J = 0$ in Gd).

INTRODUCTION

The Mössbauer spectroscopy is a highly useful tool for the study of the magnetic materials. It measures the microscopic magnetic and structural properties of the resonant ion. In our Mössbauer study of $R_2Fe_{14}B$ compounds we have used two rare earths : Gd and Dy.

Gadolinium

Gadolinium is normally found as the trivalent ion which forms the S-state ($L = 0$). Therefore the usually large 4f-orbital contribution to the magnetic hyperfine field vanishes as well as the 4f contribution to the electric field gradient (EFG). Thus, the EFG observed is caused only by the distribution of charges in the environment of the ion. Mostly used is the 85 keV resonance in ^{155}Gd. The source $^{155}Eu : SmPd_3$ is kept at 4.2K.

Dysprosium

Dysprosium is a typical rare earth with large orbital momentum. The orbital term ($B_{orb} \simeq 600$ Tesla) is dominant in the hyperfine field. The 25.65 keV resonance on ^{161}Dy give absorption spectra with a very good resolution. In addition, the dysprosium offers the possibility of a clear separation of magnetic and quadrupolar interactions. The source $^{161}GdF_3$ is produced by thermal neutron irradiation.

^{155}Gd MOSSBAUER STUDY OF $Gd_2Fe_{14}B$:

We have performed our experiments on powder and single crystal absorbers (for this last study, three platelets of 0.3 mm thickness,with the crystallographic c-axis perpendicular to the face of the platelets, were cut and assembled together to cover a total surface of 1 cm^2).

In figure 1(a), we display the Mössbauer spectrum of the powder absorber and in figures 1(b) and (c) we show the spectra of the single crystal absorber respectively in zero field and in a field of 8 T, applied along the c-axis, i.e parallel to the γ-rays. Obviously, the single-crystal spectra exhibit much more structure than the spectrum of the powdered sample and lead to considerably improved precision in the hyperfine interaction parameter values and in estimates for the second order crystal field parameters.

The spectra were fitted to the transmission integral. In accordance with the crystal structure, two components of equal intensity but with different hyperfine interactions, corresponding to the two rare earth sites, were assumed. Positions of the resonance absorption lines were determined by numerical diagonalisation of the hyperfine Hamiltonians.

312

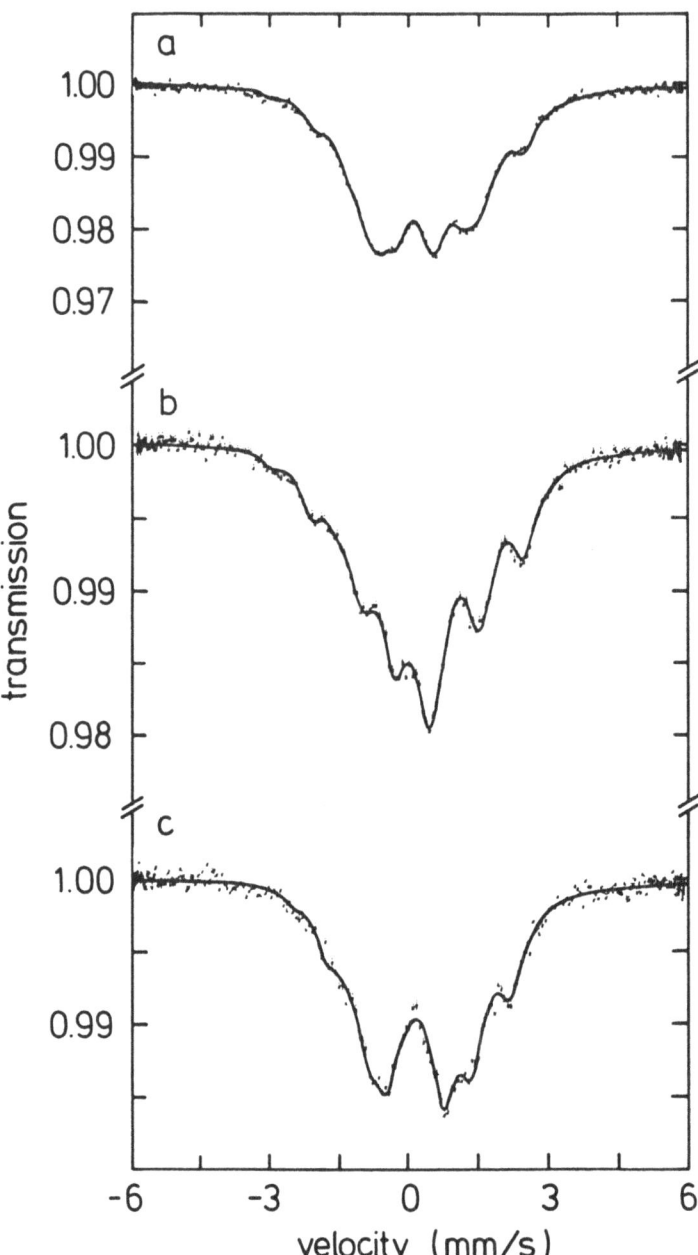

Fig. 1 ^{155}Gd Mössbauer spectra of $Gd_2Fe_{14}B$: (a) powder absorber, (b) single crystal in zero field, (c) single crystal in a field of 8T. Full curves show transmission curves resulting from least-squares fits.

Magnetisation measurements have shown the Gd moments in $Gd_2Fe_{14}B$ to be oriented parallel to the crystallographic c-axis. Since the a-b plane is a mirror plane in the point symmetry group for both rare-earth sites the c axis is also one of the principal axes of the EFG tensor. Similarly the other two principal axes are [110] and [1$\bar{1}$0]. Among the three possibilities for the polar coordinates (θ, φ) of the direction of the Gd moments with respect to the coordinates defined by the principal axes (x,y,z) of the EFG tensor [namely $(\frac{\pi}{2}, 0)$, $(\frac{\pi}{2}, \frac{\pi}{2})$ and $(0, 0)$], the best fits to the Mössbauer spectra is obtained with polar angles $(0, 0)$ for component 1 (c axis $\nparallel$ z axis) and $(\frac{\pi}{2}, \frac{\pi}{2})$ for component 2 (c axis $\nparallel$ y axis). This is in agreement with point charge calculations (Givord et al. 1984, Cadogan and Coey 1984) if we ascribe component 1 in the Mössbauer spectra to the crystallographic site 4f (GdI) and component 2 to site 4g (GdII).

Results for hyperfine interaction parameters deduced from fits to powder and single-crystal spectra along with weighted averages are summarised in table 1.

TABLE 1 Results of ^{155}Gd Mössbauer spectra for $Gd_2Fe_{14}B$. Isomer shifts with respect to the ^{155}Eu:$SmPd_3$ source. Errors are given in parentheses (unit : last digit of parameter value given).

B_{appl} (T)	Sample	Site	δ_{IS} (mm s^{-1})	B_{hf} (T)	V_{zz} (10^{21} V m^{-2})	η
0	Powder	1	0.224(5)	14.7(4)	−7.99(18)	0.44(14)
		2	0.166(7)	27.9(5)	11.01(9)	0.34(5)
	Single	1	0.225(6)	15.4(2)	−7.59(16)	0.66(8)
	crystal	2	0.172(10)	27.5(2)	11.41(9)	0.35(3)
	Average	1	0.224(5)	15.3(2)	−7.77(20)	0.61(7)
		2	0.169(7)	27.6(2)	11.21(20)	0.35(3)
8	Single	1	0.217(9)	9.4(3)	−7.75(11)	0.61†
	crystal	2	0.204(13)	20.1(3)	11.37(11)	0.35†

† Parameter was not varied in fitting.

The values for the parameters V_{zz} and $\eta = \dfrac{V_{xx} - V_{yy}}{V_{zz}}$ describing the EFG

tensor yield estimates for the crystal field parameters B_2^0 and B_2^{2c} . Separating the shielding factor $(1-\sigma_2)$ and those factors which vary from one rare-earth ion to another (Stevens factor α_J and $\langle r^2 \rangle_{4f}$), we define universal parameters A_2^m by

$$B_2^m = (1-\sigma_2)\ \alpha_J\ \langle r^2 \rangle_{4f}\ A_2^m$$

The parameters A_2^m are applicable approximately for all rare-earth ions in the compounds $R_2 Fe_{14} B$.[For the conversion from $(V_{zz}$, $\eta)$ to A_2^m see Bogé et al. 1986].

The values for A_2^m deduced from the average results for V_{zz} and η are listed in table 2 along with values resulting from point charge calculations (Givord et al. 1984).

TABLE 2 Crystal field parameters A_2^m as defined in the text, in units K/a_0^2. Calculated values from Givord et al.

Site		A_2^0	$\left| A_2^2 \right|$
1 (4f)	from MS[+]	680(17)	414(49)
	calculated	1900	500
2 (4g)	from MS[+]	661(19)	1298(27)
	calculated	1400	1600

+ MS : Mössbauer spectroscopy.

The agreement between values deduced from Mössbauer spectroscopy and values calculated is satisfactory since the first ones represent under estimates of the quantities A_2^m (Bogé et al. 1986).

The positive sign of A_2^0 implies a preferred magnetisation parallel to the c axis for those rare-earth ions for which $\alpha_J < 0$, such as Nd, Pr, Dy for example. The sign of A_2^2, on the other hand is totally irrelevant.

The difference between the measured hyperfine fields and the free-ion field for Gd^{3+}, about $-34T$ (Freeman and Watson 1962), amounts to $+49T$ for site 4f and to $+62T$ for site 4g. This difference arises from conduction electron polarisation by the local 4f moment (self-polarisation field B_s) and from transferred polarisation due to the surrounding moments. The transferred polarisation from the rather distant Gd neighbours is likely to be negligible. The transfer then is primarily due to Fe moments. From NMR data by Berthier et al. (1985) for compounds with non-magnetic Y and Lu, for which B_s vanishes, we derive an estimate of 53T for the transferred field at Gd nuclei, using the appropriate ratios of s-electron hyperfine constants calculated by Campbell (1969). This value is close to those deduced from experiment. Thus, B_s does not contribute significantly to the Gd hyperfine fields.

The difference between the hyperfine fields at the two Gd sites is consistent with results obtained for other compounds $R_2Fe_{14}B$ (R = Nd, Dy, Lu and Y) investigated by N.M.R. (Berthier et al. (1986), and Berthier, this conference).

Finally, a comparison of the entries in table 1 for the results obtained in zero field and in a field of 8T shows that the magnetic hyperfine field at both sites decreases in the field. The difference expected is

$$\delta B_{hf} = B_{appl} - 4\pi DM$$

which quantity equals 7.1T for the saturation magnetisation reported by Bogé et al. (1985) and D = 1 (flat absorber discs).

The difference actually found for site 2 agrees well with this expectation ; the result for site 1 $|\delta B|$ = 6.0 ± 0.4T, is somewhat small, but in view of the uncertainty associated with these differences we do not believe this to be significant.

For the sign of the hyperfine field we adopt the definition that it is positive if the hyperfine field points in the same direction as the local moment of the electronic shell of the probe nucleus, in our case the Gd moment. In the ferrimagnetic compound $Gd_2Fe_{14}B$, the bulk magnetisation is dominated by the Fe moments, and the Gd moments point in the opposite direction to the applied magnetic field. Thus, the results deduced from the spectra imply positive hyperfine fields at Gd nuclei on both sites.

In summary, [155]Gd Mössbauer spectra of a $Gd_2Fe_{14}B$ single crystal have yielded estimates for the crystal field parameters A_2^m in reasonable agreement with results of point charge calculations. The sign of A_2^o favours the crystallographic c-axis as the easy axis of magnetisation for both Nd sites in $Nd_2Fe_{14}B$. The magnetic hyperfine field is positive at both Gd sites. This is primarily caused by a strong transferred polarisation due to Fe moments leading to a transferred hyperfine field of about +50T at Gd nuclei.

[161]Dy MOSSBAUER STUDY OF $Dy_2Fe_{14}B$

We have performed our experiments on a polycrystalline powder sample. Figure 2 shows spectra between 4.2 K and 373 K. Due to the particular parameters of the Dy resonance one obtains at 4.2 K the hyperfine magnetic

316

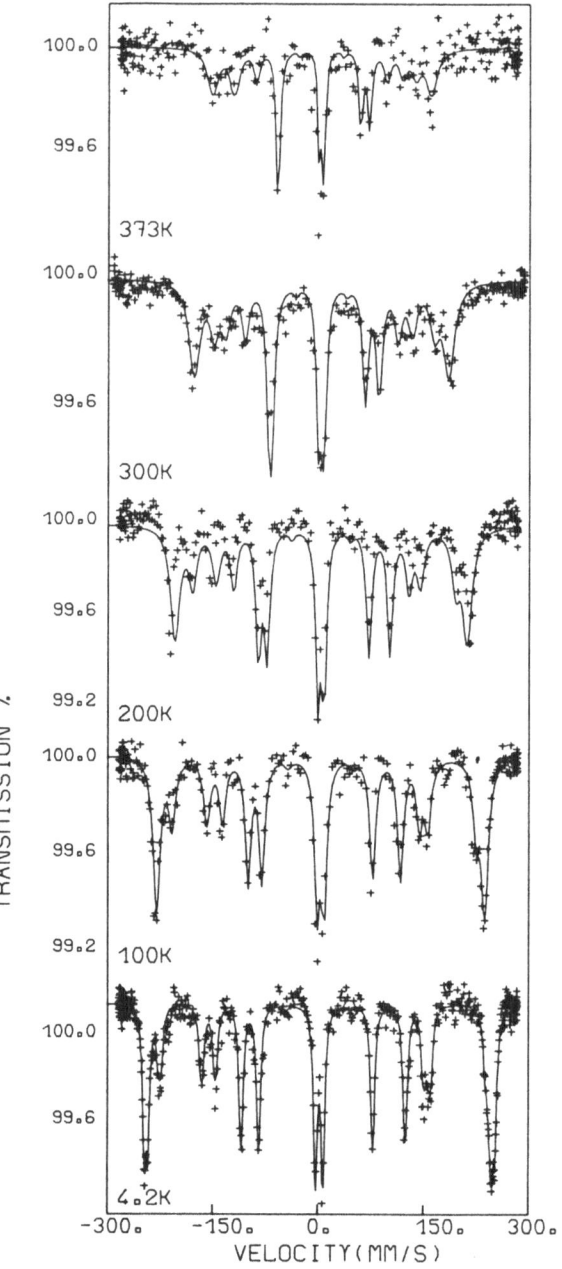

Fig. 2 ^{161}Dy Mössbauer spectra of $Dy_2Fe_{14}B$ at various temperatures.

field B_{hf} simply and accurately from the separation of the intense, outermost lines. Between 4.2 K and 100 K one can determine B_{hf} and the quadrupole interaction $e^2q\,Q$ with a relative accuracy of respectively 0.5% and 1%. Above 100 K, one observes the superposition of a paramagnetic relaxation spectra, a well known phenomenon in some Dy compounds.

Clearly, the spectrum at 4.2 K analysed with a least-squares fitting program, requires the superposition of two subspectra corresponding to two inequivalent sites R_1 and R_2 with different values of the hyperfine interaction parameters (Table 3). This is in agreement with the known crystal structure of the $R_2Fe_{14}B$ compounds.

Up to 100 K the relaxation phenomenon does not appear. We are in a static regime. The fitting program determines easily the hyperfine parameters. Above 100 K, we have used the two level formalism in the effective field approximation (Wickman, 1968) to fit these spectra. At 373 K, the two sites are not distinguished on account of high level relaxation rate.

TABLE 3   Hyperfine parameters for the two Dy sites of $Dy_2Fe_{14}B$.

T (K)	B_{hf} (Tesla)		$e^2q\,Q$ (mm/s)		Relaxation rate (MHz)
	site R_1	site R_2	site R_1	site R_2	
4.2	620	632	116	116	
50	610	624	113	111	
100	586	598	96	99	static limit
200	535	545	65	67	8×10^5
300	458	476	38	39	10×10^5
373	390		25		14×10^5

The hyperfine field acting on Dy nuclei is essentially composed of two terms. The main contribution arises from the local 4f ground state moment which is close to the saturation value, $10\mu_B$ ($B^{sat}_{4f} = 588$ T). The differences ΔB between the measured B_{hf} at 4.2K, and the calculated B^{sat}_{4f} are comparable to those observed for the Gd compound (Table 4). This field ΔB is the sum of the self polarization field B_s and of the hyperfine field B_N transferred from Fe moments.

TABLE 4   Hyperfine fields in $R_2Fe_{14}B$

Compounds	B_{hf} (Tesla) T = 4.2K		B_{4f}^{sat} (Tesla)	$\triangle B$ (Tesla)	
	Site R_1	Site R_2		Site R_1	Site R_2
Gd	14.7	27.9	- 34	48.7	62.
Dy	620.	632.	588	32.	44.7

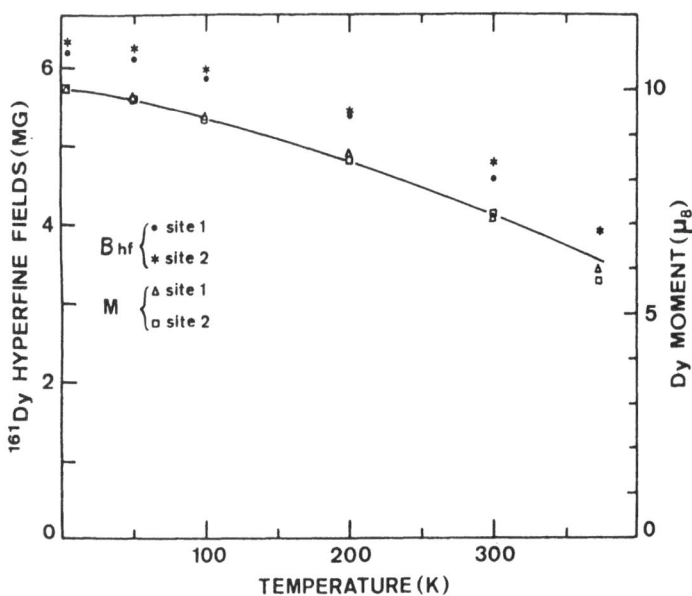

Fig. 3   Experimental temperature dependence of the ^{161}Dy hyperfine field B_{hf} in $Dy_2Fe_{14}B$ and temperature dependence deduced for the Dy moments M.

Assuming that the temperature dependence of B_N can be neglected below 400 K, and starting from the value $M^{Dy} = 10\mu_B$ at 4.2K the variations of the Dy moments with temperature are deduced from B_{hf} (T) (Figure 3). The continuous curve in this figure corresponds to the Brillouin function $M^{Dy}(T) = M^{Dy}(0) B_{15/2}(2S\mu_B B_{ex}^{Dy}/kT)$, with the value of the exchange field B_{ex} assumed to be proportional to the non-4f magnetic moment. In the figure 3 we see the excellent agreement between the calculated curve and the experimental data. Then, the crystal field effect can be neglected.

[155]Gd MOSSBAUER INVESTIGATION ON $(Gd_x Nd_{1-x})_{1+\epsilon} Fe_4 B_4$ $(0.2 \leqslant x \leqslant 1)$ AND ON $(Gd_{0.8} Sm_{0.2})_{1+\epsilon} Fe_4 B_4$.

Ternary rare earth-iron-boron compounds of the type $R_{1+\epsilon} Fe_4 B_4$ have received attention due to their presence as minor constituents in $Nd_2 Fe_{14} B$-type magnets (Kostikas 1985, Buschow et al. 1986). Their crystal structure has been determined (Bezinge et al. 1985, Givord 1985 and 1986) as consisting of two interpenetrating sublattices of tetragonal symmetry, containing respectively rare earth atoms and Fe + B atoms and having either incommensurate or very long periodicities along the c axis. From magnetisation and [57]Fe Mössbauer measurements, it was found that most alloys in this serie order magnetically at low temperatures, and that the Fe atoms carry no magnetic moments (Givord 1985, Niarchos 1986, Rechenberg 1986 and 1987). The easy magnetisation axis was determined for some alloys [R = Nd (Givord 1985)), Dy and Sm (Rechenberg and Sanchez 1987)] and found to correlate with the sign of the second-order crystalline electric field (CEF) parameter B_2^0. In order to further investigate CEF interactions at the rare-earth sites, we have undertaken [155]Gd Mössbauer measurements on $(Gd_x Nd_{1-x})_{1+\epsilon} Fe_4 B_4$, and $(Gd_{0.8} Sm_{0.2})_{1+\epsilon} Fe_4 B_4$ both above and below the magnetic order temperatures.

The alloys were prepared by R.F. melting in a cold crucible. Besides the main phase (hereafter denoted as 1-4-4), X-ray analysis revealed a significant amount of $GdFeB_4$ (crystallographic study of this compound has been made by Sobczak and Rogl, 1979) for x>0.5. As shown below, however, the contributions of both phases to the [155]Gd Mössbauer spectra could be clearly separated.

Mössbauer spectra were taken and analysed as mentioned previously for the $Gd_2 Fe_{14} B$ compound.

Fig. 4 shows [155]Gd Mössbauer spectra taken above and below the magnetic ordering temperature on a) $GdFe_{1+\epsilon} F_4 B_4$, b) $(Gd_{0.4} Nd_{0.6})_{1+\epsilon} Fe_4 B_4$, c) $(Gd_{0.8} Nd_{0.2})_{1+\epsilon} Fe_4 B_4$ and d) $(Gd_{0.8} Sm_{0.2})_{1+\epsilon} Fe_4 B_4$

As indicated above, to fit the spectra where x>0.5, it is necessary to introduce two subspectra, corresponding to two kinds of Gd atoms. One of them corresponds to the compound $GdFeB_4$. The ordering temperature of $GdFeB_4$ is not known, but our Mössbauer data clearly indicate that it lies between 4.2 K and 25 K. Hyperfine parameters for this phase are : quadrupolar interaction $e^2 q\, Q = 1.70\ (2)$ mm/s, isomer shift IS = 0.42 (1) mm/s, hyperfine magnetic field $|B_{hf}| = 33.6(3)$T at 4.2 K and angle between V_{zz} and H_{hf}, $\theta = 60°$. The other subspectrum, present for all the x values corresponds to the 1-4-4 phase. A very good fit of the Mössbauer spectra investigated leads to the results summarized in table 5. These results correspond to the outer lines of the spectra presented in Fig 4.

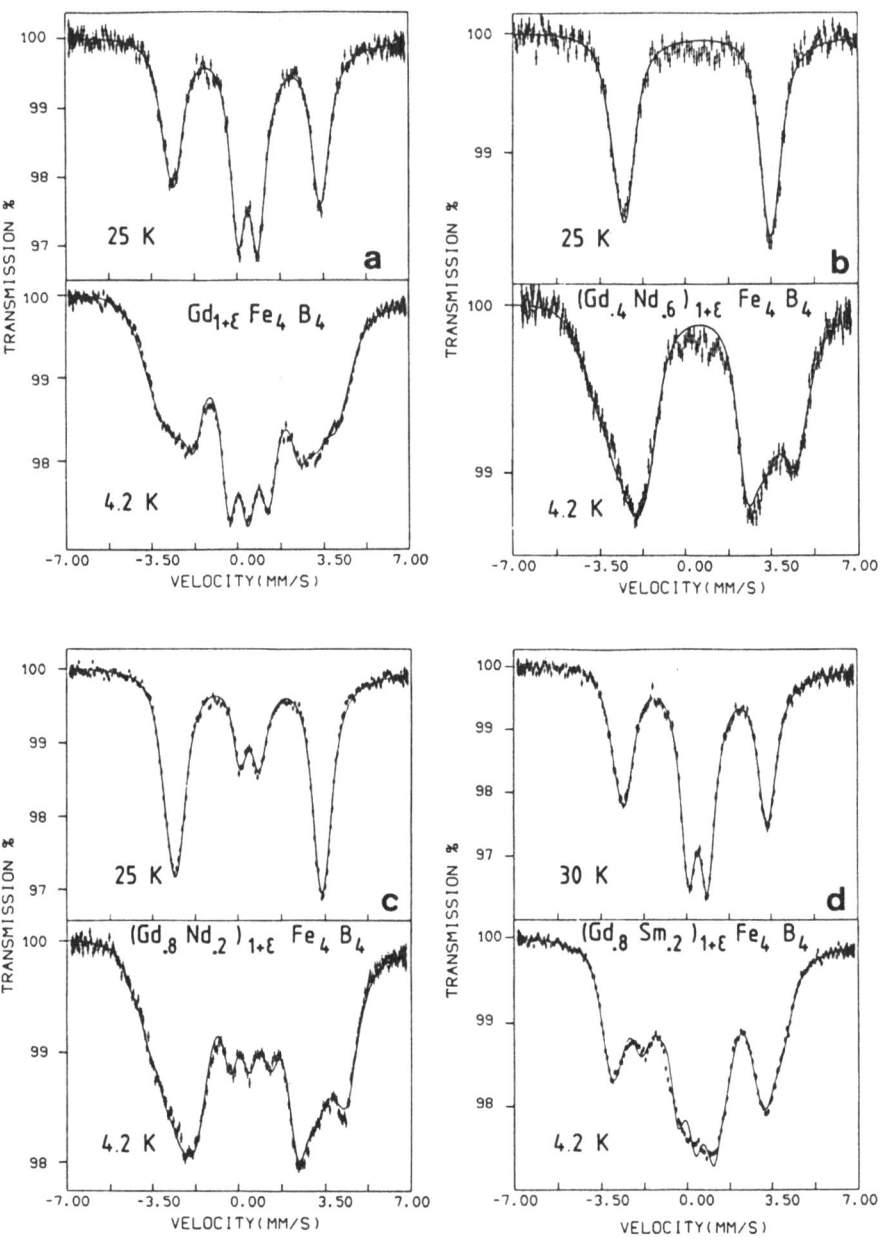

Fig.4 ^{155}Gd Mössbauer spectra of a) $Gd_{1+\varepsilon} Fe_4 B_4$, b) $(Gd_{0.4} Nd_{0.6})_{1+\varepsilon} Fe_4 B_4$,
c) $(Gd_{0.8} Nd_{0.2})_{1+\varepsilon} Fe_4 B_4$, d) $(Gd_{0.8} Sm_{0.2})_{1+\varepsilon} Fe_4 B_4$

TABLE 5   Results derived from ^{155}Gd Mössbauer investigation.
I.S. is the isomer shift given with respect to the SmPd$_3$ source. e^2qQ is
the average value and σ the width of the Gaussian distribution of the
quadrupole interaction. θ is the angle between V_{zz} and B_{hf}. Errors are
given parentheses in units corresponding to the last digit of the
parameter value quoted.

	x	I.S. (mm/s)	e^2qQ (mm/s)	σ (mm/s)	$\|B_{hf}\|$ (T)	θ (°)
GdFe$_4$B$_4$ (1a)	–	.32(1)	12.67(5)	.42(8)	23.4(3)	?
Gd$_x$Nd$_{1-x}$Fe$_4$B$_4$	.8	.33(1)	12.60(5)	.62(4)	24.9(2)	81(2)
	.6	.32(1)	12.68(5)	.73(4)	24.8(2)	82(2)
(1b and 1c)	.4	.32(1)	12.70(5)	.88(6)	25.6(5)	90(1)
	.2	.33(1)	12.83(5)	.88(6)	25.6(5)	90(1)
Gd$_x$Sm$_{1-x}$Fe$_4$B$_4$	.8	.32(1)	12.6(1)	.72(6)	21.8(5)	3(4)

A Gaussian distribution (of width σ) of the quadrupolar interactions
has to be used to ensure a good fit of the 1-4-4 spectra. This is a
consequence of the quasi incommensurate structure of this phase.
Nevertheless, it is not necessary to consider such a distribution of the
hyperfine field below T_c.

Values of isomer shifts, quadrupolar interactions and hyperfine
fields seem to be nearly independent of x, but the width σ of the gaussian
distribution increases significantly when x decreases. It is noteworthy
that when Nd is substituted to Gd the hyperfine field is perpendicular or
nearly so to V_{zz} ($\theta = 90°$), but no unique value of θ permits to obtain a
good fit of the spectrum for the non substituted compound. The fit shown
in Fig 1a was obtained by assuming an isotropic θ distribution. This
result suggests that Gd$_{1+\epsilon}$Fe$_4$B$_4$ does neither show a collinear structure
nor a helicoïdal structure about the c-axis. Upon Nd substitution, its
CEF-induced anisotropy favors a collinear structure of the Gd moments in a
plane perpendicular to the c-axis (Fig 4b and 4c). Upon Sm substitution
(Fig 4d), its CEF-induced anisotropy favors a collinear structure of the
Gd moments parallel to the c-axis. It is known that the second-order
Stevens factor α_J is of opposite sign in Nd and Sm (whereas $\alpha_J = 0$ for

Gd). This seems to imply a difference in the preferred direction of Nd and Sm moments and this direction is imposed to the Gd moments.

The ^{155}Gd quadrupolar splitting $e^2qQ \simeq 12.7$ mm/s in the 1-4-4 compounds is, to our knowledge, the largest ever observed in Gd metallic compounds. This value yields the A_2^0 term in the crystalline electric field Hamiltonian to be -2450 ∓ 50 K/a$_0{}^2$ (instead of 670 K/a$_0{}^2$ inGd$_2$Fe$_{14}$B). Detailed point charge calculations, in good agreement with the experimental values, are presented elsewhere (Rechenberg, 1987).

REFERENCES

Barton W A and Cashion J D, J.Phys. C : Solid State Phys. **12**, 2897 (1979)

Berthier Y, Bogé M, Czjzek G, Givord D, Jeandey C, Li H S and Oddou J L, J.Magn.Magn.Mater. **54-57**, 589 (1986)

Bezinge A, Braun H F, Muller J and Yvon K, Solid State Commun. **55**, 131 (1985)

Bogé M, Coey J M D, Czjzek G, Givord D, Jeandey C, Li H S and Oddou J L, Solid State Commun. **55**, 295 (1985)

Bogé M, Czjzek G, Givord D, Jeandey C, Li H S, Oddou J L, J.Phys F : Met.Phys. **16**, L67 (1986)

Buschow K H J, Materials Sci. Rep. **1**, 1 (1986)

Cadogan J M and Coey J M D, Phys.Rev. **B30**, 7326 (1984)

Campbell I A, J.Phys. C :Solid State Phys. **2**, 1338 (1969)

Freeman A J and Watson R E, Phys.Rev. **127**, 2058 (1962)

Givord D, Li H S and Perrier de la Bâthie R, Solid State Commun. **50**, 497 (1984)

Givord D, Moreau J M and Tenaud P, Solid State Commun. **55**, 303 (1985)

Givord D, Moreau J M and Tenaud P, J.Less-Comm.Met. **123**, 109 (1986)

Kostikas A, Papaefthymiou V, Simopoulos A and Hadjipanayis G C, J.Phys F : Met.Phys **15**, L129 (1985)

Niarchos D, Zouganelis G, Kostikas A and Simopoulos A, Solid State Commun. **59** 389 (1986)

Rechenberg H R, Bogé M, Jeandey C, Oddou J L, Sanchez J P and Tenaud P, Solid State Commun. **64**, 277 (1987)

Rechenberg H R, Paduan-Filho A, Missell F P, Deppe P and Rosenberg M, Solid State Commun.**59**, 541 (1986)

Rechenberg H R and Sanchez J P, Solid State Commun. **62**, 461 (1987)

Sobczak R and Rogl P, J.Solid St. Chem. **27**, 343 (1979)

Tanaka Y, Laubacher D B, Steffen R M, Shera E B, Wohlfarth H D and Hoehn M V, Phys.Lett. **108B**, 8 (1982)

Tomala K, Czjzek G, Fink J and Schmidt H, Solid State Commun. **24**, 857 (1977)

Wickman H H and Wertheim G K, in "Chemical Applications of Mössbauer Spectroscopy" ed. Goldanskii V I and Herber R H (Ac.Pr.N.Y.1968) p.546

DEVELOPMENT OF ^{157}Gd AND ^{145}Nd MÖSSBAUER RESONANCES

G. Czjzek and H.-J. Bornemann

Kernforschungszentrum Karlsruhe
Institut für Nukleare Festkörperphysik
P.O. Box 3640, D-7500 Karlsruhe
Federal Republic of Germany

ABSTRACT

A ^{145}Pm($t_{1/2}$=17.7y) radiation source intended for ^{145}Nd Mössbauer spectroscopy has been produced by the reaction ^{146}Nd(p,2n)^{145}Pm. Concurrent reactions lead to a very high background radiation level in the energy range of the Mössbauer line at 67.1 keV. The source will only be usable for Mössbauer spectroscopy after several years when the nuclei causing the background have decayed. ^{155}Gd Mössbauer spectra were obtained for a $Gd_2Fe_{14}B$ single crystal absorber in the temperature range 4.2 K to 120 K. The results indicate different variations of Gd moments at the 2 inequivalent Gd sites with temperature. A programme has been written for simultaneous fits to a series of Mössbauer spectra taken with an absorber at several temperatures in terms of crystal and exchange fields acting on the 4f electrons. The programme has been applied to the evaluation of ^{174}Yb Mössbauer spectra of $Yb_2Fe_{14}B$ in the temperature range 4.2 K to 125 K.

INTRODUCTION

The magnetic anisotropy of rare-earth transition-metal compounds is caused primarily by the single-ion anisotropy of the rare-earth ions due to the crystal field acting on the 4f electrons. Measurements of magnetization curves of single crystals with fields applied along different crystallographic directions can be employed for determination of the crystal field parameters B_n^m (Givord et al., 1988). An alternative possibility is given by determination of hyperfine interactions at rare-earth nuclei by NMR or Mössbauer spectroscopy (Ofer et al., 1968).

For S-state ions such as Gd^{3+} or Eu^{2+}, the 4f contribution to the electric field gradient (EFG) interacting with the nuclear quadrupole moment is negligible. The EFG observed in Mössbauer spectra is due to the arrangement of charges in the crystal lattice. It is directly related to the crystal field parameters B_2^o and B_2^2 (M. Bogé et al., 1986) which are expected to play a dominant role for the magnetocrystalline anisotropy.

For non-S-state ions, on the other hand, the 4f contributions dominate all hyperfine interactions. Both the EFG and the magnetic hyperfine field depend sensitively on the nature of the 4f states as determined by the crystal field. The variation of hyperfine interactions with temperature is due to the thermal occupation of excited crystal field states. Hence, for rare-earth nuclei, the analysis of hyperfine interactions and their temperature dependence can be employed in determining crystal field parameters.

For permanent-magnet materials such as $Nd_2Fe_{14}B$, information about crystal field parameters would be of particular interest for Nd. In the following section, a report is given about an attempt to produce sources for high-resolution Mössbauer spectroscopy with ^{145}Nd.

Then we report on the variation of Gd moments in $Gd_2Fe_{14}B$ in the temperature range 4.2 K $\leq$ T $\leq$ 120 K, deduced from ^{155}Gd Mössbauer spectra of a $Gd_2Fe_{14}B$ single crystal.

Finally, a programme for evaluating a series of Mössbauer spectra taken for an absorber at several temperatures in terms of crystal and exchange fields acting on the 4f electrons of a non-S-state rare-earth is described. This programme has been applied for the analysis of ^{174}Yb Mössbauer spectra of $Yb_2Fe_{14}B$.

SOURCE FOR ^{145}Nd MÖSSBAUER SPECTROSCOPY

Two excited levels of ^{145}Nd are suitable for Mössbauer spectroscopy. The first level at 67.1 keV with $t_{1/2}$ = $(29.4 \pm 1.0) \cdot 10^{-9}$ sec should give rise to a very narrow resonance whereas the second level at 72.5 keV with $t_{1/2}$ = $(0.72 \pm 0.05) \cdot 10^{-9}$ sec gives spectra of rather poor resolution (Kaindl and Mössbauer, 1968). Technical difficulties associated with detector resolution and with source preparation have precluded the exploitation of the high-resolution resonance in the past.

Sources have been produced in a 2-step process by capture of thermal neutrons in ^{144}Sm. The resulting unstable nucleus ^{145}Sm decays to ^{145}Pm with a half-life close to 1 year. The decay of ^{145}Pm ($t_{1/2}$ = 17.7y) to ^{145}Nd is accompanied by emission of γ-rays at 67.1 or at 72.5 keV employed in Mössbauer spectroscopy. Thus, preparation of sources useful for ^{145}Nd Mössbauer spectroscopy involved a long waiting time after the neutron irradiation of ^{144}Sm until a sufficient number of ^{145}Pm nuclei had accumulated.

We hoped to arrive at a useful source much faster by the reaction $^{146}Nd(p,2n)^{145}Pm$. Several test irradiations of natural Nd and of enriched ^{146}Nd showed the optimum proton energy for this reaction to be near 15 MeV. A test source gave a γ-spectrum with a signal-to-background ratio near

3:1 when the source was suspended freely above the detector. Insertion of a $NdPd_3$ absorber, of the absorber chamber and the cryostat walls, however, reduced this ratio to a value below 1:10. As this implies a reduction of the observable Mössbauer effect by an order of magnitude, the source is not usable for Mössbauer spectroscopy at short terms.

The reason for the problems lies in high-energy γ-activities (~400 to 700 keV) due to concurrent reactions $^{146}Nd(p,n)^{146}Pm$ $(t_{1/2} = 5.5y)$ and $^{146}Nd(p,3n)^{144}Pm$ $(t_{1/2} = 1y)$. At the optimum proton energy, the cross-sections for these reactions are smaller than that for the reaction $^{146}Nd(p,2n)^{145}Pm$ by an order of magnitude. This gain, however, is compensated by the shorter half-lives of ^{144}Pm and of ^{146}Pm. In addition, unfavourable branching ratios further increase the high-energy activities compared to the 67.1 keV line of interest. Multiple scattering in all materials finally leads to the background obscuring the Mössbauer line. Thus, it will be necessary to wait for a sufficient time, i.e. for a few years, until the nuclei causing the background have decayed.

TEMPERATURE DEPENDENCE OF Gd-MOMENTS IN $Gd_2Fe_{14}B$

Mössbauer spectra of ^{155}Gd in a $Gd_2Fe_{14}B$ single crystal were measured in the temperature range $4.2 K \leq T \leq 120 K$. Analysis of the spectra was performed in the same way as described by Bogé et al. (1986) and the results derived from the 4.2 K spectrum agreed with those reported earlier (Bogé et al., 1986). Differences in the center shifts δ can be ascribed to the different sources employed in the 2 investigations (^{155}Eu in $SmPd_3$ by Bogé et al., 1986; ^{155}Eu diffused into Pd in the present work). Variations of center shifts δ, of the magnetic hyperfine fields B_{hf} and of the absorber linewidth Γ_A with temperature are summarized in table 1.

The temperature variation of the center shifts is as expected for second-order Doppler shifts. The linewidth Γ_A has the natural value $\Gamma_{nat} = 0.25$ mm/s at 4.2 K. With increasing temperature a systematic increase of Γ_A was found. This indicates the occurrence of rather slow relaxation processes as they have been found for ^{161}Dy in $Dy_2Fe_{14}B$ for $T \gtrsim 100$ K (M. Bogé, private communication).

For both Gd sites, Table 1 shows increasing values of the hyperfine field B_{hf} with increasing temperatures. This apparently paradoxical result is understandable if we separate the contributions to B_{hf} caused by Gd moments (B_{Gd}) and by Fe moments (B_{Fe}).

TABLE 1   Temperature dependence of absorber linewidth Γ_A, of center
           shifts δ and of hyperfine fields B_{hf} at the 2 Gd sites in
           $Gd_2Fe_{14}B$ deduced from [155]Gd Mössbauer spectra. Standard
           deviations in parentheses.

| T(K) | Γ_A(mm/s) | site 4f | | site 4g | |
		δ(mm/s)	B_{hf}(tesla)	δ(mm/s)	B_{hf}(tesla)
4.2	0.249(6)	0.257(3)	14.45(8)	0.226(4)	27.52(11)
60	0.259(6)	0.251(3)	14.86(9)	0.207(4)	27.46(11)
100	0.287(12)	0.229(7)	15.74(18)	0.203(8)	27.60(25)
120	0.303(15)	0.226(8)	15.80(22)	0.205(9)	27.97(31)

The contribution B_{Gd} due to Gd moments, the resultant of the core
polarization field $B_{CF} \sim -(5\pm0.5)\cdot\mu_{Gd}$ tesla (μ_{Gd} in μ_B) and the conduction
electron polarization due to the local Gd moment ("own" polarization),
$B_{OP} \sim +(2\pm1)\cdot\mu_{Gd}$ tesla, is expected to be negative, $B_{Gd} \sim -(3\pm1)\cdot\mu_{Gd}$ tesla.
The transferred field due to Fe moments was found to be positive in all
intermetallic Gd-Fe compounds (Tomala et al., 1977). For Gd nuclei in
$Gd_2Fe_{14}B$ where $B_{hf} = B_{Gd}+B_{Fe}$ is positive at both Gd sites (Bogé et al.,
1986), obviously $|B_{Fe}| > |B_{Gd}|$. An increase of B_{hf} with temperature then
indicates a more rapid decrease of Gd moments compared to Fe moments. This
is in accordance with the conclusions deduced by Bogé et al. (1985) from a
comparison of the temperature dependence of the saturation magnetization
of $Gd_2Fe_{14}B$ with that of $Y_2Fe_{14}B$. An additional result evident from the
entries in Table 1 is a significantly smaller variation of B_{hf} at 4g sites
compared to 4f sites. This indicates stronger Gd-Fe exchange interactions
for Gd in 4g sites than for Gd in 4f sites.

CRYSTAL AND EXCHANGE FIELDS IN $Yb_2Fe_{14}B$

Hyperfine interactions at nuclei of non-S-state rare-earth ions are
dominated by the contributions due to the local 4f electrons. Hence,
information about 4f states and the fields determining them can be
deduced from measurements of hyperfine interactions by NMR or Mössbauer
spectroscopy.

We have developed a programme by which a series of Mössbauer spectra
is fitted simultaneously in terms of crystal field parameters and exchange
fields acting on 4f electrons. The programme is appropriate for compounds

between rare-earths and magnetic 3d transition metals in which 3d-3d exchange interactions are dominant. It is based on the assumptions that

- changes of crystal field parameters (for example due to structural changes) are negligible in the temperature range under study, and

- exchange fields acting on 4f electrons are dominated by 3d-4f interactions such that a self-consistent determination of exchange fields due to 4f moments is not necessary.

Furthermore, magnetostrictive terms are not included in the present version of the programme.

Finally, the temperature range covered by the spectra must extend to an upper limit which corresponds to an energy value at least close to the energy of the first excited crystal field level. Otherwise, information is obtained on the crystal field ground state only, not sufficient for the determination of more than one or two parameter values.

The spectra obtained with ^{174}Yb in $Yb_2Fe_{14}B$ in the temperature range 4.2 to 125 K, shown in the report of CEAM group 1.08, are being used for testing the programme. A problem is presented in this case by impurity phases Yb_2O_3 and Yb_2Fe_{17} whose contributions to the Mössbauer spectra must be accounted for. This leads to a considerable number of additional unknown variables to be determined in the fitting procedure.

In order to avoid an excessive number of fitting variables, the following restrictions were introduced:

- the ratios B_2^2/B_2^o, B_4^4/B_4^o and B_6^4/B_6^o were restrained to the averages of the values determined by Givord et al. (1988) for several $R_2Fe_{14}B$ compounds; B_2^o was varied independently for the 2 Yb sites, for B_4^o and B_6^o, the same value was assumed at both Yb sites;

- the temperature dependence of the exchange field was assumed to be given by $B_{ex}(T) = B_{ex}(0) \cdot (1-\beta T^2)$; $B_{ex}(0)$ was varied independently for the 2 Yb sites, β was set equal for both sites;

- the direction of the exchange field was confined to [100] for $T \leq 110$ K, in accordance with the results by Burlet et al. (1986)

- the temperature dependence of the f-factors (assumed to have the same values for both sites) was approximated by $f(T) = f(0) \cdot e^{-\alpha T^2}$.

Preliminary results are presented in Table 2. Whereas the values obtained for $B_{ex}(0)$ agree quite well with the result of calculations by Radwański and Franse (1987), the values of the crystal field parameters are substantially larger than those expected on the basis of magnetization measurements (Givord et al., 1988).

TABLE 2   Preliminary results deduced from ^{174}Yb Mössbauer spectra of $Yb_2Fe_{14}B$ in the temperature range 4.2 to 125 K. The meaning of entries is explained in the text.

site	$\mu_B \cdot B_{ex}(0)$ (K)	$\beta \cdot 10^6$ (K^{-2})	B_2^0 (K)	$B_4^0 \times 10$ (K)	$B_6^0 \times 10^3$ (K)
4f	26.8(0.5)	8(2)	25.0(0.6)	2.25(0.06)	5.0(0.5)
4g	30.7(0.6)	8(2)	15.0(0.3)		

constant ratios according to Givord et al. (1988)

site	B_2^2/B_2^0	B_4^4/B_4^0	B_6^4/B_6^0
4f	0.65	2.8	-12.6
4g	2.0	2.8	5.0

For the spectrum obtained at 125 K, above the spin reorientation temperature determined by Burlet et al. (1986), a good fit could not be obtained if the exchange field was assumed to point in the [001] direction for both Yb sites. The best result was obtained with an exchange field close to [001] for the 4f site, but close to the a-b plane for the 4g site.

REFERENCES

Bogé, M., Coey, J.M.D., Czjzek, G., Givord, D., Jeandey, C., Li, H.S. and Oddou, J.L. 1985. 3d-4f magnetic interactions and crystalline electric field in the $R_2Fe_{14}B$ compounds: magnetization measurements and Mössbauer study of $Gd_2Fe_{14}B$. Solid State Commun. 55, 295-298.

Bogé, M., Czjzek, G., Givord, D., Jeandey, C., Li, H.S. and Oddou, J.L. 1986. A ^{155}Gd Mössbauer study of a $Gd_2Fe_{14}B$ single crystal. J. Phys. F: Metal Phys. 16, L67-L72.

Burlet, P., Coey, J.M.D., Gavigan, J.P., Givord, D. and Meyer C. 1986. A note on exchange and crystal field interactions in $R_2Fe_{14}B$ compounds: $Yb_2Fe_{14}B$. Solid State Commun. 60, 723-727.

Givord, D., Li, H.S., Cadogan, J.M., Coey, J.M.D., Gavigan, J.P., Yamada, O., Maruyama, H., Sagawa, M. and Hirosawa, S. 1988. Analysis of high field magnetization measurements on $R_2Fe_{14}B$ single crystals (R = Tb,Dy,Ho,Er and Tm) to be published in J. Appl. Phys.

Kaindl, G. and Mössbauer, R.L. 1968. Recoilless nuclear resonance absorption in ^{145}Nd. Phys. Lett. 26B, 386-387.

Ofer, S., Nowik, I. and Cohen, S.G. 1968. The Mössbauer effect in rare-earths and their compounds. In "Chemical Applications of Mössbauer Spectroscopy" (Ed. V.I. Goldanskii and R.H. Herber) (Academic Press, New York and London) pp. 427-503.

Radwański, R.J. and Franse, J.J.M. 1987. Rare earth contribution to magnetocrystalline anisotropy energy in $R_2Fe_{14}B$. Phys. Rev. B 36, 8616-8621.

Tomala, K., Czjzek, G., Fink, J. and Schmidt, H. 1977. Hyperfine interactions in intermetallic compounds between Gd and 3d transition metals. Solid State Commun. 24, 857-861.

SUMMARY OF DISCUSSIONS :

ATOMIC SCALE MAGNETISM

G. Czjzek, D. Niarchos

Information on atomic scale magnetism has proved to be very valuable for the C.E.A.M. project. A better understanding of the local magnetic moments, exchange interactions and their contribution to the magnetic performance of known materials such as $Nd_2Fe_{14}B$, as well as other new phases, was very helpful in explaining experimental data and developing new theories.

The techniques used for studies of magnetism on the atomic scale were neutron diffraction, Mössbauer, NMR and muon spectroscopies.

From neutron diffraction data it was possible to measure the magnetic moments on each site, both of the transition metal and of the rare earth. These values were compared with those obtained from Mössbauer and NMR data. Neutron diffraction was also used to determine the magnetic structure and detect changes with temperature or with hydrogen absorption, notably in $Ho_2Fe_{14}B$.

Mössbauer spectroscopy, whether on ^{57}Fe or on conventional rare-earths (^{161}Dy, ^{155}Gd) or an exotic one (Yb) has been useful for detecting the spin reorientation as a function of temperature in pure compounds and solid solutions, as well as providing the values of the key crystal field parameters, A_2^0 and A_2^2.

The NMR technique on the other hand has given independent values of the hyperfine fields at the six Fe sites in the $Nd_2Fe_{14}B$ compound as well as the transferred hyperfine fields. The ^{11}B, ^{57}Fe, ^{59}Co and some rare earths have all been exploited resonances.

In conclusion these powerful techniques for looking at magnetism on the atomic scale have proven to be complementary to bulk techniques in the course of the study of the $R_2M_{14}B$ series of compounds.

SECTION III

— MATERIALS

CHAPTER 4

PHASE RELATIONS

AND MICROSTRUCTURE

PHASE EQUILIBRIA IN Fe-Nd-B AND RELATED SYSTEMS AND MICROSTRUCTURE OF SINTERED Fe-Nd-B MAGNETS

Gerhard Schneider, Ernst-Theo Henig, Bernd Grieb, Gerhard Knoch

Max-Planck-Institut für Metallforschung, Institut für Werk-
stoffwissenschaften, Heisenbergstr. 5, D-7000 Stuttgart 80

ABSTRACT

Phase relations in the systems Fe-Nd, Fe-Nd-B, Fe-Nd-O, Fe-Dy-B, Fe-Tb-B and Fe-Nd-Dy(Tb)-B were investigated. The knowledge of these phase relations is used for a better under-standing and an enhancement of the microstructure of the Fe-Nd-B-based sintered magnets. The important microstructural parameters: phase distribution, grain-size distribution, con-tiguity, porosity and microstructure of the intergranular regions and their influence on the coercivity were investi-gated. The effect of additions like Al and oxides (e.g. Al_2O_3) was determined. Al stabilizes an additional phase and improves the wetting behaviour between the Nd-rich liquid and Φ.

1 INTRODUCTION

The magnetic properties of Fe-Nd-B-based magnets are determined both by the intrinsic physical properties of the occurring phases (H_A, M_S, T_C) and the extrinsic properties of the microstructure. Consequently both of these properties must be improved for an enhancement of the material. Therefore the knowledge of the phase relations and the development of the microstructure due to certain sintering parameters must be available for the manufacturers.

Conflicting knowledge of the Fe-Nd-B ternary system already exist, but details of the solidification path and the equilib-

ria at temperatures close to the sintering temperature have
not been available. In addition there is a lack in the phase
relations in higher order systems as needed to survey effects of partial substitution of the transition elements and
the rare earth metals, and that of other additives.

Oxygen as an additional alloying element has to be taken into
account because it is always present as a contamination element. Oxygen containing phases have been found by TEM investigations in grain boundaries and grain junctions in the magnet
/87Fid/. Furthermore oxygen was reported to affect the corrosion resistance of the Fe-Nd-B magnets. More information about
the phases and the phase equilibria in the quaternary system
Fe-Nd-B-O is desirable.

For a refinement of the magnetic properties of sintered magnets the morphology of the microstructure is of fundamental
importance. Therefore the microstructural features as phase
distribution, grain size, grain shape and contiguity have to
be studied in detail. The influence of the master alloy composition, the powder characteristics, the sinter parameters and
the following heat treatment on the microstructure and the
consequences for the magnetic properties have to be correlated
for a systematic improvement. In addition, the influence of
additives and contaminations in the grain boundary areas is
known to have a major importance.

2 EXPERIMENTAL

The alloys were prepared from materials of the following
purities, in mass %: Fe, 99.8, Nd, Dy, Tb, 99.9 and a ferro-
boron master alloy, Fe-18.5B. The specimens were arc melted
under argon and remelted 5 times to assure homogeneity. For
the differential thermal analyses the Al_2O_3-crucibles were
coated with a slurry of Nd_2O_3 to avoid a reaction with the
crucible material. The heating and cooling rates for the DTA
were either 5 or 10 K min^{-1}.

The powder processing for the magnets was done in glove boxes under purified argon atmosphere. The ingots were crushed to a coarse powder and then ground in a vibration ball mill. To avoid particles larger than 20 μm the powder was sieved. The powder, which was aligned in a field of 0.5 T perpendicular to the pressing direction, was compacted with 500 MPa. The compacts were sealed in tubes of fused silica, sintered 1 h and annealed.

The magnetic properties were measured in a vibrating sample magnetometer.

3 CONSTITUTION

3.1 The Binary System Fe-Nd

In the microstructure of Fe-Nd alloys two intermetallic compounds are observed. The $Fe_{17}Nd_2$ phase is peritectically formed at 1210 ±5°C. The second phase, referred to as ϵ, is never observed as a primary solidification product but only as a constituent of the Nd-rich eutectic. Two different eutectic microstructures are visible, (Nd+$Fe_{17}Nd_2$) and (Nd+ϵ). Whether (Nd+ϵ) also contains $Fe_{17}Nd_2$ and is therefore ternary is not certain. Microprobe analysis of ϵ places its Nd content below 33 at.%. In larger lamellae of ϵ a domain pattern is visible in polarized light, suggesting a uniaxial magnetocrystalline anisotropy. The volume fraction of ϵ is always higher in specimens with a higher oxygen contamination (DTA specimens). This suggests ϵ to be a ternary compound in the system Fe-Nd-O. Thermal analysis of alloys with Nd contents $\geq$ 20 at.% consistently show a pair of peaks at 685 and 675°C, corresponding to the two eutectics observed in the microstructure. The Fe_2Nd-phase reported by Terekhova et al. /65Ter/ and synthesised at high temperatures and high pressure by Cannon et al. /72Can/, could not be found in the Fe-Nd binary system as a stable phase at normal pressure. A phase Fe_7Nd of the $CaCu_5$ structure type is only obtained by splat cooling and is metastable /86Sta/. The phase diagram shown in Fig. 1 was calculated on the basis of thermodynamic data and the DTA experimental results /87aSch/.

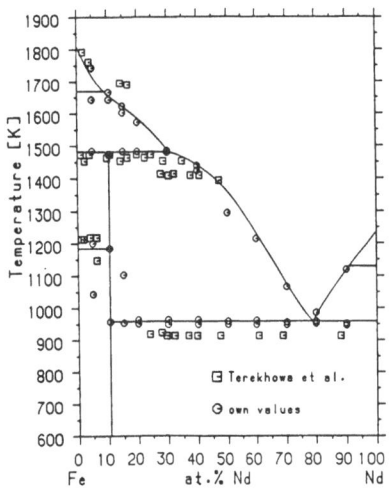

Fig. 1: Computed Fe-Nd
binary phase dia-
gram /87aSch/.

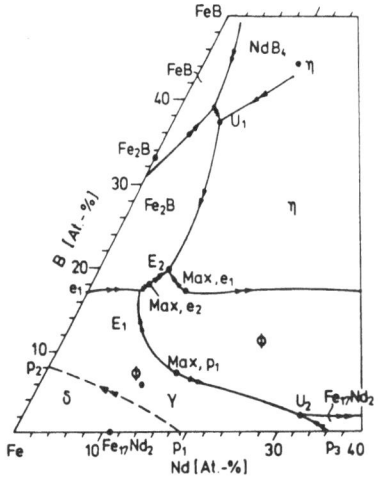

Fig. 2: Fe-Nd-B, liquidus
projection.

3.2 The Ternary System Fe-Nd-B

In the following the phase $Fe_{14}Nd_2B$ is referred to as Φ and Fe_4NdB_4 as η. The solidification in the Fe-rich corner of the Fe-Nd-B system is shown by the <u>liquidus projection</u> (Fig. 2) and the <u>reaction scheme</u> (Fig. 3). The critical tie line L = Φ + Fe_2B and a ternary eutectic L = Fe + Φ + Fe_2B represent stable reactions though they contradict earlier reports in which Fe coexists with η /84Sta,85Mat/.
The <u>isothermal section</u> at 1000°C is shown in Fig. 4. A two-phase field (Φ + Nd-rich liquid) exists between the maximum temperature of Φ formation (1180°C) and the temperature of the ternary eutectic, at which the last liquid solidifies (655°C). Four <u>vertical sections</u> have been constructed with the aid of DTA data. The section Nd:B=2:1 is shown in Fig. 5a. In specimens with a composition near Φ, metastable solidification was observed if the specimens were superheated. The short range order of Φ-clusters in the liquid is destroyed at a certain

temperature above the decomposition temperature of Φ. Cooling
down from this temperature, the crystallization of Φ is
suppressed for kinetic reasons. Then the primary crystallization field of Fe is expanded to lower Fe compositions and
to lower temperatures. A secondary metastable crystallization
of a high temperature phase (χ) was observed. This phase
decomposes at 1105°C into Fe and Φ. In the related system Fe-
Dy(Tb)-B quenching of this phase from high temperatures
succeeded /87Gri/. It was found to be of the 17/2 type, containing some B. The further solidification is analogous to the
stable system. The vertical section at a ratio Nd:B=2:1 of the
metastable system is shown in Fig. 5b.

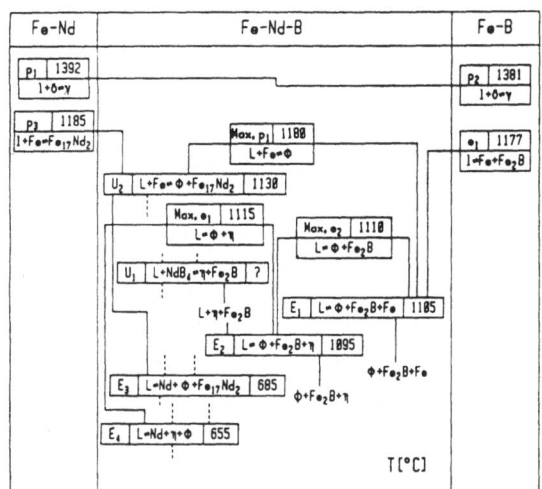

Fig. 3: Fe-Nd-B, reac-
tion scheme
/86Sch/. Tempe-
ratures in °C.

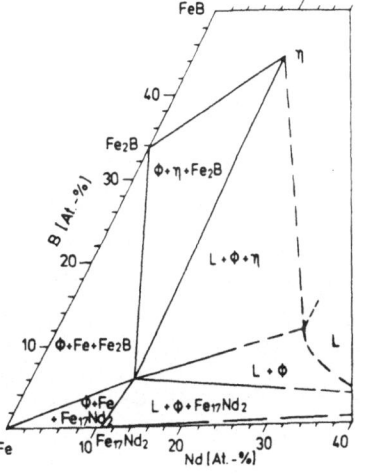

Fig. 4: Fe-Nd-B, 1000°C
isothermal sec-
tion /86Sch/.

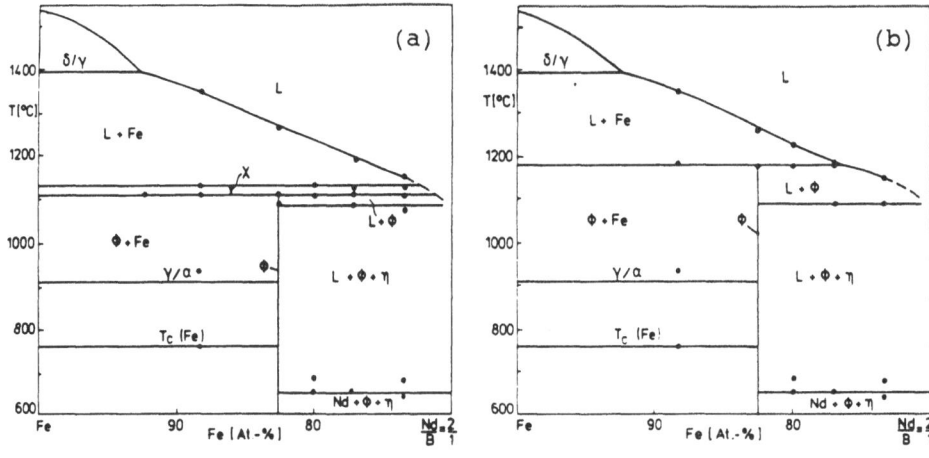

Fig. 5: Fe-Nd-B, vertical sections, Nd:B=2:1, (a) stable,
(b) metastable.

Under some experimental conditions one can expect to find the
four-phase reactions according to Matsuura et al. /85Mat/,
Sagawa et al. /85Sag/, or Zhang and Luo /87Zha/ which are L +
Fe₂B = Fe + η and L = Fe + η + Φ. On the other hand we obtain
L = Fe + Φ + Fe₂B and L = Fe₂B + η + Φ as stable reactions,
which are not compatible with the ones of Matsuura et al..
This has led to conflicting versions of the liquidus projection in the literature. The comparison of the reaction temperatures, composition of invariant reactions, superheating of
the melt and the microstructures /87Hen/ show that the liquidus projection with the reactions L = Fe + Φ + Fe₂B and L =
Fe₂B + n + Φ is the stable one and that of Matsuura et al.
/85Mat/ is metastable. It is suggested, that oxygen impurities
support the metastability in Fe-Nd-B and we have indeed verified by oxygen analysis that the alloys that solidify in the
metastable mode have consistently higher oxygen content. This
also explains the fact that (Fe + n), though metastable,
continues to exist after annealing at 600°C as shown by Chaban
et al. /79Cha/ and 900°C as found by Stadelmaier et al.
/84Sta/ and Buschow et al./85Bus/.

3.3 The Systems Fe-Dy-B, Fe-Tb-B, Fe-Nd-Dy(Tb)-B

3.3.1 Fe-Dy-B and Fe-Tb-B

Cooling from temperatures slightly above the Φ formation temperature gives a stable peritectic formation of Φ out of $Fe_{17}RE_2$ (plus $\approx$ 3 at.% B) and liquid at 1214°C in the Fe-Dy-B system and 1218°C in the Fe-Tb-B system. Cooling from temperatures significantly above the liquidus, however, the formation of the Φ phase is suppressed by as much as 60 K below the stable formation temperature.

3.3.2 Fe-Nd-Dy-B and Fe-Nd-Tb-B

Lattice Parameters:
Along the sections $Fe_{14}(Nd_{1-x}(Tb$ or $Dy)_x)_2B$ ($0 \leq x \leq 1$) the lattice parameters of the tetragonal unit cell decrease continiously. No decomposition into two Φ phases could be observed, at least above 800°C, the temperature at which the samples were annealed /87Gri/.

Constitution along $Fe_{14}(Nd_{1-x}(Tb$ or $Dy)_x)_2B$ ($0 \leq x \leq 1$):
In Fig. 6 the stable and the metastable formation temperatures of Φ along $Fe_{14}(Nd_{1-x}Dy_x)_2B$ are shown. The results in the Fe-Nd-Tb-B system are quite similar.
Iron is the primary phase on the Nd-rich side. $Fe_{17}RE_2$ is the primary phase on the Dy-rich and secondary on the Nd-rich side of the system.
The two-phase region of $L + Fe_{17}RE_2$ exists throughout the vertical section. The decreasing temperature range of the two-phase field $L + Fe_{17}RE_2$ with increasing Nd content clearly demonstrates that also for the ternary Nd-Fe-B the metastable formation of Φ follows the reaction $L + Fe_{17}Nd_2 \rightarrow \Phi$.
In the curve of the stable system in Fig. 6, two ranges of different slopes can be distinguished. (i) The temperature differences from $Nd_2Fe_{14}B$ up to a substitution degree of 20% Nd are about 20 K and (ii) for the residual 80% substitution

20 K also. In the stable systems the formation mechanism of Φ apparently has to change from L + Fe -> Φ on the $Fe_{14}Nd_2B$ side to L + $Fe_{17}Dy_2$ -> Φ on the $Fe_{14}Dy_2B$ side. The change in slope with composition of the Φ formation temperatures as well as the intersection of the formation temperature curves of Φ and $Fe_{17}Dy_2$ indicate that this transition takes place at about 20% Dy substitution for Nd.

The extrapolation of the phase field boundary between L + Fe and L + $Fe_{17}RE_2$ (dashed line) marks the metastable conditions for the $Fe_{17}RE_2$ formation on the Nd-rich side when superheating suppresses the formation of Φ.

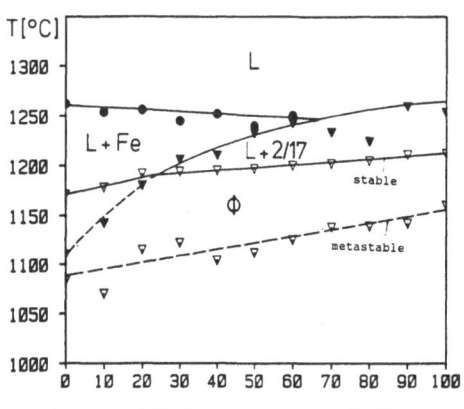

Fig. 6: Fe-Nd-Dy-B, superposition of the stable and metastable systems along $Fe_{14}(Nd_{1-x}Dy_x)_2B$.

3.4 The System Fe-Nd-O

From the literature three ternary phases are known, two of them being oxygen rich. This is the orthorhombic perovskite phase $FeNdO_3$ /70Mar/ and a hypothetical cubic garnet phase $Fe_5Nd_3O_{12}$ /62Esp/. Dariel et al. /76Dar/ reported a metallic phase $Fe_{16}R_6O$ to be stable with R = Y, Gd, Tb, Dy, Ho, Er. The structure is a cubic Ag_8Ca_3 or $Ni_6Si_2R_3$ type. This phase is apparently not formed with lighter rare earth elements.

We have succeeded in synthesising three intermetallic compounds in the oxygen-poor region of the Fe-Nd-O system. The composition of the most Fe-rich phase, designated as β, was determined by microprobe analysis with an Nd content of 12-14 at.% and a very small oxygen content. β is non-cubic, ferromagnetic with a uniaxial anisotropy. The domain width of surfaces perpendicular to the unique axis was measured to be 0.92 ± 0.05 μm. A crystal of this phase, surrounded by an Nd-rich eutectic, is shown in Fig. 7.

The second phase, designated to as π, has an Nd content of about 20 at.% and a slightly higher oxygen content than β. π is non-cubic and ferromagnetic with a Curie temperature of 230°C.

The third phase, designated as τ, is richer in Nd and has the highest oxygen content of these three phases. τ is also a non cubic phase, as indicated by its optical activity.

The existence of a miscibility gap, reported by Stadelmaier et al. /85Sta/, can be confirmed. The crystallization product of the oxygen rich liquid is mainly the perovskite, whereas the metal rich liquid solidifies to form the metallic binary and ternary phases (Fig. 8).

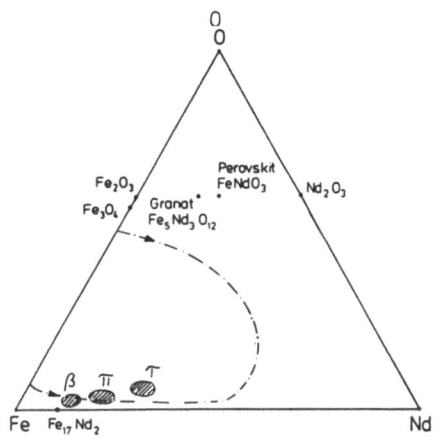

Fig. 7: Fe-55Nd-5O, pol. light. β-crystal in Nd-rich eutectic.

Fig. 8: Fe-Nd-O, ternary phases and schematic shape of the miscibility gap.

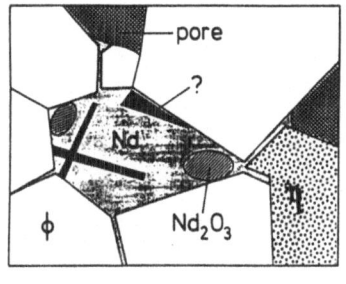

(a)

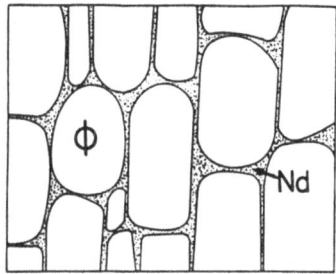

(b)

there exists a two-phase field L + Φ at a sintering temperature of about 1050°C. If the magnet composition is located in
this two-phase field, only Φ grains surrounded by the Nd-rich
liquid phase at the sintering temperature exist. Large η
grains do not occur. Upon cooling, some n particles crystallize out of the liquid and finally as part of the eutectic.
This amounts to less than 0.5 vol.% of very small η particles.

Figs. 10a,b show the difference in the phase distribution
between a commercial magnet with the composition $Fe_{77}Nd_{15}B_8$
and the "two-phase" magnet with the composition $Fe_{75}Nd_{18.5}B_{6.5}$.
No n phase can be seen in the "two-phase" magnet.

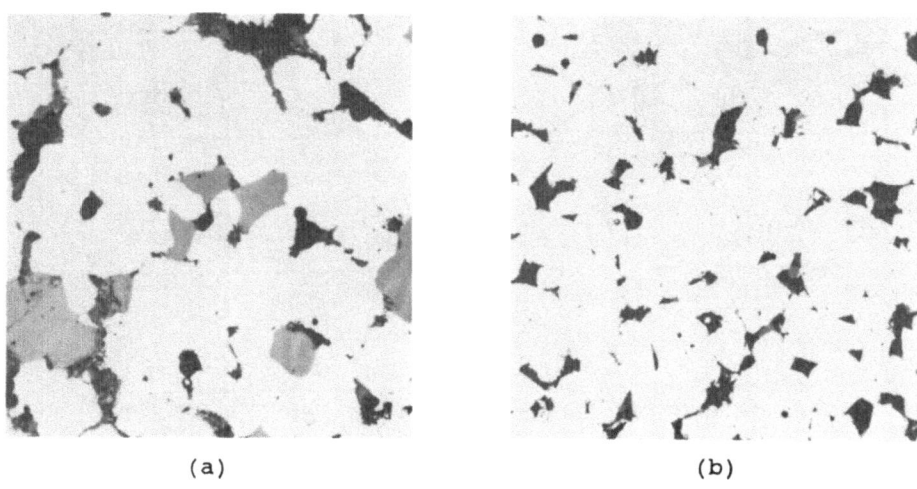

(a)   (b)

Fig 10: Microstructures of sintered Fe-Nd-B magnets.
      (a) Fe-15Nd-8B; three phases: Φ + η + Nd.
      (b) Fe-18.5Nd-6.5B; two phases: Φ + Nd.

Two effects are attained with this optimized magnet.
1. The magnetic properties of the sintered magnet have been
   improved. The improved temperature dependence compared with
   a "three-phase" magnet can be seen in Fig. 11.
2. The corrosion resistance of the "two-phase" magnet is superior to that of the "three-phase" magnet /87Men/.

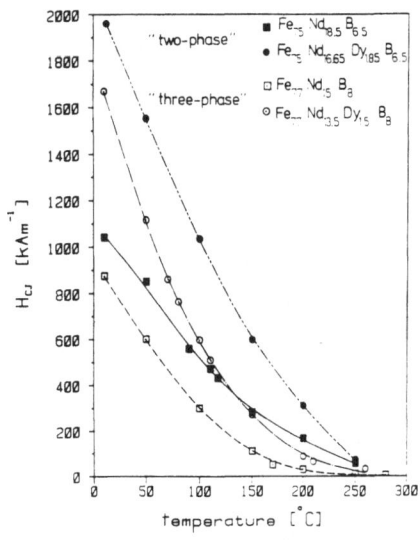

Fig. 11: Temperature dependence of the coercive field of different magnet compositions.

4.2 The Grain-Size Distribution in Sintered Fe-Nd-B Magnets

Two physical reasons exist, why a small grain-size is desirable to attain high coercivity.
1. Local demagnetising strayfields at edges increase with increasing grain size.
2. The probability of the existence of inhomogeneities in the grains which can nucleate the demagnetization is higher in large grains. If the reversal of the magnetization starts, in smaller grains it is stopped earlier.

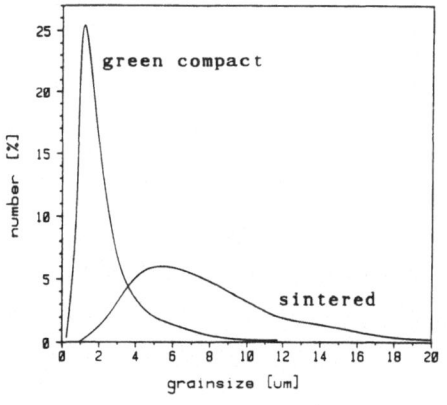

Fig. 12: Grain-size distribution of an Fe-18.5Nd-6.5B powder (30 min milling) and of the sintered body (1050°C/60min).

Fig. 12 shows a typical grain-size distribution of the Fe-Nd-B powder and of the sintered magnet. The powder-grain size can be varied both by the milling time and by influencing the brittleness of the masteralloy (e.g. hydrogen decrepitation).

The large difference in the grain-size distribution between the powder and the sintered body illustrates the strong grain growth during liquid phase sintering. Mainly the sintering temperature and time influence this grain growth.
Variation of the sintering temperature in the range of 1000 to 1100°C shows, that the highest $_jH_c$ can be achieved with sintering temperatures of 1000 to 1050°C (sintering time 60 min). Above 1060°C the coercive field strongly decreases (Fig. 13a). The accelerated grain growth above 1075°C can be explained by the large amount of liquid phase. In these magnets some grains > 50 μm were observed.
Fig. 13b illustrates that a short sintering time is advantageous for a high coercivity (sintering temperature 1050°C). The growth of the mean grain diameter can be described by

$$D = D_o + kt^{0.25}$$

with D_o = 1.9 μm and k = 2.0.

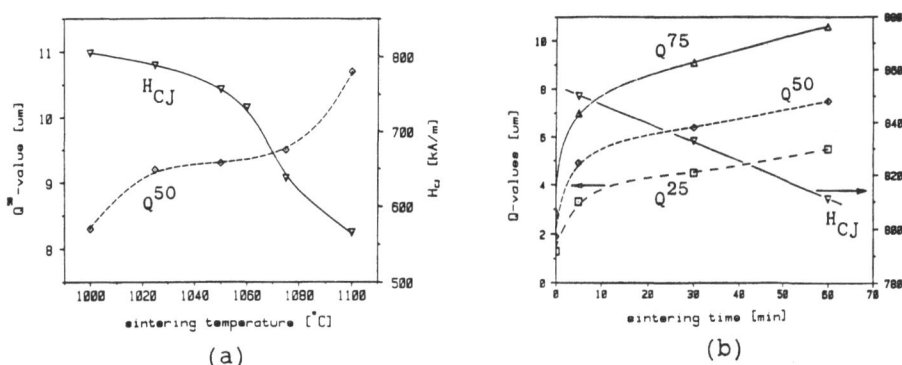

(a)   (b)

Fig. 13: Coercivity and grain-size of Fe-18.5Nd-6.5B sintered magnets, (a) vs. sintering temperature, (b) vs. sintering time.

Besides sintering temperature and time the heating rate profile, the volume fraction of liquid phase and dispersions,
which hinder the grain boundary migration influence the grain-
size distribution.
From Fig. 14 it can be seen, that it is important to have in
sintered Fe-Nd-B magnets a mean grain diameter < 12 μm (this
is different in Al$_2$O$_3$-doped magnets). It is supposed, that the
few grains > 50 μm are responsible for the deterioriation of
the coercivity in the magnets with a mean grain size > 12 μm.
Efforts for decreasing the grain size below 5 μm seem not to
be promising considering Fig. 14. However, Ma et al. /87Ma/
reported a considerable increase of $_jH_c$ in Fe-Nd-B sintermagnets with a mean grain size < 4 μm.

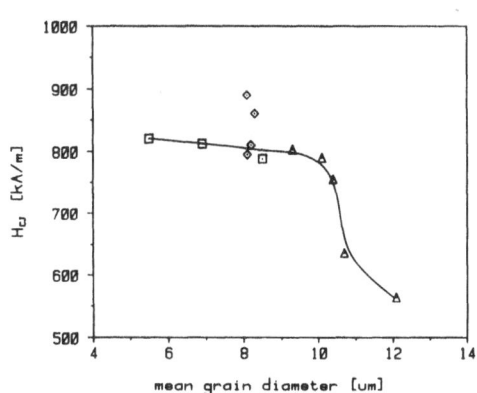

Fig. 14: Coercivity of
Fe-18.5Nd-6.5B
sintered magnets
vs. mean grain-
size.

4.3 The Contiguity

The contiguity C_{SL} is a measure for the fraction of grain
boundaries which are filled with, at sintering temperature,
liquid phase.

$$C_{SL} = \frac{S_{SL}}{S_{SL} + S_{SS}}$$

S_{SL} = interface solid-liquid
S_{SS} = interface solid-solid.

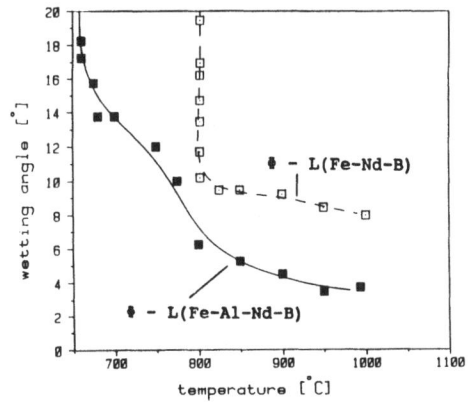

Fig. 16: Temperature dependence of the wetting angle Φ - L(Fe-Nd-B) and Φ - L(Fe-Al-Nd-B).

4.4 Microstructure of Grain Junctions and Grain Boundaries

At the sintering temperature the grain junctions and grain boundaries in the Fe-Nd-B sintered magnet are filled with a liquid Nd-rich phase. This liquid phase plays an important role in the Fe-Nd-B magnet.

1. It aids the densification by liquid phase sintering.
2. The non-ferromagnetic Nd phase in the grain boundaries of Φ causes magnetic decoupling of neighbouring Φ grains. 3. The Nd phase is a critical factor in the corrosion behaviour.
4. The crystallization products of the Nd-rich liquid may affect the remagnetization mechanism of the magnet.

Nd + Φ + η is not the only crystallization product of the Nd-rich liquid. It is well known that oxygen contamination during magnet processing can introduce 0.2 to 0.9 mass % oxygen in the sintered magnet. Because of the oxygen enrichment in the liquid phase, it must be considered quaternary, belonging to the system Fe-Nd-B-O /85Sta/.
From our investigations in the systems Fe-Nd, Fe-Nd-B and Fe-Nd-O, we must assume that ferromagnetic phases of metallic character are present in the grain junctions of the sintered magnet (see Chapter 3.4).
In /87bSch/ a microstructure of an Fe-Nd-B magnet (0.17 mass %

oxygen) is shown, which has been annealed for 10 days at 600°C for grain coarsening. Platelets of an optically active phase are visible, which does not belong to the system Fe-Nd-B. Notice that this magnet does not contain Al, so these phases are not identical to those reported in Chapter 5.1 for Al-doped magnets. The optical properties suggest that this is a phase with metallic character belonging to the system Fe-Nd-O.

When considering the demagnetization mechanisms of the Fe-Nd-B magnet the effect of these additional ferromagnetic phases must be included. They could act either as pinning centres or as nucleation sites for reversed domains (Fig. 17).

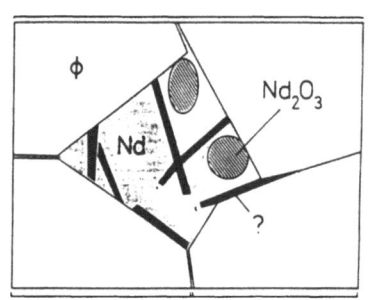

Fig. 17: Sketch of a grain junction in a real Fe-Nd-B sintered magnet.

5 INFLUENCE OF ADDITIONS

5.1 Al

Al which replaces the Fe atoms in the Φ cell, lowers the intrinsic properties of Φ, the Curie temperature, the saturation magnetization and the crystal anisotropy. So the positive effect of Al additions on the coercive field $_JH_c$ of sintered Fe-Nd-B magnets must be explained as a microstructural one. Fig. 18 shows the coarsened microstructure of an Al-containing sintermagnet due to a long annealing time. An additional Al-stabilized non-cubic platelet phase can be observed. The composition was analyzed to be $Fe_{60}Nd_{24}Al_{16}$. This phase is softmag-

netic and has a Curie temperature of 190°C. The analysis of
the magnetic properties of the sintered magnet shows, that
this Al-stabilized phase can not be responsible for the increase of coercivity /87bDur/. As shown in Chapter 4.4 Al
improves the wetting angle of the liquid phase, leading to a
higher fraction of Nd-filled grain boundaries. This is the
present explanation of the positive effect of Al on the coercive field.

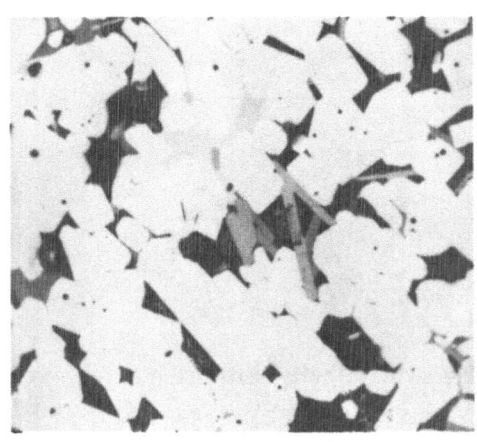

Fig. 18: Fe-2.5Al-20Nd-
6.5B sintered
magnet. Al-sta-
bilized phase in
grain junctions.

5.2 Dispersions

Dispersions, which are added to the Fe-Nd-B powder during milling, affect the coercivity of sintered magnets. We have
investigated the influence of Nd_2O_3, Fe_2O_3, Al_2O_3, AlN and BN.
Al_2O_3 powder yields the strongest increase of the coercive
field $_jH_c$ (Fig. 19).
In Nd_2O_3-doped magnets a hindrance of the grain growth by
these dispersions due to a grain boundary pinning was observed, leading to a small increase in $_jH_c$. In Al_2O_3 containing
magnets this cannot be used to explain the effect, because not
a hindrance but an accelleration of the grain growth was
observed. High coercive fields > 1500 kAm^{-1} have been measured
in magnets with a mean grain diameter of about 12 μm. Up to
now the effect of Al_2O_3 is not clarified.

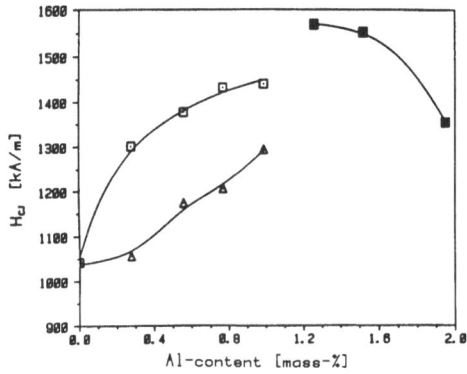

Fig. 19: (73.5-x)Fe-xAl-20Nd-6.5B, coercive field vs. Al-content.
($\triangle$) Al-addition during melting of the master alloy, ($\square$) Al_2O_3-addition during milling.

REFERENCES

62Esp G.P.Espinosa, "Crystal Chemical Study of the Rare-Earth Iron Garnets", J. Chem. Phys. 37 (1962) 2344-2347

65Ter V.F.Terekhova, E.V.Maslova, Y.M.Savitskiy, "Iron - Neodymium Equilibrium Diagram", Russ. Metall. 6 (1965) 50-52

70Mar M.Marezio, J.P.Remeika, P.D.Dernier, "The Crystal Chemistry of the Rare Earth Orthoferrites", Acta Cryst. B26 (1970) 2008-2022

72Can J.F.Cannon, D.L.Robertson, H.T.Hall, "Synthesis of Lanthanide-Iron Laves Phases at High Pressures and Temperatures", Mat. Res. Bull. 7 (1972) 5-12

76Dar M.P.Dariel, M.Malekzadeh, M.R.Pickus, Proc. 21st Ann. Conf. Magn. Magn. Materials (1976) 583

79Cha N.F. Chaban, Y.B. Kuz'ma, N.S. Bilonizhko, O.O. Kachmar, N.V. Petriv, "Ternary {Nd,Sm,Gd}-Fe-B Systems", Dopov. Akad. Nauk URSR Ser. A. : Fiz.-Mat. Tekh. Nauki 10 (1979) 875-877

84Sag M.Sagawa, S.Fujimara, N.Togawa, H.Yamamoto, Y.Matsuura, "New Material for Permanent Magnets on Base of Nd and Fe", J. Appl. Phys. 55 (1984) 2083-2087

84Sta H.H.Stadelmaier, N.A.Elmasry, N.C.Liu, S.F.Cheng, "The Metallurgy of the Iron-Neodymium-Boron Permanent Magnet System", Mat. Lett. 2 (1984) 411-415

85Bez A.Bezinge, H.F.Braun, J.Muller, K.Yvon, "Tetragonal Rare Earth (R) Iron Borides, $R_{1+\epsilon}Fe_4B_4$ ($\epsilon \approx 0.1$), With Incommensurate Rare Earth and Iron Substructures", Solid State Com. 55 (1985) 131-135

85Bus K.H.J.Buschow, D.B.De Mooij, H.M.Van Noort, "The Fe-Rich Isothermal Section of Nd-Fe-B at 900°C", Philips J. Res. 40 (1985) 227-238

85Liv J.D. Livingston, "Iron-Rare Earth Permanent Magnets", 8th Int.Workshop on Rare Earth Magnets and their Applications, Dayton, Ohio (1985) 423-440

85Mat Y.Matsuura, S.Hirosawa, H.Yamamoto, S.Fujimura, M.Sagawa, K.Osamura, "Phase Diagram of the Nd-Fe-B Ternary System", Japan. J. Appl. Phys. 24 (1985) L635

85aSta H.H.Stadelmaier, N.A.ElMasry, "Understanding Rare Earth Permanent Magnet Alloys: Crystal Chemistry and Problems Inherent in Phase Constitution", Proceedings of the 4th Int. Symp. on Magn. Anisotropy and Coercivity in Rare Earth - Transition Metal Alloys, Dayton, OH, (1985) 613-633

86bSch G. Schneider, E.-Th. Henig, G. Petzow, and H.H Stadelmaier, "Phase Relations in the System Fe-Nd-B", Z. Metallkde. 77 (1986) 755

86Sta H.H.Stadelmaier, G.Schneider, M.Ellner, "A $CaCu_5$-Type Iron-Neodymium Phase stabilized by Rapid Solidification", J. Less-Common Metals 115 (1986) L11-L14

87aDur K.-D.Durst, H.Kronmüller, "The Coercive Field of Sintered and Melt-Spun NdFeB Magnets", J. Magn. Magn. Mat. 68 (1987) 63-75

87bDur K.-Durst, H.Kronmüller, G.Schneider, "Magnetic Hardening Mechanisms in Fe-Nd-B Type Permanent Magnets",
Proceedings of the 5th Int. Symp. on Magn. Anisotropy
and Coercivity in Rare Earth - Transition Metal Alloys,
Bad Soden, FRG (1987) 209-225

87Fid J.Fidler, "The Role of the Microstructure on the Coercivity of Nd-Fe-B Sintered Magnets", Proceedings of
the 5th Int. Symp. on Magn. Anisotropy and Coercivity
in Rare Earth - Transition Metal Alloys, Bad Soden, FRG
(1987) 363-377

87Gri B.Grieb, E.-Th.Henig, G.Schneider, G.Petzow, "Homogeneity Ranges and Phase Equilibria Around $(Nd,RE)_2Fe_{14}B$
with RE=Tb,Dy", Proceedings of the 5th Int. Symp. on
Magn. Anisotropy and Coercivity in Rare Earth - Transition Metal Alloys, Bad Soden, FRG (1987) 395-402

87Hen E.-Th.Henig, G.Schneider, H.H.Stadelmaier, "Metastable
Solidification of Fe-Rich Iron-Neodymium-Boron Alloys",
Z. Metallkde. 78 (1987) 818-820

87Hoc S.Hock, H.Kronmüller, "Intrinsic Magnetic Properties of
$Fe_{14}Nd_2B$ Single Crystals", Proceedings of the 5th Int.
Symp. on Magn. Anisotropy and Coercivity in Rare Earth
- Transition Metal Alloys, Bad Soden, FRG (1987) 275-
282

87Ma B.M.Ma, R.F.Krause, "Microstructure and Magnetic Properties of Sintered NdDyFeB Magnets", Proceedings of
the 5th Int. Symp. on Magn. Anisotropy and Coercivity
in Rare Earth - Transition Metal Alloys, Bad Soden, FRG
(1987) 141-148

87Men R.van Mens, G.W.Turk, M.Brouha, "Magnetic and Corrosion
Properties of Mechanically Alloyed $Nd_2Fe_{14}B$ with a Rare
Earth Hydride", Proceedings of the 9th Int. Workshop on
Rare Earth Magnets and their Applications, Bad Soden,
FRG (1987) 311-316

87aSch G.Schneider, E.Th.Henig, G.Petzow, H.H.Stadelmaier,
"The Binary System Iron-Neodymium", Z. Metallkde. 78
(1987) 694-696

87bSch G.Schneider, E.-Th.Henig, H.H.Stadelmaier, G.Petzow,
        "The Phase Diagram of Fe-Nd-B and the Optimization of
        the Microstructure of Sintered Magnets", Proceedings of
        the 5th Int. Symp. on Magn. Anisotropy and Coercivity
        in Rare Earth - Transition Metal Alloys, Bad Soden, FRG
        (1987) 347-362

87Zha   N.Zhang, Y.Luo, "Two Important Sections in the Nd-Fe-B
        Ternary Phase Diagram", Proceedings of the 5th Int.
        Symp. on Magn. Anisotropy and Coercivity in Rare Earth
        - Transition Metal Alloys, Bad Soden, FRG (1987) 453-
        460

88Sch   G.Schneider, "Konstitution und Sinterverhalten von
        Hartmagnetwerkstoffen auf Fe-Nd-B-Basis", Dissertation,
        Universität Stuttgart (1988)

EFFECT OF SOME ADDITION ELEMENTS ON THE PHASE
EQUILIBRIA AND MICROSTRUCTURE OF Nd-Fe-B ALLOYS

C.H. ALLIBERT

Laboratoire de Thermodynamique et Physico-Chimie Métallurgiques
Institut National Polytechnique de Grenoble
38402 Saint Martin d'Hères, France

ABSTRACT

For Nd-Fe-B alloys with magnet composition, the modification of the phase equilibria due to small additions of Cu, Nb, Zr has been studied. Some information on the influence of Mo, Co, Dy has been obtained. At sintering temperature, Cu, Nb, Zr and Mo do not dissolve in significant amount in $Nd_2Fe_{14}B$. Cu enters the Nd rich liquid and forms Nd_xCu_y phases on cooling. Nb, Zr and Mo concentrate in solid precipitates. Co and Dy substitutes for Fe and Nd in the major phases but also change the constitution of the binder. Concerning the microstructure, Nb and Zr are expected to change the hard phase grain size. The experimentation shows that Cu modifies the composition and fraction of binder and increases the $Nd_2Fe_{14}B$ grain size. The drastic decrease of coercivity due to Cu appears mainly related to the change of binder composition and decrease of grain boundary area.

INTRODUCTION

Sintered permanent magnets based on a phase presenting a magneto crystalline anisotropy have two kinds of properties : the intrinsic properties (magnetization Js, anisotropy field Ha, Curie Temperature Tc) that are determined only by the composition and crystal structure of the hard magnetic phase and the coercivity iHc that depends strongly on the microstructure. When addition elements are introduced in such materials, they can dissolve in the hard phase and modify its intrinsic properties. They can also form secondary phases. If these phases are present inside the hard phase - as coherent precipitates with suitable size and composition - they can originate a magnetic hardening by pinning the domain walls. When the secondary phases are distributed around the hard phase grains, they change the microstructure of the alloy (grain size, contiguity ...) and affect the coercivity.

Finally, the influence of addition elements on the magnetic characteristics of a given alloy can be understood only if the effect of these elements on the phase equilibria of the alloy is determined.

Consequently, the purpose of the present work was to study the effect of several addition elements on the phase equilibria of the Nd-Fe-B system.

First, the influence of the substitution for Fe of 3-10 at% Cu in alloys with compositions close to the magnets was determined at temperatures corresponding to the sintering and heat treatment. Then the modification of the microstructure due to 0-4 at% Cu was studied and related to the coercivity of sintered specimens. This part was carried out in collaboration with W.A.RODEWALD (Vacuum Schmelze).

Secondly the effect of the substitution for Nd of 1 at% Nb or Zr on the phases present in Nd15Fe78B7 was determined. This literature results on the magnetic properties of the Nb, Zr containing magnets were analysed using the metallurgical data.

Finally, some information was obtained on Nd-Fe-B alloys containing Mo, Co and Dy in order to estimate the contribution of such elements to the magnetic properties.

MATERIALS AND TECHNIQUES

The starting materials were a master alloy Nd-25Fe, electrolytic Fe(99.9), Cu ingot (99.95) and B powder (99.99). The alloys were prepared by induction melting of the elements contained in a boron nitride crucible, then casting in a watercooled mold. The alloys were submitted either to DTA followed by slow cooling 300°C/h or to annealings in vacuum sealed silica tubes at 1000°C (40h), 900°C (150h) and 600°C (800h). The phases present in the specimens were characterized by scanning electron-microscopy, electronmicroprobe analysis and X-Ray diffraction. The micro-structure was studied by image analysis of SEM or optical micrographs.

RESULTS

Effect of Cu on the phase equilibria

The studied compositions, shown in Table 1, are located around the composition of Nd2Fe14B. In these alloys, 3 to 10 at% Cu are introduced. The phases present in such alloys are identified after either annealing or slow cooling at 300°C/h.

TABLE 1   Compositions and treatments of the studied alloys.

Material Content(at%)	1	2	3	4	5	6
Nd	8.4	8.4	12	10	16	13.8
Cu	9.8	5	4	5	4	3.3
B	8.4	8.4	17	6.5	8	7.8
Treatments :	Slow cooling (300°C/h)			900°C	1000°C 600°C	1000°C 600°C

The phases identified in the specimens are grouped in Table 2.

TABLE 2   Phases present in the specimens.

Material	Phases
1	Fe + Fe2B + Nd2Fe14B + NdCu2 + Nd(Cu1-xFex) + "Nd2Cu1-yFey"
2	Fe + Fe2B + Nd2Fe14B + NdCu2 + Nd(Cu1-xFex) + Nd
3	Fe + Fe2B + Nd2Fe14B + NdCu2 + Nd
4	Fe2B + Nd2Fe14B + Nd1.1Fe4B4 + NdCu2 + Nd(Cu1-xFex)
5	Nd2Fe14B + Nd1.1Fe4B4 + NdCu2 + Nd(Cu1-xFex)+Nd
6	Nd2Fe14B + Nd1.1Fe4B4 + NdCu2 + Nd(Cu1-xFex)

The examination of the composition values leads to the following
remarks :

At high temperature, Cu is present in the liquid phase but does not
dissolve in significant amount in the solid compounds : about 1 at% Cu
substitutes for Fe in Nd2Fe14B while less than 0.5% Cu is detected in
Nd1.1Fe4B4. The solidification forms first the phases present in the
ternary section Nd-Fe-B while Cu concentrates in the liquid phase. The
present observations are consistent with the solidification path determined by Schneider et al (1986). Then the Cu containing liquid freezes by
forming the subsequent NdxCuy compounds and likely some Nd1.1Fe4B4. The
solubility of Fe in NdCu2 is very small (# 1 at%) but can reach some
10 at% in NdCu. Some analyses give compositions around "Nd2(Fe,Cu)" :
they could correspond to the overall composition of a ternary eutectic
Nd-NdCu-Nd2Fe17, according to the phase equilibria plotted by
Terechkhova et al. (1965) and Carnasciali et al. (1983) for the Nd-Fe and
Nd-Cu systems.

Annealings at 1000 and 600°C provide results that confirm the solidification investigation. The Cu for Fe substitution rate in Nd2Fe14B slightly decreases with temperature (# 1 at% at 1000°C, 0.7 at% at 600°C).

From the view point of the magnetic properties, the present results underline the following points :

Cu cannot induce a significant modification of the intrinsic properties (Tc, Js, Ha) of Nd2Fe14B.

Cu does not provide the precipitation of a coherent secondary phase in the Nd2Fe14B matrix. Consequently no magnetic hardening effect due to pinning has to be expected from Cu additions.

At sintering temperature, Cu changes the composition and volume fraction of the liquid phase surrounding the Nd2Fe14B grains. On cooling, Cu induces the precipitation of NdxCuy phases. Thus, Cu can produce a modification of the microstructure of the sintered alloys : such a modification will be analysed in the last part of the report.

Effect of Nb and Zr on the phase equilibria

A small amount (1 at%) of Nb or Zr was substituted for Nd in the Nd15Fe78B7 alloy. The phases present in the specimens were analysed after annealing at 1000°C and 600°C. The table 3 groups the main results.

TABLE 3   Main phases present in the alloys.

	Nd14Nb1Fe78B7	Nd14Zr1Fe78B7
1000°C	Nd2Fe14B + Nb3Fe3B4 + "liquid"	Nd2Fe14B + ZrB2 + "liquid"
600°C	Nd2Fe14B + Nb3Fe3B4 + Nd1.1Fe4B4 + Nd	Nd2Fe14B + ZrB2 + Nd1.1Fe4B4 + Nd

The compositions of the different phases indicate that :

Nb and Zr are present in very small amounts in Nd2Fe14B (0.3at%). They concentrates in ternary or binary compounds, the morphology of which (Fig. 1a,b) suggests that they are solid at sintering temperature. The compositions of these compounds correspond to Nb3Fe3B4 and ZrB2 : these formulae have still to be checked by crystal structure analysis, specially for the Nb compounds. Effectively, three compounds with close

1a N d 1 4 F e 7 8 N b 1 B 7

———· N b 3 F e 3 B 4

· ———· N d 2 F e 1 4 B   m a t r i x

1b N d 1 4 Z r 1 F e 7 8 B 7

Z r B 2

N d 2 F e 1 4 B   m a t r i x

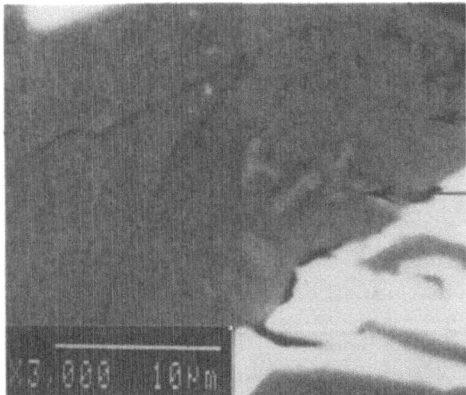

1c N d 2 4 M o 1 F e 6 8 B 7

· — — N d 2 F e 1 4 B   m a t r i x

——— M o   r i c h   p h a s e

Fig. 1   Microstructure and phase compositions of Nd-Fe-B-X alloys
annealed at 600°C.
1a : Nd14Nb1Fe78B7   1b : Nd14Zr1Fe78B7   1c : Nd24Mo1Fe68B7

compositions (NbFeB, Nb3Fe3B4 and Nb2FeB2) were found by Kuzma et al. (1968) in the Nb-Fe-B system and might form in the Nd-Fe-B-Nb alloys depending on their composition.

From these results, it appears that, in equilibrium conditions, Nb and Zr do not dissolve in Nd2Fe14B and consequently do not change its intrinsic magnetic properties. However, the formation of Nb or Zr rich precipitates, that likely occurs in the sintering temperature range of the magnetic alloys, is expected to affect their microstructure.

Effect of Mo, Co, Dy on the phases present in Nd-Fe-B alloys

The characterization of samples with compositions Nd24Mo1Fe68B7 and Nd16(Fe0.8Co0.2)76B8 after annealing at 1000 and 600°C shows that :

Mo very slightly dissolves in Nd2Fe14B(# 1 at%) and also forms small Mo, Fe rich precipitates (Fig. 1c).

Co substitutes for Fe in the major phases Nd2Fe14B and NdFe4B4 but also induces the precipitation of phases such as Nd(Fe,Co)2, Nd3Co in the Nd rich binder (Fig. 2a).

Only qualitative information is obtained on the Nd13.8Dy2.2Fe76B8 alloy because of the difficult analysis of the specimen (Fe Kα and Dy Lα peaks interference). However it indicates that Dy substitutes for Nd in all the phases.

The overall results indicate that Mo will not significantly modify the intrinsic magnetic properties of the hard phase but is able to slightly change the microstructure. For Co and Dy, it is already well-known that the Fe, Nd for Co, Dy substitutions in Nd2Fe14B strongly affect Tc, Ha, Js. The present findings emphasize that the minor phases are also changed by Co, Dy (composition, volume fraction) and this can produce a microstructural change.

Effect of the addition elements on the microstructure of the sintered magnets

No experimentation has been carried out with Nb, Zr, however the analysis of the equilibria results enables some predictions. By forming Nb3Fe3B4 or ZrB2, Nb or Zr consumes Fe, B or B and the composition of the Nd-Fe-B mixture evolves towards the limit of the 3 phase domain Nd2Fe14B+Nd2Fe17+liquid : this slightly decreases the binder fraction and

Nd2(Fe.85Co.15)14B

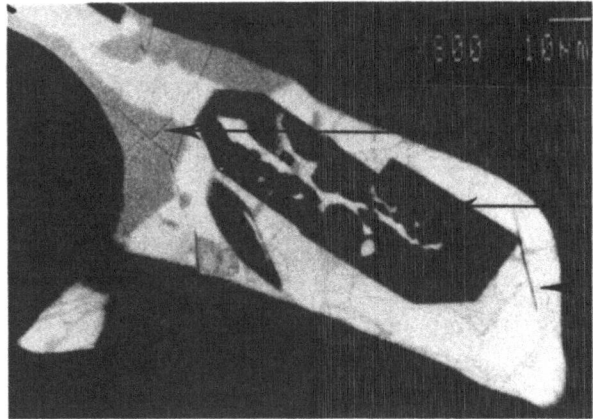

Nd(Fe.5Co.5)2

Nd(Fe.8Co.2)4B4

·Nd3Co

2 a

·Nd

**NdCu2
+
Nd(Fe,Cu)**

·Nd2Fe14B matrix

2 b

Fig. 2   Microstructure and phase compositions of Nd-Fe-B-X alloys
annealed at 600°C.
2a : Nd16(Fe0.8Co0.2)76B8.   2b : Nd15Fe75Cu3B7.

increases its Nd content. When an excess of Nb, Zr is added, the 3 phase domain boundary can be overpassed and some Nd2Fe17 can form. Such a behaviour is consistent with the observations of Xiao et al. (1987), and easily explains the coercivity decrease. For small additions of Nb, Zr, only a change of microstructure is expected. According to the general knowledge of the liquid sintering process, the growth of the Nd2Fe14B grains is expected to increase with the liquid fraction decrease and to be slowered by the small solid precipitates. At the moment, the experimentation is still necessary to analyse the overall effect of Nb, Zr on the microstructure and relate it to the coercivity enhancement mentioned by several authors.

The effect of Cu has been studied in collaboration with W.A.Rodewald at Vacuumschmelze. Five specimens containing 0-4 at% Cu were prepared by mixing powdered alloys with atomic compositions Nd15.7Fe77.2B6.9Al0.2 and Nd15.15Fe75Cu4B5.85. The compacted rods were sintered at 1090°C for 1h then annealed at 630°C for 1h.

As predicted by the equilibrium results, the Cu richest specimen (4 at% Cu) consists of a packing of Nd2Fe14B grains surrounded by a Nd rich binder containing Cu and Fe with different ratios (Fig. 2b). Some grains of Nd1.1Fe4B4 are detected. When the overall Cu content of the alloys decreases, the only composition change is a decrease of the Cu content in the binder.

The microstructures of the sintered specimens, shown on Fig. 3, evidence differences which were analysed by comparing the volume fractions of binder, the grain sizes of Nd2Fe14B phase and the Cu contents. The corresponding values and the intrinsic coercivity iHc are grouped in Table 4.

TABLE 4   Main characteristics of the Nd(Fe-Cu)-B alloys.

Material	1	2	4	5
Cu content at%	4	2	0.5	0
Binder Fraction %	16	16	13	9
Porosity %	5	5	3	2
Grain size μm	12-30		6-30	6-10
iHc KA/m		312	736	710

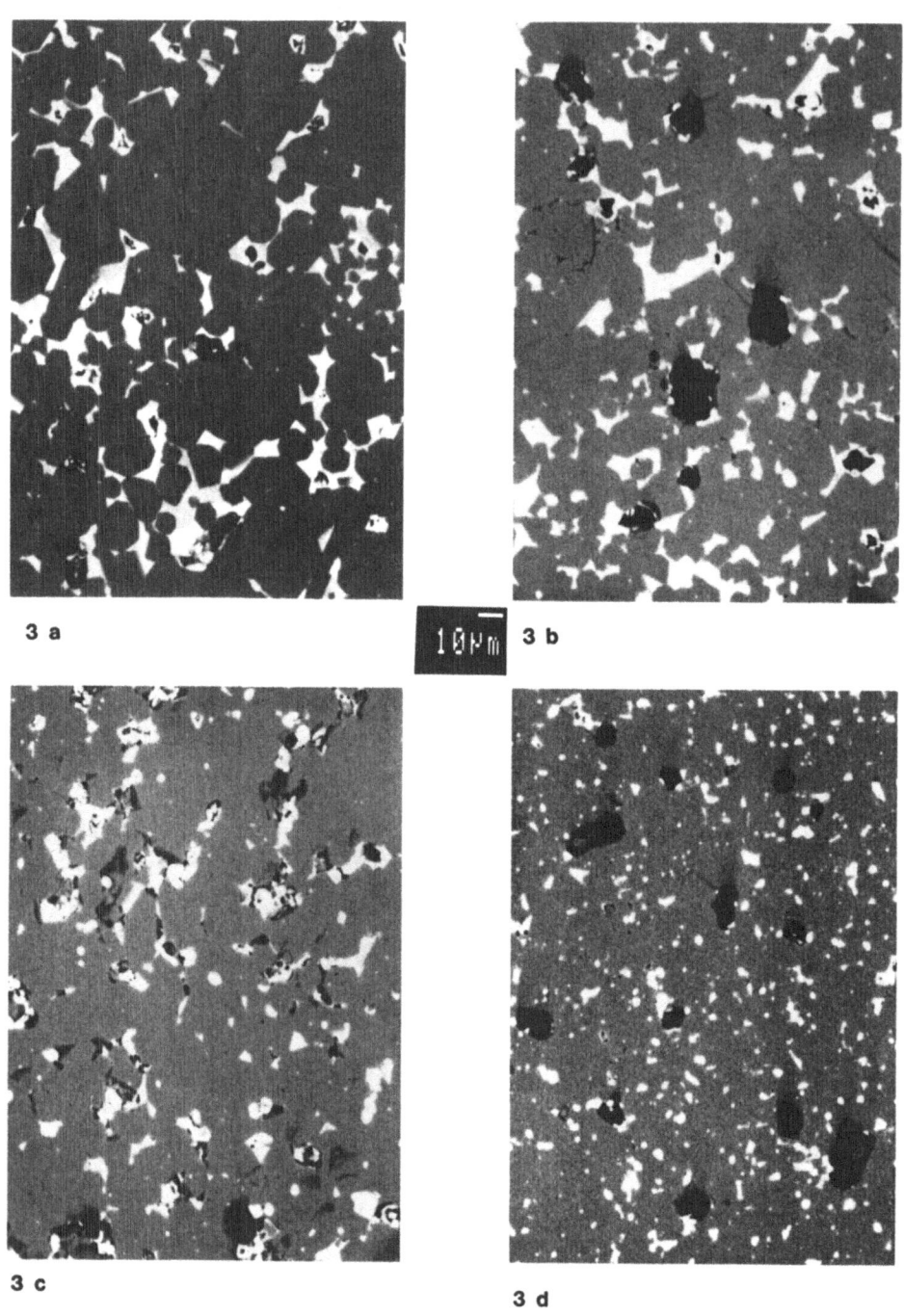

3 a   10μm 3 b

3 c

3 d

Fig. 3   Microstructure of Nd-Fe-Cu-B sintered alloys.
3a : 4 at% Cu   3b : 2 at% Cu   3c : 0.5 at% Cu   3d : 0 at% Cu.

The results indicate that Cu increases slightly the volume fraction of binder and significantly the grain size of the hard phase. Cu also produces a slightly higher porosity and finally induces a drastic decrease of coercivity.

From the previous data published by Sagawa et al. (1984), an increase of binder fraction increases iHc in the Nd-Fe-B alloys. Moreover, iHc values in the same range are obtained for different grain sizes (15 μm after Sagawa, 6-10 μm in the present specimen). By comparison the decrease of coercivity due to Cu appears related mainly to the binder composition. However, the measurement of other microstructural features that appear very different on the micrographs of etched samples (size distribution, interface and grain boundary areas) is still necessary to really understand the detrimental effect of Cu on the Nd-Fe-B magnetic properties.

CONCLUSION

The present work emphasizes the very different behaviour of some addition elements in the Nd-Fe-B alloys :

Nb, Zr and Mo form small precipitates that could slowered the grain growth in the sintered specimens.

Co and Dy distribute in the main phases $Nd_2Fe_{14}B$ and $NdFe_4B_4$ but also induce the formation of compounds in the Nd rich binder. Such a change could produce a microstructural modification. This modification might be taken into account besides the strong change of intrinsic magnetic properties already well-known, in order to analyse the effect of Co and Dy on the coercivity.

Cu dissolves in the Nd rich binder and changes its volume fraction and composition. Cu also affects the microstructural features (grain size). A complementary, but very difficult, investigation of the microstructure is still required to enable the complete analysis of the detrimental effect of Cu on the coercivity. Such a characterization would be of interest to really understand the coercivity mechanism in the Nd-Fe-B alloys.

REFERENCES

Carnasciali, M.M., Costa, G.A. and Franceschi, E.A. 1983. J. Less Common Met., 92, 97-103.

Kuzma, F.B., Tsolkovsky, T.J. and Baburova, O.P. 1968. Izv. Akad Nauk. Neorg Mat. 4, 1081-1085.

Sagawa, M., Fujimura, S., Togawa, N., Yamamoto, M. and Matsuura, Y. 1984. J. Appl. Phys., 55, 2083-2087.

Schneider, G., Henig, T.E., Petzow, G. and Stadelmaier, H.H. 1986. Zeitsch. Metallk., 77, 755-761.

Terechkova, V.F., Maslova, E.V. and Savitsky, Y.M. 1965. Russian Metallurgy (Met)., 3, 50-52.

Xiao, Y., Strnat, K.J., Mildrum, M.F. and Ray, A.E. 1987. 5th Int. Workshopon Rare-earth Mag. (Ed. Herget C. and Poerschke R.), pp 467-476.

THE Nd₁₅Fe₇₇B₈ MICROSTRUCTURE: SOME EFFECTS OF OXYGEN FOR DIFFERENT SOLIDIFICATION CONDITIONS

D. Cochet-Muchy and S. Païdassi

Centre d'Etudes Nucléaires de Grenoble
Département de Métallurgie
85X, 38041 Grenoble Cedex, France

ABSTRACT

Some microstructural effects of oxygen in cast and rapidly quenched Nd15Fe77B8 alloys have been observed. Starting from as-cast materials, we have studied the effects of an oxidation in air near 300°C, followed by electron beam surface remelting. During the oxidation process, oxygen promotes, along the diffusion paths, the growth of glassy regions containing strongly textured α-Fe precipitates. After surface remelting of these oxidised samples, various rapid solidification microstructures are obtained. Oxygen contamination effects have also been studied in Nd15Fe77B8 ribbons directly elaborated by the planar flow casting process. For partially amorphous ribbons, the study of the devitrification mechanisms reveals the formation of very finely crystallized two phase (α-Fe + C-Nd2O3) regions, in addition to the main phase (Nd2Fe14B) crystallization.

INTRODUCTION

Recently, the role of oxygen on the stabilization of different minor phases which influence the bulk properties of Nd-Fe-B magnets has received attention. For instance, in the Nd-rich binder of the sintered magnets, phases not belonging to the Nd-Fe-B phase diagram have been observed: Ramesh et al. (1987) have found a fcc Nd-O type oxide with NaCl structure and a=0.52 nm. In a recent study, Schneider et al. (1987) also mentioned the oxygen influence on the corrosion resistance. Their study is concentrated upon the phase equilibria in the systems Nd-Fe, Nd-Fe-B and Nd-Fe-O, as a basis for understanding the actual system Nd-Fe-B-O. Here, we give complementary informations about some phases observed in the quaternary system, using TEM and EPMA results. Various oxygen concentration and solidification conditions have been explored in Nd15Fe77B8 samples: as-cast, oxidized, oxidized with the surface remelted by electron beam, devitrified flow-cast ribbons.

AS-CAST SAMPLES

Induction-melted ingots of the nominal atomic composition Nd15Fe77B8 have been obtained from commercially available products (Nd,Fe 99.9%,B 99%). As reported by Fidler *et al.* (1985), the as-cast microstructure shows primary α-Fe dendrites, surrounded by the Nd2Fe14B (T1) matrix phase. Plates of the boride phase of approximate composition NdFe4B4 (T2) are also found. The remaining regions are Nd-rich overall and present a complex microstructure (Fig. 1). They are interpreted as the result of the last solidification step near 655°C, which could be written L -> T1 + T2 + Nd in a true ternary Nd-Fe-B system (Schneider *et al.*, 1986). In these regions, one can distinguish two bright phases of non-cubic character, with different morphologies. Probably due to its preferred graingrowth perpendicularly to the c-axis, T1 (brighter) forms lamellae with a thickness ranging from 5 µm to less than 0.5 µm. T2 is associated with the more equiaxed grains. The remaining material, which is cubic and not metallic Nd, presents two well-defined level of grey; the compositions estimated by EPMA are around 80-85 at.% Nd, 2-3 at.% Fe, balance O, for the medium grey phase (A), and around 37 at.% Nd, balance O, for the dark grey phase (B)

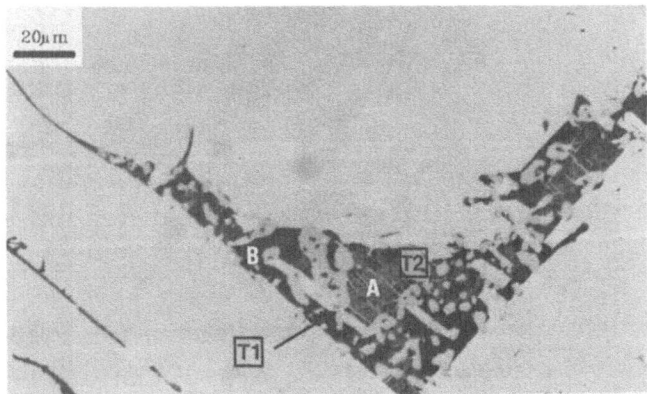

Fig. 1   Optical micrograph of a typical Nd-rich region in as-cast Nd15Fe77B8.

(boron contents unknown). The phase richer in neodymium is in the range of compositions announced for the fcc Nd-O type oxide with a=0.52 nm (Ramesh *et al.*, 1987; Sagawa *et al.*, 1984). The second phase corresponds to C-Nd2O3, also cubic with a=1.108 nm. We have observed by TEM that these phases are heavily faulted and contain T1 lamellae and T2 grains inside, but, the diffraction lines of the two oxides being very close, it is difficult to distinguish them from the electron diffraction patterns. Nevertheless, the C-Nd2O3 phase can be identified unambiguously because forbidden reflections due to multiple electron scattering appear sometimes (reflections of the (110) type for instance). From another EPMA result, Nd(OH)3 is expected to form at the surface of these oxides, but not in the bulk material.

These observations clearly show that, in the Nd15Fe77B8 alloy conventionnally prepared, the Nd-rich liquid corresponding to the last solidification step is quarternary. Stadelmaier *et al.* (1985) have suggested the existence of a miscibility gap in the liquid, separating the melt in an oxygen rich part and an oxygen poor part. Phases of metallic character issued from the oxygen poor part, such as the Fe-rich phases reported by Schneider *et al.* (1987) in the Nd-Fe-O system, have not yet been identified.

OXIDIZED SAMPLES

An heat treatment in air, 100 hours at 270°C, has been performed on the previous Nd15Fe77B8 samples. These values of time and temperature correspond to an upper limit beyond which the samples are considerably fracturing. Volume changes due to the solid state microstructural modifications reported below are probably responsible for this result.

Figure 2 is an optical micrograph, near the surface, of the alloy after such a treatment. Modifications appear in the Nd-rich regions and in the T1 phase. In the Nd-rich regions, it seems that the proportion of the Nd-O type oxide has decreased, in favor of C-Nd2O3. In the T1 phase, new elongated zones (X) are present. The X zones are growing from the surface, the pores (P), and also the Nd-rich regions, where the oxygen seems to diffuse at first. Then, they follow the oxygen diffusion paths in the T1 phase (a dark line is clearly visible in the middle of most of them), probably grain boundaries or structure defects. Their typical thickness is 10 µm, but large variations are observed. The overall chemical concentrations measured by EPMA in these zones are around 7 at.% Nd, 56 at.% Fe, 31 at.% O (boron content unknown). The median line is enriched in O, depleted in Fe and Nd, but it is not thick enough to allow a quantitative analysis. No modification of the T2 phase is found. Contrary to previously reported results (Christodoulou *et al.*, 1987), the T2 phase seems so to be the more oxidation resistant.

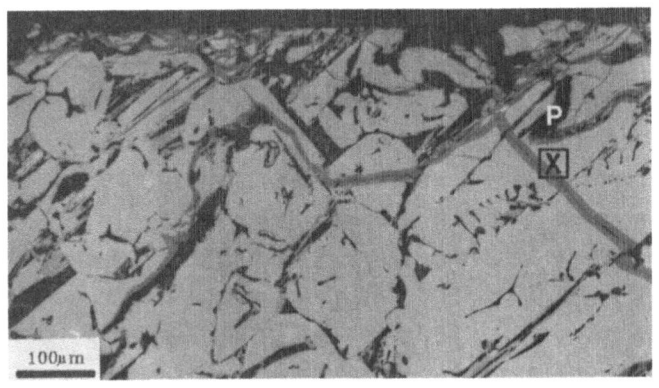

Fig. 2   Optical micrograph of Nd15Fe77B8 annealed 100 hours in air at 270°C, near the surface of the sample.

We have succeeded in imaging the X zones by TEM: they are constituted by an amorphous matrix phase, containing strongly textured α-Fe precipitates in the nanometer range. Figure 3 is a dark-field image (DF), corresponding to (110)α-Fe reflections, showing such a zone. The other features appearing on the diffraction pattern are an amorphous ring at d=0.28 nm, and some spots originating from the T2 phase defining the bottom of the zone. In this case, an orientation relationship between the T2 crystal and the texture of the precipitates is found. A part of the X zone has been damaged during the ion milling of the foil (the top half approximately), indicating a poorer resistance to the ion milling. Just below the hole, the median line is visible. Finally, it must be emphasized that, considering size, morphology and texture, such a zone has nothing in common with the famous "bcc phase" layers (Hiraga *et al.*, 1985), interpreted as an α-Fe precipitation due to the ion-milling (Ramesh *et al.*, 1987).

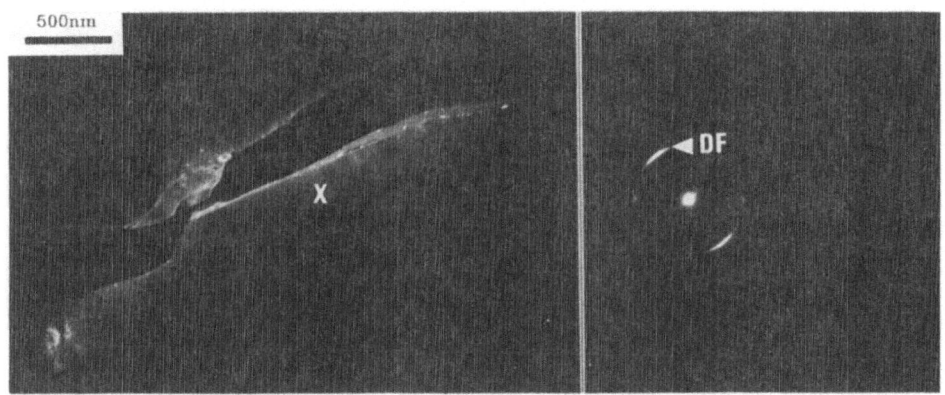

Fig. 3   Dark-field electron micrograph of a X zone and corresponding diffraction pattern.

The X zones growth can be interpreted as the result of the T1 phase oxidation process in the following way: oxygen extracts Nd from Nd2Fe14B, leading to elemental Fe precipitation. In agreement with Stadelmaier *et al.* (1985), oxide glass formation in the system Nd-B-O occurs.

SURFACE REMELTING BY ELECTRON BEAM

Remelting and subsequent rapid quenching of the previous oxidized samples have been produced by electron beam surface treatment, using linear beam scannings with different velocity, power and focusing. Considering the initial microstructure (Fig. 2), a melted depth in the mm range is necessary to insure a nearly constant composition of the treated zone. Figure 4 shows such a zone, resulting from three sucessive scannings at 5 cm/s, with a focused 800 W beam. The region with a fine grained microstructure (3-10 μm) defines the melted depth for the T1 matrix phase. Inside this region,

phases with a higher melting point sometimes appear unmelted:
these are Fe dendrites and X zones. On the contrary, the
melting front of the Nd-rich regions propagates outside, the
limit generally being underlined by cracks opening (C). The
corresponding resolidified regions, containing T1 and T2
precipitates, are observed to be free of C-Nd2O3, and look
like the Nd-O type oxide. The concentrations detected by EPMA
for this oxide phase are now around 55 at.% Nd, 25 at.% Fe, 20
at.% O, thus indicating a large composition range for various
solidification conditions. The SEM observation of the rapid
solidification microstructure in the completely remelted zone
has revealed no fundamental novel features: T2 and Nd-rich
minor phases appear at T1 grain boudaries. No Fe has been
detected. The increased oxygen concentration leads to various
size C-Nd2O3 segregations (1 to 100 μm)(S). Their localization
and shape (as a prolongation of the non-remelted X zones)
unfortunately indicate that a complete homogenization has not
been attained during the electron beam treatment, even after
five successive scannings.

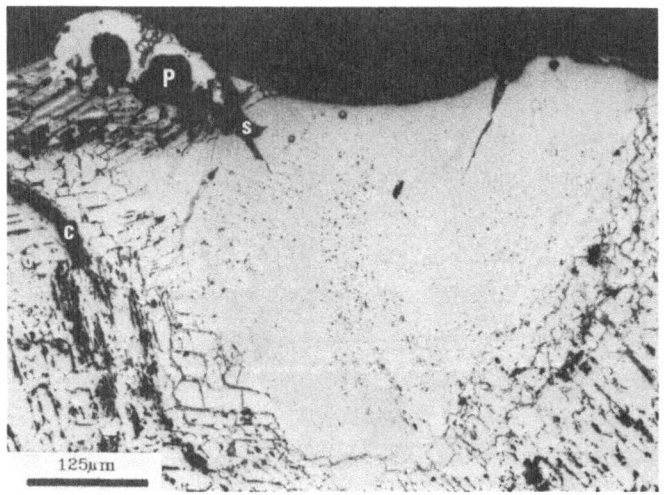

Fig. 4   Optical micrograph of an electron beam treated
zone in oxidized Nd15Fe77B8.

DEVITRIFIED FLOW-CAST RIBBONS

Ribbons of the Nd15Fe77B8 composition have been obtained by using the planar flow casting process under vacuum (wheel speed=60 m/s). The Nd and Fe concentrations have been checked by EPMA, indicating a less than 0.2 wt.% shift. An oxygen level in the 10^3 ppm range has been determined by chemical analysis. TEM has revealed an amorphous part on the wheel side of the 15-20 μm thick ribbons. Small α-Fe precipitates (10-20 nm) are heterogeneously distributed in this amorphous matrix. Such a nucleation seems to be related to locally bad contacts between the wheel and the ribbon, and to the subsequent lower quenching rate. When the TEM observations are made in the middle or near the free side of the ribbons, only Tl grains are found. DTA of such ribbons shows a Curie temperature near 320°C for this phase, and the crystallization peak of the amorphous part into Tl near 600°C (heating rate=20 K/mn).

Heat treatments of the as-quenched ribbons have been performed under vacuum (P<10^{-4} mbar) to obtain the fully crystallized material, and the resulting microstructures have been observed by TEM. Figure 5 shows nearly equiaxed Tl grains (50-100 nm) resulting from the crystallization of the amorphous part after a 5 minutes annealing at 650°C. Some contrasts appear inside the grains, but they have not been interpreted. This microstructure is different from that observed in the initially crystallized regions, where the graingrowth leads to defect-free and randomly faceted grains (25-100 nm). Another type of region has been observed as a result of the crystallization of the amorphous part during such a treatment: they are constituted by a finely crystallized (20 nm) mixture of phases showed in Fig. 5. The corresponding diffraction pattern has been indexed on the basis of α-Fe + C-Nd2O3. Other rare-earth compounds, such as borates, could be also present, but their identification is difficult due to the great structural similarity of all these compounds, as pointed out by Stadelmaier et al. (1985). These TEM observations do not allow us to quantify the proportion of each type of regions (Tl and α-Fe + C-Nd2O3). Also, at present, we have not established the possible relationship between the different Fe crystallites present before and after the heat treatment. Nevertheless, it is now possible to explain the increase of the α-Fe amount during the heat treament of overquenched ribbons (Becker, 1984) as an effect of the dissolved oxygen, which promotes the simultaneous formation of rare-earth oxides.

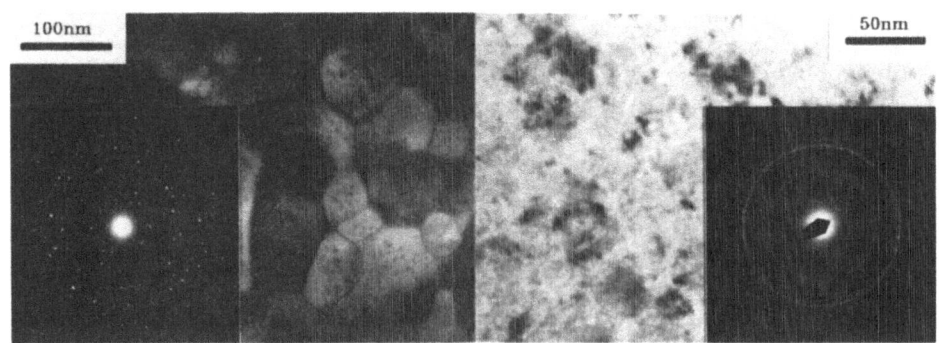

Fig. 5   Electron micrographs of devitrified Nd15Fe77B8 (5 mn at 650°C): T1 phase (left) and   -Fe + C-Nd2O3 (right) regions, and corresponding diffraction patterns.

REFERENCES

Becker J.J., 1984, J. Appl. Phys., 55, 2067-2072.
Christodoulou C.N., Schlup J., Hadjipanayis G.C., 1987, J. Appl. Phys., 61, 3760-3762.
Fidler J., Yang L., 1985, 4th Int. Symp. on Magnetic Anisotropy and Coercivity in Rare-Earth Transition Metal Alloys, Dayton (Proceedings Book by: University of Dayton, KL-365, Dayton, Ohio 45469, USA).
Hiraga K., Hirabayashi M., Sagawa M., Matsuura Y., 1985, Jpn. J. Appl. Phys., 24, 699-703.
Ramesh R., Chen J.K., Thomas G., 1987, J. Appl. Phys., 61, 2993-2998.
Sagawa M., Fujimura S., Yamamoto H., Matsuura Y., 1984, IEEE Trans. Mag., MAG-20, 1584-1589.
Schneider G., Henig E.T., Petzow G., Stadelmaier H.H., 1986, Z. Metallkde., 77, 755-761.
Schneider G., Henig E.T., Stadelmaier H.H., Petzow   G., 1987, 5th Int. Symp. on Magnetic Anisotropy and Coercivity in Rare-Earth Transition Metal Alloys, Bad Soden, (Proceedings Book by: D.P.G., D-5340 Bad Honnef 1, FRG).
Stadelmaier H.H., El-Masry N.A., 1985, 4th Int. Symp. on Magnetic Anisotropy and Coercivity in Rare-Earth Transition Metal Alloys, Dayton.

SUMMARY OF DISCUSSIONS :

PHASE RELATIONS AND MICROSTRUCTURE

C. Allibert, G. Schneider

Results on phase equilibria in Nd-Fe-B and related systems have been presented. Systems which were investigated over wide composition and temperature ranges are Fe-Nd, Fe-Nd-B, Fe-Dy-B and Fe-Tb-B (Max Planck Institut). Information over special temperature ranges or for compositions of technical importance was also given on the systems Gd-Fe-Ti (Centre d'Etudes Nucléaires de Grenoble), Fe-Nd-O, Fe-Nd-Dy(Tb)-B (M.P.I.) and on the effect of addition elements Zr, Nb, Mo, Cu, Co, Dy in Nd-Fe-B alloys (Institut National Polytechnique de Grenoble). It was emphasized that the additions can change the intrinsic properties of the hard magnetic phase, and also influence the microstructure by forming additional phases or changing the binder-phase structure and behaviour.

The study of the addition elements (I.N.P.G., Allibert) effect showed that Cu, Nb, Zr, Mo do not dissolve in $Nd_2Fe_{14}B$ (≤ 1 at %) and cannot therefore change its intrinsic properties. Nb, Zr and Mo form precipitates and can lead to the formation of Nd_2Fe_{17} if added in excess of few atom per cent. Cu is concentrated in the Nd-rich binder forming $NdCu_2$ and Nd(Fe,Cu). Cu significantly changes the microstructure and decreases the coercivity.

Oxygen effects (C.E.N.G., Cochet-Muchy) were observed in as-cast $Nd_{15}Fe_{77}B_8$ heat treated in air : different oxide type phases and amorphous zones containing textured α-Fe precipitates are evidenced. In melt-spun ribbons, heat treatment promotes the crystallization of a mixture of α-Fe and oxide phases beside $Nd_2Fe_{14}B$.

In Gd-Fe-Ti, two compounds were identified : one with ThMn$_{12}$ structure, with approximate composition Gd$_{6.5}$Fe$_{81}$Ti$_{12.5}$ and another one with the CeMn$_6$Ni$_5$-type.

The importance of basic knowledge of the phase relations to guide of the refinement microstructure was illustrated at the M.P.I. (Schneider). The effect of other microstructural parameters including grain size and wetting behaviour between the Nd-rich liquid and the Nd$_2$Fe$_{14}$B grains were also discussed. A decrease of the wetting angle is proposed as the micro-structural origin of an improvement in intrinsic coercivity.

SECTION III

MATERIALS

CHAPTER 5

MICROSTRUCTURE AND COERCIVITY

COERCIVITY MECHANISMS IN Nd-Fe-B AND OTHER SINTERED MAGNETS

D. Givord, P. Tenaud, T. Viadieu

Laboratoire Louis Néel
C.N.R.S.
166 X, 38042 Grenoble Cedex, France

ABSTRACT

A phenomenological model for coercivity mechanisms is proposed, which considers that magnetization reversal is initiated in a volume equal to the activation volume and is determined by the formation of a domain wall. From magnetic viscosity measurements, the activation volume is found to be proportional to the cube of the domain wall width, δ. The temperature dependence of the coercive field is interpreted by considering both the energy lost in the formation of a domain wall and magnetostatic interactions at a local scale. The observed angular dependence of the coercive field reveals that, in the activation volume, the anisotropy is much larger than the coercive field and is not strongly reduced with respect to the bulk. If the coercivity is determined by true nucleation in a fully saturated sample, this is unlike the usual assumption that the magnetocrystalline anisotropy is strongly reduced in the volume of the nucleus.

INTRODUCTION

In Nd-Fe-B permanent magnets, as well as in ferrite and $SmCo_5$ magnets, coercivity originates from magnetocrystalline anisotropy in compounds of uniaxial symmetry. In such systems, the theoretical value of the coercive field H_c is that of the anisotropy field, H_A. However, it is well known that the coercive field is actually much weaker than the anisotropy field and its value is dependent on the size of the magnetic grains. In fact, the physical processes which determine coercivity are not well understood. Schematically magnetization reversal requires an energy barrier E to be overcome. A non-saturated magnetization state may be thermally excited, extending over a certain volume which depends on the temperature and on the applied magnetic field. In fields $H < H_c$, this non-saturated state collapses once formed. As H reaches H_c, reversal of the whole magnetization follows the formation of a non-saturated magnetic state of volume equal to that of the activation volume, v.

In recent years, several attempts towards theoretical models of coercivity have been developed (Gaunt, 1983; Kronmüller, 1987) but few experimental studies of coercivity mechanisms appeared in the literature. We describe in this paper results of magnetic viscosity measurements, temperature dependence of the coercive field, angular dependence of the coercive field on Nd-Fe-B, $SmCo_5$ and ferrite magnets. Prior to the description of the results, the model developed to analyse them is presented.

MODEL FOR RESULT ANALYSIS

The energy barrier involved in magnetization reversal may be expressed as :

$$E = E_0 + H \, \partial E/\partial H \tag{1}$$

where E_0 is the field independent part of E and $\partial E/\partial H$ the field derivative of E. Let us assume that magnetization reversal is initiated by the formation of a small nucleus (the activation volume) of reversed magnetization. Schematically, two terms are expected to contribute to the energy barrier E_0 :

$$E_0 = E_p - \alpha 4\pi \, M_s^2 v \tag{2}$$

The first term represents the energy lost in the formation of a domain wall of energy per unit area γ and surface s :

$$E_p = \gamma s \tag{3}$$

with $\gamma = 4\sqrt{AK}$ where A and K represent the exchange interactions per unit length and the anisotropy energy per unit volume respectively. The second term in relation (2) is associated with stray fields representing the magnetostatic interactions in the magnet ; M_s is the spontaneous magnetization, v, the activation volume, α a phenomenological parameter.

From relations (2) and (3), it results that the intrinsic coercive field H_0 is :

$$H_0 = - \gamma s/v M_s + \alpha \, 4\pi \, M_s \tag{4}$$

The second term in Eq. 1 involves the field derivative of E which may be determined from magnetic viscosity measurements. This is related to the activation volume (Néel, 1950),

$$v = (\partial E / \partial H) / M_s \tag{5}$$

Thus, the experimental coercive field H_c and the intrinsic coercive field H_0 are related through,

$$H_c = H_0 - H_f \tag{6}$$

H_f being the fluctuation field (Néel, 1950),

$$H_f = E_{activ} / v M_s \tag{7}$$

where E_{activ} is the energy brought by thermal activation (see next section).

MAGNETIC VISCOSITY MEASUREMENTS

The time dependence of the magnetization under different applied magnetic fields measured on NdFeB magnets at 25 K and 300 K is shown in figure 1. As temperature decreases, and in particular below 77 K, the change of magnetization with time becomes weaker. At 4.2 K, the time dependence effects were very weak and no accurate data could be obtained. At all temperatures, it was found that the magnetization varies approximately as the logarithm of time. The viscosity $S = - \dfrac{dM}{d \ln t}$ can thus be deduced from the data. The field dependence of S is shown in figure 2.

A logarithmic dependence of the magnetization is often observed in permanent magnets and spin glass systems (Barbier, 1954; Préjean and Souletie, 1980). This means that the relaxation of the system is determined by activation energies E with a broad distribution f(E) about a mean value $\bar{E}$ (Street and Wooley, 1949). The relaxation time associated with a given value of E is

$$\tau = \tau_0 \exp E/kT, \tag{8}$$

where τ_0 is typically 10^{-11} s . In the time t of a measurement, it can be assumed to a very good approximation that energy barriers smaller than

$$E = kT \ln t/\tau_0 \simeq 25 \ kT = E_{activ} \tag{9}$$

are overcome, and those larger than 25 kT are not.

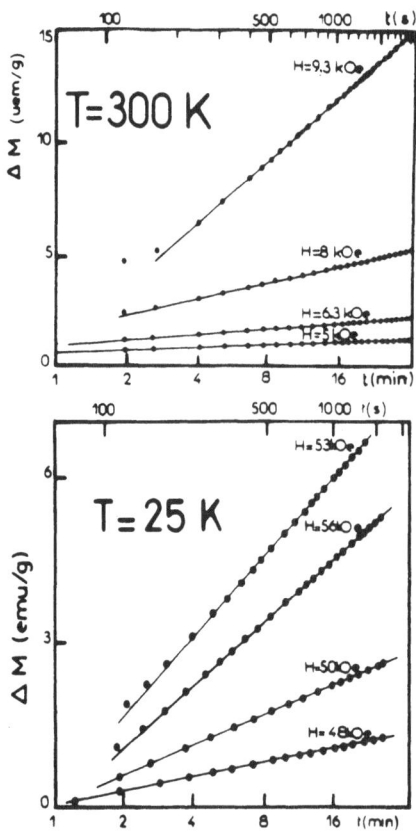

Fig. 1   Time dependence of magnetization

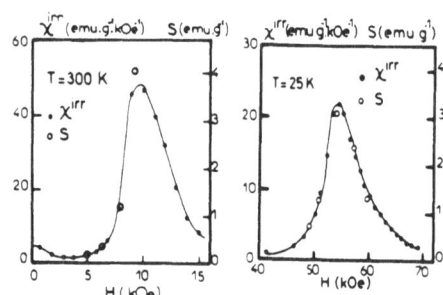

Fig. 2   Field dependence of the magnetic viscosity S and irreversible susceptibility χ^{irr}

Within this hypothesis, it is derived (Gaunt, 1976) that

$$S = 2M_s \, kT \, f(E) \qquad (10)$$

At a given temperature, the viscosity can be related to the irreversible susceptibility χ^{irr} deduced from the hysteresis loop of the magnet. Indeed, it can be shown (Gaunt, 1976) that

$$\chi^{irr} = 2 \, M_s \, f(E) \, (\partial E/\partial H) \qquad (11)$$

Thus

$$S = \frac{kT}{(\partial E/\partial H)} \, \chi^{irr} \qquad (12)$$

S and χ^{irr} are found to be proportional (figure 2). Such behavior was observed at all temperatures. One can therefore write that :

$$S = S_v \chi^{irr}, \qquad (13)$$

where

$$S_v = \frac{kT}{(\partial E/\partial H)} .$$ (14)

S_v, which has the dimension of a magnetic field, is the amplitude of the effective fluctuation field H_f. S_v is sample-shape independent, unlike S and χ^{irr}. From the values of S and χ^{irr} obtained at different temperatures, the temperature dependence of S_v was deduced. A rapid increase in S_v occurs from 4.2 K to 77 K, though at higher temperatures it decreases smoothly. The maximum value of S_v is 200 Oe. The values obtained at room temperature are in excellent agreement with those of Rodewald (1985).

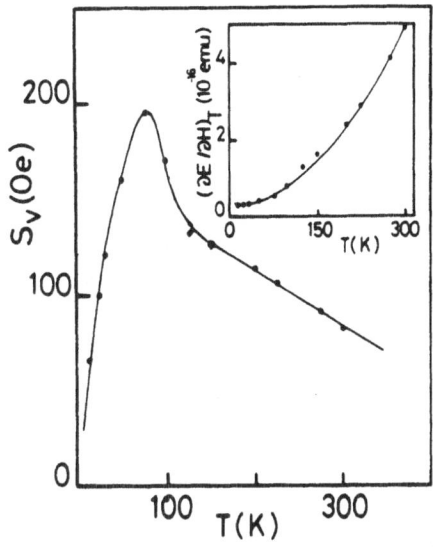

Fig. 3   Temperature dependence of the coefficient $S_v = S/\chi^{irr}$. Inset, temperature dependence of $(\partial E/\partial H)$.

The peak observed in the temperature dependence of S_v results simply from the dominant effect of the kT factor in relation (14) at low temperatures. This is a straight forward illustration of the very rapid increase of thermal activation in this temperature range. On the contrary, $\partial E/\partial H$ increases monotonically with temperature (fig. 3 inset). The value of the activation volume v is thus derived using relation (5).

It is meaningful to relate v to the wall width δ. In an uniaxial system where exchange interactions and anisotropy are well known, the calculation of Y is straight-forward. In $Nd_2Fe_{14}B$, the value of exchange interactions (A = $45 \ 10^{-7}$ erg/cm at 0 K) was simply deduced from the value of the Curie temperature (T_c = 593 K). The anisotropy constants K_1 and K_2 were deduced from measurements on a $Nd_2Fe_{14}B$ single crystal (Cadogan et al,

1988), they are in good agreement with those of Durst and Kronmüller (1986). In this system, because of the magnetization reorientation below 130 K, it is found numerically that the wall width and energy are both reduced by about 30 % compared to the values obtained in a uniaxial system. It is thus derived that experimental activation volume v is proportional to δ^3 (fig. 4) (Givord et al, 1987). This suggests that the activation volume corresponds to the volume of non-uniform magnetization which can be formed without loosing too much exchange or anisotropy energy.

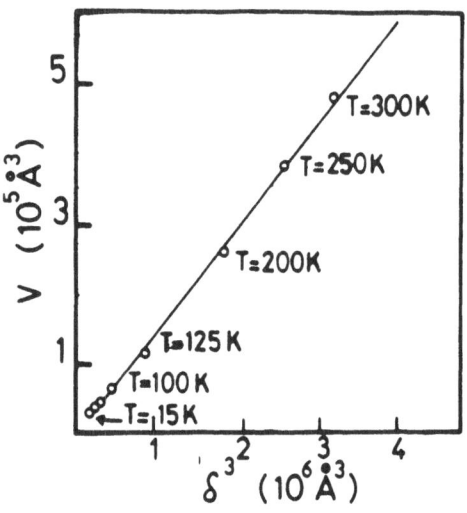

Fig. 4   Activation volume v versus the cube of the domain-wall width δ.

TEMPERATURE DEPENDENCE OF THE COERCIVE FIELD

The temperature dependence of the coercive field H_c in NdFeB and ferrite magnets are shown in figures 5 and 6 respectively. It is striking that the coercive field decreases in NdFeB magnets as temperature is increased, whereas in ferrite magnets, the coercive field increases with temperature. The activation volume being known from measurements of the magnetic viscosity, the intrinsic coercive field H_0 may be deduced using relations (7) and (6). The temperature dependences of H_0 were fitted to relation (4) with some further simplification.

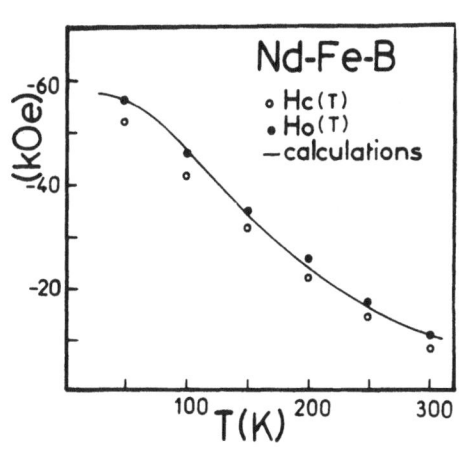
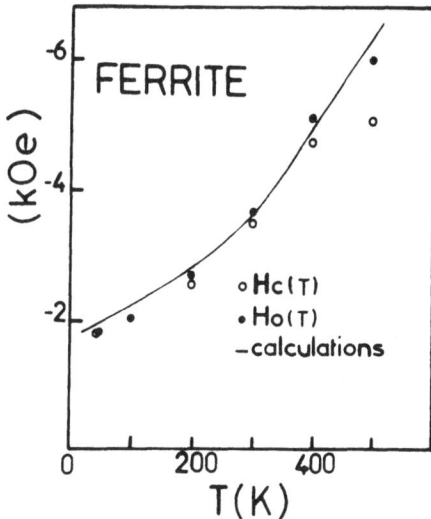

Fig. 5   Temperature dependence of the experimental coercive field H_c in the Nd-Fe-B magnet. Deduced temperature dependence of the intrinsic coercive field H_0. Calculated temperature dependence of H_0 (see text).

Fig. 6   Temperature dependence of the experimental coercive field H_c in the Sr-ferrite magnet. Deduced temperature dependence of the intrinsic coercive field H_0. Calculated temperature dependence of H_0 (see text).

The surface of the activation volume was assumed to be related to v through :

$$s/v = \beta \, v^{-1/3} \tag{15}$$

thus

$$H_0 = - \beta Y/v^{1/3} + \alpha \, 4\pi \, M_s \tag{16}$$

The values of the parameters associated with the best fit (solid lines in the figures) are respectively $\alpha = 0.7$, $\beta = 0.4$ for the Nd-Fe-B magnet (figure 5), $\alpha = 1.36$, $\beta = 0.85$ for the Sr-ferrite magnet (figure 6). Similar analysis proposed by other authors to interpret the temperature dependence of the coercive field in ferrite (Kools, 1985) and Nd-Fe-B

(Hirosawa et al, 1986) magnets led to values of the parameters in fair agreement with those determined in this study.

For the Nd-Fe-B magnet, the first term in relation (16) being dominant, determines the temperature dependence of the coercive field. As it is found experimentally that v is proportional to δ^3, H_0 decreases approximately as γ/vM_s, i.e. as the anisotropy field H_A. In the ferrite magnet, the two terms in relation (16) are of the same order of magnitude, but of apposite sing. The activation volume is proportional to δ^3 as in Nd-Fe-B magnets (Givord et al., 1988a). The first term is thus proportional to the anisotropy field H_A which is nearly constant between 4.2 K and 500 K (Jahn et al., 1969). The increase of H_0 as temperature increases is due to the decrease of the term $\alpha \, 4\pi \, M_s$ in relation (16), associated with the decrease of the spontaneous magnetization.

ANGULAR DEPENDENCE OF THE COERCIVE FIELD

The angular dependence of coercivity was measured in NdFeB, $SmCo_5$ and ferrite magnets.

Experimental procedure

After saturation in a positive field of 76 kOe applied along the $\vec{z}$-axis, the sample was turned and the field H applied at different angles θ_H from $\vec{z}$ ($\theta_H = 0°$ corresponds to a field antiparallel to the saturation field) (figure 7). For each θ_H, the associated coercive field H_c^θ was defined as the field corresponding to the maximum of χ_{irr}. It is associated with the largest number of grains reversing their magnetization towards the field. The measured angular dependence of the coercive field H_c^θ/H_c^0 for NdFeB and $SmCo_5$ magnets is shown in figure 8 and 9. It is systematically observed that H_c^θ increases as θ_H increases from 0° to about 90°. In NdFeB magnets the results are in agreement with those reported earlier by Kronmüller et al (1988). Jahn et al (1987) have also observed an increase of H_c with θ_H.

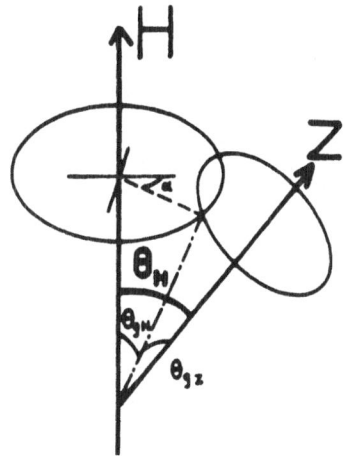

Fig. 7   Scheme of the definition of the angles involved in the determination of the angular dependence of the coercive field.

Analysis

Following the analysis developed in the previous sections, the intrinsic energy barrier in a grain g of coercive field h_c is :

$$E_0 = -v \vec{M}_s \vec{h}_c + 25 \text{ kT} \tag{17}$$

Let us consider that the easy axis of the grain under examination makes an angle θ_{gH} with the applied field ($\theta_{gH} = 0$ when the field is anti-parallel to the magnetization) (figure 7). As the coercive field is much weaker than the anisotropy field, the magnetization in the bulk remains very close to the easy axis. If it is assumed, with no further justification, that the anisotropy in the activation volume is not significantly reduced, the magnetization in this volume remains also very close to the easy axis and the magnetostatic energy can be written as :

$$-v\vec{M}_s\vec{H} = -vM_sH \cos\theta_{gH} \tag{18}$$

where H is < 0 when it is antiparallel to the magnetization. The above assumption holds to consider that the physical system is not modified by changing the orientation of the easy axis as regard to the field, i.e. E_0 is independent of θ_{gH}. It follows, taking into account (17) and (18), that the coercive field h_c^{θ} of such a grain is :

$$h_c^{\theta} = -\frac{E_0 - 25 \text{ kT}}{vM_s \cos\theta_{gH}} = \frac{h_c^0}{\cos\theta_{gH}} \tag{19}$$

where h_c^0 represents the coercive field for $\theta_{gH} = 0$.

In order to calculate the angular dependence of the coercive field in a magnet, which is an assembly of crystallites, the distribution of grain orientations must be taken into account (Givord et al., 1986 b). The calculated angular dependence of the coercive field in NdFeB and SmCo$_5$ magnets at 300 K is then in good agreement with experiment (figure 8).

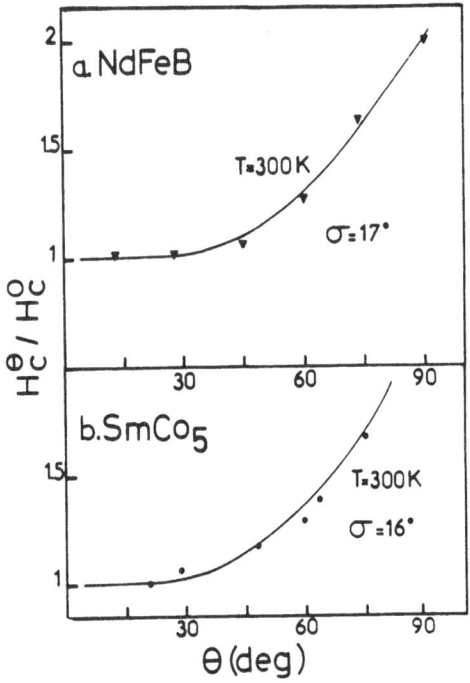

Fig. 8 Angular dependence of the coercive field in Nd-Fe-B and SmCo$_5$ magnets. Full lines = calculations (see text).

In Nd-Fe-B magnets at 200 K, the experimental dependence of the coercive field is not as large as the one calculated within the above model (full line in figure 9). The discrepancy increases for large θ_H. In fact in these systems, for an applied magnetic field of the order of H_c^θ, the magnetostatic energy of the moments in the applied field is of the same order of magnitude as the anisotropy energy. A coherent and reversible rotation of the magnetization from the easy axis towards the field occurs. This phenomenon modifies the energy barrier E_0 governing the reversal of magnetization in the activation volume. In particular, the domain wall energy γ is reduced. A quantitative analysis of the angular dependence of coercivity must then be based on a more microscopic understanding of magnetization reversal (Givord et al., 1988b). The calculated angular dependence of the coercive field becomes then in good agreement with experiment (figure 9).

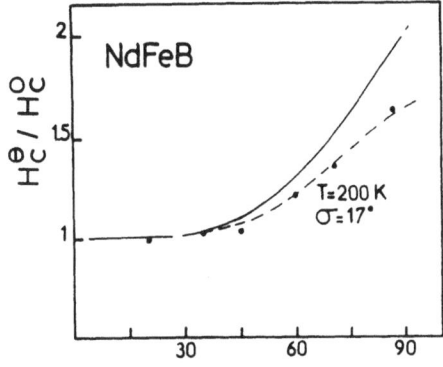

Fig. 9   Angular dependence of the coercive field in Nd-Fe-B magnet at 200 K. Full line : calculation as in figure 8. Dashed line : calculation taking into account the reversible rotation of the moments (see text).

An important conclusion may be drawn from the above results. In the calculation of the angular dependence of coercivity, the anisotropy in the activation volume was taken as equal to that of the bulk. The good agreement observed between calculated and experimental curves means that the anisotropy in the activation volume is not significantly reduced compared to that of the bulk. If magnetization reversal is determined by "true nucleation", i.e. by nucleation from a fully saturated state, this is unlike the usual assumption that magnetization reversal is first initiated in a region of the sample in which large anisotropy defects are present (Kronmüller, 1988; Durst et al., 1987).

REFERENCES

Barbier, J.C. 1950. Ann. Phys. Paris, 9, 84.
Cadogan, J.M., Gavigan, J.P., Givord, D. and Li, H.S. 1988. J. Phys. F (Met. Phys.)
Durst, K., Kronmüller, H. and Schneider, G. 1987. Proc. 5th Intern. Symp. on Anisotropy and Coercivity in RE-TM Alloys, Ed. Herget, C., Kronmüller, H., Poerschke, R., 209.
Durst, K. and Kronmüller, H. 1986. J. Magn. Magn. Mat., 59, 86.
Gaunt, P. 1976. Phil. Mag., 34, 775.
Gaunt, P. 1983. Phil. Mag., 48, 261.
Givord, D., Lienard, A., Tenaud, P. and Viadieu, T. 1987. J. Magn. Magn. Mat., 67, L281.
Givord, D., Tenaud, P., Viadieu, T. 1988. I.E.E.E. Trans. Mag.
Givord, D., Tenaud, P., Viadieu, T. 1988. J.Magn. Magn. Mat.
Hirosawa, S., Tokuhara, K., Matsuura, Y., Yamamoto, H., Fujimura, S. and Sagawa, M. 1986. J. Magn. Magn. Mat., 61, 363.
Jahn, J., Elk, K. and Schumann, R. 1987. J. Magn. Magn. Mat., 68, 335.
Kool, F. 1985. J. de Phys., C6, 349.
Kronmüller, H. 1987. Phys. Stat. Sol. (b), 144, 352.
Kronmüller, H., Durst, K. and Martinek, G. 1988. J. Magn. Magn. Mat.
Néel, L. 1950. J. Phys. Rad., 11, 49.
Préjean, J.J. and Souletie, J. 1980. J. de Phys., 41, 1335.
Rodewald, W. 1985. Proc. 4th Inter. Symp. on Magn. Anisotropy and Coercivity in RE-TM alloys, 737.
Street, R. and Wooley, J.C. 1949. Proc. Phys. Soc., 62A, 562.

MAGNETIC HARDENING MECHANISM IN RE-FE-B PERMANENT MAGNETS

H. Kronmüller and K.-D. Durst

Max-Planck-Institut für Metallforschung, Institut für Physik,
7000 Stuttgart 80, F.R. Germany

ABSTRACT

High coercive fields of RE-Fe-B magnets in general are interpreted either by pinning or nucleation mechanisms. In both cases the magnetic properties of phase boundaries between the grains are suggested to determine the actual value of H_c. In this paper a detailed analysis of the deleterious effects of the microstructure on H_c is given and the conditions for optimizing H_c are outlined. It is shown that remarkable improvements of H_c are possible by tailoring the microstructure of sintermagnets. From an analysis of the temperature dependence of H_c it is concluded that the nucleation mechanism leads to a far more coherent interpretation of the relevant properties of H_c than pinning models.

INTRODUCTION

Permanent magnets (pms) based on rare-earth transition metal intermetallic compounds exceed the properties of conventional pms in general by one or two orders of magnitude. These improvements are directly connected with a high magnetocrystalline anisotropy energy of uniaxial crystal systems, as, e.g., the hexagonal lattice of Co_5Sm, $Co_{17}Sm_2$ and the tetragonal lattice in the case of $Fe_{14}Nd_2B$. This latter compound has become of special interest because in comparison to the CoSm compounds Fe and Nd are much cheaper raw materials and $Fe_{14}Nd_2B$ has a saturation magnetization of $\mu_0 M_S \sim 1.6$ Tesla as compared to ~ 1.0 Tesla of Co_5Sm. The high coercive fields of the RE-transition metal compounds are based on their large magnetocrystalline anisotropy as clearly demonstrated by Fig. 1 where coercive fields of some prominent pms are presented as a function of the anisotropy constant K_1. Fig. 1 also contains the theoretically predicted coercive field which according to the pioneering work of W.F. Brown (1945) corresponds to the so-called nucleation field and is given by

$$H_c = \frac{2K_1}{M_s} - N \cdot M_s \quad . \tag{1}$$

Eq. (1) in general predicts coercive fields which are a factor of 5 to 10 larger than those actually realized in technical materials. This discrepancy between theory and experiment is known as Brown's paradox (1945). The importance of this paradox is demonstrated in Fig. 2 where the ratio H_c^{exp}/H_N of experimental coercive fields, obtained technically or in the laboratory are compared with the theoretical nucleation field. Up to the beginning of this decade only 10 to 20% of the theoretical nucleation field could be realized. In some new alloys as $(Co,Fe,Cu)_{17}$ $(Zr,Sm)_2$ and $Fe_{14}Nd_2B$-based alloys it was possible to achieve 40% of the theoretical nucleation field (Kronmüller et al., 1984, K.-D. Durst et al. 1987a, 1987b). According to Fig. 1 the steady increase of H_c during the last four decades is mainly due to the increase of K_1 of these new compounds. In principle a further increase of H_c should be possible by eliminating all sources which lead to a reduction of the nucleation field given by eq. (1)

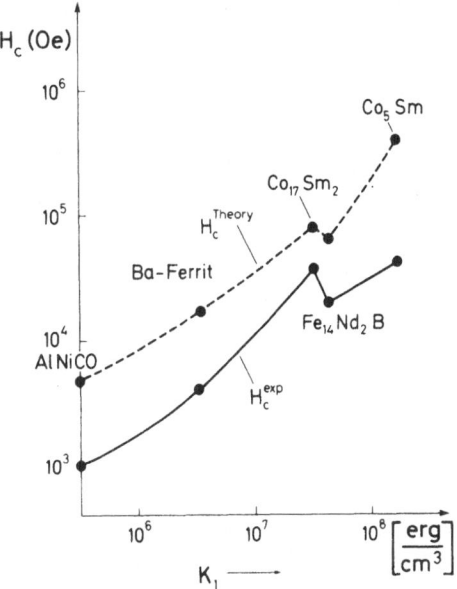

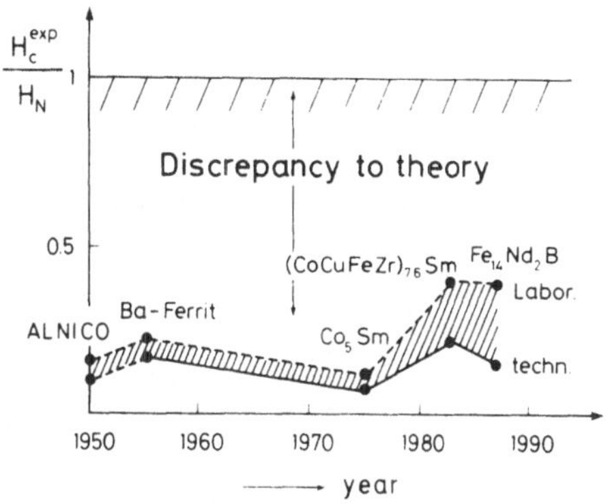

Fig. 2. Comparison between the measured coercive field, H_c^{exp}, and the theoretical nucleation field H_N of pms developed during the last 4 decades (- ● --- ● - = laboratory magnets, —●—●— = technical magnets).

Fig. 1. Coercive field of prominent pms as a function of K_1 and comparison with theoretical nucleation fields ($H_c^{th} = H_N$).

THE ROLE OF MAGNETIC INHOMOGENEITIES IN SINTER MAGNETS

The excellent hard magnetic properties of sinter magnets are based on the properties of the grains which behave approximately as single domain particles. In the case of an ideal sinter magnet only two phases should

constitute the pm: A ferromagnetic phase of high uniaxial magnetic anisotropy, corresponding to the single domain grains, and a nonmagnetic phase,
isolating the ferromagnetic grains perfectly from each other by a thin
layer. The transition region between the ferromagnetic and the nonmagnetic
phase should take place within an atomic layer. These rather strict requirements in practice cannot be realized in technical sinter magnets. As
shown in Fig. 3.1-3.4 in general four different types of magnetic inhomogeneities lead to a deterioration of the coercive field.

1. Nonmagnetic phases **2. Misoriented grains** **3. Incomplete liquid ph.b.**

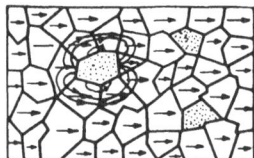 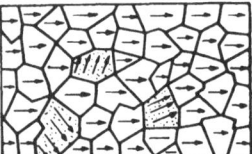 

4. Microstucture of grain boundaries

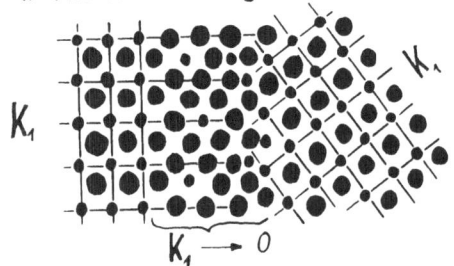

Fig. 3. Illustration of magnetic
inhomogeneities in pms. The pointed
grains represent either nonmagnetic,
misaligned or magnetically decoupled
grains.

1. <u>Polyhedral nonmagnetic grains and holes</u>: In general it is difficult to
   produce ideal two-phase sintermagnets, because this requires precise
   atmospheric, stoichiometric and thermal sintering conditions. According to Fig. 3.1 nonmagnetic grains, e.g. Fe_4NdB_4, produce local demagnetization fields oriented opposite to $\underline{M}_s$ within the neighbouring
   grains, thus reducing the nucleation field.

2. <u>Misoriented grains</u>: Misoriented grains produce also local demagnetizing fields and in addition the nucleation field is reduced intrinsic-
   ly (Fig. 3.2)

3. <u>Incomplete nonmagnetic grain boundaries</u>: If some neighbouring grains
   are not completely magnetically isolated reversal of magnetization in
   one of these grains leads to a cascade of demagnetization processes in
   neighbouring grains (Fig. 3.3).

4. Extended grain and phase boundaries: In general the grain boundaries do not correspond to an abrupt change of the magnetic properties. In this case nucleation of reversed domains may occur by reduced nucleation fields in the transition region between the ferromagnetic phase and the isolating grain boundary phase.

In the following it will be shown that the coercive field due to these different deteriorating effects may be written as (Durst and Kronmüller 1987b, Hirosawa and Sagawa 1987)

$$H_c(T) = \alpha(T) \frac{2K_1}{M_s} - N_{eff} M_s , \qquad (2)$$

where $\alpha(T)$ corresponds to a temperature dependent microstructural parameter and N_{eff} takes care of the demagnetizing fields.

MICROMAGNETIC BACKGROUND OF HARDENING MECHANISMS

1. Pinning of domain walls by planar inhomogeneities

Planar magnetic inhomogeneities may act as rather strong pinning centres for domain walls. Well known examples are the Cu-rich Co_5Sm precipitations in a $Co_{17}Sm_2$ matrix (Nagel, 1979). Also grain or phase boundaries may be considered as planar pinning centres. On the basis of the atomistic Heisenberg model the coercive field due to planar narrow inhomogeneities (Hilzinger and Kronmüller, 1975), of width $r_0 < \delta_B$ is found to be

$$H_c = \frac{1}{3 \cdot \sqrt{3}} \frac{2K_1}{M_s} \frac{\pi r_0}{\delta_B} \left[\frac{A}{A'} - \frac{K_1'}{K_1} \right] - N_{eff} M_s , \qquad (3)$$

where the material parameters with an upper prime refer to the inhomogeneity. For wide inhomogeneities $r_0 > \delta_B$ we use $K_1(z) = K_1 - \Delta K/ch^2(z/r_0)$ where ΔK denotes the change of K_1 at the centre of the inhomogeneities. The coercive field being due to the varying wall energy for $\Delta K = K_1$ and $A = A'$ is given by

$$H_c = \frac{2K_1}{M_s} \left(\frac{2\delta_B}{3\pi r_0} \right) - N_{eff} M_s . \qquad (4)$$

Here the wall width of the matrix $\delta_B = \pi\sqrt{A/K_1}$ has been introduced, where A means the exchange energy of the matrix. The term $N_{eff} M_s$ takes care of the demagnetization field resulting from grain surfaces and volume charges The α-parameters for pinning $\alpha_K^{p\,in}$ introduced by eq. (2) may be easily determined from eq. (3) and eq. (4). Fig. 4 shows, $\alpha_K^{p\,in}$, as a linear function for $r_0/\delta_B < 0.6$ and as an $1/r_0$-decrease for $r_0/\delta_B > 0.6$. In the transition region between the narrow and wide inhomogeneity the maximum α-parameter for pinning is obtained, being given by $\alpha_{max}^{pin} = 0.3$.

2. Nucleation of reversed domains in planar inhomogeneities

The nucleation fields of planar inhomogeneous regions have been calculated previously on the basis of the theory of micromagnetism (Kronmüller, 1978, 1987). The basic micromagnetic equation is obtained from a linearization of the variation of the magnetic Gibbs free enthalpy and is given by

$$2A\ d^2\varphi/dz^2 - [K(z) - M_s(H_{ext} - H_d + 2\pi M_s)]\ \varphi = 0. \qquad (5)$$

Here φ denotes the angle between $\underline{M}_s$ and the easy direction within the nucleus shown in Fig. 5. H_{ext} denotes the external field and H_d a local demagnetizing field resulting from nonmagnetic inclusions, misoriented grains and external surfaces. $K_1(z)$ corresponds to the inhomogeneous first crystalline anisotropy constant within the inhomogeneity. A suitable Ansatz for $K_1(z)$ writes

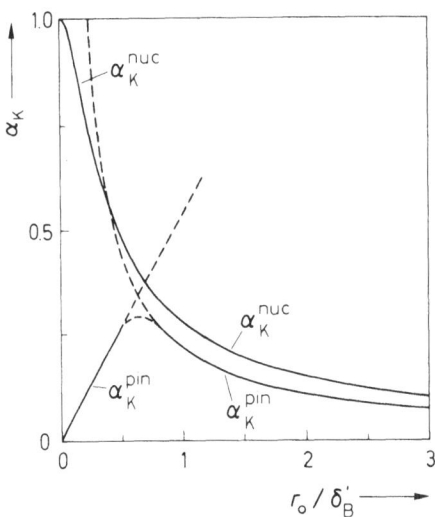

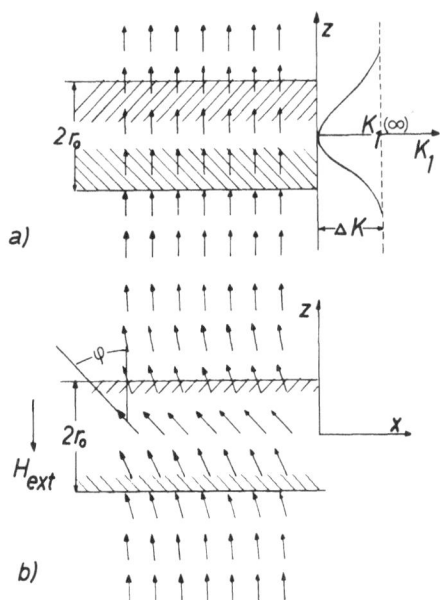

Fig. 4. The α_K parameter for pinning (α_K^{pin}) and nucleation (α_K^{nuc}) as a function of r_0/δ_B. α_K^{nuc} has been determined for $\Delta K = K_1$. α_K^{pin} has been determined for $A = A'$ and $K_1' = 0$.

Fig. 5. Model of a planar nucleus within a planar inhomogeneity. The upper part shows the variation of distribution of a quasi-uniform rotation process.

$$K_1(z) = K_1(\infty) - \frac{\Delta K}{ch^2(z/r_0)} \quad . \tag{6}$$

$K_1(\infty)$ = magnetocrystalline constant within the ideal matrix, ΔK = change of K_1 at the centre of the inhomogeneity. The nucleation field follows from (5) as an Eigenvalue of H_{ext} and is given by

$$H_c = \frac{2K_1}{M_s} \alpha_K^{nuc} - N_{eff} M_s \quad , \tag{7}$$

with the α-parameter for nucleation

$$\alpha_K^{nuc} = 1 - \frac{1}{4\pi^2} \frac{\delta_B'^2}{r_0^2} \left[-1 + \sqrt{1 + \frac{4\Delta K r_0^2}{A}} \right]^2 \quad , \tag{8}$$

and $\delta_B' = \pi\sqrt{A/K_1(\infty)}$ denoting a fictitious wall width of a uniaxial crystal. Fig. 4 gives a comparison of α_K^{pin} and α_K^{nuc}, clearly showing that throughout $\alpha_K^{pin} < \alpha_K^{nuc}$ holds.

3. Effect of misaligned grains

In real sinter magnets the grains orientation shows an angular distribution with a main standard deviation of $20°$ (Givord et al., 1987, Durst and Kronmüller, 1986). The effect of these misalignments may be taken into account by an additional microstructural parameter, α_ψ, now leading to a structural reduction parameter, $\alpha = \alpha_K^{nuc} \cdot \alpha_\psi$, where α_ψ is given by (Kronmüller et al., 1987)

$$\alpha_\psi^{nuc} = \frac{1}{\cos\psi_0} \frac{1}{(1+(tg\psi_0)^{2/3})^{3/2}} \left\{ 1 + \frac{2K_2}{K_1} \frac{(tg\psi_0)^{2/3}}{1+(tg\psi_0)^{2/3}} \right\} \quad . \tag{9}$$

Fig. 6 shows the angular dependence of α_ψ for room temperature with K_2 = second anisotropy constant, and ψ_0 denoting the angle between the applied field and the easy axis of the grain. Fig. 6 also includes the angular dependence of H_c for a pinning mechanism which is given by $\alpha_\psi = 1/\cos\psi_0$.

For a quantitative analysis of H_c we have to take care of the angular distribution of the grain's easy axes. The type of averaging of ψ_0 in eq. (9) depends on the magnetic correlation between neighbouring grains. Here we have to consider two limiting cases:

1. Uncoupled grains: In this case each grain with $H_c(\psi_0) \lesssim H_{ext}$ reverses its magnetization and α_ψ has to be replaced by a certain average $\langle\alpha_\psi\rangle$ as outlined by Kronmüller et al. (1987).

2. Strongly coupled grains: If the grains show strong magnetic coupling (e.g. incomplete, nonmagnetic grain boundaries) we may assume that the grains with the lowest coercive field, i.e., the minimum value of α_ψ, determines the bulk coercive field. In this case it is assumed that a reversed grain induces also reversal of the magnetization in the neighbouring grains.

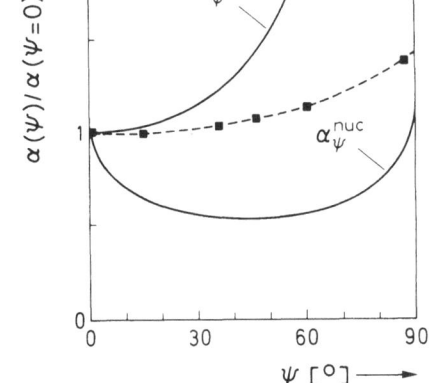

Fig. 6. Angular dependence of the theoretical α_ψ-parameters for pinning and nucleation. Experimental points (-■-■-) denote the angular dependence of the relative critical field, $H_{crit}(\psi_0)/H_{crit}(0)$ of $Fe_{71}Nd_{20}Al_2B_7$.

According to these considerations the lower bound of H_c for nucleation is given by

$$H_c = \frac{2K_1}{M_s}\, \alpha_K^{nuc} \cdot \alpha_\psi^{min} - N_{eff}\, M_s \ . \tag{10}$$

Also for pinning such a relation holds, with α_K^{nuc} replaced by α_K^{pin} given by eq. (3) and (4). Here it is noteworthy that α_ψ^{min} for pinning is given by $\alpha_\psi^{min} = 1$ whereas according to eq. (9) for nucleation α_ψ^{min} depends on temperature through K_1 and K_2 and at RT $\alpha_\psi^{min} = 0.7$ holds.

EXPERIMENTAL RESULTS

1. Coercive field of minor hysteresis loops.

The coercive field of minor hysteresis loops depends in a characteristic manner on the maximum applied field. As an example Fig. 7 shows the results for $Fe_{71}Nd_{20}Al_2B_7$. Similar results were obtained by Hirosawa and Sagawa (1987). According to Fig. 7 $H_c(H)$ increases rather moderately below $\mu_0 H_{ext} = 0.5$ Tesla and then increases steeply approaching a saturation coercive field of $\mu_0 H_c = 2$ Tesla at a field of 1.0 Tesla.

2. Temperature dependence of H_c.

Fig. 8 shows the temperature dependence of H_c of a series of FeNdB magnets. A monotonic decrease of H_c is observed for all oriented sintered pms over the whole temperature range. Sintermagnets produced at the Max-Planck-Institut with compositions leading to two-phase materials ($Fe_{14}Nd_2B$ + Nd-rich boundary phase) above room temperature show a much better temperature stability of H_c as compared with technical pms composed at least of three phases including the nonmagnetic Fe_4NdB_4 phase.

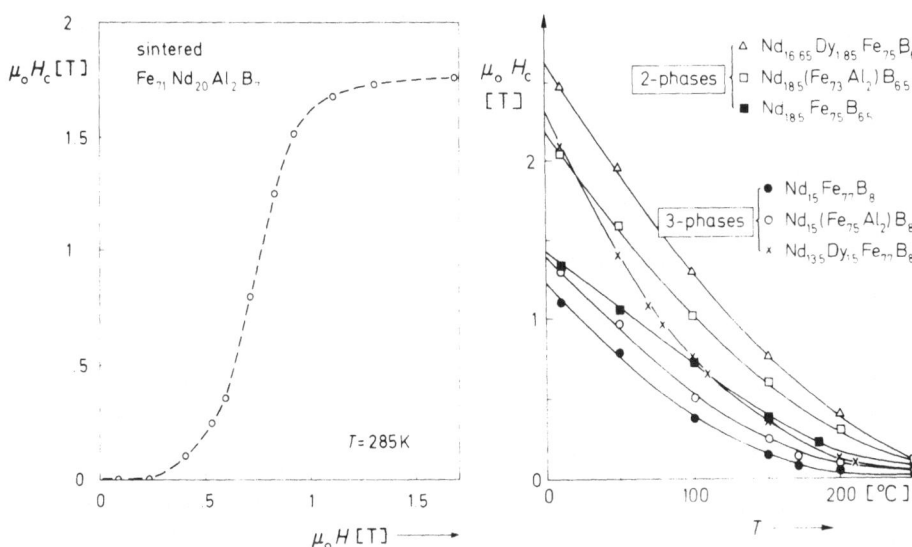

Fig. 7. Field dependence of $H_c(H)$ of minor hysteresis loops of a two-phase $Fe_{71}Nd_{20}Al_2B_7$ magnet.

Fig. 8. Temperature dependence of H_c of two-phase and three-phase FeNdB-type pms.

3. Angular dependence of the critical field, H_{crit}.

Fig. 6 in addition to the theoretical angular dependence of α_ψ also contains the angular dependence of H_{crit} as measured for $Fe_{71}Nd_{20} Al_2B_7$. H_{crit} is defined as the point of inflection of the hysteresis loop in the second or third quadrant. This point corresponds to the maximum of the irreversible susceptibility being narrowly connected to the coercive field. At small angles ψ_0 the critical field is nearly independent of ψ_0 and then increases up to $\psi_0 \lesssim \pi/2$ by a factor of 1.4. Even at $\psi =$

$\pi/2$ a maximum of the irreversible susceptibility can be observed.

4. Domain patterns.

Magnetooptical Kerrtechnique has been used to study domain structures in sinter pms. It has been shown (Pastushenkov et al., 1987) that under the influence of a reversed field in saturated grains reversed domains nucleate at the grain boundaries, preferentially in the neighbourhood of nonmagnetic phases and if the $Fe_{14}Nd_2B$-grain shows a misorientation.

ANALYSIS AND DISCUSSION OF RESULTS

A satisfactory model of the hardening mechanism in FeNdB-type pms has to explain a large spectrum of experimental results: Temperature dependence of H_c, angular dependence of H_c, field dependence of $H_c(H)$ of minor hysteresis loops, nucleation of reversed domains at grain boundaries. The last two of these experimental facts support the nucleation model. In particular the field dependence of H_c clearly shows that the long-range displacement of domain walls takes place at magnetic fields which are only 1/4 - 1/3 of the coercive field (Hirosawa and Sagawa, 1987; Durst and Kronmüller, 1987b). Accordingly the barriers for domain wall displacements are much smaller than the nucleation barrier. If by a large magnetic field all domain walls are eliminated from the sintered grains a reversal of the magnetization may only take place by the nucleation of domains. Such grains with angular deviations from the applied field are characterized by smaller nucleation fields as shown by Fig. 6. Magnetooptical Kerreffect domain patterns by Pastushenkov et al. (1987) have shown that the reversion of a grain induces also reversion of neighbouring grains due to magnetic couplings between the grains. Insofar it seems to be justified to assume that the bulk coercive field is determined by the strongly misaligned grains. Since there exists always an angular distribution of the easy axes of the grains this implies that the decrease of H_c predicted for the nucleation model (see Fig. 6) possibly cannot be observed experimentally. In fact the angular dependence presented in Fig. 6 fits neither the pinning nor the nucleation model. As shown by Kronmüller et al. (1987), however, an averaged value of α_ψ^{nuc} agrees rather well with the experimental results. In conclusion also the angular dependence of H_c agrees sufficiently well with the nucleation model.

For a discussion of the temperature dependence of H_c we have to consider eq. (10) which for pinning as well as nucleation models holds. The differences in the temperature dependences of both models are due to the different expressions of α_K and α_ψ valid for both cases. In the case of the pinning mechanism a test of the temperature dependence of H_c is possible by plotting ($A = A'$, $K_1' = 0$)

$$\frac{H_c}{M_s} \quad \text{vs.} \quad \frac{2\pi}{3\cdot\sqrt{3}} \frac{K_1}{M_s^2} \frac{1}{\delta_B'} \quad , \quad (r_0 < \delta_B') \qquad (11)$$

For a calculation of the numerical factor of eq. (11) we have used $A = A'$ and $K_1' = 0$. For plotting eq. (11) we have used experimental results for K_1, M_s and δ_B' obtained just recently by Hock (1988) and Hock and Kron-müller (1987). If eq. (11) would be valid a linear relationship between the experimental left side and the theoretical right side has to be expected. According to Fig. 9 however only a moderate linear behaviour is observed for three different pms. In particular at lower temperatures considerable deviations from linearity are observed, in agreement with Hirosawa et al. (1986). If the $\alpha_K^{p,in}$ parameters and the widths, r_0, of the inhomogeneity are determined from the slopes in Fig. 9 it turns out that the parameters are incompatible with the pinning model. The $\alpha_K^{p,in}$-parameters

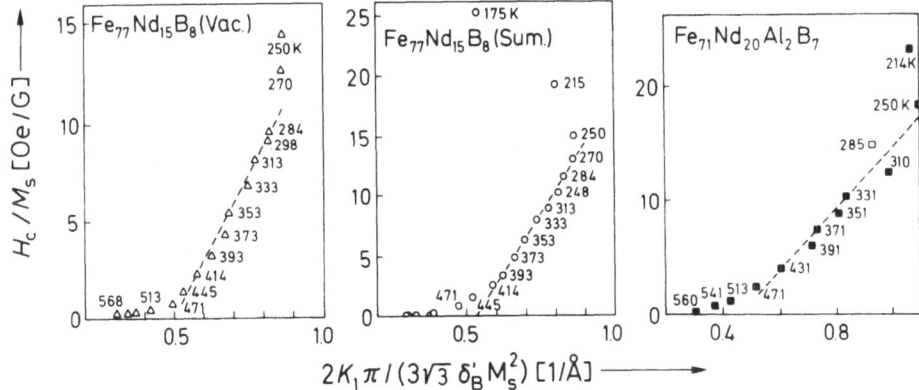

Fig. 9. Test of the temperature dependence of H_c for pinning of domain walls by narrow inhomogeneities. The pms considered are those of Vacuumschmelze, Sumitomo (Hirosawa and Sagawa, 1987) and of the Max-Planck-Institut.

throughout are found to be larger than 0.25 in contrast to the pinning theory which predicts $\alpha_K^{p,in} < 0.25$ for narrow inhomogeneities. A test of the temperature dependence by the pinning formula derived for wide domains (δ_B'

< r_0) leads to even more drastic discrepancies.

A test of the nucleation mechanism starts from eq. (10) by inserting for α_ψ the minimum values as determined from eq. (9) (Kronmüller et al., 1988). A further fitting parameter then is the width, r_0, of the nucleation region defined by the $1/\mathrm{ch}^2(z/r_0)$ inhomogeneity introduced by eq. (6). It turns out that excellent linear behaviour of the experimental results is obtained by plotting

$$ H_c/M_s \quad \text{vs.} \quad \frac{2K_1}{M_s^2} \, \alpha_\psi^{min} \, \alpha_K^{nuc} \quad . \tag{12} $$

where α_ψ^{min} follows from eq. (9) and α_K^{nuc} from eq. (8). Linear behaviour is found for r_0-parameters varying between 4 Å < r_0 < 14 Å as shown in Fig. 10. Concerning these rather small r_0-parameters it should be noted that the actual width of the nucleation region corresponds to $2r_0$.

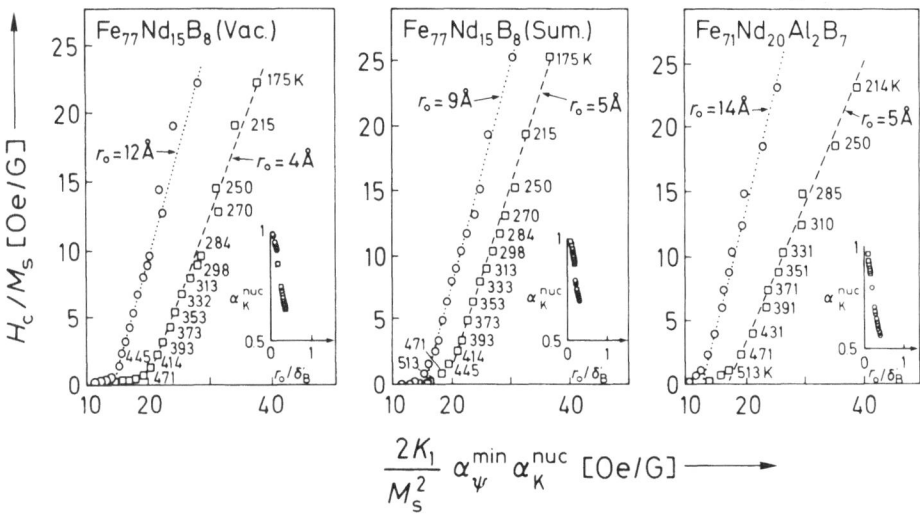

Fig. 10. Test of the nucleation mechanism by plotting H_c/M_s vs $(2K_1/M_s^2)\alpha_K^{nuc}\,\alpha_\psi^{min}$ for the pms of Vacuumschmelze, Sumitomo and of the Max-Planck-Institut. The insets show the variation of α_K^{nuc}.

From the slopes of Fig. 10 an experimental α-parameter of 0.6 - 0.7 is obtained which according to $\alpha = \alpha_\psi^{min} \, \alpha_\psi^{nuc}$ is composed of $\alpha_\psi^{min} \sim 0.8$ and $\alpha_K^{nuc} \sim 0.9$. From the extrapolated linear regions of Fig. 10 we obtain the effective demagnetization factors. For the Vac- and the Sumitomo magnet we find $N_{eff} = 7.2 \, \pi$ and for the Max-Planck-FeNdAlB magnet $N_{eff} = 6.4 \, \pi$ is obtained. Such large local demagnetization factors are expected near nonmagnetic inclusions and sharp edges of the grains. Recently is has been shown that values of $N_{eff} = 20\pi/3$ may exist near nonmagnetic precipitations (Kronmüller, 1987, Adler and Hamann, 1985).

Summarizing the results of this section it is concluded that the nucleation hardening mechanism determines the coercive field in FeNdB pms. The following results support this conclusion:

1. Technical saturation of H_c is achieved by fields much smaller than H_c.
2. The temperature dependence of $H_c(T)$ is far better described by the nucleation field of misaligned grains and modifications due to inhomogeneities of reduced anisotropy energy.
3. Considerable deteriorations of H_c are due to local demagnetization fields which may be as large as 2.5 Tesla.
4. The numerical results obtained for the widths of the nucleation regions and for the local demagnetization fields are compatible with our present knowledge of the microstructure of pms.
5. Further improvements of sinter magnets require a more perfect alignment of the individual grains, a reduction of the magnetic coupling between the grains and the elimination of extended nonmagnetic precipitations including holes.

REFERENCES

Adler, E., and Hamann, P. 1985, Proc. 8th Intern. Workshop on RE Magnets (Ed. K. Strnat) (University of Dayton, Ohio 1985), p. 747.

Aharoni, A., 1962, Rev. Mod. Phys. 34, 227.

Brown, W.F., 1945, Rev. Mod. Phys. 17, 15.

Durst, K.-D., and Kronmüller, H., 1986, J. Magn. Magn. Mat. 59, 86.

Durst, K.-D., Kronmüller, H., and Schneider, G., 1987a, in: 5th Intern. Symp. on Magn. Anisotropy and Coercivity in RE-Transition Metal Alloys (Eds. C. Herget, H. Kronmüller and R. Poerschke). (DPG-GmbH, Bad Honnef) p. 209.

Durst, K.-D., and Kronmüller, H., 1987b, J. Magn. Magn. Mat. 68, 63.

Givord, private communications, 1987.

Hilzinger, H.-R., and Kronmüller, H., 1975, Phys. Lett. 51A, 59.

Hirosawa, S., and Sagawa, M., 1987, J. Magn. Magn. Mat. 71, L1.

Hirosawa, S., Tokuhara, K., Matsuura, Y., Yamamoto, H., Fujimura, S., and Sagawa, M., 1986, J. Magn. Magn. Mat. 61, 363.

Hock, S., Dr.rer.nat. Thesis, 1988, University of Stuttgart.

Hock, S., Kronmüller, H. 1987, in: 5th Intern. Symp. on Magn. Anisotropy and Coercivity in RE-Transition Metal Alloys (Eds. C. Herget, H. Kronmüller and R. Poerschke). (DPG-GmbH, Bad Honnef) p. 275.

Kronmüller, H., 1978, J. Magn. Magn. Mat. 7, 341.

Kronmüller, H., Durst, K.-D., Ervens, W., and Fernengel, W., 1984, IEEE Trans. Magnetics 20, 1569.

Kronmüller, H., Durst, K.-D., and Martinek, G., 1987, J. Magn. Magn. Mat. 69, 149.

Kronmüller, H., 1987, phys. stat. sol. (b) 144, 385.

Kronmüller, H., Durst, K.-D., and Sagawa, M., 1987, J. Magn. Magn. Mat., to be published.

Nagel, H. 1979, J. Appl. Phys. 50, 1026.

Pastushenkov, J., Durst, K.-D., and Kronmüller, H., 1987, phys. stat. sol. (a) 104, 487.

THE MICROSTRUCTURE AND EXTRINSIC MAGNETIC PROPERTIES OF NdFeB-BASED MATERIALS

P J Grundy, D G Lord, S F H Parker and R J Pollard

Department of Pure and Applied Physics,
University of Salford,
Salford M5 4WT, U.K.

ABSTRACT

The microstructure of as-cast material, sintered magnets and melt-spun ribbon samples prepared from ternary NdFeB alloys and quaternary alloys containing additions of refractory elements, such as Nb, Mo and Zr, and also Al and Ga has been examined using electron microscopy, mainly transmission electron microscopy.

We have defined the microstucture of the ternary materials and agree in general with the findings of other workers. However, particular attention has been given to the quaternary materials, and our microstructural findings have been correlated with magnetic measurements on these systems where possible.

Our investigations show that additional precipitate phases are formed in some of the quaternary materials and that this precipitation could be associated with the changes in coercivity and reduced irreversible losses often observed in sintered magnets containing such additions (Tokunaga et al, 1987a,b). Increases in the coercivity of ribbons seem to be definitely related to the precipitation. The possibility of a contribution to coercivity from a suitably controlled, and sufficiently fine precipitate structure in NdFeB and other types of rare earth-iron magnets is of some interest and importance.

INTRODUCTION

The aim of the research has been to make a contribution to the knowledge of the microstructure and related magnetic properties of magnets, and materials for magnets, based on the generic composition $Nd_{15}Fe_{77}B_8$. The objectives have included electron optical and magnetic characterisations of as-cast alloys, sintered magnets and rapidly solidified material with an emphasis on the latter two material types. Electron microscopy has concentrated on S/TEM studies in conventional, X-ray microanalysis and Lorentz modes.

Attempts to produce and characterise a small scale precipitate in the magnetically hard $Nd_2Fe_{14}B$ phase by making additions to the start material has been one of the main and most successful themes of our work. We have shown that it may be possible to produce a second and new contribution to coercivity in NdFeB materials in this way, i.e. an identifiable domain wall pinning contribution to augment any nucleation

process. Our results have been published in several publications (Parker
et al, 1987a; Parker et al, 1987b; Pollard et al, 1987).

METHODS AND TECHNIQUES

As-cast specimens have been prepared from both commercially
available 99.5% pure cast materials and from some 99.99% purity vertically
zoned material. The 99.5% material was a master alloy prepared by Rare
Earth Products Ltd. of composition $Nd_{16.5}Fe_{75.5}B_8$. In making substituted
alloys a fourth element - Nb, Mo, Zr, Al, Ga, or Dy was added at the 1 to
3% level to the master alloy by re-melting in an argon arc furnace.

Sintered magnets were prepared using a small batch facility
at Mullard Magnetic Components Ltd. This enabled us to fabricate magnets
from 25g batches of alloy under controlled conditions involving: hydrogen
decrepitation, ball milling, alignment in a 1600 kA/m field and isostatic
pressing prior to sintering and annealing. Typical sintering and annealing
temperatures were 1070°C and 630°C.

The same start materials were melt-spun in an argon atmosphere
over a range of cooling rates to produce a variety of microstructures
giving a corresponding variation in magnetic properties.

Specimens for transmission electron microscopy (S/TEM) were
prepared by electrochemical and ion beam thinning and examined in a JEOL
200CX instrument equipped with a fully quantitative X-ray microanalysis
attachment (EDS). Specimens for optical and scanning
electron microscopy (OM, SEM) were prepared by conventional metallographic
techniques.

Magnetic hysteresis parameters were measured in a permeameter
and a vibrating sample magnetometer (VSM). Some thermo-magneto-gravimetric
(TGA) analyses were obtained with a Mettler TA3000 system.

RESULTS AND DISCUSSION

As-cast Alloys

We have examined as-cast and large grained zoned materials using OM,
SEM, S/TEM and EDS. Main impurities were the rare earths Pr, Sm, La and
Ce. In the ternary NdFeB alloy (and sintered magnets) we have, in common
with other workers (e.g. Sagawa et al, 1984; Fidler, 1985), identified four
phases, Fig. 1, including the main $Nd_2Fe_{14}B$ tetragonal compound (a = 0.879
nm, c = 1.22 nm). Surrounding about half of the crystals of the main phase

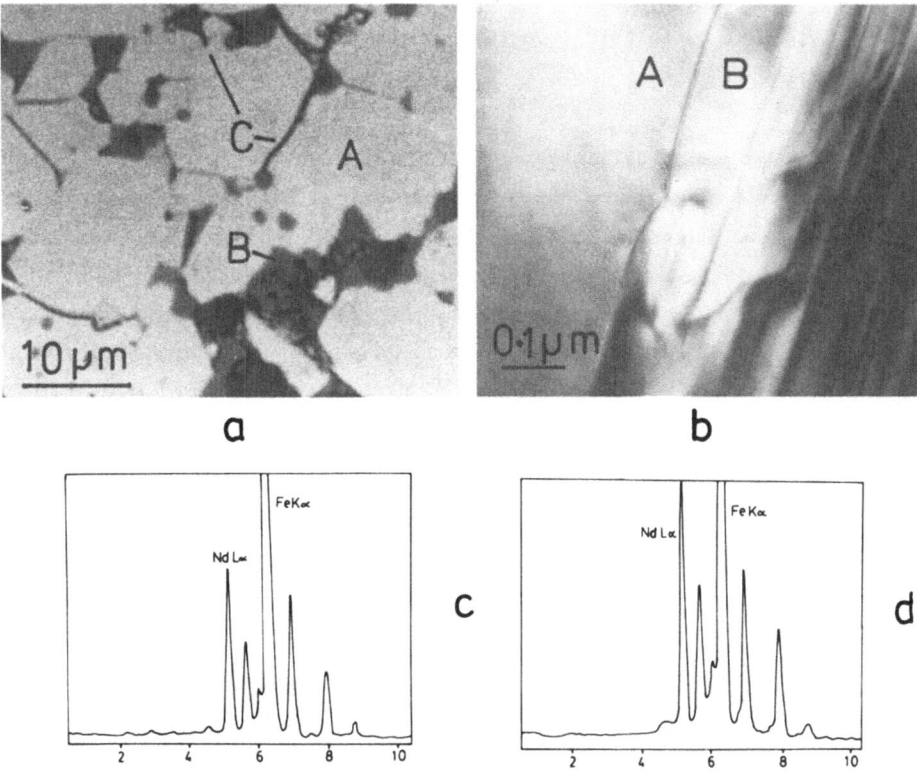

Figure 1(a) and (b) - Optical and TEM micrographs showing the $Nd_2Fe_{14}B$ phase (A), the $NdFe_4B_4$ phase (B) and the Nd-rich grain boundary phase (C) (c) and (d) - X-ray microanalyses of A and B.

are two non-magnetic phases; a tetragonal $Nd_{1+\epsilon}Fe_4B_4$ compound and a Nd-rich phase, occupying 10% and 5% respectively of the volume of the as-cast alloy. SEM observations (Parker et al, 1987b) showed the presence of fairly large inclusions of α-iron in the as-cast (and sintered and melt-spun) samples and the presence of ferromagnetic iron was also suggested by TGA, as discuused below.

Magnetic measurements on sintered magnets have shown a dependence on the cooling rate at which the original cast alloy is reduced from the melt. This is probably related to a cooling rate dependence of the relative volumes of the various phases. Indeed microstructure may have some dependance on cooling rate (Ramesh et al, 1987). The subsequent sintering and aging treatments would have little effect on these proportions because of the relatively low temperatures employed (1045-1085°C and 600-650°C). It was therefore desirable to make magnets under near identical conditions and with reproducible properties from the same start material in order to

determine the effects of any additions to the ternary compositions. It has
proved difficult to do this from the small batch facility given the time
scale available.

Sintered Magnets - Ternary Alloys

SEM and OM of sintered magnets prepared from the ternary compositions
showed the same phases as found in the as-cast alloys but with a smaller
grain size (5-10 μm) in optimally (magnetic) prepared samples. Kerr
magneto-optic images showed the demagnetised $Nd_2Fe_{14}B$ grains to be multi-
domained, Fig.2, and to retain a large number of reverse domains in fully
magnetised samples. This is surprising, as the demagnetising field of the
sample should be well below its coercivity and the hysteresis loop usually
shows a high squareness. These common observations indicate that surface
domains are not a good indication of the magnetisation processes occuring
in the bulk of the material, due probably to easy nucleation sites at
surface irregularities and oxide inclusions.

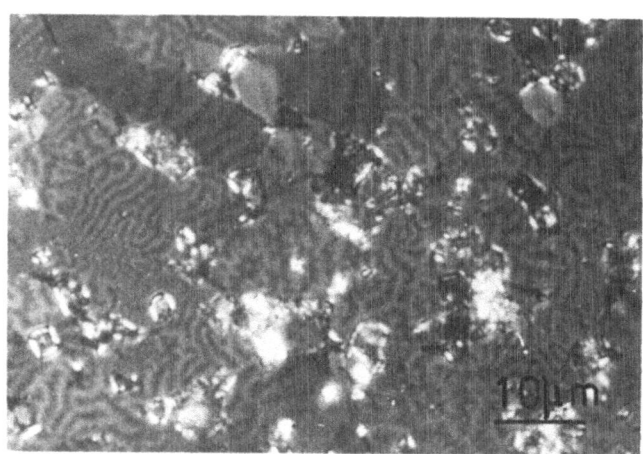

Figure 2 - Kerr magneto-optical micrograph of surface domains in a sintered
magnet.

Optimum coercivity is obtained in sintered magnets after a heat
treatment at about 630°C (e.g. a change from 700 to 1100 kA/m). The
coercivity after sintering has been shown to vary with quench rate (Ramesh
et al, 1987) and may be brought to a common optimum value by heating at
630°C. As suggested above, we have found that these improvements involve
no obvious changes in microstructure. We have not observed the additional
trans-grain boundary b.c.c. phase claimed by Hiraga et al (1985) which
supposedly dissolves under the correct heat treatment thus reducing domain

wall nucleation with a corresponding increase in coercivity. We have, however, observed the presence of segragated α-iron in SEM of sinter specimens and thermo- momagneto-gravimetric analysis did show that further iron was probably precipitated during heat treatment following sintering, Fig. 3.

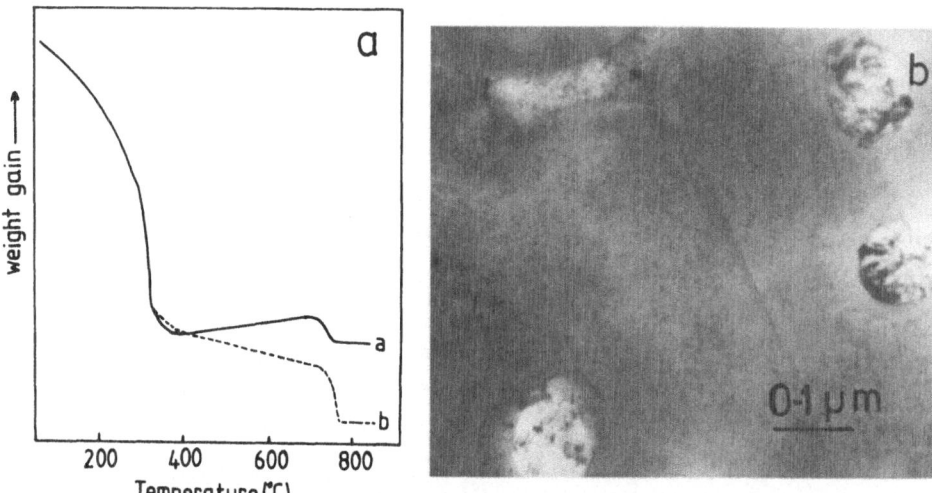

Figure 3(a) – Thermo-magneto-gravimetric analysis of sintered magnet showing evolution of magnetic iron (curve a) and reduced effect on second run (curve b). Note Curie temperatures of the hard magnetic phase and of iron.
(b) – TEM micrograph of segragated iron in melt-spun ribbon.

Sintered Magnets – Quaternary alloys

The application of NdFeB-based magnets has been limited to some extent by the relatively low Curie temperatue (~300°C) of the hard magnetic phase. The useful temperature range over which these magnets can be employed may be increased by reducing losses and the temperature coefficient of coercivity, or by increasing the room temperature coercivity so that reasonable properties are still maintained at elevated temperatures. The addition of Dy, Ga and Al can increase the room temperature coercivity by a factor of two (Endoh et al, 1987) whilst reducing remanence and energy product. Reduced losses have also been observed with Nb additions (Tokunaga et al, 1987a).

We have made a high resolution TEM and microanalysis study of magnets containing such additions in order to investigate the possibility of microstructural contributions to the reported changes in magnetic properties. Some of our measurements of coercivity are shown in Table 1.

TABLE 1   Effect of Additions on Coercivity

Composition	Coercivity (kA/m)	Change
$Nd_{16.5}Fe_{75.6}B_{7.9}$	1100	---
+1% Al	1175	+ 75
+3% Al	1210	+110
+1% Ga	1300	+200
+3% Ga	850	-250
+1% Nb	1075	- 25
+1% Mo	1075	- 25
$Nd_{15}Dy_{1.5}Fe_{75.5}B_8$	1032	---
+1% Nb	1280	+248

In general, apart from the (Nd,Dy) composition, the observed changes are more modest than most reported in the literature and can indeed show a reduction in value.   However, the situation in melt-spun ribbons is quite different, as discussed below.

In the case of Nb-containing magnets we have found two additional phases to those in the ternary systems, Fig. 4.   One is a Fe_2Nb Laves

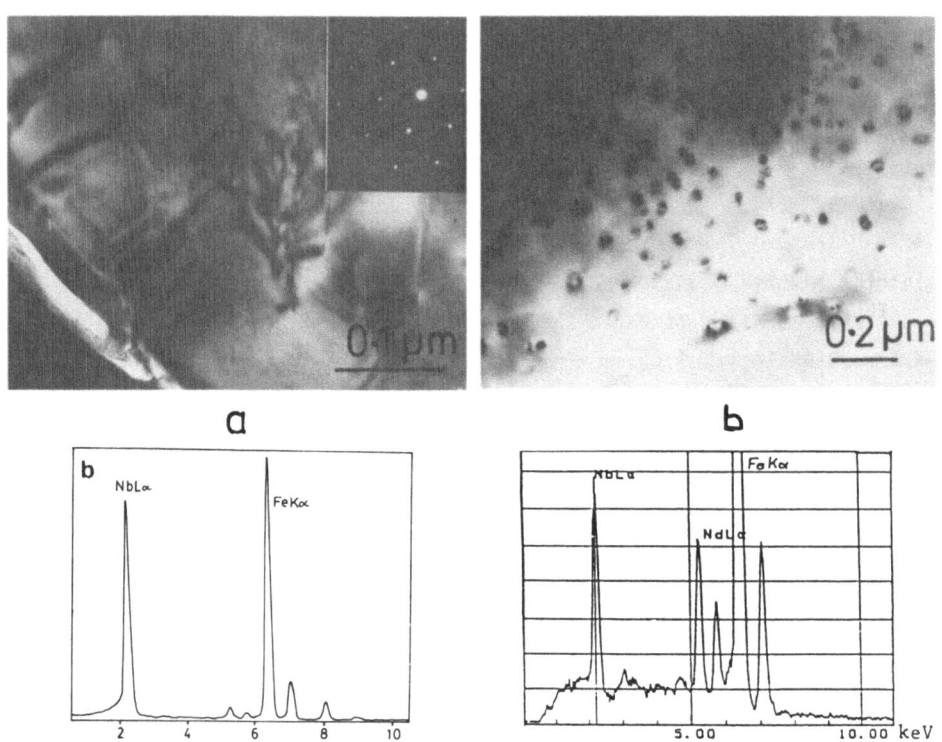

Figure 4(a) - TEM micrograph and X-ray microanalysis of Fe_2Nb Laves compound (with inset diffraction pattern) in sintered quaternary magnet.
        (b) - TEM micrograph and X-ray microanalysis of small Nb-containing precipitates

compound with a hexagonal $MgZn_2$ structure (a = 0.482 nm, c = 0.787 nm) found in the form of micrometre-size inclusions inside grains of the $Nd_2Fe_{14}B$ phase.   The second addition takes the form of small coherent precipitates, typically 20-50 nm in size at a density of $10^{21}/m^3$, within grains of the hard phase.   Apart from a small lattice rotation across the coherent interface the particles would appear to have essentially the same structure and lattice parameter of the matrix. Compositional analysis of these particles suggests a composition near to that of the matrix but with about 15% of the iron substituted by Nb.   SEM and optical microscope observations of the Nb-containing magnets revealed no obvious microstructural changes apart from the visible Fe_2Nb inclusions.

Lorentz TEM micrographs of the magnetic grains, Fig. 5, show domain walls changing direction as they pass near or through these small precipitates.   These particles may therefore act as pinning sites although it is not possible to say wether they contribute significantly to the coercivity of the magnet.   Indeed, their relatively large size as compared to the domain wall width, estimated at about 4 nm by Durst and Kronmuller (1986), may cause domain nucleation.   In melt-spun alloys the precipitate size is much smaller and of the same order as the domain wall width (Parker et al, 1987b).

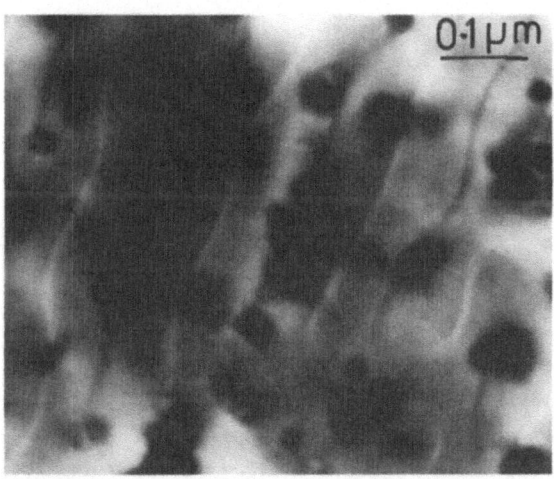

Figure 5 - Lorentz TEM micrograph of domain walls and Nb-containing precipitates in sintered magnet.

Zirconium additions (1%) result in two Zr-rich regions and a precipitation of small coherent Zr-containing particles in the hard phase (Pollard et al, 1987).   The coherent precipitates are generally of the form observed in the Nb-containing system but in some cases they can take on an acicular morphology with a needle width of about 5 nm, Fig. 6. This dimension is significant in that it approaches the width of the domain wall in the $Nd_2Fe_{14}B$ phase. Clearly, a dense precipitation field of such particles could provide a considerable barrier to domain wall motion. Again, molybdenum additions give a small coherent precipitate.

Aluminium and gallium additions do not appear to produce any detectable fine precipitation. Indeed, X-ray microanalysis suggests that these elements are concentrated in some of the grain boundaries and are not substituted in the hard phase. These latter observations agree with measurements by Grossinger (1987) which show little change in the anisotropy field in Al-containing magnets. Improvements in coercivity are therefore probably due to some subtle refinements of the microstructure at the grain boundaries.

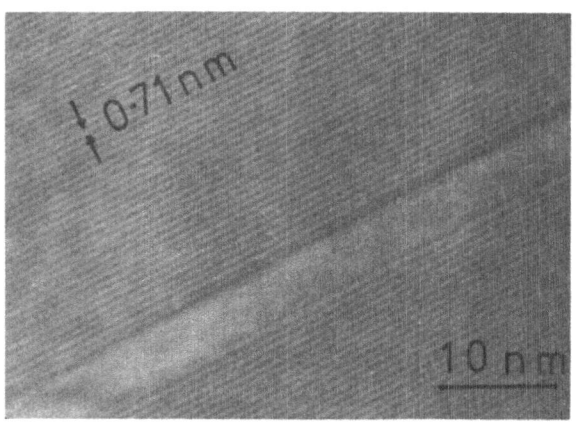

Figure 6 - TEM micrograph showing acicular morphology of coherent Zr-containing precipitate in sintered magnet.

Melt Spun Ribbons - Ternary Alloys

In agreement with other workers we have found that the coercivity of the ternary alloys is dependent on the cooling rate (≃ wheel speed) in rapid solidification. Some results for a $Nd_{16.5}Fe_{75.5}B_8$ alloy are given in Figure 7. Each experimental point represents the average of several closely placed measurements. In the apparatus used by us the optimum wheel speed is about 12 m/s which corresponds to a single domain grain size of less than 50 nm, Fig. 8. High wheel speeds produce an amorphous ribbon with low coercivity, and underquenched material with multi-domain grains formed at low speeds also shows relatively low coercivity.

However, unlike previous workers we observe an apparently anomalous result at ≃8.6 m/s, where the coercivity drops to very small values of the order of 8 kA/m (100 Oe). Now, a better indicator of the quench rate than wheel speed is the ribbon thickness. Fig. 7 shows that the ribbon thickness decreases monotonically with increasing wheel speed and that the minimum in coercivity is not produced by a faulty experiment.

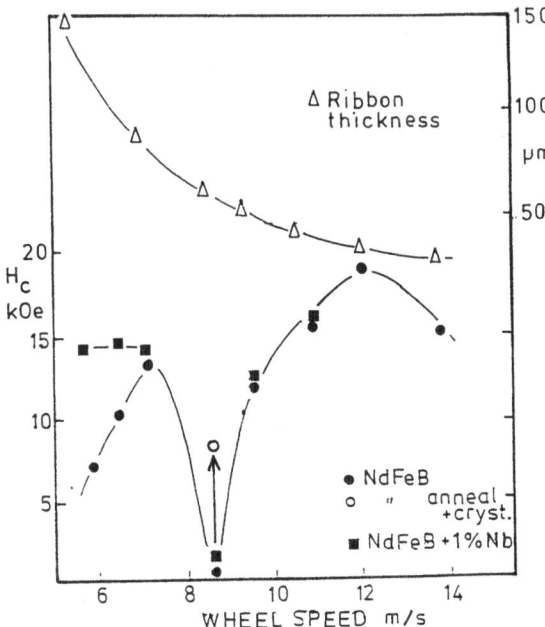

Figure 7 — Coercivity and ribbon thickness as a function of wheel speed in melt-spun ternary and quaternary (Nb) alloy ribbons.

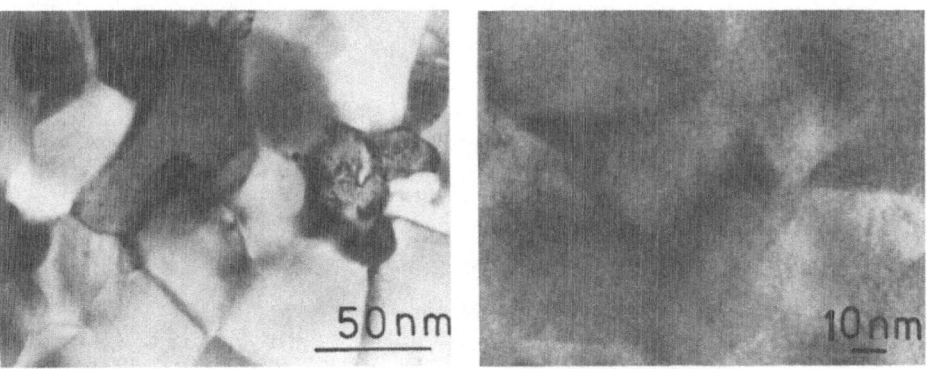

Figure 8 – TEM micrograph showing grain structure of optimally quenched ternary alloy ribbon.

TEM observations reveal a change in microstructure in the very low coercivity material; grains of the hard magnetic $Nd_2Fe_{14}B$ phase are surrounded by amorphous boundary region, Fig. 9b.   Magnetisation measurements show that this amorphous material is magnetic and domain boundaries have been observed to traverse some grains and the grain boundary region.   This microstructure can be compared to those on either side of the anomalous quench rate, Fig. 9a and c, which show crystalline grain boundary regions.

The lack of coercivity in the "anomalous" ribbon highlights the importance of the grain boundary region for the pinning of domain walls or the prevention of nucleation of reverse domains.   A short anneal of several minutes at 600°C was sufficient to crystallize the grain boundary region, causing a reduction of magnetisation, without otherwise changing the microstructure of the ribbon.   This treatment magnetically "isolated" the grains of the hard phase and significant coercivity was recovered – see the open circle symbol in Fig. 7.

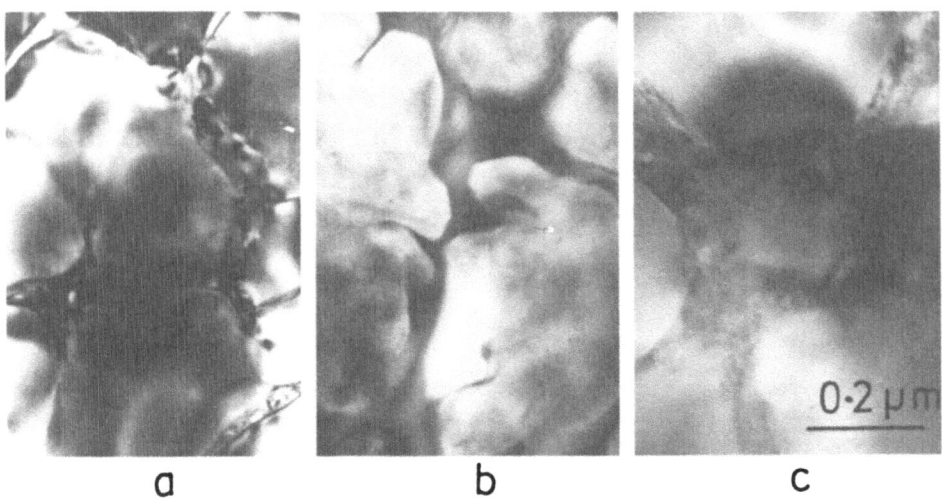

Figure 9 – TEM micrographs showing grain and grain boundary structures in ternary alloy ribbons formed at (a) 7m/s, (b) 8.6 m/s and (c) 9.5 m/s.

Melt-Spun Ribbons – Quaternary Alloy

Samples containing 1% and more Nb show a similar dependence of coercivity on wheel speed, as shown by the square symbols for 1% Nb in Fig. 7.   Again there is a marked drop in coercivity at 8.6 m/s.   However, the value is significantly higher than in the ternary alloy at about 150 kA/m (1-2000 Oe).   Amorphous grain boundary regions were again observed in TEM of this ribbon.

A further interesting and important result is that increased
coercivity over the ternary alloy is obtained in ribbons quenched at wheel
speeds below the "anomalous" rate. It can be seen in Fig. 7 that
coercivities more than twice that measured in the ternary ribbons are
obtained at wheel speeds below ≈ 7 m/s. In this range of cooling rate the
ribbons show a fine and dense precipitation of niobium containing
particles, Fig. 10. These cuboidal particles are extremely small,
typically 5 nm on edge, and are oriented along [100] matrix directions
(Parker et al, 1987b). The increase in coercivity is attributed to pinning
effects in these particle arrays. In ribbons that cool at a faster rate
the niobium presumably remains in solution as no precipitation is observed
and coercivity values are not significantly different from those of the
ternary alloy.

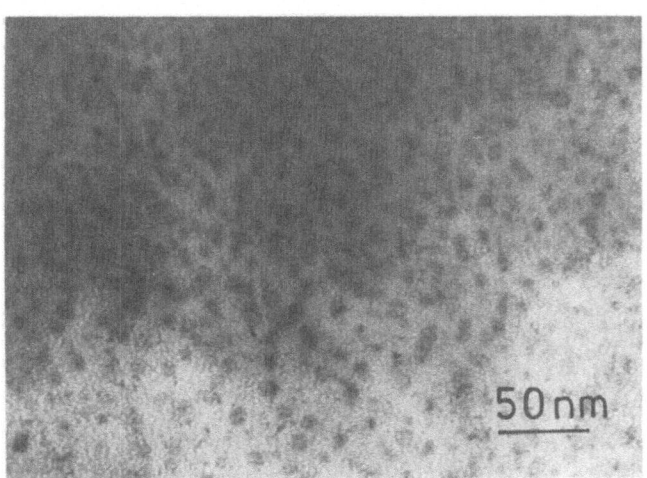

Figure 10 - TEM micrograph revealing a dense and fine precipitaion of
coherent Nb-containing precipitates in quaternary alloy ribbon spun at
7 m/s.

SUMMARY AND CONCLUSIONS

We have carried out a microstructural and magnetic investigation of
NdFeB-based alloys and magnets, including material containing refractory
and other metallic additions. Emphasis has been laid on transmission
electron microscope studies and the identification of the important phases
in as-cast alloys, magnets and rapidly solidified materials.

New and potentially important possibilities have been revealed with
the observation of fine scale precipitation in the hard magnetic $Nd_2Fe_{14}B$

phase of the materials containing additions such as Nb and other refractories. The link between this precipitation and the often claimed increases in coercivity and decreases in magnetic losses in sintered magnets is not clear. This is probably due to the relatively large precipitate size in sintered material, which is usually many times the calculated or measured domain wall width. However, investigations of melt-spun ribbons where cooling rates are much faster have shown that the much finer precipitation of the same order of size of the wall width can be associated with increases in coercivity.

The importance of grain boundary structures and compositions has also been highlighted by compositional analysis of Al and Ga containing magnets and the melt spun-alloys. In the first, improved magnetic properties appear to be associated with concentration of the additions at the grain boundaries and in the ribbons the amorphous or crystalline nature of the grain boundaries seems crucial. Further investigations of the effects of elemental additions and heat treatments on sintered and melt-spun materials are called for.

ACKNOWLEDGEMENTS

The aothors would like to thank many colleagues for their collaboration in this research - J. Ormerod and E. Rozendaal for the facilities for magnet fabrication, H. A. Davies and R. Greenhough for melt spinning, S. Abell for zoned material, J. Fidler for early microanalysis and D. Kennedy for master alloy material. Support from the U.K. Science and Engineering Research Council is also acknowledged.

REFERENCES

Durst K. D. and Kronmuller H., J. Magn. Magn. Mat., 59 (1986) 86.
Endoh M., Tokunaga M. and Harada H., IEEE Trans. Magn., MAG-23 (1987) 2290
Fidler J., IEEE Trans. Magn., MAG-21 (1985) 1955.
Grossinger R., Haslinger F., Zang Shougong, Eiber R., Liu Yingle, Schneider J. and Kirchmayr H. R., paper AA06, EMMA Conference, Salford 1987, to be published in IEEE Trans. Magn., March 1988.
Hiraga K., Hirabayashi M., Sagawa M. and Matsuura Y., Jap. J. Appl. Phys., 24 (1985) 699.
Parker S. F. H., Grundy P. J. and Fidler J., J. Magn. Magn. Mat., 66 (1987a) 74
Parker S. F. H., Pollard R. J., Lord D. G. and Grundy P. J., IEEE Trans. Magn. MAG-23 (1987b) 2103.
Pollard R. J., Grundy P. J., Parker S. F. H. and Lord D. G., paper AA05, EMMA Conference, Salford 1987, to be published in IEEE Trans. Magn. March 1988.

Ramesh R., Thomas G., Okada M. and Homma M., paper AC-07 INTERMAG
  Conference, Tokyo 1987.
Sagawa M., Fujimura S., Yamamato H., Matsuura J. and Hiraga K., IEEE Trans.
  Magn., MAG-20 (1984) 1584.
Tokunaga M., Harada H. and Trout S. R., IEEE Trans. Magn., MAG-23 (1987a)
  2284.
Tokunaga M., Kogure H., Endoh M. and Harada H., IEEE Trans. Magn., MAG-23
  (1987b) 2287.

Magnetic Hardening Studies in Nd-Fe-B Magnets

G.C. Hadjipanayis[*]

Research Center of Crete
Heraklion, Crete, GREECE

ABSTRACT

The origin of coercivity has been examined with SEM and TEM in a series of Nd-Fe-B magnets with different Dy and Al concentrations. No obvious Al segregation has been observed at grain boundaries while Dy is often found there. The SEM data indicate that Al substitutions may change the structural morphology of the samples. Clear domain wall pinning has been observed at grain boundaries and Nd-rich inclusions.

*Currently at the Department of Physics, Kansas State University.

INTRODUCTION

Large coercive fields (Hadjipanayis et al., 1983) and energy products exceeding 40 MGOe (Sagawa et al., 1984) have been obtained in Nd-Fe-B based magnets. These outstanding hard magnetic properties are attributed to the presence of the highly anisotropic tetragonal phase $Nd_2Fe_{14}B$ (Herbst et al., 1984). However, the details of magnetization reversal are not yet clear (Hadjipanayis, 1987). It is generally accepted that the domain walls move easily inside the grains and are pinned (or nucleated) at grain boundaries. This is known as localized domain wall pinning or otherwise referred to as "nucleation." It is also observed that Dy and Al substitutions tend to increase the coercivity substantially. It has been reported (Kim, 1988) that small additions of Al combined with Dy significantly increase the coercivity with a minimum sacrifice of Curie temperature and thus substantially extend the application temperature range of the magnets.

In the present study we have examined the origin of coercivity and the effect of Al and Dy substitutions by correlating the magnetic properties with the microstructure and magnetic domain structure.

EXPERIMENTAL

Sintered magnets of Nd-Dy-Fe-B with and without Al have been obtained from IG Technologies. The grain structure and chemical composition were determined with a scanning electron microscope equipped with an energy dispersive x-ray analysis detector. The microstructure and magnetic domain structure were obtained with a JEOL 100C transmission electron microscope. The magnetic measurements were made with a vibrating sample magnetometer in the temperature range of 10-400 K and in fields up to 75 kOe.

RESULTS AND DISCUSSION

SEM Studies

A systematic study of grain structure, morphology and chemical composition has been made with scanning electron microscopy. Several measurements have been made on commercial magnets with different amounts of Al and Dy and the results are reported.

In all the samples a main phase which we call the dark phase is present in which the ratio of Fe to rare-earth is ~7:1 (Fig. 1). This is the 2:14:1 phase. Also present in all the samples are areas rich in rare-earth which we call the light phase. These rare-earth rich areas may not be just one phase because their compositions vary from grain to grain. In addition, a small amount of a "grey" phase is observed with Fe/R~4:1, presumably the RFe_4B_4 phase.

Small changes in Al, Dy and Nd content seem to leave the dark phase alone while they change the light phase. Each effect is discussed below.

The Effect of Aluminum: The main effect of Al substitution was to decrease the size of the Nd-rich grains, from 2.5 μm to 1.8 μm. A secondary effect was to increase the number of light phase grains from 6 per 1000 μm^2 to 10 per 1000 μm^2. This led to a decrease in H_c from 15.3 to 14.9 kOe while B_r remained the same at 11.3 kG.

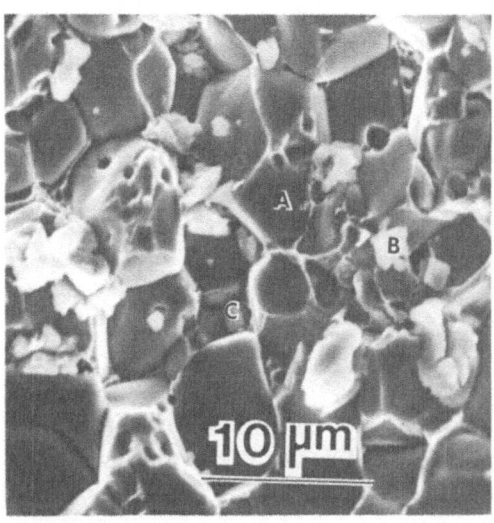

Fig. 1 Grain structure of a Nd(Dy)-Fe(Al)-B magnet. A 2:14:1, B Nd-rich, C 1:4:4 phase.

Aluminum was detected most often in the dark (2:14:1) phase. In the light (Nd-rich) phase it was rarely, if ever, detected. In the boundaries between the light and dark phases, alumimum was not detected as often as it was between the dark to dark phase boundaries. This leads to the conclusion, that the aluminum is concentrated in the dark (2:14:1) phase. This was consistent in all the samples. The amount of Al detected in the dark phase was slighty above the value recorded in the original composition.

The Effect of Dysprosium: Increasing the Dy content was found to increase the number of Nd rich grains from 10 per 1000 μm^2 to 15 per 1000 μm^2. A secondary effect was to increase the size of the Nd-rich grains from 1.8 μm to 2.4 μm. This changed the H_c from 14.9 to 16.6 kOe while B_r went down slightly from 11.3 to 11.0 kG. The concentration of Dy is more inhomogenous. In general there is much more Dy in the light phase grains. Because of peaks overlapping (with Fe) the error in Dy analysis is about 20%. The SEM studies indicate that Dy substitutions may change the structure morphology of the samples and can change the coercivity substantially.

Microstructure Studies

The phases found by the SEM studies have been verified by transmission electron microscopy (Fig. 2) (Hadjipanayis, et al. 1986). The Nd-rich phase are not found to be distributed uniformly along the grain boundaries; they are often found at intersections of several 2:14:1 grains and they are also observed as small isolated regions (Fig. 3).

Lorentz microscopy shows clearly domain walls interacting with both grain boundaries (Fig. 4) and Nd-rich inclusions (Fig. 5). The domain wall pinning in Figure 6 is probably by a fine inclusion which cannot be seen with the magnification used. By tilting the specimen and varying the objective field current, it is found that pinning at grain boundaries is much stronger than pinning at Nd-rich inclusions.

CONCLUSIONS

In summary, the SEM data do not indicate any obvious Al segregation at grain broundaries. However, more Dy is often found there. These substitutions probably alter the grain structure morphology of the samples. Lorentz microscopy shows clearly domain wall pinning at grain boundaries and a Nd-rich inclusions with the former pinning being the stronger.

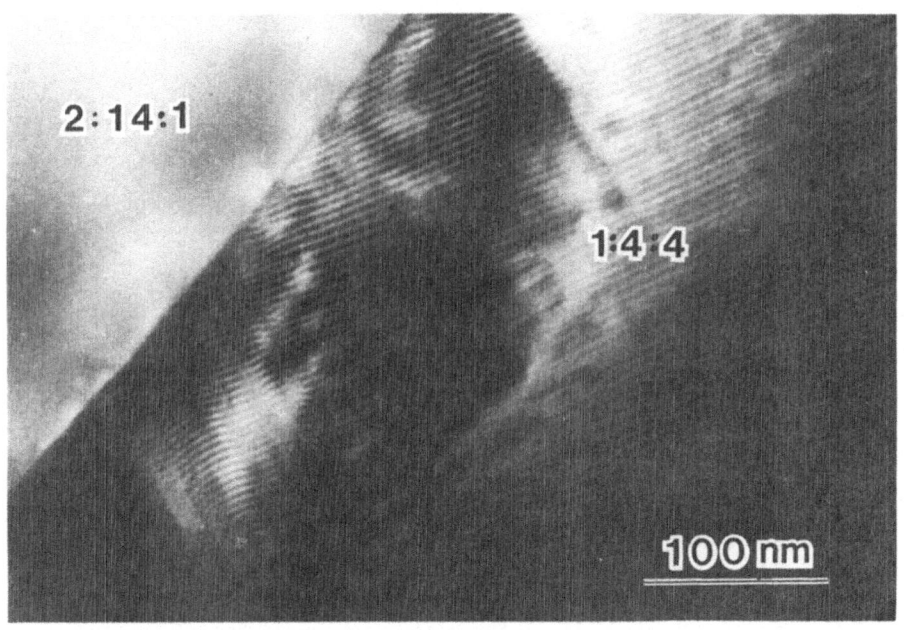

Fig. 2 Bright field micrograph showing the 2:14:1 and 1:4:4 phase.

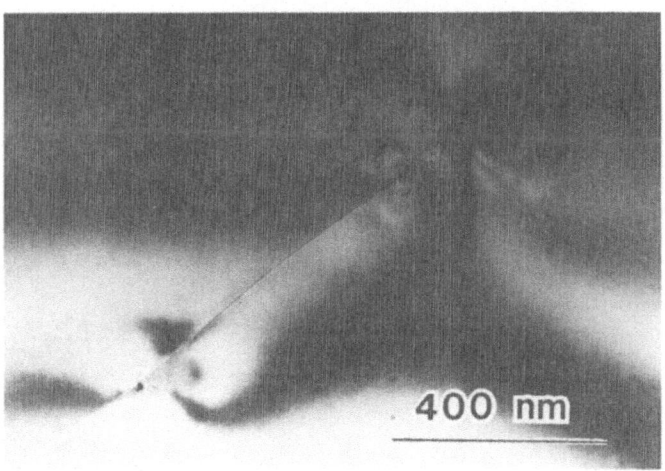

Fig. 3 Microstructure showing isolated Nd-rich inclusions in a Nd(Dy)-Fe(Al)-B.

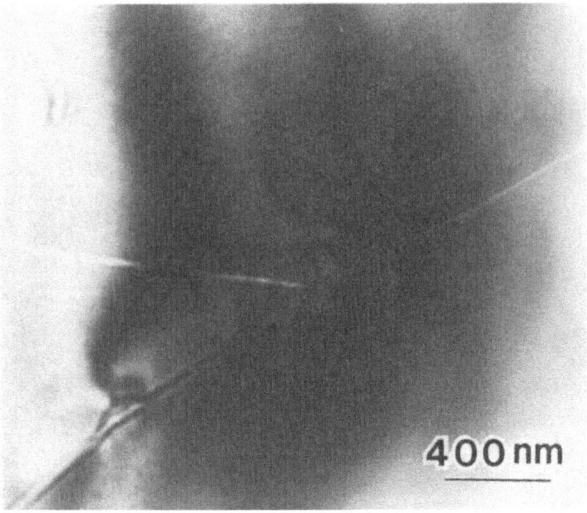

Fig. 4 Domain walls interacting with grain boundaries and Nd-rich inclusions.

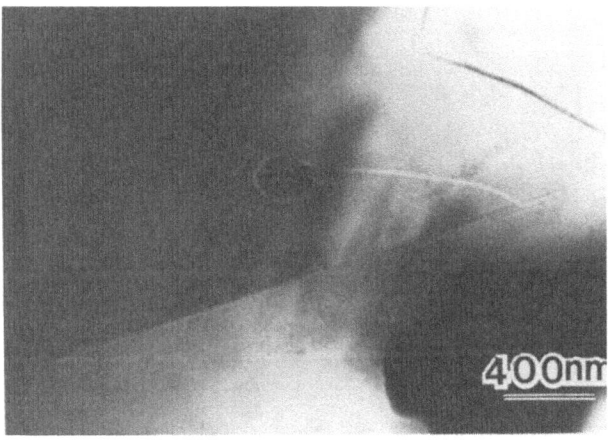

Fig. 5 Domain wall pinning by a Nd-rich inclusion.

Fig. 6 Domain wall pinning by a fine inhomogeneity which cannot be seen with the magnification used.

REFERENCES

Herbst, J.F., Croat, J.J., Pinkerton, J.F. and Yelon, W. 1984. Phys. Rev. B29, 4176.

Kim, A. IG. Technologies (private communication).

Hadjipanayis, G.C., Hazelton, R.C. and Lawless, K.R. 1983. Appl. Phys. Lett. 43, 797.

Hadjipanayis, G.C., Dickenson, R.C. and Lawless, K.R. 1986. J. Magn. Magn. Mater. 54-57, 559.

Hadjipanayis, G.C. 1987. 3M Conference, Chicago.

Sagawa, M., Fujimura, S., Yamamoto, H., Matsuura, Y. and Hiraga, K. 1984. IEEE Trans. Magn. MAG-20, 1584.

MAGNETIZATION, MAGNETOCRYSTALLINE ANISOTROPY, DOMAIN WALL ENERGIES AND THICKNESSES IN $R_2Fe_{14}B$ MATERIALS WITH R = Nd,Gd,Dy AND Ho

W.D. Corner and M.J. Hawton

Physics Department, University of Durham, South Road,
Durham, DH1 3LE, England

ABSTRACT

Measurements of magnetization and magnetocrystalline anisotropy have been made in three compounds (R = Gd,Dy and Ho) over a range of temperatures from 4.2 to 350K. Observations of domain structure have been made on these compounds and also on the corresponding Nd compound. From measurements on the basal plane domain patterns the domain wall energies and thicknesses have been calculated using the measured magnetization and anisotropy. Good agreement is found between values calculated in this way and values calculated using the basic magnetic parameters except in the case of $Gd_2Fe_{14}B$. It is suggested that the model used is inappropriate in this case owing to the weak anisotropy. Changes of domain structure in a $Nd_{15}Fe_{77}B_8$ ingot have been observed around part of the hysteresis cycle. Saturation can be achieved in grains with their c-axis parallel to the applied field at fields low compared to the coercivity. This indicates the importance of the correct processing of material with good intrinsic properties to enhance the coercivity.

INTRODUCTION

A good permanent magnet material is one in which reversal of magnetization is impeded. This may occur because of the intrinsic properties of the material, such as high magnetocrystalline anisotropy, or be due to other factors which depend on processing, such as shape anisotropy. Reversal of magnetization usually occurs, starting from the saturated state, by the nucleation of small domains of reverse magnetization and their growth by movement of domain boundary walls. A knowledge of the magnitude of the magnetocrystalline anisotropy and an understanding of its origin are needed if any potential permanent magnet material is to be assessed. In addition information about the energy and thickness of domain walls and how they move under applied fields contributes to an understanding of the magnetization reversal process. Magnetocrystalline anisotropy between 42K and 350K has been measured in materials with R=Gd,Dy and Ho. Domain wall energies and thicknesses at room temperature have been obtained for these materials and for $Nd_2Fe_{14}B$. Changes of domain structure under applied fields have also been studied. In order to obtain values of anisotropy constants and wall energies from torque magnetometer readings and measurement of domain wall spacing a knowledge of the magnetization is required. This has been measured in the same materials and over the same temperature range as the anisotropy.

MATERIALS

Single crystals of $Gd_2Fe_{14}B$ and $Dy_2Fe_{14}B$ were provided by Dr. D. Givord. These were cut by electro-spark erosion to produce suitably oriented discs for magnetic measurements and surfaces were polished using diamond paste down to 1/4 μm grade for domain observation. Similar specimens of $Ho_2Fe_{14}B$ were cut from an ingot with a high degree of grain orientation. In addition discs for magnetic measurements were made using powder derived from the same ingot and set in epoxy resin with an applied field of 0.7T. Measurements of domain wall spacing on $Nd_2Fe_{14}B$ were made on suitably cut surfaces of a large grained ingot. Further details of the anisotropy measurements described here are given by Hawton and Corner (1988).

MAGNETIZATION

Measurements of magnetization were made with a VSM operating in a superconducting solenoid with a maximum field of 13T. Values of saturation magnetization as a function of temperature for the three compounds with R=Gd,Dy and Ho are shown in Fig. 1. For the Gd compound there was little difference in values measured in the c-axis direction or in the basal plane, the anisotropy being comparatively weak. Measurements on the Dy and Ho compounds were made on ingot material with the field along an easy direction. Measurements were also made on aligned powder samples of the Ho compound. Comparison of the results from the powdered samples and those from ingot material gave a method of estimating the fraction of the sample volume occupied by the powder and this was used in the analysis of torque measurements made on the same powder samples. For both the Ho and the Dy compounds the magnetization was found to increase steadily with temperature in the range investigated. The Gd compound showed a slight decrease. Comparison with the results of other workers at isolated temperatures are also shown in Fig. 1.

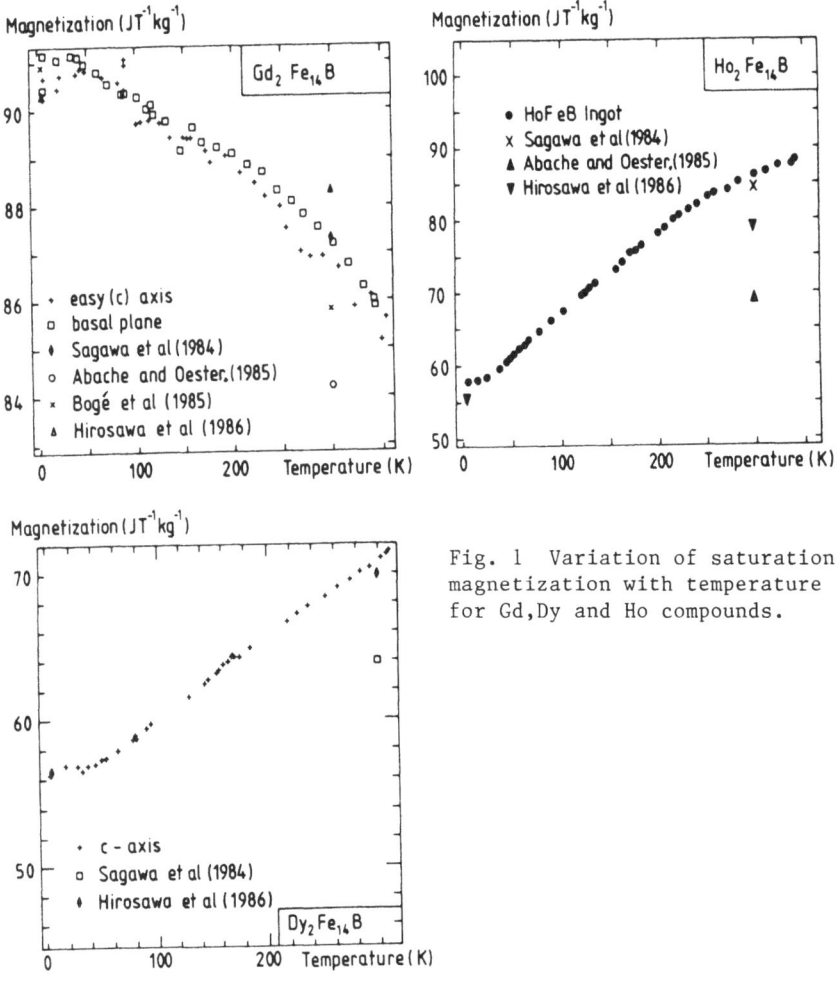

Fig. 1 Variation of saturation magnetization with temperature for Gd,Dy and Ho compounds.

MAGNETOCRYSTALLINE ANISOTROPY

Measuring Technique

It is usual to represent the magnetocrystalline anisotropy energy density F_K by an empirical expression involving a series of anisotropy constants K_i and functions of the direction of the magnetization M_s relative to the crystal axes. All the compounds studied have a tetragonal structure and an appropriate expression is,

$$F_K = K_1\sin^2\theta + K_2\sin^4\theta + K_3\sin^4\theta\cos4\phi \tag{1}$$

where θ and ϕ are the angles between M_s and c- & a-axes respectively. By using disc-shaped samples cut in appropriate crystal planes it is possible to measure the values of the K_i's using a torque magnetometer as described by Hawton and Corner (1987) in a field of up to 13T. With this instrument torque/field angle curves were obtained and corrected to give torque/ magnetization angle curves. A further correction was applied to allow for the disc not being positioned exactly on the rotation axis. The resulting curves were Fourier analysed, the K_i's being derived from the Fourier co-efficients. The number of constants was chosen to give a good fit, while not including spurious data.

Anisotropy of $Gd_2Fe_{14}B$

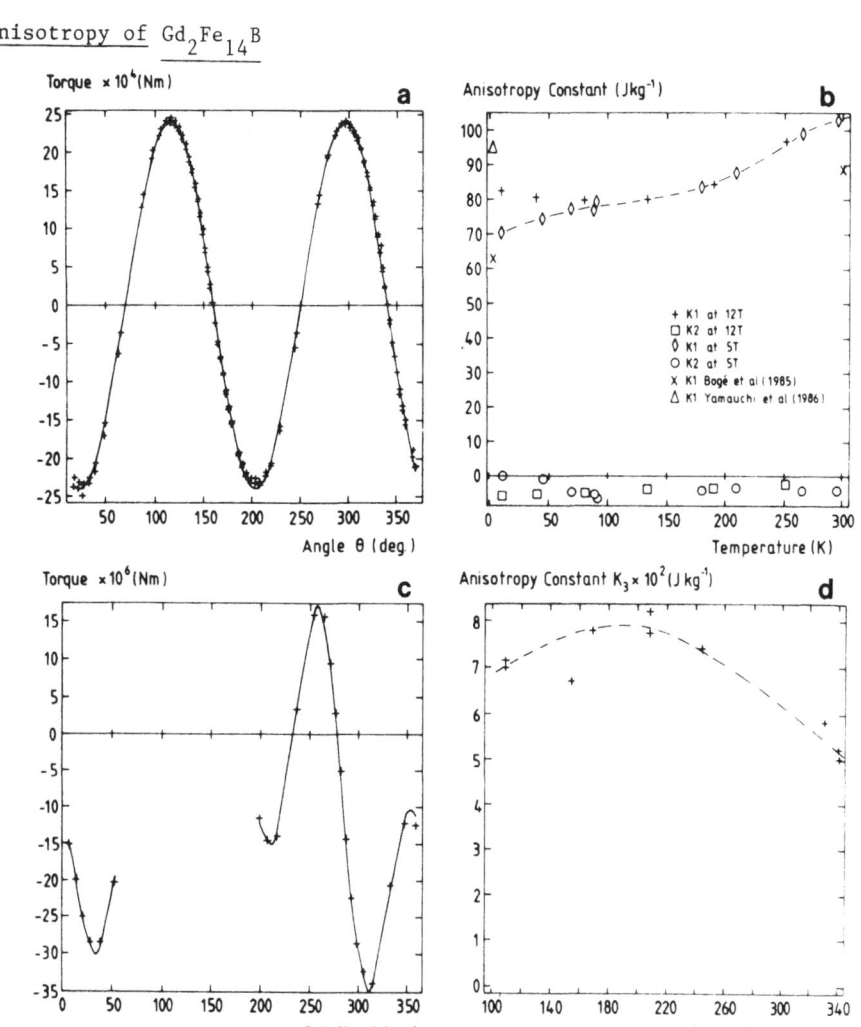

Fig. 2 $Gd_2Fe_{14}B$, (a) Typical torque curve for disc with c-axis in disc plane at 180K and 5T, (b) Temperature variation of K_1 and K_2, (c) Typical torque curve for disc with c-axis normal to disc plane at 245K and 12T. Line indicates fitted function using values of K_1 from Fig. 2(a) and K_3 from Fig. 2(d)., (d) Variation of K_3 with temperature at 5T.

With a disc containing the c-axis torque curves were of sinusoidal form (Fig. 2(a)). Attempts to fit with two constants gave a slightly better representation than with K_1 alone, though K_2 was comparatively small. The variation of K_1 and K_2 with temperature are shown in Fig. 2(b). Values with a field of 12 T show an increase in K_1 as the temperature falls below 80K which may be due to a separation of the coupling as found by Franse et al (1987) in ferrimagnetically aligned compounds of similar structure. For this reason the values at 5 T are probably more reliable, the aniso-tropy being comparatively low. The basal plane constant K_3 was found from measurement on a disc cut perpendicular to the c-axis. Here a further correction was needed to allow for a small error in the cutting of the disc which could introduce a contribution from the uniaxial anisotropy. In the disc used this added a further term to the expression for torque as a function of angle of magnetization with a 2ϕ periodicity. Using the measured value of K_1 the amplitude of this term showed that the disc had been cut to within half of a degree of the (001) plane. The basal plane anisotropy is small and the magnetometer was operating close to its lowest sensitivity limit, this being set by mechanical hysteresis effects in the connecting wires. After correction portions of the torque curves could be recovered from the data as shown in Fig. 2(c). From such data the temperature variation of K_3 shown in Fig. 2(d) was derived. Below 100 K reliable measurments of K_3 could not be obtained.

Anisotropy of $Dy_2Fe_{14}B$

The anisotropy of this compound is much higher than that of the Gd com-pound and the magnetization is smaller. These facts combine to make it difficult to rotate the magnetization through large angles from the easy (c-axis) direction. Even at 12 T a rotation of only about 35° could be achieved. Curves of torque on a disc cut with its plane containing the c-axis were as shown in Fig. 3(a). The fit was not improved appreciably by the inclusion of a second constant K_2 and the results shown in Fig. 3(b) represent an expression involving K_1 only.

Anisotropy of $Ho_2Fe_{14}B$

Torque curves measured at higher temperatures on a disc of $Ho_2Fe_{14}B$ were similar to those obtained with $Dy_2Fe_{14}B$ and could be analysed in a similar way. Below 70 K, however, a kink developed around the easy direction corresponding to the onset of the spin-reorientation process and this became acute enough to represent an easy cone below 42 K. A typical curve

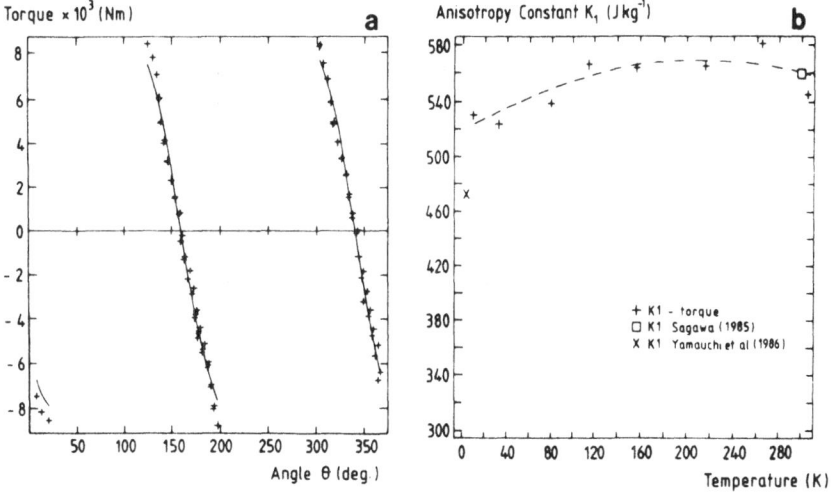

Fig. 3 $Dy_2Fe_{14}B$ (a) Typical torque curve for disc with c-axis in plane at 83 K and 12.5 T, line indicates fit with K_1 only, (b) Variation of K_1 with temperature at 12.5 T.

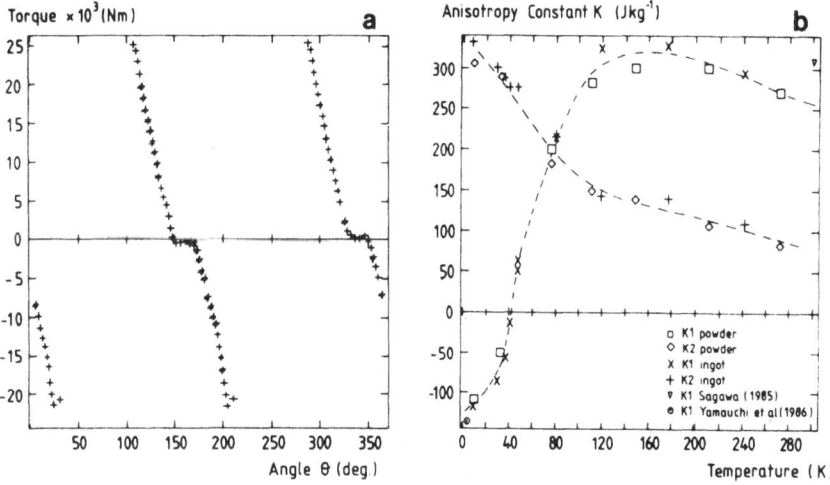

Fig. 4 $Ho_2Fe_{14}B$ (a) Torque curve for disc with c-axis in plane at 41 K and 12.5 T, (b) Variation of K_1 and K_2 with temperature at 12.5 T

is shown in Fig. 4(a). The cone angle was found to increase to about 20° when the temperature fell to 4.2K. Values of K_1 and K_2 obtained by the analysis of such curves are shown in Fig. 4(b). Lack of perfect alignment in the multigrain samples may have reduced the magnitude of the constants below those appropriate for single crystals, but there is good agreement with the results obtained using an aligned powder when corrections had been applied for the fact that only part of the volume consisted of active material. Nevertheless the results lie somewhat lower than those at 4.2K and 300K obtained by Yamauchi et al (1986) and Sagawa et al (1984) respectively.

Discussion of Anisotropy Results

As expected the Gd compound shows the weakest anisotropy, the rare earth as having a $^8S_{7/2}$ ground state. This sets an upper limit on the contribution of the iron ions to the anisotropy. Single ion interactions of the Gd ion would be expected to be weak and most of its anisotropy to originate in two-ion effects which would be difficult to distinguish from the anisotropy due to the iron lattice. The increase with temperature of K_1 could be due to the presence of a two-ion contribution which falls more rapidly with temperature than that of the iron lattice. The overall rise in anisotropy could then be explained in terms of the competition between the 3d anistropy and the two-ion 4f anisotropy.

Both the Dy and the Ho compounds showed an easy c-axis, not surprising since Dy, Ho and Nd all have negative Stevens α_J factors. Detailed comparison of measured anisotropy constants with those derived from crystal field theory must await more reliable calculations of the latter. It would clearly be desirable to extend the anisotropy measurements to give more complete values of K_2 and K_3 over the entire ferromagnetic range of temperatures.

DOMAIN WALL ENERGIES AND THICKNESSES

Technique

After polishing the surface of a crystal domain patterns were made visible by the use of a fine-grained ferrofluid. This consisted of particles of magnetite less than 50 nm in diameter suspended in petroleum ether with Solsperse 3000 as a surfactant. A drop of this was placed on the specimen surface and covered by a 5mm dia. glass disc. The patterns could then be viewed using a metallurgical microscope. Using a specially constructed stage it was possible to apply magnetic fields of up to 1.3T to the specimen during observation of the patterns. All measurements were made at room temperature, 290K.

Domain Wall Energies

The observed patterns were all of the same type. On surfaces containing the c-axis the structure was as shown in Fig. 5 where for compari-

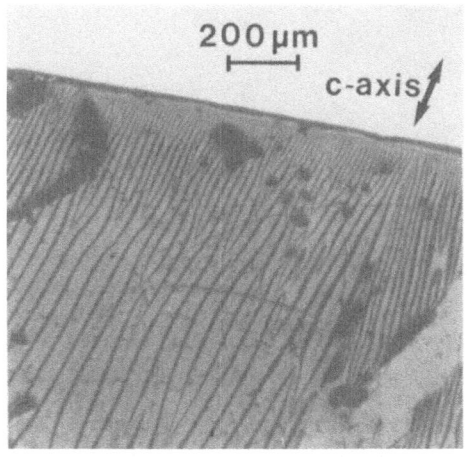 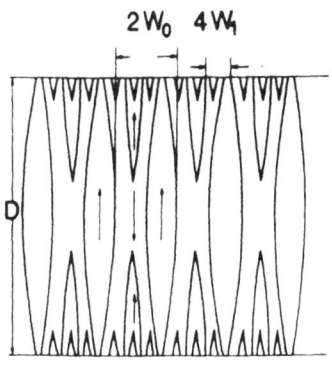

Fig. 5.   Domain patterns on $Dy_2Fe_{14}B$ crystal compared with model of Bodenberger and Hubert (1977).

son is shown that proposed by Bodenberger and Hubert (1977) for domains in $SmCo_5$.   The patterns on the basal plane were similar to that shown in Fig. 6 and represent the ends of the main domains and the reverse spikes illustrated in Fig. 5.   It was shown by Bodenberger and Hubert that the domain wall energy per unit area γ may be calculated from the mean wall

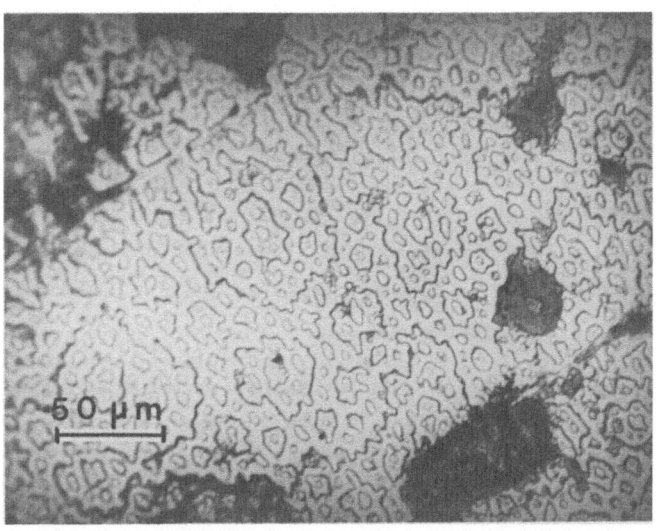

Fig. 6.   Domain pattern on basal plane of $Dy_2Fe_{14}B$.

spacing W_1 on the basal plane using,

$$\gamma = \frac{\mu_o W_1}{\beta} \left(\frac{M_s}{4\pi}\right)^2 \qquad (2)$$

M_s is the magnetization and β is a factor whose size depends on the type of surface structure. They found that for $SmCo_5$ crystals a value of $\beta = 0.31$ was appropriate and since the structures on the present materials were similar the same value has been used in this work. The value of W_1 was obtained by ruling lines in random directions on a photograph of the patterns on the basal plane and counting the number of intersections per unit length of line. To obtain W_1 the mean spacing of intersections was multiplied by $2/\pi$ as suggested by Bodenberger and Hubert. Using this and measured values of M_s, values of wall energy could be calculated for four compounds as shown in Table 1.

TABLE 1. Domain wall energies calculated from W_1 and M_s

R.E. Ion	W_1 [μm]	M_s [$MJT^{-1}m^{-3}$]	γ [$J\ m^{-2}$]
Nd	0.78	1.283	0.033
Gd	2.64	0.686	0.032
Dy	4.35	0.564	0.036
Ho	1.50	0.694	0.019

For comparison the wall energy may be calculated using the expression proposed by Träuble et al (1986),

$$\gamma = 2(AK_1)^{\frac{1}{2}}\left[1 + \left[\frac{(1+K)}{K^{\frac{1}{2}}}\right]\sin^{-1}\left(\frac{K}{K+1}\right)^{\frac{1}{2}}\right] = 2(AK_1)^{\frac{1}{2}}G \qquad (3)$$

where $K = K_2/K_1$ and A is the exchange parameter. A value of A may be obtained from the relation $A = kT_c/a$ where T_c is the Curie temperature and a the lattice parameter. Table 2 shows results obtained using this expression and comparison with the figures in Table 1 shows a good measure of agreement except for the Gd compound.

TABLE 2. Domain wall energies calculated from magnetic parameters

R.E. Ion	K_1 [MJm^{-3}]	K_2 [MJm^{-3}]	G	a [nm]*	T_c [K]+	A [Jm^{-1}]	γ [Jm^{-2}]
Nd	5.0+	0.66+	2.082	0.879	588	9.2×10^{-12}	0.028
Gd	0.80	-0.04	(2)	0.878	660	10.3×10^{-12}	0.012
Dy	4.59	0.44	2.063	0.876	593	9.3×10^{-12}	0.027
Ho	2.29	0.61	2.171	0.875	574	9.1×10^{-12}	0.020

Numerical values are from our own measurements except where otherwise marked; + from Buschow (1986) and * from Sinnema et al (1984).

Domain Wall Thickness

Values of wall thickness δ may be calculated using the much quoted expression

$$\delta = \pi \left[\frac{A}{K_1 + K_2} \right]^{\frac{1}{2}} \qquad (4)$$

Alternatively, by eliminating A from Eqns (3) and (4) the wall thickness can be calculated from the wall energy using,

$$\delta = \frac{\pi\gamma}{2G} \left[\frac{1}{K_1(K_1 + K_2)} \right]^{\frac{1}{2}} \qquad (5)$$

Results of the application of Eqn (4) and of using Eqn (5) along with the values of γ shown in Table 1 are compared in Table 3. Again the agreement is good with the exception of the Gd compound.

TABLE 3. Domain wall thicknesses by two different methods.

R.E. Ion	δ [nm] by Eqn (4)	δ [nm] by Eqn (5)
Nd	4.1	4.9
Gd	11.6	32.1
Dy	4.3	5.6
Ho	5.6	5.2

Discussion of Wall Energy and Thickness Results

The Bodenberger and Hubert model seems to be very satisfactory for the analysis of the results except for the Gd compound. Here, to obtain agreement with the value of γ obtained using Eqn (3) a value for β of 0.83 would be required. Careful study of the patterns obtained on a surface of $Gd_2Fe_{14}B$ containing the c-axis (Fig. 7) shows that the formation of reverse daggers is not so frequent as in the other compounds. There are very few such structures at the free surface though some do appear attached to inclusions. This is presumably due to the low anisotropy of the Gd compound. Though no flux closure domains can be seen on Fig. 7 it is possible that these exist, but are masked by the heavy ferrofluid deposit at the free surface. In the absence of an exact knowledge of the form of any such closure structure it is not possible to calculate the wall energy from the mean domain wall width measured on Fig. 7.

Values of wall energy for $Nd_2Fe_{14}B$ may be compared with those of Durst and Kronmüller (1986) who found $\gamma = 0.024$ Jm^{-2} corresponding to $A = 7.7 \times 10^{-12}$ Jm^{-1}. Givord et al (1987) found $\gamma = 0.025$ Jm^{-2} but took a much larger value of $A = 45 \times 10^{-12}$ obtained from $T_c = 593$ K and $a = 0.182$nm.

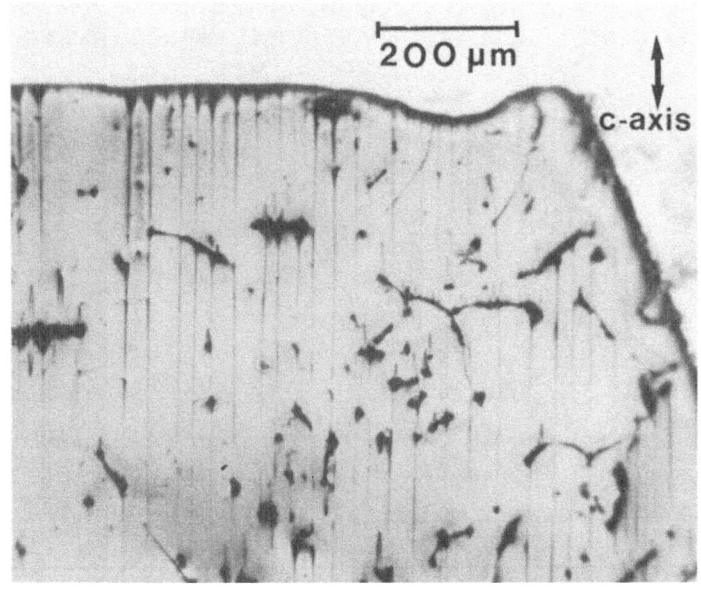

Fig. 7   Domain pattern on $Gd_2Fe_{14}B$ crystal.

The   lattice   parameter they used is the spacing of (110) planes
containing R ions. Using similar A values for the four compounds studied
here, Eqn (3) yields energies which are rather more than twice those
given in Table 2.

Magnetization Processes in $Nd_{15}Fe_{77}B_8$ Ingot

Using fields up to 1.3 T the changes in domain structure in a coarse
grained ingot were observed. In this grains were oriented in various
directions and the field was applied parallel to the easy c-axis in one
set of grains. At a field between 0.34 and 0.51 T those grains had been
completely saturated, though a domain structure could still be seen in
less favourably oriented grains up to the maximum field. On reducing the
field reverse domains began to nucleate at grain boundaries at about 0.3T
and to spread rapidly as the field was reduced to zero. When zero field
was reached the reverse domains covered about 50% of the area of the
grains. Increase of the field in the reverse direction showed that
saturation was again achieved in the grains with c-axis parallel to the
field soon after this reached 0.3 T.

The fields quoted are applied fields and those acting on the grains must have been somewhat reduced due to demagnetizing effects. Nevertheless saturation in favourably oriented grains was achieved at applied fields which were appreciably lower than the values of coercivity quoted for magnet material (0.6 to 1.0 T or 500 to 800 kAm^{-1}). This indicates the importance of the techniques used in production of magnet material in enhancing the coercivity in materials based on compounds with high anisotropy and narrow domain walls. Further details of the domain observations reported here are given by Corner and Hawton (1988).

Acknowledgements.

It is a pleasure to acknowledge the ready cooperation of various other members of the CEAM project and particularly the help given by Dr. D. Givord, Dr. I.R. Harris and Dr. J.S. Abell.

REFERENCES

Abache, C. and Oesterreicher, H. 1985. J. Appl. Phys. 57, 4112.
Bogé, M., Coey, J.M.D., Czjzek, G., Givord, D., Jeandry, C., Li, H.S. and Odden, J.L. 1985. Solid State Commun. 55, 295.
Bodenberger, R. and Hubert, A. 1977. Phys. Stat. Sol. (a) 44, K7.
Buschow, K.H.J. 1986. Mater. Sci. Rep. 1, 1.
Corner, W.D. and Hawton, M.J. 1988. J. Magn. Magn. Mat. 72, 59.
Durst, K.D. and Kronmüller, H. 1986. J. Magn. Magn. Mat. 59, 86.
Franse, J.J.M., 1987. Presented at CEAM meeting Dublin.
Givord, D., Lienard, A., Tenaud, P. and Viadieu, T. 1987. J. Magn. Magn. Mat. 67, L281.
Hawton, M.J. and Corner, W.D. 1987. J. Phys. E: Sci. Inst. 20, 406.
Hawton, M.J. and Corner, W.D. 1988. J. Magn. Magn. Mat. 72, 52.
Hirosawa, S., Matsuura, Y., Yamamoto, H., Fujimura, S., Sagawa, M. and Yamauchi, H. 1986. J. Appl. Phys. 59, 873.
Sagawa, M. 1985. J. Magn. Soc. Japan, 9, 25.
Sagawa, M., Fujimura, S., Yamamoto, H., Matsuura, Y. and Hiraga, K. 1984, IEEE Trans. Magn. MAG-20, 1584.
Sinnema, S., Radwanski, R.J., Franse, J.J.M., de Mooij, D.B. and Buschow, K.H.J. 1984. J. Magn. Magn. Mat. 44, 333.
Träuble, H., Boser, O., Kronmüller, H. and Seeger, A. 1966. Phys. Stat. Sol. 10, 283.
Yamauchi, H., Yamada, M., Yamaguchi, Y. and Yamamoto, H. 1986. J. Magn. Magn. Mat. 54-57, 575.

NdFeB MAGNETS BY MELT SPINNING

A. Chamberod, F. Vanoni,

Centre d'Etudes Nucléaires de Grenoble
DRF/SPh/Métallurgie Physique
85X - 38041 Grenoble Cédex, France.

ABSTRACT

A study of the different parameters acting on the performances of NdFeB melt spun magnets was developped. Samples were prepared either directly by compacting flakes obtained by rapid quenching on a rotating wheel, or heat treatment : "hot pressing" or "die upsetting". In each case, we explored first the concentration range of the major elements Fe, Nd and B, and considered the role of some adding elements like Al or Dy. On the other hand different roll quenching parameters were investigated.

Moreover it was established that the growth axis of a $Nd_2Fe_{14}B$ crystal corresponds to an "a axis" of its tetragonal structure, perpendicular to the "c" magnetic axis.

INTRODUCTION

Two ways have been followed successfully to prepare NdFeB magnets. The first technology is based on a powder metallurgy followed by powder sintering (Sagawa, 1984) and was first developped by Sumitomo (1984). The second one starts from flakes produced by rapid quench (Croat, 1984 ; Herbst, 1985), and was proposed by General Motors (1984). In the framework of CEAM, our activity was devoted to the production and characterization of NdFeB magnets by rapid quench : melt spinning or planar flow casting.

There are many parameters to explore. First of all the composition of the alloy is of major importance, not only considering the main elements Nd, Fe and B, but the adding of some other elements. Secondly, there are many roll quenching parameters to be precised : wheel diameter, wheel speed rotation, nature of the material of this wheel, nature and pressure of the gas in the device, ejection pressure (ie pressure difference in the crucible and in the device), distance of the crucible nose to the wheel and relative orientation, heating rate up to the melt and melting duration before quench.

The flakes obtained by rapid quench can be used directly or with subsequent heat treatment : "hot pressing" or "die upsetting". Pressure and temperature conditions during these treatments are very critical and must be carefully determined.

MATERIALS

Alloy preparation

The alloy ingots were prepared using good purity constituents : 99.9

% Nd, Fe and B. Fe and B were first melt together, and Nd was further added. Melting is realised in a copper reactor with water cooled sectors, under pure argon atmosphere, to avoid pollution to the best. The resulting load is quite spherical, 60g weight.

Quench characteristics

Loads are HF heated in the roll-quenching device, in a copper reactor with water-cooled sectors, under pure argon pressure of typically 7.10^4 Pa (less than the atmospheric pressure). Just above the melting point, an overpressure of about $2,5.10^4$ Pa is applied at the top of the crucible, pushing away the melt material through the hole (or the slit) of the nose at the bottom of the crucible. The distance nose-wheel is typically 0.3mm. Then the melt is thrown on the rotating wheel. Usually the wheel is made of copper (some trials were made with stainless steel), its diameter is 20cm (trials were made with 15 or 32 cm). One obtains ribbon fragments (flakes) a few mm length, 20 to 70 µm thickness, very friable. But in certain conditions, the ribbon gets out full and breaks only when bumping against the device wall.

Magnet forming

- type 1 : By a coarse crushing, the ribbon or ribbon fragments are broken to obtain flat particles (surface 1 to 4 mm^2). These particles are linked with a polymer, hardened and densified by Υ-irradiation or by chemical reaction at room temperature under pressure (3.10^8Pa). The density obtained is about 6, corresponding to 80% of that of the massive material and the shape is cylinders ($\Phi \sim$ 8mm, h $\sim$ 10mm).

- type 2 : the small particles (see type 1) are heated up typically for 3min to 640°C under a pressure of 3.10^8Pa. Different temperatures were tested between 620 and 720°C, with no significant difference. 640°C is the best for the mechanical behaviour of the magnet.

- type 3 : The samples treated "type 2" are heated again between 700 and 745°C for 1 to 4min (typically : 2min). A pressure of 5.10^7Pa is applied during the isothermal stay. In the opposite of that happens for type 2, the choice of temperature and pressure for this "die-upset" treatment is very sharp and must be adapted to the alloy considered.

MEASUREMENTS

Sample characterization is made from several measurements : X-ray diffractograms were made directly on the ribbon fragments, or on sections of magnets.

Differential calorimetry measurements were made with a DSC-2C Perkin-Elmer calorimeter.

Mössbauer effect measurements were performed with a conventional spectrometer and a Co^{57} source.

Magnetic measurements consist on measuring the induction B as a function of the applied field for increasing fields, and decreasing fields from the highest value possible. This value, 26 kOe, is not sufficient to saturate the magnet, but is enough to give curves representative of the properties of the magnet in the first and second quadrant of the hysterisis loop.

RESULTS

Study of the as-prepared pure magnetic phase $Nd_2Fe_{14}B$

This phase is the basis for FeNdB magnets, both sintered or melt spun. It is responsible for the high value of magnetization but it is not possible to reach a correct value of the coercitivity and therefore a high value of (BH)max with this phase alone. However there is an interest to understand its behaviour as a function of different parameters.

Samples of $Nd_2Fe_{14}B \equiv Nd_{11.77}Fe_{82.35}B_{5.88}$ were prepared by planar flow casting with different peripheric wheel speeds : 18, 20, 22.5, 25 and 30 m/s.

An amorphous phase can be retained by quench, depending on the wheel speed. The presence of such an amorphous phase can be detected just looking at the demagnetizing curve, which shows, if any, an anormalous flat shape. Calorimetry measurements (fig. 1) allow to determine the crystallization temperature, at around 875 K, not depending on the speed. A Curie point is observed at 575 K.

If we link as-quenched particles with a polymer at room temperature, the energy product $(BH)_{max}$ is maximum (5 MGOe) for a wheel speed of 22.5 m/s. The remanent induction B r is near 0.6 Tesla, and the coercitive field is rather low (6.4 kOe).

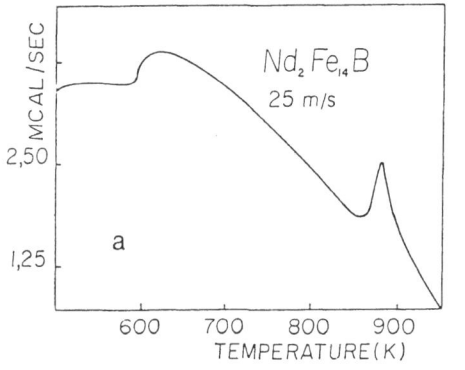
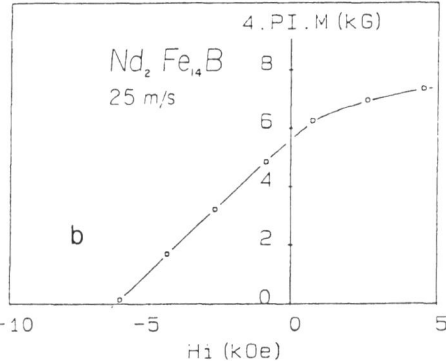

Fig. 1 $Nd_2Fe_{14}B$ melt spun (25/s) : a) calorimetry curves b) magnetization curves.

Effect of addition of elements

The addition of 1 % Aluminium lowers significantly the coercivity, and $(BH)_{max}$ (7.44 MGOe), while the optimal speed shifts towards 20 m/s or less.

It appears from the calorimetry measurements that the optimal speed is that one which is the limit for the formation of an amorphous phase : in the $Nd_{11.77}Fe_{82.35}B_{5.88}+ 1$ % Al alloy the amorphous phase is detected for speeds higher than 20 m/s which corresponds to the maximum of (BH)max (see fig. 2a,b,c,d).

The addition of both 1 % Aluminium and 1 % Dysprosium (substituted to 1 % Neodymium), so that the formula becomes $Nd_{10.77}Dy_1Fe_{82.35}B_{5.88}$, or $Nd_{1.83}Fe_{14}Dy_{0.17}B$, gives a good result (5.6 MGOe) at a speed between 20 and 25 m/s.

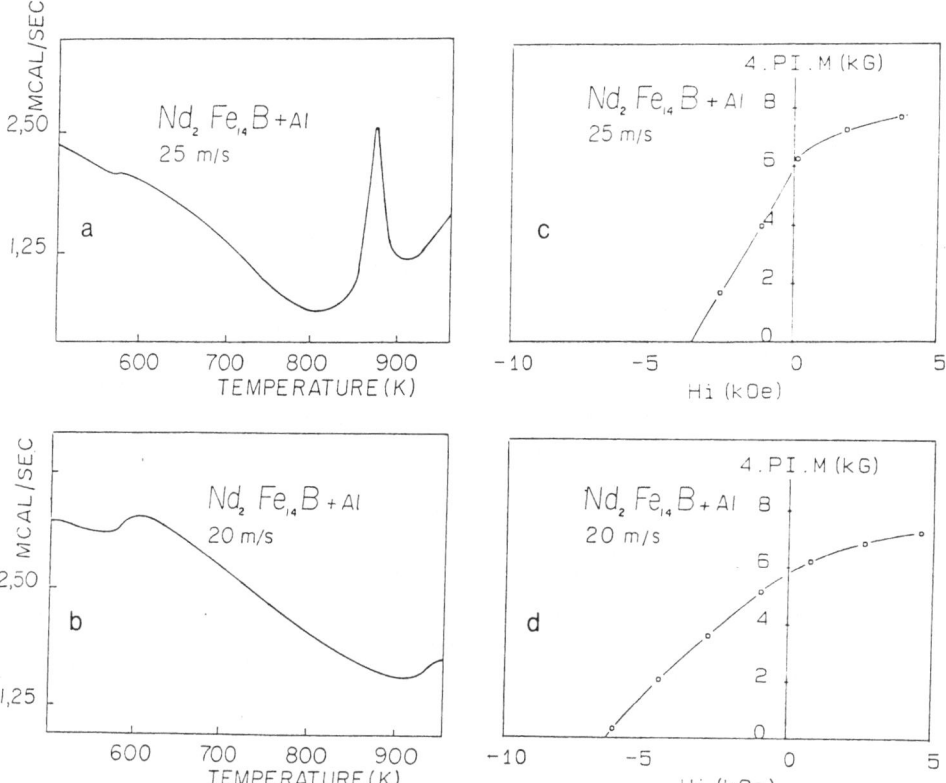

Fig. 2 Nd_2Fe_{14} + Al melt spun. Calorimetry curves a)wheel speed 25m/s b) 20m/s. Magnetization curves c) wheel speed 25 m/s d) 20m/s.

Effect of annealing

We have heated such samples under pressure (10^2MPa) i) for 5 mn at 900K, just above the crystallization temperature and ii) for 5 mn at 985 K which is the temperature required by GM. For both annealings, $(BH)_{max}$

at the optimum speed is a little higher (8.5 to 9 MGOe) than in the as-prepared state. This improvement results from an increase of both Br and Tc. The increase of $(BH)_{max}$ is more pronounced for the other speeds, smaller or faster than 22.5 m/s, but this speed remains the optimal one. After such annealings, no more amorphous phase is present in the sample, and the Curie point has increased to 585 K.

We think that the increase of Br observed is due to an increase of density by pressing and also an anneal of "defects" (deformation, stresses), which would explain the increase of the Curie point from 575 to 585 K, too. Concerning Hc, following Mishra (1985) we propose that it is optimal when the magnetic grains of $Nd_2Fe_{14}B$ are single domains, and the domain walls are pinned at the boundaries. In the as-prepared state, such boundaries should be made of the amorphous phase. After heating, both the magnetic grains may have grown to the right size, as observed in the overquenched samples (wheel speed higher than 22.5m/s), and the crystallised phase formed around these grains may be more favourable to a wall pinning. No major difference seems to come from the difference of treatment temperature, 900 or 985K. 900K seems to be slightly more efficient for a better $(BH)_{max}$, maybe because it corresponds to the righter grain size.

Considering the alloy with 1 % Aluminium, it appears that after anneal the alloy quenched with speeds of 25 to 30 m/s is better than the alloy quenched at 20 m/s. A marked difference is observed between 985 and 900K, the latter corresponding to the best $(BH)_{max}$ (9 MGOe instead of 7.3), with an increasing value of Hc (4.5 kOe). Thus Aluminium plays a role on the coercitivity essentially during annealing.

Study of other compositions

As still reminded, correct performances for magnets can be reached only if the $Nd_2Fe_{14}B$ phase is embedded in one or several other phases responsible for coercivity. Then we have analysed the behaviour of alloys around the composition $Nd_2Fe_{14}B$ ($\equiv Nd_{11.77}Fe_{82.35}B_{5.88}$) considering that any alloy is made of $(Nd_2Fe_{14}B)$ + addition of (NdFe) or (NdB) or (FeB). In all the cases 1 % at Aluminium was added, which is known to contribute to good properties.

Several observations have been made :

For a given alloy composition, the most favourable wheel speed changes according as one considers magnets of type 1, 2 or 3. Thus in the case of alloy $Nd_{13}Fe_{82}B_5$ + Al (of GM), the best $(BH)_{max}$ is obtained for a speed of 22.5 m/s if the alloy is bounded with polymer (type 1) while the speed should be 25 m/s if the alloy is heated to 630°C (type 2) and $\gtrsim$ 30 m/s for die upset samples (type 3).

On the other hand, the maximum of $(BH)_{max}$ for a given treatment (type 1,2 or 3) and a given wheel speed depends on the composition of the alloy. Fig. 3 shows the variations of Hi, Br and $(BH)_{max}$ as a function of the ratio Nd:Fe of 1,2% Nd_xFe_{100-x} alloy added to the $Nd_2Fe_{14}B$ phase in the case of a treatment type 2 (hot pressing). Clearly Br has a stronger influence than Hc on $(BH)_{max}$, but the shape of the magnetization curve plays an important role.

Concerning the heat treatment for types 2 and 3, it appears that they are very critical. Heating rate, maximum temperature, treatment duratyon, characteristics of the applied pressure must be precised accurately.

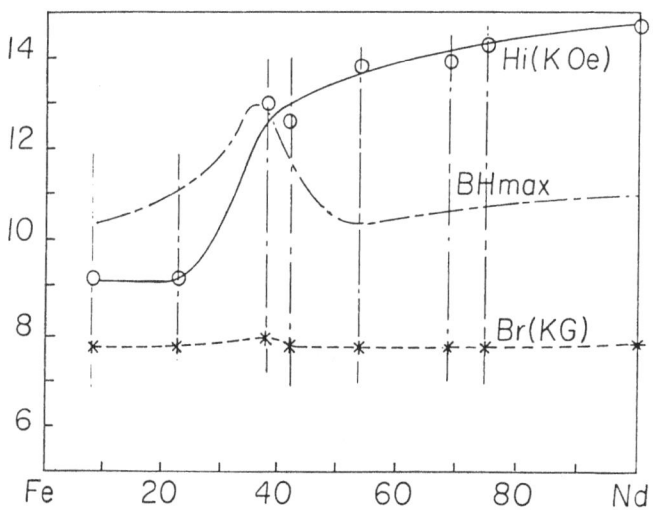

Fig. 3   Hi, Br and $(BH)_{max}$ as a function of the ratio Nd : Fe of the alloy $Nd_2Fe_{14}B + 0.012 \; Nd_xFe_{100-x}$.

Choice of the different materials

As a conclusion of the study of these different parameters, we propose the following materials and treatments :

- type 1 : $Nd_{10.77}Fe_{82.35}Dy_1B_{5.88}$ (+ Al 1%) 25 m/s (fig. 4a) which is the magnetic phase $Nd_2Fe_{14}B$, ie $Nd_{11.77}Fe_{82.35}B_{5.88}$, where 1 % Nd was substituted by 1 % Dy. Its $(BH)_{max}$ is 5.6 MGOe. The alloy $Nd_{12}Dy_1Fe_{82}B_5$ 25 m/s shows a $(BH)_{max} = 5.9$ MGOe but the coercitivity is good : 13 kOe.

- type 2 : $Nd_{13.5}Fe_{81}B_{5.5}$ (+ Al 1%) 22.5 m/s (fig. 4b) which can be considered as the magnetic phase $Nd_{11.77}Fe_{82.35}B_{5.88} + 0.012$ $(Nd_{38}Fe_{62})$. Its $(BH)_{max}$ is 12.9 MGOe. The alloy $Nd_{15.5}Fe_{79}B_{5.5}$ (magnetic phase + 0.012 ($Nd_{69}Fe_{31}$)) shows a $(BH)_{max}$ a little lower (11.6 MGOe) but a coercitivity equal to 15 kOe.

Scanning microscopy observations show that the grain size varies from 150 to 700 Å. The grains are all crystallised. The Neodymium is rather concentrated at the grain boundaries.

- type 3 : $Nd_{13.5}Fe_{81}B_{5.5}$ (+ Al 1 %) (cf. type 2) (fig. 4c)

The best treatment corresponds to 700°C with a height reduction of 50%.

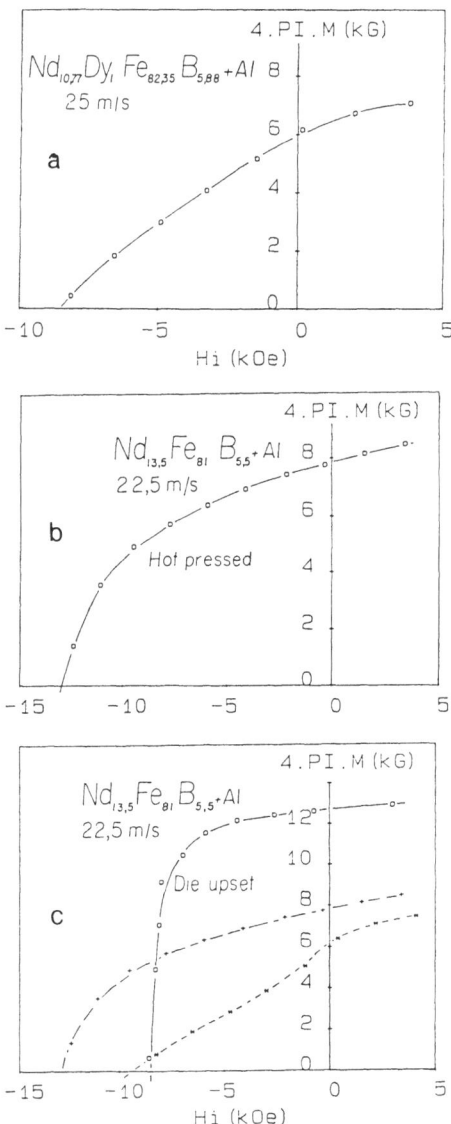

Fig. 4   Magnetization curve for a) the best "type 1" b) the best "type 2" c) the best "type 3".

$(BH)_{max} = 36.9$ MGOe with $Br\!\!/ = 12,7$ $Br\perp = 4,6kG$ and $Hc\!\!/ = 8,3$ $Hc\perp = 10kOe$. Let us recall that the parameters defined here are very sensitive depending on the concentration.

DETERMINATION OF THE EASY GROWTHS AXIS OFR A $Nd_2Fe_{14}B$ SINGLE CRYSTAL

The studies related above and others works in literature show up that "anisotropic magnets" like the die upset ones, are by far better than the "isotropic" ones. For example, an isotropic magnet (the magnetization axis for each grain is isotopically distributed) has an maximum energy product $(BH)_{max} = 8$ to 12 MGOe with a remanent induction $Br = 6$ to 8 kG ; but an anisotropic magnet (the magnetizations of all the grains are parallel) has $(BH)_{max} = 30$ to 45 MGOe with $Br = 11$ to 13kG. Therefore the interest of orientating the grains into the magnet is clear. As a contribution to the understanding of the orientation mechanism, we have undertaken to determine the axis of easy growth of a $Nd_2Fe_{14}B$ crystal.

Experiments

We used the CHZOCHRALSKI method to extract a crystal from a melt of $Nd_2Fe_{14}B$ in a crucible under Argon, with no seed and a high rotation speed. Then a "several crystal" sample is obtained (at least 100 single crystals optically detected) with a preferential growing direction. A cylinder ($l = 8mm$, $\Phi = 6mm$) was cut with the axis along the drawing direction).

X-ray measurements :

Fig. 5 shows X-ray diffractograms of the section of our sample, obtained with the Co radiation ($\lambda = 1.7903$ Å). The peaks observed are easily indexed with the well-known tetragonal structure of $Nd_2Fe_{14}B$ (Herbst 1984, Givord 1984). A comparison with a powder shows that the reflections $h^2+k^2 \ll l^2$ are very weak (ex 105) while some with $h^2+k^2 \gg l^2$ (ex 400) are very intense. Qualitatively, these results indicate that the growth axis is roughly in the basal plane of the cylinder. More quantitatively we can plot the ratio $I_{hkl}^{sample}/I_{hkl}^{powder}$ as a function of the angle Φ between the direction (hkl) and the reference axis "a" of the tetragonal structure (fig. 6). The curve obtained can be fitted to a half gaussian $\exp(-\Phi^2/2\sigma^2)$ with $\sigma = 10°$, to be compared to $\sigma = 20°$ usually observed in sintered magnets (Givord 1985) or usual SmCo magnets. Thus the orientation effect is significant and better than in usual "anisotropic" materials.

Magnetic measurements.

Magnetization measurements performed along directions in, or perpendicular to, the basal plane of the cylinder, show that the sample

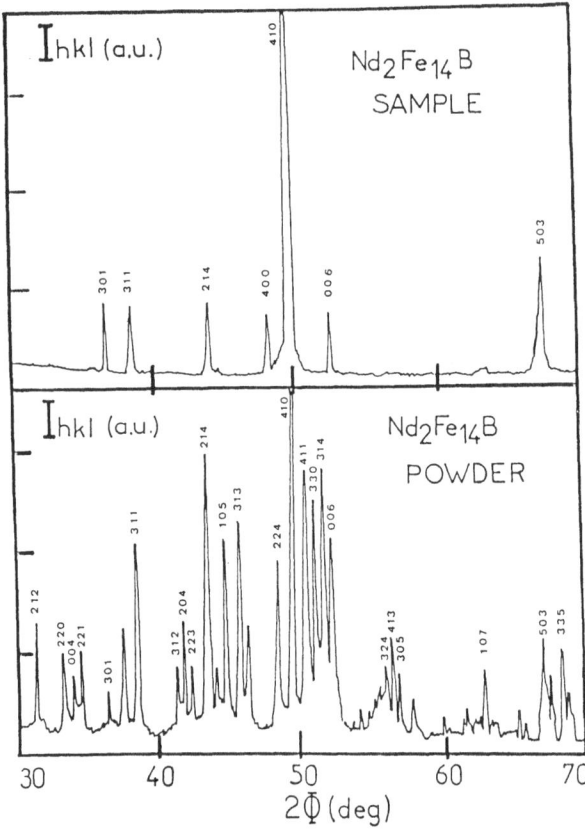

Fig. 5   X-ray   diffractogram   of   a Nd$_2$Fe$_{14}$B   "several   crystal"
compared to that one of a Nd$_2$Fe$_{14}$B powder.

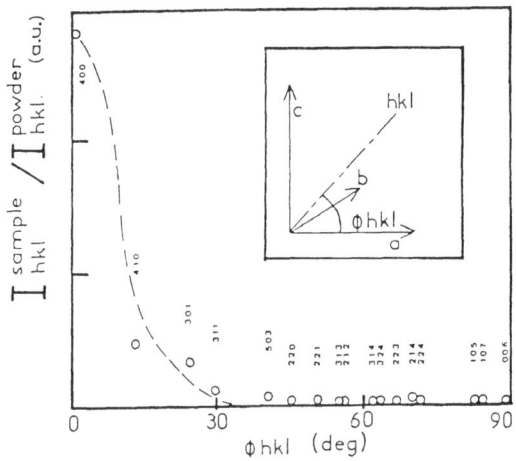

Fig. 6   Nd$_2$Fe$_{14}$B   "several   crystal" :   ratio $I_{hkl}^{sample}/I_{hkl}^{powder}$ as a
function   of   the   angle Φ between   the   [hkl]   direction   and the
reference "a" axis of the tetragonal structure.

is quite isotropic in the plane, while the magnetization curve (corrected from the demagnetizing factor) is much lower in the perpendicular direction (fig. 7). This indicates that the easy magnetization axis lies in or near the basal plane of the cylinder.

Remembering that the easy magnetisation axis of a $Nd_2 Fe_{14} B$ crystal is the c axis, we can calculate magnetisation curves for a field applied along, or perpendicular to, the cylinder axis, assuming an anisotropic distribution of C axis in the plane of the cylinder. These curves agree well with the experimental ones.

Thus X-ray and magnetization measurements give coherent results : the growth axis of a $Nd_2 Fe_{14} B$ single crystal is the "a" axis of the tetragonal structure, perpendicular to the easy magnetization axis, which is the "c" axis.

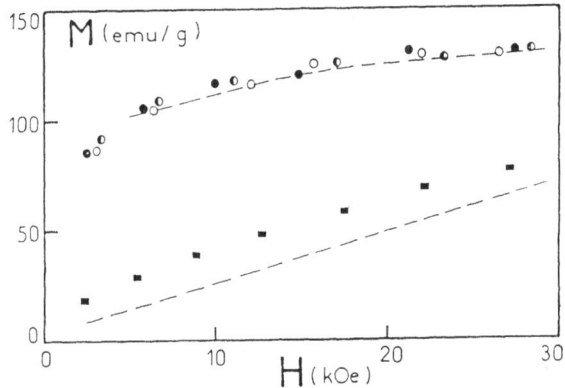

Fig. 7 Magnetization curves of a $Nd_2 Fe_{14} B$ "several crystal" measured in the plane of the cylinder "o" or perpendicular to this plane □.

Application to the structure of NeNdB permanent magnets.
Sintered magnets.

The volume contraction associated to sintering of NdFeB magnets is observed to be anisotropic : $\Delta l/l = 21$ % and 8 % along, and perpendicular to, the preferred axis of the magnet. In the light of the result above, we can consider that the single crystal grains have their c axis parallel to the preferred axis of the magnet. Sintering favours a growth of the magnet along a "a" axis, which is perpendicular to the preferred axis of the magnet. As a result, one observes a lower contraction perpendicular to this preferred axis.

Melt spun magnets (Tenaud 1987)

The effect of die upsetting (for type 3 magnets) can be understood as follows : the high temperature treatment favours the grain growth. But due to pressing, only the grains with "a" and "b" axis perpendicular to the press direction are allowed to grow, at the expense of the others.

Fig. 8 shows the X-ray diagram of the die-upset alloy $Nd_{13.5}Fe_{81}B_{5.5}$ (+ Al 1%), which is the best of the "type-3 magnets" that we have obtained. The plot of the ratio $I_{hkl}^{sample}/I_{hkl}^{powder}$ as a function of the angle Φ between the direction [hkl] and the reference axis "c" of the tetragonal structure is represented fig. 9. From this curve we can deduce

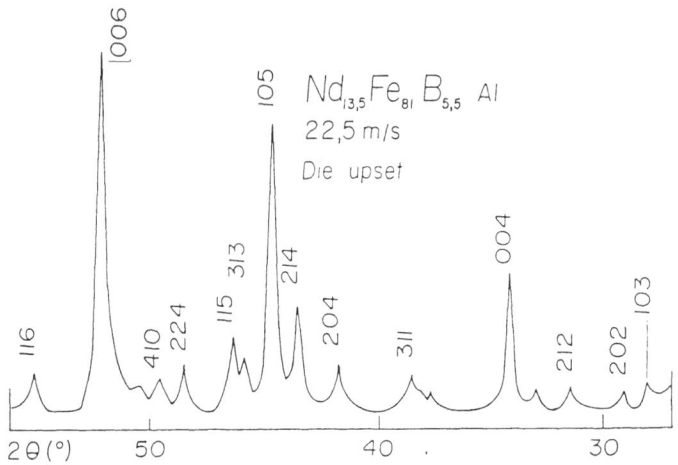

Fig. 8   X-ray diffractogram of the best "type 3" material.

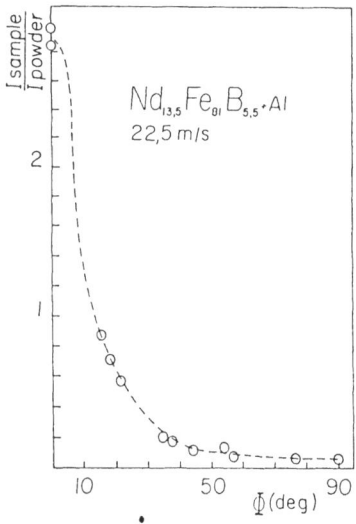

Fig. 9   Ratio $I_{hkl}^{sample}/I_{hkl}^{powder}$ as a function of the angle Φ between the hkl direction and the reference axis of the tetragonal structure, for the best "type 3" material.

several data. Assuming that I^{sample}/I^{powder} is made of two parts, the first one corresponding to the grains randomly distributed, the second one corresponding to the grains preferentially oriented with a gaussian distribution of angles, we can write :

$$f(\Phi) = \frac{I^s}{I^P} = A + B \exp\left(-\frac{\Phi^2}{2\sigma_o^2}\right)$$

A is experimentally drawn from the curve. B and σ_o are deduced from the straight line obtained by plotting $\text{Log}\left(\frac{I^s}{I^P} - A\right)$ as a function of Φ^2.

Then the number of grains "at random" is

$$N_1 = \int_o^{\frac{\pi}{2}} A \sin \Phi$$

and the number of grains "oriented" is

$$N_2 = \int_o^{\frac{\pi}{2}} B \exp\left(\frac{-\Phi^2}{2\sigma_o^2}\right) \sin \Phi \, d\Phi,$$

so that the ratio of grains "at random" is $x = \dfrac{N_1}{N_1 + N_2}$

We obtain here $x = 0.35$ and $\sigma_o = 17.5°$.

Br can be calculated remembering that $Br_{max} = 16$ kG, and $Br_i = \dfrac{Br_{max}}{2} = 8kG$ $\dfrac{Br}{2}max = 8kG$ for grains "at random". Then :

$$Br = x \, Br_i + (1-x) \frac{\displaystyle\int_o^{\frac{\pi}{2}} \cos \Phi \sin \Phi \exp\left(\frac{-\Phi^2}{2\sigma_o^2}\right) d\Phi}{\displaystyle\int_o^{\frac{\pi}{2}} \sin \Phi \exp\left(\frac{-\Phi^2}{2\sigma_o^2}\right) d\Phi} \cdot Br_{max}$$

The calculated value $Br = 12.3$ kG fits very well to the experimental one $Br = 12.7$ kG.

Therefore it appears that all the grains are not oriented, and it would be profitable to improve the orientation. That is probably possible by optimizing temperature and pressure. The grain size obtained after the die upsetting treatment defined here is $(2000 \times 2000 \times 300) \text{Å}^3$. In the opposite to the sintered magnets, there is no change of density during this operation.

REFERENCES

Croat, J.J., Herbst, J.F., Lee R.W., Pinkerton F.E. 1984. High energy product NdFeB permanent magnets. Appl. Phys. Lett. $\underline{44}$, 148.

General Motors Corporation, European Patent application 0125752 A2, 1984. Bonded Rare Earth-Iron magnets.

Givord, D., Li, H.S., Moreau, J.M. 1984. Magnetic properties and crystal structure of $Nd_2Fe_{14}B$. Solid State Com. $\underline{50}$, 497.

Givord, D., Lienard, A., Perrier de la Bathie, R., Tenaud, P., Viadieu, T. 1985. J. Physique C6, 313.

Herbst J.F., Croat, J.J., Yelon, W.B. 1985. Structural and magnetic properties of $N_2Fe_{14}B$. J. Appl. Phys. $\underline{57}$, 4086.

Herbst, J.F., Croat, J.J., Pinkerton, F.E., Yelon, W.B. 1984. Phys. Rev. $\underline{B29}$, 4176.

Mishra, R.K. 1985. Research publications General Motors Research Laboratories. Microstructure of melt spun NdFeB Magnequench Magnets.

Sagawa, M., Fujimara, S., Yamamoto, H., Matsuura Y., Hiraga, K. 1984. IEEE Trans. Magn. Mag. $\underline{20}$, 1584.

Sumitomo Special Metals Co, Osaka. European patent application 0101552 A2, 1983. Magnetic Materials and Permanent Magnets.

Tenaud, P., Chamberod, A., Vanoni F. 1987. Texture in Nd-Fe-B magnets analysed on the basis of the determination of $Nd_2Fe_{14}B$ single crystals easy growth axis. Solid State Com. $\underline{63}$, 303.

COERCIVITY DISTRIBUTION AND MAGNETIC STABILITY
OF Nd-Fe-B ALLOYS

S.Stieler[1], K.Zeibig, M.Kemper[2], W.Kurtz, A.Höhler, R.Beranek[3], C.Heiden

University of Gießen, 6300 Gießen, FRG

ABSTRACT

The investigations were carried out towards two aims: 1. Analysis of magnetic stability of rare earth permanent magnets and 2. study of the magnetization behaviour of small Nd-Fe-B particles.

1. Thermal aftereffect and magnetization creep during repeated minor hysteresis cycles were studied on samples prepared from sintered magnets of $SmCo_5$, Sm_2Co_{17} and $Nd_2Fe_{14}B$. The temperature dependence of the magnetic viscosity constant S_v and the field dependence of the creep effect were investigated at different operating points in the vicinity of remanent magnetization.

2. While small particles from crushed Nd-Fe-B magnets (sintered and pressed melt spun) show strong variation in their magnetic properties, measurements on the starting material, the melt spun ribbons, provide an explanation for this complicated behaviour.

PART ONE : MAGNETIC STABILITY

INTRODUCTION

The first part of our work is concerned with the magnetic stability of Nd-Fe-B (and -for comparison- that of Sm-Co) magnets with respect to small changes ΔH of the external field ($\Delta H \approx 10 \ldots 500\,A/cm \ll H_c$). Irreversible variations of the sample magnetization due to such small field changes influence the applicability of a magnet, for instance in precision measuring equipment. Two effects can be observed: magnetization creep and thermal aftereffect.

MATERIALS AND METHODS

The samples were spheres of 2.9mm diameter, grinded from commercial sintered magnets (Sumitomo). After being magnetized, the sample is in the sheared remanent state M_r. With a counter field pulse an operating point M_{op} was adjusted, lying on the shearing line $\mu_o \cdot H_{int} = -M/3$. M_{op} was varied between $+0.95\,M_r$ and $-0.95\,M_r$. Starting from this operating point the creep and aftereffect measurements were done.

---

present adress:
[1] Hoechst AG, D-6000 Frankfurt
[2] Thyssen Edelstahlwerke AG, Magnetfabrik Dortmund, D-4600 Dortmund
[3] Siemens AG, D-8000 München

Figure 1 shows the adjustment of the working point and both types of instabilities: creep and aftereffect. The magnetisation creep is defined to be the magnetization change at the endpoints of asymmetrical hysteresis cycles as function of the number n of complete cycles. The magnetization M_n at each endpoint behaves as $M_n = M_1 + M^* \cdot x_n$, where M^* is the creep constant and x_n the creep parameter which approximately increases with n as $\sqrt{\ln n}$.

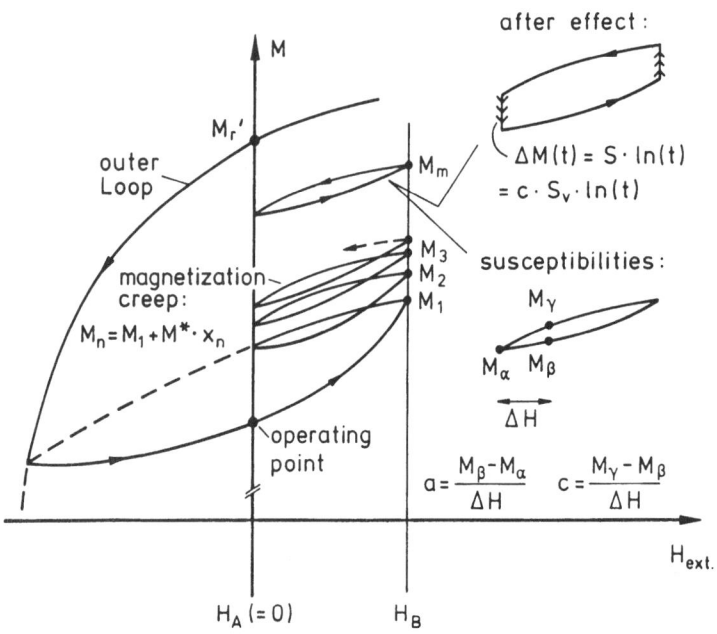

Figure 1 : Starting from a sheared remanent state M_r, the operating point is adjusted by a counter field pulse. Periodic field sweeps between H_A and H_B cause the magnetization to creep upwards. After a large number of cycles the minor loop becomes quasistationary (endpoint M_m). At such a loop a) the aftereffect can be measured during waiting periods at the endpoints by observing the time dependence of the magnetization. The corresponding reversible susceptibility a and the irreversible susceptibility c can be obtained by determining the indicated quantities (ΔH is small).

Results for $SmCo_5$, Sm_2Co_{17} and $Nd_2Fe_{14}B$ with $M_{op} \approx 0.9 \cdot M_r$ and T=300K were given at the Amsterdam meeting in March '86 [1]. Experiments were now extended to other operating points between -0.95 and 0.95 [2].

For an unsaturated ferromagnet a jump of the external field produces a corresponding jump of the magnetization. Aftereffect is the change M(t) of the magnetization with time (at constant external field), following this magnetization jump. After Néel [3] and Street and Woolley [4] the time dependence of M within a limited time intervall can be described by $M(t) = S \cdot \ln(t) = c \cdot S_v \cdot \ln(t)$. S as well as S_v are called aftereffect constants; c is the irreversible susceptibily, which is defined in figure 1. We have measured the aftereffect of SmCo and NdFeB samples at the endpoints of small minor hysteresis cycles, after the magnetisation creep had become negligible by applying 300 field sweeps (c.f. fig.1, loop no. m). The field excursion was about 90 A/cm, the temperature was varied between 4K and 300K and M_{op}/M_r between -0.95 and 0.95. The observed magnetization changes due to both effects are small and of the order of 10ppm of M_r ($\approx 10\mu T$), so that a very sensitive magnetometer is needed for the measurements. We used a microprocessor controlled high slew rate and low drift microwave SQUID system [5].

The specimen was inserted into a special sample chamber, which is surrounded by liquid helium, required for SQUID operation. Specimen temperature can be varied between 4K and 300K by electric heating. The temperature dependence of the magnetization creep requires a precision temperature controller: at T=300K for instance, temperature has to be constant within about 1mK to restrict the drift of the magnetization to 1ppm ($\approx 1\mu T$) [6].

RESULTS

The main results of the aftereffect measurements at $M_{op} \approx 0.9 \cdot M_r$ are (c.f. fig.2):

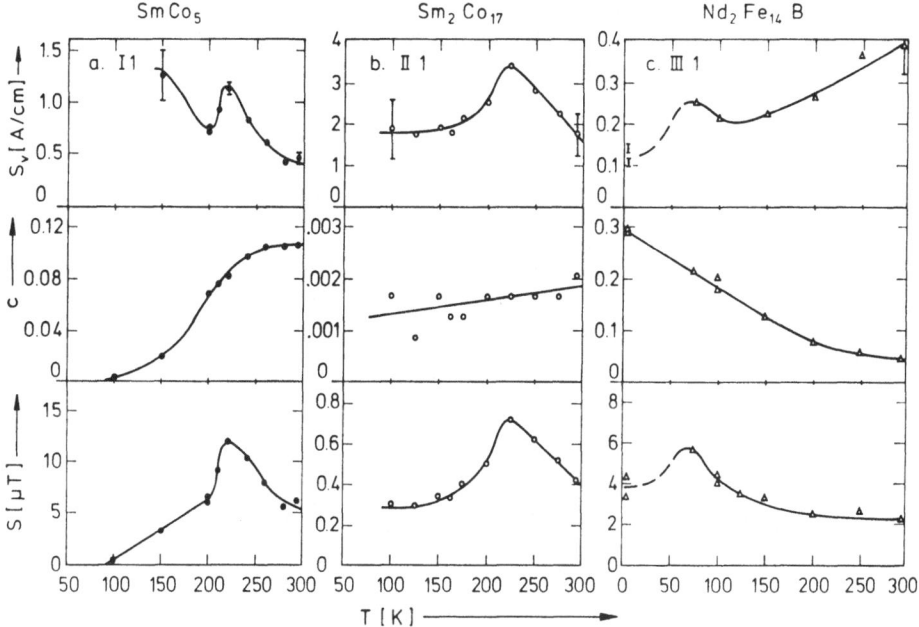

Figure 2 : Temperature variation of the aftereffect constants S and $S_v = S/c$ and of the irrevesible susceptibility c for $SmCo_5$, Sm_2Co_{17} and $Nd_2Fe_{14}B$ samples. Field excursion H = 90 A/cm, operating point $M_{op} \approx 0.9 \cdot M_r$.

a.) At room temperature the aftereffect (described by the constant S) is largest for $SmCo_5$, smaller for $Nd_2Fe_{14}B$ and smallest for Sm_2Co_{17}. This result has been confirmed with other samples.

b.) The aftereffect constant S_v at room temperature is approximately equal for the nucleation type magnets NdFeB and $SmCo_5$; it is about ten times larger for the pinning type magnet Sm_2Co_{17}.

c.) The room temperature values for the irreversible susceptibility c are of the same order of magnitude for the nucleation type magnets, but almost two orders of magnitude larger than for Sm_2Co_{17}.

d.) The temperature variation of S and S_v is complicated for all three types of magnets and not in accordance with any existing theory of the thermal aftereffect.

e.) For $SmCo_5$ and Sm_2Co_{17} the irreversible susceptibility c increases with temperature as expected, for NdFeB c decreases with T.

Figure 3 shows the dependence of creep, aftereffect and the susceptibilities a and c on the operating point for NdFeB.

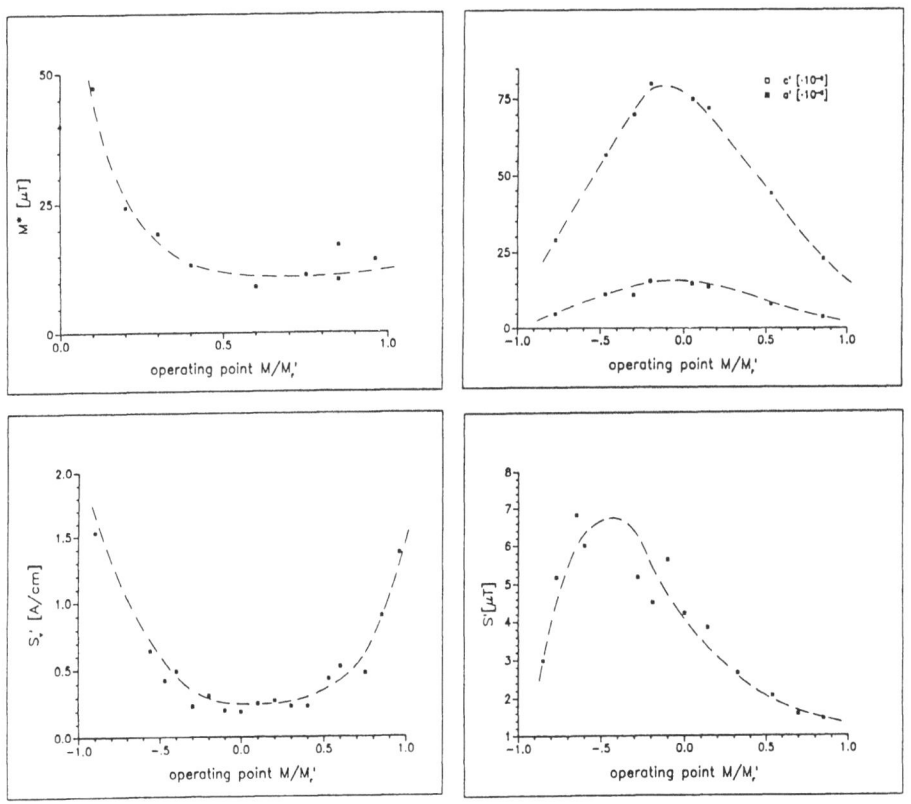

Figure 3 : Dependence of creep constant M*, aftereffect constants S and S_v and of the susceptibilities a and c on the operating point M_{op} for a sintered NdFeB sample. All measurements were performed at T = 300K with a field excursion of 90 Å/cm.

a.) Magnetization creep, aftereffect and the susceptibilities decrease, as the operating point approaches the sheared remanence, because the number of reversed domains decreases.

b.) The aftereffect constant S_v increases with M_{op}. This might be explained by the decrease of the reversible susceptibility a, because S_v after Street and Woolley [4] is proportional to $1/v$, where v is the average volume, that a wall passes during its reversible movement before each Barkhausen jump.

PART TWO : COERCIVITY DISTRIBUTION OF SMALL NdFeB-PARTICLES

INTRODUCTION

In the second part of our work we investigated the behaviour of small NdFeB particles mainly from melt spun ribbons, the starting material for bonded magnets. Homogeneity of the starting material is the most important condition for producing good magnets. Inhomogenious material can lead to a pronounced degradation of the magnet and also to wrong conclusions regarding the magnetic behaviour.

MATERIALS AND METHODS

We studied the magnetization behaviour of small NdFeB particles with dimensions of the order of $10\mu m$ to $100\mu m$, prepared from crushed magnets or directly from melt spun ribbons. Magnetization curves were measured using a SQUID based sensitive vibrating sample magnetometer in fields up to 5 Tesla [7].
The small particles ($50\mu m$ average diameter) were prepared
a.) by crushing a sintered magnet (Sumitomo),
b.) by crushing a pressed magnet made from $Fe_{82}Nd_{13}B_5$ melt spun ribbons,
c.) by cutting melt spun ribbons of $Fe_{82.3}Nd_{11.8}B_{5.9}Al_1$ compound.
Crushing, particle selection and mounting in the sample holder was done under helium atmosphere to avoid oxidation.

RESULTS

Concentrating on coercivity we obtained the following results:
Sample type a): The average coercivity is reduced by a factor of ca. 4 with regard to a sintered bulk specimen of 3mm diameter. There is no much spread in the H_c-values of individual particles ($H_{c\,max} / H_{c\,min} \approx 1.6$), see table 1.

Table 1 : Distribution of coercivity $\mu_o \cdot H_c$ (T) of sintered NdFeB particles

No.	3ooK	15OK	4.2K
1	0.231	0.460	0.543
2	0.221	0.494	1.150
3	0.221	0.494	0.970
4	0.193	0.425	1.120
5	0.198	0.488	1.200
6	0.198	0.485	1.250
7	0.159	0.459	1.090
bulk	0.956	3.600	----

Sample type b): The average coercivity is reduced by a factor of ca. 1.4 with regard to the pressed bulk specimen. There is a considerable spread in the H_c-values of individual particles ($H_{c\,max}$ / $H_{c\,min} \approx 36$, measured on 20 samples), see table 2.

Table 2 : Distribution of coercivity $\mu_o \cdot H_c$ (T) in small pressed melt spun samples

No.	300K	150K	4.2K
1	0.918	1.090	0.921
2	0.206	0.202	0.350
3	0.680	0.685	1.000
4	0.189	0.261	0.325
5	1.080	1.270	0.923
6	0.646	0.563	0.584
7	0.194	0.271	0.220
8	0.940	1.300	1.150
9	0.660	0.968	0.904
10	0.110	0.131	0.151
11	0.275	0.521	0.354
12	0.030	0.031	0.067
13	0.677	0.791	0.821
14	0.489	0.564	0.614
15	0.485	0.618	0.656
16	0.897	1.080	0.940
17	0.049	0.111	0.249
18	0.276	0.471	0.337
19	0.100	0.119	0.116
20	0.816	0.838	0.849

Clearly the particles from the sintered magnet are more homogeneious than those from the pressed unsintered specimen. The reduction of the average coercivity may be attributed to surface effects.

Sample type c): Melt spun ribbons are the starting material for samples like type b). Homogeneity investigations (more than 150 samples of one charge have been measured) should give an explanation for the complicated behaviour shown in table 2.

The investigations were performed on melt spun ribbon-pieces of one charge with a maximal size of 2mm x 12mm and a thickness of 40μm in average. To obtain a first information on the average H_c-distribution of the whole charge we prepared 56 samples (platelets of 1mm x 1mm) each taken from a different ribbon piece. The corresponding H_c-distribution, which exhibits two accumulations of H_c-values at low (0.2 Tesla) and high (0.7 ... 1.5 Tesla) coercivity is given in fig. 4.

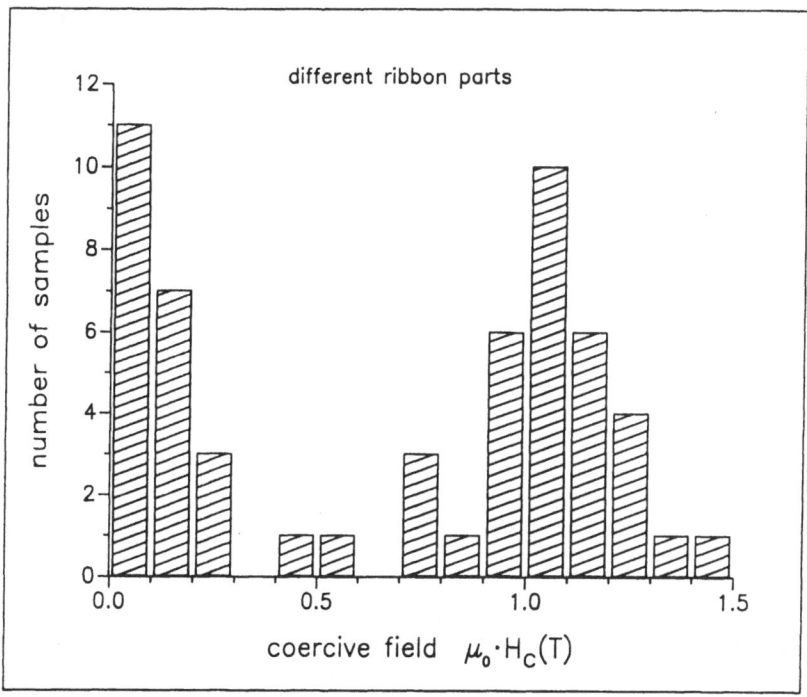

Figure 4 : Coercivity distribution of 56 samples each taken from different ribbon parts

The reason for this behaviour: Some samples of the investigated material consists of soft- and hardmagnetic volume fractions. Typical magnetization curves are shown in figure 5. a)-c). Figure 5.a) shows an ideal magnetization curve of a sample consisting only of hardmagnetic volume parts (V_H), figure 5.b) and 5.c) show samples with softmagnetic parts (V_S).

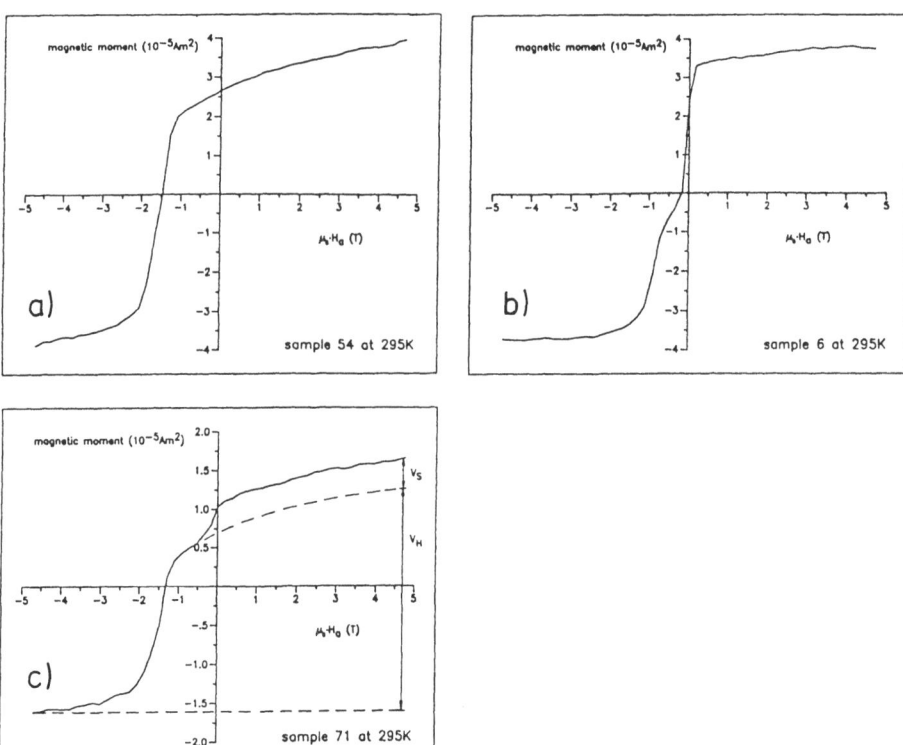

Figure 5 : Typical magnetization curves of samples each taken from different ribbon parts.

To get an answer to the question on what local scale a noticable dispersion of coercivity can be found, we divided some ribbon-pieces into 20 ... 30 samples, noted their positions and measured the magnetization curves at room temperature.

Figure 6.a)-c) shows the corresponding H_c-distribution and the spatial arrangement of coercivity of a "good" (high H_c) ribbon (ribbon A, samples 76 to 96) and magnetization curves from neighboured samples of different H_c-cluster (H_c-values in a common intervall). All samples of this ribbon consist of pure hardmagnetic parts.

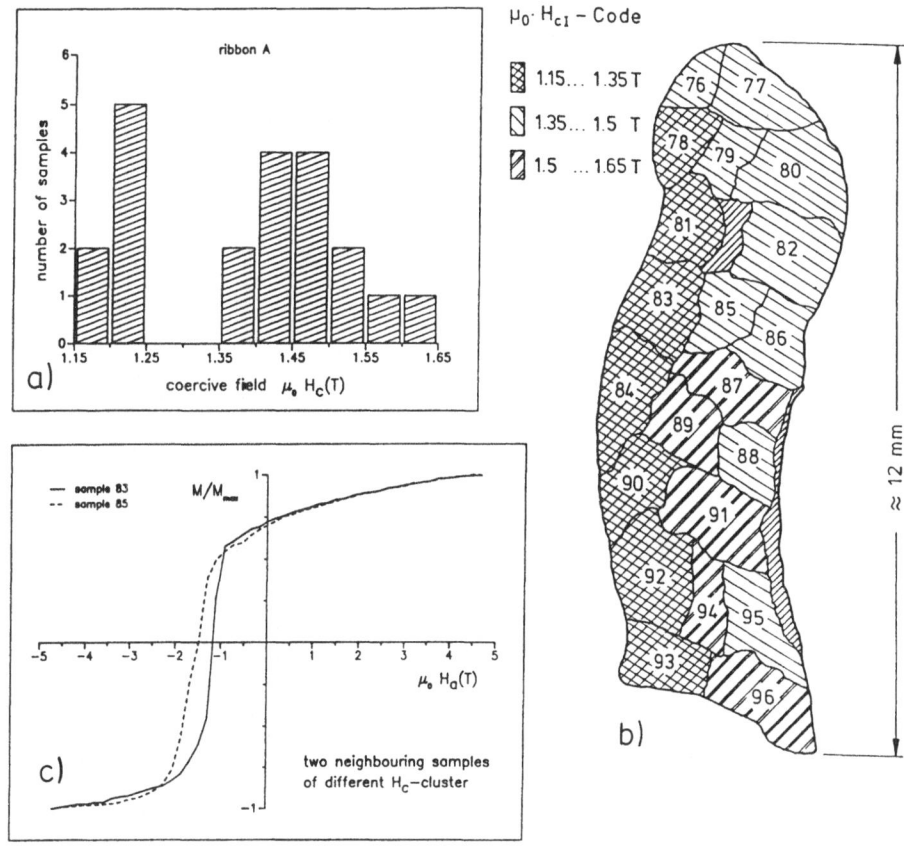

Figure 6 : a) Distribution of coercivity at 295K in ribbon A, b) Topography of corresponding H_c-values, c) Magnetization curves of two neighboured samples (No. 83 and 85) of different H_c-cluster.

Figure 7.a)-c) shows the behaviour of a "bad" ribbon (ribbon B, sample 120 to 148). The topography can be divided into one part of high (left) and one of low (right) coercivity. The magnetization curves of neighboured samples of these different H_c-cluster show a strong discrepancy.

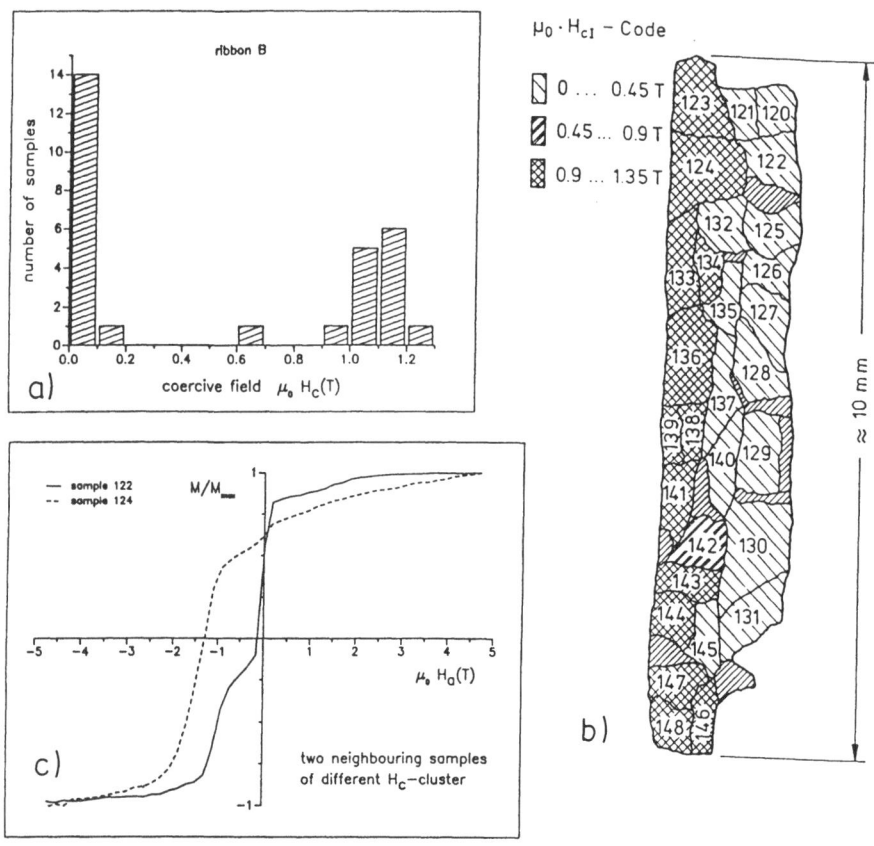

Figure 7 : a) Distribution of coercivity at 295K in ribbon B, b) Topography of corresponding H_c-values, c) Magnetization curves of two neighboured samples (No. 122 and 124) of different H_c-cluster.

These measurements can be used with advantage to select samples with ideal properties for other experiments, for example the temperature dependence of coercivity. Figure 9 shows the $H_c(T)$ dependence of three ideal hardmagnetic samples, figure 8 the magnetization curve of one of these samples at 4.2K.

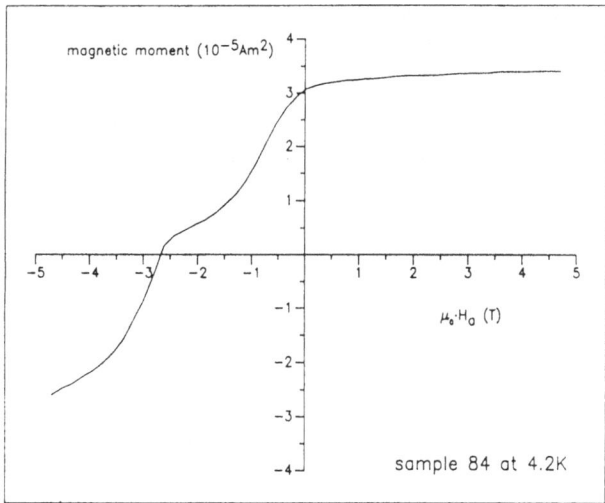

Figure 8 : Magnetization curve of a pure hardmagnetic sample at 4.2K. At room temperature there is no shoulder in the magnetization curve, which is due to the spin reorientation. The room temperature curve is similar to the one in fig.5 a).

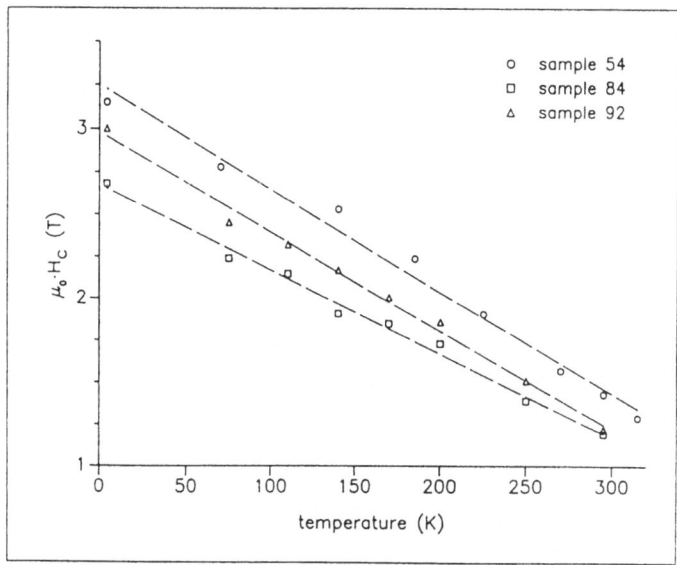

Figure 9 : Temperature dependence of coercivity of three pure hardmagnetic samples.

We found an increasing coercivity with decreasing temperature in contrast to some results in the literature given by [8] and [9].

It is clear that softmagnetic parts can reduce the magnetization at the low temperature H_c-values in addition to the spin reorientation (see fig.8); the coercivity can move to lower values.

It is possible to simulate an inhomogeneous sample by superposition of measured curves of samples with soft- and hardmagnetic parts. We assume that to first order approximation the magnetization curve of a larger sample can be constructed by linear superposition of the magnetization curves of the individual particles. The example in figure 10 shows the sum curve of 13 ideal hardmagnetic particles. Replacing selected curves by those of samples with soft magnetic parts different ratios of V_S / V_H can be created.

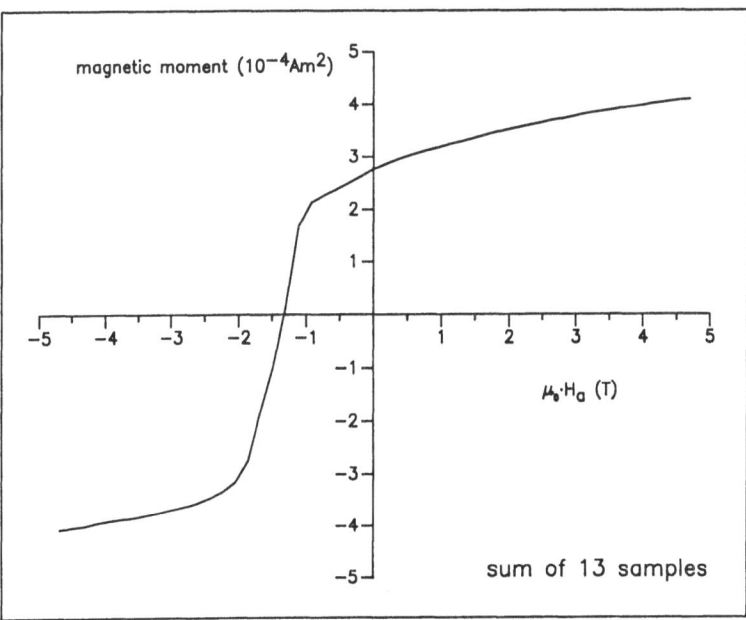

Figure 10 : Sum magnetization curve of 13 individual samples.

First information about the temperature dependence of coercivity gives the ratio $H_c(4.2K) / H_c(295K)$. Figure 11 shows this ratio for different mixtures V_S / V_H.

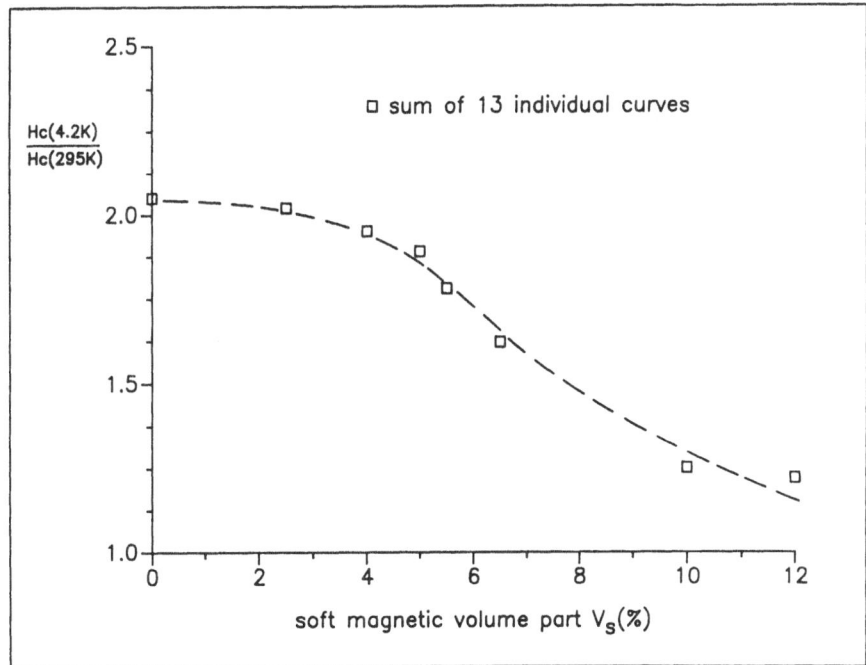

Figure 11 : Ratio H_c(4.2K) / H_c(295K) depending on soft magnetic volume parts.

With few volume percent of soft magnetic material the global H_c-value measured at low temperature can differ strongly from the local H_c of the hardmagnetic component. This is the explanation for the observed complicated behaviour of particles prepared from pressed samples given in table 2 [10].

Further development of our work will concentrate on microstructure investigations of some selected samples, to answer the question: why are "good" samples good and "bad" samples bad.

ACKNOWLEDGEMENT

We highly appriciate the collaboration with Dr.A.Chamberod, who provided the investigated samples.

REFFERENCES

[1]   CEAM Poster presentation, Amsterdam march 1986
[2]   D.Givord, C.Heiden, A.Höhler, P.Tenaud, T.Vadieu and K.Zeibig ,
      Proc.EMMA-Conference, Salford, 1987
[3]   L.Néel, J.Phys.Rad. 11, 49 (1950)
[4]   R.Street and J.Woolley, Proc. Phys. Soc. (A) 62, 562 (1949)
[5]   C.Heiden and H.Rogalla, J.M.M.M. 19, 240 (1980)

[6] S.Stieler and C.Heiden, Proc. 9. Intern.Cyrogenic Eng. Con., Y.Yasukochi, A.Nagano, eds., Butterworth, 551 (1982),

[7] R.Beranek and C.Heiden, J.M.M.M. 41, 247 (1984)

[8] K.D.Durst, H.Kronmüller and G.Schneider, Proc. 9. Int.Workshop on rare earth magnets, 209, (1987)

[9] G.Hilscher, R.Grössinger, S.Heisz, H.Sassik and G.Wiesinger, J.M.M.M. 54-57, 577 (1986)

[10] M.Kemper, W.Kurtz and C.Heiden, to be published

SUMMARY OF DISCUSSIONS :
MICROSTRUCTURE AND COERCIVITY

J. Fidler, P. Grundy

The main themes of the session were the origins of coercivity in Nd-Fe-B materials and studies of the microstructure using electron microscopy.

Both Givord (C.N.R.S., Grenoble) and Martinek (M.P.I., Stuttgart) presented results from detailed theoretical and experimental investigations of coercivity. Angular dependence and temperature dependence of the intrinsic coercivity were compared with models derived from magnetization reversal considerations. In general, pure pinning or nucleation modes do not agree with experiments and, as yet, conditions in real materials, such as grain alignment effects and microstructural variations complicate such a comparison. However, the investigation of viscosity effects is particularly rewarding and Givord showed the importance of the concept of activation volume in defining the initiation of magnetization reversal by a nucleus which is of the order of the domain wall width in size.

Corner presented details of the temperature dependence of magnetization and anisotropy of several $R_2Fe_{14}B$ compounds. These results extended considerably the few and limited measurements reported in the literature. Investigations of domain structure in these materials also gave values of domain wall energy and width in good agreement with other investigations. The possible inhomogeneities in "real" NdFeB materials were revealed by sensitive magnetic measurements of small regions of melt-spun ribbons. Heiden compared the demagnetization curves of such specimens and showed that wide variations in coercivity can be obtained, even from adjacent regions.

The complications and variations in the microstructure of materials for magnets were highlighted in the reports on electron microscope investigations. Grundy discussed the various phases present in sintered magnets, and showed that certain additions to the ternary compositions can produce precipitation effects. As yet, evidence for such a contribution to coercivity in sintered materials is uncertain, but large increases in intrinsic coercivity of melt-spun ribbons containing niobium can be associated with the small precipitates, similar in size to the domain wall width, observed in underquenched materials. Hadjipanayis and Fidler presented detailed results on the microstructure of sintered and melt-spun samples. Hadjipanayis discussed the importance of microstructure for coercivity and showed evidence, by Lorentz microscopy, of apparent pinning of domain walls at intergranular phases and grain boundaries. The absence or presence of coercivity in melt-spun materials with similar microstructures is a particular puzzle. Fidler showed that a wide variety of compositions can be observed in grain boundary phases and discussed the role of these phases (including oxides) in providing coercivity.

The presentations in this session, and the subsequent discussion, underscored the very strong relationship between microstructure and coercivity. But real materials are complex and model structures are simple. Microscopic investigations focus on a localised area while magnetic measurements give the average properties. Nevertheless further investigations are essential if the structure of real materials is to be tailored to provide optimium properties.

SECTION III

— MAGNET PROCESSING

CHAPTER 6

INGOT MATERIALS :

CHARACTERISATION,

POWDER PROCESSING

AND MELT SPINNING

CAST RARE EARTH ALLOYS FOR MAGNETIC

MATERIALS RESEARCH

D.Kennedy, D.W.A.Murphy

Rare Earth Products
Waterloo Road,
Widnes,
Cheshire.
WA8 0QH
U.K.

ABSTRACT

The materials cast, analysed and supplied to research groups forming the Concerted European Action on Magnets (CEAM) are described in this paper. In the early stages of the research two compositions were selected by the processing group, these were supplied in quantities sufficient to manufacture magnets for comparative purposes. Alloys have also been produced under various casting conditions enabling micro-structural investigations of changing cooling rates by members of the processing group. During the course of collaboration with CEAM researchers, a multiplicity of alloy compositions have been supplied, culminating in production of samarium-iron alloys for wide distribution. The information presented in this paper covers input materials, the nominal compositions of alloys, reported analytical results on major and impurity elements, and the ingot morphologies used in the cooling rate experiments.

INTRODUCTION

The objectives of the work and products described in this paper were to supply known or new materials for research into permanent magnets. The expertise and flexibility of manufacturing were available as a resource to facilitate study of processing methods and new materials.

Prior to the commencement of the CEAM programme the production route for neodymium-boron-iron had been investigated (Kennedy, 1985) which enabled supply of research compositions at the beginning of CEAM in October 1985.

INPUT MATERIALS

Though the cast alloys are regarded as the raw material in a research and development programme on magnetic materials, the source of the constituent materials requires explanation. Iron, ferro-boron and other common or low concentration additions have low supply sensitivity or contribution to alloy cost. The rare-earth content comprises a large part of cost and requires strategic purchasing to minimise risk to production and profits. The substitution or reduction in concentration of strategically problematic constituents improves the industrial potential of the magnetic material, hence the attraction of neodymium which is six times more plentiful than samarium, and iron which is more abundant than cobalt.

The alloys investigated chiefly contain neodymium and samarium; as described in a previous publication (Kennedy, 1985) the abundant and freely available purity of neodymium oxide is of a 95% pure grade which is converted to a metallic form of the following specification:

Neodymium 95% (min.)

Typical Analysis	97.3%	Nd
	1.4%	Pr
	0.3%	La
	<0.1%	Sm
	0.2%	Ca
	<0.05%	Mg
	<0.01%	C
	200 ppm	Oxygen

The samarium is converted to metal through an alternative route which produces a purer metal:

Samarium 99.5 - 99.9%

Typical Analysis	99.9%	Sm
	0.025%	Nd
	0.03%	Mg
	0.025%	Ca
	0.015%	Mn
	100-200 ppm	Oxygen

These metals are melted in refractory crucibles with the necessary alloying elements. The chief elements used in the research programme have been iron and boron. The form of these metals has been electrolytic iron (min. 99.9% Fe) and ferro-boron (17-20% boron, impurities 1.5% Al and 0.5% Si; balance Fe).

ALLOY COMPOSITIONS

During the course of the programme a wide range of compositions have been supplied and are listed in Table 1.

Materials of particular note in Table 1 are: BM1557 - BM1561 (controlled cooling experiments); IS259 and IS265 (used in forming magnets by commercial techniques); BM1803 - BM1810 (SmFe based alloys circulated to 10 laboratories researching new alloys).

ALLOY ANALYSIS

The principal constituents of alloys are listed in Table 2.

Impurities are introduced into alloys from the ingredient metals and are also caused in the process of melting and casting (i.e. from crucibles and furnace atmosphere). On some occasions purification of a melt is effected, as for NdFeB where the Ca or Mg in the primary Nd(95%) is driven from the liquid. The impurities present in the NdFeB alloys are typically:

Al 0.1% - 0.15%; Ca <0.05%; Mg <0.05%

The principle impurity, aluminium, is derived from the ferro-boron additive, the aluminium impurity level is directly related to the quantity of boron added in a ratio of approximately 10 : 1.

The oxygen content of the cast alloys has been determined by vacuum fusion technique for a number of batches:

	e.g.	BM1822	100 ppm
		BM1823	508 ppm
		BM1824	120 ppm
		BM1825	100 ppm
		BM1826	60 ppm

mean value 148 ppm (wt) oxygen.

In these results the high value would appear erroneous, possibly due to analysis of a sample with surface oxidation or oxide inclusions. The validity of inclusion of this result is debatable, as in industrial terms the entire ingot will be crushed and this oxygen contained will be incorporated in the magnet product. In research terms the general level of 100 ppm indicates that oxygen has low solubility in rare earth transition metal melts and that oxides are immiscible. In practice, oxides and other dross tend to deposit on crucible walls though some inclusions from oxide floating on the liquid alloy can become incorporated in the cast ingot.

Following observation by some workers, of high concentrations of halides (especially chlorine) in rare earth rich areas of the microstructure, a test for halide content has been made. The technique used was a turbidity test where a substantial sample of alloy (greater than 10 grams) is dissolved in nitric acid, addition of silver nitrate will cause precipitation with halides (Cl, Br, I) excluding fluorine. The results of the analysis showed much less than 50 ppm of halides. This result and results produced by other groups indicate that halides (incl. fluorine) are not present in the alloys described in this report. Products from other suppliers are known to contain chlorides derived from alternative production routes, though it is also possible that during metallographic preparation the samples have come into contact with halogenated hydrocarbons which have reacted with the rare earth content of the castings. It is known that rare earth alloys can react explosively with halogenated hydrocarbons, this hazard is worth repeating to researchers and industry working with chemicals normally considered inert.

The analytical methods used to determine compositions and impurity levels were:

Element	Method	Accuracy
Total Rare Earth (TRE) (used for alloys containing Nd 95%)	Gravimetric	1% relative (i.e. Nd nominally 30% then ± 0.3%)
Rare earth mixtures (e.g. Nd/Dy)	XRF	1% relative
Samarium	Gravimetric	1% relative
Boron	Inductively Coupled Plasma	± 0.1%
Iron (major element)	Gravimetric	Normally quoted as balance.
Other transition metals (and silicon)	Techniques as applicable	± 0.1% for minor additions.
Al, Ca, Mg	Atomic Absorption	± 0.01%
Oxygen	Vacuum Fusion	Variable

CONTROLLED COOLING

In association with a group of collaborators a set of castings were produced for microstructural investigation of the effects of cooling rate. Five castings were made, each of nominal composition: Nd 35% (wt)

Fe 63.7%

B   1.3%

In each case the melts were soaked at 1420°C and poured at 1400°C under the following conditions:

	BM1559	15mm thick water-cooled copper mould
Slower cooling rates	BM1557	40mm thick water-cooled copper mould
	BM1560	30mm thick water-cooled copper mould
	BM1558	100mm dia. copper mould (uncooled)
	BM1561	Crucible cooled, 100mm dia., hemispherical base.

Fellow workers have shown coarsening of structures and a tendency to allow precipitation of primary iron with slower cooling rates. Faster cooling rates have suppressed primary iron which can cause problems in subsequent processing.

SAMARIUM-IRON-ALLOYS

Following a presentation made to CEAM and a publication (Buschow 1987), interest was expressed by members of CEAM for a set of Sm-Fe alloys with some substitution of Fe by vanadium, silicon and titanium. To help understand the microstructure and to predict melting behaviour, phase diagrams were studied (figs.1-4) (Smithells 1978). During melting the silicon containing and titanium containing alloys were formed according to the chemical compositions required, but considerable difficulty was experienced with the vanadium containing alloy which did not easily dissolve vanadium, consequently these alloy products are low in Sm and V.

The alloys were supplied in pairs, nominally stoichiometric and a second composition enriched with 3 wt% Sm. The purpose of the enrichment is to counter the loss of Sm due to oxidation in the course of magnet preparation by fine milling and sintering.

Preliminary metallographic investigations have been made of three of these alloys:

BM1803, $SmFe_{10}V_2$ as illustrated in fig.5 (50 x mag.)
and fig.6 (400 x mag.)

BM1809, $SmFe_{10}Si_2$ as illustrated in fig.7 (50 x mag.)
and fig.8 (400 x mag.)

BM1810, $SmFe_{10}Si_2$ (+ 3wt% Sm) as illustrated in fig.9 (50 x mag.)
and fig.10 (400 x mag.)

Variation in microstructure throughout the ingots cast has not been studied and may be considerable, therefore interpretation of the micrographs should be made with due consideration for potentially non-representative samples.

The most striking feature of the microstructures is the comparison between BM1809 and BM1810, where the increased Sm content appears to have drastically changed the form of the phases.

DISCUSSION

This paper describes 52 alloys supplied to more than 16 laboratories for research and development of magnetic materials. These have been characterised in terms of major components, impurities and microstructure. The results listed are essential data to be used by magnet processors or researchers into fundamental properties of materials.

The compositions were selected by research groups and have enabled refinement of the alloy production route and led to production of a new range of alloys with potential for investigation. The new $SmFe_{12}$ type alloys may not be suitable for full economic exploitation due to a magnetization lower than NdFeB, similar Curie temperature to NdFeB and, most significantly, their utilisation of comparatively rare samarium which is strategically problematic.

REFERENCES

Kennedy, D. 1985. Nd-Fe Permanent Magnets – Their Present and Future Applications. I.V.Mitchell, Ed. Elsevier, London.
Buschow, K.H.J. et al. 1988. Magnetic Properties of Ternary Fe-rich Rare Earth Intermetallic Compounds. IEEE Transactions on Magnetics.
Smithells, C.J. 1978. Metals Reference Book, 5th Edition, Butterworths, London.

TABLE 1. Alloys supplied to research groups.

Batch Number		Nominal Composition	Remarks
BM1419		$Nd_{15}Fe_{77}B_8$	
BM1482		$Nd_{13.5}Fe_{81.75}B_{4.75}$	
BM1483		$Nd_{14}Fe_{78.26}B_{7.74}$	
BM1546		$Nd_{10}Fe_{53}B_3$	
BM1557	)		
Bm1558	)		Controlled
Bm1559	)	$Nd(35\%)Fe(Bal)B(1.3\%)$	cooling
Bm1560	)		experiments
BM1561	)		
IS259		$Nd(35\%)Fe(Bal)B(1.3\%)$	Large batch
IS265		$Nd_{14.5}Dy_{1.5}Fe_{76}Nb_1B_7$	Large batch
BM1585		$La(4.66\%)Ce(11.11\%)Nd(10.23\%)$ $Fe(74\%)$	99.9% pure input
BM1630		$Nd_2Fe_{14}B$	
IS274		$Nd(35\%)Fe(Bal)B(1.3\%)$	
IS324		$Nd(35\%)Fe(Bal)B(1.3\%)$	
BM1652	)		
BM1653	)		
BM1655	)	See analysis Table 2.	Grain refining
BM1656	)		trials
BM1658		$Nd(13.1\%)La(7.4\%)Ce(7.25\%)Ni(1.25\%)$ $Fe(70.9\%)B(0.15\%)$	
BM1659		$Nd_{15}Fe_{62.5}B_{5.5}Co_{16}Al_1$	
BM1673		$Nd(35.5\%)Fe(73.2\%)B(1.3\%)$	
BM1676		$Nd_2Fe_{14}B$	
BM1677		$Nd_{15}Fe_{77}B_8$	
BM1696		$Nd_{20}B_{6.26}Fe_{73.75}$	
BM1697		$Nd_{27}B_6Fe_{67}$	
BM1698		$Nd_{13}B_{6.5}Fe_{80.5}$	
BM1699		$Nd_2Fe_{14}B$	
BM1700		$Fe_{76}Nd_{16}B_7Nb_1$	
BM1702		$Fe_{76}Nd_{14.5}B_7Dy_{1.5}Nb_1$	
BM1703		$Fe_{79.6}Nd_{13.2}Si_{1.2}B_6$	
BM1704		$Fe_{77.8}Nd_{13.9}Si_{2.3}B_6$	
IS348		$Nd_2Fe_{14}B$	
BM1718		$MM_2Fe_{14}B$	
BM1732	)		
BM1733	)		
BM1734	)	$Nd(35\%)Fe(63.7\%)B(1.3\%)$	
BM1735	)		
BM1756		$Nd(23.27\%)Fe(68.74)Ce(7.12)B(0.87)$	
BM1780		$SmFe_{10}V_2$	
BM1785		$SmFe_{11}Ti_1$	
BM1803		$SmFe_{10}V_2$	
BM1804		$SmFe_{10}V_2(+3\%Sm)$	
BM1806		$SmFe_{10.8}Ti_{1.2}$	
BM1807		$SmFe_{10.8}Ti_{1.2}(+3\%Sm)$	
BM1809		$SmFe_{10}Si_2$	
BM1810		$SmFe_{10}Si_2(+3\%Sm)$	
BM1822		$Nd_{11.77}Fe_{82.35}B_{5.88}$	
BM1823		$Nd_{10.27}Fe_{82.35}Dy_{1.5}B_{5.88}$	
BM1824		$Nd_{11.77}Fe_{80.35}Al_2B_{5.88}$	
BM1825		$Nd_{11.77}Fe_{80.35}Si_2B_{5.88}$	
BM1826		$Nd_{10.27}Fe_{80.35}Dy_{1.5}Si_2B_{5.88}$	

TABLE 2. Analytical results for alloys used in research (see table 1 for nominal compositions).

Batch No.	Nd as TRE*	Dy	Sm	La	Ce	Fe	B	Co	Al	Si	Nb	Ni	V	Ti
BM1419	33.7					Bal	1.3							
BM1482	31.8					Bal	1.3							
BM1483	28.9					Bal	0.7							
BM1546	32.8					66.7	0.6							
BM1557	34.6					64.3	1.2							
BM1558	35.0					64.1	1.2							
BM1559	35.1					64.0	1.3							
BM1560	34.6					64.0	1.2							
BM1561	34.4					64.1	1.3							
IS259	34.4					64.0	1.3							
IS265	31.0	4.0				63.8	0.9				1.4			
BM1585	10.7			4.9	10.8	Bal								
BM1630	26.5					Bal	0.9							
IS274	34.9					Bal	1.3							
IS324	34.6					Bal	1.2							
BM1652	26.9					Bal	0.1					1.1		
BM1653	26.7					Bal	0.1					1.1		
BM1655	32.7					Bal	1.3							
BM1656	32.3					Bal	1.3							Zr 1.6
BM1658	13.4			7.0	6.6	Bal	0.1					1.2		
BM1659	32.3					Bal	0.8	14.4	0.5					
BM1673	35.5					63.0	1.3							
BM1676	26.5					72.1	0.9							
BM1677	32.6					65.7	1.3							
BM1696	40.5					58.5	0.9							
BM1697	49.9					49.4	0.8							
BM1698	29.6					69.4	1.0							
BM1699	27.2					71.8								
BM1700	33.4					64.2	1.4							
BM1702	30.3	3.5				63.1	1.0				1.4			
BM1703	29.1					69.5	1.0			0.5				
BM1704	30.7					67.4	0.9			1.0				
IS348	26.8					72.1	1.0							
BM1718	(MM)26.6					72.0	0.8							
BM1732	34.8					63.6	1.2							
BM1733	35.1					63.7	1.2							
BM1734	35.3					63.3	1.2							
BM1735	35.0					62.8	1.3							
BM1756	23.2				7.2	68.6	0.8							
BM1780			17.6			76.7								5.8
BM1785			16.5			72.3							10.9	
BM1803			18.3			68.7							12.8	
BM1804			14.1			72.0							13.7	
BM1806			18.1			75.5								7.3
BM1807			20.7			72.3								7.1
BM1809			18.6			74.4				6.7				
BM1810			22.7			70.3				7.0				
BM1822	26.4					Bal	0.9							
BM1823	23.7	3.5				Bal	0.9							
BM1824	26.1					Bal	0.9			0.7				
BM1825	26.6					Bal	0.8				0.8			
BM1826	22.9	3.6				Bal	0.9				0.9			

* Nd values quoted are total rare earth (TRE) content of alloys using 95% pure Nd metal.

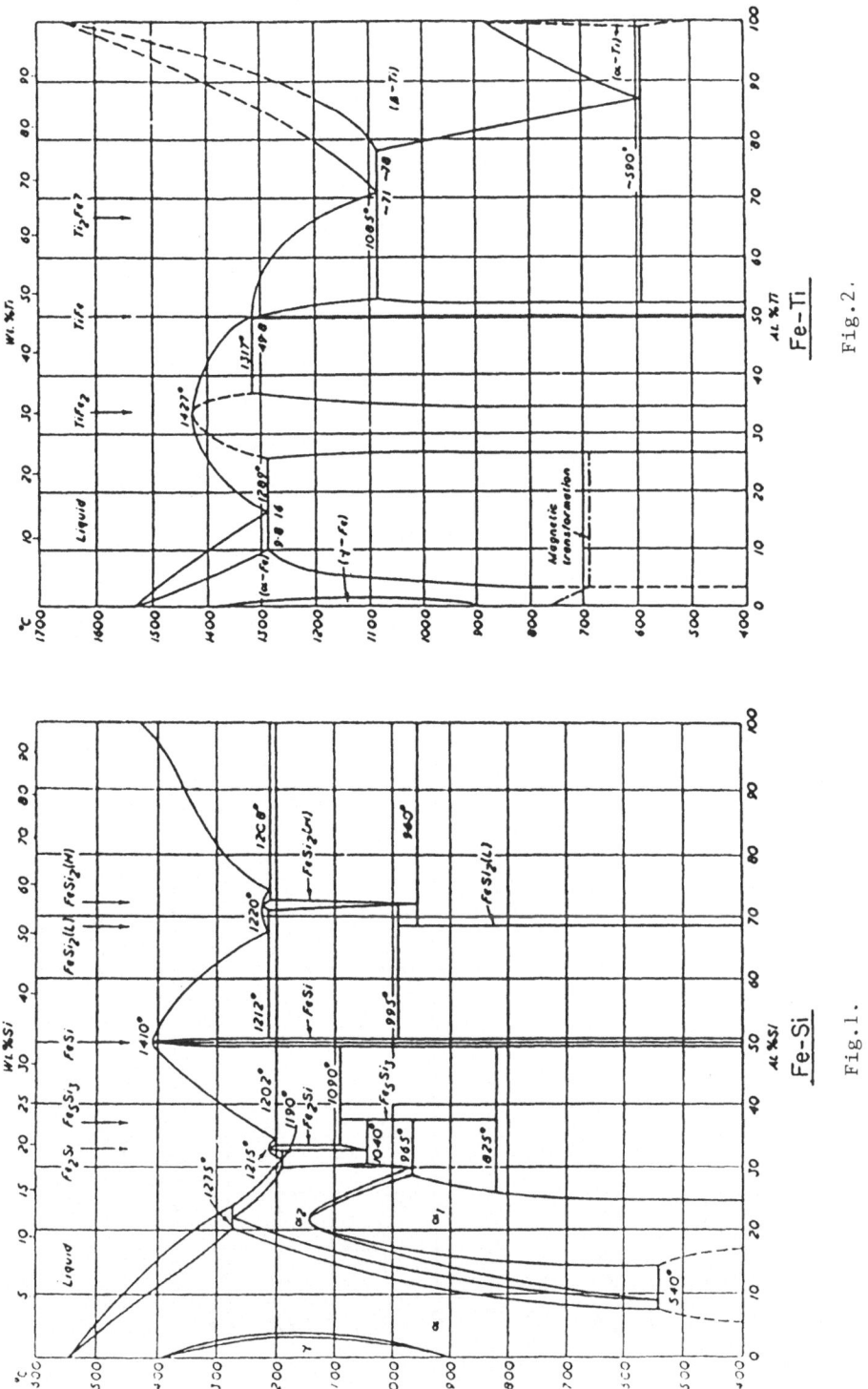

Fig. 2.
Fe-Ti

Fig. 1.
Fe-Si

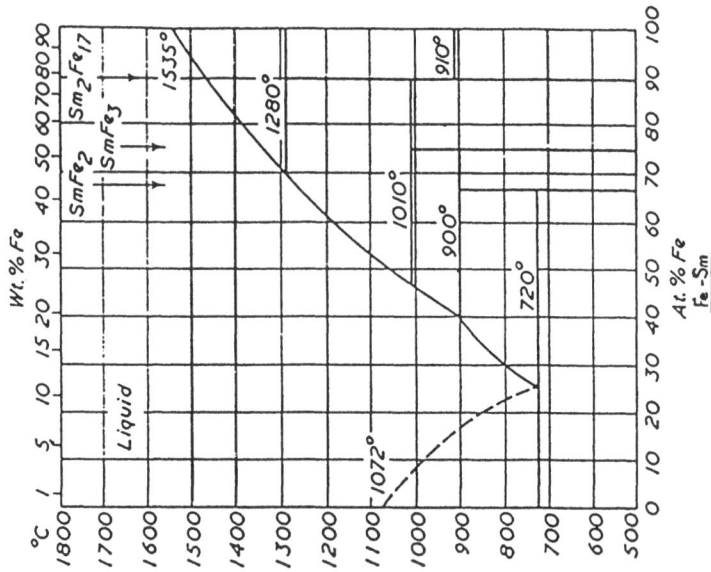

Fig.4.

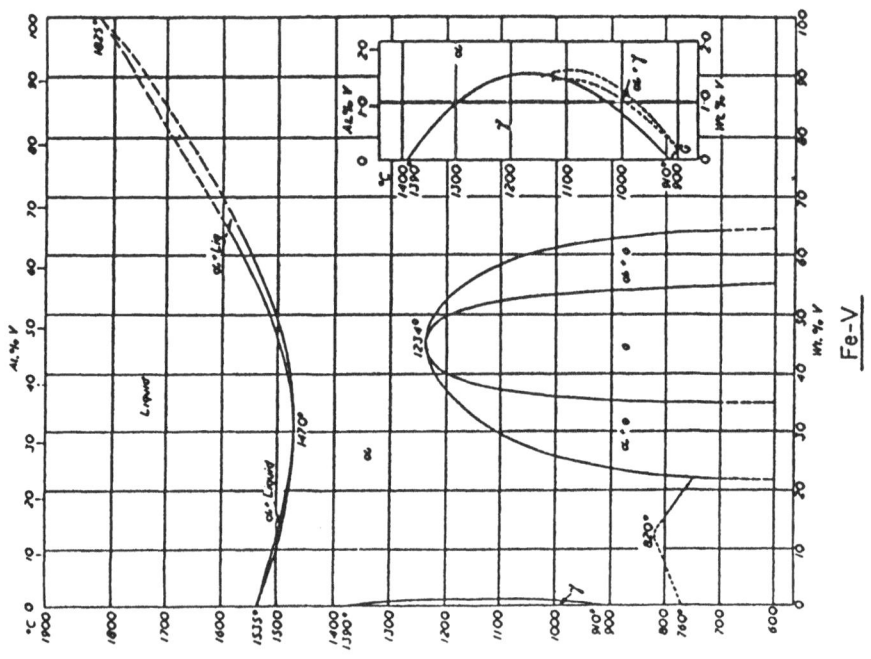

Fig.3.

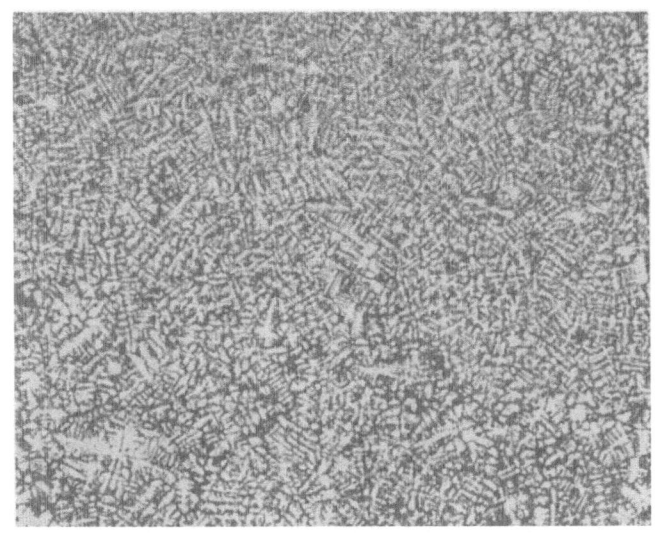

Fig.5. BM1803 (50 x mag.)

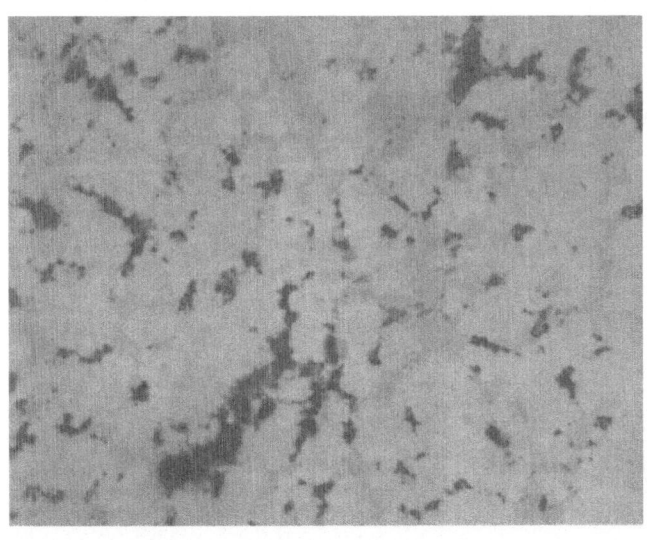

Fig.6. BM1803 (400 x mag.)

Fig.7. BM1809 (50 x mag.)

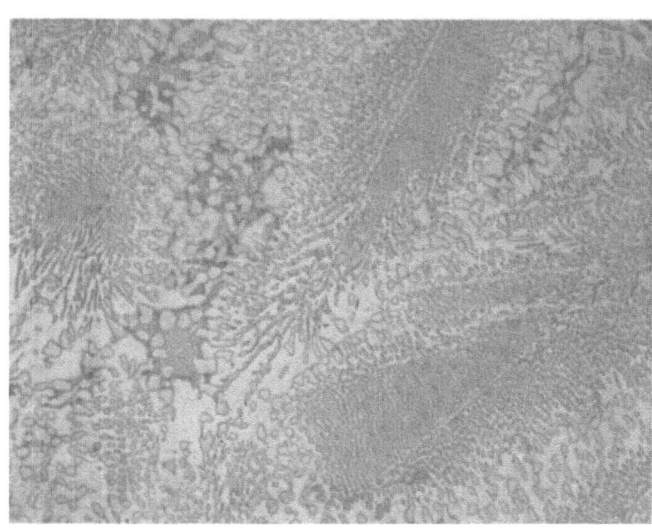

Fig.8. BM1809 (400 x mag.)

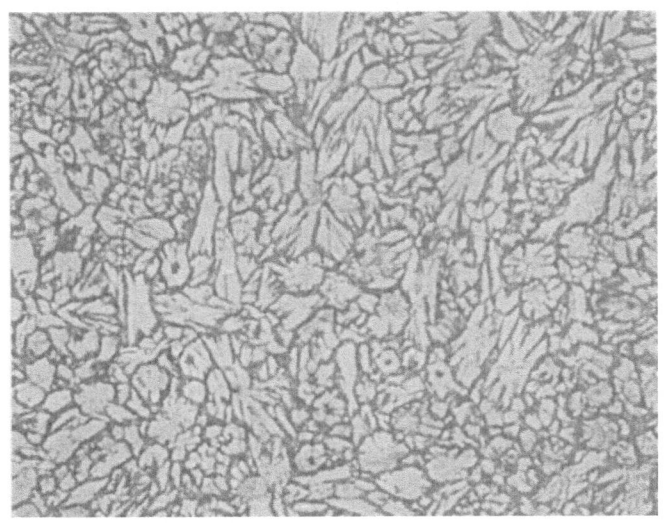

Fig.9. BM1810 (50 x mag.)

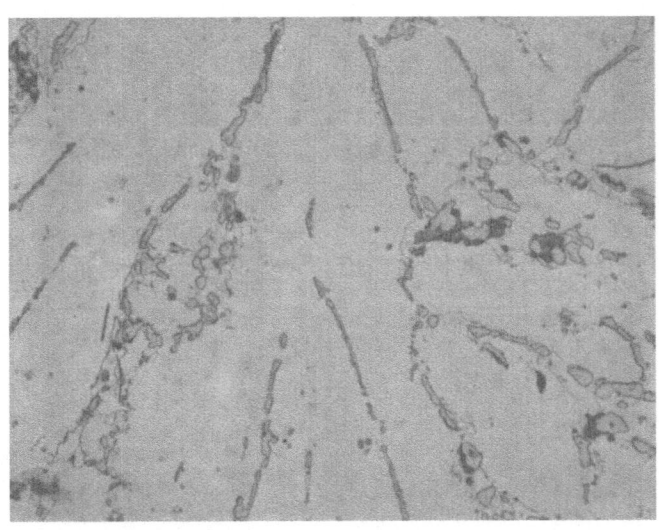

Fig.10. BM1810 (400 x mag.)

THE PREPARATION OF NEODYMIUM-METAL

AND NdFeB-ALLOYS

S. Sattelberger, R. Hähn

GfE Gesellschaft für Elektrometallurgie m. b. H.
Höfenerstrasse 45
D-8500 Nürnberg, Germany

ABSTRACT

The various possible methods of processing neodymium-metal by metallothermic reduction of neodymium-alloys were theoretically investigated. Thermodynamic calculations as well as reflections on feasibility lead to the conclusion that calciothermic reduction of neodymium fluoride presents an ideal solution to the problem. Based upon the theoretical investigations a process for production of neodymium-metal was worked out and developed up to a technical scale.

In parallel various possible methods of production of NdFeB-based magnetic alloys were tested. The theoretical investigations pointed towards advantages of the Vacuum-Induction-Melting-Process (VIM). Based upon this process a technical scale production process was worked out and the parameters for crucible material, cooling rate, superheating/pouring temperature and the control of impurities optimised. The process was developed up to a commercial scale.

Within the framework of the CEAM-cooperation a process for technical production of NdFeB-hydrides was developed. The hydride is produced in a two-stage-process:

- Alloy production
- Hydrogenation

The results show that a controlled hydrogenation yields a hydride with contents of about 0,3 - 0,4 % of homogeneously distributed hydrogen. The production process was also developed up to a commercial scale.

PRODUCTION OF NEODYMIUM METAL

Choice of the Most Advantageous Processing Method

The production of Neodymium is effected on a commercial scale
by two basically different processing methods

- electrolysis of fused electrolytes
- metallothermic reduction of Nd-alloys.

Within the framework of the CEAM-cooperation the metallothermic reduction of Nd-alloys was investigated.

Thermodynamical Considerations

For metallothermic reductions we use reactive metals, for
example

    Lithium (Li)   Magnesium (Mg)   Sodium (Na)
    Calcium (Ca)   Potassium (K)   Aluminium (Al)
    and Silicon (Si).

The metallothermic reduction of Nd-compounds proceeds by the
general reaction formula:

$$Nd_2O_3 + 6\ M \longrightarrow 2\ Nd + 3\ M_2O$$

$$2\ NdR_3 + 6\ M \longrightarrow 2\ Nd + 6\ MR$$

$$Nd_2O_3 + 3\ Me \longrightarrow 2\ Nd + 3\ MeO$$

$$2\ NdR_3 + 3\ Me \longrightarrow 2\ Nd + 3\ MeR_2$$

R = F, Cl
M = monovalent
Me = bivalent

The choice of the optimal reduction method was made according
to three criteria:

- the applicability of the various metals
- the feasibility of various metallothermic techniques
- the cost comparison.

The thermodynamical calculation of the ΔG-value leads to the
following order of the Δ G-values:

    - Oxide System
      $CaO \gg Nd_2O_3 > MgO > Al_2O_3 > SiO_2 > Na_2O > K_2O$

    - Fluoride System
      $CaF_2 > NdF_3 > LiF > NaF > MgF_2 \gg AlF_3$

    - Chloride System
      $KCl \sim NaCl \sim LiCl \sim CaCl_2 > NdCl_3 > MgCl_2 > AlCl_3$

Reflections on Feasibility

Metallothermic reactions:

- proceed quickly and give high yields

- give compact reaction products, the metal as an ingot and the slag as a well separated layer to avoid additional separation steps

- give good quality metals

- and materials should be available for the high temperature reaction vessels or crucibles.

From this point of view and in consideration of the thermodynamics, the best process for producing Nd-metal should be the reduction of $NdCl_3$ with Ca, Li, Na or Potassium.

However, there are some disadvantages:

- Li and K are very expensive.

- Na has a relatively low boiling point; it can vaporize in considerable quantities during the reaction.

- The commercial form of $NdCl_3$ is the hexahydrate chloride. It is not easy to get waterfree products, which are necessary to get high-quality Nd-metal.

Considering the fluoride system it is to be seen that only Calcium can reduce NdF_3. The disadvantage of the calciothermic reduction is the high melting point of the CaF_3-slag.

However, this problem can easily be solved by addition of $CaCl_2$ as a flux.

The investigations within the framework of the CEAM-cooperation lead to the development of an optimal process for production of Nd-metal by calciothermic reduction of NdF_3. The flow-sheet is shown in Fig. 1.

The specification of the Nd-metal is as follows:

Nd	min	95	%
Pr) Ce) La)	max	3	%
Mg	max	0,02	%
Ca	max	0,1	%
O	max	0,5	%
Al	max	0,1	%
Fe	max	0,5	%

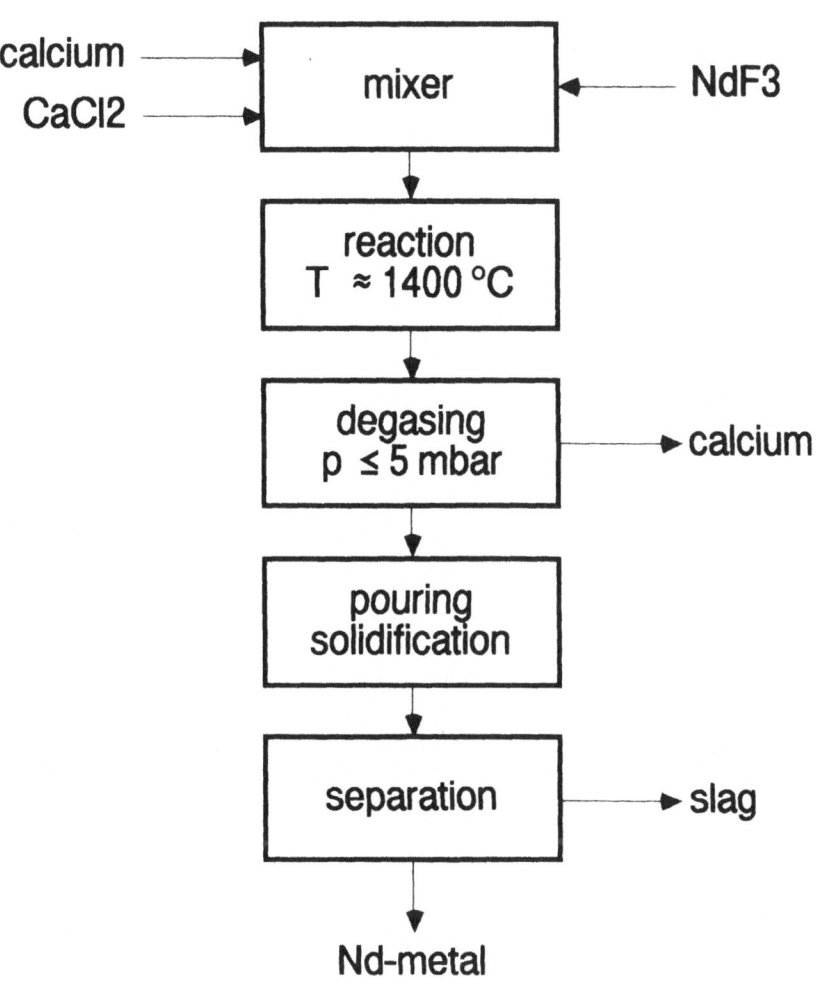

<u>Fig. 1 :</u>   Preparation of Neodymium Metal

PRODUCTION OF NdFeB-MAGNETIC ALLOYS

The production of NdFeB-alloys on a commercial scale is accomplished mainly by two different processing methods:

- by the vacuum-induction-melting-process (VIM-process)
- by the Co-reduction-process

Within the framework of the CEAM cooperation the VIM-process was investigated thoroughly. The following parameters were investigated and optimised:

- crucible materials
- cooling rate
- superheating/pouring temperature
- control of impurities

The following results were obtained:

Crucible material

From a technological and economical point of view a ceramic, based on the spinel-system $MgO \cdot Al_2O_3$ proved to be most suitable. The durability of this material was better than that of pure Al_2O_3 resp. MgO.

Cooling rate

During production of NdFeB alloys it is not only important to keep a determined chemical analysis. It is also necessary to obtain exactly defined crystal structures and phase compounds.

In order to achieve a high portion of the magnetic phase $Nd_2Fe_{14}B$ as well as to avoid a precipitation of α-iron and other disturbing phases, a special cooling technique is necessary. Experiments have shown that the phase composition can be controlled exactly over the layer thickness of the decanted ingots. The pouring of the melt into water-cooled copper-moulds with a layer thickness of about 15 mm proved to be an optimal process. The macrostructure of a typical as-cast ingot is shown in Fig. 2.

Fig. 2: Macrostructure of typical as-cast NdFeB-Ingot

Superheating / pouring temperature

Together with REP, various samples were prepared with different superheating and pouring temperatures to investigate the
influence of these parameters on the phase compositions. The
results are shown in the report of Messrs. Mullard, Southport.

Control of impurities

The purity of the alloys is influenced by the following factors:

- purity of the raw materials
- choice of the crucible material
- crushing technology

Investigations have shown that it will be possible to produce alloys in sufficient purity if the following factors are considered:

- application of

  Nd- metall with the above mentioned specification

  electrolytic iron

  special carbothermically reduced ferroboron with < 0,1 % C.

- Use of spinel type crucible ceramic

- Consequent avoidance of ingress of air during crushing and packaging by the working under safety gas (argon). By these means it is possible to produce a material with the following specifications:

| Nd | 26 | - | 36 | % ± 0,3 % |
| B | 0,8 | - | 1,3 | % ± 0,1 % |

Pr	max.	1	%
Ce	max.	0,2	%
La	max.	0,3	%
other RE	max.	0,5	%

C	max.	0,05	%
Ca	max.	0,01	%
Al	max.	0,15	%
Si	max.	0,05	%
Mg	max.	0,01	%
O	max.	0,1	%

The systematic addition of further alloying elements such as Dy, Al, Co, Nb is possible with a tolerance of +/- 0,1 - 0,3 % depending on the contents.

The flow-sheet of the VIM-Process is shown in Fig. 3.

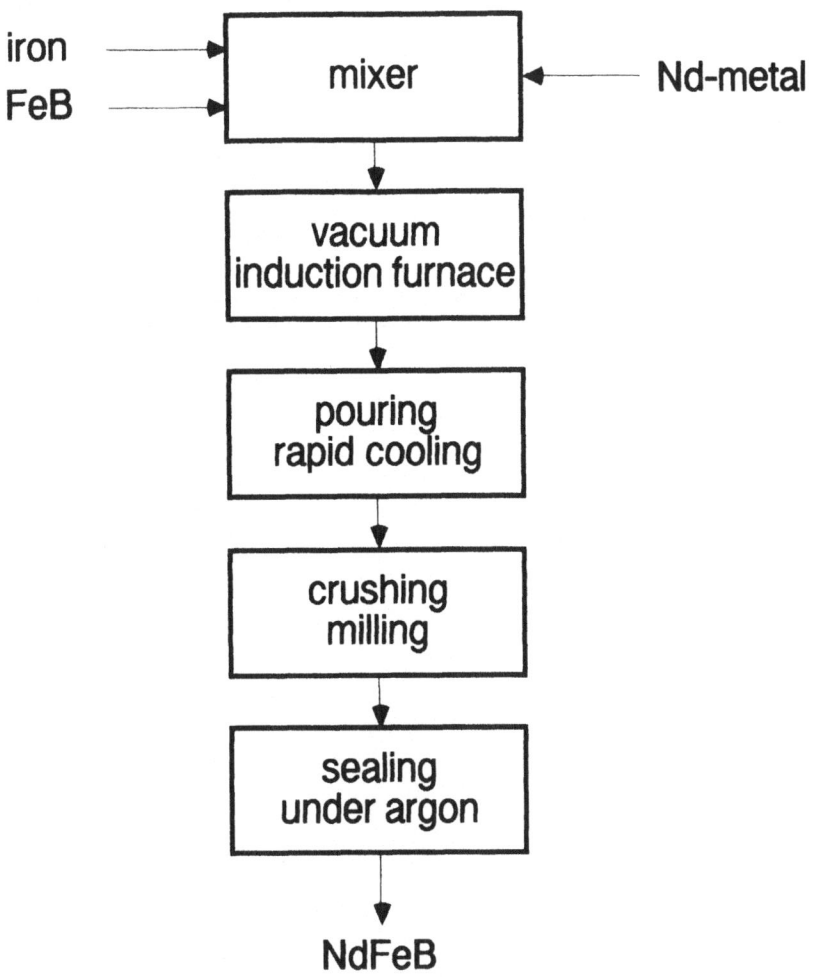

<u>Fig. 3:</u>   Preparation of NdFeB-Alloys

PRODUCTION OF NdFeB-HYDRIDES

Within the framework of the CEAM-cooperation a process for technical production of NdFeB-hydrides has been developed. The hydride is produced in a two-stage-process:

- Alloy production
- Hydrogenation

The results show that a controlled hydrogenation at about 100 - 150 C in the bulk material yields a hydride with contents of about 0,3 - 0,4 % homogenously distributed hydrogen.

The material is rather brittle and thus easily crushable. However, a marked tendency towards oxygen pick-up was noted.

MICROSTRUCTURAL AND HYDROGEN ABSORPTION STUDIES ON NEODYMIUM-IRON-BORON MATERIALS.

I.R. Harris, J.S. Abell, P.J. McGuiness, E.J. Devlin, P.A. Withey.

DEPARTMENT OF METALLURGY AND MATERIALS,
UNIVERSITY OF BIRMINGHAM,
BIRMINGHAM B15 2TT,
UK.

ABSTRACT

Hydrogen decrepitation has been developed into a highly successful industrial process. The hydrogen absorption/desorbtion behaviour of alloys based on $Nd_{15}Fe_{77}B_8$ (Neomax composition) and the stoichiometric phase $Nd_2Fe_{14}B$ have been examined so as to relate these effects to the microstructure. Microstructural characterisation was carried out with a variety of techniques with a special emphasis on the detection of boron. Variations in the hardness of the material were investigated as a function of ageing time over a range of temperatures. Microstructural observations on ingots of NdFeB alloys solidified by horizontal and vertical float zoning techniques have allowed the phase distribution, grain size distribution and the nature of the solid-liquid growth interface to be studied and compared with the as-cast microstructure. The nature of hydrogen absorption as observed by DTA is complex but may prove to be a useful phase analysis aid as well as a decrepitation route to produce single crystals from large grained material.

INTRODUCTION

This report details the research conducted at Birmingham University as part of the CEAM collaboration in the period Oct. 1983 - April 1988. This is not intended to be an exhaustive account, rather a description of the major discoveries and successes. These include hydrogen absorption/desorption experiments on a range of NdFeB alloys and the transference of the Hydrogen Decrepitation Process, a method of pre-milling NdFeB alloys, from the laboratory to an industrial process. An investigation into light element analysis techniques for the characterisation of boron in these alloys, and the influence of ageing times and temperatures on the micro-hardness of these materials. Hydrogen absorption was studied as a possible phase analysis technique and ultimately as a means of selectively separating out individual grains from large grained material. An investigation was also carried out into the usefulness of zone refining and directional solidification as a technique for producing single crystals. A spin-off from this work yielded information on phases present in the grain boundary, particularly oxygen stabilised phases which may have important effects on the coercivity of sintered magnets.

MATERIALS

The alloys investigated in this work contained neodymium of 95% purity, and the alloys were produced by induction melting the components (Nd, iron and ferroboron) by Rare Earth Products plc.

The differential thermal analysis (DTA) measurements were obtained by means of a specially constructed high pressure unit which could be operated over a wide range of hydrogen pressures (0 to 40 bar) and temperatures (77 to 1073 K). The system incorporated sheathed chromel-alumel thermocouples and the samples were in the form of small lumps with a total weight of around 200 mg.

METALLOGRAPHIC STUDIES

In the as-cast condition the $Nd_{11.8}Fe_{82.4}B_{5.8}$ ($Nd_2Fe_{14}B$) alloy is very inhomogenous and the central regions of each grain consists of free iron. This is consistent with the reported [1] peritectic nature of this phase. The alloy was then homogenised by annealing at 950°C for 5 days to produce a single phase material.

The "as-cast" microstructure of the $Nd_{16}Fe_{76}B_8$ alloy, Fig. 1 shows details of the grain boundary phase which are very complex; the description of the grain boundaries as uniformly Nd-rich is clearly an over simplification.

GRAVIMETRIC STUDIES

The $Nd_{16}Fe_{76}B_8$ alloy was exposed to hydrogen (at 1 bar) in a Sartorius microbalance at room temperature and the resultant weight gain was monitored as a function of time. This is shown in Fig. 2 and there is some evidence of a two stage absorption process consistent with earlier statements [2] on the nature of the absorption process in this alloy.

The gravimetric measurements give a total hydrogen content of $Nd_{16}Fe_{76}B_8H_{27.8}$ (0.42±0.02 w/o) which is in good agreement with the values given in Table 1. The vacuum desorption behaviour is consistent with a two stage desorption process [2], first from the matrix $Nd_2Fe_{14}B$ phase and then from grain boundaries.

The hydrogen absorption behaviour of the $Nd_2Fe_{14}B$ sample is shown in Fig. 3 and in this case it was necessary to heat to 160°C in 1 bar of hydrogen before absorption took place. This is in marked contrast to the $Nd_{16}Fe_{76}B_8$ alloy indicating the importance of the grain boundary phases in the activation process. On cooling the sample to room temperature a composition of $Nd_2Fe_{14}BH_{2.7}$ was obtained, in good agreement with previous observations [6,8-10].

Reheating the sample under hydrogen (1 bar) shows a slight desorption of hydrogen until $720^{\circ}C$ when disproportionation took place.

Table 1

Ref.	Stoichiometry of Hydride	% wt Increase on forming the hydride
3	$Nd_{15}Fe_{77}B_8$	-
4	$Nd_{15}Fe_{77}B_8H_{29}$	-
5	-	0.35($\pm$2)
6	$Nd_{15}Fe_{77}B_8H_{25}$	-
7	$Nd_{16}Fe_{76}B_8H_{28}$	0.42($\pm$0.02)

DTA AND DSC STUDIES

The DTA measurements show that the $Nd_{16}Fe_{76}B_8$ alloy readily absorbs hydrogen at room temperature with an exothermic reaction or reactions. It was not possible to obtain exactly the same behaviour on each absorption and the ΔT vs time plots exhibited the range of behaviours shown in Fig. 4.

Possible explanations for the various DTA behaviours shown in Fig.4 are:

-The multiple peaks are genuine evidence of the multi-stage nature of the hydriding process.

-Different parts of the samples activate at different times.

-There are variations of the microstructure between samples.

-There are variations in the degree of oxidation of the grain boundary material between samples.

It is probable that the observed behaviour is a combination of all four factors.

Quantitative DSC measurements made on a range of NdFeB boron alloys also revealled the importance of the intergranular Nd-rich phase in the reaction with hydrogen. Table 2 shows how the more Nd-rich materials have a greater heat of formation than those closer to the stoichiometric $Nd_2Fe_{14}B$ material.

Table 2

Alloy	$\Delta H(cal/g)$
$Nd_2Fe_{14}B$ (homog)	17.4
$Nd_{15}Fe_{77}B_8$	41.8
$Nd_{16}Fe_{76}B_8$	45.4
$Nd_{17}Fe_{75}B_8$	49.6
$Nd_{27}Fe_{67}B_6$	104.8

The activation time and the time taken for the reaction to reach completion are both very dependent upon the hydrogen pressure and the dependence of the activation time on the hydrogen pressure is shown in Fig. 5. From a practical viewpoint, this plot shows that the time taken to begin the hydrogen decrepitation process can be considerably shortened by the application of high hydrogen pressures.

If, instead of immediate transfer to the DTA unit, the crushed material is left in air for 48 h then it was found that the sample could not be activated at room temperature, on subsequent exposure to hydrogen at a pressure of 1bar. In this case, activation could only be achieved at elevated temperatures and this observation is consistent with the oxidation of the Nd-rich material at the grain boundaries which is no longer available for the activation of the sample. The exotherm at elevated temperature then corresponds with the absorption of hydrogen into the matrix phase.

The hydrogen absorption behaviour of the $Nd_2Fe_{14}B$ alloy (homogenised to remove the free iron) was very different from that of the freshly crushed $Nd_{16}Fe_{76}B_8$ alloy in that there was no hydrogen absorption reaction at room temperature even up to hydrogen pressures of 25 bar. The sample had to be heated to 150-160°C in a hydrogen pressure of 1 bar in order to achieve hydrogen absorption. In this case a single exothermic peak was always observed (Fig. 6). These observations are in agreement with the gravimetric measurements referred to earlier and again emphasise the importance of the Nd-rich grain boundary material in achieving room temperature absorption.

It should be noted that in the uncoated condition, sintered

magnets cannot be exposed to even low hydrogen pressures during their lifetime as such exposure would lead to their complete disintegration.

MASS SPECTROSCOPY STUDIES

The $Nd_{16}Fe_{76}B_8$ alloy was hydrided in-situ (at a pressure of 1 bar) and then vacuum degassed by heating at 5°C per minute to 1000°C. The desorbed hydrogen was monitored by means of a mass spectrometer and the variation of partial pressure of the hydrogen signal with temperature is shown in Fig. 7(a). There are two distinct desorption steps which is in agreement with earlier gravimetric observations on the vacuum degassing of this alloy in its hydrided condition.

The first peak corresponds to the loss of hydrogen from the matrix $Nd_2Fe_{14}B$ phase (confirmed by X-ray diffraction measurements after this stage) and the second peak corresponds to the loss of hydrogen from the grain boundaries. Gravimetric measurements have shown that all the hydrogen is removed from the sample at the end of the second stage and these observations confirm that no hydrogen remains in the material after the vacuum sintering stage of a magnet produced using the HD process.

The areas under the two peaks in Fig. 7a are a measure of the relative amounts of hydrogen desorbed at the two stages and the ratio of stage 1 to stage 2 is: 1 : 1.05. This indicates, that despite accounting for only 15 vol% of the alloy the Nd rich phase absorbs as much hydrogen as the matrix phase.

Similar experiments were carried out on the desorption of hydrogen from the hydrided $Nd_2Fe_{14}B$ alloy (see Fig. 7(b)). In this case only a single desorption peak was noted over the same range of temperature (20 to 1000°C). This peak occurred over the same temperature range and was of the same shape as the first desorption peak in the degassing of the $Nd_{16}Fe_{76}B_8$ alloy offering further evidence that the first peak corresponds to desorption from the matrix phase of the alloy.

PRODUCTION OF PERMANENT MAGNETS

The use of hydrogen in the processing of rare-earth magnets has been studied in the Department of Metallurgy and Materials, University of Birmingham, since 1978. These experiments have been reviewed in references [11,12].

Work on the production of a NdFeB magnet using hydrogen decrepitation as a method of premilling the powder was initiated by Harris et al. [2], and a comprehensive study has been carried out in conjunction with Philips (Mullard) Components and Lucas Research Centre [13]. A bulk ingot of a NdFeB alloy was powdered by a combination of hydrogen decrepitation and attritor milling (HD/AM). The powder was aligned and pressed in the hydrided condition and the green compact sintered at 1080°C for 1 hr. The magnets produced using this process (HD/AM) showed good sintered densities and energy products of around 250 kJm^{-3}.

The production of high energy product permanent magnets using a combination of hydrogen decrepitation and jet milling has been carried out in a collaboration with Philips (Mullard) Components [14] and is described elsewhere in this proceedings.

CHARACTERISATION OF HYDROGEN DECREPITATED POWDER

The detailed morphology of the HD powder can be seen in Fig(8). The powder consists of large pieces of ~50-100μm with fine debris on the surface which is approximately 1-10μm in size. Chemical analysis carried out on the material as a whole <1> (see Table 3) shows, as expected, that the Nd/Fe ratio was almost equal to 16:76. Slightly less neodymium-rich values were obtained from spot analysis on the large 50-100μm pieces denoted by <2> and <3>, indicating that their composition is somewhere between that of the $Nd_{16}Fe_{76}B_8$ starting alloy and the $Nd_2Fe_{14}B$ matrix phase.

The small disc shaped debris, denoted by <4>,<5> on the surface of the large pieces were found to have very high high neodymium contents up to a Nd/Fe ratio 6.7/1 indicating that they are formed from the grain boundary Nd-rich phase. In addition a small number of pieces of ~10μm denoted by <6> were found which had a Nd/Fe ratio indicating that they consisted primarily of the Nd_4Fe_4B boride phase.

An oxygen analysis of the HD powder indicated a value of 0.15 $^w/_o$.

Table 3

Area	Nd/Fe ratio	Atomic ratio
Whole sample<1>	0.228	17.3/76
Point<2> on large grain	0.198	15.1/76
Point <3> on large grain	0.195	14.8/76
Debris<4>	6.69	6.69/1
Debris<5>	2.70	2.70/1
Debris<6>	0.26	1/3.9

AGEING EXPERIMENTS ON NdFeB ALLOYS

The behaviour of the $Nd_{15}Fe_{77}B_8$ alloy during heat treatment is not fully understood as no obvious change in the microstructure of the alloy takes place during the post sintering heat treatment at 630°C. In an effort to understand the changes which occur the microhardness of the matrix phase ($Nd_2Fe_{14}B$) was measured as a function of ageing time and temperature.

This investigation is the continuation of similar studies on $Nd_{15}Fe_{77}B_8$ (15) and $Nd_{16}Fe_{76}B_8$ (16). The alloy used in these studies was $Nd_{15}Fe_{77}B_8$ supplied by GfE of Nurnberg. The alloy was solid solution treated (SST) at 1100°C and then quenched to room temperature, this heat treatment gave the alloy a large grain size .

The alloys were then heat treated at various temperatures for periods of 5 and 10 minutes with rapid heating and cooling to ensure the accuracy of the timing. The microhardness measurements of the alloy was obtained on the polished surface of the bulk alloy by means of a "Leitz" microhardness tester. Special care had to be taken to avoid the production of microcracks during the indentation tests and also to avoid the grain boundary regions as these areas are much softer and cause anomalously large indentations in regions close to the boundary. The hardness of the matrix phase of the solid solution treated alloys was found to be 872 (40) VHN .

The temperatures chosen to age the NdFeB alloy were 350°C, 400°C, 500°C and 600°C. The results can be found in Fig.9. By taking these peak times and assuming an Arrhenius type behaviour, namely;

$$\ln(rate) = \ln A - Q/RT$$

where A is a constant and R is the molar gas constant, a value for the activation energy, Q, can be calculated. This variation is shown in Fig.10. and it can be seen that a typical Arrhenius plot is obtained. From these results the value for the activation energy was found to be 41 (±3) kJ/mol.

The microstructure of the material and the domain structure does not noticeably alter during the ageing trials.

The process giving rise to the hardness behaviour could be the redistribution of the boron atoms in the matrix phase after quenching the alloy from 1100°C. It is possible that the atoms are displaced from their normal lattice sites to a metastable site at high temperatures. Thus they move to their normal lattice positions on ageing and give the variation in the hardness observed here. This is supported by work on the diffusion of boron through crystal lattices, for example γ-iron, where the activation energy is found to be 86.1

kJ/mol (17).

The information gained in this study may aid the investigation into suitable systems to produce precipitates within the matrix phase with the possibility of removing the need for a reactive grain boundary phase in NdFeB based magnets.

LIGHT ELEMENT ANALYSIS

Analysis of boron in the NdFeB alloys and magnets is a difficult problem and microanalytical work has therefore been concentrated in this direction. Techniques for analysis of boron with a spatial resolution of a few microns or better are as follows.

TECHNIQUE	RESOLUTION	COMMENTS
SIMS/LIMA	1 μm	Destructive technique, Imaging not good.
Autoradiography	1μm	Rather specialised technique.
EPMA,SEM WDX	1μm	Accurate boron analysis
STEM + EDX	10nm	High resolution, difficult sample preparation, EELS boron capabilty
STEM + EELS	10nm	limited by neodymium.

SIMS / LIMA

These methods have the built-in advantage that they use mass-spectroscopy but the average spatial resolution is only 1 μm. Nor is this method quantitative for the phases we are inspecting with relatively high concentrations of boron present. This is because of the strong effect the matrix has on the rate of sputtering of the ions. This method is generally more useful for doped systems where small changes in the dopant concentration can be observed and measured as there will be negligible changes in the matrix. It is useful for generating linescans across grains. This technique did reveal the presence of aluminium in the $Nd_2Fe_{14}B$ phase. This substitutes for iron in this phase but not in the $NdFe_4B_4$ phase where the iron sites are smaller.

WDX / EPMA

This technique has a similar spatial resolution to SIMS but gives excellent compositional reproducibility, correctly analysing the concentration of boron in the $Nd_2Fe_{14}B$ phase. The accuracy of measurements may be improved by using a lower excitation voltage for the incident electron beam, thereby lowering the volume of sample generating x-rays (improved spatial resolution) and reducing the amount of bremsstrahlung (improved signal to noise ratio of the low energy x-ray peaks).

The low spatial resolution of these methods limits their application to alloys. To obtain information on magnets with a much finer microstructure (grain size $=10\mu$m approx.) one of the transmission electron techniques must be used.

STEM + EDX

Normal EDX detectors are fitted with beryllium windows about 10 μm thick. These windows exclude the x-rays of elements lighter than sodium. Thin window or windowless dectors must therefore be used. Using an ultrathin window, LINK EDX detector in conjunction with the JEOL 4000 FX 400 KV TEM we have been unable to detect boron in the $Nd_2Fe_{14}B$ phase in NdFeB magnets. The same system readily detects boron in boron nitride, BN. We believe the reason for this absence of boron x-rays is the absorption of the light x-rays in the relatively heavy matrix. Boron can be detected in the more boron rich phases such as $NdFe_4B_4$ and NdB_6, the latter being a phase found in some samples made up for the purpose of testing and calibrating the various boron detection systems.

STEM + EELS

The competing TEM technique for light element detection is electron energy loss spectroscopy, EELS. The boron in $Nd_2Fe_{14}B$ is right on the edge of detectability. A thin specimen is essential. The high beam voltage of the microscope used was a tremendous help. Parallel instead of serial EELS detection would improve the system but was not available to us. Quantifying the spectra is difficult because the highest energy Nd edge is an N edge and EEL cross-sections for an N edge are not currently available. Other non-neodymium containing samples such as Fe_2B yielded excellent spectra which also gave accurate quantitative results.

To summarise, SIMS is an extremely sensitive tool though it does not give accurate analysis for our samples nor is the quality of the imaging good. EPMA provides excellent boron analysis (using WDX) down to a resolution of 1 μm. Below this, EELS will detect boron with a minimum limit of 5-10 at% in Nd containing materials.

EFFECT OF CONTROLLED SOLIDIFICATION ON MICROSTRUCTURE

Controlled solidification is being investigated as a possible means of modifying the phase distribution and grain size of NdFeB alloys in order to study the influence of these factors on the magnetic properties. This study is also a necessary precursor to attempts to grow large single crystals of the matrix $Nd_2Fe_{14}B$ phase for physical property measurements, by identifying appropriate compositions which minimise additional phases, and determining suitable growth parameters to produce a significant grain growth. Hydrogen absorption is being studied as a possible phase analysis technique [7] and ultimately as a means of selectively separating out individual grins frm large grained material.

Alloys based on the stoichiometric phase $Nd_2Fe_{14}B$ have been directionally solidified by two techniques, one vertical the other horizontal.

The horizontal mode involved passing a solid-liquid interface along a bar of material held in a water-cooled boat, while the vertical process was performed on a rod contained in a silica tube. In both cases solidification was carried out under an atmosphere of purified Argon at rates typically of 2-5 cm/hr. In the horizontal mode the interface was not traversed right to the end of the bar; this end was quenched in order to study the nature of the interface. The bars were typically 8 mm diameter by 100 mm in length. Sections cut from the bars were mounted, polished and examined by standard metallographic techniques, including SEM-BSE and EDX modes. Magnetic domain observation was achieved by polarised light, ferrofluid and directly in the SEM. Hydrogen absorption and desorption as observed by DTA has been monitored in a specially constructed high pressure system described elsewhere [7].

The complex grain boundary phases of the as cast material, were imaged and identified as the now well established $NdFe_4B_4$ and Nd-rich areas with evidence for the presence of a Nd oxide. The morphology of the domain patterns revealed by polarised light and ferrofluid indicate that the columnar grains grow with their major axis predominantly in the basal plane rather than along the c-axis.

HORIZONTAL SOLIDIFICATION OF NdFeB ALLOYS

Sections cut from various positions down the length of the horizontally solidified bar revealed evidence of grain growth and phase redistribution as a result of the controlled passage of the solid-liquid interface. The first end to solidify exhibits a similar microstructure to the as-cast material but with more regular equi-axed grains rather than the columnar structure typical of the as-cast alloy; this is clearly a result of the controlled rather than rapid solidification. However, the presence of the cold boat tends to generate a small chill zone at the base of the bar where the solidification front is affected primarily by vertical rather than horizontal freezing and thus modifies the shape of the solidification front. Associated with this chill zone there occurs a relatively high density of $NdFe_4B_4$ particles, their globular morphology distinguishing them from the more commonly observed grain boundary phase. In the majority of the bar, however, there occurs a redistribution of the $NdFe_4B_4$ phase which is swept along the bar coarsening in nature as it goes. A similar redistribution occurs for the other major complex additional phase region, the multi-phase Nd-rich area. The culmination of this microstructural modification is clearly depicted in Fig. 11 taken from the quenched end of the bar. Clear evidence of significant grain growth is visible in this figure with the grains being typically 200-500µm in extent in the proximity of the solidification front. It is also interesting to note that the grain boundaries are relatively devoid of the usual additional phases; this is presumably associated with the high density of complex lower melting point material observed at the top of this figure corresponding to material quenched in advance of the interface. The extra phases tend to be swept along ahead of the solidification front and subsequently infill between the primary needles of the matrix phase during the sudden freezing. Some coarse $NdFe_4B_4$ particles are also seen close to the faceted growth front. Domain patterns indicate that the grains forming the faceted front exhibit no preferred orientation and that in some cases the rapidly formed primary needles are direct extensions of the already crystallised grain. Detailed analysis of the quenched region indicated a variety of compositions including borides and several NdFe alloys suggesting a complex multi-phase structure. However many grains had no such inclusions and exhibited clean grain boundaries and well defined domain patterns.

VERTICAL SOLIDIFICATION OF NdFeB ALLOYS

The sections examined from the vertically solidified bar all showed evidence of a surface reacted layer (~0.5 mm thick) adjacent to the silica tube, typified by Fig. 12 taken from the lower end of the rod, the first to solidify. A sharp interface between the reacted layer and the matrix was a feature of all the sections cut from this rod indicating that the layer served to passify the surface against further reactions. XRD and SEM-EDX analysis of this layer showed it to be Nd_2O_3.

Grain growth was not quite so evident in this bar. The microstructure of the lower section was not radically different to the as-cast grain boundary region. However, in the top half of the rod the grain boundary region was indeed different. Discrete $NdFe_4B_4$ was still observed but the predominant extra phase was eutectic in nature, the two components being the $NdFe_4B_4$ and a Fe-rich phase. The domain patterns revealed by polarised light, ferrofluid and by secondary electron detection in the SEM do not indicate domains in the grain boundary phase. $NdFe_4B_4$ is known to be non-ferromagnetic but if the Fe-rich component is free Fe then the domains may be distorted by surface strains; however, reference to the phase diagram suggests it is more likely to be a binary boride of Fe such as Fe_2B.

HYDROGEN ABSORPTION

Samples taken from the clean, large grained region of the horizontally solidified bar shown in Fig.11 absorbed no detectable hydrogen when subjected to a pressure of 1 bar at room temperature in the DTA equipment. This is consistent with previous findings [7] where it was established that the presence of the Nd-rich phase is a necessary precursor for hydrogen absorption. A sample from the quenched multi-phase structure however, showed a large exothermic reaction associated with Nd-rich material shown in Fig. 13(a). This was unlike any other reaction previously seen on as-cast alloys which usually show single or sometimes unresolved double peaks typified by that shown in Fig. 13(b) recorded on the as-received GfE alloy. A sample from the vertically solidified rod containing little Nd-rich grain boundary phase showed no absorption at room temperature but a small reaction at elevated temperature as seen in Fig. 13(c), consistent with hydrogen absorption by the matrix phase [7].

The possibility of inducing grain growth and phase redistribution in these alloys by controlled solidification is clearly demonstrated in these experiments. The largest grains appear at the end of the bar last to freeze where the intergranular phases have been swept away; this is consistent with the idea of Stadelmaier et al. [18] that grain boundary phases inhibit recrystallisation. The eutectic inclusions observed within grains are somewhat unexpected as this is a low melting point phase. This suggests either grain growth subsequent to solidification or liquid immiscibility involving the formation of an impervious oxide layer as in the case of the silica enclosed rod. The relatively low incidence of grain growth in this rod may be associated with the change in overall composition due to the reaction with the silica.

The observation of the binary $NdFe_4B_4$ plus Fe-rich grain boundary phase in the vertically processed rod together with the absence of the usual Nd-rich phase is clearly connected with the composition .change associated with the formation of the Nd_2O_3

reacted layer. A similar intergranular phase has recently been observed by Schneider et al. [19] in slowly cooled samples, who stated that the lamellar structure was Fe and $NdFe_4B_4$ but did not conclude whether it was a eutectic reaction, as suggested by Matsuura et al. [1] or a solid state reaction. Establishing the composition as Fe or Fe_2B is difficult in this case but the alternative suggestion by Schneider that this reaction could involve contamination by a fourth component such as oxygen appears to have some relevance in the present case.

The possibility of oxygen stabilised phases has recently been invoked by Fidler [20] and Schneider [21] during investigations of cast alloys and sintered magnets, and our observations of the complex nature of the quenched end of the bar are consistent with this interpretation. In Table 4 the Nd/Fe ratios measured in the present work are compared with those reported by Fidler, where the similarity is clear.

Table 4

Nd/Fe	Fidler [20]	Present Work
a	1.2 - 1.4 : 1	1.3 : 1
b	2.0 - 2.3 : 1	1.8 : 1
c	3.5 - 4.4 : 1	
d	>7 : 1	

The hydrogen absorption data is complex; the large exotherm is associated with the Nd-rich region and that observed at elevated temperature in single phase material is clearly the matrix phase, with the multi-phase as-cast alloy often exhibiting more than one reaction. More work is required before this can be fully characterised; hydrogen absorption might then be a useful phase analysis aid. Information on the kinetics of absorption might subsequently allow the parameters to be chosen to enable controlled decrepitation of large grained material to produce individual single crystals of sufficient size for physical property determinations.

ACKNOLEDGEMENTS

We would like to thank all those connected with the CEAM project.

REFERENCES

[1]   Matsuura, Y., Hirosawa, S., Yamamoto, H., Fujimura, S., Sagawa, M. and Osamura, K., Jap. J. Appl. Phys. 24, (1986)L635.

[2]   Harris, I.R., Noble, C. and Bailey, T., J. Less-Common Metals, 106, (1985)L1.

[3]   Cadogan, J.M. and Coey, J.M.D., Appl. Phys. Lett. 48, (6), 442 (1986).

[4]   Wiesinger, G. Hilscher, G. and Grossinger, R., J. Less-Common Metals, 131, (1987) 409.

[5]   Scholz, U.D., Report to CEAM Group, (Sept. 1986).

[6]   Pollard, R.J. and Oesterreicher, H., IEEE Trans. Mag. Mag., 22, (1986) 735.

[7]   Harris I.R., McGuiness P.J., Jones D.G.R., and Abell J.S., Physica Scripta. Vol. T19, (1987) 435.

[8]   Abache, C. and Oesterreicher, H., J. Appl. Phys. 57, (1984) 4112.

[9]   Oesterreicher, H., and Abache, C., J. de Phys. C6, 45 (1985).

[10] Oesterreicher, K. and Oesterreicher, H., Phys. Status Solidi (a) 85, K61, (1984).

[11] Harris I.R., J. Less-Common Met. 131(1987) 245.

[12] Harris, I.R. Paper No.52, 9th Int. Workshop on Rare Earth Magnets and their Applications, Bad Soden, FRG, August 31st-September 2nd, (1987) 267.

[13] McGuiness, P.J., Harris, I.R., Rozendaal, E., Ormerod, J. and Ward, M., J. Mat. Sci. 21 (1986) 4107.

[14] McGuiness, P.J., Harris, I.R., Rozendaal, E., Ormerod,. J. To be published in J.Mat.Sci.

[15] I.R. Harris and T. Bailey. J. Mat. Sci. Lett. 4 (1985) 151.

[16] M.A. Hixon. Ph.D. Thesis, University of Birmingham (1987).

[17] P.E. Busby, M.E. Warga and C. Wells. J. Metals. 5 (1953) 1463.

[18] H.H. Stadelmaier, N.A. Elmasry, N.C. Liu, S. Cheng, Materials Lett., 2 (1984) 411.

[19] G. Schneider, E.-T. Henig, G. Petzow and H.H. Stadelmaier, Z. Metallkde, 77 (1986) 755.

[20] J. Fidler, 5th Int. Symp. Magnetic Anisotropy and Coercivity in RE-TEM Alloys, Bad Soden (1987) 363.

[21] G. Schneider, E.-T. Henig, H.H. Stadelmaier and G. Petzow, ibid. 347.

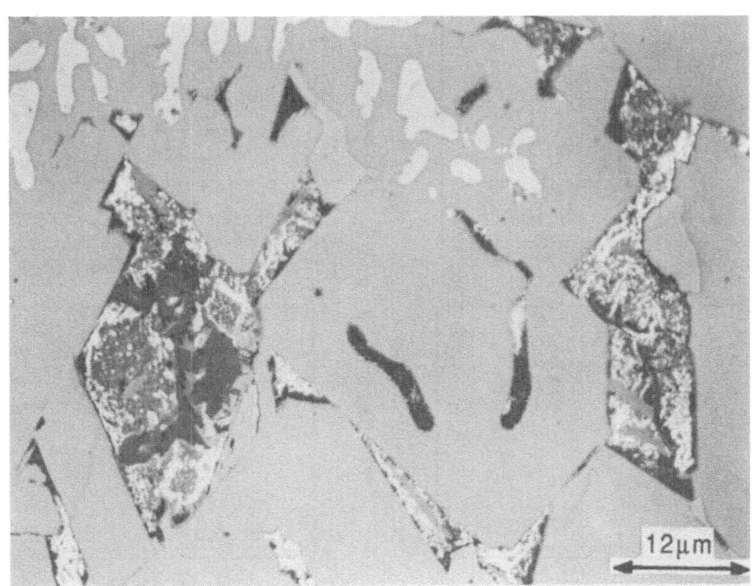

Fig 1.   As cast $Nd_{16}Fe_{76}B_8$

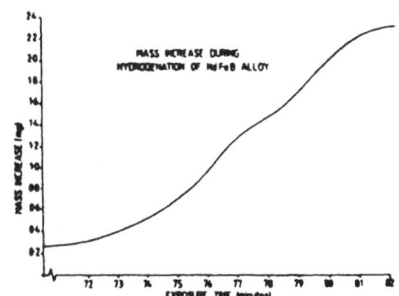

Fig 2. Two stage absorption of hydrogen by $Nd_{16}Fe_{76}B_8$

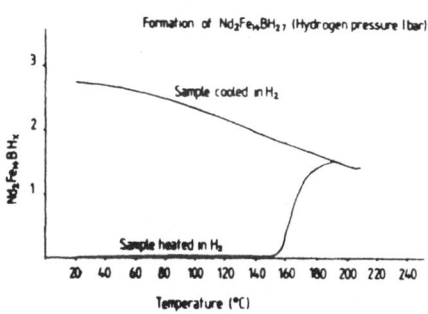

Fig 3. H_2 absorption in $Nd_2Fe_{14}B$

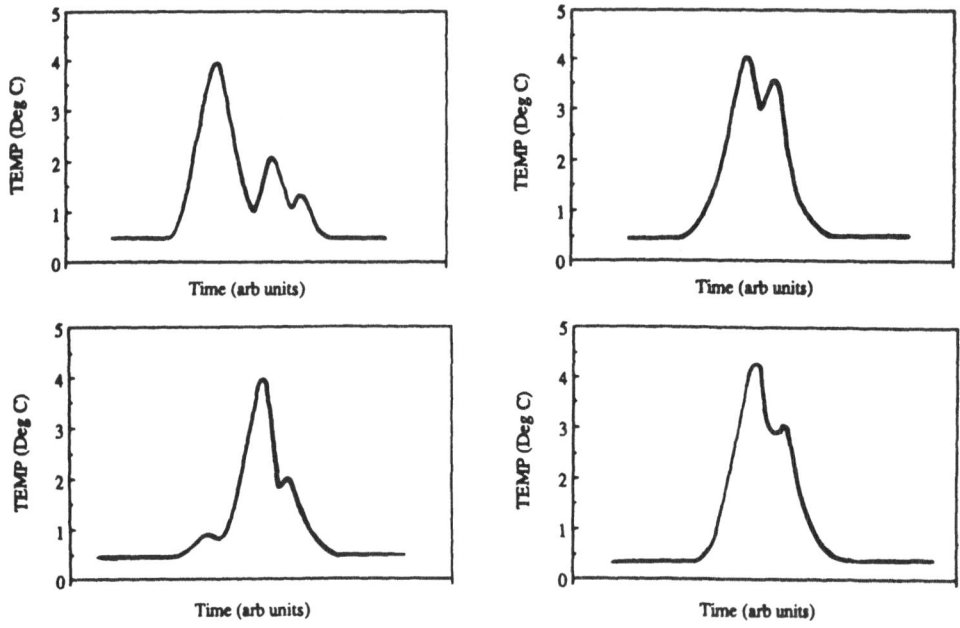

Fig 4.   ΔT vs Time traces for DTA
         hydrogen   absorption   in
         $Nd_{16}Fe_{76}B_8$

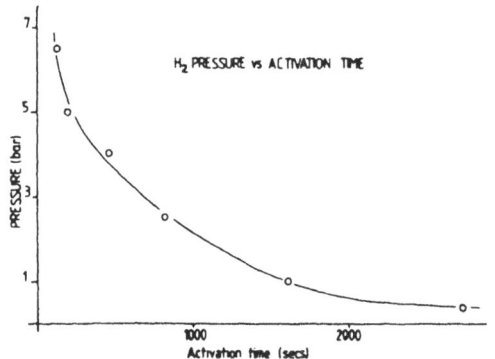

Fig 5.   The variation in activation
         time   with   hydrogen
         pressure   for   the
         $Nd_{16}Fe_{76}B_8$ alloy

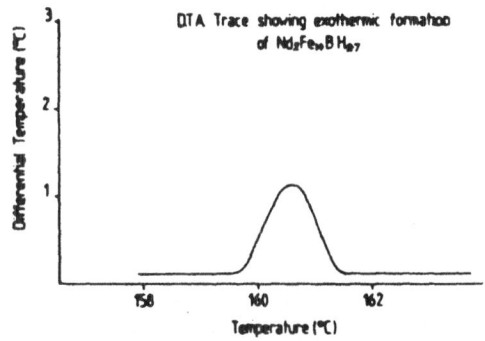

Fig 6.

ΔT vs Temp trace for the $Nd_2Fe_{14}B$ sample. Heated at a rate of 4^0C per minute in hydrogen (1bar)

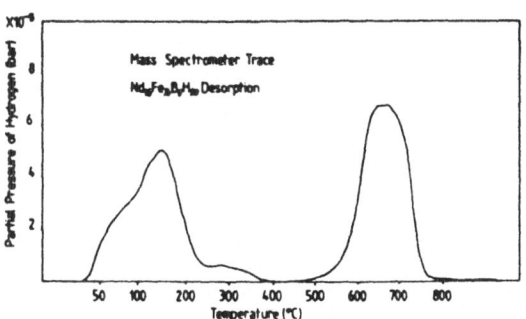

Fig 7a. Mass spectrometer hydrogen signal vs temperature during the vacuum degassing of the $Nd_{16}Fe_{76}B_8$-Hydride

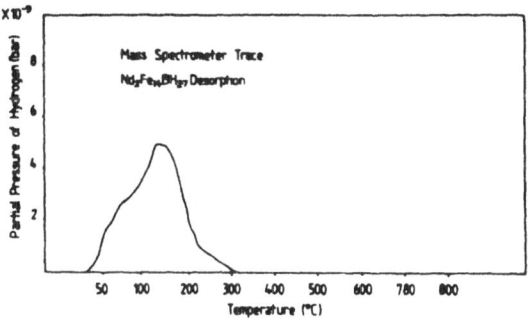

Fig7b. Mass spectrometer hydrogen signal versus temperature during the vacuum degassing of a $Nd_2Fe_{14}B$ homogenised alloy

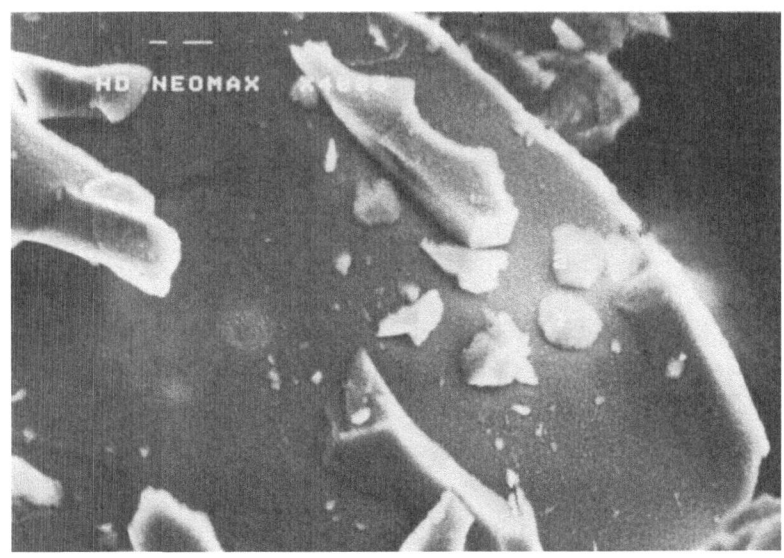

Fig.8   HD Neomax powder particle
with Nd-rich particles on
the surface

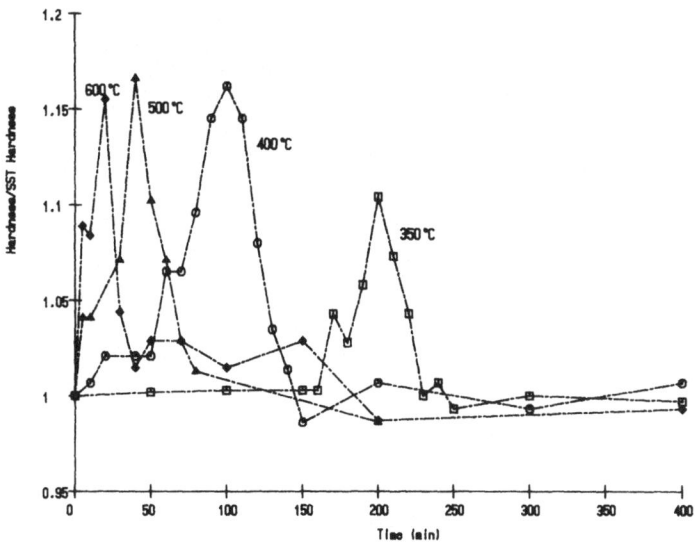

Fig.9   Variation in microhardness
         of $Nd_{16}Fe_{76}B_8$ aged at 350,
         400, 500, 600 ºC

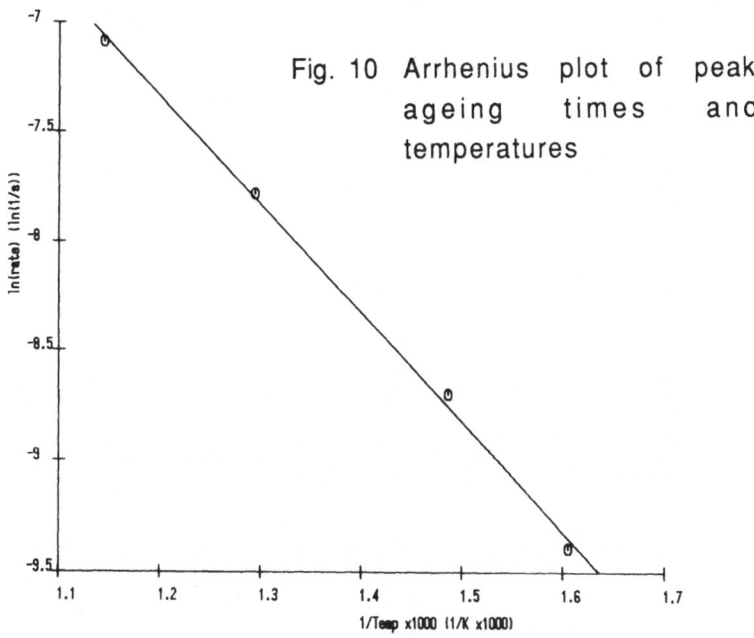

Fig. 10 Arrhenius plot of peak
         ageing times and
         temperatures

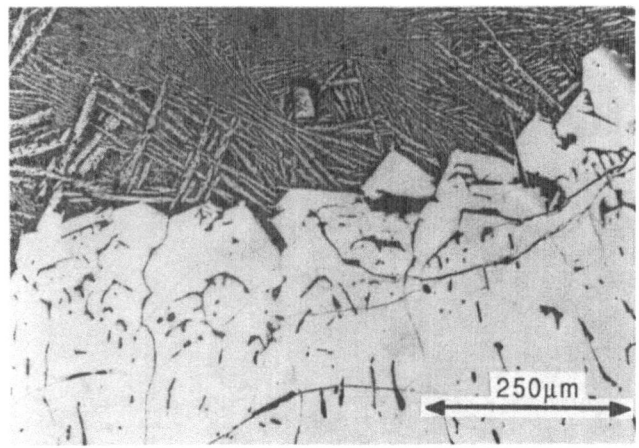

Fig. 11 Part of the solidification
front taken from a
longitudinal section from
the horizontal bar

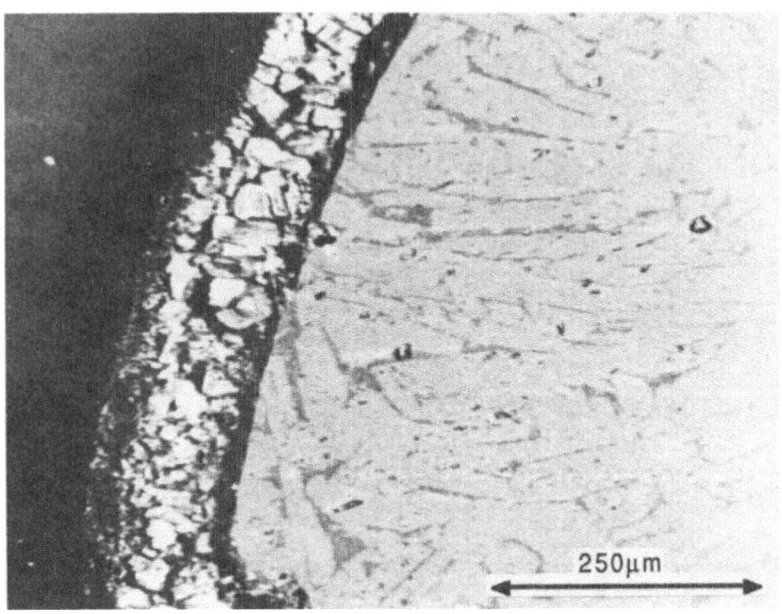

Fig. 12 Transverse section of the
vertical rod showing the
reaction layer

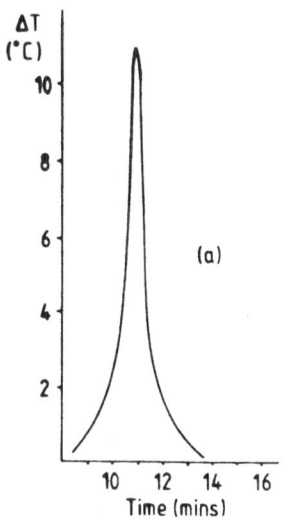
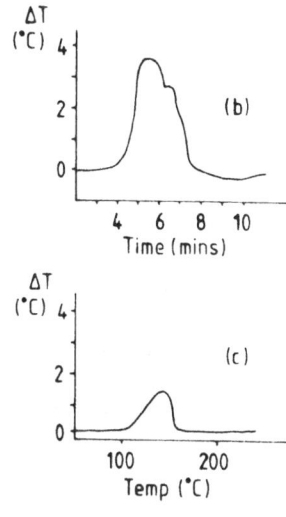

Fig. 13 Hydrogen absorption exotherms in (a) the quenched eutectic region of Fig. 11, (b) the as-cast GfE alloy and (c) the upper section of the vertical rod.

A STUDY OF Nd-Fe-B MAGNETS PRODUCED USING A COMBINATION OF HYDROGEN DECREPITATION AND JET MILLING

E Rozendaal and J Ormerod,
Philips Components Limited Southport,
Balmoral Drive, Southport, PR9 8PZ, U.K.

P.J. McGuiness and I.R.Harris,
Department of Metallurgy and Materials,
University of Birmingham, Edgbaston, B15 2TT, U.K.

ABSTRACT

A combination of hydrogen decrepitation (HD) and jet milling (JM) has been used to produce powder for the processing of permanent magnets. The procedure has proved to be very successful for both NdFeB ("Neomax") alloys and the NdDyFeNbB high coercivity alloys. The magnets produced by the HD/JM process showed excellent coercivities when sintered between 980°C and 1040°C, at higher temperatures, excessive grain growth reduced the coercivity values significantly.

INTRODUCTION

The use of hydrogen in the processing of rare-earth magnets has been studied in the Department of Metallurgy and Materials, University of Birmingham, since 1978. (Harris, 1987; Harris, 1987). Early work concerned hydrogen decrepitation of the intermetallic $SmCo_5$ and sintered and polymer bonded magnets were produced from hydrogen decrepitated powder. (Harris, 1987; Harris, 1987; Harris et al, 1979). In the case of $SmCo_5$, hydrogen could be absorbed and desorbed at room temperature prior to the magnetic alignment and compaction procedure.

The HD process was extended to $Sm_2(Co,Fe,Cu,Zr)_{17}$-type magnets and for these materials the combination of high hydrogen pressure and a temperture of 200°C was required to complete the process. (Kianvash and Harris, 1984; Kianvash and Harris, 1985). Sintered magnets were produced by aligning the hydrided powder and the hydrogen was subsequently removed during the vacuum sintering operation.

Soon after the announcement of the NdFeB "Neomax" magnetic alloy the effect of exposing the alloy to hydrogen at room temperature was reported. (Sagawa et al. 1984; Harris et al, 1985). These studies showed that the material reacted readily with hydrogen at moderate pressures with a strongly exothermic reaction. Work on the desorption of hydrogen from the $Nd_{16}Fe_{76}B_8$ alloy indicated that vacuum degassing consisted of two stages whereby hydrogen was first desorbed from the ($Nd_2Fe_{14}B$) matrix phase below 300°C with the remainder being evolved from the neodymium rich phase between 350 and 650°C. (Harris et al., 1985).

The formation of two hydrides is consistent with the decrepitation behaviour where the initial activation process corresponds with the hydriding of the intergranular, neodymium-rich material, followed by the hydriding of the matrix phase with the attendant transcrystalline cracking of the individual crystallites.

Work on the production of a NdFeB magnet using hydrogen decrepitation as a method of premilling the powder was initiated by Harris et al. and a comprehensive study has been carried out in conjunction with Birmingham University and Lucas Research Centre. (Harris et al., 1985; McGuiness et al., 1986). A bulk ingot of a Nd-Fe-B alloy was powdered by a combination of hydrogen decrepitation and attritor milling (HD/AM). The powder was aligned and pressed in the hydrided condition and the green compact sintered at 1080°C for 1 hr. The magnets produced using this process (HD/AM) showed good sintered densities and energy products of around 250 kJm^{-3}. In a recent paper the hydrogen absorption/desorption behaviours of the alloy $Nd_{16}Fe_{76}B_8$ and $Nd_2Fe_{14}B$ have been examined. (Harris et al., 1987). These studies reveal clearly the essential role of Nd-rich grain boundaries in the $Nd_{16}Fe_{76}B_8$ alloy in the hydrogen activation process.

Report of other groups carrying out similar studies are scarce. Cadogan and Coey have produced results on the hydrogen absorption and desorption properties of $Nd_{15}Fe_{77}B_8$ alloys but magnets were not produced from the resulting powders. (Cardogan and Coey, 1986). Pollard et al. have also reported on hydriding of "Neomax" type alloys, and have proposed a route for producing fine powder direct from the hydrogen decrepitation process. (Pollard et al. 1986) However, no results of the production magnets from this powder have been published.

In our previous paper on the HD/AM process we were of the opinion that "the HD process would be more effective if combined with jet milling". (McGuiness et al., 1986). Thus in this paper we report on the application of the hydrogen decrepitation process in conjunction with jet milling (HD/JM), to the production of NdFeB magnets from bulk alloys with composition $Nd_{16}Fe_{76}B_8$ and $Nd_{14.5}DY_{1.5}Fe_{76}Nb_1 B_7$.

PROCESSING

In order to produce the fine powder necessary for the production of sintered magnets the following procedure was adopted:

  i) Bulk alloy castings were placed in a stainless steel hydrogenation vessel which was then evacuated to backing pump pressure.

  ii) Hydrogen was then introduced into the vessel to a pressure of ~1 bar.

  iii) The hydrogen absorption process occurred after a short incubation period and was accompanied by audible clicks and a rise in temperature. The former is consistent with the decrepitation of the bulk material and the latter with the high exothermicity of the process. (Harris et al., 1987; Pollard et al., 1986, Oesterreicher and Oesterreicher, 1984).

  iv) The material produced during the hydrogen decrepitation process was transferred in 10 kg batches to the mill hopper. The milling process was carried out in a spiral type jet mill, using a nitrogen atmosphere to produce the material suitable for pressing.
  Significantly shorter milling times were required than those needed to mill the conventional pre-

milled material.

v) The hydrogen decrepitated/jet milled (HD/JM)
powder was then pressed into an aligned green
compact using a pressing force of between 1000-2500
kgcm^{-2} and a perpendicular alignment field of
1000 kAm^{-1}.

vi) The compacts were sintered over a range of
temperatures from 960°C to 1080°C, in a
commercial vacuum furnace. A marked degradation in
the vacuum at quite low temperatures (~200°C)
was noted, due to the evolution of hydrogen from the
$Nd_2Fe_{14}B$ matrix phase as the compacts were heated
to the sintering temperature. This is in agreement
with the previous observations on the
hydrogen desorption behaviour of this material.
(Harris et al., 1985; McGuiness et al., 1986; Harris
et al., 1987; Cadogan and Coey, 1986).

The samples were maintained at the sintering temperature
for 1 hr. and then slowly cooled to room temperature.

The sintered compacts were magnetised in a magnetic field
of 2400 kAm^{-1} prior to the determination of the second
quadrant demagnetisation loops.

RESULTS AND DISCUSSION

1) Characterisation of the HD powder

The morphology of the HD powder was investigated with a
SEM. The HD powder consists of large pieces of ~50-100 um
with fine debris on the surface which is approximately 1-10 um
in size.

The small disc shaped debris were found to have very high
neodymium contents of up to a Nd/Fe ratio of 6.7/1 indicating
that they are formed from the grain boundary Nd-rich phase.
In addition a small number of pieces of ~10 um were found
which had a Nd/Fe ratio indicating that they consisted
primarily of the $NdFe_4B_4$-boride phase.

The composition of the large 50-100 um pieces are analised
somewhere between that of the $Nd_{16}Fe_{76}B_8$ starting alloy

and the $Nd_2Fe_{14}B$ matrix phase.

An oxygen analysis of the HD powder indicates a value of 0.15 ^{W}/o.

2) Characterisation of the HD/JM powder

Typical HD/JM powder is shown in Fig. 1 for comparison, Fig. 2 shows typical powder produced by the HD/AM route. It is evident that the HD/JM powder exhibits an extremely uniform size whereas a much wider range of particle size is obtained in the case of HD/AM material. Fisher sub-sieve measurements on the HD/JM powder indicate a value of 2.3, comparable with that obtained for the HD/AM powder. An oxygen analysis of the HD/JM powder gives a value of 0.5 ^{W}/o.

3) Characterisation of green compacts

The density of green compacts produced from HD/JM material was found to be significantly lower than similar compacts produced from standard attritor milled and HD/AM material. It was found that to achieve a green compact density of 4 gcm^{-3} the HD/JM powder required a pressing force of 1500 $kgcm^{-2}$ as opposed to the force of 1000 $kgcm^{-2}$ needed for the HD/AM powder. This difficulty in achieving compact density has been attributed to the very narrow particle size distribution and the difficulty in packing such material.

Due to the effect hydrogen has on the magnetic anisotropy of the $ND_2Fe_{14}B$ hard magnetic phase the green compacts exhibited a much reduced coercivity in comparison with standard milled material. (Oesterreicher and Oesterreicher, 1984). In previous studies the alignment characteristics of HD/AM material were determined and these studies established that hydrogen did not effect the uniaxial nature of the anisotropy so that HD/AM material could be fully aligned in a magnetic field. (McGuiness et al., 1986). Similarly, no problems were encountered in achieving full alignment while processing the HD/JM material.

4) Characterisation of sintered magnets

In order to establish the optimum sintering conditions, a

515

FIG. 1   HD/JM POWDER

FIG. 2   HD/AM POWDER

study of the relationship between the properties of the HD/JM magnets and the sintering temperature has been carried out for the $Nd_{16}Fe_{76}B_8$ alloy and the results are shown in Figures 3,4, and 5. The most striking feature of these graphs is the variation of Hc_i with sintering temperature. Little more than half the full intrinsic coercivity is realised by sintering at the usual sintering temperature of 1080°C and to achieve 800 kAm^{-1} the temperature must be reduced to below 1040°C for a 1 hour sinter. The Br-values of the material are not so dependent upon the sintering temperature except for compacts sintered below 1000°C, where the lack of a full sinter is reflected in the sharp drop in density from the fully dense value.

Metallographic examination indicates that the low coercivity, which is observed in magnets sintered above 1040°C, is due to excessive grain growth.

A quantitative examination of the sintered micro structures was undertaken using a VIDS II system which combines the video output from a T.V. camera and microscope with the graphics display of an Apple computer so that measurements may be made directly from the T.V. image. The VIDS II software is able to provide information on the grain size distribution as well as the mean grain size.

Fig. 6 shows how the grain size is affected by the sintering temperature. An increase in both average grain size and grain size distribution is observed, with material sintered at 1080°C having a mean size of greater than 20 um. Measurements made on the coercivity of the sintered magnets (Fig. 4) indicate that an average grain size of <14 um is necessary to achieve good coercivity from this particular alloy composition. Fig. 7 which shows how the spread of grain size varies with sintering temperature, indicates how important it is to maintain a narrow grain size if good coercivities are to be obtained. The sharp increase in the range of grain sizes corresponds to the point where the coercivity begins to decrease.

The degree of grain growth (at 1080°C), not normally observed in conventional processing, has been attributed to

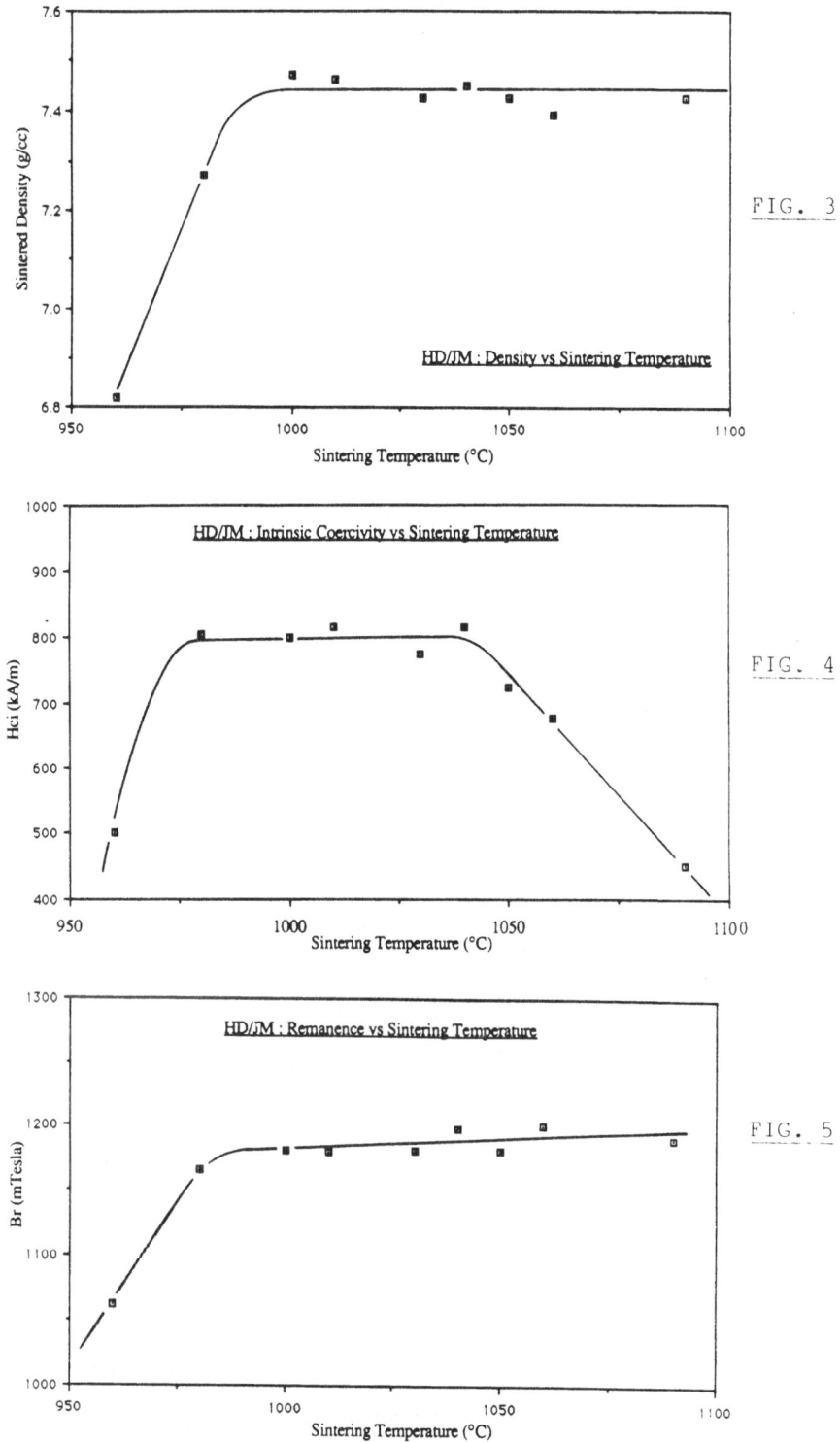

FIG. 3

HD/JM : Density vs Sintering Temperature

FIG. 4

HD/JM : Intrinsic Coercivity vs Sintering Temperature

FIG. 5

HD/JM : Remanence vs Sintering Temperature

518

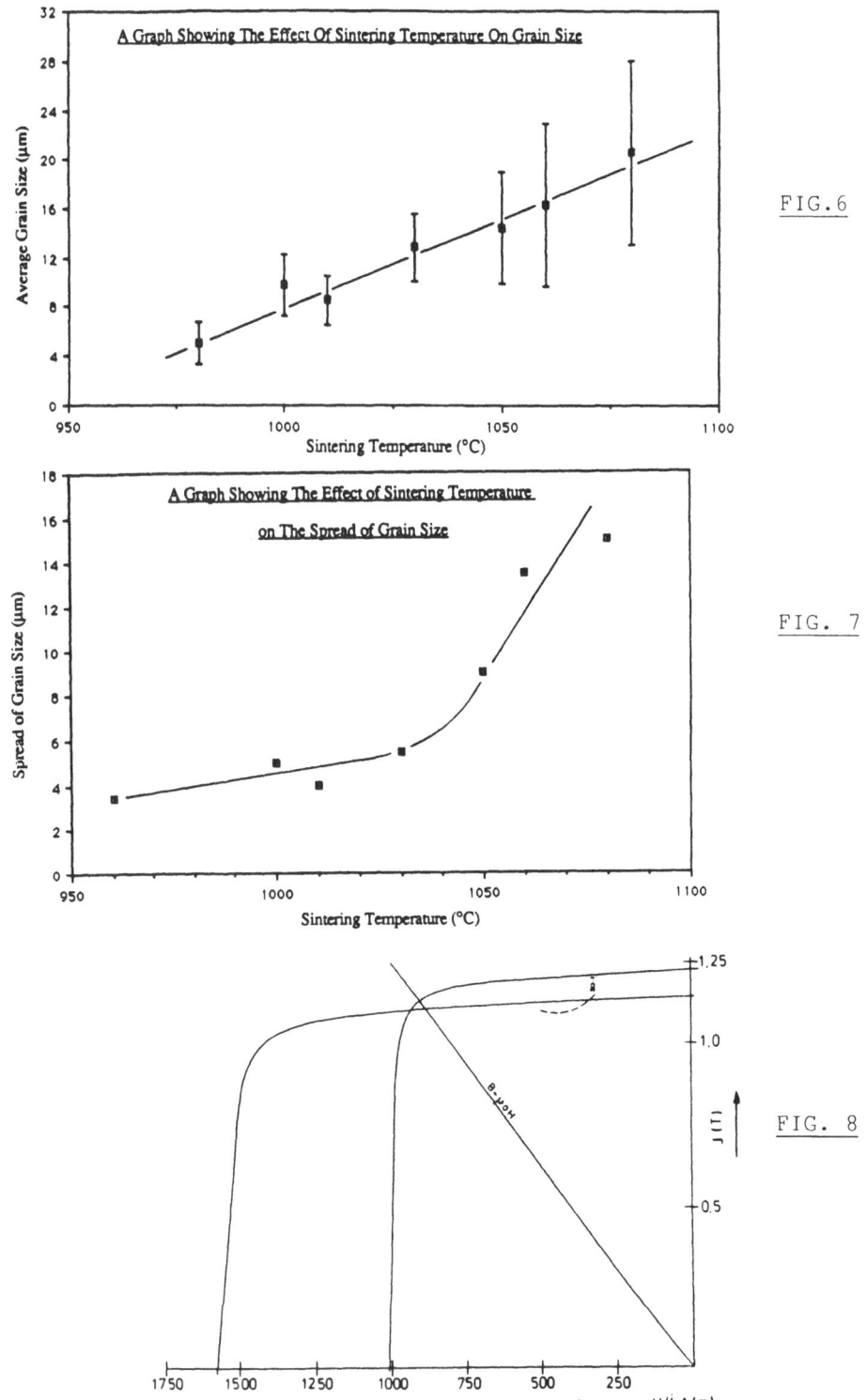

A Graph Showing The Effect Of Sintering Temperature On Grain Size

FIG.6

A Graph Showing The Effect of Sintering Temperature
on The Spread of Grain Size

FIG. 7

FIG. 8

the clean, relatively debris free, material (Fig. 1) produced
by the HD/JM process.

Typical (B-H) against H curves for sintered and slowly
cooled magnets made from HD/JM powder of the $Nd_{16}Fe_{76}B_8$
and $Nd_{14.5}Dy_{1.5}Fe_{76}B_7$ alloys are shown in Fig. 8. The
values of B_r Hc_B and Hc_i and $(BH)_{max}$ are summarised in
Table I together with typical values for magnets produced by
the HD/AM process. (McGuiness et al., 1986)

Table 1

	Nd16Fe76B8		Nd14.5Dy1.5Fe76B7
	HD/AM	HD/JM	HD/JM
Br (mT)	1175 ± 5	1240 ± 5	1145 ± 5
Hc_B (kA/m)	690 ± 5	915 ± 5	875 ± 5
Hc_i (kA/m)	740 ± 5	1010 ± 5	1570 ± 5
$(BH)_{max}$ (kJ/m^3)	250 ± 1	305 ± 1	260 ± 1

CONCLUSION

The combination of hydrogen decrepitation and jet milling
has introduced several significant advantages over the
conventional milling route in the production of permanent
magnets.

(1) Hydrogen decrepitation overcomes the problems of breaking
     up ingots, which can be extremely tough if they contain
     significant amounts of free iron.
(2) Hydrogen decrepitation provides an inexpensive and
     straigh forward method of producing large quantities of
     material suitable for either attritor or jet milling.
(3) The extremely friable nature of the hydride enables it to
     be milled for a shorter time, in the case of
     attritor milling, or at a higher feed rate in the case of
     jet milling, when compared to conventionally pre-milled
     material.
(4) The HD/JM material produced can be sintered at a
     temperature significantly below 1080°C and still
     achieve excellent properties.

REFERENCES

Harris, I.R., 1987. J. Less-Common Metals, 131 245.
Harris, I.R., 1987. Paper No. 52, 9th Int. Workshop on Rare
    Earth Magnets an their Applications, Bad Soden, FRG, Aug.
    31-sep. 2, 267.
Harris, I.R., Evans, J., Nyholm, P.S. 1979. British Patent 1
    554 384.
Kianvash, A. and Harris, I.R. 1984. J. Mater. Sci. 19, 353.
Kianvash, A. and Harris, I.R. 1985 J. Mater. Sci. 20, 682.
Bailey, T. and Harris, I.R.
Sagawa, M., Fujimura, S., Togawa, N., Yamamoto, H. and
    Matsuura, Y. Appl. Phys. 55 2083.
Harris, I.R., Noble, C. and Bailey, 1985 T. J. Less-Common
    Metals 106 L1.
McGuiness, P.J., Harris, I.R., Rozendaal, E., Ormerod, J.
    Ward, J. 1986 J. Mater. Sci. 21 4107-4110.
Harris, I.R., McGuiness, P.J., Jones, D.G.R. and Abell, J.S.
    1987. Scripta Physica. T19, 435.
Cadogan, J.M. and Coey, J.M.D. 1986 Appl. Phys. Lett. 48 (6)
    442.
Pollard, R.J. and Oesterreicher, H. 1986. IEEE Transactions on
    Magnetics, Sept. Vol. Mag. 22, No. 5, 735.
Oesterreicher, K. and Oesterreicher, H. 1984. Phys. Stat. Sol.
    (a) 85 K1.

A CONTRIBUTION TO THE MECHANISM
OF THE HYDROGENĀTION OF Nd-Fe-B-ALLOYS
AND THE USE OF HYDROGENATED ALLOY
FOR PERMANENT MAGNET PRODUCTION

U.D. Scholz, H. Nagel

Thyssen Edelstahlwerke AG
Magnetfabrik dortmund
D-4600 Dortmund 41, FRG

ABSTRACT

The relationship of Nd-content and hydrogen absorption behaviour of Nd-Fe-B permanent magnetic alloy was studied. Measurements showed that the higher the Nd-content of the alloy the more hydrogen is absorbed. With increasing Nd-content the activation time for hydrogen absorption decreases. In the pressure range between 10^{-3} and 1200 mbar no plateau pressure typical for stable hydrides could be found. The influence of the hydrogenation of Nd-Fe-B alloys on the production of permanent magnets is described especially for the milling step.

INTRODUCTION

Like most rare-earth transition metal compounds Nd-Fe-B alloys absorb hydrogen. First investigations showed that the Nd-Fe-B permanent magnetic alloy and the tetragonal hard magnetic phase could be hydrogenated at pressures between 20 and 50 bars (l'Héritier et al., 1984; Harris et al., 1985; Oesterreicher et al., 1984; Pollard et al., 1986; McGuiness et al., 1986). Recently it was shown that hydrogenation of $Nd_2Fe_{14}B$ is also possible at ambient pressure (Cadogan et al., 1986; Harris, 1987; Rozendaal et al., 1987; Scholz et al., 1987). The tetragonal phase $Nd_2Fe_{14}B$ could be hydrogenated up to the stoichiometry $Nd_2Fe_{14}B\ H_5$ within the range of ambient pressure and 20 bar (l'Héritier et al., 1984). During hydrogen absorption the alloy decrepitates as referred elsewhere (Harris, 1987 and references cited therein). This decrepitation process offers a possible alternative for the production of permanent magnets based on Nd-Fe-B. In this work we consider the hydrogen absorption with respect to the production of permanent magnets.

EXPERIMENTS ON THE ABSORPTION OF HYDROGEN

The reaction of hydrogen with standard Nd-Fe-B bulk alloy (35 w% Nd, 1.3 w% B, balance Fe) below ambient pressure was carried out in a stainless steel vessel using 5N6 hydrogen. The amount of absorbed hydrogen was always about 0.4 weight %. Absorption took place within a time of 30 min. The absorption velocity of an alloy in the as-cast condition and after subjecting it to an absorption/desorption cycle respectively is shown in figure 1. Hydrogen was desorbed by vacuum degassing at 800 °C. As can be seen in figure 1, the maximum absorption velocity increases by a factor of 3 after subjection to an activation cycle, i.e. 0.9 l H_2/s compared to 0.3 l H_2/s in the as-cast state. Furthermore the reaction of activated material starts immediately, whereas in case of the "virgin" alloy hydrogen absorption starts after an activation time of about 6 minutes. During the hydrogenation process temperatures up to 350 °C were detected on the surface of the alloy depending on the mass of the hydrogenated alloy. This fairly high temperature is due to the reaction enthalpy of the hydrogenation process. The reaction enthalpy was calculated to be ΔH_R = -57.2(8) kJ/mol H_2 by means of isobare DTA/TG experiments. Figure 2 shows a typical DTA/TG diagram of the reaction of bulk Nd-Fe-B alloy with H_2 (1 bar). The first peak corresponds to the hydrogenation of the tetragonal hard magnetic phase. In accordance with other authors (Cadogan et al., 1986) we found the second exothermic peak corresponds to the irreversible decomposition of the tetragonal hard magnetic phase. The heat of reaction for the decomposition was determined to be ΔH_R = - 53.3 (6) kJ/mol H_2. For the standard alloy used the volume expansion of the tetragonal compound was always found to be in the same order of magnitude (expansion of the tetragonal $Nd_2Fe_{14}B$ unit cell from 0.9449 nm³ to 0.9713 nm³, dV/V = 0.028). Two different experiments were carried out to get a better understanding of the hydrogenation mechanism.

The first experiment deals with the question whether

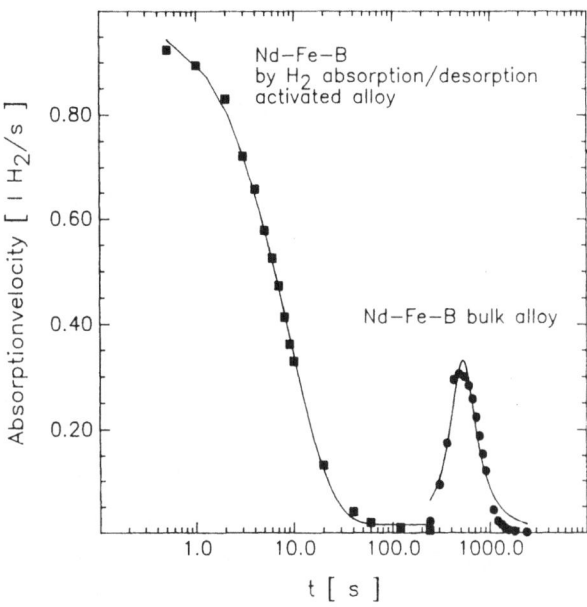

Fig. 1   Comparison of the absorption velocity of
Nd-Fe-B bulk alloy and by H_2 absorption/
desorption activated alloy

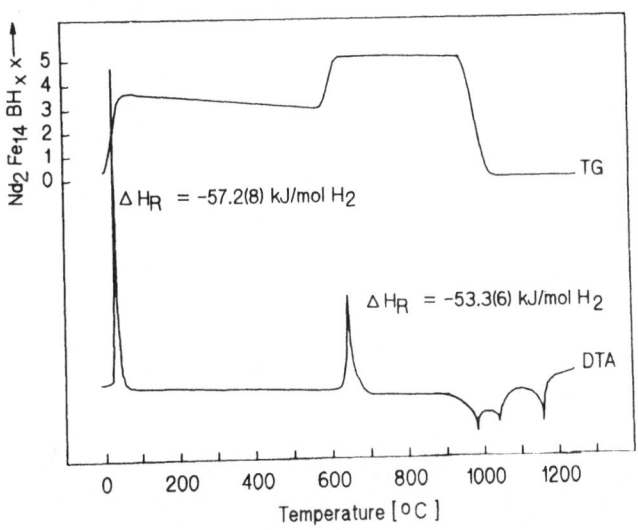

Fig. 2   Typical DTA/TG diagram of an isobare reaction
of Nd-Fe-B with 1 bar H_2

there exists a stable hydride in the technically interesting
pressure range or not. Therefore the isothermal absorption
and desorption of hydrogen was measured. Measurements of the
c-p isotherms within the pressure range of $3.10^{-3} < p < 1200$ mbar
of four different temperatures (323K, 363K, 423K, 453K) were
carried out. Neither an indication to an existing plateau
typical for a stable hydride nor a difference between desorption and absorption curves could be observed. Therefore in
figure 3 only the desorption curves are plotted. This leads
to the conclusion that no stable hydride exists in the investigated range. An extension of the investigation to smaller
hydrogen pressures is difficult because of the long time needed to get equilibrium conditions. During the time used to get
equilibrium the leak rate of the equipment became a determina-
tive factor for the result of measurement. The fact that
there is no stable hydride in the technically interesting pressure range ($p > 10^{-3}$ mbar) makes it obvious that cyclic charg-
ing/decharging prosesses do not result in an extensive decrepitation of the alloy (figure 4). This is in contrast to other
alloys like $SmCo_5$ or $LaNi_5$ (Kuipers, 1973).

To get information regarding the influence of the Nd-rich
phase on the hydrogenation mechanism absorption measurements
on Nd-Fe-B alloys with different Nd-content were carried out.
The phase distribution of the investigated alloys is shown in
figure 5 (McGuiness et al., 1988). McGuiness et al. report a-
bout the measurements of the relation between reaction enthalpy and Nd-content. Due to their results the reaction enthalpy
increases nearly linear with the Nd-content. The determination
of the maximum absorbed hydrogen as a function of temperature
and Nd-content at 1 bar equilibrium pressure shows a linear
decreasing hydrogen concentration for increasing temperature
(figure 6). The higher the Nd-content of the alloy the higher
the maximum absorbed hydrogen. This connexion is fairly linear
for constant temperature (figure 7). The real absorption velocity of hydrogen is not strongly influenced by the Nd-content
(figure 8), but a very strong influence of the Nd-content
on the activation time of the reaction was found. The maxi-

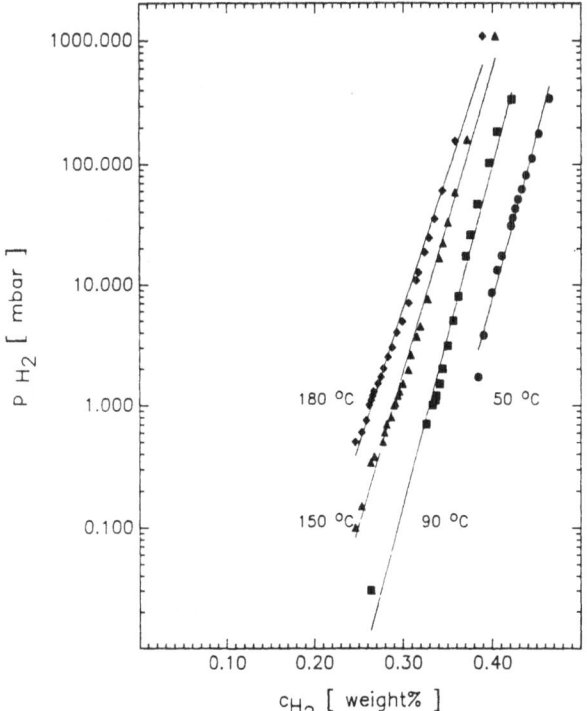

Fig. 3   c-p isotherms of Nd-Fe-B-H

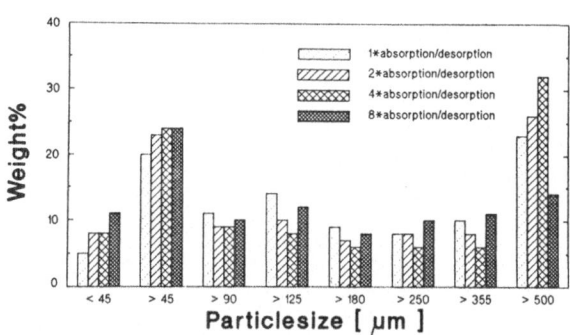

Fig. 4   Particle size distribution of Nd-Fe-B-H
          as a function of charging / discharging cycles

526

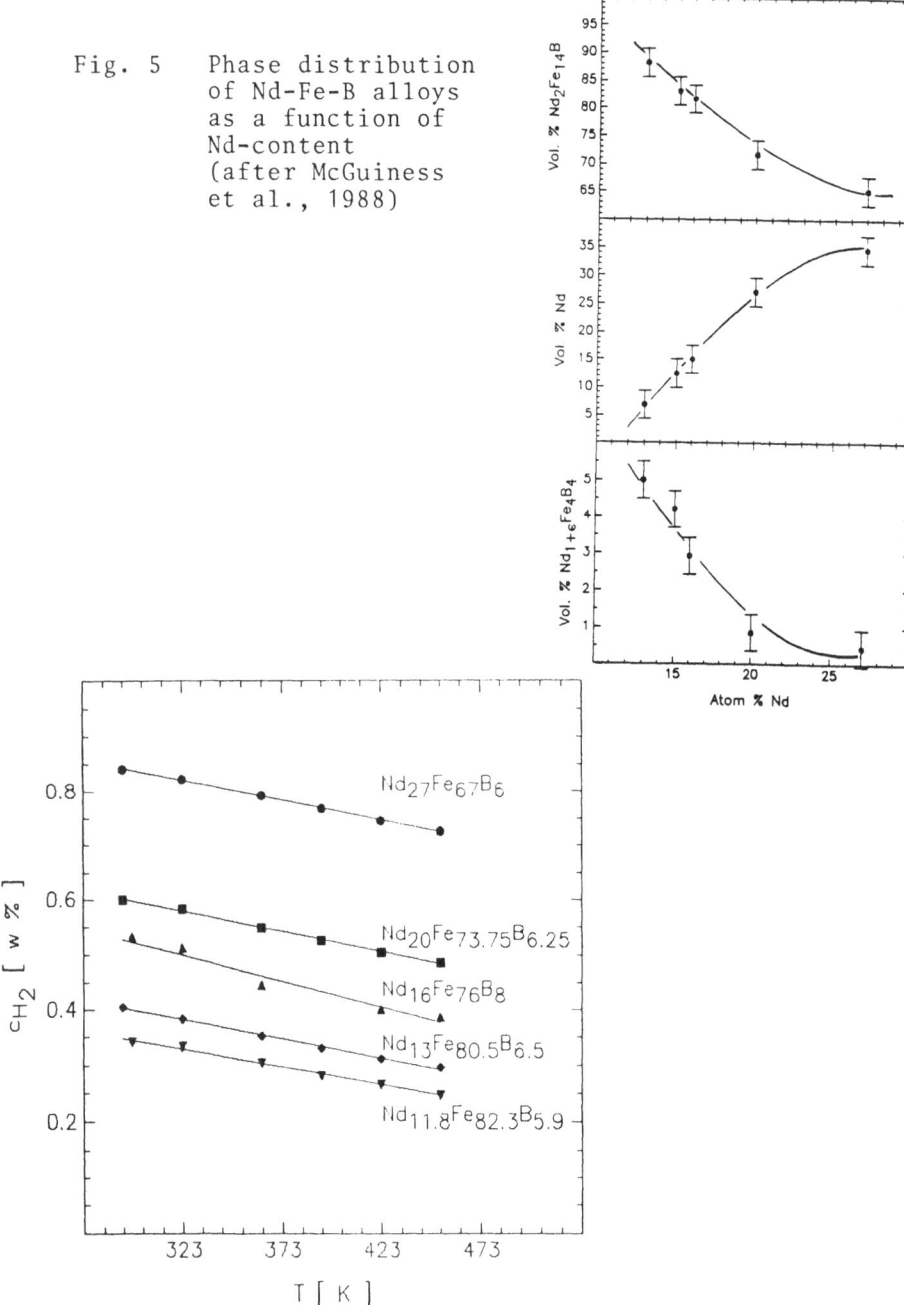

Fig. 5   Phase distribution
          of Nd-Fe-B alloys
          as a function of
          Nd-content
          (after McGuiness
          et al., 1988)

Fig. 6   Relation of maximum H-concentration in
          Nd-Fe-B between Nd-content and T at 1 bar
          equilibrium pressure

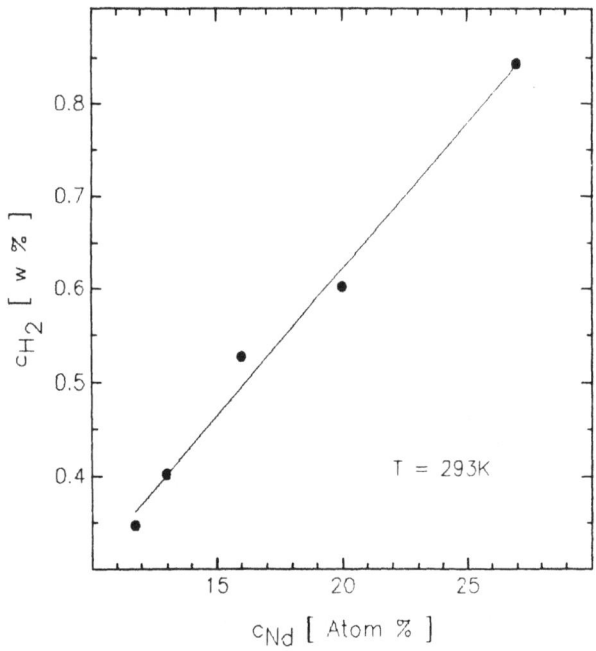

Fig. 7   Relation between maximum absorbed H- and
Nd-content at 1 bar H_2 and 293K

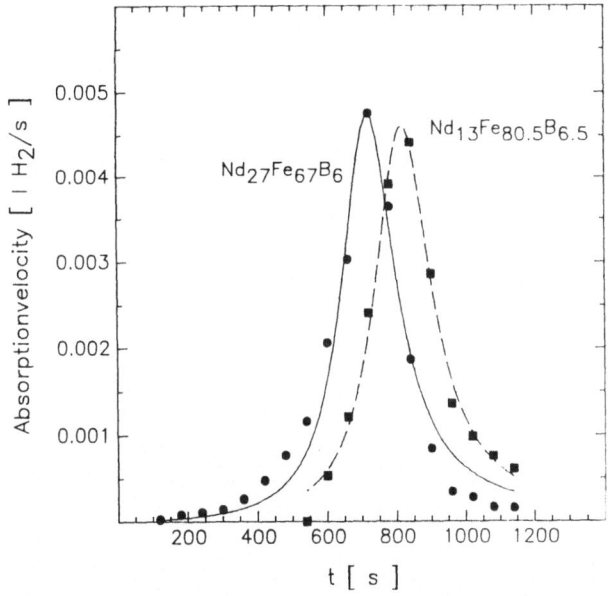

Fig. 8   Comparison of the H_2-absorption velocity of
Nd-Fe-B alloys with different Nd-content

mum of the absorption velocity of an alloy with high Nd-content is shifted to shorter activation times. These experiments showed that the amount of Nd-rich phase present in the alloy has a strong influence on the hydrogen absorption behaviour of Nd-Fe-B alloys.

INFLUENCE OF HYDROGEN IN Nd-Fe-B ALLOYS ON THE PROCESSING OF MAGNETS

The particle size distribution of hydrogenated Nd-Fe-B may be compared with the particle size distribution of conventionally premilled material. For both the maximum is at 125 μm (figure 9), whereas the average of the particle size distribution of hydrogenated Nd-Fe-B measured by FSSS was at 16 - 20 μm. This fact explains why the feed rate of the following jetmill step was increased by a factor between 2 and 4, depending on the particle size of the final powder. The hydrogenated material is cracked so that the milling energy is decreased. That makes the final powder finer than conventionally treated material. The hydrogen in the Nd-Fe-B alloy has a positive effect on the stability of the milled material. It seems that hydrogenated alloy is more stable against oxygen than not hydrogenated alloy.

The absorbed hydrogen is desorbed from the green compacts during vacuum sintering (figure 10). In accordance to other authors (Harris, 1987) we found two desorption stages during the sinter-process. The first appears at a temperature of about 400 °C and the second at about 600 °C. The sinter temperature for Nd-Fe-B magnets made from hydrogenated alloy is decreased by 50 - 80 °C compared to magnets made from not hydrogenated alloys. This may be an effect of the lower oxygen content in the powder and the better thermal contact of the grains in the green compacts. The hydrogen content of the final magnets is in the order of magnitude of 10 - 20 ppm. This is the same for magnets made from not hydrogenated alloy. Characteristic magnetic properties of magnets made from hydrogenated Nd-Fe-B alloy are shown in the table.

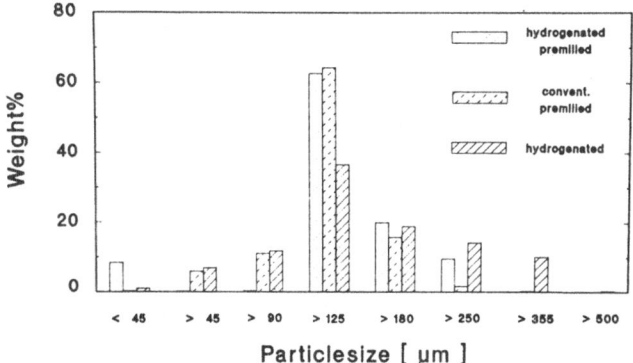

Fig. 9   Particle size distribution of differently
          treated Nd-Fe-B alloy

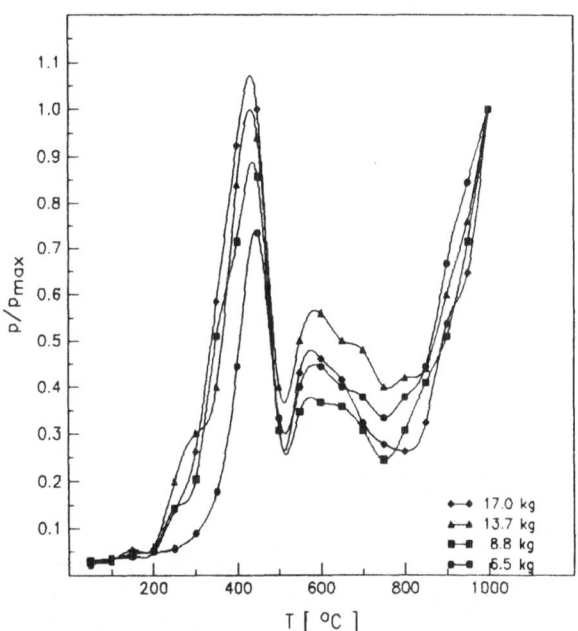

Fig. 10 H_2-desorption of Nd-Fe-B-H green compacts
           as a function of the increasing temperature

TABLE: Characteristic magnetic properties of magnets made
by combination of hydrogen decrepitation and jet-
milling technique.
The powders had particle sizes of 2.0 µm (FSSS)

	Nd-Fe-B 35 w% Nd 1.3 w% B balance Fe	(Nd,Dy)-Fe-B 31.5 w% Nd 3.5 w% Dy 1.3 w% B balance Fe
$_iH_c$	12.1 kOe	21.3 kOe
$_bH_c$	10.8 kOe	11.6 kOe
B_r	1.25 T	1.17 T
$(BH)_{max}$	35.0 MGOe	33.0 MGOe
density	7.57 g/cm³	7.64 g/cm³
H_2-content	20 ppm	10 ppm

ACKNOWLEDGEMENTS

We are indepted to M.A. Friedrich for the measurements
of the c-p-isotherms. The competent technical assistance of
V. Marczylo and R.H. Nitschke is greatly appreciated.

REFERENCES

Cadogan, J.M. and Coey, J.M.D. 1986. Hydrogen absorption and
desorption in $Nd_2Fe_{14}B$. Appl. Phys. Lett. 48, 442.
Harris, I.R. 1987. The potential of the HD-Process in perma-
ment magnet production. 9th Int. Workshop on Rare-Earth
Magnets and Their Appl., FRG, paper W 5.2.
Harris, I.R.; Noble, C. and Bailey, T. 1985. The hydrogen
decrepitation of an $Nd_{15}Fe_{77}B_8$ magnetic alloy. J. Less-
Comm. Met. 106, L 1.
Kuipers, F.A., 1973. $R-Co_5$-H and related systems. Thesis,
Technical University, Delft.
l'Héritier, P.; Chaudouët, P.; Madar, R.; Rouault, A.;
Sénateur, P. and Fruchart, R. 1984. MAGNÉTISME - Une
nouvelle série d'hydrures métalliques ferromagnétiques
de type $Nd_2Fe_{14}B H_x$ $(0 < x < 5)$
C.R. Acad. Sc. Paris, t. 299, Série II, n° 13, 849.

McGuiness, P.J.; Harris, I.R.; Rozendaal, E.; Ormerod, J. and Ward, M. 1986. The production of a Nd-Fe-B permanent magnet by a hydrogen decrepitation / attritor milling route. J. Mat. Sci. $\underline{21}$, 4107.

McGuiness, P.J. and Harris, I.R., 1988. Paper presented at the CEAM-meeting Madrid, and private communication.

Oesterreicher, R. and Oesterreicher H. 1984. Structure and Magnetic Properties of $Nd_2Fe_{14}B\ H_{2.7}$. phys. stat. Sol. (a) $\underline{85}$, K61.

Pollard R.J. and Oesterreicher H. 1986. Novel recording media: Fe-RE-B particles, Paper DB-11 Intermag 1986, Phoenix.

Rozendaal, E., Ormerod, J., McGuiness, P.J. and Harris, I.R., 1987. The production of Nd-Fe-B permanent magnets by a hydrogen decrepitation/jet milling step. 9th Int. Workshop on Rare-Earth magnets and their applications. 1987 Bad Soden F.R.G. Paper W5.4.

Scholz, U.D., Krönert, W.E. and Nagel, H., 1987. The influence of Nd-Fe-B alloy hydrogen absorption on permanent magnet production. 9th Int. Workshop on Rare-Earth magnets and their applications. 1987 Bad Soden F.R.G. Paper W5.3.

Oxidation sizing experiments on neodymium-iron-boron

magnet powders

by

M Stewart M G Gee and B Roebuck

National Physical Laboratory
Division of Materials Applications
Teddington
Middlesex, UK

ABSTRACT

Rare earth (NdFeB-based) permanent magnets can be manufactured by a powder processing route. Reliable methods for characterisation of the size and size distribution of the powders are thus required. Oxidation experiments on NdFeB powders showed that the rate of oxidation was dependent on powder size. Fine powders oxidised more quickly than coarse powders. The oxidation process was monitored by measurement of change in mass in a thermogravimetric analyser (TGA) and by X-ray diffraction experiments on oxidised and partially-oxidised powders. A theoretical expression for the change in mass during oxidation was developed to compare with experimental measurements.

1. INTRODUCTION

Because of their larger surface area per unit mass fine powders oxidise more quickly than coarse powders. Thus, in principle, measurement of the rates of oxidation of powders with different particle sizes should enable a particular batch of powder to be discriminated from others. The technique has been established in experiments on WC and other transition metal carbide powders[1,2]. The oxidation process is monitored by measurements of change in mass in a thermogravimetric analyser. A theory was developed[1] which allowed values of particle size and size distribution to be calculated and compared with experiment. Two kinds of experiment can be performed, isothermal and isochronal. The latter refers to experiments in which a powder is heated from room temperature at a specific linear rate. Isochronal experiments can usually be performed more quickly than isothermal experiments and are consequently more convenient. However, to perform the theoretical analysis for isochronal experiments requires an activation energy for the oxidation process. Also, it is important to know whether the oxidation process is linear or parabolic. In the case of transition metal carbides one of the oxidation products is CO_2 and the evolution of the gas at the carbide/oxide interface continually disrupts the latter resulting in a linear oxidation rate as fresh surface is always available for reaction. However, in the case of NdFeB powers there are no gaseous products and it is likely that the process is parabolic, governed by diffusion through the oxide layer. Support for this assumption has been obtained in stability experiments[3] on NdFeB powders where it was found that in the temperature range 373-623 K oxidation was parabolic with an activation energy of about 24 kJ/mol.

The thermogravimetric experiments were performed on two separate NdFeB powders. The powders were prepared from NdFeB ingots, one by jet milling and the other by a hydrogen decrepitation process followed by ball milling. The latter powder was prepared from a homogenised ingot to yield material which had the nominal stoichiometric composition of $Nd_2Fe_{14}B$. Experiments were also carried out to examine the effects of changes in the sample heating rate and sample mass. This was partly to generate data for comparison with theory but also to identify a combination of powder mass and heating rate which could be recommended as the most sensitive to changes in particle size and yet remain convenient to perform. For example, to compete with other methods of particle size measurement it is important that an isochronal experiment on a given batch of powder can be performed in not more than a few hours.

In order to compare the results of the oxidation experiments with the theoretical analysis samples of the hydrogen decrepitated powder were also examined in a scanning electron microscope. Micrographs were obtained and values for the particle size and size distribution were determined by direct measurement. Also a few X-ray diffraction experiments were performed on fully and partially oxidised powders, in order to examine the products of oxidation.

2. THEORY

2.1 Basic equations

For linear oxidation of spherical powders in isothermal experiments it has been shown[1] that

$$
\frac{M_t}{M_I} = \frac{\int_{\alpha t}^{\infty} (r_i - \alpha t)^3 N(r_i) dr_i}{\int_{0}^{\infty} r_i^3 N(r_i) dr_i} \tag{1}
$$

where M_t was the mass of unoxidised powder at time t, M_I was the initial mass, $N(r_i)$ was the number distribution density and α was a reaction rate constant. The integration limits were between αt and infinity to take account of the fact that particles with sizes smaller than αt were completely oxidised after time t. Equation (1) was derived from the assumption that for linear oxidation of a single spherical particle of initial radius, r_o

$$
r = r_o - \alpha t \tag{2}
$$

However for parabolic oxidation and assuming the oxide directly replaces the original particle, the radius r of the remaining non-oxide is given by

$$
r = r_o - \alpha t^{\frac{1}{2}} \tag{3}
$$

and equation (1) has to be modified accordingly.

A computer program was written to calculate values of the mass ratio $\frac{M_t}{M_I}$ as a function of time for comparison with experiment. The program allows

(a) a single value of radius can be chosen. Thus for isothermal experiments

$$\frac{M_t}{M_I} = \frac{(r_o - at^n)^3}{r_o{}^3} \qquad at^n \le r_o \qquad (4)$$

where n = 1 for linear and n = 0.5 for parabolic oxidation. For isochronal experiments it is necessary to account for the change in α with time and temperature since

$$\alpha = \alpha_o \exp \left[-\frac{Q}{RT} \right] \qquad (5)$$

where α_o is a pre-exponential factor, Q is an activation energy, R is the gas constant and T is the temperature. For linear heating rates

$$T = T_o + Ht \qquad (6)$$

where A is the heating rate and T_o is the starting temperature of the experiment. Equations (5) and (6) are incorporated in (4) and $\frac{M_t}{M_I}$ is evaluated numerically in steps of constant increments of time.

This is equivalent to a consecutive series of isothermal experiments and α is re-evaluated from (5) and (6) for each step. Thus the theoretical plots of $\frac{M_t}{M_I}$ versus time (or temperature) for isochronal experiments converge as the chosen size of time increment becomes smaller. It was found that a convenient size of the time increment was about 1/200 of the time taken for the mass ratio to change from 1 to 0 (start to finish of the experiment).

(b) mixed fractions of particles of single size can be chosen, since in this case

$$\frac{M_t}{M_I} = \frac{\sum_j N_j{}'(r_j - at^n)^3}{\sum_j N_j{}' r_j{}^3} \qquad (7)$$

for j fraction populations, $N_j{}'$, of initial radius r_j. In practice this option is a summation of case (a) for each size of particle.

Thus the ratio $\frac{M_t}{M_I}$ for isochronal experiments is calculated numerically in the same way as for the single particle size distribution in case (a).

3.   ISOCHRONAL EXPERIMENTS

Isochronal oxidation expiments were performed on the powders in a
Stanton Redcroft 780 TGA. Weighed samples were placed in a platinum
crucible and heated, in still air, at a linear rate from room
temperature to 1100 $^{\circ}$C. Temperature measurements were made using a
thermocouple next to the sample and were found to differ markedly from
the nominal heating rates.

3.1   Effect of sample mass and heating rate

To examine the effect of heating rate and powder mass on the oxidation
behaviour, experiments were performed on the H_2 decrepitated powder,
using nominal heating rates of 2.5, 5, 10 and 20 $^{\circ}$C/min, and varying
masses of 20, 40, 80 and 160 mg. Fig 1 shows the relative mass change
for several different initial masses with a constant heating rate of
10 $^{\circ}$C/min. The 20 mg and 40 mg samples were almost identical; the 80 mg
and 160 mg samples initially exhibited similar behaviour but then the
reaction slowed down. For a given mass this change occurred at a fixed
temperature for all heating rates, and increasing the mass decreased the
temperature of the change. For the larger masses, 80 mg and 160 mg, at
the highest heating rate 20 $^{\circ}$C/min, the reaction became pyrophoric and
the heating rate was no longer linear. The apparent reduction in the
oxidation rate of the larger samples was probably due to the formation
of a protective crust, effectively sealing the unoxidised material from
the atmosphere. Since these effects were an experimental artefact the
curves from these experiments were unsuitable for particle size
analysis. However the coincidence of the 20 mg and 40 mg curves
indicated that for these values of sample mass the oxidation was
proceeding under equilibrium conditions. It was found that for a sample
mass of 20 mg, heating rates up to 20 $^{\circ}$C/min could be used for the
particle size analysis experiments. Fig 2 shows typical curves for the
H_2 decrepitated and the jet milled powders.

3.2   Particle Sizing

It is evident from Fig 2 that the technique can be used directly as a
method for "fingerprinting" a particular batch of powder and thus could
be of use as it stands for quality control. However the method would be
more useful if particle size distributions could be obtained from the

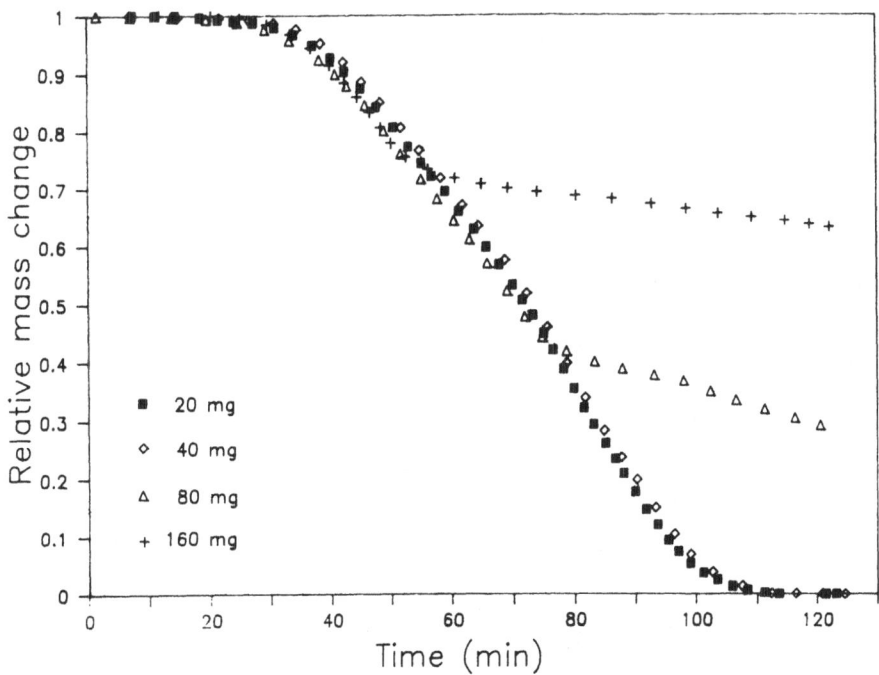

Fig 1 - Isochronal curves, 10 $^{\circ}$C/min;
effect of different sample masses

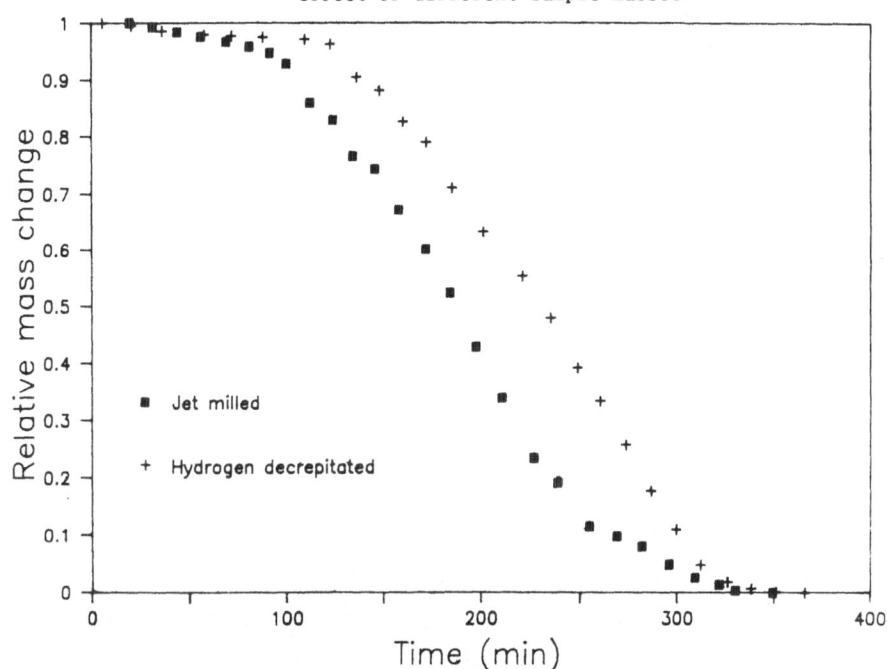

Fig 2 - Isochronal curves for jet milled and
H$_2$ decrepitated powder; ~ 2 $^{\circ}$C/min

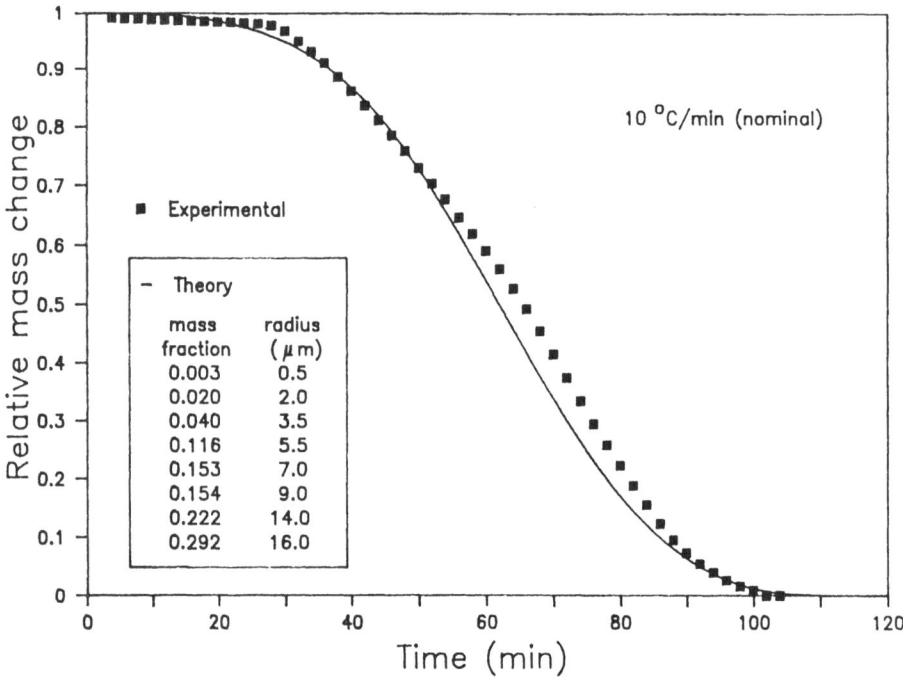

Fig 3 - Comparison of theory and experiment
for H$_2$ decrepitated powder

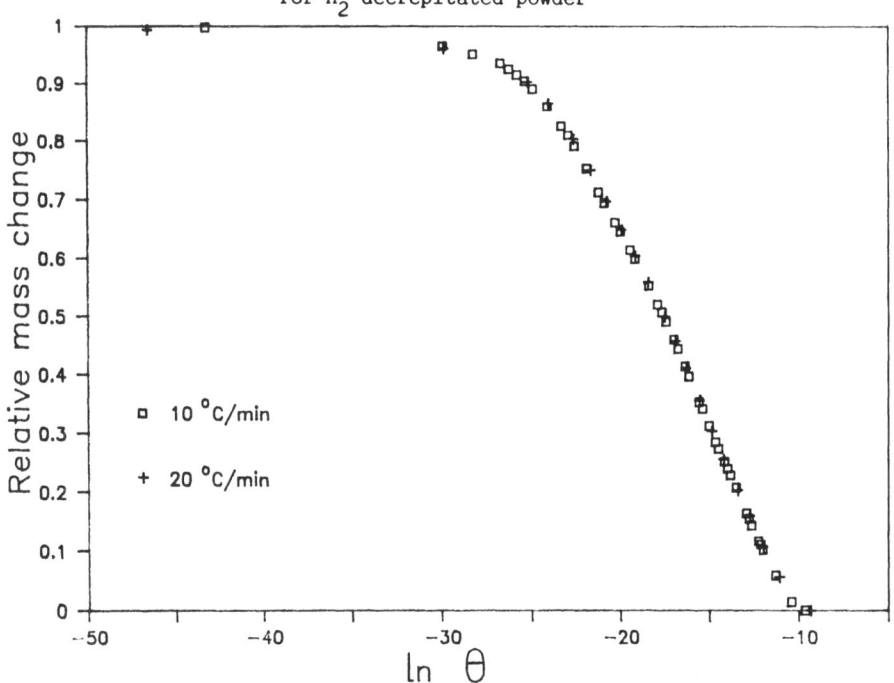

Fig 4 - Comparison of relative mass fraction and θ
for an activation energy of 120,000 J/mol at
two heating rates

isochronal curves. The two unknown parameters in the theoretical equations are the activation energy Q and pre-exponential factor α_o. In order to obtain values for Q and α_o theoretical plots of $^M t/M_I$ were derived using the particle sizes measured by scanning electron microscopy for the hydrogen decrepitated powder and compared with the experiment isochronal curve. The best fit, Fig 3, was obtained with a value of 35 kJ/mol for Q and 330 μm^2/min for α_o.

In order to check these values a more direct method was attempted using first of all isothermal experiments on small blocks. However this proved difficult because of the friability of the material. It is, however, possible to calculate the activation energy from two isochronal experiments on the same powder. For two isochronal experiments at heating rates H1 and H2, when the relative mass fractions are equal, at time t1 and t2, equation (4) (incorporating equations (5) and (6)) is equivalent for both experiments. Since it is the same powder the r_o and α_o terms cancel, leaving:-

$$\int_0^{t1} \exp\left[-Q/R(H1t+T_o)\right].dt = \int_0^{t2} \exp\left[-Q/R(H2t+T_o)\right].dt = \theta \quad (8)$$

Thus a value for Q can be obtained by comparing curves of mass fraction versus equation (8). It can be seen from Fig 4 that a value of 120 kJ/mol for Q gave a satisfactory fit for the H_2 decrepitated powder.

In order to determine the pre-exponential factor, α_o, oxidation experiments were carried out on sieved powders. Sieving the powders established the maximum particle size in the sample, by passing through finer sieves until there was a detectable difference in the curve. The largest particles were the last to fully oxidise and this was where the curve met the x-axis. Thus knowing Q and the maximum particle size a value for α_o could be determined from (7). This is represented graphically in Fig 5 for the H_2 decrepitated powder, for an α_o of 2×10^7 μm^2/min.

Thus it can be seen that the two methods used for obtaining values for Q and α_o gave distinctly different values. Further work is therefore needed to obtain the correct values for Q and α_o, probably by a direct experimental method, before the technique can be used reliably for particle size measurements.

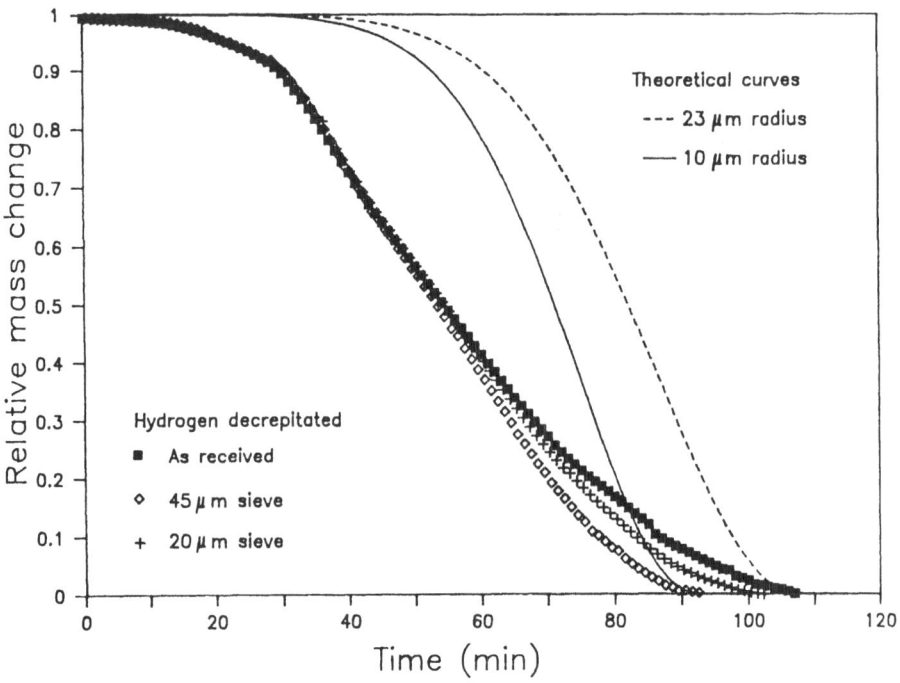

Fig 5 - Effects of sieving on isochronal curves compared
with theoretical plots for single particle sized powders

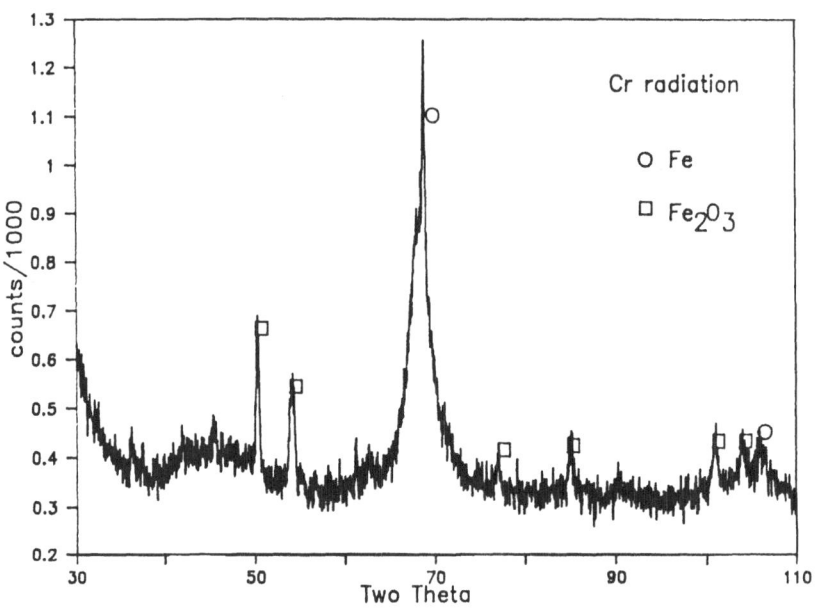

Fig 6 - XRD showing iron peak

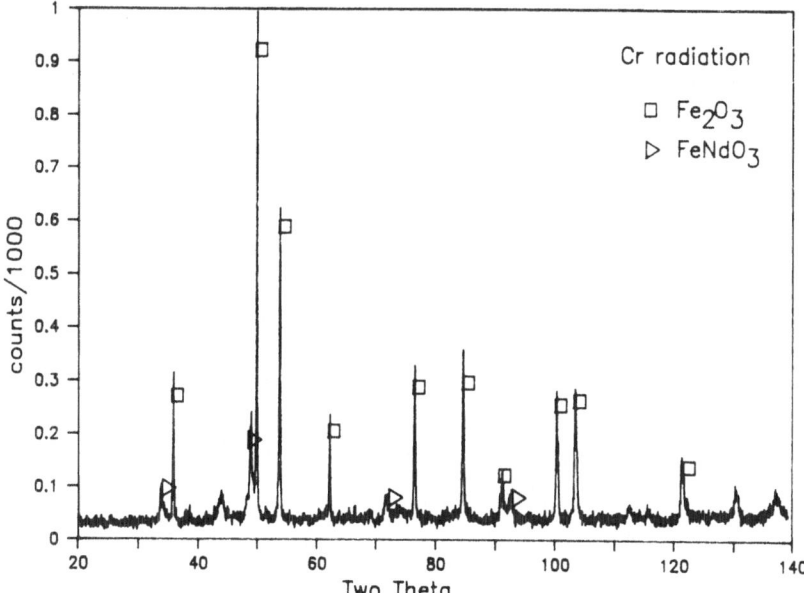

Fig 7 - XRD showing final products of oxidation

4.   X-RAY DIFFRACTION

The progress of the reaction was also monitored by performing XRD on
partially and fully oxidised samples. Fig 6 shows that initially the
NdFeB structure disappeared and a broad iron peak appeared, this was
consistent with other investigations[4]. The final oxidation products were
mostly Fe$_2$O$_3$ and FeNdO$_3$, (Fig 7), with some Fe$_3$O$_4$ forming at
intermediate stages. The total mass gain of the samples, approximately
35%, was consistent with the formation of oxides, with two metal atoms
for every three of oxygen.

Although the complexity of the reactions revealed by XRD indicated that
the use of a single activation energy was an over simplification, the
particle size analysis appeared to fit for different heating rates.
Since in the early stages of oxidation NdFeB transforms to iron its
oxidation behaviour might be similar to that of iron; preliminary tests
on iron powders did result in a similar value for the activation energy
as that obtained on the NdFeB powders.

5. SUMMARY

In was found that the oxidation rate of a powder sample of NdFeB magnet material was dependent on the average size and size distribution of the powder particles. Isochronal oxidation measurements, which are convenient to perform and can be done in about 1 h, can be used to characterise a particular batch of powder. A theoretical analysis, which was derived for experiments on transition metal carbide powders, was applied to the results in order to obtain values for the particle size and size distribution. The analysis requires further work before it can be used with confidence because accurate values for the constants of the activation process could not be obtained. However, the technique should be particularly suitable for identifying the presence of larger particles in a particular batch of powder because the results depend on measurements of changes in mass.

6. REFERENCES

1   B ROEBUĆK, E G BENNETT, E A ALMOND and M G Gee, J Mater Sci, 21, 1986, 2033-2042.

2   B ROEBUCK, E A ALMOND and J L KELLIE, PM66-EPMF, Dusseldorf, FDR, July 1986, Horizons of Powder Metallurgy, Eds W A Kaysser and W J Huppman, Pt 1, 123-126.

3   B E HIGGINS and H OESTERREICHER, IEEE TRANS. MAGNETICS, 23-1, (Jan 87), 92-93.

4   P SCHREY, IEEE TRANS. MAGNETICS, 22-5, (Sept 86), 913-915.

COMPOSITIONAL AND PROCESS EFFECTS ON STRUCTURES AND PROPERTIES OF Fe-Nd-B-BASED RIBBONS AND MAGNETS PRODUCED BY THE MELT SPINNING ROUTE

H. A. Davies, K. J. A. Mawella[*], R. A. Buckley, G. E. Carr[**], A. Manaf and A. Jha

School of Materials, University of Sheffield
Mappin St., Sheffield S1 3JD, U.K.
* Present Address: R.A.R.D.E., Fort Halstead, Kent
** Present Address: Rolls Royce Associates, Derby.

ABSTRACT

The effects of alloy composition and of process variables such as roll material and velocity, ribbon thickness, annealing temperature and pressing conditions on the microstructure and magnetic properties of chill-block melt spun FeNdB-based ribbons and of magnets derived by hot-pressing the ribbons have been studied. Measurements by X-ray diffraction line broadening analysis indicate that a very small grain size (typically 30-80nm) plays a major role in promoting high coercivity $_iH_c$ and energy product $(BH)_{max}$ but they also suggest that other, less identifiable, factors also play a role. Substitution of Fe by Co leads to generally improved properties at elevated temperatures for hot pressed magnets as well as for the as-spun ribbon. Small additions of niobium and of silicon can lead to substantial improvements in coercivity and to remanence respectively, although the effect of the silicon appears to be particularly sensitive to the melt spinning parameters. The morphology of the microcrystallites is also, correspondingly, changed markedly by the silicon additions but the role of this in the enhancement of energy product requires further clarification.

INTRODUCTION

The development by General Motors of iron-neodymium-boron magnets produced by a chill-block melt spinning route (Croat et al 1984a, Lee 1985) represented a radical departure from conventional rare earth magnet technology. Although the use of melt spinning for producing amorphous and microcrystalline alloys was already an established technology, for instance, in producing novel soft magnetic alloys in ribbon form, its application to hard magnetic alloys had not been forseen. The melt spinning to ribbon typically 15-60μm thick induces ultra-rapid solidification, reliably estimated to be in the range 10^5-$10^6 Ks^{-1}$, depending on thickness and other factors (Davies 1983). Two major effects of such rapid quenching are substantial refinement of the microstructure, notably the grain size, and, if the cooling rate exceeds a critical value, broadly characteristic of the particular alloy, avoidance of crystallisation and solidification to an amorphous or glassy phase. The latter may be subsequently devitrified to a

microcrystalline structure. Both of these effects can be utilised in the
melt spinning route for the production of FeNdB magnets and it is control
of the final grain size that is crucial in the development of the hard
magnetic properties.

The broad aims of this study were to investigate some of the effects
of composition and melt spin process variables such as roll material and
ribbon thickness on the microstructure and properties of as-cast ribbon
and to study the effects of composition and annealing parameters on the
microstructure and properties of initially amorphous, overquenched ribbon.
In particular, the role of cobalt in enhancing the thermal stability of the
magnetic properties and of other elements such as niobium and silicon in
modifying the properties were also considered.

EXPERIMENTAL

A sealed atmosphere chill-block melt spinning unit, based on a large
evacuable stainless steel chamber having a ferro-fluidic rotating seal,
was constructed in our laboratory under the C.E.A.M. programme. This can
be evacuated to better than 10^{-5} torr, prior to filling with argon and
facilitates the spinning of very clean ribbon. Since this was not fully
commissioned until well into the project period, during the earlier phase
ribbon was produced in an existing smaller and less sophisticated chamber.

The chamber was designed to spin up to 200g of ribbon but experimental
samples of alloy were generally spun in amounts less than 30g. The sample
was melted in each case by r.f. induction in a quartz crucible and propel-
led by argon overpressure through a nozzle typically 0.7mm in diameter onto
a rapidly rotating roll. Copper, steel and chromium surfaced rolls of
nominal diameter 150mm were available. The ribbon thickness was controlled
within the range $\sim$ 15-60μm by varying the roll substrate velocity between
7 and 40ms^{-1}.

A hand operated hydraulic press was adapted for hot pressing of magnets of diameters 5-7mm from finely crushed and sieved ($<$ 150μm) ribbon.
High density graphite dies and pistons were employed and these were contained, using O-ring end seals in a flowing argon environment within a
quartz tube. The design was similar to that described by Gwan et al
(1987). Subsequently, using an enlarged die and pistons, some of the hot-
pressed magnets were upset-forged to half their original length to induce
anisotropy.

Most of the magnetic measurements on ribbon were made with a Princeton Applied Research vibrating sample magnetometer (VSM) in Dr. Grundy's laboratory at Salford University (Group I member) with additional measurements being made at Durham University (Dr. D. Corner-Group I member). Permeameter measurements on magnets were made at Sheffield University (Dr. D. Howe-Group III member) and at Birmingham University (Dr. I. R. Harris - Group II member).

Determinations of the Curie temperature T_c and of the crystallization parameters and kinetics for 'overquenched' ribbon were made on Perkin Elmer DSC2 and DSC7 differential scanning calorimeters in the authors' laboratories. X-ray diffraction phase analysis and particle size line broadening analysis were made with a Philips automatic diffractometer.

All the scanning electron microscopy and most of the t.e.m. and micro-analysis studies on ribbon and magnet samples were performed, respectively, with Philips PSEM 500 and STEM 400 electron optical instruments at Sheffield. Some additional observations were made with the AEI 1000 keV electron microscope at B.S.C. Swinden Research Labs., Rotherham and with a JEOL 4000FX instrument at Birmingham University in collaboration with Dr. I. P. Jones. However, attempts at boron analysis by EELS proved difficult because of interference from the tail of the Nd signal.

RESULTS AND DISCUSSION

a) Curie temperature enhancement

The value of T_c for the basic ternary alloy in melt spun form was shown by D.S.C. to be 310C, identical to the starting ingot material, and within 2K of that of the sinter route magnets of similar composition. Although various alloying additions were considered and investigated for enhancing T_c, the most effective was found to be Co in substitution for Fe. Substitutions up to x = 25 wt% in $Fe_{78.26-x} Co_x Nd_{14} B_{7.74}$ alloys were made and T_c determined for both ingot and melt spun ribbon samples. The data were in good agreement with those of Matsuura et al (1985) and of Arai and Shibata (1985) for sinter route material and lent confidence to the alloy preparation technique (compositional adjustment, by argon arc melting, to master alloys obtained from Rare Earth Products Ltd.). T_c increased almost linearly with Co content from $\sim$ 583K for x = 0 to 820K for x = 25 (Fig. 1). Magnet alloys containing $\sim$ 15 at % Co were likely, in principle, to provide a good compromise between improved properties and cost. On the basis of T_c alone, alloys containing $\lesssim$ 20 wt % Co would be unlikely to

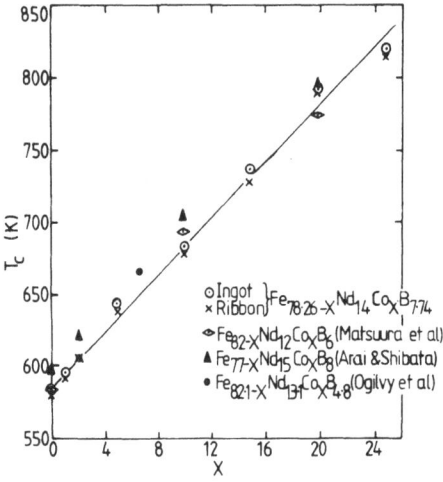

Fig. 1 Effect of Co on Curie temperature T_C of ingot and melt spun FeNdB, with previous data for sinter-route magnets.

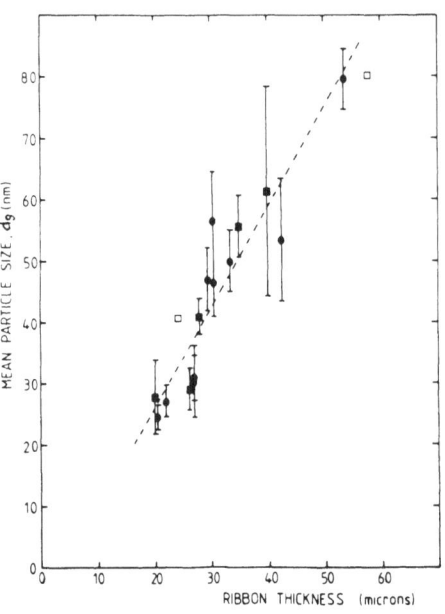

Fig. 3 Mean X-ray grain size d_g of $Fe_{14}Nd_2B$ phase in as-spun $Fe_{82.6}Nd_{13.1}B_{4.2}$ and $Fe_{77.9}Nd_{14.3}B_{7.8}$ ribbon on Cu and Cr rolls.

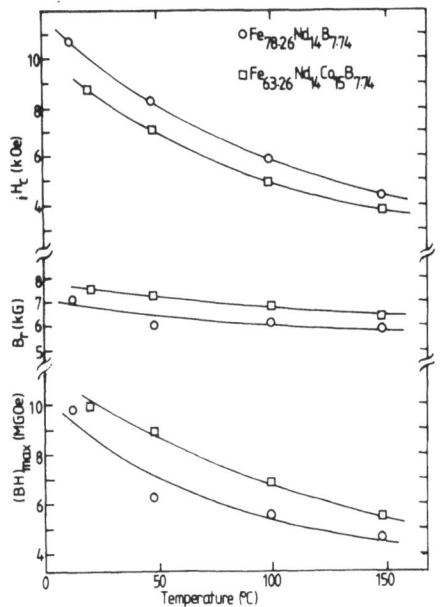

Fig. 2 Temperature dependence of $_iH_c$, B_r and $(BH)_{max}$ for FeNdB and FeCoNdB hot pressed magnets.

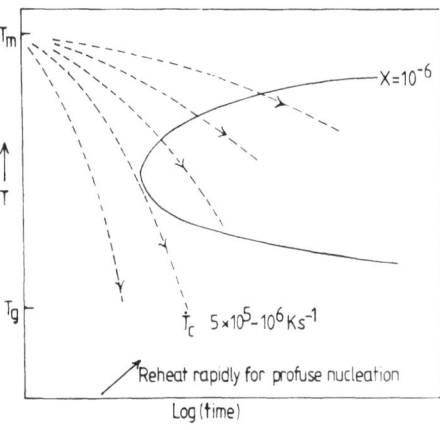

Fig. 4 Schematic C-C-T curve for crystallisation of molten $Fe_{14}Nd_2B$ alloy and curves representing different cooling rates.

meet the stringent temperature specifications for aerospace applications.

To test the real effectiveness of the Co additions it was necessary to determine the temperature dependence of the magnetic properties. Facilities were not available for such tests on ribbon samples by v.s.m. but permeameter measurements were performed up to $150^{\circ}C$ on hot pressed magnets fabricated from the ribbon (the hot pressing is described in a later section). This was fitted with a heater system around the magnet sample positioned between the field magnet poles. The results are summarised in Fig. 2 together with the corresponding data for the base ternary alloy. Although, as expected, B_r was increased by Co substitution, $_iH_c$ was reduced, so that the increase in $(BH)_{max}$ at any given temperature within the range was smaller than expected from the T_c data. This was thought to be due to the Co substitutions inducing a reduction in the magneto-crystalline anisotropy field for the tetragonal 14/2/1 phase. However, it should be borne in mind that the ternary and quaternary magnets were not pressed under similar conditions (2 min at $750 \pm 10^{\circ}C$) and although these were close to optimum conditions for the ternary alloy this need not necessarily be the case for the Co-containing quaternary. Thus, some improvement in $_iH_c$ and its temperature dependence might be expected as a result of process optimisation.

b) X-ray diffraction studies of crystallite size

i) As-spun ribbon

The pioneering work on melt spun Fe-Nd-B ribbons by the General Motors group in the U.S.A. (Croat et al 1984 a,b) had shown that directly quenched ribbon of optimal $_iH_c$ and $(BH)_{max}$ had an average grain size of the order of 50nm. This, they indicated, resulted from melt spinning at a critical roll speed V_s^c and they concluded that the high $_iH_c$ was due to the ultra-fine grains. Both $_iH_c$ and $(BH)_{max}$ were shown to increase steeply with increasing roll speed V_s, due it was presumed to a progressively decreasing grain size resulting from increasing cooling rate ↑, up to a maximum at V_s^c where vitrification of the alloy begins. Further increase in V_s led to an equally sharp decrease in the properties as the proportion of amorphous phase increased. In this study, we have preferred to take ribbon thickness t_r rather than V_s as a measure of ↑ since, unless a constant flow rate of melt is employed, from one study to another, t_r and thus ↑ will not necessarily correlate directly with V_s. Nevertheless even t_r is not totally satisfactory since the interfacial heat transfer coefficient h can also be influenced independently by changes in V_s.

In the absence of any previous systematic study of the grain size d_g, we determined the correlation between d_g and t_r for as-cast ribbons over a range of t_r from $\sim 60\mu m$ down to $20\mu m$ (Carr et al 1988). The mean particle size, which was equated with d_g, was determined by X-ray line broadening analysis for the $Fe_{14}Nd_2B$ phase using the Scherrer formula with correction for instrumental broadening using an annealed Cu powder sample having $d_g \gg 100nm$. Six peaks for the $Fe_{14}Nd_2B$ phase were used: (214),(105),(313),(224), (205) and (411) and d_g taken as the arithmetic mean. d_g decreased about linearly from 80nm for $t_r = 60\mu m$ to 23nm at $t_r = 20\mu m$ (Fig. 3). For $t_r \lesssim 27\mu m$ the structure became increasingly amorphous so that d_g then relates to crystallites within a glassy matrix. This critical t_r also corresponds roughly to the maximum in $(BH)_{max}$ reported by Croat et al (1984 b).

The effect of $\dot{T}$ on the microstructure can be explained by reference to the C-C-T curve for the start of the liquid to crystal transformation, shown diagramatically in Fig. 4. As $\dot{T}$ increases, the temperature at which crystallisation begins decreases (i.e. undercooling increases) until, at a critical value $\dot{T}_c$, crystallisation is avoided and the liquid is cooled to a glass (Davies 1983).

No systematic difference in d_g was observed between ribbons cast on solid Cu and on Cr-plated rolls and this is consistent with our previous observation that peak properties occurred at a similar ribbon thickness for the two roll materials (Ogilvy et al 1985). Measurements were made on low boron (4.2 at %) and high boron (7.8 at %) alloys; no significant effect of B content on d_g was observed, which is surprising, since it has been shown to influence significantly the magnitude of, and the t_r corresponding to, the maximum $_iH_c$ for as-spun ribbon (Croat et al 1984 a). Thus, although the data indicate that the small d_g plays a major role in promoting $_iH_c$ in the melt spun ribbon, they also suggest that other factors are also important.

ii) Overquenched and annealed ribbon

Clearly a more controllable and preferred means of obtaining the fine microcrystalline structure is to overquench the ribbons to an amorphous or partly-amorphous structure and subsequently anneal at an appropriate temperature and time. Accordingly, we also studied the evolution of the microstructure during the annealing of such overquenched ribbons.

The mean d_g, derived by X-ray line broadening analysis, as a function of annealing time at 600°C and 700°C for initially overquenched $Fe_{77.9}$ $Nd_{14.3}B_{7.8}$ ribbon is given in Fig. 5. Interestingly, after the initial increase, corresponding to the devitrification process, d_g remains almost independent of anneal time, at least up to $\sim$ 30 mins. This was also found to be the case for the Co-substituted ribbons (Fig. 6). In this case, the effect of the magnitude of the prior quench rate was also investigated. 'Well over-quenched' ribbon (a higher $\dot{T}$ - $t_r \sim$ 20μm) yielded a smaller d_g than 'slightly overquenched' ribbon ($\sim$ 25μm), evidently indicating pre-existing nuclei formed during the initial quench and thus a higher nuclea-tion frequency during the quench in the former case.

In contrast to d_g, the magnetic properties of the ribbons reach a maximum at annealing times in the range 3-5 mins (corresponding, we believe, to complete devitrification) and then decrease. (Fig. 7 gives examples for the Co-substituted alloys). Clearly, since d_g plateaus approximately beyond 5 mins anneal time, then the high $_iH_c$ must derive from factor(s) in addition to the small grain size.

Figure 7 also indicates that, of the Co compositions investigated, the 15 wt% alloy has the broadest and highest maximum for both $_iH_c$ and $(BH)_{max}$. Thus, it offers the greatest flexibility for this process route and is least sensitive to variations in annealing parameters.

We have also investigated the kinetics of crystallisation of the amorphous phase in overquenched ribbon for a low boron alloy ($Fe_{81.75}$ $Nd_{13.5}B_{4.75}$) using D.S.C. The Kissinger method was used whereby the peak temperature T_p of the crystallisation exotherm is determined as a function of the heating rate β and the activation energy for crystallisation E_x derived from the slope of a linear plot of log $[\beta/T_p^2]$ vs. $[1/T_p]$. Measurements were made for 3 different mean ribbon thicknesses 16, 18 and 22μm and values of E_x were 293, 257 and 220 kJ mol^{-1}, respectively. The value of $\sim$ 290 kJ mol^{-1} for the 16μm thick ribbon is broadly consistent with other data for metal-metalloid glasses but the dependence of E_x on thickness is of particular interest. It may indicate that incipient crystals were present in the thicker samples and that, in such cases, the activation energy for crystal growth is substantially smaller. Data of this type are of relevance with regard to processing parameters for hot pressing of overquenched ribbons.

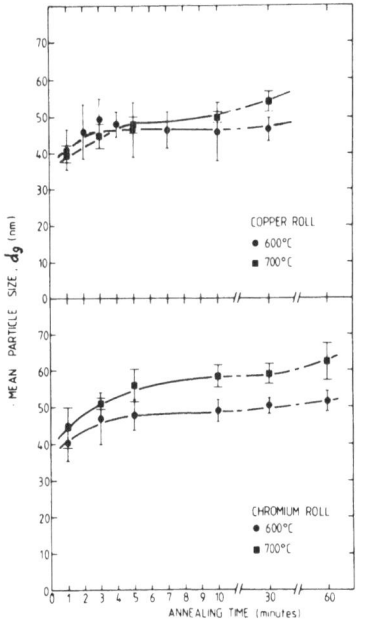

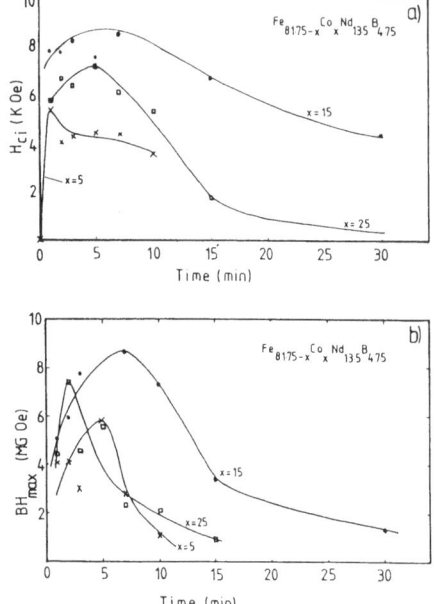

Fig. 5 Effects of annealing temperature and time on d_g for overquenched $Fe_{77.9}Nd_{14.3}B_{7.8}$.

Fig. 7 Effect of annealing time at 600^oC on (a) $_iH_c$ and (b) $(BH)_{max}$ for $Fe_{81.75-x}Co_xNd_{13.5}B_{4.75}$ overquenched ribbon.

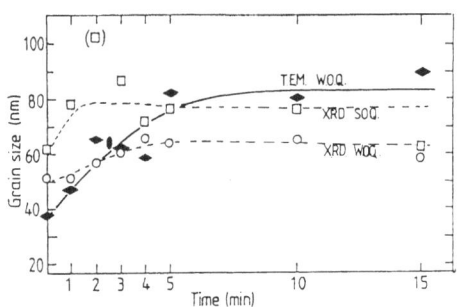

Fig. 6 Effect of time at 700^oC on X-ray and t.e.m. d_g for overquenched $Fe_{63.26}Co_{15}Nd_{14}B_{7.74}$ ribbon. ⬥ indicates value for hot pressed magnet .

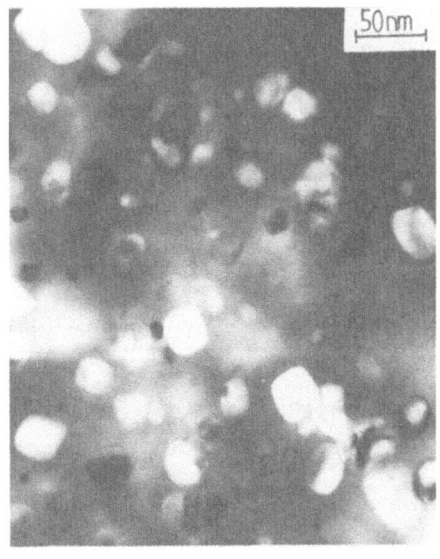

Fig. 8 Overquenched and annealed $Fe_{78}Nd_{14.3}B_{7.7}$ ribbon in partly devitrified state.

(iii) <u>Transmission electron microscopy</u>

For the 'well overquenched' $Fe_{63.26}Co_{15}Nd_{14}B_{7.74}$ ribbon, d_g was also measured directly by t.e.m. on foils ion-beam thinned from ribbon samples (Fig. 6). There is a broad measure of agreement with the X-ray line broadening data. Of course, the latter relate to mean grain dimensions normal to the plane of the ribbon, whilst the t.e.m. data relate to dimensions in the plane of the ribbon (since the foils were derived by thinning down normal to the plane). We have not yet prepared foils by thinning in the ribbon plane and we cannot be certain that the grains are equiaxed along three orthogonal axes.

The microstructure in the very early stages of annealing, in which the growing 14/2/1 crystallites are surrounded by untransformed amorphous matrix, is exemplified by Fig. 8. The typical microstructure on complete devitrification, in this case after 4 mins at $600^{\circ}C$, is shown in Fig. 9; the mean crystallite diameter is about 50nm and this is typical of the high $(BH)_{max}$ condition. After a long (30 mins) anneal, the same ribbon specimen shows some grain growth (Fig. 10) but more obvious is the tendency for the 14/2/1 crystals to spheroidise and for the development of intergranular blocky areas of the Fe_4B_4Nd phase. Clearly, microchemical changes are occurring for long anneals which can account for the decreasing property values, as summarised in Fig. 7.

Of greater interest, however, is the reason for the decreases in $(BH)_{max}$ and $_iH_c$ beyond the peak values at 3-5 mins and also the phase constitution corresponding to peak properties. The existence of a thin layer of an amorphous Nd-rich phase around the $Fe_{14}Nd_2B$ grain boundaries has been proposed (Mishra 1986); it is supposed that this plays a major role in promoting a high coercivity. Microscopy at very high resolutions (e.g. Fig. 11) has failed to reveal such a grain-boundary phase in the case of our ribbon samples. In any case, it is not clear that this would be a <u>metallic</u> phase, as implied in previous studies, since, being supposedly Nd-rich, its crystallisation temperature should be lower than that of the amorphous $Fe_{14}Nd_2B$ phase. Thus, this leaves the possibility that, if such a phase exists, it may be a Nd-rich oxide. There might be a tendency, for example, for accelerated oxidation at grain boundaries in thin foils of an active rare-earth alloy such as this, either during preparation or during subsequent storage, handling and study. One could not then be certain that a grain boundary phase observed by t.e.m. was not an artefact. Even so, it would still be unclear why we have not observed this phase in our present

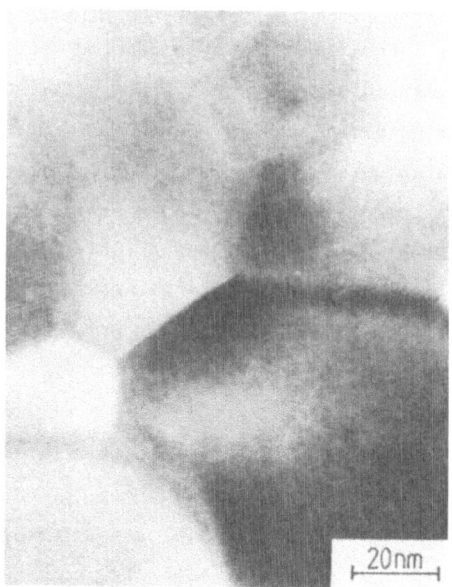

Fig. 9 Overquenched and annealed (4 mins at 600°C) $Fe_{78}Nd_{14.3}B_{7.7}$ ribbon after complete devitrification.

Fig. 11 High resolution electron micrograph of overquenched $Fe_{78}Nd_{14.3}$ $B_{7.7}$ after anneal at 700°C for 4 mins.

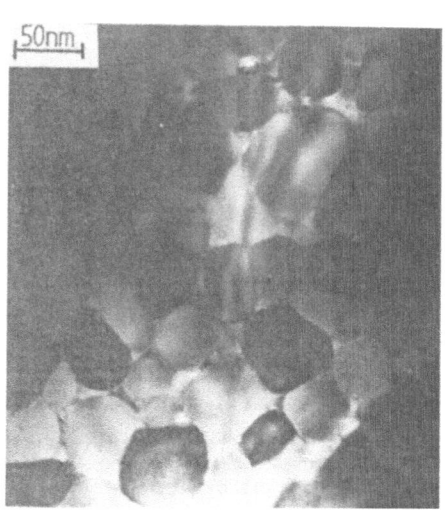

Fig. 10 Overquenched $Fe_{78}Nd_{14.3}B_{7.7}$ ribbon after prolonged anneal (30 mins at 600°C) showing grain spheroidisation and grain boundary phase.

Fig. 12 Overquenched $Fe_{78}Nd_{14.3}B_{7.7}$ ribbon cast onto Cu roll and annealed 10 mins. at 600°C.

studies.

We also observed occasionally very large grains associated with the characteristically fine microcrystallites (Fig. 12) in the over-quenched and annealed ribbon samples. We believe these to have been the result of locally reduced cooling rate and under-cooling, probably adjacent to air pockets on the underside (i.e. roll-contact surface) of the ribbon. Melt spinning onto a Cr-plated roll proved to be beneficial in avoiding this phenomenon due, we believe, to better and more uniform physical and thermal contact with the FeNdB melt.

c) Alloying investigations

i) Niobium and dysprosium additions

The presence of small concentrations of Nb and/or Dy is sintered FeNdB-based magnets has been shown to considerably enhance the coercivity, the Dy largely through a significant increase in the anisotropy field of the 14/2/1 phase on partial substitution of Nd. Small amounts of $Fe_{76}Nd_{14.5}$ $B_7Dy_{1.5}Nb_1$ and $Fe_{75.1}Nd_{16.0}B_{7.8}Dy_{1.1}$ were chill block melt spun in argon for Dr. Grundy at Salford. The ribbon samples of the Nb containing alloy were shown to have intrinsic coercivity $_iH_c > 16kOe$ considerably in excess of that of the equivalent ternary alloy. This was ascribed to domain wall interaction with a fine coherent Nb-rich precipitate. (Parker et al 1987).

Samples of three alloys containing Nb or Nb + Dy were melt spun into an overquenched state and the ribbon in each case subsequently hot-pressed to fully dense magnets at Sheffield. These also had substantially enhanced $_iH_c$ compared with all the ternary and Co-containing alloys investigated (see section d)).

ii) Silicon

Although the melt spun FeNdB ribbon derives its high coercivity from the ultra fine grain size, the energy product $(BH)_{max}$ is limited by the fact that this grain structure is crystallographically isotropic. Since the grain size in the optimum state is so extremely fine, it is not realistic to pulverise the ribbon down to single grain particles and so facilitate magnetic alignment, as is practiced for sinter route materials. An obvious means of enhancing the remanence Br in the appropriate direction, and thus of improving $(BH)_{max}$, is to induce crystallographic texture. The method developed by General Motors is to die-upset forge the hot-pressed magnet. This causes the material to flow laterally and results in substantial c-axis alignment parallel to the press direction. $(BH)_{max}$ values approximating to those typical for sinter route magnets ($\sim$ 35MGOe) have been achieved. This is, however, a two-stage process after melt spinning and methods of improving the properties in the as-spun ribbon have been sought.

However, substantially enhanced properties in as-cast ribbon have been reported by workers at Energy Conversion Devices in the U.S.A. (European Patent Application 0195219) for alloys containing small concentrations of silicon. Moreover, these were claimed to be magnetically isotropic, even though $J_r \gg 0.5 J_{sat}$. To investigate these, we melt spun two of the alloys of interest $Fe_{79.6}Nd_{13.2}B_6Si_{1.2}$ and $Fe_{77.8}Nd_{13.9}B_6Si_{2.3}$. The former especially was found for isolated ribbon samples to have substantially increased B_r and $(BH)_{max}$ over the ternary FeNdB alloys in the as-spun condition (up to 10.7kG and 21.8MGOe, respectively measured by v.s.m. longitudinally, in the plane of the ribbon (Fig. 13)). However, the values varied greatly from sample to sample even when these were nominally of constant thickness, and it has not yet been possible to establish any clear correlation with other parameters.

High resolution electron microscopy on thin foils parallel to the ribbon plane (a typical micrograph is shown in Fig. 14) indicated a very different microstructure from that of the ternary alloys with the grains, though comparable in size with the ternary, being columnar in appearance, at least in the plane of the ribbon. It appears also that the grains are delineated by a very thin film of an, as yet, unidentified phase and it is possible that this plays a role in promoting the enhanced properties. Electron diffraction rings for the 14/2/1 phase indicated uniform intensity around the spotty Debye rings and thus no preferred orientation in any direction within the ribbon plane, in spite of the structure being columnar in this plane. We have not yet succeeded in characterising the grain morphology normal to the ribbon plane. Measurements of the properties through the ribbon thickness are not yet conclusive because of difficulties associated with correction for the demagnetising factor. Similarly, although there is clearly preferred orientation of the (001) planes normal to the foil in thicker ribbons cast at lower roll velocities, as manifest by X-ray diffraction, it is not yet clear whether such texture is present at the critical velocity and thickness.

These particular alloys appear to add a new dimension to the rapid

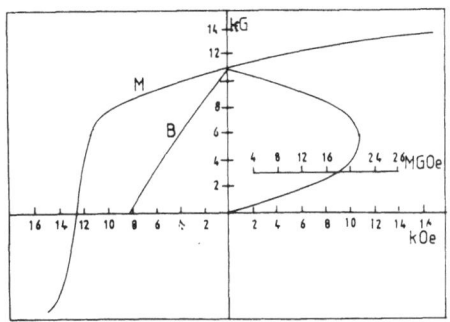

Fig. 13 Second quadrant M–H and B–H curves and B.H for as-spun Fe$_{79.6}$ Nd$_{13.2}$B$_6$Si$_{1.2}$ ribbon.

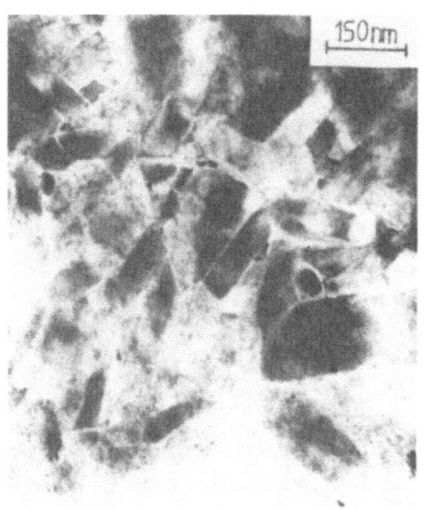

Fig. 14 Electron micrograph showing columnar grains in ribbon plane for as-spun Fe$_{79}$Nd$_{13.2}$B$_6$Si$_{1.2}$.

solidification route for FeNdB magnets and suggest that polymer bonded magnets, based on as-cast ribbon, having (BH)$_{max}$ well in excess of 12 MGOe are feasible (compared with 8MGOe at present). However, clearly, further work is required to fully characterise and understand the material.

d) Fully dense magnets

Investigations of the hot pressing process showed that a pressure of $\sim$ 150MPa and a pressing temperature of 750 $\pm$ 10°C (measured with a sheathed thermocouple placed in the die) for $\sim$ 2 minutes resulted in practically fully dense (> 99%) magnets whereas temperatures $\lesssim$ 725°C gave inconsistent densities. A typical microstructure is shown in Fig. 15; the ribbon fragments are seen to be in a layered configuration resulting from the uniaxial pressing. It is doubtful, however, that the particles are fully sintered together since the mechanical properties of this type of magnet are inferior to those of the conventional powder sinter-route magnets.

Initial trials were performed with powdered GM Magnequench and this gave satisfactory results. Subsequently, trials were performed with ternary FeNdB ribbon spun at Sheffield, based on both low-boron 4.75 wt% B (GM) and high-boron 7.75 wt% B (Sumitomo) compositions. Typical property values for the former after pulse magnetising at 45kOe were $_iH_c$ = 10.8kOe,

Br = 7.5kG and (BH)$_{max}$ = 9.0MGOe while typical corresponding values for the latter were 10.8kOe, 7.1kG and 9.9MGOe, respectively. Addition of Al (2wt%) to the high boron alloy did not improve the coercivity significantly in spite of the fact that it has been reported to give an improvement for sinter-route magnets. More recently, we have concentrated on the Fe$_{63.26}$Co$_{15}$Nd$_{14}$B$_{7.74}$ alloy to assess the influence of Co on the thermal

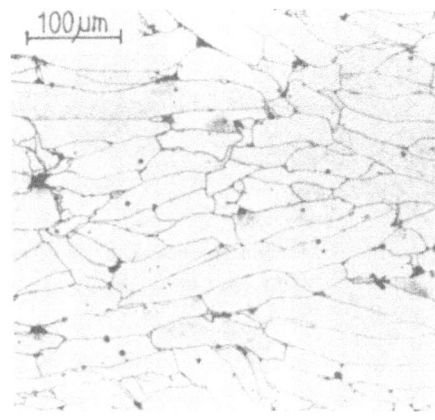

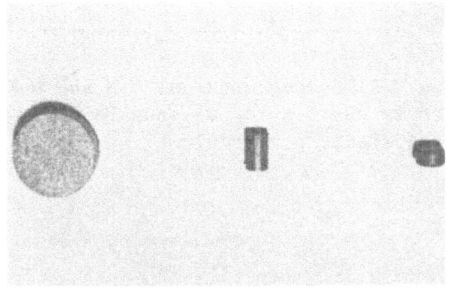

Fig. 16 Magnets produced by (a) polymer bonding (b) hot pressing and (c) die upset forging of FeCoNdB melt spun ribbon.

Fig. 15 Microstructure of hot pressed Fe$_{63.26}$Co$_{15}$Nd$_{14}$B$_{7.74}$ magnet.

stability of the magnets (Fig. 16). The magnetic data are summarized in Fig. 2. Small additions of Nb and of combinations of Nb and Dy, however, resulted in dramatic improvements in coercivity in the hot pressed magnets. The three compositions investigated, Fe$_{76}$Nd$_{16}$B$_7$Nb$_1$, Fe$_{77}$Nd$_{14.5}$B$_6$Dy$_{1.5}$Nb$_1$ and Fe$_{76}$Nd$_{14.5}$B$_7$Dy$_{1.5}$Nb$_1$ all had $_i$H$_c$ outside the range of measurement of the permeameter (> 16kOe). It is not possible from these data to assess the relative effectiveness of the Nb and Dy but work is in progress to study the role of the presumed fine Nb precipitate in pinning domain walls as was reported by Parker et al (1987) for ribbon. With such small additions of these two elements, alloys of this type have commercial potential, particularly if B$_r$ could also be enhanced by suitable further alloying.

Some of the FeCoNdB hot pressed magnets were subsequently upset forged in larger graphite dies. By performing the upsetting slowly it was possible to reduce the length by 50% without barrel cracking (Fig. 16). The properties of these magnets at room and elevated temperatures are currently under investigation.

SUMMARY AND CONCLUSIONS

The grain size for as-cast melt spun FeNdB ribbon, deduced from X-ray line broadening analysis,has been shown to decrease approximately linearly with decreasing ribbon thickness. Below a critical thickness, corresponding to a critical cooling rate of $\sim$ 5 x 10^5 Ks^{-1}, the ribbon becomes increasingly amorphous. On annealing at temperatures in the range 600-700°C the structure completely devitrifies in 2-4 minutes attaining a grain size of about 50 nm which resists coarsening for times up to $\sim$ 30 mins. In contrast $_iH_c$ and $(BH)_{max}$ are shown to decrease beyond $\sim$ 5 mins. which may be associated with grain spheroidisation and the formation of grain boundary phase(s) that was observed.

Substitution of Fe by Co was shown to lead to increased T_c for the ribbon and to enhanced B_r and $(BH)_{max}$ and improved temperature stability for hot-pressed magnets produced from the ribbon.

It was shown that small additions of Si could dramatically affect the microstructure and enhance the energy product of as-cast ribbon although the reproducibility was not good. Further work is needed to understand this. Similarly, small additions of Nb were shown to substantially increase the coercivity of magnets hot pressed from overquenched ribbon, possibly through additional domain pinning by finer, coherent Nb-rich precipitate.

REFERENCES

Arai, S. and Shibata, T. 1985. IEEE Trans. Magnetics MAG-21, 1952-4.
Carr, G. E., Davies, H. A. and Buckley, R. A. 1988. Mater. Sci. Eng. 99, 147-151.
Croat, J. J., Herbst, J. F., Lee, R. W. and Pinkerton, F. E. 1984a. Appl. Phys. Lett. 44, 148-9.
Idem. 1984b. J. Appl. Phys. 55, 2078-82.
Davies, H. A. 1983. "Amorphous Metallic Alloys" (Ed. F. E. Luborsky). (Butterworths, London) pp. 8-25.
Gwan, P. B., Scully, J. P., Bingham, D., Cook, J. S., Day, R. K., Dunlop, J. B. and Heydon, R. G. 1987. "Proc 9th Intl. Workshop on Rare-Earth Magnets" (Eds. C. Herget and R. Poerschke). (D.P.G.-GmbH, Bad Honef) pp. 295-9.
Lee, R. W. 1985. Appl. Phys. Lett. 46, 790-1.
Matsuura, Y., Hirosawa, S., Yamamoto, H., Fujimura, S. and Sagawa, M. 1985. Appl. Phys. Lett. 46, 308-10.
Mishra, R. K. 1986. J. Mag. Mater. 54-7, 450.
Ogilvy, A. J. W., Gregan, G. P. and Davies, H. A. 1984. "Nd-Fe Permanent Magnets - Their Present and Future Applications".(Ed. I. V. Mitchell). (C.E.C. Brussels) pp. 93-8.
Parker, S. F. H., Pollard, R. J., Lord, D. G. and Grundy, P. J. 1987. IEEE Trans. Magnetics MAG-23, 2103-5.

PREFERENTIAL CRYSTALLITE ORIENTATION IN Nd-Fe-B MELT SPUN FLAKES

R. Coehoorn and J.P.W.B. Duchateau

Philips Research Laboratories
P.O.Box 80.000, 5600 JA Eindhoven, The Netherlands

ABSTRACT

Textured melt spun flakes containing the intermetallic compound $Nd_2Fe_{14}B$ were prepared using different alloy compositions and quench rates. The c-axis is oriented preferentially normal to the surface of the flakes. X-ray diffraction shows that the degree of preferential orientation is much stronger at the free side of the flakes than at the wheel side. For the optimum composition and quench rate the energy product $(BH)_{max}$ is 120 kJm^{-3}, which is 10-15 % better than in isotropic meltspun flakes.

1.INTRODUCTION

Generally the crystallite orientation in melt spun Nd-Fe-B flakes, obtained either directly by quenching at the optimum quench rate or after annealing overquenched amorphous flakes, is random (Croat et al., 1984). Assuming that the system can be considered as an assembly of noninteracting magnetically hard crystallites with uniaxial anisotropy, the remanence is then only half of the saturation magnetization and the maximum energy product is only one quarter of the theoretically attainable maximum of aligned magnets. Lee (1985) has shown that a high degree of preferential orientation can be obtained by the die upsetting process, in which the sample flows plastically normal to the press direction. However, in view of the critical press parameters and the restricted shaping possiblities of these magnets, one would prefer to produce anisotropic flakes directly during the quench process. In this report we show that under certain spinning condition the as spun flakes indeed show preferred orientation, which leads to improved magnetic properties.

2.PREPARATION AND TEXTURE DETERMINATION

The preparation of the flakes has been described in detail by Coehoorn and Duchateau (1988). A stable quench process was obtained by ejecting the liquid through a slit of dimension 10 mm x 0.4 mm, mounted 200 μm above a copper wheel which was 60 cm in diameter. In fig. 1 the X-ray diffraction patterns of the free side, wheel side and powdered $Nd_{13.5}Fe_{79.6}B_{6.9}$ flakes, quenched at a wheel speed of 9 ms^{-1}, are shown. The high intensity of the (0 0 l) peaks shows that the c-axis is oriented preferentially normal to the flake. The degree of preferential orientation is higher at the free side than at the wheel side. The orientation distribution of the crystallites that are oriented is very sharp: the full width at half maximum is $2 - 4°$ at the free side and $8 - 15°$ at the wheel side. Apart from a fraction of crystallites which are oriented sharply, another fraction has a random orientation. From the enhancement of the (0.0.l) peaks with respect to the other X-ray peaks and from the width of the orientation distribution function (also determined by X-ray diffraction), the fraction x of oriented crystallites can be determined (Coehoorn and Duchateau, 1988). For the flakes of which the X-ray diagrams were shown in fig. 1, we found x = 0.37 at the free side and x = 0.08 at the wheel side.

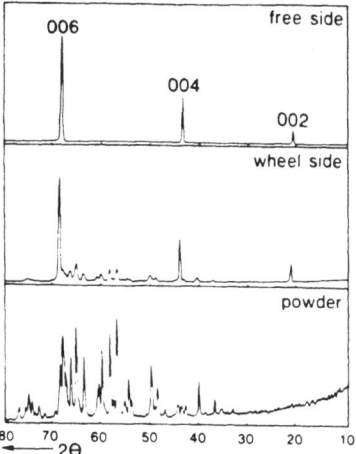

Fig. 1. Cr- Kα X-ray diffraction pattern of $Nd_{13.5}Fe_{79.6}B_{6.9}$ *flakes, spun at 9 ms^{-1} (a) free side, (b) wheel side, (c) powdered flake.*

The average fraction of oriented crystallites ($\bar{x}$) follows from measurements of the remanence, measured after application of a magnetizing field parallel and perpendicular to the texture axis:

$$\bar{x} = \frac{(M_{r\parallel} - M_{r\perp})}{(M_{r\parallel} + M_{r\perp})}.$$ [1]

In order to saturate the samples completely, the measements were performed in a 15 T Bitter magnet at the High Magnetic Field laboratory of the University of Nijmegen (The Netherlands). For the flakes discussed above we found that $\bar{x} = 0.14$.

3.COMPOSITIONAL VARIATIONS

The fraction $\bar{x}$ of preferentially oriented crystallites is strongly dependend on the composition (fig.2a). Flakes, spun under nominally identical process conditions with a wheel speed of 14 ms^{-1} showed a maximum fraction of oriented crystallites $\bar{x} = 15$ % for the composition $Nd_{11.8}Fe_{79.4}B_{8.8}$. However, X-ray diffraction and scanning electron microscopy (SEM) showed that in flakes of this composition the grains were quite coarse (diameter approximately 5 μm) and that the flakes contained some $\alpha -Fe$. This unfavorable microstructure leads to a relatively low remanence and coercive field in these flakes (fig. 2b and 2c, respectively).

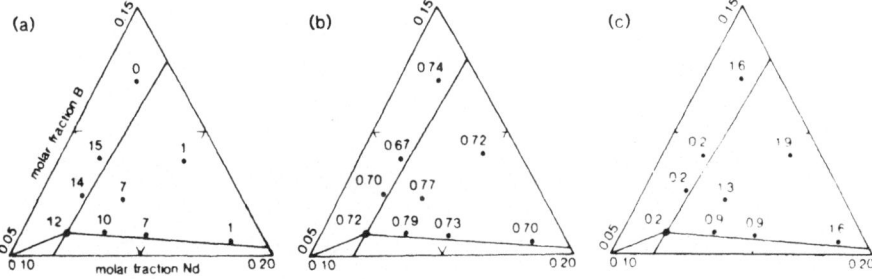

Fig. 2. Dependence on composition of: (a) average fraction of preferentially oriented crystallites, from magnetic measurements, in percents; (b) remanence $\mu_0 M_{r\parallel}$ in Tesla; (c) coercive field $\mu_0 H_c$ (average of parallel and perpendicular values) in Tesla.

4.VARIATION OF QUENCH RATE

The second step in the optimization of the texture and $(BH)_{max}$ energy product was a variation in the quench rate, which was achieved by a variation of the wheel speed. For this investigation we selected the composition $Nd_{13.5}Fe_{79.6}B_{6.9}$ for which in the investigation on the compositional dependence a favorable combination of high coercivity, remanence and texture was found (sec. 4).

The scanning electron micrographs in fig. 3 show the fracture surface of flakes, spun at different wheel speeds. The upper edge in these micrographs is the wheel side. The micrographs show that an increase in wheel speed leads to a decrease in the thickness, from 97 μm at 3 ms^{-1} to 15 μm at 28 ms^{-1}. The corresponding increase in quench rate resulted in a refinement of the grain size. In each flake the grain size at the wheel side is generally much smaller than the grain size at the free side.

The grain size decrease with increasing wheel speed leads to a pronounced wheel speed variation of the oriented fraction, the remanence and the coercive field (fig. 4). The preferential orientation is most pronounced at low wheel speeds. At 3 m/s the average fraction of oriented crystallites is approximately 25 %, as derived from $M_{r\parallel}$ and $M_{r\perp}$. At the free side, where the quench rate is lower than the average quench rate, an analysis of the X-ray spectra shows that the oriented fraction is much higher: $x = 75\%$. On increasing the wheel speed the texture becomes weaker. In spite of the high degree of preferred orientation, the slowly quenched flakes do not have the optimum energy product. The grain size in these flakes, in particular close to the free surface, is much larger than the single domain size, which is approximately 0.3 μm in $Nd_2Fe_{14}B$. This results in a relatively low coercive field (fig. 4c). The wheel speed dependence of the coercivity shows a characteristic maximum at medium wheel speeds, as already reported by other authors (Croat et al., 1984; Hilscher et al., 1986). Above 16 ms^{-1} the flakes contain an increasing fraction of amorphous material, which leads to a decrease in H_c . The low coercivity in highly textured flakes also leads to a lower average remanence than the remanence in high coercivity isotropic flakes. This can be explained by assuming that already in the remanent state there are multidomain crystallites, or single domain crystallites with a reversed magnetization.

Fig. 5 shows the low field parts of the hysteresis loops at some selected wheel speeds. The maximum in the energy product was obtained at $v = 16$ ms^{-1}: $(BH)_{max} = 120$ kJm^{-3} (15 MGOe). These flakes have a very fine microcristalline grain structure (fig. 3), in particular at the wheel side.

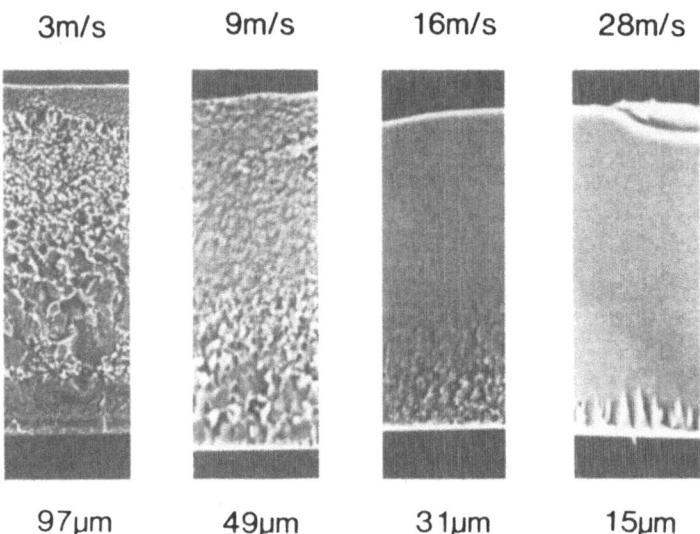

3m/s   9m/s   16m/s   28m/s

97μm   49μm   31μm   15μm

Fig. 3. Scanning electron micrographs of fracture surfaces of flakes, spun at different wheel speeds. The upper edge is the wheel side.

5.CONCLUDING REMARKS

A high degree of preferential orientation in melt spun Nd-Fe-B flakes can be obtained by using low quench rates. However, due to the large crystallite size in highly textured flakes, the coercive field is relatively small. In the more rapidly quenched optimum flakes, with $(BH)_{max} = 120$ kJm^{-3}, the average degree of orientation was 7 %, and the coercive field was $\mu_o H_c = 1.7$ T. The improvement of the energy product over that of isotropic flakes is 10-15 %, which is mainly due to the increased remanence after magnetization normal to the flake surface: $\mu_o M_r = 0.87$ T.

The same type of preferred orientation was observed for all other rare earths elements in alloys $R_{13.5}Fe_{79.6}B_{6.9}$. In a future publication we will report on these results in more detail, as well as on the effects of partial substitution of neodymium by dysprosium and of temperature variations of the melt.

Finally we remark that the rapid growth direction (c-axis) observed in our experiments is different from the rapid growth direction which is observed in castings of $R_2Fe_{14}B$ alloys, and in single crystal growth experiments, where the a-axis is the preferential growth direction (Tenaud et al., 1987). Presently no explanation of this difference is available. The observation that the texture is most pronounced at the free side, where the shear stress due to the vertical gradient in the horizontal momentum during solidification is much lower than at the wheel side, excludes the possibility of shear stress induced anisotropy. In several scanning electron micrographs we have clearly observed columnar crystallites, oriented perpendicular to the surface, and originating from an area of very fine crystalline grains. A similar microstructure was observed also by Dadon et al (1987). These observations suggest a texture formation mechanism by growth selection of crystallites that are oriented favorably with respect to the temperature gradient.

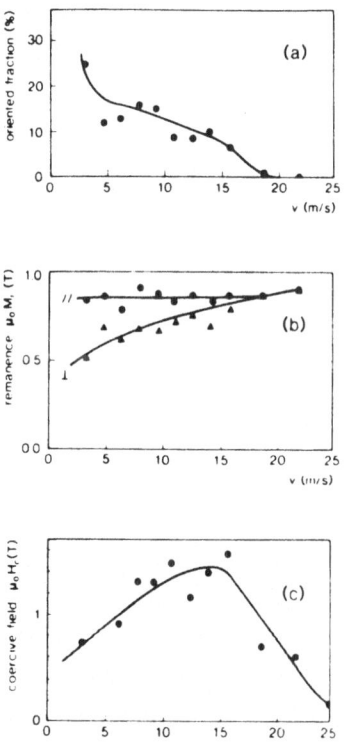

Fig. 4. Wheel speed dependence of: (a) average fraction of prefentially oriented crystallites from magnetic measurements; (b) Magnetic remanence $\mu_o M_{r\|}(o)$ and $\mu_o M_{r\perp}(\Delta)$ in Tesla; (c) coercive field $\mu_o H_c$ (average of parallel and perpendicular values) in Tesla.

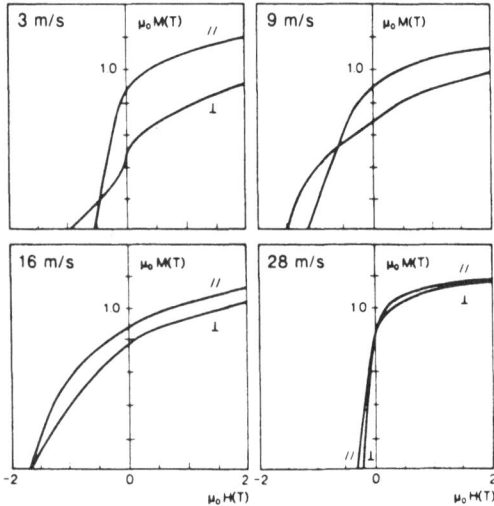

Fig. 5. Low field parts of magnetization curves of flakes, spun at different wheel speeds, for field directions parallel and perpendicular to the texture surface.

REFERENCES

R. Coehoorn and J.P.W.B Duchateau (1988), Materials Science and Engineering, to be published.

J.J. Croat, J.F. Herbst, R.W. Lee and F.E. Pinkerton (1984), Appl. Phys. Lett., **44** , 148.

D. Dadon, Y. Geffen and M.P. Dariel (1988), Proceedings of the Intermag Conference, april 14-17 1987, Tokio, to be published.

G. Hilscher, R. Grossinger, S. Heisz, H. Sassik and G. Wiesinger(1986), J. Magn. Magn. Mater., **54-57**, 577.

R.W. Lee (1985), Appl. Phys. Lett., **46** ,790.

P. Tenaud, A. Chamberod and F. Vanoni (1987), Sol. St. Comm., **63** , 303.

THE PRODUCTION OF NdFeB MAGNETS BY THE RAPID
SOLIDIFICATION PROCESSING ROUTE

J H Vincent and M-J P D Wyborn[*]

GEC Hirst Research Centre, East Lane, Wembley, UK

[*] The Micanite and Insulators Co Ltd., Trafford Park, Manchester, UK

ABSTRACT

The rapid solidification process route for NdFeB magnets has been studied. Alloy supplied by CEAM participants REP and GfE has been melt spun to yield amorphous and crystalline ribbon which was incorporated into polymer bonded magnets produced by compression moulding. A detailed study of the compression moulding process was undertaken using Magnequench ribbon. The improved metal loading which could be achieved with a floating die to minimise load transfer to the die walls was apparent. Studies of the hot processing process highlighted the difficulty of selection of die and ram materials, but demonstrated that the magnetic properties of the pressings were reasonably insensitive to the pressing procedure employed.

Corrosion protection studies on hot pressed magnets were carried out at Micanite and Insulators and on samples sent to Olivetti. It was established that satisfactory protection could be achieved and that procedures developed for coating sintered magnets were equally applicable to hot pressed compacts.

INTRODUCTION

The rapid solidification process route for the production of NdFeB based magnets is intrinsically attractive, in that a common feed stock of quenched strip can subsequently be processed by a number of different routes to yield magnets ranging in energy product from 30 to over 300 kJm^{-3}. The work undertaken at the GEC Hirst Research Centre (HRC) initially concentrated on melt spinning of NdFeB alloys and the incorporation of the cast product into compression-moulded polymer bonded magnets. The Micanite and Insulators Co (a member of the GEC group of companies) later began an investigation into the hot pressing process using rapidly quenched alloy supplied from HRC and from General Motors, and information generated from this effort was also presented within CEAM.

MELT SPINNING

At the outset of the CEAM programme, the rapid solidification facility at HRC did not have the capability of operating within a sealed enclosure. Early experiments thus concentrated on establishing whether it was possible to cast NdFeB successfully by the use of a local protective atmosphere.

Casting was performed onto either a Cu-Be or mild steel roller with a peripheral velocity in the range of 10-25 ms^{-1}. Silica crucibles were used with drilled nozzles of 0.5-1.0 mm diameter; the charge weight was typically 100 grammes. During casting, the melt temperature and ejection over-pressure were recorded, typical values being 1380-1420°C and 40-80 kPa respectively. Melt spinning of $Nd_{16}Fe_{76}B_8$ is shown in Figure 1. The first few casting runs highlighted the difficulties associated with casting without an enclosure. These were, firstly, the accelerated reaction between the melt and the silica crucibles which often resulted in partial or complete blockage of the nozzle outlets, either before or during the run. Secondly, even when the strip remained in contact with the wheel for several centimetres, and had thus cooled to below red-heat on departure from the wheel, there was invariably a degree of oxidation. Thirdly, the contact length tended to increase during a run as the wheel surface warmed and adsorbed gas was driven off, leading to improved wetting. A system was therefore developed to classify the product according to contact length, with the limited contact material being discarded.

A corollary of the reaction between the crucibles and the melt around the nozzle outlets, was that nozzles of <0.8 mm diameter were unusable and that a higher than normal ejection pressure had to be used to ensure that the melt was expelled. The resulting melt flow rate of about 2,000 mm^{3}s^{-1}, meant that, even at the higher casting speeds used, it was difficult to produce strip of less than about 50 μm thick, compared with the 40 μm thickness which had been found previously (Ogilvy et al, 1984) to be the optimum for maximising coercivity.

During the development of the melt spinning process the protective atmosphere was continuously improved to include low pressure flowing argon shielding over the melt and an argon shrouded enclosure around the

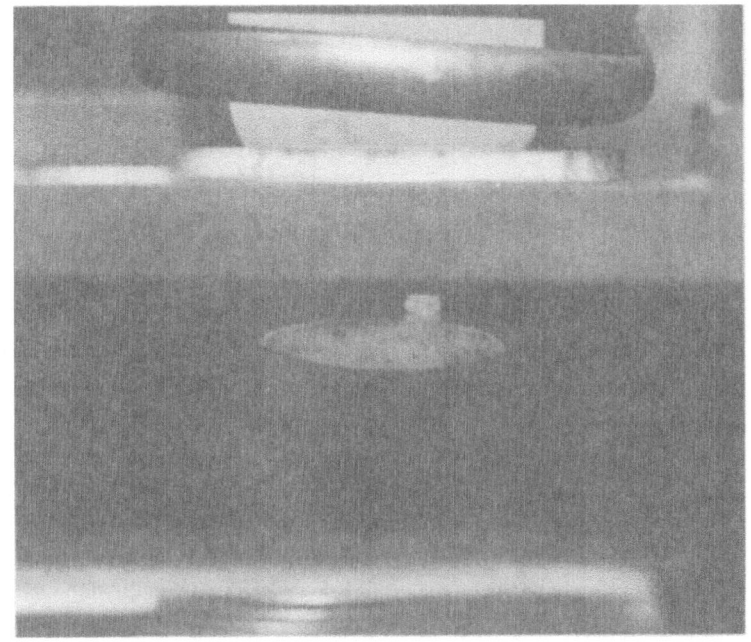

1   Photograph of melt spinning of NdFeB.

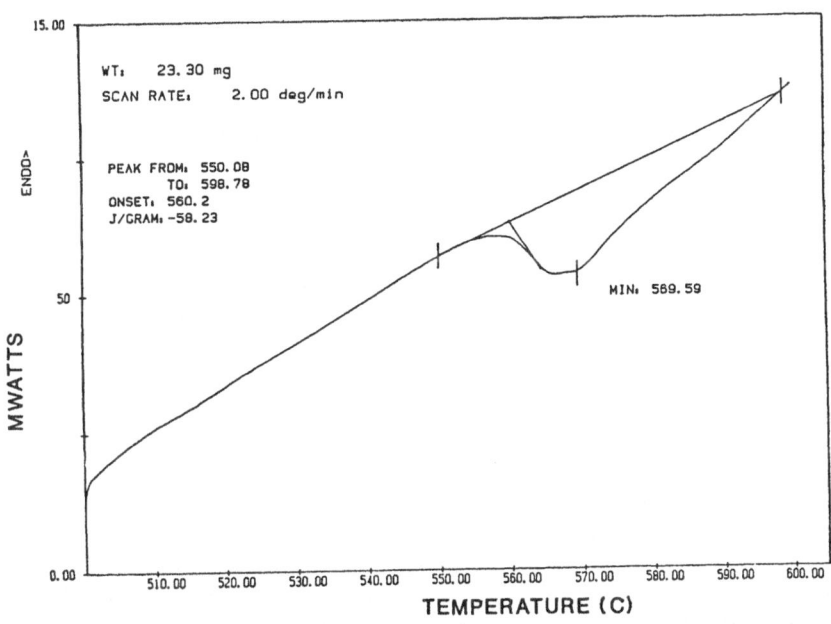

2   DSC trace of crystallisation event of amorphous NdFeB ribbon.

casting point and downstream to protect the strip while it cooled. However, these measures only served to prevent the most severe examples of oxidation observed earlier and did not cure the basic problem of oxygen ingress.

Despite the difficulties described above, a limited amount of amorphous strip about 20 μm thick was produced. Figure 2 shows the exothermic crystallisation event recorded at a heating rate of 2 Kmin^{-1} on a Differential Scanning Calorimeter (Perkin-Elmer DSC4). This strip was completely free of oxide, indicating that if the quench rate was sufficient, oxidation could be prevented. Samples were crystallised by heating within an evacuated and argon backfilled silica ampoule for 10 minutes at 600°C prior to magnetic characterisation.

Magnetic properties

Table 1 lists the measured magnetic properties typical of material described above, together with data from a Magnequench sample. The measurements were made at the Lucas Research Centre on compression moulded magnets and the results have been corrected to 100% metal density. Also included in the Table are crystallite size determinations of the samples derived from X-ray diffraction line broadening measurements.

TABLE 1: **Magnetic properties of rapidly quenched NdFeB strip**

Sample Type	Intrinsic Coercivity (kAm^{-1})	Remanence (T)	Crystallite size (nm)
As-cast, short contact length 50 μm thick	220	0.46	40 ± 10
As-cast, long contact length 50 μm thick	460	0.54	40 ± 10
Initially amorphous, crystallised at 600°C, 20 μm thick	1,000	0.78	50 ± 10
Magnequench flake, 25 μm thick	1,230	0.79	35 ± 5

It can be seen that, although the crystallite size for the as-cast strips is actually less than that for the initially amorphous material, the intrinsic coercivity and remanence are substantially lower. It was presumed that the cause of the lower properties of the as-cast strips was either oxidation of the ribbon or structural modifications, for example growth of a second phase, during the cooling of the strip after it left the wheel. It was not possible to conduct the detailed microanalytical and microstructural investigations needed to confirm this. The coercivity of the Magnequench ribbon was somewhat higher than the HRC produced equivalent, and in this case the reduced grain size of the Magnequench ribbon may be significant. This may be analogous to previous results (Croat, 1983) showing that, after crystallisation, very rapidly quenched amorphous NdFeB ribbons do not achieve the energy product of ribbons quenched only slightly faster than the critical quench rate. This was probably because the reduced number of nuclei in the former case resulted in an ultimately larger grain size.

In the final period of the CEAM programme an enclosure for the casting unit was completed which enabled casting to be carried out within an inert atmosphere. As expected, the casting process was much more reliable, and the ribbon produced was completely free of surface contamination.

COMPRESSION MOULDING

In order to measure the magnetic characteristics of the melt spun material produced at HRC it was first necessary to develop a means of bonding the ribbon together, in the absence of a facility to measure individual ribbons. This work was extended into a more general investigation of the compression moulding technique.

The basic process developed was intrinsically very simple. The as-quenched ribbon was coarse-crushed and repeatedly sieved through a 250 μm mesh. In this way the production of fines was minimised. The resulting powder was then mixed with between 10 and 20 volume percent of organic binder before loading in pre-weighed amounts into a cylindrical die cavity. The temperature of the die could be varied between room temperature and 200°C. Pressures of between 300 and 700 MPa were applied and the compact was then either ejected for post-curing of the binder or cured in-situ at 180°C for 60 minutes, under a pressure of 300 MPa. A modified procedure developed later involved maintaining the die at about 100°C and using a resin which flowed, but did not cure rapidly at this temperature. This method had the advantage that the sample did not have to be held in the die during curing, and it gave almost as good compaction.

After curing, the compacts were weighed and measured and, from the calculated density of the compact and the known density of the metal flake and the resin used, values of metal content and porosity were obtained. The results are presented below, with the standard deviations included where a statistically significant number of samples was produced.

TABLE 2: Effect of pressing technique on metal volume fraction and porosity.

Compaction technique	Metal/resin mix (vol %)	Volume fraction of alloy %(σ)	Volume fraction of pores %(σ)
Cold press, post cure	80/20 90/10	64.7 (1.9) 68.5	19.1 (2.3) 23.8
Hot press, cure in-situ	80/20 90/10	73.8 (1.8) 72.3	7.6 (1.8) 19.7
Warm press, post cure	80/20	71.5 (0.7)	10.5 (0.9)

The trends in these results are clear. There was no significant increase in the ultimate metal loading by increasing the initial metal content, instead the porosity rose and the compacts became more friable. The improvement offered by curing the sample under pressure within the die was clear, but the 1 hour cycle time was not practical, even for the production of samples for measurement. The warm press procedure required a press cycle time of only 3 minutes and enabled almost the same metal loading to be achieved, with somewhat improved reproducibility.

The main reason for the poor metal loadings found originally in the cold press route was determined to be the very uneven packing of the flake through the thickness of the compact. At the top face of a pressing (the face towards the moving ram) the volume fraction of metal was generally about 50%, rising towards the centre of the pressing. Similarly the lower face of the compacts also showed a below average metal density. This was ascribed to friction within the compact and between the compact and the die walls causing uneven load distribution and load transfer to the die walls. The lubrication provided by the fluid polymer during the warm or hot press cycles allowed better densification. However, the difficulty in ejecting the compact produced by either of these techniques was a significant disadvantage. This difficulty and the inconvenience of holding the die and tools at elevated temperature were such that a second round of experimentation was begun to optimise the cold press/post cure schedule.

Optimisation of cold press/post cure technique

In order to improve the cold press method, a modified die and ram arrangement was constructed that enabled the die to move relative to the two rams, thus minimising load transfer to the die walls. A 90/10 metal resin mix was selected and the effect of dry and wet mixing and of adding camphor as a lubricant were assessed. The effect of particle size was checked by milling a batch of powder such that the particle size distribution peaked at 80 μm compared with 140 μm for the as-received powder. The resulting compacts were measured physically as before, and were also characterised magnetically.

Figure 3 shows the metal volume fraction calculated from the density of the compacts and the measured remanence as a function of applied load during compaction. There was no effect on density of using a lubricant, wet mixing or of using the finer particle size. As can be seen in Figure 3, the remanence rises with the increase in metal fraction, although for high pressures, the remanence was slightly lower than expected for a monotonic increase. The highest volume fraction achieved at 78% was considerably in excess of the levels attained previously, and demonstrated the benefits of the floating die.

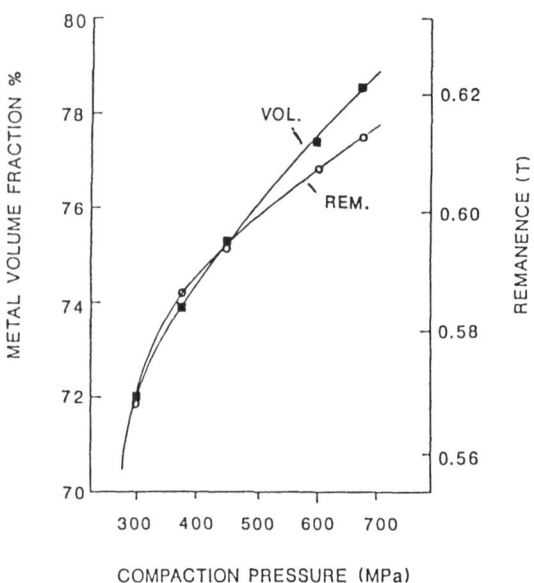

3   Effect of compaction pressure on the volume fraction of metal in the compact and on remanence.

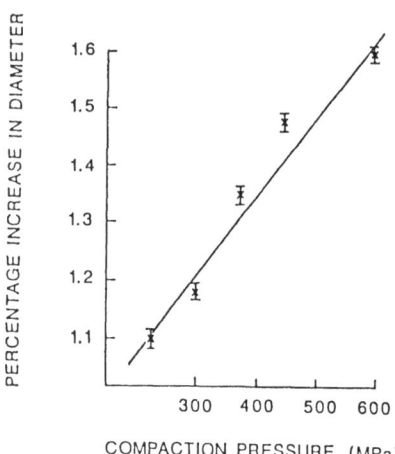

4   Effect of compaction pressure on dilation of pressed compacts.

The die diameter used was 12.70 mm and in each case the diameter of the compact was slightly greater than this value, varying between 12.84 and 12.91 mm. Figure 4 shows the percentage increase in final diameter as a function of pressing force. Again this effect was independent of the details of the sample preparation technique. The general conclusion is that, by fixing the pressing load, very accurate dimensional control can be achieved for these simple shapes.

As expected, the BH_{max} product of the magnets increased with metal fraction, to a maximum of 58 kJm^{-3}. However, the expected relationship, of BH_{max} being approximately proportional to $(B_r)^2$ was not followed, as the intrinsic coercivity of the magnets dropped linearly with increasing pressing forces from 1180 kAm^{-1} for P = 300 MPa to 1130 kAm^{-1} for P = 600 MPa. This effect is illustrated most clearly in Figure 5, which plots the average recoil permeability μ_r (taken as $B_r/\mu_0 H_c$) against pressing force. A clear distinction was evident between the coarse and fine initial powders and the increase of μ_r with pressure was presumably a result of the breakdown of particles under load. This effect has also been observed in the polymer bonding trials undertaken at SG Magnets (Ward, 1989) and is likely to be a function of the increased surface/volume ratio, and hence increased oxidation, of the finer particle sizes. This adverse influence of high pressing forces on intrinsic coercivity, and hence μ_r, would have to be eliminated if the potential properties of bonded NdFeB magnets are to be achieved.

HOT PRESSING

The maximum metal loading achieved by polymer bonding in this study was 78%, with a corresponding energy product of 58 kJm^{-3}. The ability to approach 100% of theoretical density by an apparently simple hot pressing procedure, and thereby increase the energy product to >100 kJm^{-3} was attractive, and consequently an investigation was made into the technical viability of this process.

Figure 6 is a schematic diagram of the hot press used in these experiments. The bolster was graphite and was maintained at temperature by an induction heating unit. An argon purged enclosure was fitted around the hot press apparatus. Initially a tool steel die and rams were

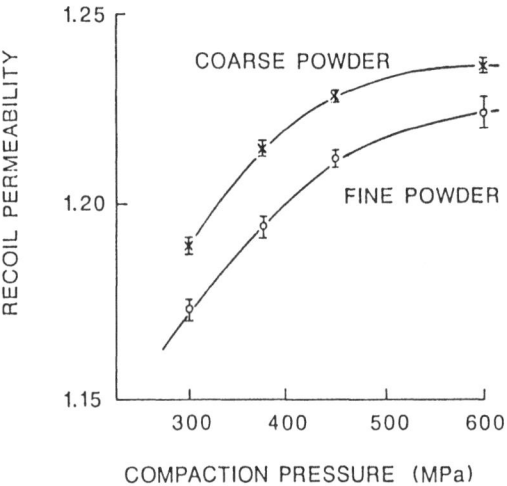

5 Effect of compaction pressure on recoil permeability for coarse and fine powder fractions.

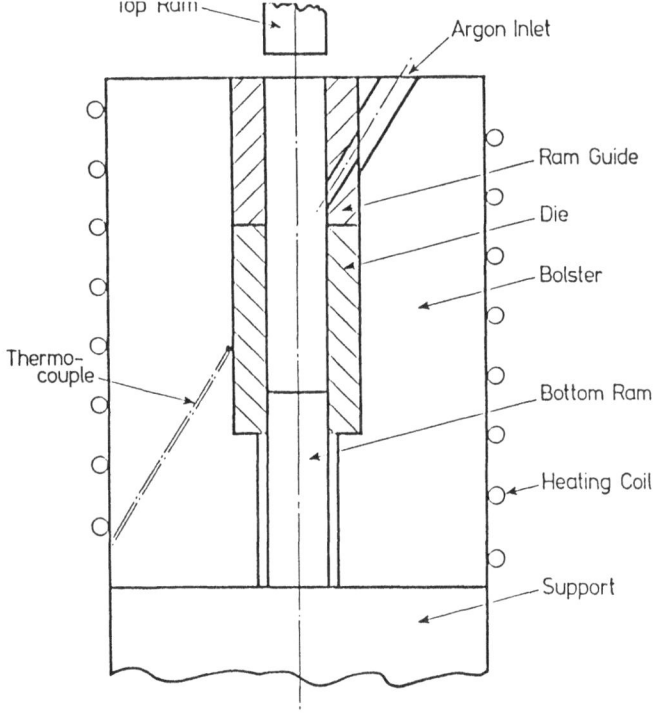

6 Schematic diagram of hot pressing die.

used with alumina paste as a release agent but this was soon replaced because of the difficulty in ejecting the compact. Subsequently a graphite die was used and graphite discs were loaded into the die above and below the powder to facilitate ejection.

The hot pressing procedure was generally to pre-heat the die to 700-800°C before loading the NdFeB powder and then to apply a pressure of 90-150 MPa for up to 3 minutes. Thermocouples protruding through the die wall confirmed that the compact reached thermal equilibrium with the die in about 60 seconds. The lower ram was then removed and the compact ejected to cool naturally. A semi-automatic handling system was developed which enabled a number of compacts to be produced before the chamber was opened and the product removed.

Compacts of >99% theoretical density could be reproducibly formed in this way and the process was very tolerant of small changes in the operating parameters, within the ranges indicated above for each type of powder used. The compacts had smooth, blackened surfaces and were not susceptible to corrosion when handled. The surface film was analysed using X-ray photo-electron spectroscopy and was found to consist entirely of carbon and oxygen.

A number of samples were characterised magnetically at the Lucas Research Centre. Figure 7 shows the effect of the initial pulse magnetising field on subsequent demagnetisation; full properties were only achieved for fields of 4 Tesla and greater. This highlights a potential practical disadvantage in that magnets of this type must be magnetised prior to assembly into components if their full properties are to be utilised.

Figure 8 shows the demagnetisation characteristics measured parallel and perpendicular to the press direction. The degree of anisotropy is consistent with GM's published data (Lee, 1985). X-ray diffraction studies at HRC of hot pressed compacts have shown enhancement of (001) base plane peaks in the surface of some samples, but the effect is neither consistent nor reproducible. If a degree of texture could be developed in a one-stage hot pressing process, rather than the two-stage process used to date, then significant increases in energy product could be achieved without extra process costs.

574

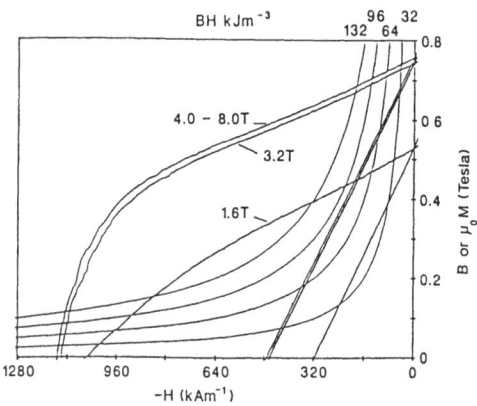

7 Effect of initial magnetising field on demagnetisation characteristics for a hot pressed magnet.

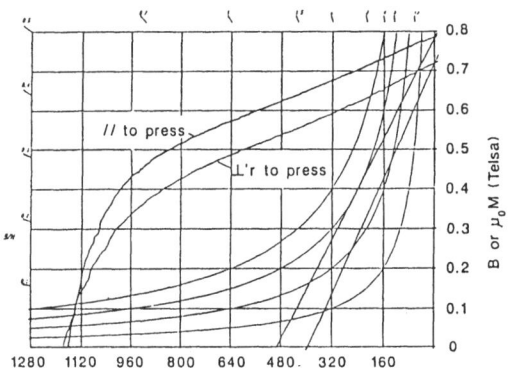

8 Demagnetisation characteristics of a hot pressed magnet measured parallel and perpendicular to the press direction.

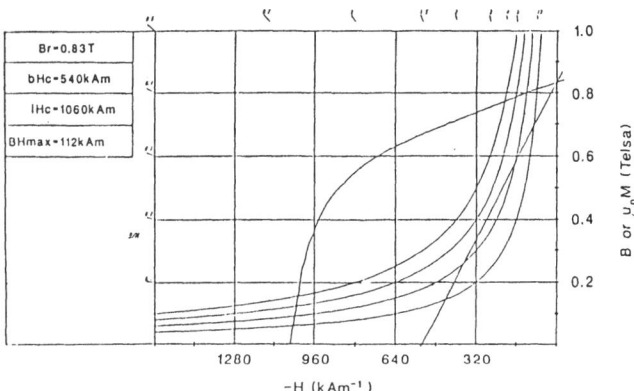

9 Demagnetisation characteristics of a hot pressed magnet with enhanced anisotropy.

Figure 9 shows the demagnetisation curve of a sample which did exhibit a degree of alignment, hence the increase of remanence of 0.83T and energy product to 112 kJm^{-3}. Subsequent work carried out on other NdFeB powders using conditions similar to those described above showed that it was possible to achieve energy products of 120 kJm^{-3} with intrinsic coercivities of up to 1,360 kAm^{-1}.

CORROSION PROTECTION

At the Micanite and Insulators Co Ltd samples of hot-pressed NdFeB magnets were electroplated with zinc, nickel or tin-nickel and subjected to two accelerated environmental tests; resistance to humidity (BS3900.f2) and resistance to salt spray (ASTM-B-1A).

The presence of graphite on the surface of the magnets and the sensitivity of the NdFeB alloy to acids, caused problems achieving satisfactory pre-treatment prior to plating. The samples were finally successfully processed by cathodic alkali treatment followed by a 2-3 second immersion in hydrochloric acid.

Only the zinc-plated and passivated samples yielded a satisfactory degree of protection on the salt spray test. Zinc also galvanically protected the magnets for some 1500 hours of the humidity test, although the samples were covered with the usual zinc corrosion products after this period. The nickel samples (barrel plated) gave reasonable results on the humidity test, but the deposits were porous at the low deposited thickness (3.5 μm) as where those of the tin-nickel samples.

Although zinc plating provided a reasonable degree of corrosion protection, it is limited in the long term because it is lost galvanically. It is considered that "barrier" type protection such as obtained from nickel will be more acceptable.

Samples of hot pressed magnets were also sent to Olivetti for anti-corrosion coating, and the results are reported in full elsewhere in this volume (Tori et al 1989). Their finding that the coating procedures employed were as effective on the hot pressed as on the sintered magnets demonstrated that initial fears that surface porosity would be deleterious were unfounded.

CONCLUSIONS

The work undertaken at GEC on the technical assessment of different aspects of the rapid solidification processing route has confirmed that the claimed magnetic properties for bonded and hot-pressed NdFeB magnets can be readily achieved.

The melt spinning process is made difficult by the reactiveness of the molten alloy, but the provision of an inert atmosphere facility does enable strip to be cast reproducibly. The compression moulding process is technically viable and simple shapes, e.g. cylinders, can be produced within tight dimensional tolerances. The adverse affect of high pressing loads on coercivity would need to be better understood if higher metal loadings and thus increased energy products were to be achieved. The hot pressing process was shown to be very promising for the production of simple shapes, and again good dimensional control could be achieved. The magnetic properties of the compacts were consistent with 100% dense material, and evidence of anisotropy in some magnets highlighted the development potential of this process.

ACKNOWLEDGEMENTS

The authors are grateful to numerous colleagues who have contributed to this work, both at HRC and at Micanite and Insulators.

REFERENCES

Croat, J. 1987. European Patent Application No. 0108474

Lee, R.W. 1985. Appl. Phys. Lett., 48(8), 790

Ogilvy, A.J.W., Gregan, G.P., Davies, H.A. 1984. 'Nd-Fe permanent magnets - their present and future applications' Ed. I.V. Mitchell (CEC, Brussels) 93-97

Tori, S. et al 1989. This volume

Ward, A.J. 1989. This volume

SUMMARY OF DISCUSSIONS :

INGOT MATERIALS - CHARACTERISATION, POWDER PROCESSING AND MELT SPINNING

E. Rozendael and H.A. Davies

The possible routes for alloy preparation were described (6.1). Already in the early stages of CEAM it was clear that 95% pure Nd could be used. The origin, levels and detection methods were reviewed for some of the impurities, such as Oxygen, Fluorine and Aluminium. The analysed composition of all castings made over the CEAM period were presented (6.1) and microstructure were shown of some new alloys. The economics of $SmFe_{11}Ti$ were shown to be doubled.

The effects of applying hydrogen decrepitation (HD) on some of the process steps were presented (6.4). Major effects were found on sinter behaviour, such as reducing the sinter temperature by 60°C and the formation of cracks on very large blocks. The presence of Fe in cast alloy was found to have no negative effect on final sintered properties when the HD method of pre-milling was used. Radially aligned thin walled rings were successfully pressed and sintered and were then built into four prototype motors (Sheffield University Applications group).

The results presented by TEW (6.5) also contributed to characterisation of the HD process. Absorption velocity and absorption energies were measured as a function of Nd-content. No plateau pressure for hydrogen was found in the chosen pressure range. Due to this, cycle absorption/desorption has no effect on the final size of the powder. The Nd content could be measured by the mass of hydrogen absorbed.

The results obtained by the National Physical Laboratory, in setting up a powder characterization technique (6.6) to measure the oxidation rate of NdFeB powder using TGA, were shown to be very promising. This technique measures the distribution of fine milled powder rather than an average size value. After determining the activation energy of NdFeB, a remarkably good fit between theoretical and experimental results were found. The current state of application area's for the various permanent magnets was highlighted (6.5). The process costs for the manufacturing of sintered NdFe magnets were shown to be higher than for the SmCo types, due to handling a more reactive powder and the necessity of coating the magnets.

It was made clear that the operating temperature in the application determines the type of magnet materials to choose from. So far, even with Dy-doping, NdFeB magnets are not magnetically stable enough above 140°C and the temperature range of 140 to 250°C is still covered by $SmCo_2Co_{17}$.

The Philip's Group reported coercivities of 650 KA/m after annealing jet milled powder in a rotary vacuum furnace. They further showed that fairly dense-isotropic spheres less than 00.5 mm were obtained which could be used as input material for a resin-bonded type magnet.

Polymer bonded magnets of the MQI type and hot pressed, fully dense magnets of the MQII type have been satisfactorily produced (6.7) using crushed ribbon material and the magnetic properties attained were comparable with those of the Magnaquench magnets. Systematic studies of the relationships between alloy composition, melt spinning process variables, microstructure and magnetic properties have been performed both on ternary FeNdB alloys and also on cobalt substituted alloys having enhanced thermal stability. Optimum cobalt concentrations giving a suitable increase of Curie temperature without an unacceptable decrease in coercivity have been identified and hot pressed magnets have been produced from this ribbon. Moreover, by systematic optimisation and melt spinning processing conditions, it has been possible to obtain significant improvement in magnetic properties of ribbon over those reported elsewhere.

SECTION III

– MAGNET PROCESSING

CHAPTER 7

MAGNETS : PRODUCTION, CHARACTERISATION AND CORROSION

PREPARATION AND CHARACTERISATION OF SINTERED NdFeB MAGNETS

A. Cartocetti

Industria Ossidi Sinterizzati SpA
Malgesso (VA), ITALY

EXTENDED ABSTRACT

The aim of the work at IOS was to acquire knowledge of the
basic   process   variables   for   the   production   of   permanent
magnets of the Nd-Fe-B family. The magnets were produced from
six alloys provided by three companies participating in the
CEAM programme. The magnetic properties were measured as
functions of changes in the process variables. Observations
were made by optical and electron microscopy in order to
evaluate grain size and the presence of different phases. The
alloys were subjected to a three stage milling in cyclohexane.
It was found that the milling stage was very important in
obtaining the best magnetic properties. The sintering trials
indicated that, for the standard $Nd_{15}Fe_{77}B_8$ alloy the optimum
sintering temperature was 1080°C   whereas for the alloy
containing Nb and Dy the optimum sintering temperature was
found to be higher at 1,100°C. A post sintering temperature
of one hour at 620°C rather than 600°C gave the best results
for the standard alloy whereas an annealing time of 1 1/2
hours gave the best results for the alloys with Dy and Nb.
Some of the NdDyFeNbB magnets that were thermally
demagnetised, when measured again, gained up to 7% in Hc_1.
This suggested that a multi-step treatment cycle might improve
the magnetic properties.

THE DEVELOPMENT AND IMPLEMENTATION OF COMMERCIALLY VIABLE PROCESSING TECHNIQUES FOR NEODYMIUM IRON BORON MAGNETS

A.J. Ward

SG Magnets Limited, Rainham, Essex, England

ABSTRACT

Commercially viable and economic techniques are described to enable the production of consistent quality anisotropic NdFeB permanent magnets by the powder metallurgy route. Throughout the development work associated with this project, particular attention has been paid to the safety aspects, whilst optimisation of process and pilot plant variables enable the fabrication of sintered magnets up to 33 MGOe (265 kJm^{-3}). The plastic injection moulding route has also been considered and has culminated in the experimental formation of isotropic, thermoplastic magnets up to 5.0 MGOe (40 kJm^{-3}).

INTRODUCTION

The European research efforts to replace samarium cobalt high strength permanent magnets with alloys of cheaper and less strategic raw materials coincided with the independent announcements of the discovery of very promising permanent magnetic properties in the $Nd_2Fe_{14}B$ ternary compound. General Motors in the USA have developed a rapid solidification/ribbon compacting technique (Croat, 1984), whilst Sumitomo Special Metals in Japan favour a conventional powder metallurgy process to produce this material (Sagawa et al, 1984). SG Magnets Limited have extensive experience of both Alnico and $SmCo_5$ magnet production by the powder metallurgy route so our initial experiments with the Neodymium Iron Boron alloys naturally concentrated on the latter system (the "Sumitomo Route"). Numerous processes have been considered to establish the most economic and technically efficient manufacturing route. Extensive evaluation of the favoured procedure, from alloy purchase to magnet despatch, has enabled consistent production of good quality magnets.

Later work has concentrated on the General Motors' ribbon and the examination of numerous methods to consolidate this material into a useable magnet. The area which has received most attention is the thermoplastic blending/injection moulding process to produce a readily marketable product using conventional polymer processing techniques and equipment.

NdFeB MAGNETS BY THE POWDER METALLURGY ROUTE

The processing steps required to produce a sintered NdFeB magnet are illustrated in Fig 1. The development of each stage is described in the following sections.

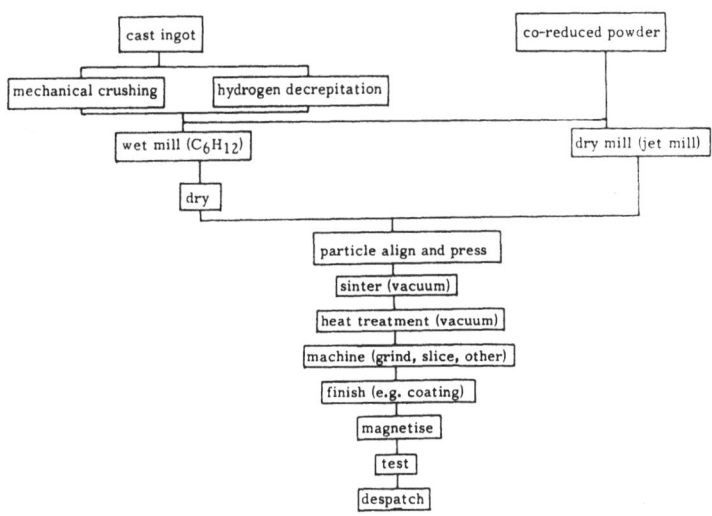

Fig 1   Process steps required to produce a sintered NdFeB magnet

Alloys

Table 1 contains a list of typical alloys studied during this work and includes sieve fractions and oxygen analyses. Lump alloy has a much lower oxygen content than the fines (as expected from surface area considerations). Goldschmidt powder had a higher oxygen content than any of the cast ingots but this did not appear to affect subsequent magnetic properties. In fact, in the milled condition this material appeared to be much "safer" and easier to handle whilst pressing so the extra oxygen may have a passivating effect on the reactivity of the powder. This supplier also provided a comprehensive data package (full chemical and sieve analyses and permeameter curve) with each alloy batch which, from a magnet manufacturer's viewpoint, is very useful.

PARTICLE SIZE REDUCTION
Mechanical Crushing

Initial size reduction of cast ingot was achieved using a specially designed mortar and pestle which contained the fragments produced. Protective atmosphere was found to be unnecessary at this stage although eye protection was required to protect the operator from occasional splinters. Most alloy suppliers now provide pre-crushed ingot (less than 5 mm diameter) which can be fed directly into a

TABLE 1 Oxygen and size analyses of some alloys studied during this programme

Supplier	Alloy	Bulk form	Size Analysis		Oxygen Content	
Trei.	35%NdFeB	Lump <10mm			8mm lump:	0.084%
					Fines <90µm:	1.30%
REP	35%NdFeB	Broken Ingot				
REP	NdDyFeBNb	Broken Ingot	3% Fines/dust:		Lump:	0.0055%
			>355µm:	71.5%		
			<355>180:	10.5%		
			<180>90:	6.5%		
			<90>45:	4.0%	<90>45µm:	1.05%
			<45µm:	7.5%	<45µm:	1.85%
GFE	35%NdFeB	Lump <10mm	9.5% Fines:		Lump:	0.0027%
			<850>355:	52.5%		
			<355>180:	22.3%		
			<180>90:	12.7%		
			<90>45:	8.5%		
			<45µm:	3.9%	<45µm:	0.37%
GFE	NdDyFeBNb	Lump <10mm	12.7% Fines:		Lump:	0.0055%
			<850>355:	50.4%		
			<355>180:	22.5%		
			<180>90:	12.4%	<180>90:	0.28%
			<90>45:	8.5%	<90>45:	0.38%
			<45µm:	6.2%	<45µm:	0.44%
Gold.	NdDyFeB	Powder, 24µm average particle size	>355µm:	4.8%		
			<355>180:	25.8%		
			<180>90:	19.9%	As received:	
			<90>45:	16.4%	0.39%	
			<45µm:	33.1%		

cross beater mill. Modifications to a standard machine were required to protect the working parts and feed/collection hoppers; an oxygen free nitrogen blanket proved effective. This method of crushing is not ideal because a very wide particle size distribution is produced (examples are given in Table 2) and despite precautions and careful working practice, the fines tend to be heavily oxidised and are best removed prior to milling.

Hydrogen Decrepitation

Cast ingot can be easily, safely and cleanly reduced to an extremely friable form by hydrogen decrepitation (Harris, 1987, McGuiness et al, 1986). Early investigations reported requirements of a pressure vessel and high purity hydrogen at 400 psi (30 bar). SGM have developed a much safer system using a flow of low pressure (7 psi, 0.5 bar) commercial grade hydrogen (dewpoint -60°C). A typical 1.5 kg batch can be processed in less than 30 minutes in a closed, water cooled steel chamber using an initial flow rate of 35 cfh $(1m^3hr^{-1})$.

The process causes both intergranular and intragranular failure of the ingot and results in a very readily oxidisable material. Surface rusting is visable within one hour exposure to air at room temperature. However, because the chamber is sealed and can be transferred directly to a glove box for subsequent handling, this does not present any problems ... risk of oxidation or fires, etc is eliminated from this part of manufacture.

TABLE 2   Size distributions (weight percent) of mechanically crushed NdFeB alloy

Sieve Fraction (μm)	Jaw Crushed Ingot	Mill Sieve 1.2mm	Mill Sieve 0.5mm	Mill Sieve 0.2mm
>1180	41.0	–	–	–
<1180>600	28.3	18.1	–	–
<600>355	11.9	36.2	–	–
<355>180	9.0	27.9	0.7	0.5
<180>90	5.2	8.2	16.2	8.8
<90>45	2.4	3.7	20.9	16.1
<45	2.3	6.7	62.4	74.6

Milling

The final stage of particle size reduction can be carried out "wet" (ball or attritor mill) or "dry" (jet mill). Milling time is much reduced for hydrogen decrepitated (compared to mechanically crushed) alloy. Co-reduced powder can be milled in the as-received condition (Herget, 1985) because of its much smaller starting size of 25 μm.

The original "wet" medium was freon (ICI Arklone 'P') but this is broken down by the neodymium and could become dangerous. Cyclohexane is believed to be inert under these conditions and is now used extensively. Dry ball milling in argon or nitrogen caused problems with the powder sticking to mill walls, making removal difficult.

A laboratory jet mill has been modified at SGM to enable initial trials by this route. Alloy feed rate, feed and drive gas line pressures (both oxygen free nitrogen) were controlled to produce powder samples between 2.3 and 4.7 μm average particle sizes. Although the arrangements were far from ideal, magnetic properties subsequently produced compared well with ball milled control magnets of equivalent particle size. Jet milling produces a much tighter particle size

TABLE 3 Selected magnetic results to illustrate the effects of process and practice improvements

Date	Alloy	Milled size μm	Sintered Density gcm^{-3}	BHmax MGOe (kJm^{-3})	jHc kOe (kAm^{-1})	Comments
3/86	REP 35% NdFeB	4.2	7.35	17.5 (140)	5.0 (400)	As sint Argon
4/86	REP 35% NdFeB	5.6	7.25	18.5 (148)	4.4 (350)	H/T Argon

* Sintering under vacuum introduced *

Date	Alloy	Milled size μm	Sintered Density gcm^{-3}	BHmax MGOe (kJm^{-3})	jHc kOe (kAm^{-1})	Comments
6/86	REP 35% NdFeB	4.1	7.20	30.0 (240)	7.8 (625)	H/T
1/87	REP 35% NdFeB	5.0	7.30	19.0 (150)	5.1 (410)	As sint Old Powder
1/87	Trei 35% NdFeB	4.3	7.35	29.0 (230)	8.4 (670)	H/T
1/87	Trei 35% NdFeB	4.0	7.45	23.0 (185)	3.1 (250)	Oversint High Br
1/87	Trei 35% NdFeB	No result: caught fire whilst pressing				

* Protective atmosphere for pressing introduced *

Date	Alloy	Milled size μm	Sintered Density gcm^{-3}	BHmax MGOe (kJm^{-3})	jHc kOe (kAm^{-1})	Comments
2/87	Gold. NdDyFeB	3.8	7.30	31.5 (250)	12.4 (990)	As sint H/T reduced jHc
2/87	Trei 35% NdFeB	3.15	6.95	28.0 (225)	10.7 (855)	H$_2$ decrep. at Lucas
2/87	Gold. NdDyFeB	3.4	7.55	32.0 (255)	8.4 (670)	H/T oversint
3/87	Trei 35% NdFeB	2.2	7.45	30.0 (240)	12.0 (960)	H$_2$ decrep. at Lucas
3/87	GFE NdDyFeBNb	3.6	7.50	31.5 (250)	13.5 (1080)	As sint H/T reduced jHc
3/87	REP NdDyFeBNb	4.1	7.55	33.5 (270)	11.2 (895)	H/T
4/87	Gold. NdDyFeB	2.8	7.60	32.5 (260)	11.6 (930)	As sint H/T reduced jHc

* Hydrogen decrepitation introduced at SGM *

Date	Alloy	Milled size μm	Sintered Density gcm^{-3}	BHmax MGOe (kJm^{-3})	jHc kOe (kAm^{-1})	Comments
4/87	Trei 35% NdFeB	1.9	7.20	26.5 (210)	8.4 (670)	As sint H/T no effect
5/87	Trei 35% NdFeB	3.1	7.40	32.0 (255)	8.3 (665)	H/T
5/87	Trei 35% NdFeB	2.7	7.40	30.5 (245)	9.3 (745)	As above, red. sint temp
6/87	Gold. NdDyFeB	3.4	7.45	33.0 (265)	11.4 (910)	H/T
6/87	Gold. NdDyFeB	3.7	7.45	32.0 (255)	12.8 (1025)	Sint/H/T run

range than other methods so magnetic properties would be further improved by more development work.

Table 3 contains a selection of magnetic results illustrating the improvements brought about by process developments during this programme.

PRESSING

All pressing has been carried out using double acting hydraulic presses designed and built at SGM for automatic control of pressure and magnetic field cycles. Both parallel and perpendicular field arrangements are available, the latter enabling an increase in remanence of approximately 5% with the associated improvement in maximum energy product. Aligning fields of $13kOe(1040kAm^{-1})$ are sufficient to produce optimum properties by this method.

Green density has been varied over a wide range with little noticeable difference in final magnetic properties, although distortion caused by excessive shrinkage for lightly-pressed compacts is undesirable from a manufacturing standpoint. Pressing load is limited by the tendency for the compact to laminate or "cap" upon ejection.

Pressing wet and dry powder has shown no effect on magnetic properties or final density. Wet powder is more difficult to handle because of reduced flow properties and greater tendency to adhere to equipment and tools.

Initially, pressing was carried out in air, but as the programme progressed powder particle sizes became smaller and this became hazardous. A nitrogen blanket (oxygen level monitored and kept below 1%) has been designed and fitted to the press tables and powder and green compacts are kept in this environment during transfer to the furnace. Since this system was installed, no serious incidents have occurred at this stage despite handling some extremely reactive powder samples (spontaneous combustion upon exposure to air).

SINTERING

Sintering of rare earth magnets at SGM is carried out in a top loading vacuum furnace with an operating capability of less than 1×10^{-5} torr. A fully programmable multi-cycle controller is fitted which can control all functions required for a process cycle including a backfilling facility (high purity argon) at any desired temperature.

The amount of outgassing depends on total charge weight (obviously) and on the method of powder preparation, for example mechanically crushed alloy outgasses much less than hydrogen decrepitated compacts. Several distinct "periods" of outgassing have been noted: at low temperatures (less than $100°C$)

the major cause is thought to be removal of residual milling medium or die lubricant. Considerable outgassing then occurs between 100 and 200°C and between 500 and 700°C, particularly with hydrogen decrepitated work.

Many experiments have been carried out to monitor the effects of sintering under vacuum (1×10^{-4} torr), under partial pressure (0.5 atmospheres high purity argon), with and without various "getter" materials, varying heating and cooling rates and time at temperature. The conditions finally arrived at compromise optimum magnetic and physical properties with ease of production and consistency of results. Heating rates are moderate (around $20°C$ min^{-1}) with allowances for outgassing, sintering time is 1 hour under "vacuum" followed by a slow (furnace) cool. Ideal temperatures vary with the alloy type and particle size but in general they agree with other workers' reported temperature ranges.

HEAT TREATMENT

Magnets which have been rapidly cooled from sintering temperature require a heat treatment cycle to realise optimum properties (Ormerod, 1984, Harris, Bailey, 1984). A time of one hour at temperatures between 600 and 680°C results in a variation in magnetic properties of less than 5% in a typical ternary alloy. Argon atmosphere can be used, but vacuum (1×10^{-4}) is preferred because any ground or sliced surfaces remain clean and do not require further finishing.

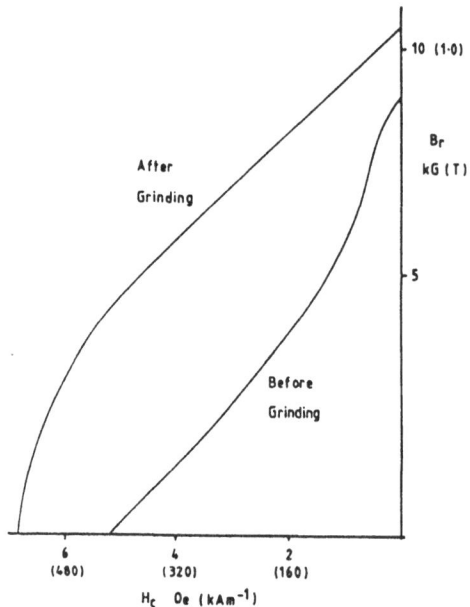

Fig 2 Permeameter curves of a heat treated (in argon) test bar before and after regrinding all faces

Impurities in the argon cause these surfaces to appear dulled and require regrinding to remove this "skin" and achieve maximum properties. An example of a small magnet (0.5 cm^3) tested before and after machining illustrates this effect clearly (Fig 2).

A separate heat treatment cycle may not be necessary if a slow (furnace) post sinter cool is allowed; indeed, further thermal processing has proved to be detrimental to the magnetic properties in several of our batches, the intrinsic coercivity exhibiting the largest reduction.

MAGNETISATION

Work done on magnetising fields agrees closely with other published information (Jubb, McCurrie, 1987). It has been found that the minimum field required for full saturation depends on the alloy composition and processing parameters, for example a higher coercivity material requires a higher saturating field than a lower coercivity counterpart. Fig 3 clearly illustrates that for the materials currently produced at SGM a magnetising field of at least 30 kOe (2400 kAm^{-1}) is required for saturation. A Hirst MC4C capacitor discharge magnetiser is used at 35-40 kOe (2800-3200 kAm^{-1}). Higher fields (up to 100 kOe, 8 MAm^{-1}) are available but have not shown any significant improvement in magnetic properties.

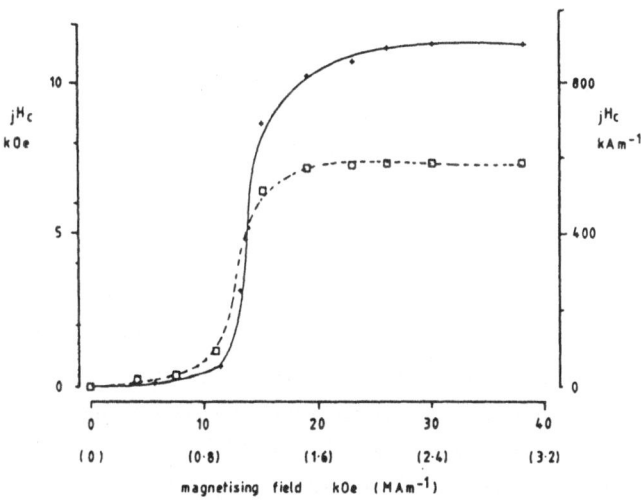

Fig 3   Magnetisation behaviour of low and high coercivity NdFeB alloys

MACHINING

Sintered NdFeB has proved to be relatively easy to machine when compared to SmCo$_5$ and Sm$_2$Co$_{17}$ materials. All cutting operations must be carried out wet and the slurry thus produced must not be allowed to dry out or serious risk of fire could result. After machining, all parts should be cleaned and dried to prevent rapid surface corrosion.

Grinding

NdFeB magnets are much less prone to chipping and more material can be removed per pass without breaking or cracking the work when compared to surface grinding SmCo$_5$. Several grades of alumina and silicon carbide wheels have been used to apparently the same effect on laboratory and production surface and centreless machines.

Slicing

Early NdFeB magnets could be sliced very easily, equivalent to SmCo$_5$ but with less risk of chipping, using diamond impregnated cutting discs. The development of high density ternary magnets and the dysprosium and dysprosium/niobium doped alloys (quaternary and quinternary compositions) with even higher density and much improved intrinsic coercivity has caused some problems. These materials appear to be much harder and tougher and slicing speeds have to be slower to maintain dimensional accuracy. The slicing machine at SGM is an ACT PACM 250A and blade widths between 0.5 and 1.5 mm have been used successfully.

Where other manufacturers' magnets have been sliced, their behaviour was similar to SGM materials.

Other Machining

Several other techniques have been tried, with varying success.

Sintered NdFeB has been drilled using a carbide tipped drill. A 4 mm diameter bore did not require a pilot hole although care must be taken to reduce breakout chipping. This is not envisaged as a suitable production route but it may be useful for producing samples or small orders.

Larger holes have been cut using either carbide or diamond tipped hole saws, the latter producing finer finishes with reduced tool wear. Again, chipping on breakout can be almost eliminated if care is taken.

Laser cutting trials produced very poor results. Despite protective atmospheres, much oxidation was evident at the cut surfaces together with excessive spatter and heat-affected zones extending up to 3 mm into the work.

The most successful methods of machining holes or more complex shapes were spark erosion (for small holes) and copper wire erosion cutting. These methods produced very good surface finishes with negligible heat-affected zones and excellent dimensional accuracy. The main disadvantage is the cost of the process, which will deter some customers although it is an excellent method for producing prototype parts.

ADDITIONS TO THE TERNARY ALLOY

Whilst NdFeB ternary alloys have produced intrinsic coercivities of 12000 Oe (960 kAm^{-1}) in the laboratory, production magnets from these are still very susceptible to temperature degradation. It has been shown by various workers (Ghandehari, 1986, Ghandehari, 1986, Tokunaga, 1986) that small additions of dysprosium (in place of neodymium) and/or niobium (in place of iron) can enhance intrinsic coercivities considerably with the associated improvements to their temperature stability. Early work involved the addition of dysprosium oxide powder directly to the milled ternary prior to pressing. This produced good results but at the expense of some remanence (caused by the addition and formation of non-magnetic oxides) and therefore maximum energy product. An extra process step was included and Dy_2O_3 powder is relatively expensive so this route is not ideal from a production viewpoint. The availability of dysprosium and dysprosium/ niobium doped quaternary and quinternary alloys from suppliers has obviated the need for the above route. Further addition of dysprosium oxide to these new alloys has a very small affect and is not necessary to produce high coercivity.

The effect on temperature stability of coercivity improvements brought about by these additions has been tested. Magnetic properties are plotted in Fig 4 and clearly illustrate that the higher coercivity materials are more stable. An interesting observation is that the isotropic magnet loses its magnetism much less quickly compared to an anisotropic sample of the same material.

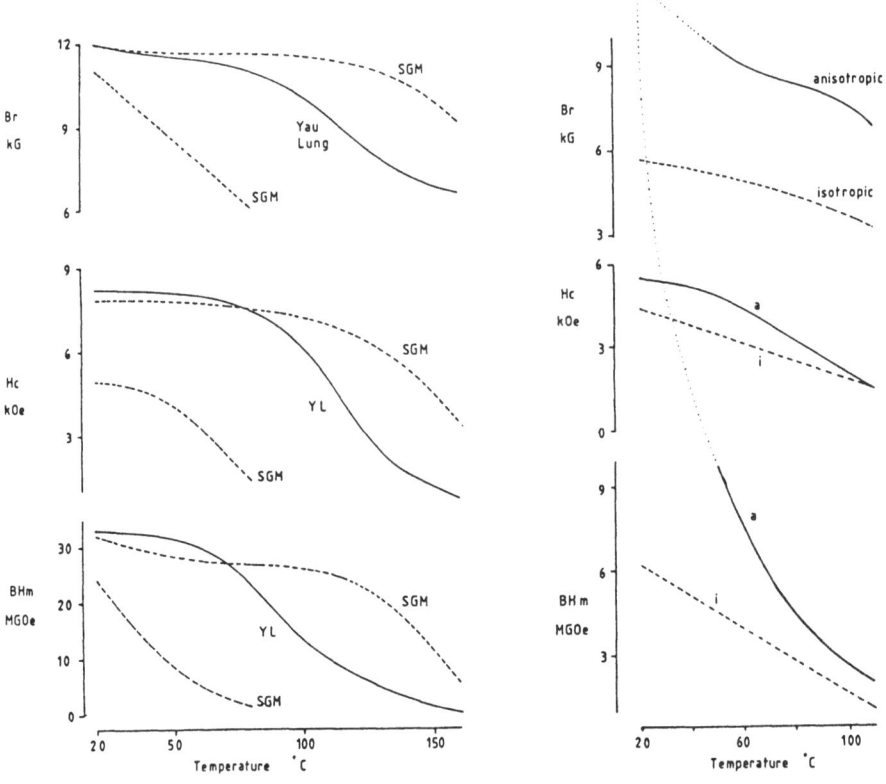

Fig 4   The effect of holding various NdFeB magnets at temperature prior to testing:

i.   SGM high jHc
ii.   Yau Lung (Chinese)
iii.   SGM low jHc

iv.   SGM anisotropic
v.   SGM isotropic
(same material)

CORROSION

During the course of the NdFeB development work it has often been noted that sintered or heat treated magnets are very susceptible to rusting if they are not thoroughly dried after machining or slicing operations. Surface corrosion under controlled test conditions is noticeable after a few hours.

Several polymeric coatings have been applied to magnets to assess the degree of corrosion protection provided. Accelerated tests to ASTM B177/6 requirements have been carried out. Inspection of the results (presented in Table 4) suggest that a thick epoxy coating is required to provide maximum environmental protection. The PTFE gave a very aesthetic appearance but, being very thin and possibly porous, did not provide the same degree of protection.

TABLE 4 ASTM B117/6 (5% salt spray at 35°C) corrosion test results for plastic coated NdFeB magnets

Coating	Time before rust easily visable	Time before coating failed
Uncoated mild steel	40 mins	
Uncoated NdFeB	<40 mins	
Black PTFE	7 hours	24 hours
Black nylon (uncured)	123 hours	430 hours
Black nylon (cured)	123 hours	430 hours
Green epoxy	123 hours	500 hours

METALLOGRAPHY

Standard techniques can be used to prepare NdFeB for optical examination, i.e. mounting in bakelite or other suitable medium, wet grinding on silicon carbide papers of successively finer grit and polishing on 6 μm and 1 μm diamond pads. When viewed in the as-polished condition, NdFeB magnets reveal a high degree of apparent porosity. This is probably due to the neodymium-rich phase being oxidised and/or removed during the grinding and polishing stages. Etching can be achieved easily with 2% Nital solution. Etching times are short (seconds) as the alloy, and particularly the Nd-rich phase, is very reactive. Because the most reactive phase is predominant at grain boundaries, the grain structure of etched magnets is easily observed.

Various magnets from other manufacturers have also been studied, for example Colt Crucible, Yau Lung, GM MQII, etc.

Macroetching of cast ingot proved to be successful using 10% hydrochloric acid aqueous solution. A deep clear etch was obtained after 1-2 minutes and revealed a typically cast ingot structure.

BONDED NdFeB

The fabrication of a permanent magnet via a bonded route has several advantages over the press-and-sinter sequence. Initial experiments have been carried out to assess the relative viabilities of numerous bonding methods. These are summarised in Fig 5 and are described below.

CRUSHED INGOT/SINTERED MAGNET

As-received cast ingot of 35% neodymium ternary and NdDyFeBNb quinternary alloys were prepared as for the sintered route. Ball milled powder was dried, blended with powdered polythene and compression moulded at 30 MPa. Mould temperature was 150°C and the mould and pellet were cooled to room temperature prior to extraction. The "magnet" thus produced exhibited no permanent magnet properties.

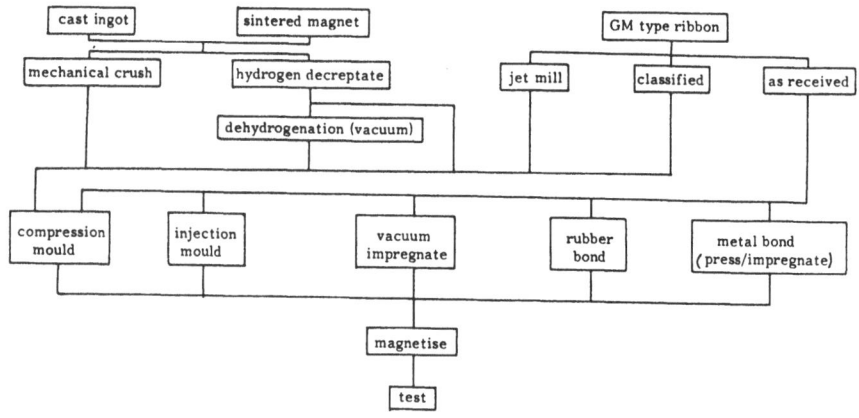

Fig 5   Process routes investigated during this study to produce a bonded NdFeB magnet

A good quality (30 MGOe, 240 kJm^{-3}) sintered magnet was similarly crushed and moulded. It failed to retain its magnetic properties.

Both cast ingot and good quality sintered magnet were broken down by hydrogen decrepitation and moulded as above. These also failed to exhibit permanent magnet properties.

The reason for the total lack of coercivity in these pellets is not fully understood. It is known that hydrogen decrepitation and the associated formation of hydrides lowers the magnetic properties, particularly maximum energy product, of the cast ingot and probably the sintered magnet also (Wiesinger, 1986). In the case of the mechanically crushed samples, some loss of properties due to oxidation would be expected, but even using larger particle sizes, for example unmilled, showed no improvement.

The conclusion is drawn that the production of a bonded magnet from the "sintered route" starting materials is not straightforward.

RAPIDLY SOLIDIFIED, I.E. "GENERAL MOTORS TYPE" MATERIAL

Most of the work on bonded NdFeB at SGM has used the GM MQI type of annealed, crushed, melt spun ribbon (General Motors, 1985). The numerous binders and processing routes considered are described below.

Sintered Ribbon

A sample of as-received ribbon was pressed at 800 MPa and sintered under high purity argon for one hour at 1050°C. The resultant pellet looked oxidised and was barely magnetic. The central portion was sliced out and tested. This appeared to be less heavily oxidised but was still very porous. Measured magnetic properties were:

Br = 2700 G, 0.27 T;   Hc = 1000 0e, 80 kAm^{-1};   BHm = 0.6 MGOe, 5 kJm^{-3}

This branch of the investigation was not pursued further.

Compression Moulding

A small hydraulic press fitted with a cylindrical electrically heated mould and water cooling system has been used to produce one-off pellets for magnetic testing. Temperature is controlled at 150°C +/- 10°C and pressures are available up to 30 MPa. Several binders have been used in this press, with varying success.

Crushed bakelite powder (<90 µm) was used initially, dry blended with the ribbon and pressed as above. This binder was far from ideal because of poor adhesive properties ... 10 percent by weight binder resulted in a very "powdery" pellet.

Acrylic powder was cleaner and easier to use and its thermoplastic nature enabled improved flow between the ribbon flakes to produce a "better" compact.

Higher magnetic properties were achieved with a two-part liquid epoxy resin. Excess resin was squeezed out of the mould and resulted in increased volume fraction of ribbon (and therefore magnetic properties), when compared to other binders. However, the compact was very difficult to remove from the die set, caused by the excess cured resin, so this route is less practical than others.

Vacuum Impregnation

Following the magnetic success of the epoxy binder, ribbon compacts were dry pressed at 130 MPa and impregnated with epoxy resin under vacuum. When cured, the pellets exhibited excellent magnetic and physical properties, equal to the commercially produced MQI magnets. During these trials, various additives were studied to assist pressing. Small quantities of an amide wax appeared to reduce the lamination problems occurring with the highest pressures, although 2 percent by weight prevented adhesion of the resin which resulted in a very weak compact. Magnetic properties were not affected by these additions.

Metal Bonded Ribbon

Various metal powders were blended with ribbon prior to cold pressing in an attempt to mechanically bind the flakes. Small quantities, for example 5 weight percent copper, can be added without affecting magnetic properties. The added powder is thought to fill the interstices so volume percent of ribbon remains unchanged. However, these quantities were insufficient to produce a robust compact and such test bars were vacuum impregnated with epoxy resin to assist handling whilst magnetic testing. High percentage additions, for example 50 weight percent copper, produced mechanically strong pellets which were fully dense, or at least sealed at the surface because such bars did not take in resin during impregnation. Magnetic properties were correspondingly lower due to reduced ribbon fill.

Iron additions proved to be disasterous. The loop shape was affected with 5 weight percent, and increased content severely affected coercivity also. Remanence was less affected and showed only a small drop with 50 weight percent, as illustrated in Fig 6. The explanation is that the soft iron powder provides efficient return paths for the flux from each ribbon flake so the net coercivity as measured outside the pellet is much less than when a non-ferromagnetic filler is used. The apparent retention of remanence is caused by the higher iron content in the overall pellet.

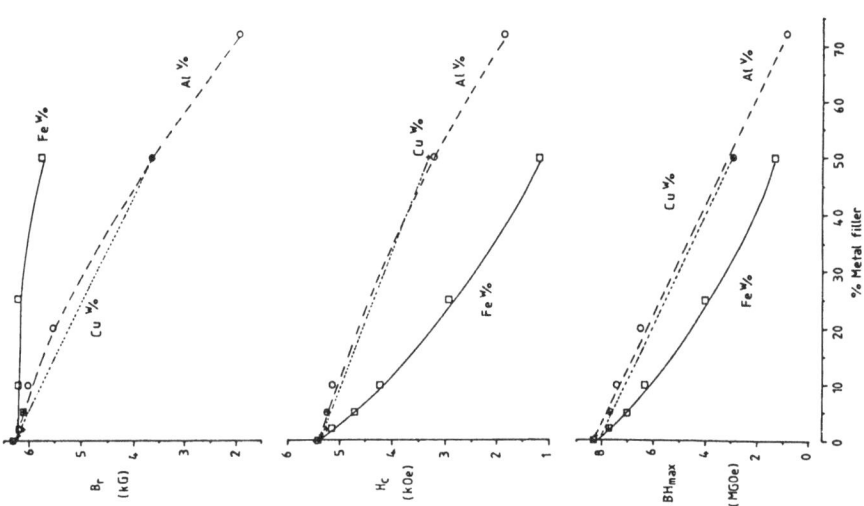

Fig 6   Magnetic properties vs binder content for metal bonded NdFeB ribbon compacts

Classified Ribbon Fractions

A sample of as-received ribbon was separated into four size fractions and a pellet of each in 10 weight percent of fine bakelite (<90 μm) was tested. The results presented in Table 5 show that the coarser fractions give higher magnetic properties. This is thought to be directly related to the degree of oxidation, i.e. the fines contain a higher percentage of oxygen due to the much greater surface area.

TABLE 5 Magnetic properties of seived melt spun ribbon fractions in 10 weight percent bakelite

Ribbon Fraction μm	Pellet Density gcm^{-3}	Br G (T)	Hc Oe (kAm^{-1})	BHm MGOe (kJm^{-3})
<355>180	4.70	4300 (0.43)	3840 (310)	4.00 (32.0)
<180>90	4.72	4200 (0.42)	3810 (305)	3.95 (31.5)
<90>45	4.70	4150 (0.42)	3700 (295)	3.80 (30.5)
<45	4.60	3850 (0.39)	3300 (265)	3.10 (25.0)
Jet milled, 4.7 μm	4.68	4100 (0.41)	3630 (290)	3.70 (29.5)

To extend this table, a sample of as-received ribbon was jet milled to 4.7 μm average particle size and a similar compact made and measured (Table 5). This result was surprising but it clearly shows that flake size is not important from a magnetic standpoint providing that precautions are taken to avoid oxidation of fine powders. This route is perhaps less suited to mass production than using coarser, as-received flake.

Injection Moulding

An injection moulded, polymer bonded NdFeB magnet was set as a target for this part of the investigation. Compounding with thermoplastic binders was carried out in a small Z-blade mixer with temperature controlled electric heating jacket.

Several polymers have been tried, for example polythene, polypropylene, polyester, several polyamides (nylons) and polyphenylene sulphone (PPS). The latter was very difficult to work with because of its high melting temperature (over 300°C). Polypropylene degraded very easily even with antioxident additives.

Nylon 6 was found to be unsuitable with a high processing temperature and poor flow characteristics. Two polymers have been identified as potentially useful to concentrate future work, namely nylon 12 and a low viscosity polythene. Most work has concentrated on a straight ribbon/polymer blend with very limited assessment of other additives, for example antioxidents, lubricants/flow promoters, coupling agents, etc. The effects of these additives will be investigated fully in due course.

Test bars from each blend were produced by either compression or injection moulding. Some results are attached in Fig 7. Nylon blends could not be compression moulded because the temperature of that mould could not be increased sufficiently.

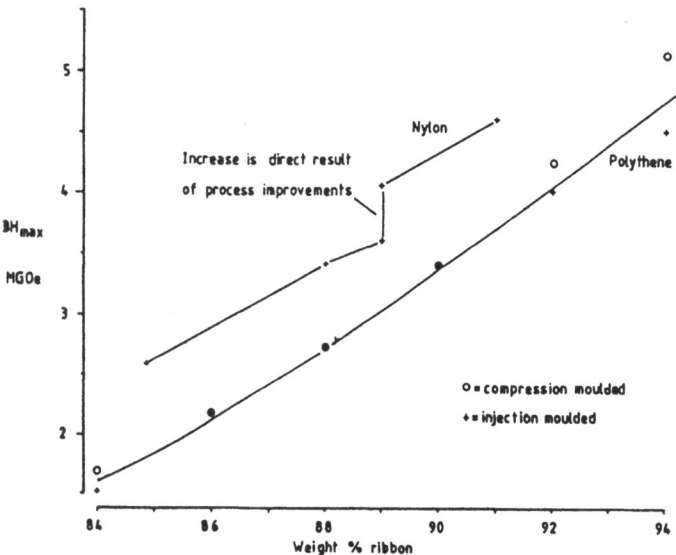

Fig 7 BH max of test bars produced by injection and compression moulding of polymer bonded NdFeB ribbon

The "injection mouldability" of different blends was assessed qualitatively with a single gate thin walled ring mould with dimensions 32 mm outside diameter, 25 mm inside diameter and 28 mm deep. The uniformity of fill was compared both visually and magnetically by field strength measurements of the magnetised ring.

CONCLUSIONS

From the work described above, the following conclusions can be drawn.

For sintered NdFeB:

i. Hydrogen decrepitation or use of co-reduced alloy is preferable to mechanical crushing of cast ingot because a more uniform particle size is produced in much reduced time.

ii. A fine, uniform particle size produces a better magnet than a coarse or wide size range powder.

iii. Heat treatment may not be necessary for optimum magnetic properties if a slow cool from sintering temperature is allowed.

iv. Dysprosium and niobium additions to the ternary alloy enhance intrinsic coercivity and therefore greatly improve temperature stability.

v. Pilot production quantities of magnets up to 33 MGOe (265 kJm^{-3}) can be produced from ternary alloy and up to 30 MGOe (240 kJm^{-3}) with enhanced intrinsic coercivity for higher temperature applications.

For bonded NdFeB:

vi. Bonding "sintered-type" starting materials does not work.

vii. Rapidly solidified, e.g. General Motors', ribbon can be successfully bonded in numerous ways.

viii. Magnetic properties are little affected by ribbon flake size, although handling coarser powders is easier and safer.

ix. Pressed/vacuum impregnated test bars have been made at over 8 MGOe (65 kJm^{-3}).

x. Injection moulded test bars have been made up to 5 MGOe (40 kJm^{-3}).

REFERENCES

Croat, J.J., Herbst, J.F., Lee, R.W. and Pinkerton, F.E. 1984. Appl. Phys. Lett. 44 (1) 148-149.

Ghandehari. M.H. 1986. Appl. Phys. Lett 48 (8) 548-550.

Ghandehari, M.H. 1986. Proc. Intermag Conf. Phoenix, Arizona.

General Motors' technical sales literature.

Harris, I.R. 1986. Personal communication.

Harris, I.R. and Bailey, T. 1984. Proc. CEC Workshop, Brussels, Oct. 1984. Ed. Mitchell, I.V. 99-103.

Herget, C. 1985. Proc. 8th Int. Workshop on R.E. Magnets, Dayton, Ohio, May 1985, Ed. Strnat K.J. 407.

Jubb, G., McCurrie, R.A. 1987. IEEE Trans. Magn. MAG 23 (2) 1801-1805.

McGuiness, P.J., Harris, I.R., Rozendaal, E., Ormerod, J. and Ward, M. 1986. J. Mat. Sci. 21 4107-4110.

Ormerod, J. 1984. Proc. CEC Workshop, Brussels, Oct. 1984. Ed. Mitchell, I.V. 69-92.

Sagawa, M., Fujimura, N., Togawa, H., Yamamoto, H. and Matsuura, Y. 1984. J. Appl. Phys. 55 (6) 2083-2087.

Togunaga, M., Meguro, N., Endoh, M., Tanigawa, S. and Harada, H. 1986. IEEE Trans. Magn MAG 22 (5) 904-909.

Wiesinger, G., Hilscher, G., Grossinger, R. and Kirchmayer, H. 1986. Hydrogen Energy Progress VI 887-892, Pergamon Press.

SOME ASPECTS OF THE STRUCTURE AND PERFORMANCE OF NdFeB INGOTS AND MAGNETS

M.Ward & J S. Taylor
Lucas Engineering & Systems Ltd.,
Lucas Research Centre,
Dog Kennel Lane,
Shirley, West Midlands,
UK

ABSTRACT

A range of commercial magnets has been characterised in terms of magnetic, mechanical and structural properties. A potentiodynamic technique has been used to study corrosion relative to several reference elements in terms of equilibrium corrosion potential and corrosion rate in a variety of corrosive media. Grain size analysis has been carried out on a number of ingots as part of a collaborative exercise related to control of grain size.

SUMMARY

Investigations have been carried out in four areas:-

(A)   Collaborative work on ingot characterisation and grain refinement.

(B)   Limited in-house processing.

(C)   Corrosion behaviour.

(D)   Characterisation of commercial magnet materials.

Main findings are:-

-Ingot grain size can be varied with cooling rate and some small benefits can be gained by producing magnets from fine grained ingots. However optimum yield of tetragonal phase only occurs at intermediate rates. Undesirable free ferrite can form at low cooling rates.

-Hydrogen decrepitation is a technically viable technique for in-house laboratory processing. Dy and Co containing alloys appear to require higher sintering temperature than $Nd_{15}Fe_{77}B_8$.

-The corrosion rate of NdFeB magnet materials is considerably higher than iron and this phenomenon is clearly linked with the presence of multiple phases. Of the materials evaluated only Zinc is sacrificial with respect to the 2/14/1 phase. The corrosion rate of bonded, nominally single-phase, rapidly solidified material is related to a high effective surface area arising from penetration of the corrosive medium into the particle/polymer interface.

-A range of commercial materials has been evaluated in terms of room temperature magnetic properties, grain size and mechanical strength. With a few exceptions measured and manufacturers quoted properties match. Grain size and microstructure show considerable variation. Anomalies in microstructure can be linked with deterioration of magnetic properties. There is a relatively strong correlation of mechanical and physical properties but only nominal correlation between magnetic and physical, chemical and microstructural properties. All sintered NdFeB materials are significantly stronger than Sm-Co$_5$ alloys.

INGOT CHARACTERISATION

Cooling Rate

Ingots of NdFeB (15/77/8) were prepared by REP Ltd using a number of mould types to vary cooling rate. Samples were distributed to all Group 2 members for evaluation. Homogenisation is known to be a slow process in NdFeB ingots therefore there are advantages in minimising grain size variation and the proportion of non-magnetic phases (eg borides).Ingots were 15, 30, 40 and 100mm thick rectangular and cylindrical (crucible cool). Results are shown in Table 1 as phase proportions and columnar grain width determined by manual point counting.

As might be expected columnar grain width was inversely proportional to cooling rate (inverse crucible width) and increased from edge to centre, particularly at the highest cooling rates. Main phases present were columnar grains of tetragonal 2/14/1 and intercolumnar boride and eutectic. Some ferrite was present in the centre of 40 and 100 mm ingots but not in the crucible cooled sample. It is possible that the low cooling rate in this sample permitted solution of the primary ferrite.

Ferrite has been included with the tetragonal phase in the analysis. Minor phases present were oxide and copper (from the moulds).

The boride content of the fast cooled sample was noticeably higher than other ingots. Slow cooling, as in ingot centres and the 100mm and crucible sample, appeared to promote the formation of high eutectic levels.

Maximum yield of tetragonal phase with minimal ferrite appeared to be possible over a narrow range of cooling rates (eg 30 mm ingot).

Other Ingots

Other ingots examined were REP 35% Nd, REP Nd$_{15}$ Fe$_{77}$ B$_5$ Dy$_2$ Nb, Triebacher Nd$_{15}$ Fe$_{77}$ B$_8$ and Triebacher Nd$_{14}$ Fe$_{75}$ B$_8$ Co$_2$ Nb. Some typical microstructures are shown in Figs. 1, 2 and 3. All samples exhibited similar microstructures, consisting of a coarse grained matrix with several intergranular phases (Nd rich, borides, oxides and other minor inclusions). Metallography and XRD showed the REP 35% Nd ingots to contain less free iron than the Triebacher 15/77/8 alloy.

Annealing of the REP Dy/Nb alloy (at 1150 deg. C for 1hr) resulted in a fine fibrous phase dispersed within the Nd-rich phase (Fig 3). Some grain growth is suggested by the results shown in Table 2. Ingots were also prepared by REP Ltd under various conditions to promote grain refinement (including Zr additions). Results of microstructural analysis were passed to Birmingham University for collation.

Magnetic Properties

Magnets were produced from two of the original ingots (BM1558, BM1561) to establish the effect of prior ingot grain size on magnetic properties. Ingot grain sizes of the samples used are shown in Table 3.

Samples were identically processed. Results, shown in Table 4, suggest some slight advantage in using a small prior ingot grain size.

MAGNET PROCESSING

Limited processing facilities have been established in-house using hydrogen decrepitation, grinding, press alignment, vacuum sintering and furnace cool. Two alloys (Nd_{14} Fe_{75} B_8Co_2 Nb and Nd_{15} Fe_{77} B_5 Dy_2 Nb), intended for higher temperature applications, have been used to produce sintered magnets. Both alloys appeared to require higher sintering temperatures than 15/77/8 alloys, possibly indicating a greater affinity for oxygen or reduced volume of liquid phase. Results for the Dy/Nb alloy are shown in Figure 4. Grain size data is given in Table 5, the large grain size reflects the in-house laboratory process used.

CORROSION
Influence on magnetic properties

In dry conditions Nd FeB magnets have shown great stability even after prolonged handling. However aqueous conditions are known to produce rapid corrosion. Fig. 5 illustrates the effect of corrosion on magnetic properties. This could obviously be a critical effect in applications which use thin segments.

Commercial sintered (Hitachi) and polymer bonded (MQ1) magnets were chosen as typical generic types and subjected to 100% humidity at 70deg. C and dry heat at 100deg. C for prolonged periods. Results are shown in Fig. 6. The sintered material clearly illustrates the difference between dry oxidation and aqueous corrosion.

Electrochemical Analysis

Corrosion data was generated in the form of electrochemical series and estimated corrosion rates for a variety of magnetic and reference materials.

Method

Potentiodynamic scans were carried out using a PARC corrosion monitor. The basic principle places a suitably prepared sample in an electrolyte with a carbon counter electrode. A standard Calomel electrode provides a reference. The test scans through a range of impressed voltages to produce positive (oxidation) and negative (reduction) overpotential relative to the equilibrium potential of the sample in the chosen medium.

When a sample is immersed in a solution it establishes a potential relative to the reference electrode under open circuit conditions (no applied potential). The initial value of equilibrium corrosion potential $E_{corr.}$ is taken after 200s. A Tafel plot of corrosion current vs potential is produced by scanning a preset voltage range. Corrosion rate is derived from i_{corr} around E_{corr} and plots can also indicate passivation effects. A constant area is assumed when estimating corrosion rate hence pitting or porosity will tend to artificially increase the apparent rate.

The polarisation which takes place during the scan is normally reversible and E_{corr} returns to a value close to the initial one. It is the final E_{corr} which has been quoted in the results.

Samples were 10mm discs, ground to 600 grit. Scan rate was 2.0mV/s, range -250mV to +1.2V. Highly corrosive media or a rapid rate tend to give a high scatter and results quoted are averages of 5 samples.

Initial results were obtained in 1wt% NaCl solution. Subsequent tests were carried out in electrolytes as follows:-

Neutral - 0.01% NaCl pH = 5.7
Acid - 0.01% NaCl + 0.01% H_2SO_4 pH = 2.5
Alkaline - 0.02% NaCl + 0.02% NaOH pH = 11.5

Results

Initial results are given in Table 6 as relative corrosion rates.

TABLE 6

Relative corrosion rates in 1Wt% NaCl solution.

Mildsteel	2/14/1 Phase (Homogenised)	Sintered Magnet	Bonded MQ1
1	5	19-39	21

Absolute corrosion rates were extremely high causing a wide scatter, however the effect of a multiphase microstructure is clearly evident. The corrosion resistance of the basic magnetic phase (2/14/1) appeared to be intrinsically lower than iron, which is itself not particularly good.

Results obtained in less aggressive neutral, acid and alkaline conditions are detailed in Table 7. Samples were:-

Reference Materials:

-Zinc

-Iron   (99.9%)

-Copper   (99.99%)

-Titanium

-2/14/1   material; as cast

-2/14/1   material; homogenised

-Samarium Cobalt (1:5)

Magnets:

-Hitachi (Hicorex 94EA)

-Sumitomo (Neomax 30)

-MQ1

-MQII

-Coated MQ1

As-cast 2/14/1 contained ferrite, this was removed by a homogenisation treatment.

The coated MQ1 had been supplied with a commercial corrosion preventative analysed to contain 90% Ba, 10% S.

Corrosion potential (E_{corr}) and corrosion rate are shown in Fig. 7.

Typical microstructures are shown in Fig. 8 as sections normal to corroded surfaces. General Observations:-

Build up of corrosion products on sample surfaces could cause significant increases in polarisation resistance. Magnet samples and the iron reference showed the formation of hydrated ferric oxide on the surface, forming from a precipitate of ferrous hydroxide in the solution. Under acid conditions Fe_2O_3 tended to form on the sample surface. Pitting varied from 'fine /uniform' to 'coarse/deep' especially in the presence of prior defects.

Observations drawn from the results in Table 7 are:-

1. E_{corr} and corrosion rates were similar for neutral and alkaline conditions. Rates for most materials were higher in acid. E_{corr} was more negative in acid for magnetic

materials and rapid gas evolution, pitting and passivation by corrosion products tended to cause scatter in this medium.

2. Coated MQ1 was stable under all conditions.

3. Only the homogenised 2/14/1 phase displayed corrosion resistance similar to Samarium Cobalt.

4. Only Zinc was electronegative with respect to NdFeB materials.

5. There is a reasonable correlation between Ecorr and corrosion rate with the exception of MQ1. In both alkaline and neutral conditions the corrosion rate for MQ1 was much higher than expected. Several explanations are possible:-

(A) Poor bonding between the polymer matrix and ribbon particles allows penetration of corrosive medium. The rate of corrosion might then increase due to

-higher effective exposed area.

-a form of crevice corrosion.

-corrosion products forcing the material apart near the sample surface.

Deep penetration by pitting was observed even under simple humidity trials (Fig. 8). Fig. 8(a) shows different stages of penetration normal to the ribbon plane. Fig. 8(b) shows penetration parallel to the ribbon plane. These should be compared with the uniform surface of MQII (Fig. 8(c)).

(B) Exposure of a high surface area by pitting will increase i_{corr}. The corrosion rate calculation assumes constant area.

6. The influence of multiphase structures can be seen in the behaviour of sintered magnets. A typical corroded surface is shown in section in Fig. 8(d).

The presence of ferrite in the 2/14/1 material caused a marked increase in both E_{corr} and corrosion rate in all solutions.

CHARACTERISATION OF COMMERCIAL MATERIALS

A continuous programme of evaluation of commercial materials has been undertaken throughout the programme. Room temperature magnetic properties have been determined on all materials obtained. Other tests, carried out on selected materials, have been

-physical and mechanical.

-microstructural and chemical.

Test Methods
Magnetic

All magnetic measurements have been carried out at ambient temperature using a permeameter. Sample flux is measured using a coil with an identical series - opposition wound compensation coil measuring air flux. Applied field is measured with a Hall Probe. A computer system is used to log data, plot normal and intrinsic characteristics and calculate single point data. Calibration is achieved with a Ni standard for B, traceable standard magnets for Hall probes and standard magnet samples to check reproducibility. (Thermal measurement equipment has now been installed but is not part of this programme).

Prior magnetisation is carried out in fields in excess of 4T.

Other analysis techniques

Phase proportions have been determined using an optical point counting technique. All grain size measurements quoted in this report have been determined with a Joyce-Loebl Image Analyser. Results are provided in the form of

-Mean grain length GL and breadth GB

-Mean grain diameter (GL+GB)/2

-Mean grain area GA

-Standard deviation of grain area and grain size distribution

Semi-quantitative analysis has been carried out using XRF techniques.

Transverse rupture strength was determined using a 3 point bend technique based on ASTM B528-83a, with a span of 20mm and sample depth of 4mm, at least 5 samples were used for each material.

Thyssen Magnet Validation

A number of closely matched magnets were circulated by Thyssen to establish some measure of traceability within the group. Results are shown in Table 8 and indicate close correlation.

TABLE 8
Magnetic measurement traceability

SAMPLE	ESTABLISHMENT	B r	bHc	iHc	BHmax
		kG	kOe	kOe	MGOe
11	LRC	11.9	10.9	11.4	33.6
	Thyssen	12.0	10.3	10.8	33.5
12	LRC	11.9	10.9	11.8	33.2
	Thyssen	12.0	10.6	11.3	33.2

Material Characterisation
Microstructure

Typical microstructures for rapidly solidified (MQ1 and MQII) and sintered magnets are shown in Fig. 9. Etching in acid ferric chloride was found to be more effective than Nital to reveal grain boundaries for grain size analysis, this is illustrated in Fig. 10.

A shading effect has been obtained in MQII after etching in acid ferric chloride as shown in Fig. 9b. No compositional variation has been detected by EPMA although boron could not be detected by this technique. However there is a marked hardness variation across the two regions. Microhardness (Hv25gm) results were as follows:-

Dark Area =   690 -   792

Light Area = 1002 - 1288

Major phases present in sintered alloys were the 2/14/1 magnetic phase, 1/4/4 boride, rare-earth rich intergranular eutectic, oxide(s) (particularly of Nd) and a number of minor phases dependent on composition as shown in Figs. 9c and 9d.

Some indication of the variation in phase proportions in commercial materials is shown in Fig. 11. These results were obtained by optical point counting and the overall proportions may be high, however they still illustrate a large variation.

Grain size was found to vary considerably as shown by the results in Table 9 and Fig. 12. Relatively narrow size distributions, conducive to high magnetic performance, were observed in commercial materials, for example Crumax, Hicorex. However this can be offset by relatively large areas of undesirable microstructure as shown in Figure 13. Such variations can be linked with varying magnetic performance (Fig. 14). The area shown in Fig. 13(a) was associated with a high oxide content. The narrow grain size range of Crumax materials could be the product of atomisation.

Magnetic and Mechanical Properties

Magnetic and mechanical properties are recorded in Table 10. Manufacturers data is either the mean of the quoted range, quoted minima, typical or nominal. Within this context measured and manufacturers data are reasonably close with a few exceptions (due to segregation as noted above). Typical demagnetisation curves are shown in Figs. 14-16.

Some variations in magnetisation behaviour were observed depending on the prior history of the material. Virgin magnetisation in Neomax 30 required a relatively low field and exhibited a single stage curve. Remagnetisation after magnetic demagnetisation required a higher field and the magnetisation curve was two stage.

Mechanical properties (Transverse Rupture Strength) of sintered magnets varied by a factor of 2, much higher than the test sample variance (Table 10). However all were higher than a typical SmCo, reflecting the better engineering properties of sintered NdFeB (Fig. 17a).

Property Correlation

The grain size and chemical analysis results shown in Table 9 were obtained from the same sample blocks as the magnetic and mechanical data given in Table 10. This was done in order to test for correlation between magnetic and mechanical properties and structural, chemical or physical features.

The extent of correlation was restricted by the limited number of samples and there was no clear link between microstructure and magnetic properties (although phase analysis is continuing). Some correlation was observed between chemical composition, particularly rare-earth content, and magnetic properties as shown in Figure 18, even considering the semi-quantitative analysis results obtained from XRF. All the samples are metallurgically similar so XRF analysis should be valid, at least in a relative sense.

Fig. 18 (a) shows the dependence of intrinsic coercivity on Dysprosium additions.Fig. 18(b) illustrates the strong correlation between Nd/Fe ratio and BH max for materials containing no Dy (hence no enhancement of iHc). In both cases there are diminishing returns as rare-earth content increases. The results of Fig. 18(b) might reflect the improved sintering behaviour with greater liquid phase present (eg. densification, grain surface smoothing). Mechanical properties correlate well with density (Figure 17) as might be expected from simple powder metallurgy principles.

Much more data is required before any other form of correlation can be confidently established.

PUBLICATIONS

'The production of a NdFeB permanent magnet by a hydrogen decrepitation/attritor milling route'.

P J McGuiness, I R Harris, E Rozendaal, J Ormerod and M Ward.

Journal of Materials Science 21 (1986) 4107-4110.

ACKNOWLEDGEMENTS

The authors are indebted to all those who have contributed to this work during its progress, in particular Kevin Watts, Jim Garrity and other members of the Materials Evaluation Group, Paul Magrath and other members of the Materials Engineering Group,and Mark Hixon of LE&S Ltd.

MOULD TYPE	POSITION	TETRAGONAL + FERRITE-area%	BORIDE area %	EUTECTIC + OXIDE-area %	COLUMNAR GRAIN WIDTH microns (MEAN)
15mm BM1559	edge	77.58	9.25	13.1	2.9
	centre	78.27	7.71	14.02	19.3
30mm BM1560	edge	80.9	5.56	13.54	16.7
	centre	79.98	4.23	15.79	33.7
40mm BM1557	edge	79.15	5.33	15.52	13.0
	centre	79.14	5.23	15.63	31.2
100mm BM1558	edge	78.95	4.38	16.67	21.1
	centre	64.99	4.08	30.93	29.7
Crucible BM1561	edge	72.64	4.92	22.44	190.8
	centre	76.36	4.29	19.35	241.5
MEAN ERROR	edge	± 2.39	± 1.34	± 2.12	-
	centre	± 2.45	± 1.26	± 2.23	-

TABLE 1 INGOT STRUCTURAL ANALYSIS

	"as cast"		1hr @ 1115°C	
	longitudinal section	transverse section	longitudinal section	transverse section
Minimum	3.20	3.96	9.073	4.66
Maximum	4.23	57.48	83.11	162.9
Mean	14.7	18.5	27.4	37.4
std. dev .	5.4	8.82	9.9	17.9

TABLE 2 EFFECT OF HEAT TREATMENT ON REP Nd15Fe77B8Dy2Nb

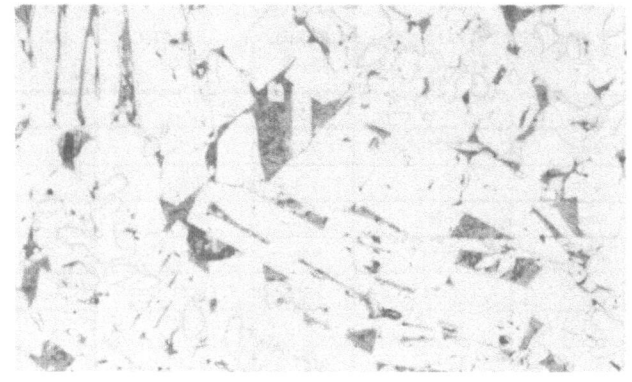

FIGURE 1 Triebacher $Nd_{15}Fe_{77}B_8$ centre of ingot x200

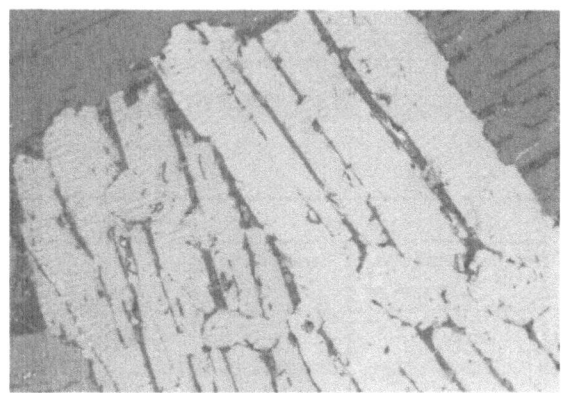

FIGURE 2 REP $Nd_{15} Fe_{77}B_5Dy_2Nb$ ingot
showing domain structure-ferrofluid x500

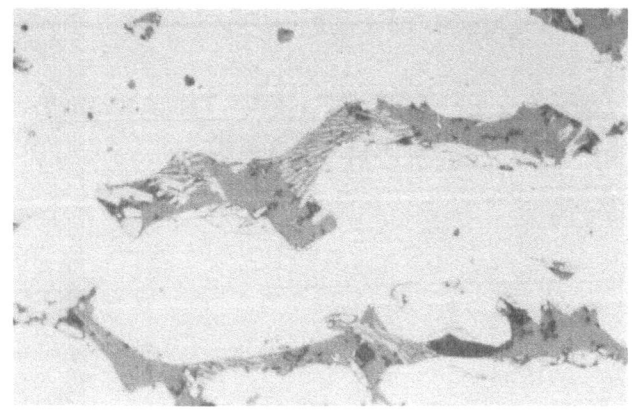

FIGURE 3 REP $Nd_{15}Fe_{77}B_5Dy_2$ Nb ingot after 1 hr at
1115 deg C x500

INGOT NUMBER	GRAIN SIZE (micron)	
	EDGE	CENTRE
BM1558	16.8	20.7
BM1561	107.4	170.4

TABLE 3   INGOTS USED FOR PRIOR GRAIN SIZE/MAGNET PROPERTY EXPERIMENT

MAGNETIC PARAMETER	BM1558	BM1561
Br (kG)	10.5	10.2
iHc (kOe)	6.7	5.4
bHc (kOe)	6.2	4.9
BHmax (MGOe)	22.0	19.2

TABLE 4   DEPENDANCE OF MAGNETIC PROPERTIES ON INGOT GRAIN SIZE

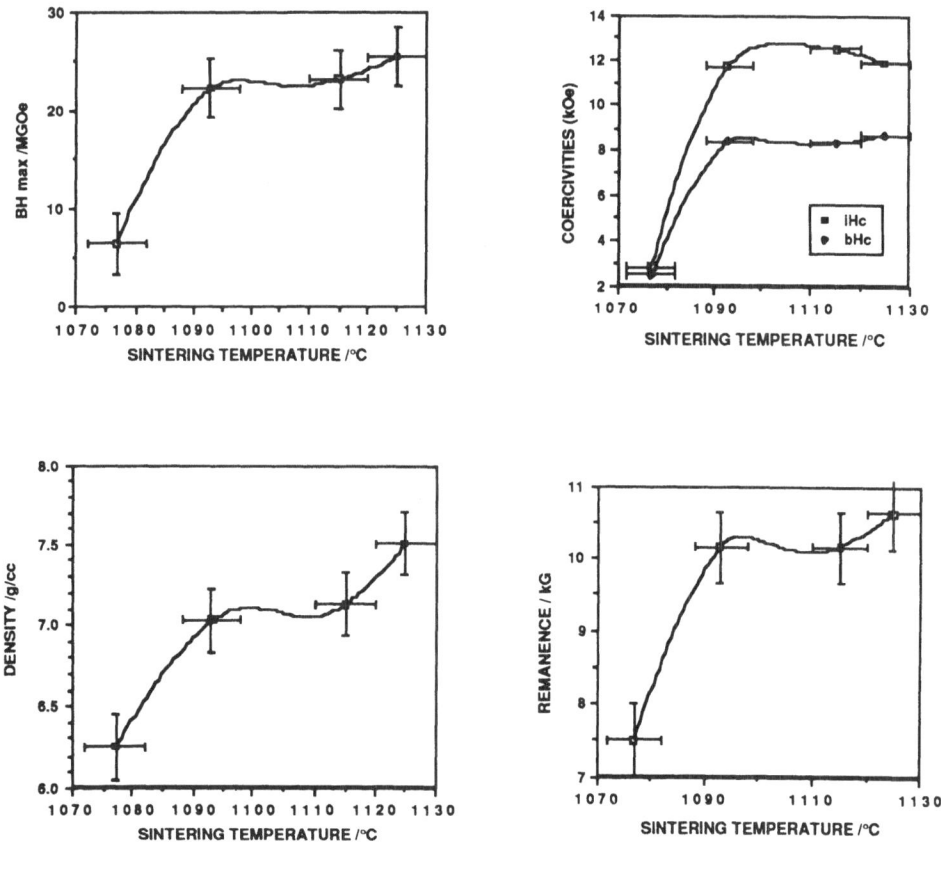

FIGURE 4   VARIATION OF MAGNETIC & PHYSI-CAL PROPERTIES OF Nd$_{15}$Fe$_{77}$B$_5$Dy$_2$Nb

	AREA sq.micron	LENGTH (micron)	BREADTH (micron)
Minimum	0.47	0.97	-
Maximum	2518	66.2	57.9
Mean	385	11.0	8.5

TABLE 5   GRAIN SIZE ANALYSIS FOR THE SAMPLES SHOWN IN FIG. 4

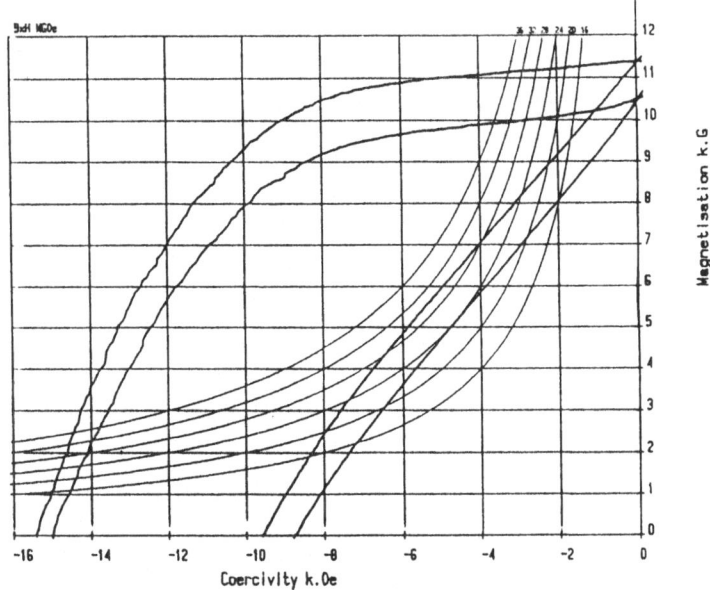

FIGURE 5   EFFECT OF CORROSION (HUMID) ON DEMAGNETISATION CHARACTERISTICS (Sintered Hitachi HX94EA-cross sectional area assumed constant-lower curve after corrosion)

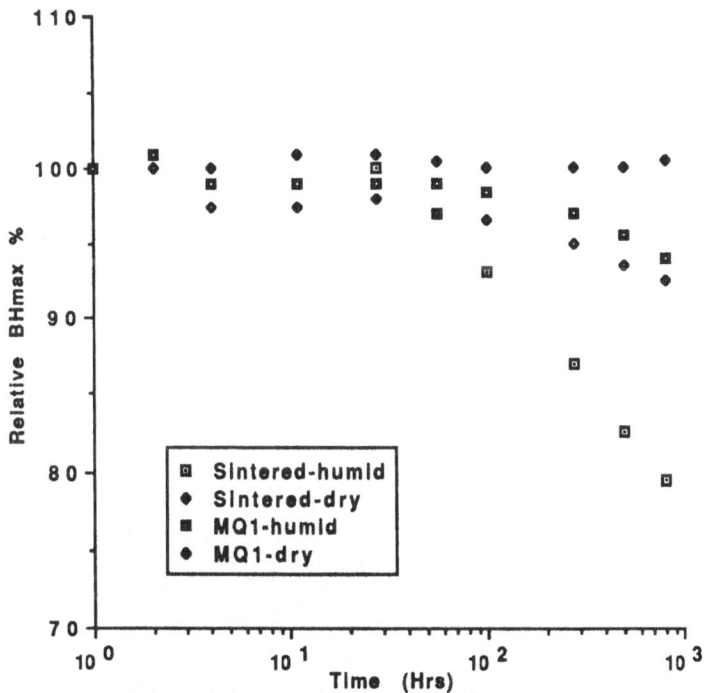

FIGURE 6 Comparison of humid & dry corrosion for sintered & bonded magnets

MATERIAL	ALKALINE pH=11.5		NEUTRAL pH=5.7		ACID pH=2.5	
	Ecorr V	Corrosion rate (milli-inches/year)	Ecorr V	Corrosion rate (milli-inches/year)	Ecorr V	Corrosion rate (milli-inches/year)
MQ1 Coated	-0.04	0.03	-0.027	0.03	-0.06	0.02
Copper	-0.093	0.9	-0.038	0.27	-0.061	0
Titanium	-0.367	0.8	-0.316	0.2	-0.232	0.7
Hitachi HX94EA	-0.402	3	-0.586	10.8	-0.708	63
Iron	-0.43	2.5	-0.37	1.2	-0.602	41.5
SmCo5	-0.49	2.7	-0.435	1.8	-0.584	36
MQ1	-0.5	17	-0.55	14.1	-0.582	22
MQ11	-0.51	4.5	-0.555	7.1	-0.76	87
2/14/1 Homogenised	-0.518	0.3	-0.53	2.7	-0.618	4.6
Neomax 30	-0.52	4.8	-0.552	10	-0.755	35.4
2/14/1 As cast	-0.575	5	-0.656	8	-0.823	31
Zinc	-0.812	7.3	-0.811	10.7	-0.979	14.2

TABLE 7   CORROSION RESULTS

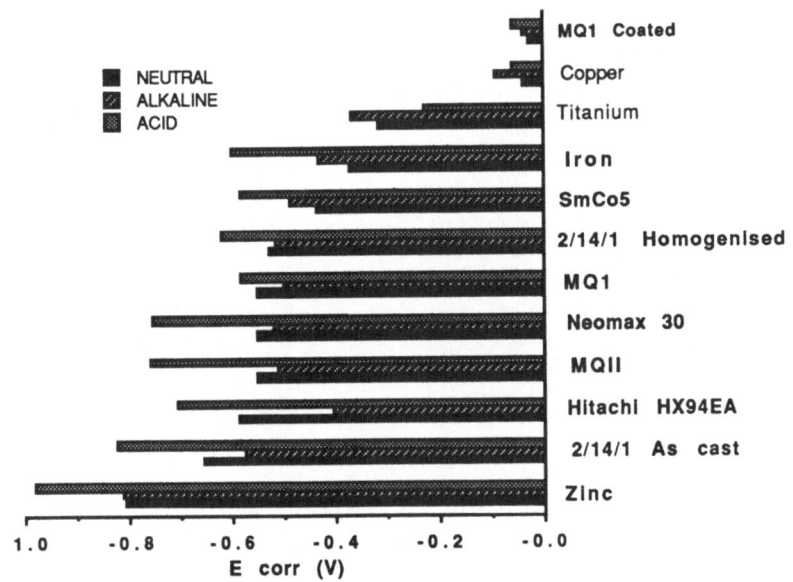

Fig. 7(a) Equilibrium corrosion potential

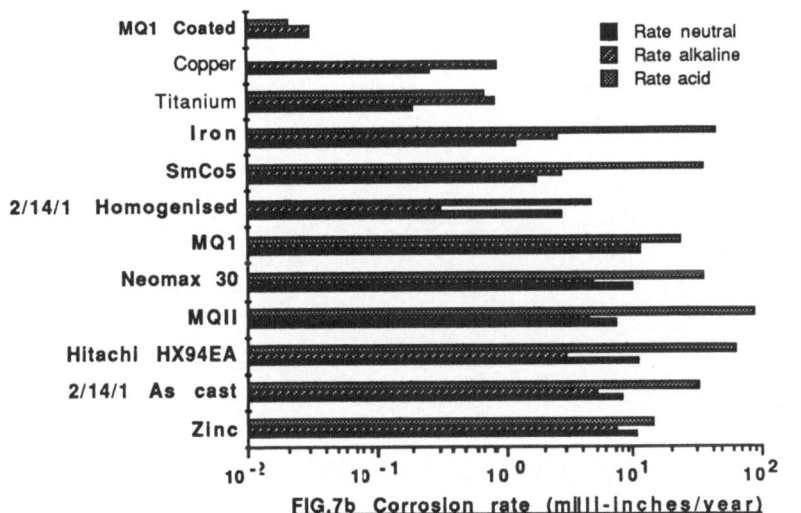

FIG.7b Corrosion rate (milli-inches/year)

FIGURE 7 CORROSION RESULTS DERIVED FROM TAFEL PLOTS

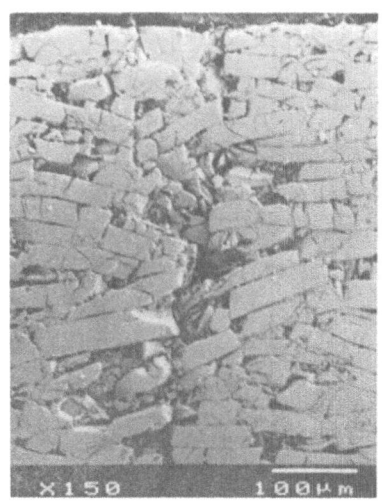

FIGURE 8(a) MQ1 showing corrosion pitting normal to the ribbon plane

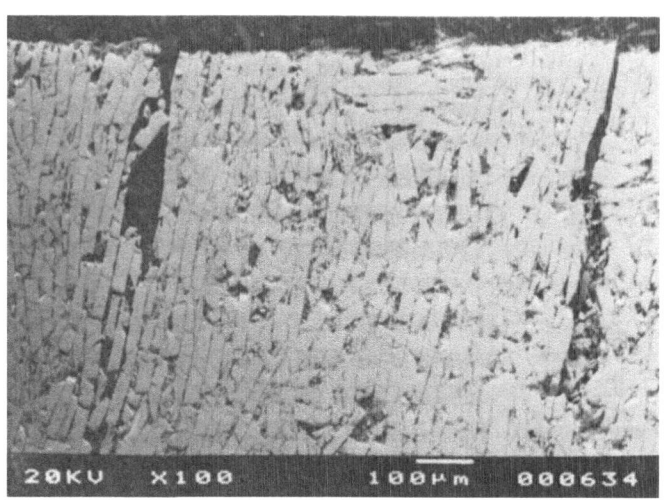

FIGURE 8(b) MQ1 showing corrosion pitting parallel to the ribbon plane

FIGURE 8 MICROSTRUCTURE OF CORRODED MAGNETS-SECTIONS NORMAL TO SAMPLE SURFACE (100% humidity)

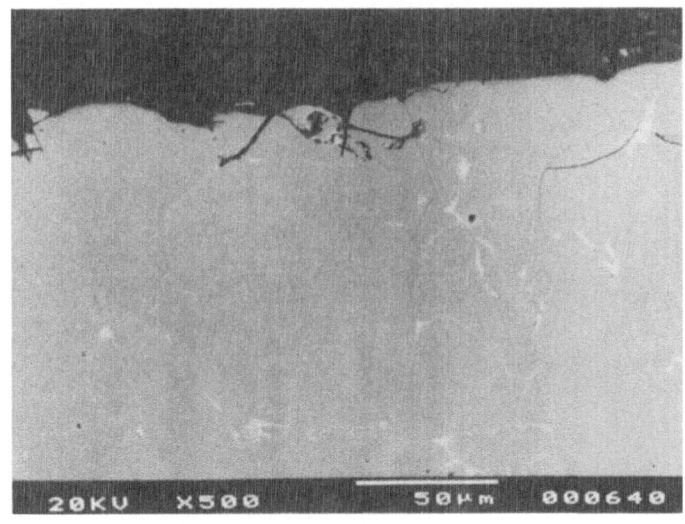

FIGURE 8(c) MQ11

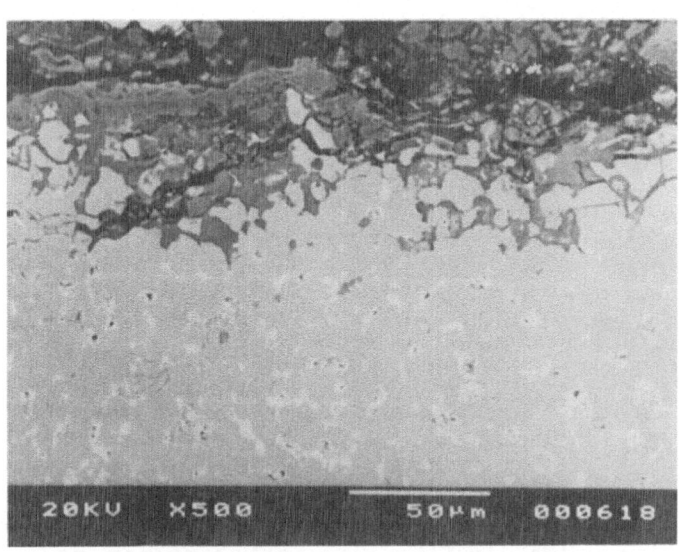

FIGURE 8(d) NeIGT 27

FIGURE 8 cont'd Microstructure of corroded magnets

9(a) MQ1 x400

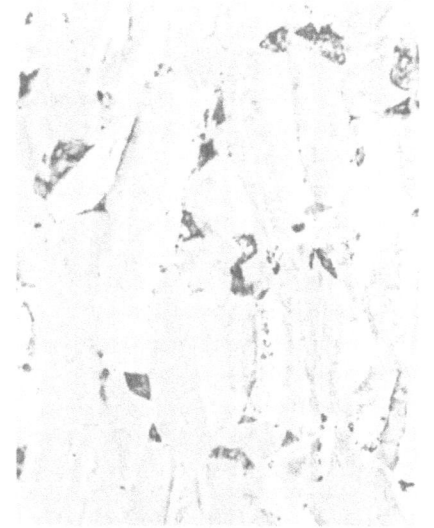

9(b) MQ11 ETCHED
ACID FeCl x400

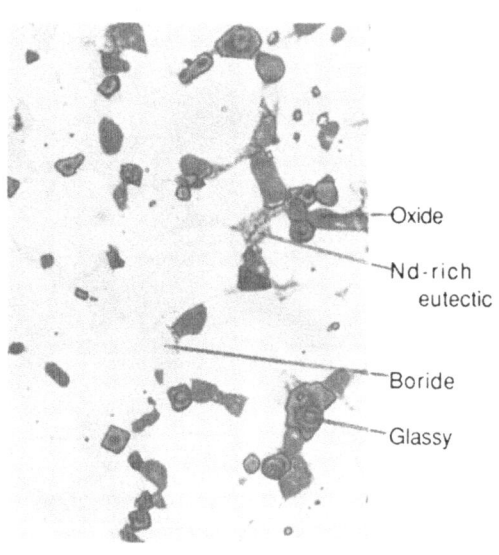

—Oxide

—Nd-rich
  eutectic

—Boride

—Glassy

9(c) NeIGT 27 UNETCHED
SHOWING VARIOUS PHASES
x1000

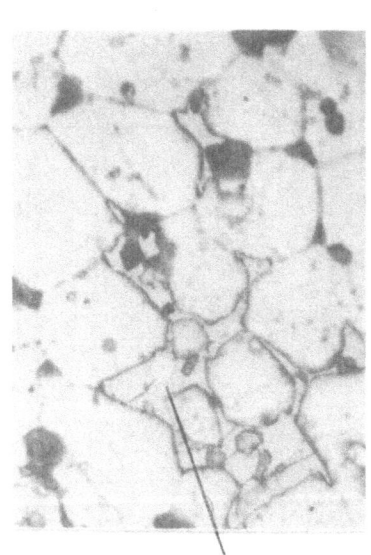

Nb rich

9(d) NEOMAX 30 etched
Nital showing Nb rich
phase x1000
phase x1000

FIGURE 9 TYPICAL MICROSTRUCTURES OF
COMMERCIAL MAGNETS

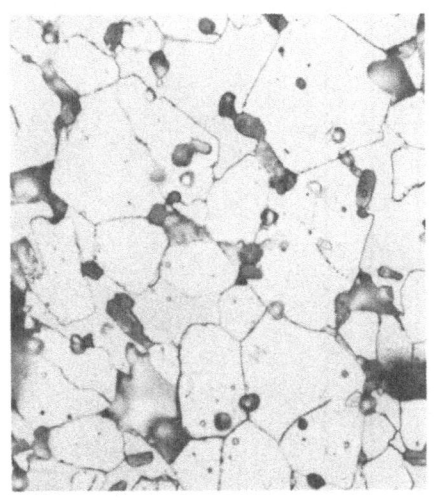 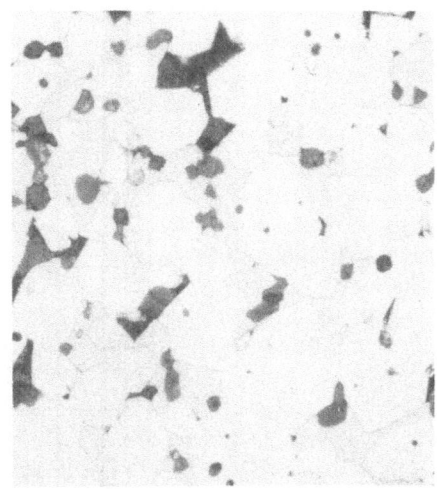

Etched   acid   **Etched   Nital**
ferric   chloride

FIGURE 10   EFFECT OF ETCHANT ON MICROSTRUCTURE
(sample= NeIGT 27   x1000

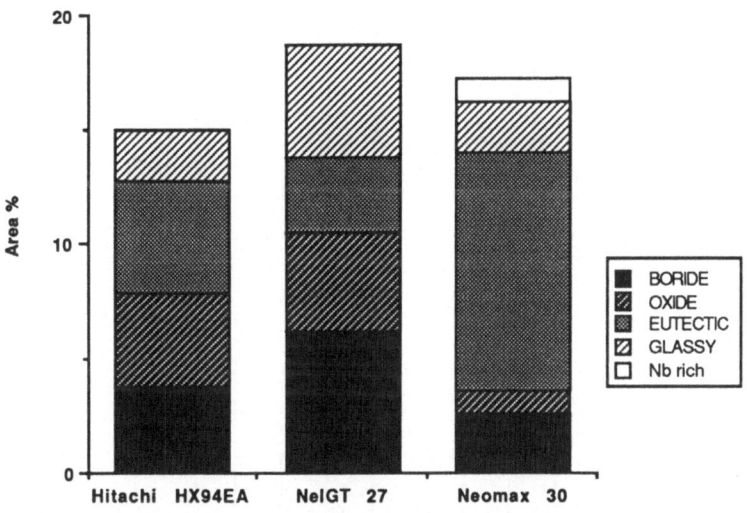

FIGURE 11   PHASE PROPORTIONS IN COMMERCIAL MAGNETS
(Manual point count)

MATERIAL	GRAIN SIZE DATA-μm					Nd:Fe	XRF ANALYSIS-approx wgt %		
	Mean GL	Mean GB	(GL+GB)/2	Mean GA	Std dev GA		Dy	Co	Si
NelGT 27	9.75	7.28	8.51	50.8	30.1	0.42	0.3	0.4	-
Hitachi HX94EA	5.67	4.4	5.03	17.18	8.37	0.37	2	0.5	-
Vacodym 370	11.8	8.92	10.4	74.9	44.3	0.38	1.8	0.3	0.4
Mullard RES270	9.3	7.3	8.3	49	28.5	0.47	-	0.3	-
Crumax 261	4.77	3.76	4.26	12.4	6.8	0.39	2	0.3	-
Crumax 355	5.3	4.21	4.75	16.1	10.6	0.45	0.4	0.2	-

TABLE 9 GRAIN SIZE DATA & XRF RESULTS

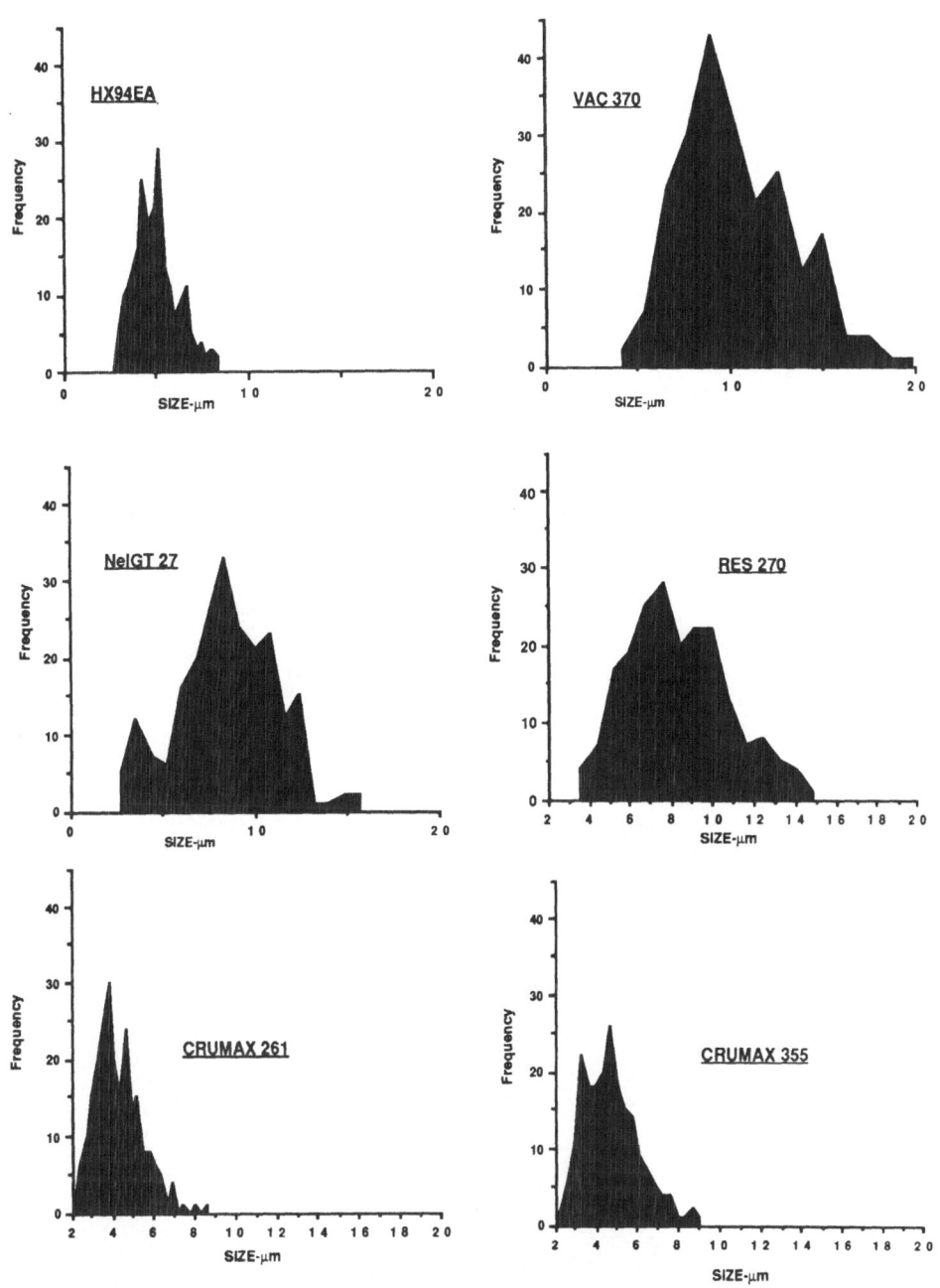

FIGURE 12 GRAIN SIZE DISTRIBUTION OF COMMERCIAL MAGNETS

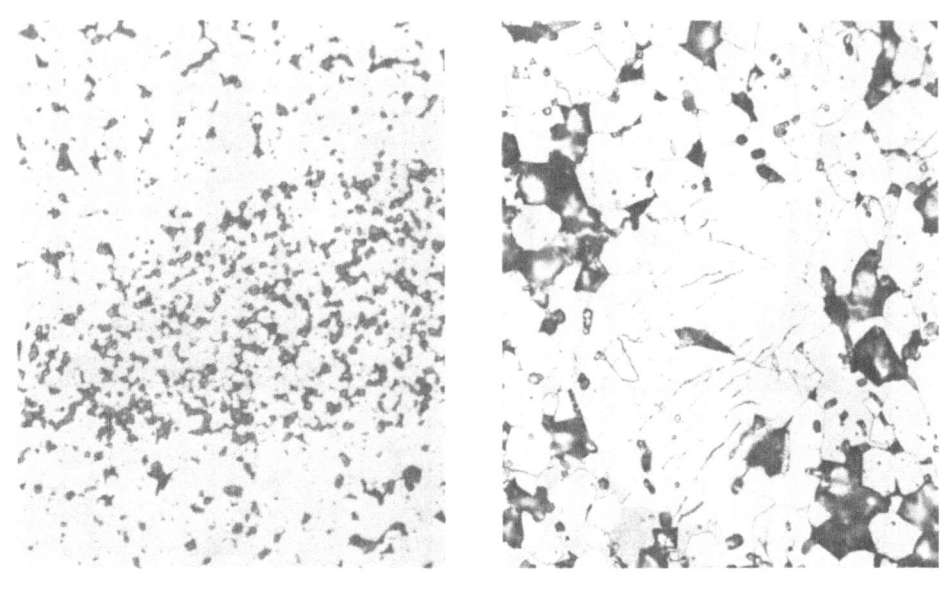

Hitachi 94EA x500   NeIGT 27   x500

FIGURE 13 SEGREGATION FEATURES IN COMMERCIAL MAGNETS

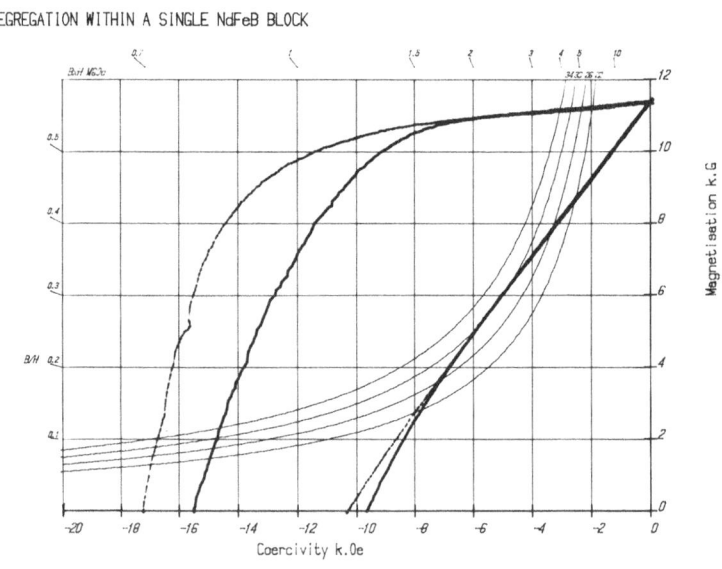

FIGURE14   EFFECT OF SEGREGATION ON MAGNETIC PROPERTIES

MATERIAL	L.R.C. MEASUREMENTS					MANUFACTURERS DATA						DENSITY	TRANSVERSE RUPTUR STRENGTH M.Pa.		
	B_r kG	$_B H_c$ kOe	$_i H_t$ kOe	BHmax MGOe	μ_{rec}	B_r kG	$_B H_c$ kOe	$_i H_c$ kOe	BHmax MGOe	μ_{rec}	kg/m³	MEAN	MIN	MAX
MQ 1	6.2	5.3	16	8.3	1.16	6.1	5.3	15	8	1.15	-	-	-	-
MQ 11	8	7	15.7	14	1.15	7.9	6.5	16	13	1.15	-	-	-	-
NelGT 27H	10.4	9.6	18	25	1.08	10.2	9.6	>17	27	1.1	7.28	218	184	245
NEOMAX 30	11.7	10.5	13.3	31.8	1.08	11.8	10.7	>12.6	33	1.05	-	-	-	-
Hitachi HX94EB	11.4	10.4	17.2	30.3	1.08	11.5	10.6	>15	31	1.05	-	-	-	-
Hitachi HX94EA	10.6	10	18.5	26.5	1.06	10.8	9.95	>15	28	1.05	7.43	332	269	369
YAO LUNG	12.2	8.6	8.9	35	1.08	-	-	-	-	-	-	-	-	-
VACODYM 370	11.1	10.3	21	29	1.07	11.75	11.3	17	31.5	-	7.15	265	255	283
VACODYM 335	12.7	9.5	10.5	35.4	1.14	12.25	8.8	10.8	33	-	7.52	-	-	-
Mullard RES270	11.1	9.8	10.7	28	1.09	11	9.4	10.5	27	1.05	7.48	395	368	415
CRUMAX 261	11.3	10.7	21	30	1.05	10.4	10	20	26	1.09	7.60	423	374	491
CRUMAX 355	12.05	11.35	13.58	34.2	1.06	12.3	11.3	14	35	1.09	7.51	431	401	454
THOMAS & SKINNER NECOOH	11.6	11.0	15.5	32	-	11.0	10.7	>20	30	-	7.48	410	399	427
THYSSEN	11.9	10.9	11.5	33.4	-	12.0	10.4	11.0	33.3	-	-	-	-	-

TABLE 10 MAGNETIC AND MECHANICAL PROPERTY DATA FOR COMMERCIAL MATERIALS

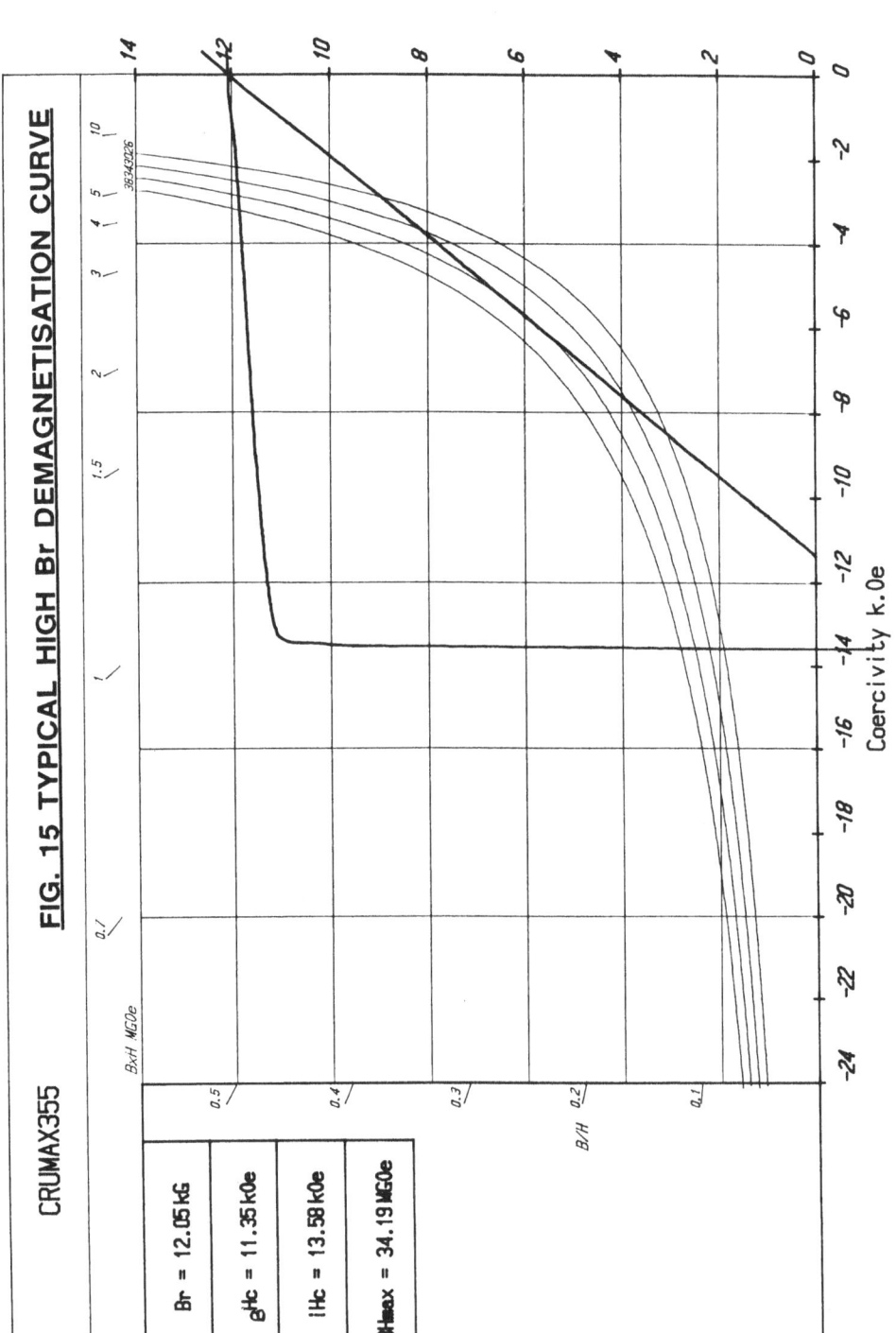

FIG. 15 TYPICAL HIGH Br DEMAGNETISATION CURVE

CRUMAX355

Br = 12.05 kG

ᵦHc = 11.35 kOe

iHc = 13.58 kOe

BHmax = 34.19 MGOe

625

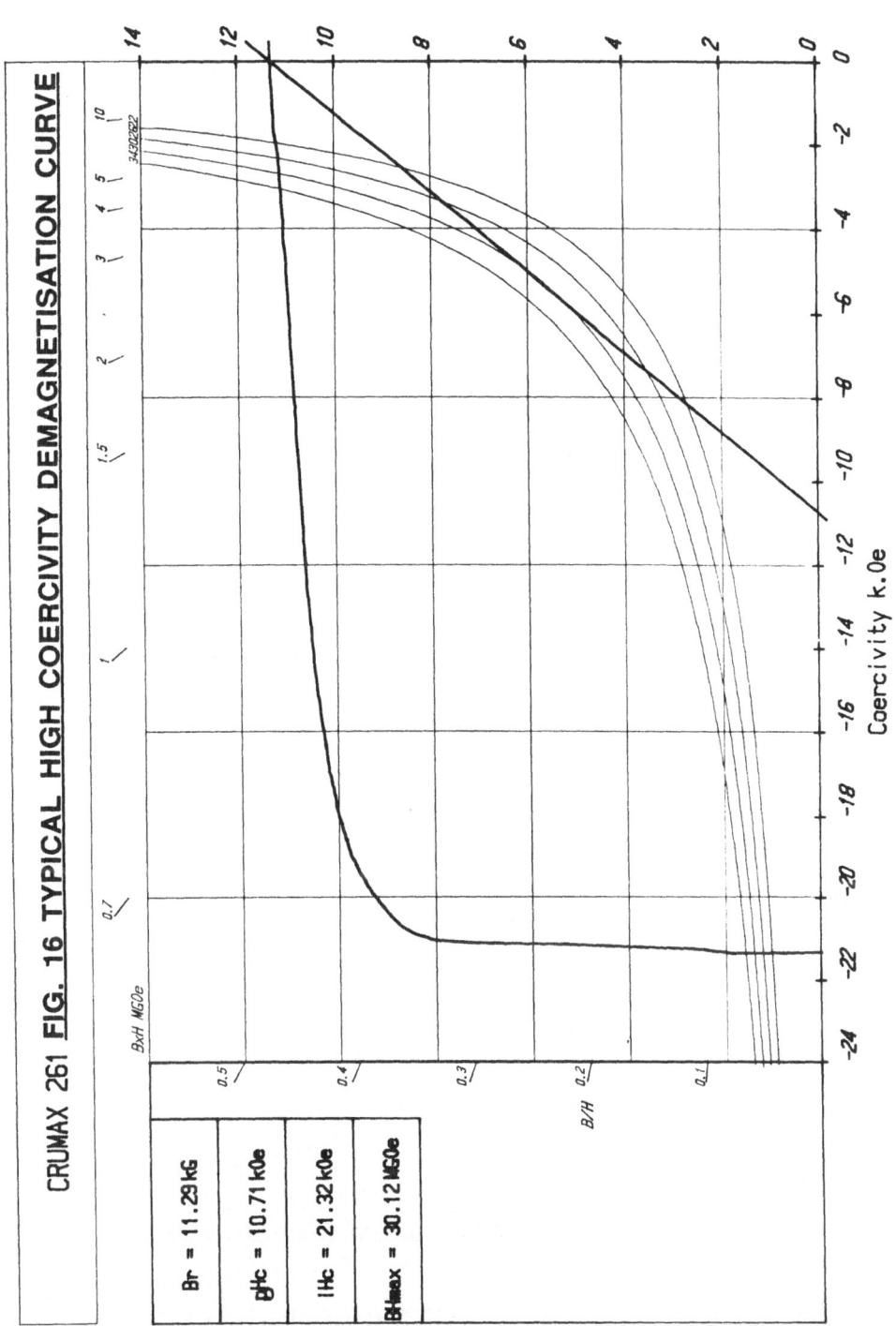

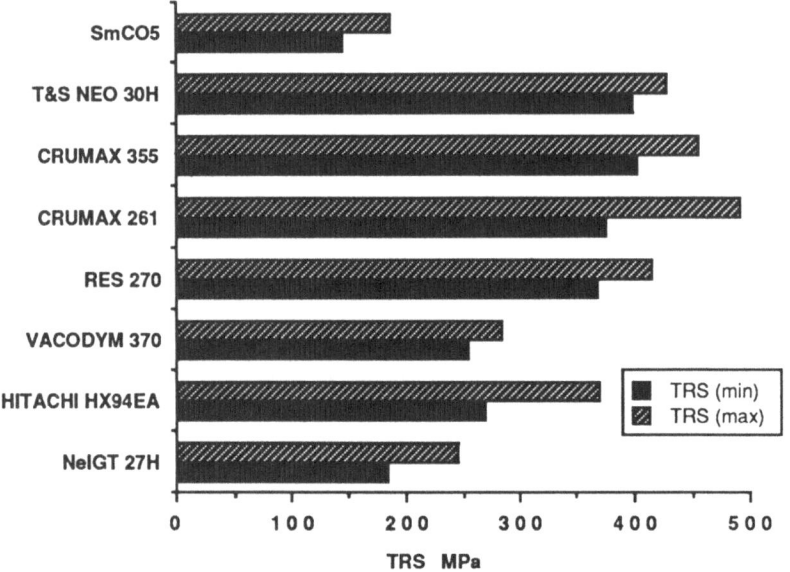

FIGURE 17(a) TRANSVERSE RUPTURE STRENGTHS

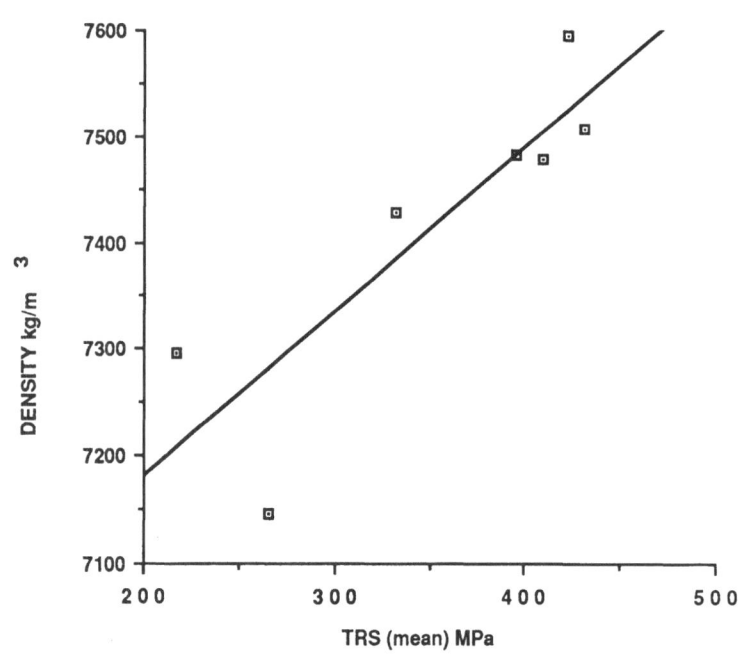

FIGURE 17(b) CORRELATION OF STRENGTH & DENSITY

FIGURE 17   STRENGTH MEASUREMENT RESULTS

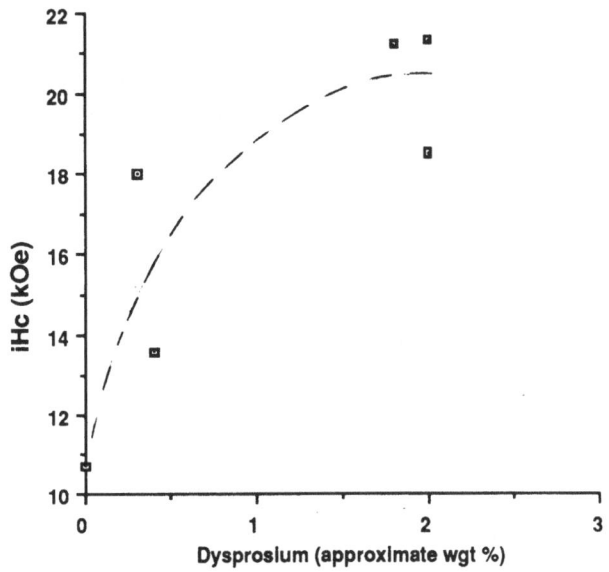

FIG 18 (a) Correlation of Dy content & iHc

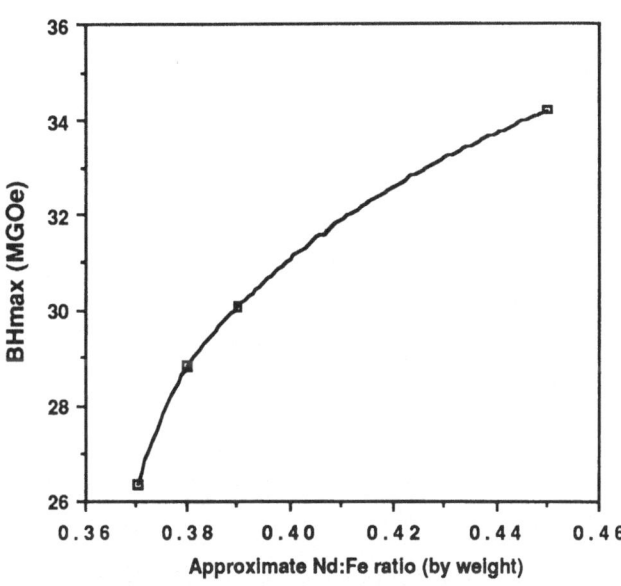

FIG 18(b) Correlation of BHmax & Nd:Fe ratio
for magnets containing no Dysprosium

FIGURE 18 CORRELATION OF MAGNETIC PROPERTIES
& COMPOSITION FOR SOME COMMERCIAL MAGNETS

FEASIBILITY STUDY OF A METHOD TO OBTAIN
CORROSION PROTECTION OF THE
NdFeB-TYPE MAGNET

S. Tori, G. Bava and co-workers

Materials and Processes Laboratory
ing.C.OLIVETTI & C. S.p.A.
10015 IVREA, ITALY

ABSTRACT

The development of a coating method for NdFeB magnets is presented in all its steps. It includes an electrolytic process of zinc deposition, with detailed description of the problem that arose with hydrogen embrittlement, either with or without a conversion coating. For extremely severe working environments, another zinc coating, not electrolytic, is presented: the DACROMET process. Alternative testing procedures have been evaluated, in a neutral atmosphere or in acidic atmospheres of SO_2 or H_2S.

INTRODUCTION

External collaboration

The work presented in this paper has been prepared using mainly magnets obtained by Olivetti via the sinter route and also magnets supplied by external collaborators in the aim to test the validity of the methods proposed on magnets produced via different routes. A list of external collaborators is presented below.

TABLE 1   List of external collaborators

Name	Company/University	Material supplied
M.Bond	S.G.Magnets Lmt.	Isostatic/sintered
H.Davies	Un.of Sheffield	Hot Pressed
J.Vincent/	G.E.C. Research Lmt./	
M.Wyborn	Micanite & Insulators	Hot pressed
Mr.Klaiber	Aimants UGIMAG S.A.	Isostatic/sintered

Requirement

The reasons for which a coating layer is needed over a NdFeB magnet depend on the application and on the working conditions of the magnets itself. Apart from the cases in which the need is obvious, such as magnets working in external environments (i.e. automotive applications) there are a number of other applications where the magnets work in an intrinsic corrosion-protected environment but need equally a coating to prevent powder loss. Examples are hard disk drives or focusing systems for optical disks, as Mullard stated during the CEAM activity. In small magnetic circuits, where miniaturization plays an important role, it is important that the thickness of the protective layer is as low as possible to avoid air gaps between the magnet and the magnetic circuit; this exigence is particularly important in "voice coil motors" for 3 1/4 inches disk drives (requests in this direction have been reported by Thyssen during the CEAM meetings).

Coating selection

Established industrial practice to solve these problems are epoxy coating and non-magnetic nickel. This study tries to follow the alternative route of zinc deposition to obtain magnet protection with a few microns, compared with the relatively thick epoxy coating, using the well established industrial process, of electrolytic zinc deposition. When the surface smoothness is not of extreme importance and working conditions are severe regarding temperature and corrosion, alternative processes based on polymer coating become interesting; in this study a commercially available process has been evaluated. This process, DACROMET^R, is commonly adopted for automotive parts directly exposed to the external world surrounding the car, such as springs, parts of the breaking system.

Test development

Another aim of the study was to develop a simple, rapid test to compare and subsequently to control, the effectiveness of the coating layer. Three substantially different methods were employed: neutral salt fog, industrial atmosphere (SO_2, H_2S) and high pressure water vapour.

ELECTROLYTIC ZINC
Standard electrodeposition of zinc

The first attempt to obtain a zinc coating was done on a standard ternary alloy, sintered at 7.4 g/cm³ density in Olivetti. Samples were cylindrical, approximately 20 mm diameter and 2 mm thickness. The total surface of every sample was ground prior to coating, in order to equalize the coating conditions from point to point. A chemical treatment to clean and "activate" the surface was done, consisting of:

- Vapour degreasing with chlorothene^R (1,1,1,trichloroethane)
- Acid pickling in HCl 25% vol., room temperature., 5 sec
- Cold rinsing in tap water, twice

The electrodeposition was carried out using the standard procedure employed in common steel parts, following the cycle:

- Zinc electoplating (acid bath pH 5.1, current density 2 A/dm², time 15 min to obtain ≈6 microns, 23 min to obtain ≈10 microns. Bath composition: Zn 30 g/l, Cl 150 g/l.
- Cold rinsing in tap water, twice.

On the zinc layer was applied the conversion coating, using the following conditions:

Yellow iridescent chromate coating, which consists of:

- HNO₃ 2% vol, 4-5 seconds
- Metapass gelb^R 25 g/l+4 cc/l HNO₃ at RT, 15-20 seconds.
- Cold rinsing in tap water, twice.
- Drying 60 ^C max temperature, hot air.

The thickness and the adherence of the layer is visible in fig.1 and fig.2, showing a coating of 6 microns and 10 microns respectively.

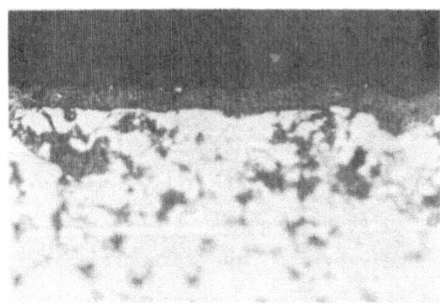 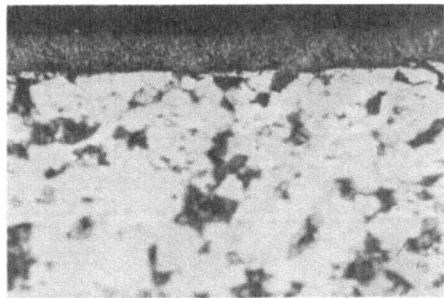

Fig.1 & 2   Micrographs of a section of a zinc layer of 6 and 10 microns, respectively.

Test method

For the evaluation of corrosion resistance the standard test used for common steel part was adopted, the neutral salt spray test, according to ISO 3768 (NaCl 50 g/l, 35 °C). This assumes a pass if no white corrosion product is visible after 72 hours of salt spray (ISO 4520, paragraph 5.3).

Improved electrolytic deposition

While the majority of the magnets treated as described above passed the test; there was spontaneous cracking after 59 hours of salt spray on a 6 microns coated sample. This was attributed to hydrogen and a new process was developed to try to completely extract hydrogen from the magnet immediately after zinc deposition. The hydrogen desorption step consisted in baking the samples in air at 200 °C for 2 hours, before the conversion coating.

Another minor change was the weakening of the pickling step, passing from a concentration of 25% HCl for 5 seconds, to 5% HCl for 25 seconds. This was done to assure a better repeatability of the process.

The corrosion resistance was again tested using salt spray and all the magnets passed the test. The first evidence of white corrosion products only appeared after 96 hours.

In conclusion, the behaviour of the coated magnets in this kind of test is completely equal to that of iron or steel, coated with a similar thickness of zinc.

Hydrogen embrittlement

Even after the hydrogen desorption step, we noticed some cases of sample cracking. A deeper examination of the problem permitted correlation of this phenomenon with the presence of macroscopic irregularities in density or, in general, with cavities in the body of the magnet. This was true for both processing routes investigated: sintered magnets, as documented in fig.3 and hot pressed magnets (processed and tested later in this study, supplied by GEC - Micanite & Insulators) as visible in fig.4.

Testing of electrolytic zinc in industrial atmosphere

The subsequent step was to evaluate if the zinc coating is effective also in an industrial environment, where other chemical reactions could increase the rate of corrosion compared with the neutral salt spray test. The environment was simulated in a chamber where it is possible to control the temperature and the relative humidity of air, as in a normal climatic

632

room, and in addition, it is possible to introduce a certain quantity of a corrosive gas and to maintain the needed concentration

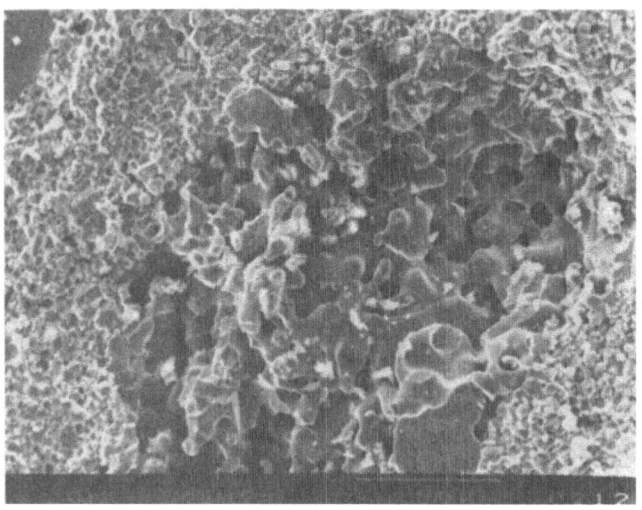

Fig.3   Cavity in fracture surface caused by hydrogen in a sintered sample (Olivetti)

Fig.4   Cavities in fracture surface caused by hydrogen in an hot pressed magnet
(G.E.C. Micanite & Insulators)   X35, ------------ 500 μm

dynamically by continuously flowing a controlled quantity of gas through the test chamber. The gases employed were SO_2 and H_2S, two commonly corrosive agents used in Olivetti to test components, parts or the whole machine, for applications that have to work in hostile environments (for example, numerical control drivers for machine tools, "post of sales", etc..). Materials evaluated were Olivetti laboratory sintered samples and certain samples supplied by S.G.Magnets Ltd. in the form of ground parallelepipeds. The zinc coating was carried out for both type of magnets, according to the improved cycle described above. The variability of the samples, derived from the utilization of magnets sintered in very different conditions, mixed with samples supplied by others, suggested a statistic method should be used to evaluate the inevitably scattered experimental results. The "factorial experiment design" was chosen and, since a method supplying a quantifiable measure of corrosion was necessary, it was decided to use the open circuit flux measure before and after the test to evaluate the influence of the corrosion on the magnetic properties of the magnets. The "factors" considered were:

1) Zn-plated vs. uncoated
2) Aged thermally vs. unaged
3) Corrosive medium vs. non corrosive atmosphere
   (SO_2 or H_2S)   (MIL 507)

The subdivision of the treatments was planned according the following standard procedure:

		SO_2						**H_2S**		
		A						**A**		
		0		1				0		1
		B		**B**				**B**		**B**
		0	1	0	1			0	1	0	1
C	0	(1)	b	a	ab	D	0	(1)	b	a	ab
	1	c	bc	ac	abc		1	d	bd	ad	abd

where: upper case letters stand for a "factor", while the numbers 0 and 1 represent the "level" of the corresponding factor. In this specific case:

A = Zn-plating, 0=without, 1=with
B = thermal aging, 0=no, 1= yes (100 °C x 24 hours)

C,D = corrosive agent, 0=none, 1=SO$_2$ in case of C, 1=H$_2$S in case of D

The identification of a "treatment", (i.e. one of all possible combinations of the 3 factors either at level 0 or 1) is made by lower case letters, with the convention that the presence of a letter means "factor at level 1" while the absence of a letter means "factor at level 0". For example: **ab** means Zn-plated, thermal aged, no corrosive agent (MIL 507). The symbol (1) has the special meaning of "all factors at level 0". The thermal aging was considered as a factor since the corrosive environments have themselves an aging effect, being conducted above room temperature, so it is necessary to separate the influence of the thermal aging alone from that of thermal stress associated with the corrosion test. The specifications for environmental conditions employed are listed below:

- REFERENCE ENVIRONMENT: A humidity test was chosen, derived from the **MIL 507** standard "Aggravated temperature-humidity cycles": Temperature: 60 °C, Relative Humidity: 95%, Time: 12 hours

Temperature: 30 °C, Relative Humidity: 85%, Time: 12 hours

Number of cycle repetitions: 7.

- SULPHUR DIOXIDE (SO$_2$): Temperature: 40 °C, Relative Humidity: 80÷90%, Time: 160^h, Concentration: 10÷20 ppm

- HYDROGEN SULPHIDE (H$_2$S): Temperature: 40 °C, Relative Humidity: 75÷80%, Time: 160^h, Concentration: 3÷5 ppm

The experimental procedure adopted was:

- The magnets under test were assigned to the treatments (3 each treatment) in a completely random sequence.

- The magnets assigned to treatment with factor A at level 1 ("plated") were Zn-plated.

- Thermal demagnetization, 320 °C in air.

- Weighing, to obtain the "init. weight".

- Magnetization in a pulse field > 45 kOe.

- Open circuit flux measure of all magnets assigned to treatment with factor B at level 1 ("aged"), in order to have the so-called "B initial", used to estimate the flux loss due to aging.

- Thermal aging of the same magnets of the previous step, in air at 100 °C x 24 hours.

- Open circuit flux measure of all magnets, to obtain the so-called "B pre", i.e. the value of B before the magnets will be subjected to the treatments.

- Execution of treatments, as specified in the experimental design.
- Open circuit flux measure of all magnets, to obtain the "B post" value.
- Thermal demagnetization in air at 320 °C.
- Weighing to obtain the "fin.weight".

NOTE: It is important to notice that two important thermal over-stresses were applied to magnets, one during thermal demagnetization and the other by the thermal aging. Those stresses were well above the thermal resistance of the conversion coating that has a maximum working temperature of about 80 °C. The mentioned stresses caused a reticulation of the surface protective layer that, in the long term, months after the completion of the test, caused a detaching of the coating from the surface of the magnets.

The results of this test could be summarized as follows:

-There is a high statistical evidence of the difference in flux loss caused by the thermal aging vs. the MIL507 test, the former being more severe. Thus in evaluating flux losses caused by the corrosion tests it is necessary to take account of the thermal history of the sample.

-Either the SO_2 and H_2S treatments caused statistically significant losses compared with the reference test.

-A different behaviour of the zinc-coating has been observed in the two corrosive mediums examined; i.e. there was statistical significance of the effectiveness of Zn-plating against corrosion promoted by an H_2S environment, while it was not possible to separate in a statistical significant way the losses caused by the SO_2 treatment from that of the reference test. The non effectiveness of Zn coating in SO_2 humid climate seems to be confirmed by the appearance of white corrosion products on the surface of the test samples.

Table 2 shows the results of the factorial experiment. A treatment causes significant differences from the reference environment if the "F ratio" value is greater than the "F limit" value. Two "F limit" values are reported: "F .05" related to 5% probability to be wrong in assuming that a certain treatment causes differences compared to the reference environment, and "F .01" related to 1% probability of error. The "treatment total" column contains the experimental results from which is possible to calculate the F ratios, following the Yate's method.

TABLE 2   Results of the factorial experiment

	SO$_2$ TREATMENT		F limit
treatment	treatment		F .05=4.7
code	total	F ratio	F .01=9.1
(1)	2692	-	
a	844	0.3	<--- Zn-plating
b	165	55.7	<--- Thermal aging
ab	1.5	0.0	
c	4151	62.4	<--- SO$_2$
ac	6519	12.3	
bc	1712	6.4	
abc	2125	7.1	

	H$_2$S TREATMENT		
treatment	treatment		F .05=4.5
code	total	F ratio	F .01=8.7
(1)	2692	-	
a	903	12.6	<--- Zn-plating
b	165	56.2	<--- Thermal aging
ab	1.5	6.5	
c	217	33.9	<--- H$_2$S
ac	392	14.6	
bc	85	29.7	
abc	-15.5	12.9	

Neutral versus acidic working environments

The above tests suggested that deeper investigations are needed to ascertain if zinc coating is applicable in cases where the magnet could work in an acidic environment without an unacceptable degradation of the corrosion resistance.

ELECTROLYTIC ZINC COATING FOR WORKING TEMPERATURE UP TO 130 °C

As mentioned before, the standard electrodeposition of zinc, with the conversion coating, is only applicable up to 80 °C. This is not a limitation in devices such as voice coil motors, where the working temperature is not far from RT, but the same is not true for the motor applications, where the working temperature are always higher. To adapt electrolytic zinc to this specifications, the conversion coating has been substituted with a thin layer of water based varnish. The materials employed in this test were supplied, other than by ourselves, also by UGIMAG (isostatic) and GEC-Micanite & Insulators (hot pressed, MQII type). In order to eliminate the surface irregularities left after grinding, some samples were impregnated using a commercially available product: LOCTITE® PMS 10E. All samples were then Zn-coated in our laboratory using the standard procedure, but without the conversion step; with the exception that in impregnated samples hydrogen was not extracted because the impregnation has a maximum working temperature of 150 °C. These samples were wet sand blasted to remove the excess product from their surfaces. Over a number of samples the protective varnish UDYLITE® AQUARES 3 was applied according to the following procedure: dilution 30%, pH 7.5, immersion time 60 sec, air drying 3 min, baking 70 °C x 15 min. The corrosion test employed was an accelerated high pressure vapour, designated AUTOCLAVE TEST. In this test the magnets were placed in a hermetically sealed container able to withstand an over pressure of several bar at 100 ÷ 160 °C. The vessel contains a small quantity of water and is heated to the desired temperature, fixed at 130 °C, max working temperature for the varnish. The vapour pressure at this temperature will be 3 bar. This provides an accelerated corrosion test, since during an overnight period (fixed as 16 hours) a corrosion level is achieved high enough to discriminate between the different coatings. After this test a corrosion level was obtained comparable to about 100 hours of neutral salt spray test; all samples presented white corrosion products, but the corrosion was less evident in varnished magnets than in those only Zn-plated. The preliminary impregnation of the samples seemed not to be important for the corrosion rate. Olivetti (sintered) and GEC (hot pressed) behaved in a similar way, while the samples supplied by UGIMAG presented less differences between varnished and only Zn-coated magnets,

638

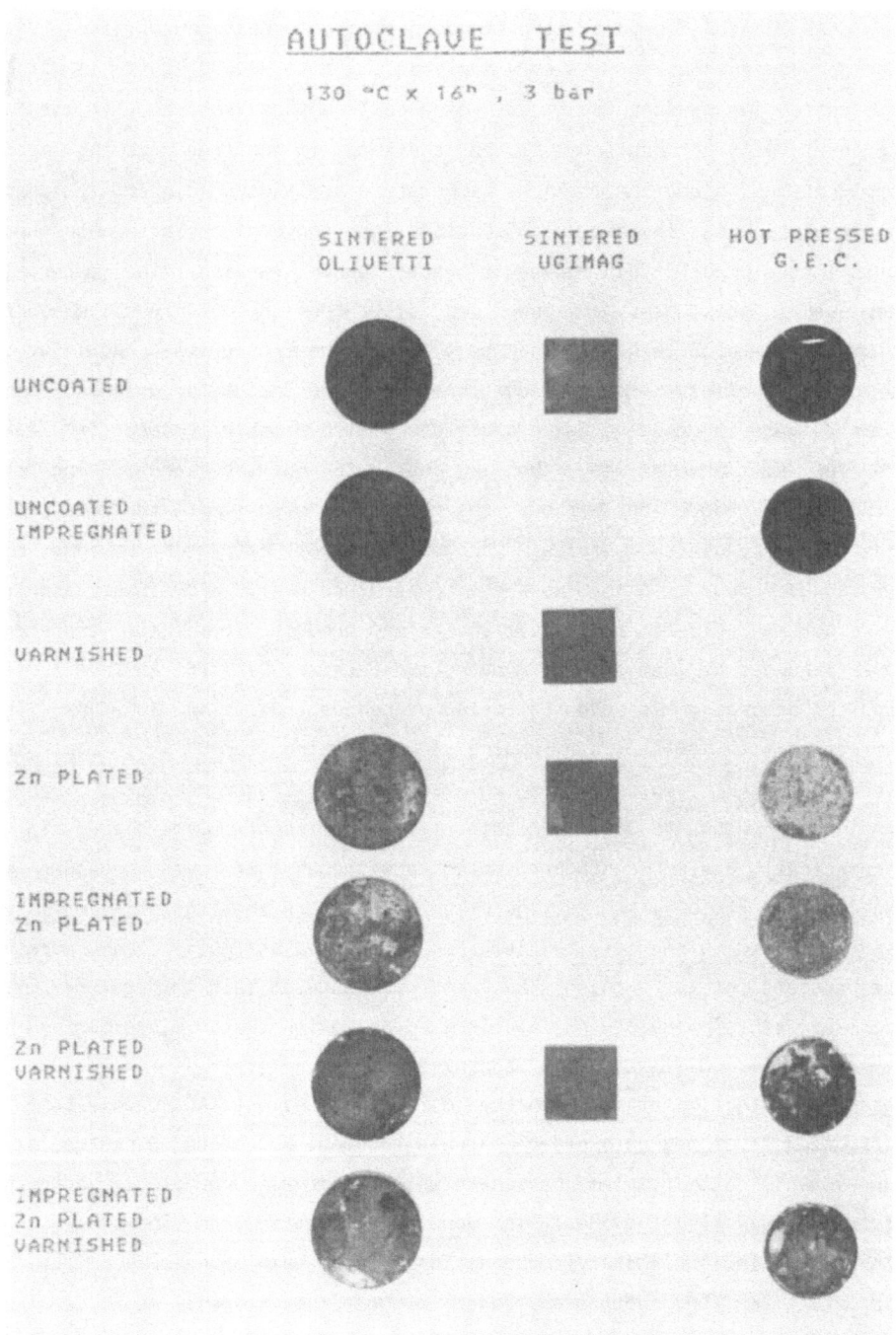

Fig.5   Synoptic table showing all the material passed through the AUTOCLAVE TEST.

even if the corrosion level of varnished was comparable to the other two categories. Fig.5 shows a synoptic table in which it is possible to see the effect of corrosion on magnets with different levels of protection, from uncoated to Zn-coated plus varnish, and to compare the three different kind of magnets examined.

EVALUATION OF THE DACROTIZING PROCESS

The deposition of a zinc layer using a method not involving electrolytic processes has been obtained using a well established industrial process, based on the DACROTIZING SYSTEM, developped by Diamond Shamrock Corp.,Metal Coating Division (USA). The product used for the test was DACROMET 320ᴿ, registered mark by DACRAL SA Europe. The process is essentially a coating in an acqueous dispersion containing zinc flakes, chromium and proprietary organics. This leaves a heat resistant layer (up to 280 °C) on the surface of the magnet. The application method consists mainly of dipping in the DACROMET bath, centrifugation to remove the surplus product, furnace drying to remove completely the water from the solution and a final baking at 300 °C for 20 min to cure the organics. The corrosion evaluation of Dacrotized magnets was carried out using a salt fog spray test. Samples were extracted after 18,24,48 and 96 hours: none of them presented red corrosion products and only little amounts of white corrosion products were evident at the end of the test.

In conclusion the DACROMET process on NdFeB magnets passed the corrosion test and seems a promising alternative in extremely severe working environments and if the roughness of the surface layer doesn't constitute a problem. Fig.6 is a SEM observation of the surface of a dacrotized magnet as it appears after the coating process; the surface presents high irregularities, with an average roughness of 1.2 microns, R_A.

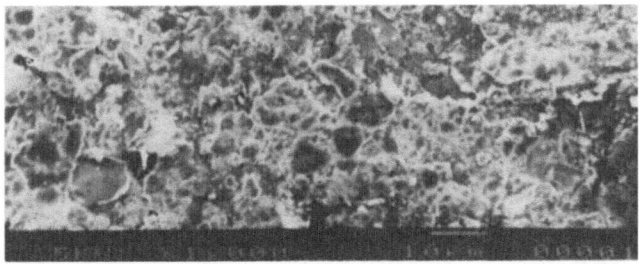

Fig.6   DACROMET 320ᴿ: Surface of the zinc layer as it appears after the curing of the polymer based protective film. SEM micrograph.

THE STABILITY OF Nd Fe B MAGNETS

A. G. Clegg, I. M. Coulson and G. Hilton

Magnet Centre, Physics Division,
Sunderland Polytechnic, Chester Road,
Sunderland, Tyne and Wear, England.

ABSTRACT

A range of materials was investigated including low and high coercivity magnets, bonded magnequench and cobalt containing alloys. The effect of temperature including irreversible and reversible changes of magnetization are discussed. Magnets with two different working points were investigated. Time effects were determined for magnets held at 100°C for up to 200 hours. It was found that the higher the coercivity H_{cM} the better is the temperature stability of the magnet. The highest coercivity materials are produced by using hydrogen decrepitation of the magnet powder.

INTRODUCTION

With permanent magnets we are always concerned to a greater or less degree about the stability of the magnet. For precision applications we need to know the amount of the change of open circuit flux with time and how this change is affected by elevated or sub-zero temperatures.

Three different effects contribute to flux changes. These are:-
1) Constitutional losses, which lead to permanent changes in the demagnetization curve. These losses are non-magnetic in nature. Typical examples are oxidation and micro-structural changes in the material. For Nd Fe B magnets the most likely source of these losses are from corrosion and other reactions with the surrounding environment. Micro-structural changes are unlikely to occur in Nd Fe B magnets because the low Curie temperature means that the magnets will not be used at temperatures high enough to cause these changes. It was thought to be possible that the binder in bonded Nd Fe B magnets would react with the alloy, but this was not found in bonded magnequench samples.
2) Irreversible losses are magnetic in nature and are losses which remain when the magnet is heated and subsequently returned to room temperature. These losses are regained when the magnet is remagnetized.
3) Reversible losses are also magnetic in nature. These are losses which take place due to a change of temperature but are regained after returning to the initial temperature.

Losses due to temperature

Irreversible losses are the greatest problem in magnet systems which are subjected to elevated temperature. The thermal energy provides the activation for domain wall movement or even domain magnetization direction reversal. The return of the magnet to room temperature will not return the domain walls to their original positions, since there is no driving force available as the magnet cools. Thus there is a change in bulk magnetization and this change is irreversible. This has been interpreted as being due to the effect of temperature on the demagnetization curve (Clegg and McCaig, 1958). The major effect of elevated temperature is to decrease the coercivity, such a decrease will then change the shape of the demagnetization curve, see Fig. 1. The modified demagnetization curve will result in a change in the working point of the magnet (McCaig and Clegg, 1987) which will consequently suffer an irreversible loss of flux.

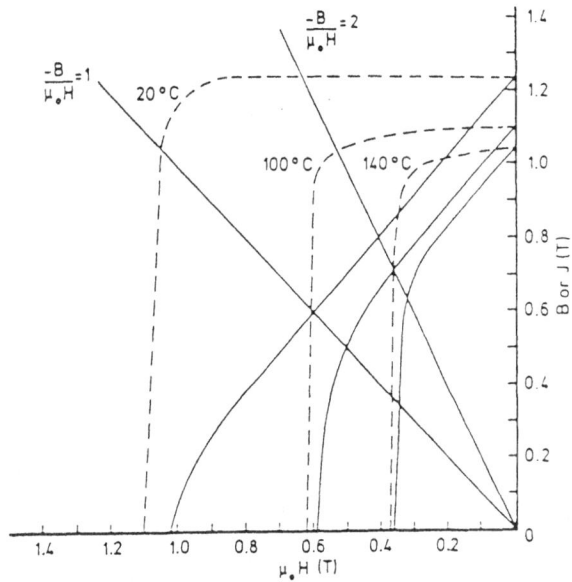

Fig. 1. *Demagnetization B, $\mu_0 H$ and intrinsic J, $\mu_0 H$ curves for NdFeB 35*

Reversible losses are due to the intrinsic magnetization of the individual domains being reduced by thermal energy. This thermal energy causes the magnetic spin moments in individual domains to depart further from the magnetization direction than at room temperature. The spin moments return to their original orientation when the temperature is returned to its original value (usually room temperature), so that if a magnet, fully magnetized at room temperature is heated for a period both irreversible and reversible changes in magnetization will occur. If the magnet is then cooled, the open circuit flux will be less due to the irreversible loss. If this magnet is reheated to the same temperature, the reversible loss will be the same. There will however, be a much lower or even negligible, irreversible loss.

Irreversible losses are temperature dependent and also dependent on the working point of the magnet. If a magnet has a small length to diameter ratio the value of $B/\mu_0 H$, the slope of the unit permeance line, will be small and the line will intersect the demagnetization curve at a point near to the coercivity point. This magnet will be subject to much greater irreversible losses than one with a large value of length to diameter ratio. Magnets can be pre-stabilised by preheating above the working temperature. After this, losses in a working environment will be very small. Reversible losses are a function of temperature and the loss is dependant on the gradient of the saturation magnetization against temperature curve. The gradient of this curve increases as the Curie point is approached and inevitably all the magnetization is lost at this point.

EXPERIMENTAL METHOD

Open circuit flux losses with temperature, both irreversible and reversible, were determined using a search coil and flux integrator. In addition irreversible open circuit flux losses with time at elevated temperatures have been determined. To determine the variation of open circuit flux accurately using a search coil in conjunction with a flux integrator, several criteria have to be met:-

(a) The search coil must always relocate to exactly the same position each time a reading is taken.

(b) The magnet test sample must always be in exactly the same position for all readings.

(c) High stability of sample temperature for the time at temperature tests. This is required because any fluctuations in open circuit flux caused by temperature variations may be much greater than those losses due to time effects.

The apparatus is described by Clegg and McCaig, 1958, but a single coil for measuring the flux changes was used rather than a differential coil system because the flux changes were relatively large.

Flux measurements were taken using an LDJ 702P digital fluxmeter. By suitable selection of the fluxmeter sensitivity flux variations could be measured to an accuracy of 1%. For determination of flux losses with temperature, the following procedure was used.

(a) The magnet was freshly pulse magnetized and located in the sample holder.

(b) A fluxmeter reading was taken at room temperature.

(c) The sample was heated to the required temperature and fluxmeter readings were taken again.

(d) The sample was allowed to cool to room temperature and new fluxmeter readings were taken.

For flux losses with time at temperature, the procedures up to stage c were the same. When the required temperature was achieved however, fluxmeter readings were then taken at suitable intervals over a period of time at this steady temperature.

DETAILS OF SAMPLES INVESTIGATED

A range of samples were tested from various suppliers and with differing magnetic properties. These were as follows:-

1) Polymer bonded magnequench, i.e. MQ1 supplied by General Motors, U.S.A.

2) Low H_{cM} Nd Fe B ⎱ both supplied by I.G. Technologies of the U.S.A.,

3) High H_{cM} Nd Fe B ⎰ and made from jet milled powder.

4) Nd Fe B supplied by S.G. Magnets Ltd., this was produced using the standard Sumitomo sintering method (Sagawa et. al., 1983).

5) Nd Fe B supplied by T.E.W. of West Germany. This was produced by pulsing a measured charge of attritor milled powder in a rubber container and isostatically pressing it. Subsequently, the green compact was sintered.

6) High H_{cM} Nd Fe B supplied by Mullard Ltd. These samples were produced using hydrogen decrepitated powder which was subsequently jet milled, pressed and sintered. This hydrogen decrepitation route (McGuiness et. al., 1986) used by Mullard allows a much shorter milling time to be used as the absorption of hydrogen at room temperature causes the Nd Fe B bulk alloy to disintegrate readily. Since this reduction process is non-mechanical there is no surface damage to the particles. During the initial stages of the sintering process the hydrogen readily desorbs from the compact and does not affect the final properties.

7) $Nd_{15}Fe_{61}B_8$ alloy samples. Here, "as sintered" samples were used as these gave the best magnetic properties.

8) A $Nd_{15}Fe_{50}Co_{27}B_8$ alloy sample. Again "as sintered" samples were used because they gave the best magnetic properties.

The magnetic properties of the above samples are given in Table 1.

TABLE 1   Room temperature magnetic properties of the test samples

Sample	B_r (T)	H_{cB} (kA/m)	H_{cM} (kA/m)	BH_{max} (kJ/m^{-3})	
1	0·575	383	1107	55	(6·9)
2	1·150	569	617	201	(25·2)
3	1·050	764	1193	215	(27·0)
4	1·150	807	923	240	(30·1)
5	1·140	875	895	250	(31·4)
6	1·120	835	1674	234	(29·4)
7	1·150	338	341	195	(24·5)
8	1·060	267	270	118	(14·8)

Note:- the C.G.S. equivalent for BH_{max} in MGOe is shown in brackets alongside the SI figure.

RESULTS OF STABILITY MEASUREMENTS

The eight samples were investigated for stability. Two different values of L/D for each sample were used, these were:

i)   L/D = 1, corresponding to a $-B/\mu_o H$ value of 2·65

ii) L/D = 0·4 corresponding to a $-B/\mu_o H$ value of 1, which is close to the BH_{max} working point on the curve.

The irreversible and reversible flux losses with temperature were measured in 25°C steps from 50°C to 175°C. The reversible losses and irreversible + reversible losses at temperatures of 60, 100 and 150°C are given in Table 2. There is a wide range of both irreversible and reversible losses with temperature over the samples which were tested.

The most significant points are:

i)   The total loss decreases as the value of the coercivity H_{cM} increases, and

ii) For the bonded magnequench (which is a different basic alloy) the irreversible loss is small.

Typical irreversible and reversible curves for sample 4 are shown in Figs. 2 and 3. This sample was in the middle of the range for losses.

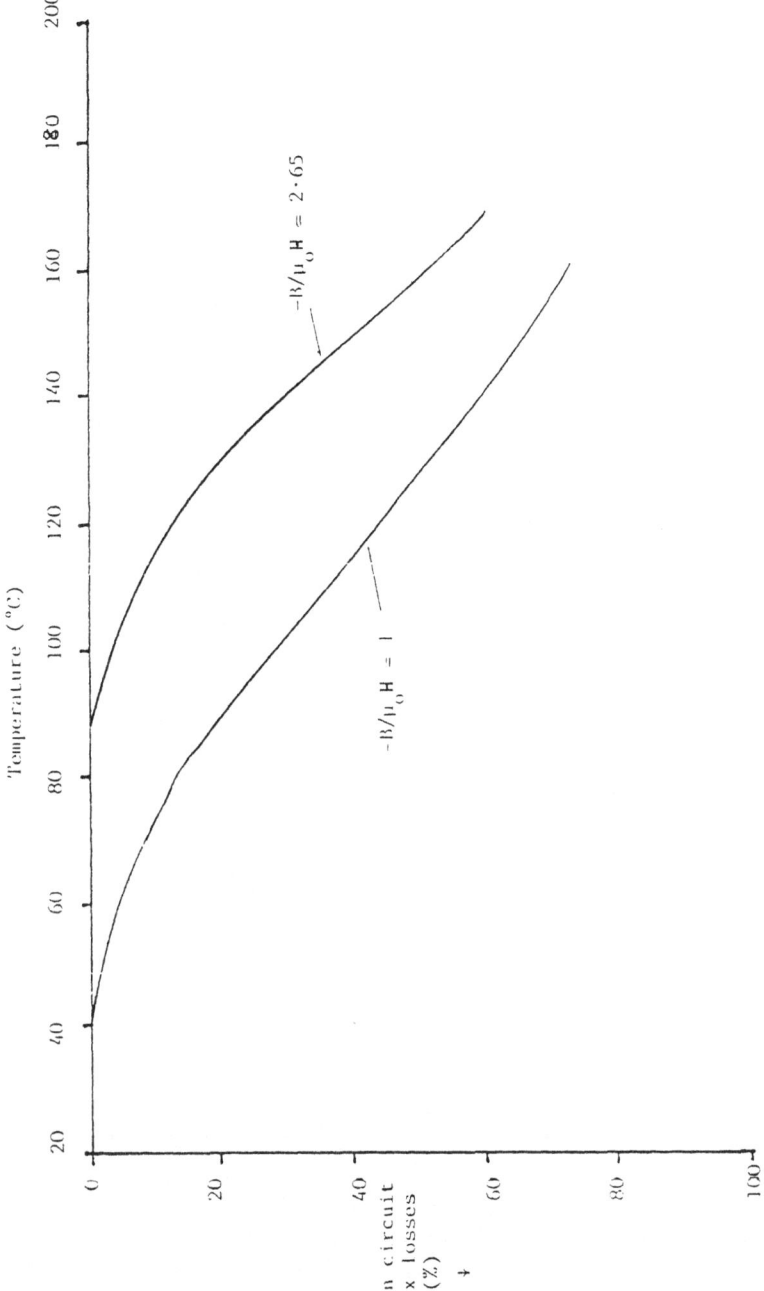

Fig. 2. Irreversible open circuit flux losses with temperature for magnet No. 4

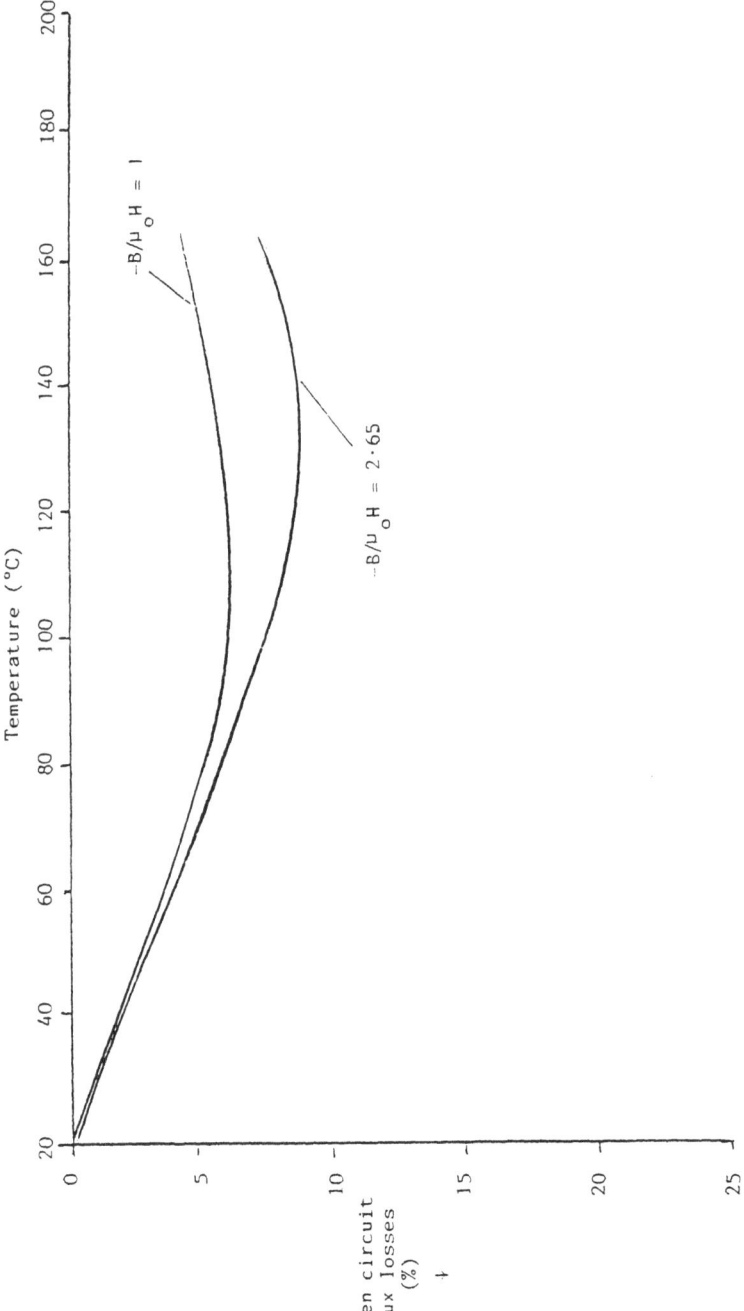

Fig 3.   Open circuit reversible flux losses with temperature for magnet No. 4

TABLE 2   **Reversible and irreversible losses with temperatures of 60°C, 100°C and 150°C**

Losses in Open Circuit Flux With Temperature

$-B/\mu_o H$	Reversible Losses						Reversible + Irreversible Losses					
	60°C		100°C		150°C		60°C		100°C		150°C	
	1	2.65	1	2.65	1	2.65	1	2.65	1	2.65	1	2.65
Sample												
1	4.4	4.0	10.7	10.0	18.7	17.8	4.4	4.0	12.0	10.8	22.0	19.8
2	4.8	2.4	7.3	8.5	7.6	12.8	5.6	2.4	27.7	10.0	64.0	36.0
3	4.0	3.2	6.3	7.5	9.5	13.9	4.0	3.2	13.8	9.7	46.7	29.5
4	3.2	3.2	6.3	7.0	5.7	8.9	7.2	3.2	34.0	10.6	70.0	49.3
5	4.4	2.4	5.9	8.5	5.4	8.6	10.3	3.17	39.0	14.1	74.0	54.2
6	2.8	2.0	7.9	7.8	15.3	14.6	2.8	2.0	8.0	7.8	20.0	17.6
7	44.5	15.0	2.4	2.9	1.9	2.2	46.5	17.5	62.0	46.0	82.3	78.0
8	26.2	13.5	1.2	1.8	0.3	3.0	27.0	15.0	57.5	42.0	76.5	68.8

For the lower coercivity magnets the irreversible losses with temperature become much greater beyond a definite temperature. This is the temperature at which the intersection of the unit permeance line with the intrinsic curve attains a field greater than that corresponding to the "knee" of this curve. This temperature is lower, as would be expected for the magnets with the smaller value of $-B/\mu_o H$. The cobalt containing samples show the greatest losses due to elevated temperature primarily because of their low coercivity.

The losses with time at a temperature of 100°C are summarised in Table 3. As can be seen these losses are quite small with the lowest values occurring in the high coercivity material and for the magnets having the higher working points determined by their greater length to diameter ratios. For the sample No. 6, with its high value of coercivity, for the magnet with $-B/\mu_o H = 2\cdot65$ a loss of only $0\cdot2\%$ occurred in a period of 200 hours. However, for several of the other samples losses of up to 4% occurred. A typical curve of loss against time is given in Figure 4. A complete set of curves for irreversible and reversible losses and losses with time for all the magnets was given by Coulson (1987).

TABLE 3   Losses in open circuit flux of sample when held at a temperature of 100°C for 200 hours

Sample	% Flux Losses	
	$-B/\mu_o H = 1$	$-B/\mu_o H = 2\cdot65$
1	0·31	0·30
2	1·30	0·85
3	1·2	0·33
4	3·11	1·43
5	5·21	1·55
6	0·23	0·20
7	3·61	3·50
8	1·83	1·30

DISCUSSION

In the high coercivity materials and the magnets with the higher load lines there is a steady increase in the reversible flux losses with temperature consistent with the decrease in saturation magnetization. The common

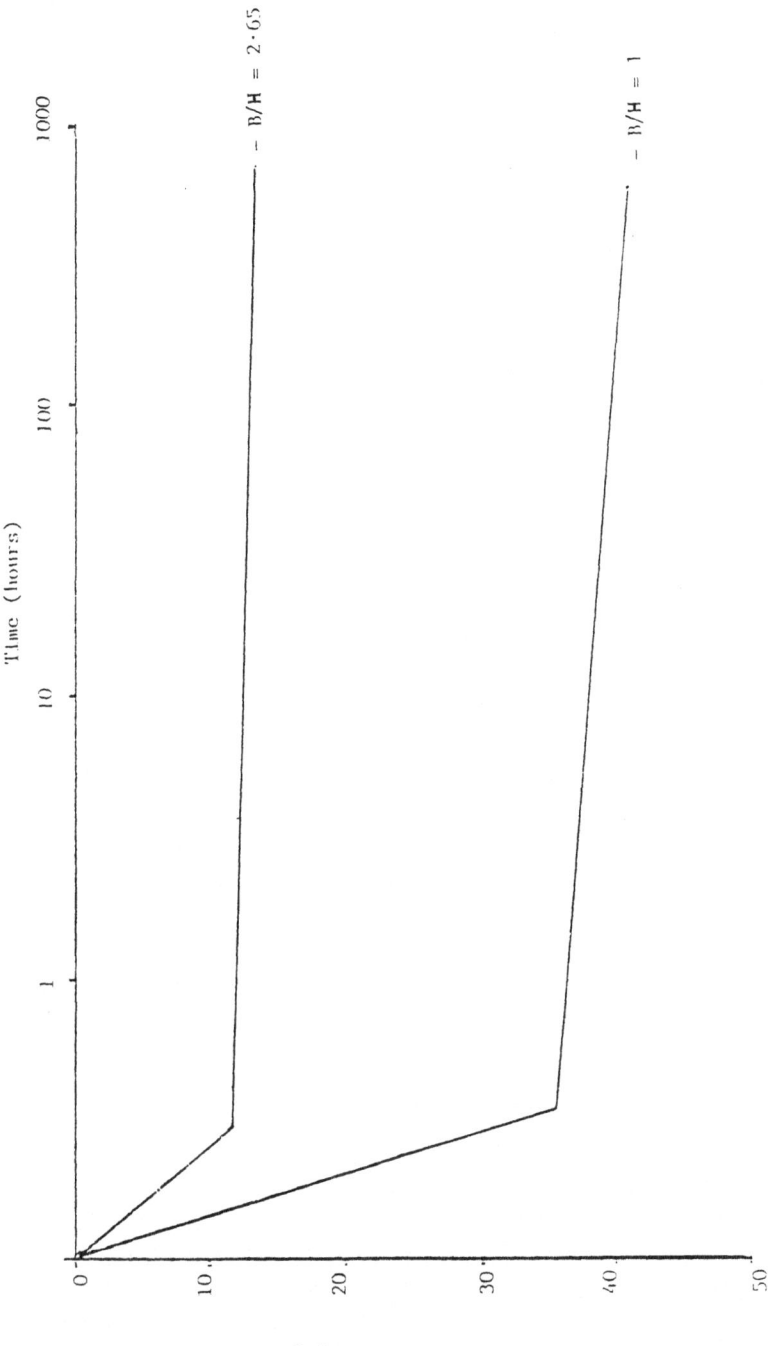

Fig. 4. Losses in open circuit flux with time for magnet No. 4 at 100°C

factor in these samples is that the working point of the magnet was such that it never approached the "knee" of the curve, a factor which normally causes the large irreversible flux losses.

This provides the information required to recommend the best properties and configuration for stability, these are a high coercivity H_{cM} and an equivalent length to diameter ratio which gives a working point well above the "knee" of the curve.

To achieve the highest coercivity it seems most convenient to use the hydrogen decrepitation technique described by McGuiness et. al. (1986). It must, however, be emphasised that for many applications it is not necessary to have the highest degree of stability and for some applications it may be advantageous to use material of lower coercivity because of difficulties in magnetizing.

Coulson (1987) discusses a number of other factors including the effect of temperature on coercivity and anisotropy and the effect of the method of powder production on the coercivity. For this information the thesis should be consulted.

REFERENCES

Clegg, A.G. and McCaig, M., 1958, The High Temperature Stability of the Nickel-iron-aluminium System. Br. J. Appl. Phys. 9, 194-199.

Coulson, I. M., 1987, The Effect of Temperature on the Properties of Permanent Magnet Alloys of the Nd Fe B Co System. Ph.D. Thesis (Sunderland Polytechnic, United Kingdom).

McCaig, M. and Clegg, A. G., 1987. Permanent Magnets in Theory and Practice, 2nd Edition (Pentech, London).

McGuinness, P., Harris, I. R. Rozendaal, E., Ormerod, J. and Ward, M., 1986. The production of a Nd-Fe-B Permanent Magnet by a Hydrogen Decrepitation/Attritor Milling Route. J. of Mat. Sci., 21, 4107-4110.

Sagawa, M., Fujimara, S., Togawa, M., Yamamoto, H., and Matsuura, Y., 1983. New Material for Permanent Magnets on a Base of Nd and Fe. J. Appl. Physics, 55, 2083-2087.

SUMMARY OF DISCUSSIONS :

MAGNETS - PRODUCTION, CHARACTERISATION AND CORROSION

H.A. Davies and E. Rozendael

A wide,range of stability measurement results on commercially available magnets were reported (7.5). It was made clear that stability can so far only be achieved with high coercive material (Nd-Fe-B-Dy alloys).

Studies on the corrosion behaviour and the coating of FeNdB magnets, both of the sintered and melt spin derived types were reported (7.2, 7.3, 7.4). This part of the CEAM programme was of crucial importance, if the more the commercial exploitation of Nd-Fe-B magnets is to proceed rapidly. Significant progress has been made within the CEAM programme in characterising the corrosion rates in different environments and manufacturers have generally established suitable coating procedures in-house but nevertheless great care will be needed in ensuring that the results of short-term tests can reliably be extrapolated to much longer periods in service.

SECTION III

— MAGNET PROCESSING

CHAPTER 8

NEW PROCESSES

ANISOTROPIC NdFeB - MAGNETS MADE BY PLASTIC DEFORMATION OF CAST SAMPLES ⟨*⟩

W. Ervens

Krupp-Widia GmbH, Essen, Germany

Introduction

Permanent magnets on the base of NdFeB are normally produced either by powder metallurgy and sintering or, starting from alloys made by melt spinning, by different compacting processes. In this paper it is shown that anisotropic NdFeB-magnets can also be achieved by hot extrusion of cast samples. This procedure is similar to the process used for the production of high grade anisotropic MnAlC-magnets. Such a process avoids the pulverisation operation and the critical handling of the powder. It could prove to be of high commercial interest if it succeeds in a reproducible production of high grade magnets.

Experimental Techniques

NdFeB ingots with compositions of 33,5 to 36% Nd, 1,1 to 1,3% B, (plus balance of Fe) were shaped to cylindrical samples of 12 mm diameter and 15 mm length. As the ingots showed a columnar structure samples were prepared with the columns parallel and perpendicular to the axis of the cylinders. These samples were encased in copper capsules, with wall thicknesses of 1,5 mm.

The capsuled NdFeB-samples were hot extruded at temperatures between 680 to 820°C. Deformation temperatures about 800°C led to the best magnetic values. The pressures which depended on the deformation temperature and alloy composition, varied from 70 to 130 kp/mm². The deformation rate referred to the sample's cross-section was ~ 80% and the deformation velocities varied from 1 to 10 mm/s.

The extruded NdFeB samples had - after removing the copper capsule - diameters of 5 to 6 mm. The samples were free of cracks and of good mechanical consistency.

Results

The extruded samples showed without exception a diametrical magnetic orientation. But there were marked differences in the flux densities, depending on the column orientation of the starting samples to the cylinder axes. Figure. 1 shows the flux distribution of an extruded sample over the cylinder surface.

⟨*⟩ Interim report of 15th April - 15th October 1986.

For the magnetic measurements, the cylinders were formed to
rectangular shapes with the preferred magnetic orientation
perpendicular to one surface. Figure. 2 shows demagnetization
curves of these magnets. Curves 2 and 3 relate to extruded
magnets, curve 1 to a sintered sample of the same chemical
composition.
The orientation of the columnar structure to the cylinder axes
before extrusion was different for the magnets of curve 2 and
3. For sample 2 it was perpendicular and for sample 3 it was
parallel. Important differences in the remanence were
evident. There is also a dependence of the magnetic values on
the extrusion velocity. With decreasing velocity from 10 mm/s
to 1 mm/s the remanence growths and the coercivity were
reduced (see Figure. 3). This dependence was different from
that found in the MnA1C-system, where B_r and $_jH_c$ grew with
decreasing pressing velocity (1).
Strong differences are apparent in the microstructure of
extruded and sintered magnets. The extruded samples show a
"rubble structure" with strong differences in particle size
whereas sintered samples have on the whole a greater
regularity.

The thermal stability of the extruded magnets was tested. For
this purpose, samples were stored at different temperatures,
starting at 80°C for one hour. After this treatment a
remarkable loss of coercivity was found. The decrease was
irreversible and could not be restored by remagnetization.
This ageing effect grew with increasing temperature and time.
To accelerate the decrease of coercivity, samples were stored
at temperatures up to 300°C. Figure. 4 shows the influence of
a treatment at 300°C for 1/2 hour. The demagnetization curve
1 was measured after deformation and curve 3 after heat
treatment. Prolonged treatments at 300°C only caused a small
evolution in the decrease of coercivity. Even at room
temperature the coercivity was not stable. This is shown by
the comparison of curve 2 and curve 1 in Figure. 4. Five
months separated the two measurements. While B_r remains
nearly constant $_jH_c$ dropped considerably.

Summary and discussion

NdFeB cast alloys can be hot extruded in the temperature range
of 700 to 800°C. By this process a diametrically magnetic
orientation is brought about in cylindrical samples. The
magnet values are distinct if, in the starting samples, the
orientation of the columns is perpendicular to the cylinder
axes. With decreasing deformation velocity there is an
increase in remanence and a strong drop in coercivity.
Extruded magnets with remanences up to 1100 mT and
coercivities up to 900 kA/m could be realised. The coercivity
stabilizing mechanism is apparently different in the extruded
NdFeB magnets from the mechanism in the sintered samples. The
$_jH_c$ of the extruded magnets showed a strong irreversible
decrease under elevated temperatures and even a reduction at
room temperature in the course of time.

The plastic deformation generates a "rubble structure" with irregular grains, lattice defects and stresses. These effects seem to be responsible for the build-up of the coercivity. However, this state is critically unstable. By recovery processes at elevated temperatures or even at room temperature the coercivity is irreversibly reduced. Whether this state can be stabilized by a special alloying process has to be investigated. The orientation-mechanism during extrusion is not yet understood.

References

(1) W. Ervens, Techn.Mitt.Krupp, Forsch.Ber. Bd.40, 1982, 117

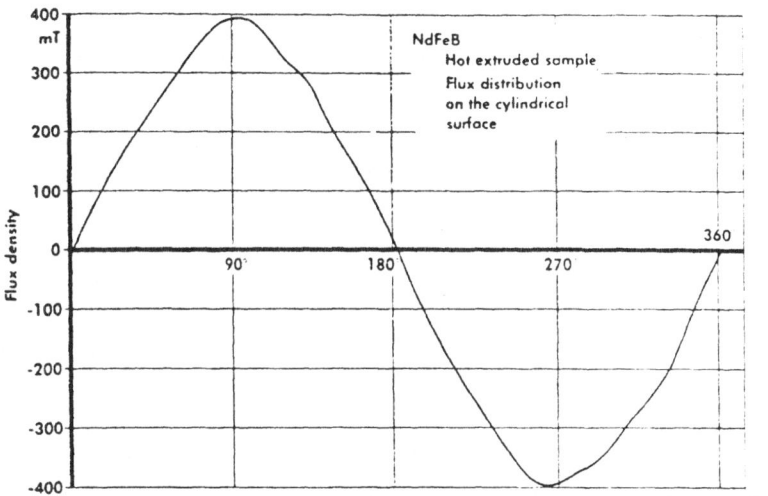

Fig. 1: Flux distribution on the cylindrical surface of an extruded sample

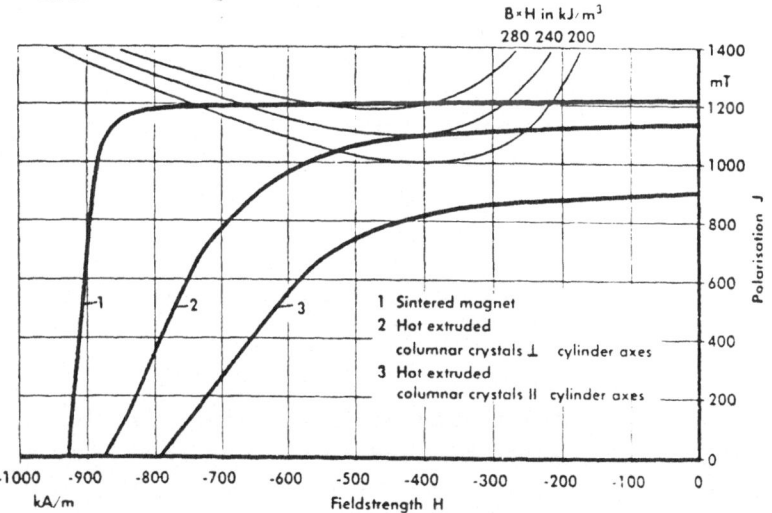

Fig. 2: Demagnetization curves of extruded NdFeB samples (2,3) and a sintered magnet (1)

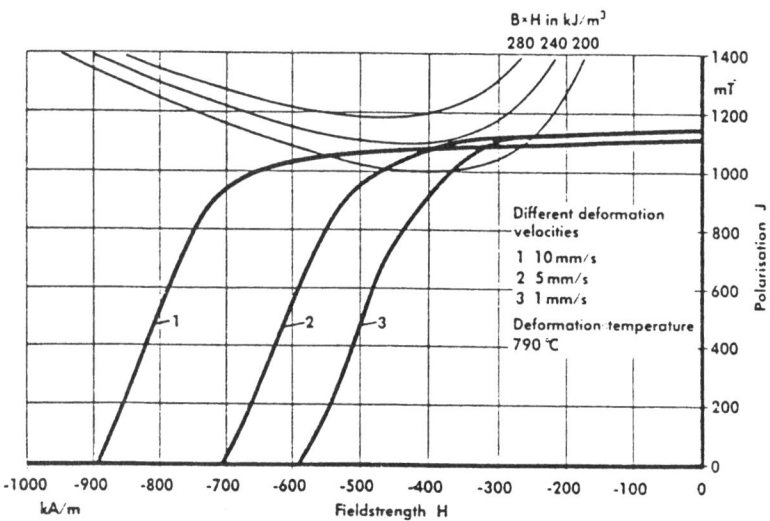

Fig. 3: Dependence of the magnetic values on the
         deformation velocities

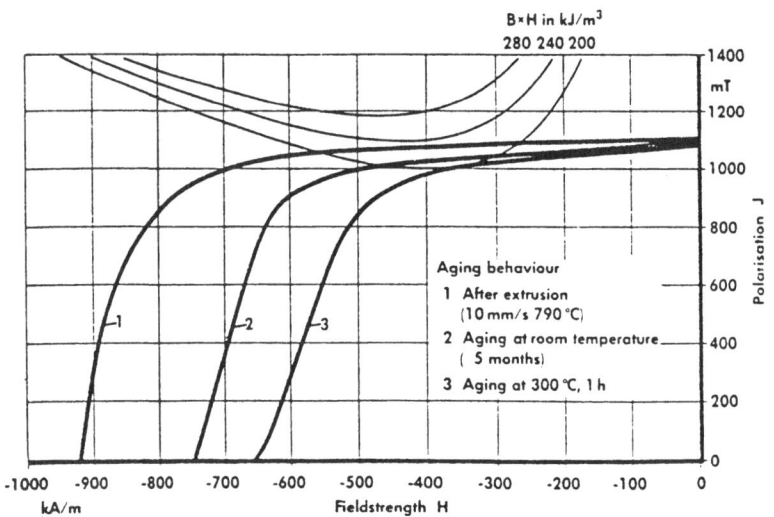

Fig. 4 : Ageing and magnetic values of extruded
         NdFeB magnets

A NOVEL PROCESS FOR RARE EARTH-IRON-BORON PERMANENT MAGNETS PREPARATION

J.P. NOZIERES, R. PERRIER de la BATHIE

Laboratoire Louis Néel

C.N.R.S.

166 X, 38042 Grenoble Cedex, France

ABSTRACT

The work described in this paper derives from research to obtain magnetically isolated grains in Rare earth-Iron-Boron alloys without passing through the divided state, which is an essential process both in powder metallurgy and melt spinning. Industrialisation of those techniques imposed serious problems : Powders obtained by cold crushing are extremely reactive and pollution may drastically reduce magnetic properties, whilst, melt spinning is a sophisticated and relatively expensive process which is difficult to apply in a continuous way.

The process presented here consists of breaking grains of as-cast alloyed ingots by an appropriate mechanical hot-working technique. To our knowledge, it is the simplest and the cheapest way to prepare Rare earth-Iron-Boron permanent magnets at the present time.

I - DETAILED DESCRIPTION

At first, the alloy is prepared by fusion of commercial products. Induction melting is the usual method, but any other process could be applied. Compositions are chosen near the standard powder metallurgy values i.e. approximately $Nd_{15.5}Fe_{78}B_6Al_{0.5}$ at. %. To improve magnetic properties, additives such as Dy or Co may also be used. It is important to notice that cooling rate is absolutely of no importance : any size or shape of grains can be obtained, and it will not influence the further treatments.

Next, as-cast ingots are hot-worked under controlled atmosphere or after sealing in an iron sheath. This step provides the optimal grain size, a partial magnetic orientation and shaping of the magnet. Any type of hot working can be applied, with variable results : extrusion (fig. 1), hot-pressing, rolling, hammering ...

Temperatures ranging from 750°C to 950°C can be used with little influence on magnetic properties. However it is important to note that at these temperatures, intergranular phases are liquid, so the material will stay homogeneous during the mechanical hot working and consequent plastic deformation, thereby preventing crushing into fragments. High deformation ratios are necessary to obtain good magnetic properties. They vary from 5 to 25.

This is the key stage of the process, starting from as-cast ingots and leading directly to permanent magnets.

A final heat treatment is applied to improve and stabilize the magnetic properties. The temperature of this treatment depends on alloy composition and previous history. No absolute rule has yet been found. However, melting of the intergranular phases as well as defect relaxation are thought to play an important role.

MICROSTRUCTURAL OBSERVATIONS

The observation of the ingot after primary crystallisation shows that the grains of the main phase ($Nd_2Fe_{14}B$) are surrounded by the segregated (heaviest) compounds, probably low melting Nd-rich eutectics involving Nd solid solution, NdAl eutectics, oxides, etc. Differential thermal analysis reveals melting points in the proximity of 700°C (fig.2)

Mechanical treatment breaks the grains below the critical size of 10 to 15 μm where coercivity appears. Typical micrography consists of a few unbroken grains of 10 to 20 μm dissiminated in a matrix formed from grains with a typical size of 0.5 to 2 μm (fig. 3). After annealing, a slight size increase of the smallest grains can be noticed reaching about 1 μm. Dynamic recrystallization does not appear to occur. It should be noticed that the effects of heat treatments were always found to be additive and reversible, except for the high temperature annealings (above 950°C) where grain size increases drastically.

MAGNETIC PROPERTIES

Typical magnetization curves at 300 K are shown by the fig.4. The $(BH)_{max}$ = 14 MGOe curve is already an industrial standard whereas $(BH)_{max}$ = 18 MGOe has been reached in laboratory experiments. More details are given in the examples described in the following section.

First magnetization curves show a high susceptibility revealing that domain walls move freely. No significant pinning has been observed, even before defect relaxation by heat treatment (fig. 5). The analysis of the coercivity vs. saturation field reveals that coercivity develops as the saturation field is increased from about to 10 to 25 kOe (fig. 6). The above mentionned properties suggest that the coercivity mechanism is the same as in sintered magnets.

The temperature dependence is also in the same range as sintered magnets (fig. 7).

Different heating cycles have been performed, with no significant variations in the magnetic properties of permanent magnets used in motor devices.

Contrary to previous studies [1], permanent magnets obtained by this process are stable. Fig. 8 shows the demagnetization curves after a 6 month period, no significant variations are observed and after appropriate annealing, this variation reduces to a negligeable amount.

ANISOTROPY

Partial anisotropy is obtained directly by mechanical hot-working which eliminates c-axes from the hot working direction. Starting from totally isotropic ingots, one should obtain all the easy-axes in a plane. The exact orientation mechanism is not yet completely understood.

However, it is possible to improve this anisotropy by using directed crystallisation : It is well known that in this system, the crystalline growth direction corresponds to a magnetic difficult axis. By favouring a unique growing direction, all easy-axes should lie in a plane. If this plane coincides with hot-working direction, one should select a unique direction in this plane, perpendicular to the hot-working direction, i.e. in the strain direction. As a result, the easy magnetization directions should all be parallel (fig. 9).

In fact, it is found that anisotropy is improved by this directed crystallisation but the magnets thus produced are still only partially anisotropic (fig.10).

EXAMPLES

This new technology enables the production of large quantities of magnets both easily and continuously. A 10 kg extrusion batch has already been prepared without difficulty and larger quantities should not pose any particuland industrial problems. Thanks to the hot working process, any type of profile can be obtained and therein lies the inherent versatility and wide range of possible applications for this process.

Fig. 11 is from a cylindrical 21,8 mm diameter extruded bar, which has been used in an electric motor. Starting from iron sheathed alloy of $Nd_{15}Fe_{18}B_6Al_{0.5}Dy_{0.5}$ at. % composition, extrusion was performed at 830°C. After annealing for one hour at 550°C, Hcj was 13,6 kOe with Br = 7.8 kG and $(BH)_{max}$ = 12.81 MGOe. Results were easily reproduced along the whole bar.

Fig. 12 is from a 28x8 mm plate which was also obtained by hot extrusion at 830°C. The billet was a 5 Kg iron sheathed $Nd_{15.5}Fe_{78}B_6Al_{0.5}$ at. % alloy. After annealing at 670°C, we obtained iH_c = 11,5 kOe with Br = 8.5 kG and $(BH)_{max}$ = 15,2 MGOe.

Fig. 13 is from a hot-pressing of a $Nd_{15.5}Fe_{78}B_6Al_4$ at. % alloy. A 200 g billet has been pressed at 750°C in a small iron can. After a 800°C annealing, iH_c = 7 kOe, Br = 9.8 kG and $(BH)_{max}$ = 19.8 MGOe.

Fig. 14 is from a hot stamping of a small sample under controlled argon atmosphere. The stamping was performed around 700°C for a few minutes. After a 580°C annealing iH_c rose to 14,5 kOe, Br to 6,4 kG for a $(BH)_{max}$ of 14 MGOe.

COMPARISON WITH OTHER NEW PROCESSES FOR MAGNET PREPARATION

Two other processes have been developed during the same period. One of them (Seiko patent [2]) consists in producing ingots of fine magnetically isolated grains using a special cooling technique, either quenching or normal cooling with certain special compositions (under stoechiometric in boron). A mechanical hot-working is involved in this process, which seems to act more like a cooling accelerator than a mechanical fracture treatment. In this process, deformation ratios are much lower. Large quantities of magnets probably cannot be produced at once, because of the fine grains required for initial crystallization. Moroever in this system, the development of coercivity results from an appropriate heat treatment applied on the fine-grained microstructure resulting from initial cooling, hot-working only improving the magnetic properties. Seiko's special compositions were tried with our own process and no significant results were observed. Magnetic properties were very low, and deteriorated on further heat treatment.

Another process, developed by General Motors [3], consists of a very special crystallization which allows one to obtain thin and long aligned platelets which are broken in a subsequent mechanical hot-working. This process does not obviate passing through the divided state, the ingot being cold-crushed or cold-cut before enclosing in an iron can. The material is then hot-worked at a low deformation ratio (around 2) and at a very well-defined temperature. No further heat treatment is applied. Because of the very stringent solidfication conditions required, this technique may have limited industrial application. Moreover, according to our experience, magnetic properties could deteriorate with time.

CONCLUSION

A novel Rare earth-Iron-Boron permanent magnet production process has been developed. It combines low cost, simplicity and adequate magnetic properties for many industrial applications and in particular for motor devices. Such magnets are now in production in Grenoble (France) by Crismatec SA, a subsidiary of La Pierre Synthétique Baïkowski.

ACKNOWLEDGMENTS

We wish to thank M. Gavinet and his team of the Laboratoire d'Elaboration et de Transformation des Matériaux, CEN Saclay for performing hot-extrusion and D.W. Taylor for a critical analysis of this paper.

REFERENCES

[1] Anisotropic NdFeB - Magnets made by plastic deformation of cast samples, W. Erwens.

[2] Seiko Epson Corporation, Brevet d'Invention FR Z 586 323 A1

[3] General Motors Corporation, European Patent EP 0231 620 AZ.

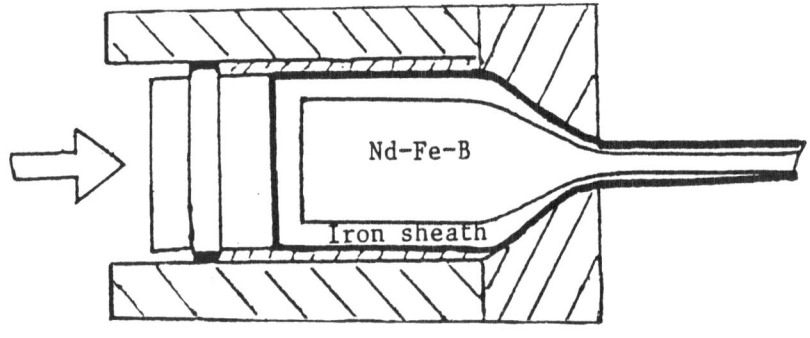

Fig. 1 :   Extrusion process

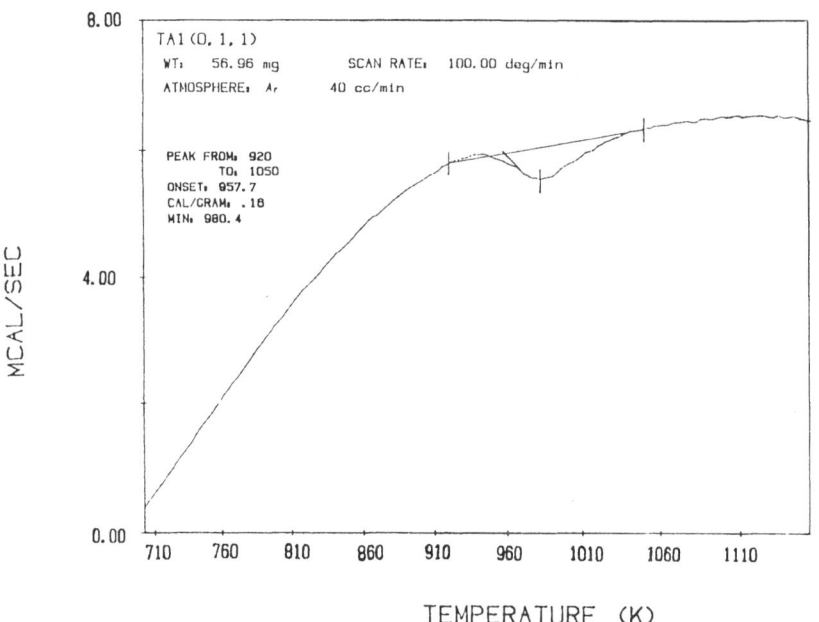

Fig. 2 :   DTA on as-cast samples

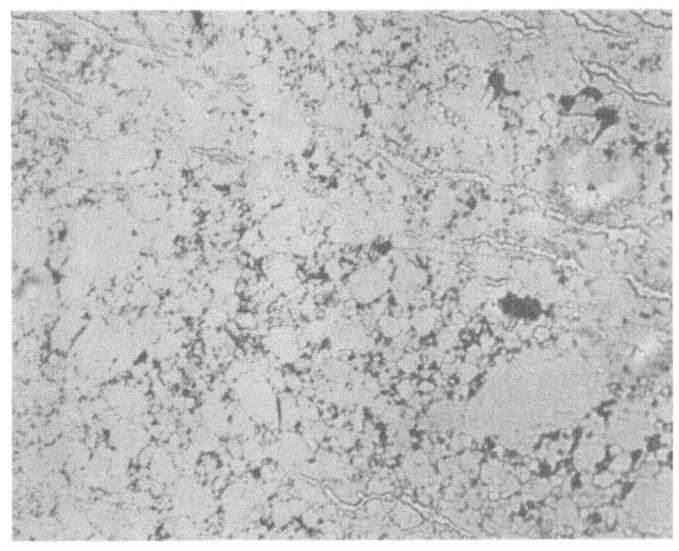

Fig. 3 : Optical micrograph of an extruded magnet (x500)

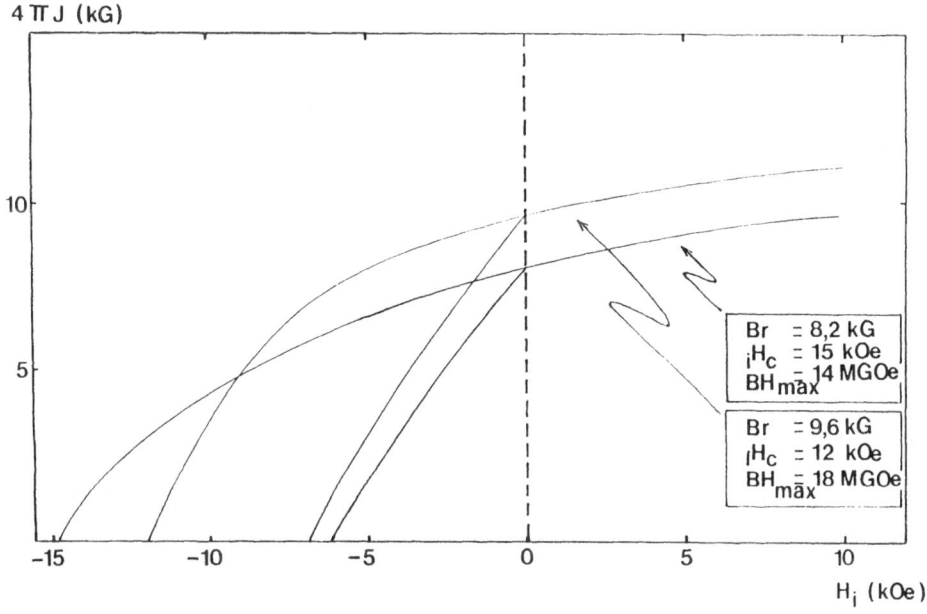

Fig. 4 : 300 K demagnetisation curves

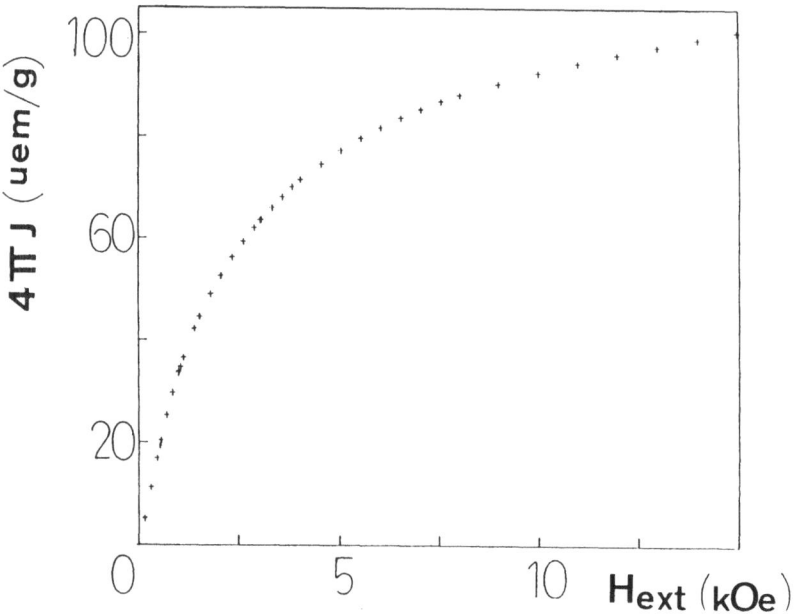

Fig. 5 : Initial magnetisation curve at 300 K

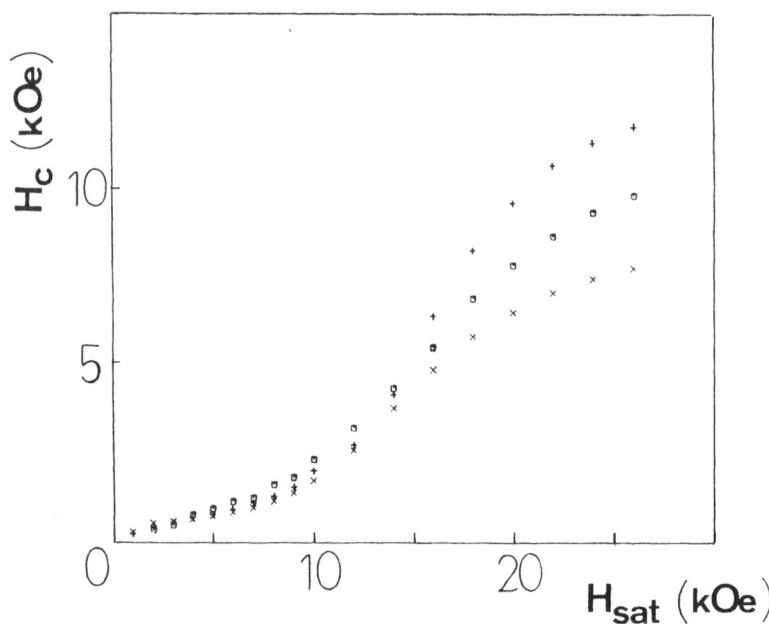

Fig. 6 : Coercivity vs. saturation field at 300 K

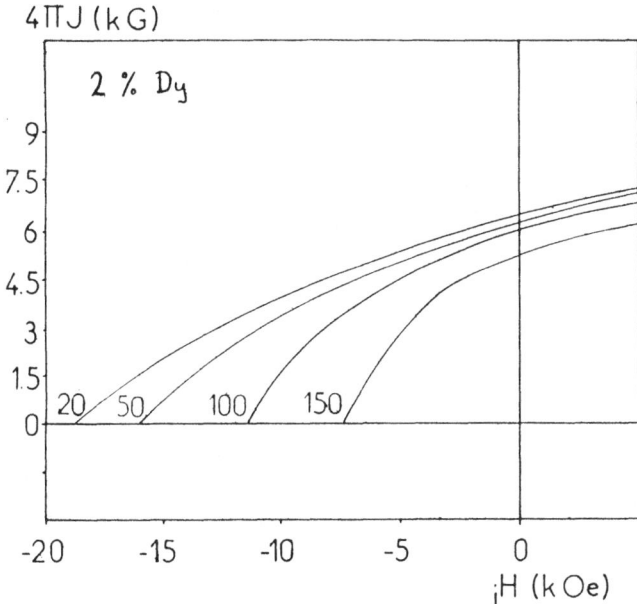

Fig. 7 :   Temperature dependence of demagnetisation curves at 300 K

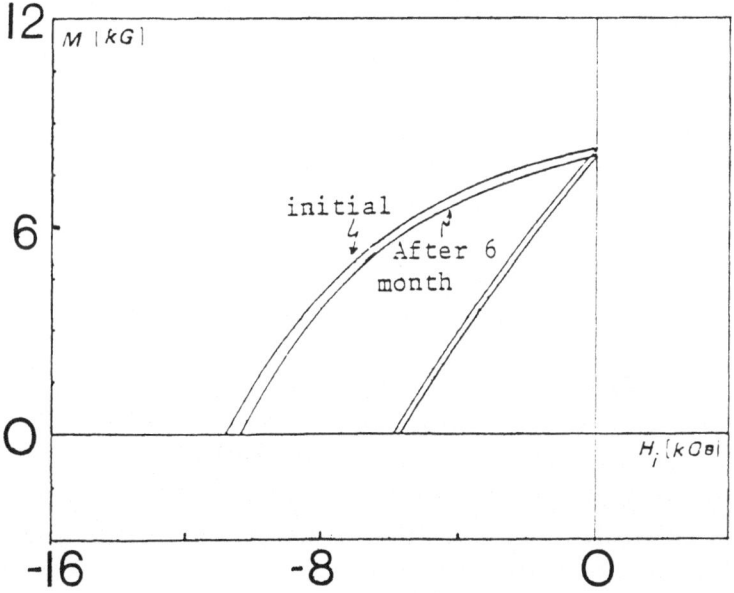

Fig. 8 :   Stability

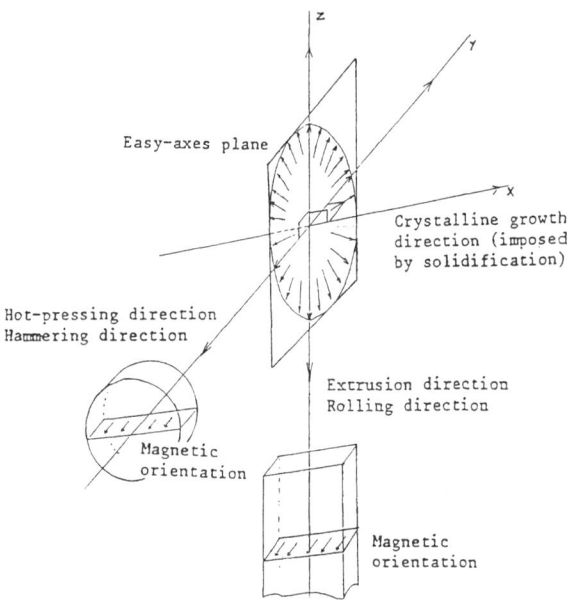

Fig. 9 : Orientation mechanism after directed crystallisation

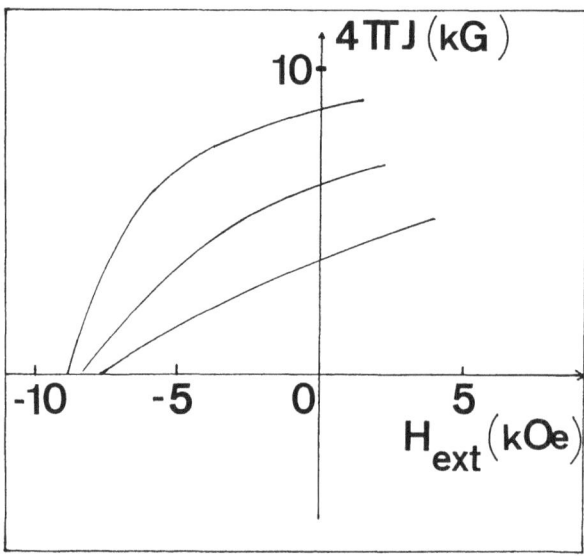

Fig. 10 : Anisotropy after directed crystallisation
(direction of measurements are refereed to fig. 9)

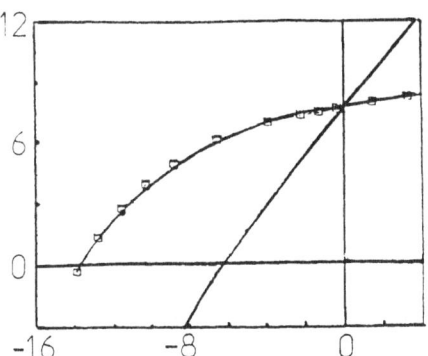

Fig. 11 : Extruded (cylindar) :

Br   = 7.8 kG

Hcj   = 13.6 kOe

Hcb   = 6.3 kOe

$(BH)_{max}$= 12.8 MGOe

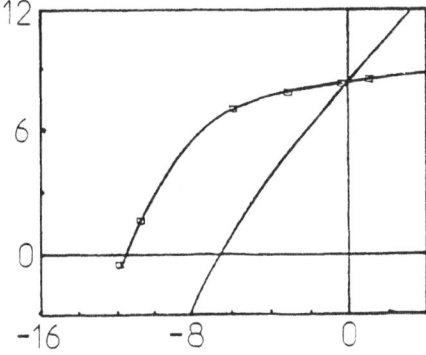

Fig. 12 : Extruded (plate) :

Br   = 8.5 kG

Hcj   = 11.5 kOe

Hcb   = 6.6 kOe

$(BH)_{max}$= 15.2 MGOe

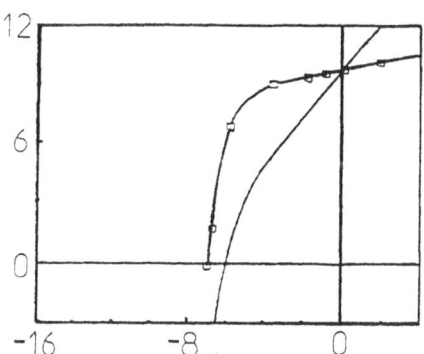

Fig. 13 : Hot-pressed :

Br   = 9.8 kG

Hcj   = 6.9 kOe

Hcb   = 6.0 kOe

$(BH)_{max}$= 19.8 MGOe

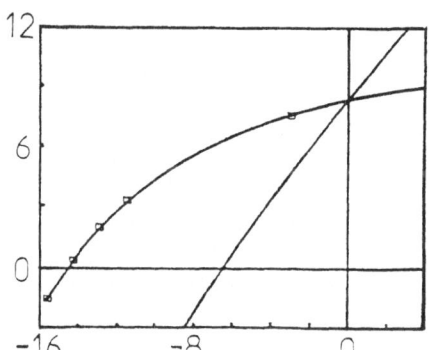

Fig. 14 : Hammered :

Br   = 8.4 kG

Hcj   = 14.5 kOe

Hcb   = 6.4 kOe

$(BH)_{max}$= 14.0 MGOe

New Developments In Bonded Nd-Fe-B Magnets

N. Rowlinson, M.M. Ashraf* and I.R. Harris,
School of Metallurgy and Materials, University of Birmingham, U.K.
*Penny and Giles Blackwood Ltd., Blackwood, Gwent, U.K.

Abstract

A new method has been developed for forming Nd-Fe-B rapidly quenched ribbons to high density magnets with energy products of up to $150kJ/m^3$. The process is carried out at room temperature using a rotary forging machine. Several soft metals (0-25% volume) Al, Zn, Sn, Cu, have been mixed prior to forging, either with anisotropic powder and then pre-aligned, or with standard Nd-Fe-B ribbon (MQ1) both supplied by GM. The densities obtained have been of the order 88-98% of the theoretical maxima. Energy products in the range $50-150kJ/m^3$ have been achieved. Magnets produced with MQ1+10% volume Al have been assessed for machinability and mechanical strength. The results of these studies are reported in this paper.

Introduction

Neodymium-iron-boron magnets are of considerable interest because of the energy products attainable at significantly reduced cost compared to those based on samarium-cobalt. There are two main processing routes used for forming Nd-Fe-B alloys into permanent magnets. One method is via conventional powder metallurgy [1,2] (i.e. mill-align/press-sinter). The other is by rapidly quenching from the melt [3,4] (giving a fine grain size and thus high coercivity). The resulting ribbons (about 30um thick and 2mm wide) are then ground to a coarse powder which can be polymer bonded to give isotropic magnets. Alternatively, the ribbon can be hot-pressed at 973K in argon to give a fully dense compact. If a further hot-press is performed in a larger die cavity which allows deformation transverse to the press direction, a preferred orientation is induced parallel to the press direction [5,6]. Furthermore, these compacts can be converted to powder suitable for forming anisotropic polymer-bonded magnets [7,8].

The subject of this paper is the formation by rotary-forging at room temperature of high-density compacts of isotropic melt-spun ribbon and aligned anisotropic powder (derived from melt-spun material) and their assessment. The binding between magnetic particles is achieved by a soft metal such as aluminium.

Experimental

Melt-spun ribbon material (supplied by GM) was mixed with various amounts of soft metal (Al, Zn, Sn, Cu) and placed in the die cavity of a rotary forging (RF) machine (see figure 1). The anisotropic material (also supplied by GM) was blended with various amounts of soft metal (Al, Zn, Sn, Cu) and pressed using a load of $4.8kN/cm^2$ in an aligning field of 1200kA/m. The preformed green compact was then placed into the RF machine die cavity. The load applied during rotary forging was varied to obtain optimum compact density. The densities were determined using a fluid displacement system.

The magnetic properties of the final compacts were assessed using a permeameter and the second quadrant loop was obtained after pulse-magnetising in a field of 5000kA/m.

Four point bending was used to assess the mechanical strength of the compacts as a function of amount and type of the soft metal binder.

Several of the aluminium-bonded magnets have been shaped using simple machining techniques such as drilling, turning and milling.

Results and Discussion

Several soft metals were investigated as binders. These were Al, Zn, Cu and Sn. The fracture strengths for 15% by volume soft metal + MQ1 ribbon are shown in figure 2. As can be seen from this data, the Al-bonded samples exhibited the highest mechanical strength i.e. 100MPa. Al was therefore chosen for more detailed study. Figure 3 shows how smaller quantities of Al give low mechanical strength, but at Al amounts over 7% by volume the mechanical strength is greatly improved. However, high amounts of Al result in significant dilution of the magnetic properties (figure 4) with the remanence falling from 743mT for no Al to 483mT for 20% by volume Al.

When larger proportions of Al were used a high relative density was obtained, up to 98% for 25% by volume Al (figure 5). Relative density is taken to be the actual density divided by the theoretical maximum and converted to a percentage, assuming no loss of Al during forging. This is consistent with the mechanical data, since a high porosity, as in the lower Al containing samples, results in a weaker binding between some particles, which gives an easy path for a fracture to follow.

Figure 6 is an optical micrograph of an Al-bonded MQ1 ribbon magnet (15% volume Al) showing that the Al has flowed and completely surrounds the ribbon particles. This encapsulation of individual ribbon particles should give excellent corrosion resistance.

The energy product obtained for 10% by volume Al-bonded MQ1 samples was $74kJ/m^3$. A typical demagnetisation curve is shown in figure 7.

Pre-aligned anisotropic compacts have been produced and the best maximum energy product achieved to date was $148kJ/m^3$. Figure 8 is a demagnetisation curve for a 7% by volume Al aligned compact. The RF process is not believed to greatly disturb the particles of the pre-aligned billet, as shown by the high energy product of the final compact.

Unlike sintered and hot-pressed Nd-Fe-B magnets, which can only be ground, the magnets produced by RF presented no problems in machining. The only precaution was to use paraffin as a lubricant.

Measurements on 100% MQ1 ribbon compacts produced via rotary forging have indicated an induced anisotropy, resulting in a significant increase in the remanence and energy product. This work will be the subject of a further publication [9].

Conclusions

The RF process gives high density magnets with energy products of up to $148kJ/m^3$ (18MGOe). These magnets have reasonable mechanical strength and good machinability. Figure 9 gives a proposed processing route for Nd-Fe-B ribbon based magnets.

Acknowledgements

We would like to thank Dr. P. Bowen and Professor W.A. Penny for technical assistance and many helpful discussions. Thanks are due to Penny and Giles for financing the work and for the provision of a research scholarship (N. Rowlinson).

References

1) M. Sagawa, S. Fujimura, N. Togawa, H. Yamamoto, Y. Matsuura and K. Hiraga, I.E.E.E. Trans. Magn. MAG20 (1984) 1584

2) M. Sagawa, S. Fujimura, N. Togawa, H. Yamamoto and Y. Matsuura, J. Appl. Phys. 55 (1984) 2083

3) J. J. Croat, J. F. Herbst, R. W. Lee and F. E. Pinkerton, Appl. Phys. Lett. 44 (1984) 148

4) J. F. Herbst, J. J. Croat, F. E. Pinkerton and W. B. Yelon, Phys. Rev. B 29 (1984) 4176

5) R. W. Lee, E. G. Brewer and N. A. Schaffel, I.E.E.E. Trans. Magn. MAG21 (1985) 1958

6) R. W. Lee, Appl. Phys. Lett. 46 (1985) 790

7) Y. Nozawa, K. Iwasaki, S. Tanigawa, M. Tokunaga and H. Harada, J. Appl. Phys. 64 (1988) 5285

8) L. J. Eshelman, K.A. Young, V. Panchanathan and J. J. Croat, J. Appl. Phys. 64 (1988) 5293

9) N. Rowlinson, M.M. Ashraf and I.R. Harris, to be published

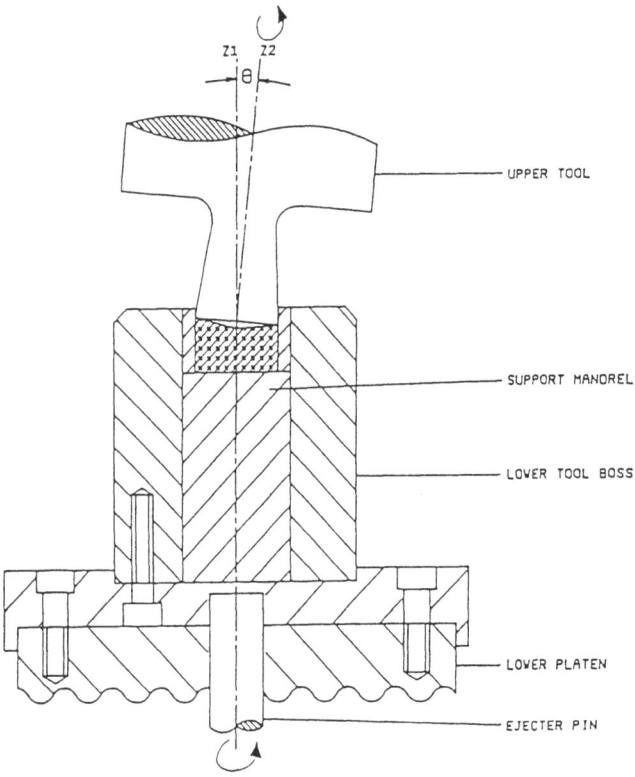

Figure 1.   Rotary forging machine principle.

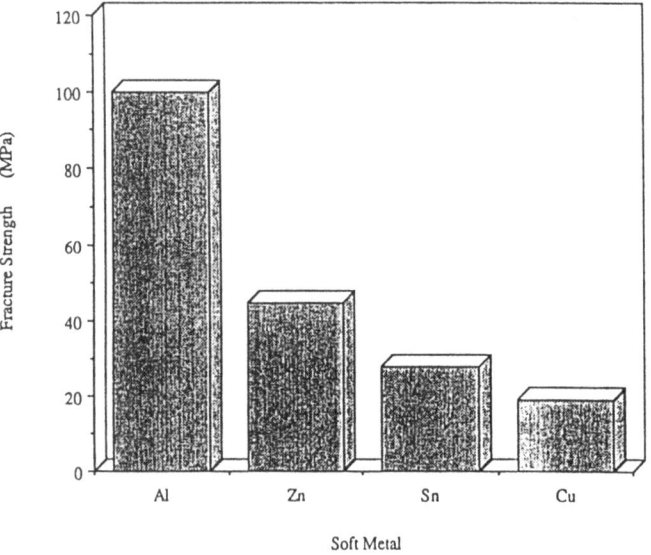

Figure 2.   Fracture strengths for 15% volume additions of various soft metals.

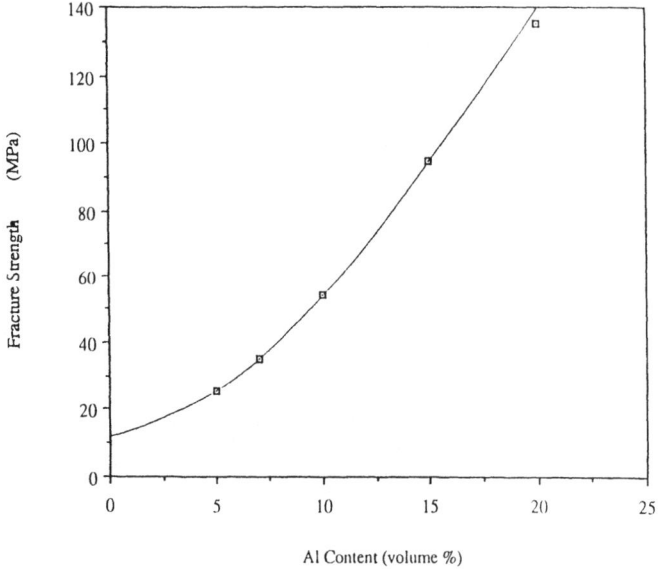

Figure 3.   Graph of fracture strength against Al content.

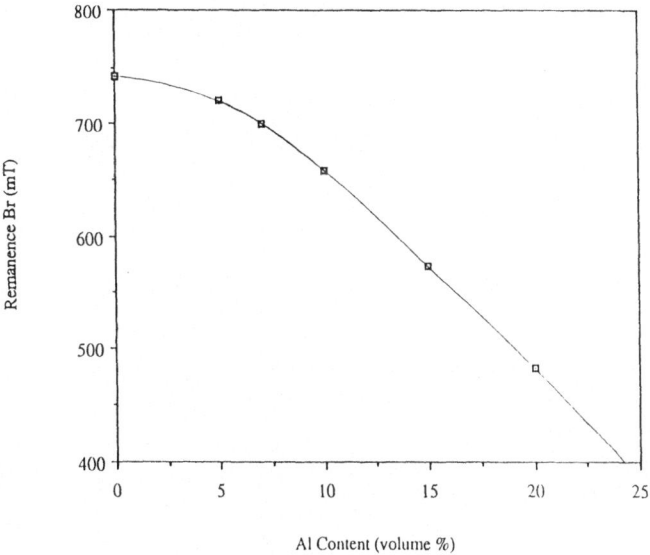

Figure 4.   Graph of remanence against Al content.

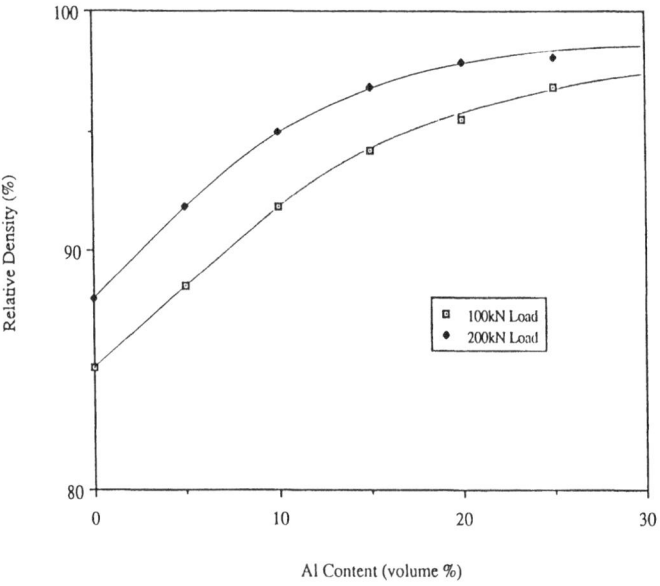

Figure 5.   Graph of relative density against Al content.

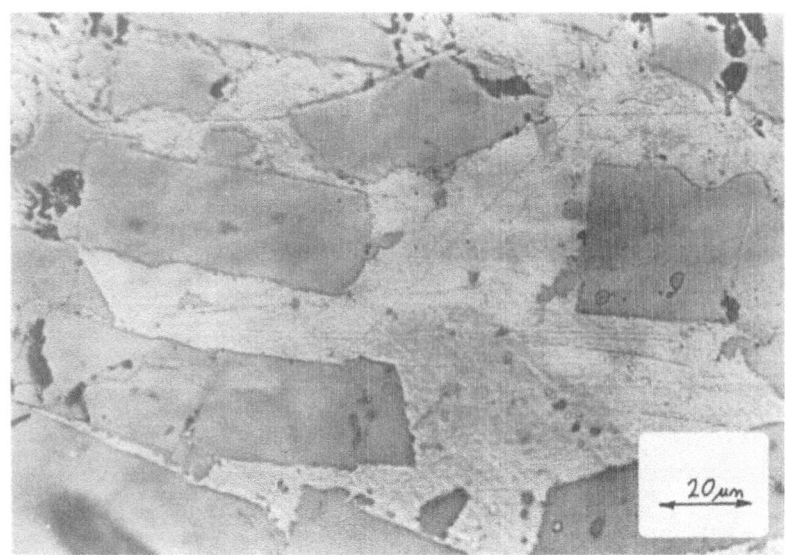

Figure 6.   Optical micrograph of Al-bonded (15% volume) MQ1 RF compact.

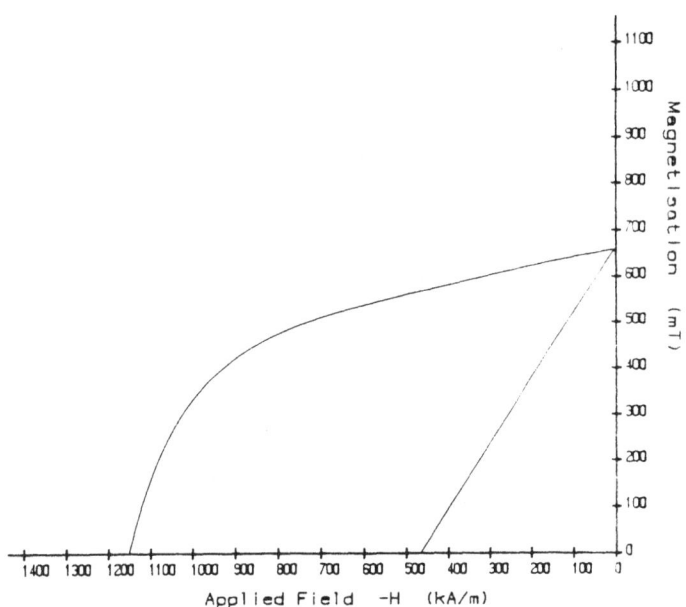

Figure 7. Demagnetisation curve for Al-bonded (10% volume) MQ1 compact

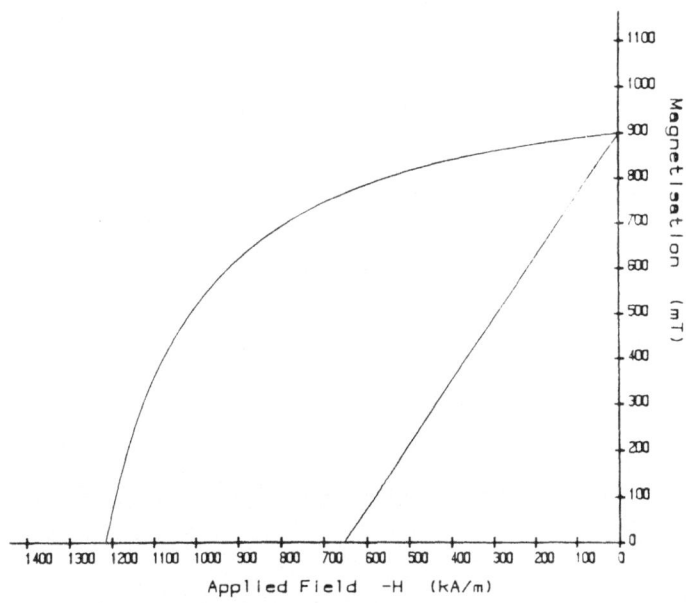

Figure 8. Demagnetisation curve for Al-bonded (7% volume) aligned anisotropic Nd-Fe-B powder.

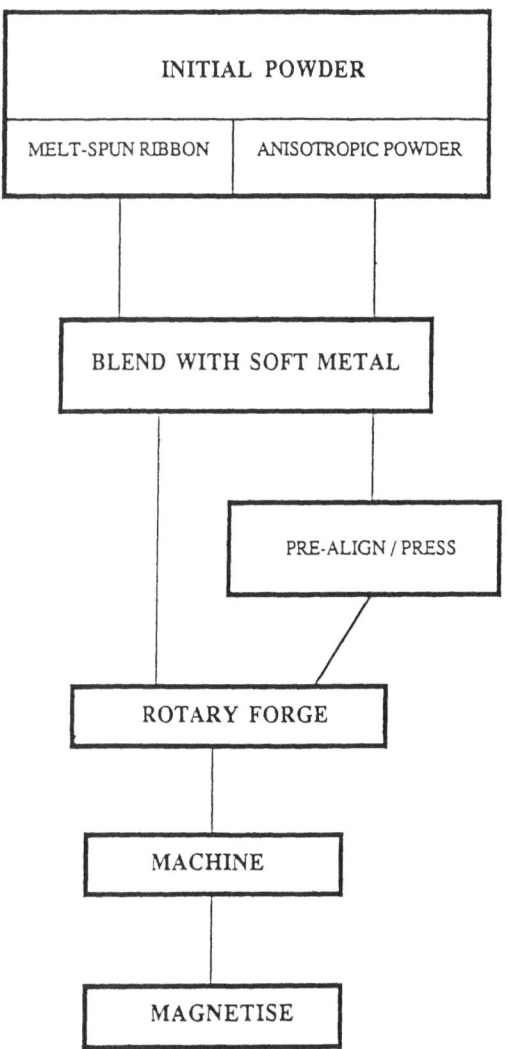

Figure 9.   A proposed process route for RF Nd-Fe-B magnets.

SUMMARY OF DISCUSSIONS :

NEW PROCESSES

J.M.D. Coey

The new hot extrusion process developed at the C.N.R.S. Grenoble (8.2) offers the prospect of partly-oriented Nd-Fe-B magnets with moderate energy product at a cost approaching that of the starting ingot. Orientation is achieved first by directional cooling of the ingot (in an iron sheath) to yield a planar c-axis texture, and then by extrusion to reduce the cross section by a factor of about 20. A fine grained microstructure is produced, which, after annealing, resembles that of a normal sintered magnet. Magnets may be produced in different shapes, and can be used directly in motors in their iron sheaths.

The hydrogen decrepitation process developed during CEAM from work done at the University of Birmingham (8.3), is now firmly established in the process route for sintered magnets. At present it greatly reduces, but does not entirely dispense with the need for jet milling to produce the fine crystallites required for orientation and sintering. There are prospects that with further developments the hydrogen decrepitation stage may avoid the need for milling entirely. Metal bonded magnets with modest energy product ($\sim$150 kJ/m^3) have been prepared in Birmingham by cold working, and powders with some coercivity have been prepared by surface treatment. The goal of a cheap high-performance bonded magnet depends on the realisation of a powder mode of crystallites that exhibit coercivity, or the achievement of a high degree of crystallite orientation in melt-spun ribbon. Significant progress in both areas has been made in CEAM, but much remains to be done.

SECTION III

— APPLICATIONS

CHAPTER 9

ROTATING MACHINES

DESIGN AND PERFORMANCE OF SMALL BRUSHLESS MACHINES
WITH Nd-Fe-B MAGNETS

R. Hanitsch, E. Hemead, A. Thoma

Institut für Elektrische Maschinen
Fachbereich Elektrotechnik
Technische Universität Berlin
1000 Berlin 10, FRG

ABSTRACT

The improved magnet properties of the new Nd-Fe-B permanent magnet material have impact on machine size and performance. For electronically controlled machines such as linear stepping motor and cup-type synchronous motor the efficiency is improved. The application of the new Nd-Fe-B material helps to improve automatic position control systems requiring fast response and high power to weight ratio. In doing the motor design a non-linear analysis with the finite element approach is applied.

INTRODUCTION

There is increased demand to have reliable and controllable small industrial drives. This will favour the use of brushless motors for a wide range of new applications. For example on vehicles in the future there will be electrical power assisted steering, electrical anti-lock brakes, electrically driven water pumps and a combined starter-generator. Servomotors should have a linear speed-torque characteristic and a high torque to inertia ratio. In addition a high power to weight ratio is required.

We concentrated on the design of a small linear stepping motor, cup-type synchronous motor and a low speed disc-type motor. For comparison we could use machines designed and built earlier making use of Samarium-Cobalt permanent magnets. It is well known that variable speed drives need a special electronic commutation circuit. In exploring system aspects relevant to Nd-Fe-B magnet motors we worked together with GEC Research Ltd. (Hirst Research Centre) on a gate array for PWM drives. The gate array is used to drive MOSFET inverters at ultrasonic carrier frequencies so that acoustic noise is re-

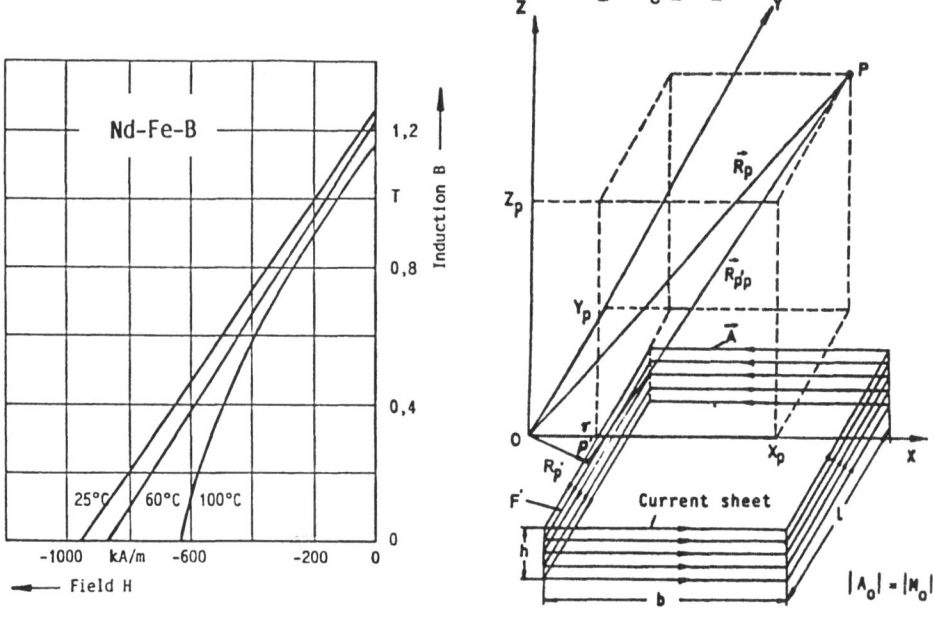

Fig. 1 Temperature influence on
demagnetization characteristic   Fig. 2 Current sheet arrangement

duced. The upper power limit for use at ultrasonic frequencies is set by the available power switching devices. Therefore a study was carried out by Behr and Hanitsch (1987) in
order to compare power MOSFET and conductivity modulated
MOSFET. However, this paper will deal mainly with the design activities with respect to small drives using Nd-Fe-B
magnets.

Nd-Fe-B PARAMETERS

Nd-Fe-B has outstanding magnetic properties at room
temperature (Fig. 1). Its behaviour at elevated temperatures
brings additional problems for the design engineer. The remanence temperature coefficient is about -0,13 %/K and the
coercivity temperature coefficient is close to -(0,5-0,6)%/K.
Because of the advantages of Nd-Fe-B compared with conventional permanent magnet material, higher airgap flux density and energy or lower magnet volume and weight, this ma-

terial helps to improve the permanent magnet motors. In co-operation with the MPI (Max-Planck-Institut) in Stuttgart some of the material parameters were measured. Table 1 gives only some of the measured data.

TABLE 1   Nd-Fe-B parameters

characteristics	data
density (g/cm³)	7,36 – 7,57
specific resistivity (cm)	130 – 160
specific heat capacity (J/kgK)	400 – 420
thermal conductivity (W/mK)	7 – 8,9
tensile strength (N/m²)	$74 \cdot 10^6$
bending strength (N/m²)	$2,5 \cdot 10^8$

MACHINE DESIGN WORK

Anisotropic sintered forms of Nd-Fe-B have now been produced having a maximum energy product of nearly 360 kJ/m³, whilst bonded isotropic forms have an energy product of some 40 kJ/m³. Especially the bonded material is suitable for in-jection moulding which will improve the cost effectiveness of motors. The finite element method of field computation makes it possible to perform detailed magnetic field ana-lysis. For a large number of field topologies a two-dimen-sional field solution will suffice. However for the prede-sign we used the current sheet approach (Fig. 2).

Linear stepping motor

We designed and built a motor with a four pole slider, which consists of four permanent magnets. The magnet height was h_m = 5 mm and the pole pitch 20 mm. The double sided stator had 32 slots. From Fig. 3 can be taken that the new Nd-Fe-B material results in an increased flux density which will give higher thrust. The effect of variation of the magnet height is demonstrated in Fig. 4. Because of the cost of the permanent magnet material it is essential to optimize the dimensions of the magnet. The measured flux

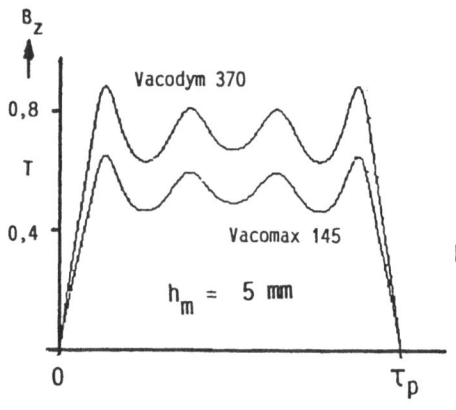

Fig. 3 Flux density in the air gap
        of linear-stepping motor

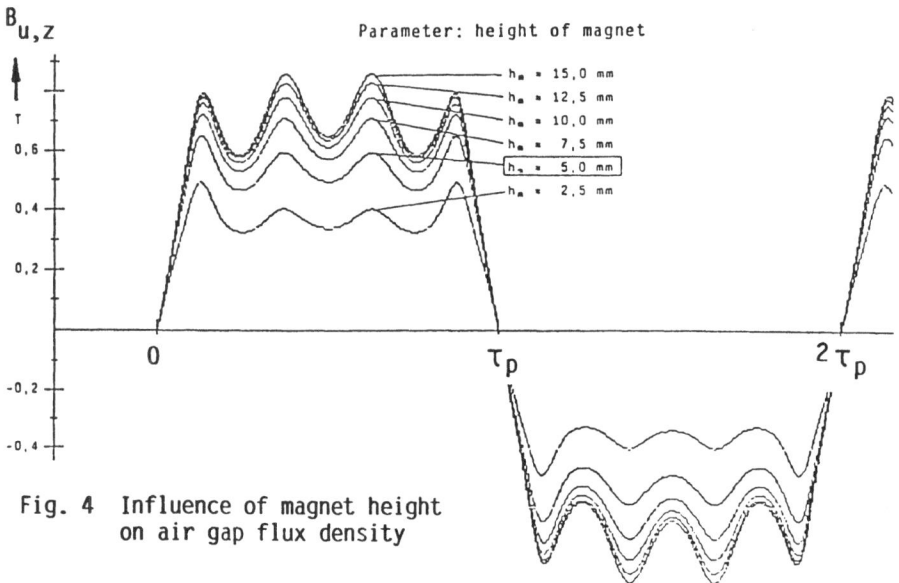

Fig. 4 Influence of magnet height
        on air gap flux density

density is shown in Fig. 5 where the effects of the stator
slots can clearly be seen. The average flux density distribution for the four-pole slider is shown in Fig. 6 in which
calculated figures are plotted. From Fig. 7, which shows
the thrust vs. speed, it is obvious that the Nd-Fe-B slider
gives a higher thrust. A recent publication by Hanitsch and
Neubauer , 1987 gives more details about this linear stepping motor.

A detailed analysis of the stepping motor was done by Neubauer (1987) who also designed the electronic control circuit for the linear drive system.

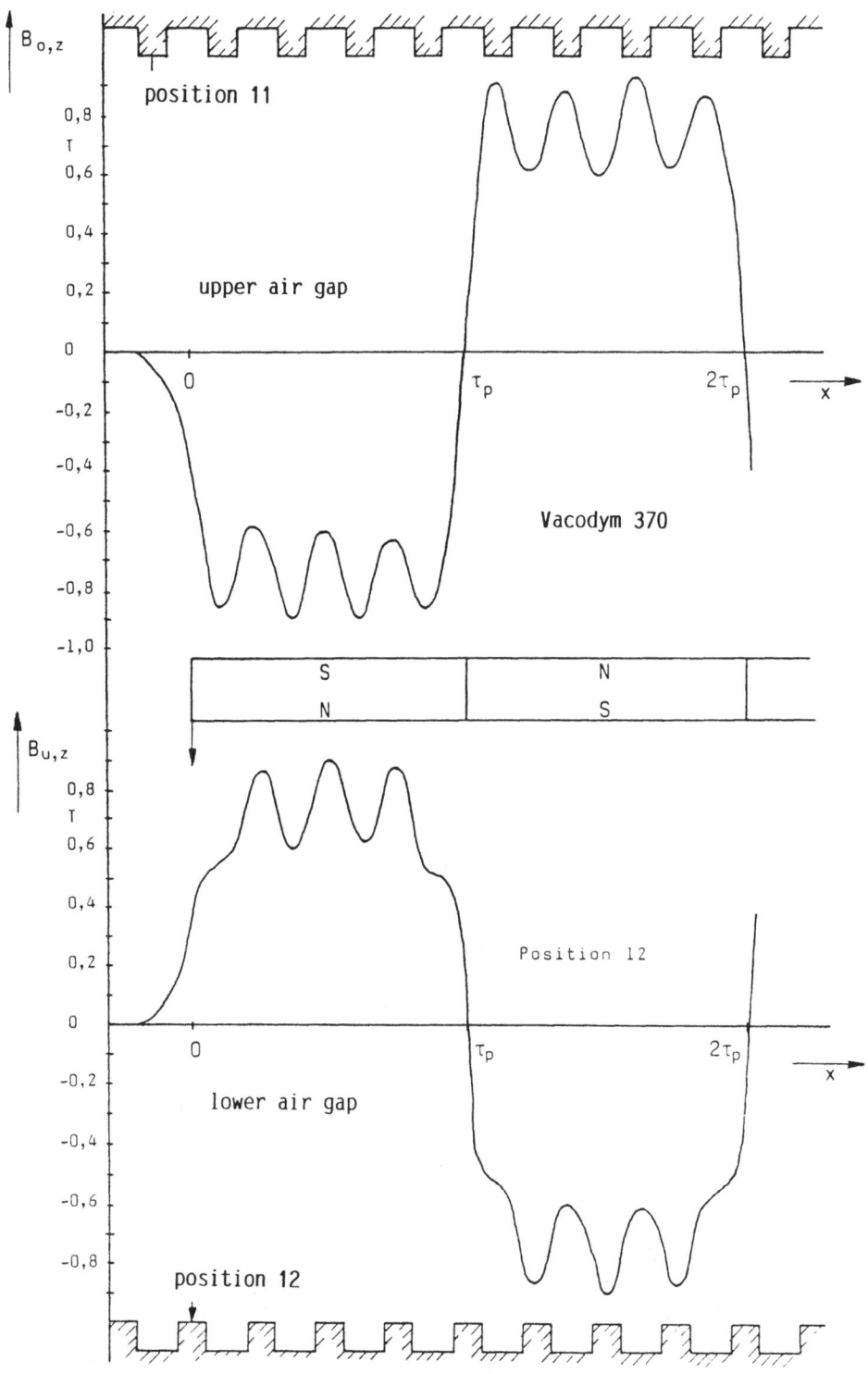

Fig. 5 Measured flux density in
upper and lower air gap

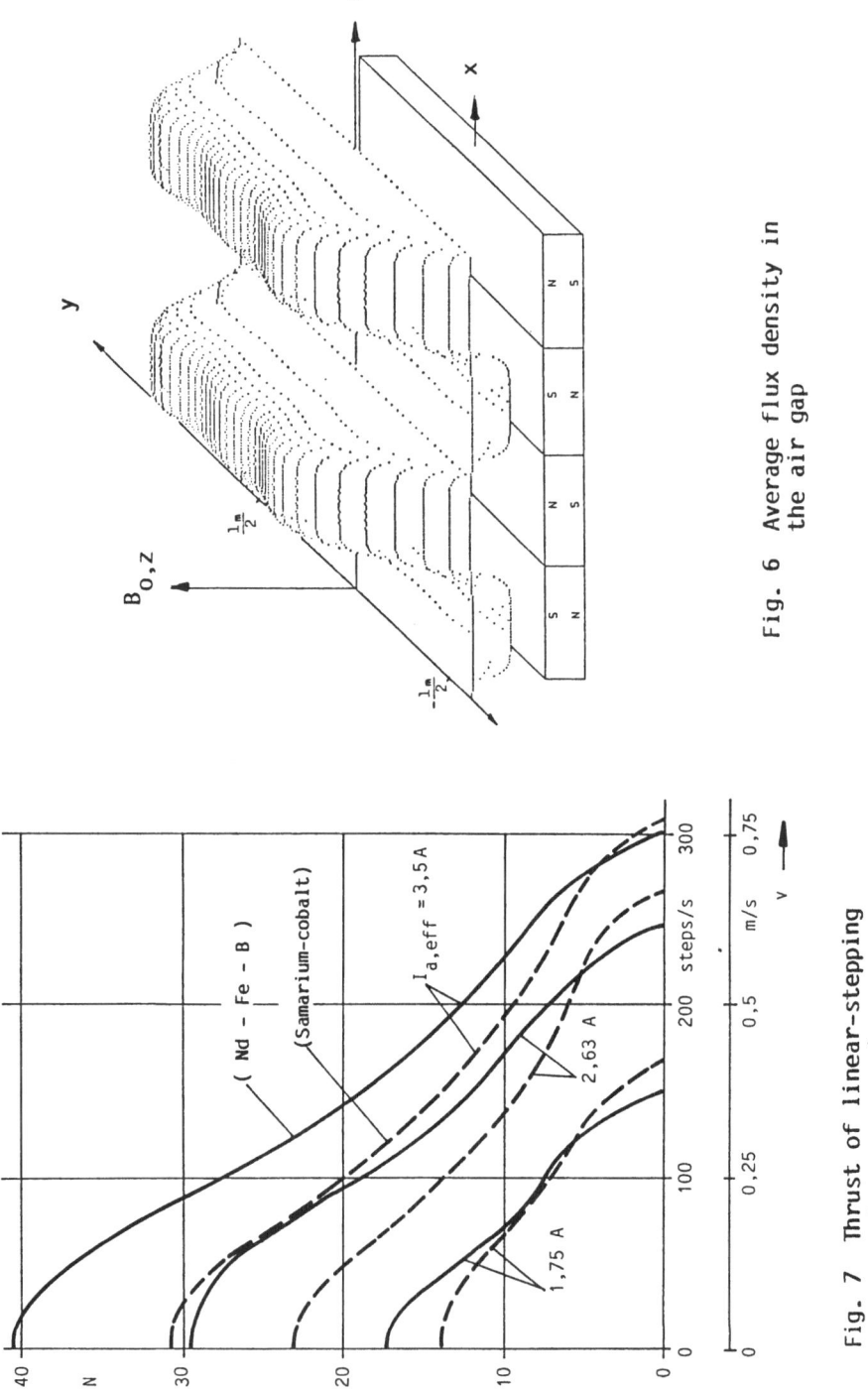

Fig. 6 Average flux density in the air gap

Fig. 7 Thrust of linear-stepping motor vs. speed

Brushless cup-type machines

The concept of a brushless DC motor with Sm-Co-magnets was realized first. A brushless DC system consists of a stationary winding, permanent magnet rotor, rotor position sensor and power inverter that produces the same linear speed-torque characteristic as a conventional DC motor. The design of the prototype can be taken from Fig. 8. The cup-type armature is a multiphase winding arrangement. The four phases were arranged in such a way that an unipolar and a bipolar circuit could be connected to the motor. The field excitation is provided by a four-pole system using prefabricated rare earth magnet material glued to a mild steel ring and the armature is constructed by nesting preformed wirewound coils to form a cylinder shape air gap winding (Fig. 9). The parameters given in Fig. 9 are used to define a specific winding parameter λ.

$$\lambda = l_m/d_m$$

With $V_m = \pi d_m^2 l_m/4$ we can express the resistor of a winding element using the specific winding parameter λ :

$$R_{w,e} = R_a + R_c = \frac{l_m + \tau_p}{\varkappa A_T}$$

$$R_{w,e} = c_o \sqrt[3]{\frac{1}{\lambda}} \; (\lambda + \frac{\pi}{2p}) \quad \text{with}$$

$$c_o = \sqrt[3]{\frac{4Vm}{\pi}} \; / \varkappa A_T$$

For $\lambda_{opt} = \pi/4p$ we obtain the smallest resistor for the winding. Different dimensions of the length l_m will result in a higher resistor. Based on this approach we calculated the efficiency for the cup-type motor. The result is shown in Fig. 10. For a more detailed information it is necessary to use finite element software as suggested by Smith (1986), Hanitsch et al. (1988) and Jabbar (1987). For the ICEM '88 a joint paper was prepared by UMIST/Manchester and TU Berlin dealing with the calculation of the flux distribution and the torque. The earlier brushless DC motor with Sm-Co magnets is compared with a new brushless type using the Nd-Fe-B magnets. The flux pattern of this new machine is shown in Fig. 11 while the winding arrangement for this three phase air gap winding can be seen in Fig. 12 .

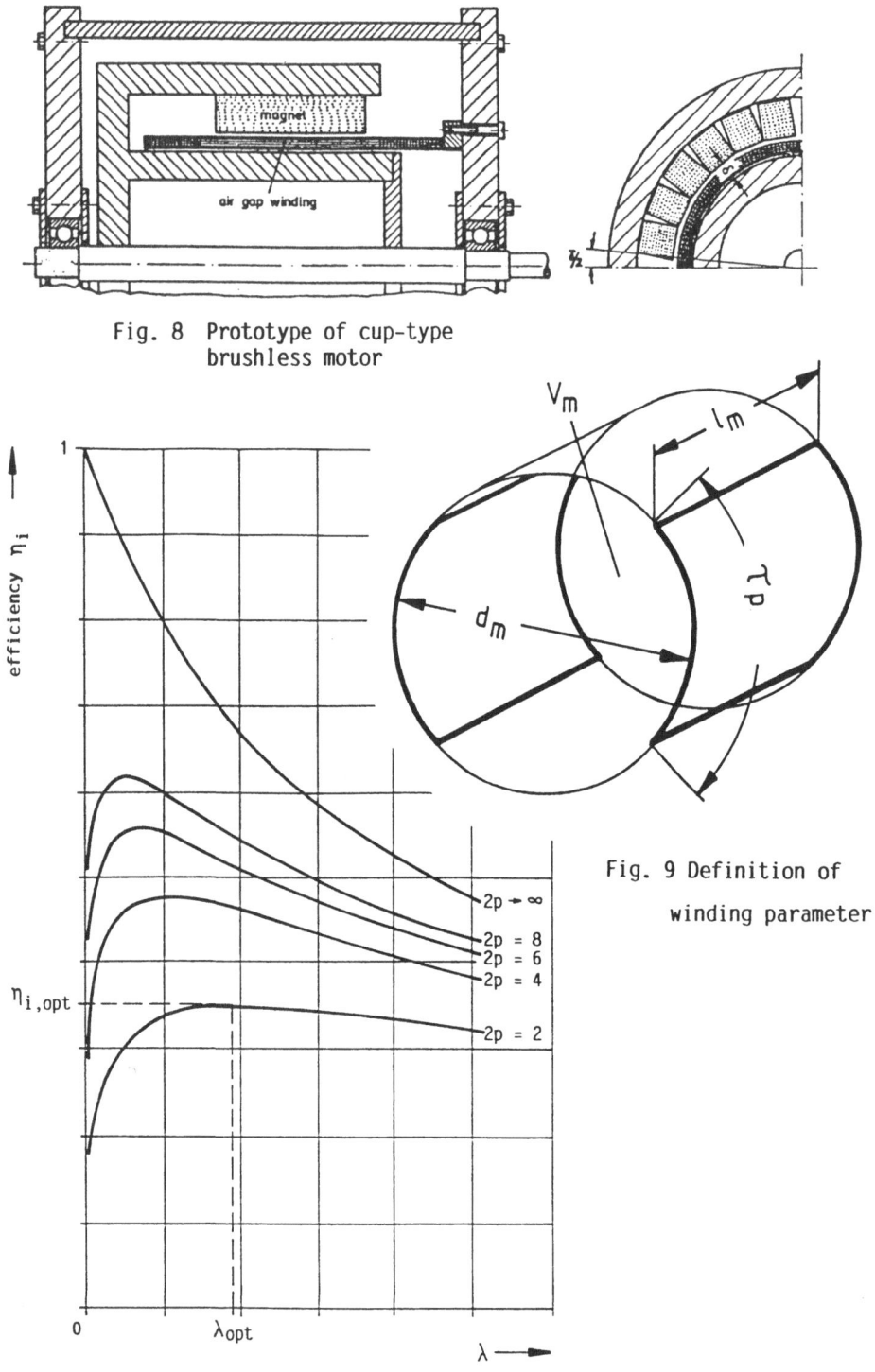

Fig. 8 Prototype of cup-type
brushless motor

Fig. 9 Definition of
winding parameter

Fig. 10 Efficiency vs. winding parameter

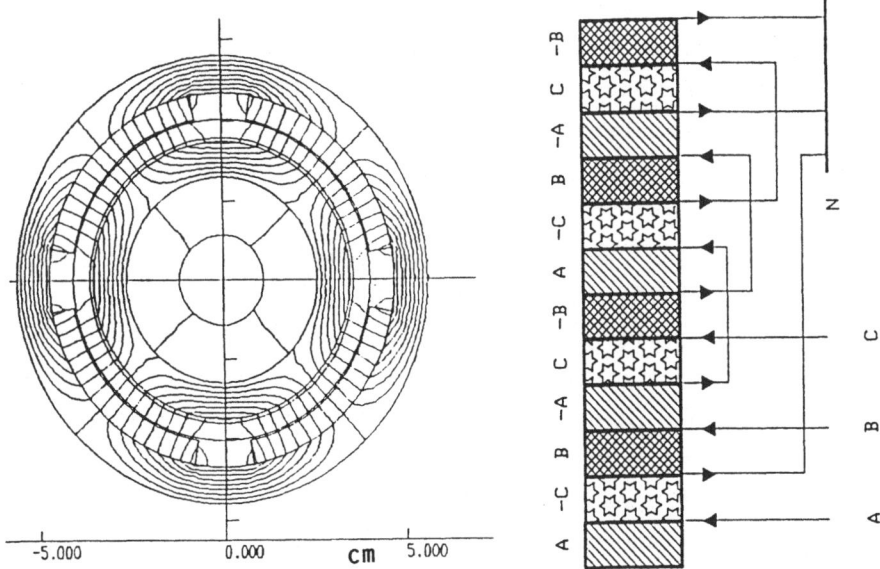

Fig. 11 Flux pattern in synchronous machine

Fig. 12 Winding arrangement

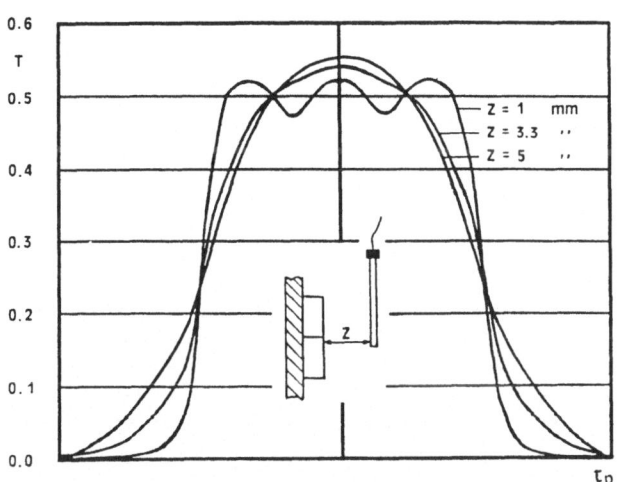

Fig. 13 Air gap flux density

Again prefabricated cubic magnets were selected for this four pole machine. As can be taken from Fig. 13 only three magnet elements were used along the circumference in this design. With the new Nd-Fe-B material about 0,5 T flux density can be achieved in the air gap. More details about the design and performance can be found in the joint publication mentioned above. Although the flux density distribution is not fully sinusoidal the induced line voltage is almost sinusoidal (Fig. 14). It is known that the type of machine we designed can operate as a DC motor or a synchronous machine. We decided to use the Nd-Fe-B machine as a synchronous motor. The comparison (Fig. 15) of the two types of machines clearly demonstrates the benefits obtained by using the new high energy permanent magnets.

With the introduction of Nd-Fe-B into the machine the air gap flux density will increase by about 25 % and in addition we have a reduction in magnet volume and weight.

ACKNOWLEDGEMENT

The authors acknowledge the support of Mr. J. Balderama, Mr. J. Holzapfel, Mr. T. Kindervater and Mr. J. Voss. Their assistance with the measurements and machine design was extremely helpful.

REFERENCES

Behr, E.-K. and Hanitsch, R. 1987. Contribution to the comparison of power MOSFET and conductivity modulated MOSFET. Proc. of the second European Conference on Power Electronics and Applications, Vol. I, Grenoble, 99-104

Hanitsch, R. and Neubauer, L. 1987. Permanent magnet motor with Neodymium-Iron-Boron magnets. Proc. of Beijing International Conference on Electrical Machines, Beijing, 746-749

Hanitsch, R. 1987. Disc-type motor with high energy permanent magnets. Proc. of third international conference on electrical machines and drives, London, No. 282, 255-259

Hanitsch, R. 1988. Beitrag zur Entwicklung von Kleinmotoren mit Hochenergie-Permanent-Magneten. Intl.wissenschaftlich-technische Tagung, Dresden, 112-117

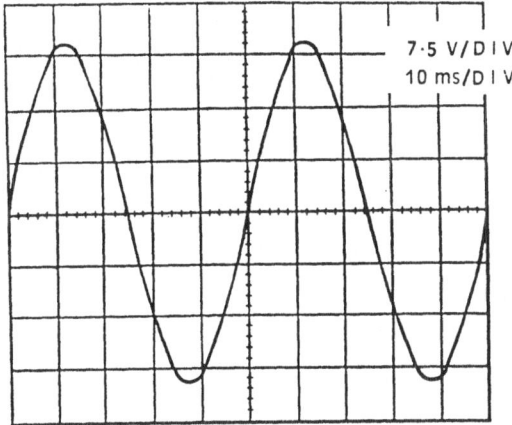

Fig. 14   Induced voltage (line to line)

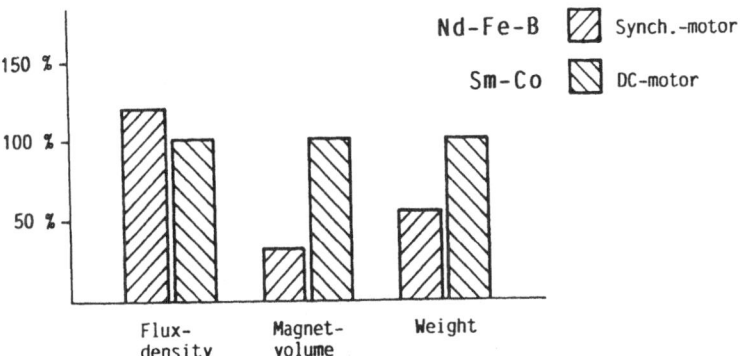

Fig. 15   Comparison of machine features

Hanitsch, R. and Chang, J.C. 1986. A cup-armature brushless
    d.c. motor. Proc. of universities power engineering
    conference, Imperial College, London, 371-374
Hanitsch, R.; Sitzia, A.M.; Chalmers, B.; Hemead, E. 1988.
    Improved brushless motor with cup-type windings and
    Nd-Fe-B magnets. ICEM '88, Proc. of the International
    Conference on Electrical Machines, Pisa
Neubauer, L. 1987. Theoretische und experimentelle Unter-
    suchungen an einem linearen, permanenterregten Schritt-
    motor. Diss. TU Berlin D 83
Smith, P. 1986. Development and verification of finite ele-
    ment software for the design of permanent magnet ma-
    chines. Proc. of 21st Universities power engineering
    Conference Imperial College, London, 328-331

694

Jabbar, M.A. 1987. Mains voltage permanent magnet motors for appliances; Aspects of design and development. Electric energy conference, Adelaide, Proceedings, 272-277

LIST OF MAIN SYMBOLS

A_T...   cross section of wire

d_m...   rotor diameter

l_m...   rotor length

R ...   resistor

V_m...   volume of rotor

λ ...   winding parameter

$\varkappa$ ...   conductivity

τ_p...   pole pitch

THE POTENTIAL FOR NdFeB MAGNETS IN ELECTRICAL MACHINES AND OTHER PERMANENT MAGNET EXCITED DEVICES

D Howe, T S Birch, K J Mitchell

Department of Electronic & Electrical Engineering
University of Sheffield, Mappin Street,
Sheffield S1 3JD, England.

ABSTRACT

The potential for NdFeB magnets in various categories and topologies of permanent magnet excited machine, for applications ranging from spindle-drives for machine tools to the electrodynamic braking of high-speed railway vehicles, as well as in other magnetic devices, has been examined. Design strategies and CAD procedures have been established for brushed dc motors, brushless dc and ac motors, line-start synchronous motors, hybrid stepper motors, and generators. In addition, finite element procedures have been implemented on an IBM PC to aid the design of permanent magnet devices in general. Theoretical designs have been validated against tests on a number of prototype units, some of which are being developed further to be launched as commercial products. The work undertaken under CEAM has indicated that there is little doubt that NdFeB will become the material of choice for many future designs.

INTRODUCTION

NdFeB magnets offer tantalising new design possibilities, and opportunities for improved performance from electrical machines and other forms of permanent magnet excited devices. However, these are not achieved simply by incorporating the new material into existing designs. Indeed, in most instances this is likely to prove deleterious. To extract the maximum leverage from NdFeB a complete redesign is necessary, with a consideration of all the facets at both motor/actuator and system levels.

CAD has been a major theme of the research undertaken by the Machines and Drives Group at the University of Sheffield. The Group have developed improved design procedures for different classifications of permanent magnet excited machine, and other devices, and, in collaboration with various industrial organisations, have undertaken numerous design studies. This report outlines the salient features of the work. It highlights some of the design considerations, and quantifies some of the improved performance factors which can be achieved by basing designs around NdFeB, thereby enhancing product value.

GENERAL DESIGN CONSIDERATIONS

Because of the outstanding energy product offered by NdFeB magnets they serve the quest for miniaturisation, since the minimum magnet volume required to establish a specified flux density in a working airgap is inversely proportional to $(BH)_{max}$. This allows the magnetic circuit to be designed more compactly, and usually permits the magnet to be moved towards the reaction space, thereby minimising flux leakage. Its high remanence means NdFeB can support a higher gap flux density, which is conducive to increasing the efficiency of electrical machines, which also benefit, in particular, from the high coercivity, since thinner magnets are required to withstand the demagnetising effect of the electromagnetic field. As a consequence, when the magnets are mounted on the rotor they do not limit the rotor inertia, as may be the case with ferrites. A result is that the torque-to-inertia is increased. If the magnets are mounted on the stator a larger diameter rotor can be accommodated for a given overall diameter. This results in a more than corresponding reduction in active length, and an improvement in torque-to-weight ratio.

Since the characteristics of NdFeB alter significantly with increasing temperature, in particular the second quadrant characteristic becomes non-linear, it is prone to irreversible demagnetisation, recoverable only by remagnetisation. However, provided the load-line of the magnetic circuit intersects the linear section of the magnet characteristics, the temperature induced change of flux will be reversible, and since the reversible temperature coefficient is comparable to the temperature coefficient for copper and aluminium it should be possible to accommodate such reversible effects in most applications.

BRUSHED DC MOTORS (1,2,3)

The pm dc motor is widely accepted both for general purpose drives as well as for servo drives. A CAD package has been developed which permits the design optimisation of commutator motors to meet a specified electrical performance within an allowable temperature rise. It combines analytically based synthesis procedures and finite element analysis, which allow the motor to be dimensioned and subsequently analysed, with due account of material non-linearities. Fig. 1 shows the results of a sequence of fe calculations on a motor equipped with ferrite segments. It shows that following a heavy overload the magnets are par-

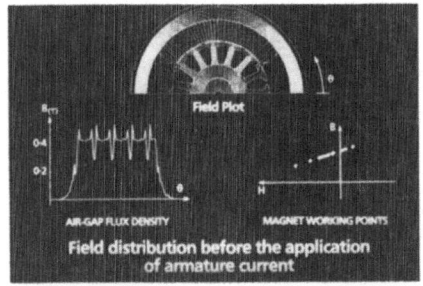

Field distribution before the application of armature current

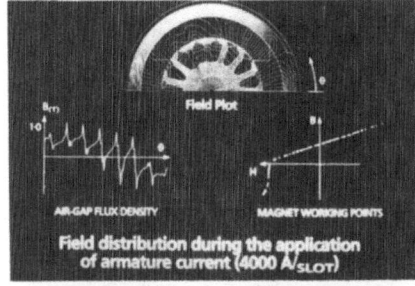

Field distribution during the application of armature current (4000 A/$_{SLOT}$)

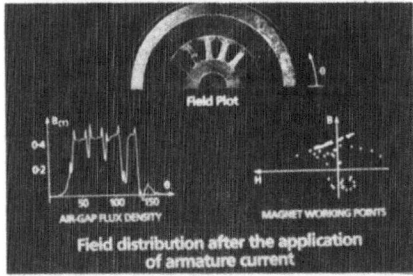

Field distribution after the application of armature current

Fig. 1 - FE Analysis of Ferrite Motor

tially irreversibly demagnetised. The ability to perform such calculations is critical, since the magnets must be sized to withstand a specified overload, typically 5XFL, without demagnetisation. FE analysis makes it possible to account for the fact that the frame is normally longer than the magnets, and, in the case of ferrites, the magnets overhang the armature, and to predict the reactance voltage induced in the coils undergoing commutation.

In order to demonstrate the potential for NdFeB over ferrite, motors have been designed to meet the specification in Table 1. The motors were required to have the same OD and a similar quality of commutation. In the case of NdFeB it was predicted that acceptable commutation could be obtained only by using a moderate grade of magnet, since the higher grades lead to thinner magnets and an excessive reactance voltage. Further, in order to realise a significant advantage, the NdFeB motor had to be designed to a higher tooth flux density. Also, unlike the ferrite motor, it did not feature magnet overhang.

Full-load output power	750W
Full-load speed	2000 rpm
Overload capacity	5 × F.L.
Supply voltage	180V
Insulation/temperature rise	Class F/100°C
Enclosure	Totally enclosed

Table 1

Parameter	Ferrite Motor	NdFeB Motor
D_a	94.3 mm	110 mm
L_a	73.0 mm	50.4 mm
L_m	88.0 mm	50.4 mm
ℓ_m	17.2 mm	5.8 mm
Magnet volume	347 cm³	72 cm³
Copper volume	200 cm³	162 cm³
Lamination volume	260 cm³	290 cm³
Magnetic loading	0.27T	0.47T
Electric loading	13 kA/m	8 kA/m
Magnet energy product (20°C)	26 kJ/m³	216 kJ/m³

Table 2

The results of the study are summarised in Table 2, which highlights a number of interesting features, the most striking of which is the reduction in the thickness and volume of magnet when NdFeB is used. This allows a larger diameter armature, and hence a significant reduction in active length of the motor. However, due to the need to accommodate end shields, commutator/brushgear etc, this does not bring about a proportionate reduction in total weight.

Fig. 2 shows the prototype motors which were constructed to the above designs. Their electrical and thermal performance bear out the theoretical predictions to a high degree of accuracy.

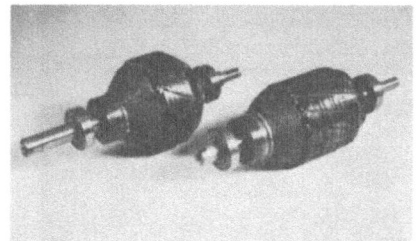

Fig. 2 - Prototype Brushed
      DC Motors

BRUSHLESS MOTORS (4)

The brushless pm motor is arguably the most important category of machine for a wide spectrum of applications. A CAD package has been developed to cater for alternative configurations and for both sinusoidal and trapezoidal induced emf waveforms. Examples of typical output from the programme are shown in Fig. 3 and relate to motors which employ a radially anisotropic NdFeB ring magnet developed by Mullards Southport.

Fig. 4 shows four alternative topologies of a 4-pole motor which use the same ring magnet, and which are supplied from a linear switchmode drive IC. For practical reasons the magnet thickness of 1.5mm is considerably in excess of that required simply for demagnetisation

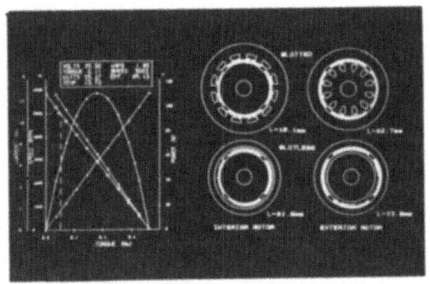 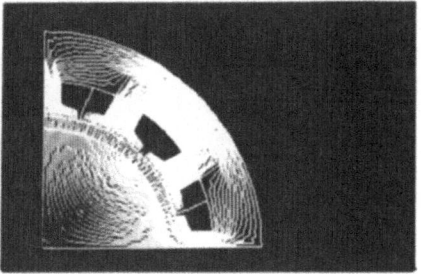

Fig. 3 - CAD of Brushless Motor

protection. The slotted designs both have the same magnetic loading, but since slot area is limited in the exterior rotor type it has a lower electric loading and hence a longer axial length. Although the slotless designs provide more space for the windings, they present a longer effective airgap. Hence their magnetic loading is reduced, and their axial length is longer still. A particular advantage of the slotless topologies is the absence of cogging torque, although this could be largely eliminated in the slotted designs by skew magnetisation, for example. Table 3 compares the measured and predicted emf constants. The errors are much larger than for any of the other machines described in this report, a consequence of the magnets not being fully dense and aligned and incompletely magnetised. However this form of magnet has immense potential for use in machines and actuators.

The adoption of 270V dc as the industry standard for aircraft of the future will allow electrical actuation powers of several kW to be provided at manageable currents, and permit the electromechanical actuation of flight control surfaces which are currently operated by hydraulic actuators. Fig 5 shows a prototype 4-pole brushless dc motor intended as a stabiliser actuator. It has an extremely high torque-to-weight ratio, 6Nm continuous at 6230 rpm on a 270V dc supply, and a short time stall torque capability of 22Nm. Heat dissipation is primarily via the flange to a substantial heat sink. Both stator and rotor are instrumented with thermocouples to validate predictions from the lumped-parameter thermal model of Fig.6, and also finite element analysis being undertaken at the University of Leuven. The measured electrical and thermal performance is in close agreement with design predictions.

The effects of stator slot skew on cogging torque has been examined

TYPE 1 - SLOTTED ARMATURE INTERNAL ROTOR

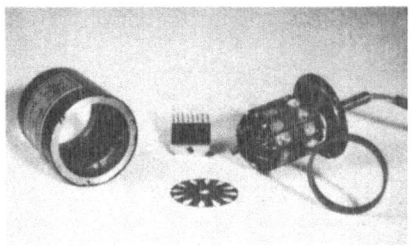

TYPE 2 - SLOTTED ARMATURE EXTERNAL ROTOR

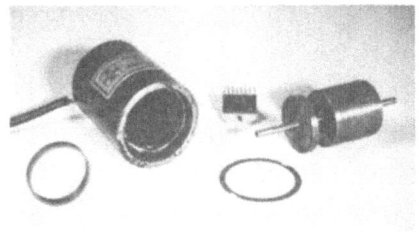

TYPE 3 - SLOTLESS ARMATURE INTERNAL ROTOR

TYPE 4 - SLOTLESS ARMATURE EXTERNAL ROTOR

Fig. 4 - Alternative Topologies of Brushless dc Motor

Motor	Predicted from Lumped Circuit (V/rad/sec)	Predicted from Finite-Element (V/rad/sec)	Measured (V/rad/sec)
Type 1	0.0278	0.0298	0.0255*
Type 2	0.0298	0.0335	0.0277*
Type 3	0.0279	0.0287	0.0230*
Type 4	0.0286	0.0277	0.0280**

* Uses magnets magnetised in solid jig.
** Uses magnets magnetised in laminated jig.

Table 3 - Measured and Predicted emf Constants

both theoretically and experimentally, using 3-1kW brushless dc motors designed with different degrees of skew. It has been shown that in order to achieve the required trapezoidal induced emf waveform the angle of skew must be less than the slot pitch necessary to eliminate cogging. In order to obtain lower levels of torque ripple, which is an important consideration in servodrives at low speed, for feed and robot applications for example, motors which produce a sinusoidal emf waveform for operation with a high accuracy encoder or resolver and a sine-wave amplifier, have been designed.

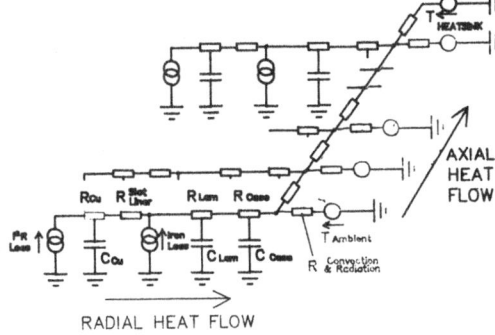

RADIAL HEAT FLOW

Fig. 5 - Prototype Aerospace
        Brushless dc motor

Fig. 6 - Lumped-Parameter Thermal
        Network

HYBRID STEPPERS (5,6)

Despite the competition from servo systems, stepper motors remain the preferred drive for many applications, particularly where they may be operated on open loop.

The potential for NdFeB has been demonstrated by design studies on an industrial range of hybrid steppers, up to 3kW rating, most of which currently use alnico. An improved design procedure, which accounts for the change of magnet working point with excitation as well as for saturation, has been developed. It enables the magnet to be dimensioned such that maximum torque per amp of excitation is developed from a given motor from no-load to full-load.

Fig. 7 compares two rotors which yield identical holding torque characteristics and give max. torque/amp over the complete excitation range. It will be seen that the NdFeB design not only requires a much lower volume of magnet than the equivalent alnico design but allows a modified rotor construction and a reduction in inertia.

Fig. 7 - Hybrid Stepper Motor Rotors

LINE-START SYNCHRONOUS MOTORS (7,8,9)

Permanent magnet excited synchronous motors offer a high power density, high efficiency, high power factor alternative to the induction motor. For line-starting the rotor is equipped with a squirrel cage, below which the permanent magnets are embedded. Fig. 8 shows two practical forms of rotor lamination which have been employed on prototype motors.

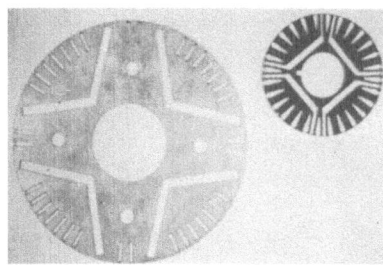
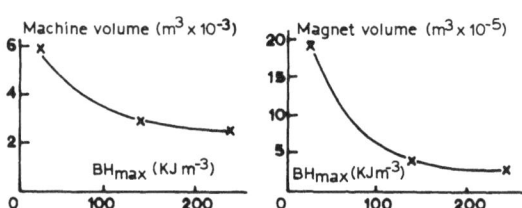

Fig. 8 - Rotor Laminations for
Line-Start Synchronous
Motor

Fig. 9 - Influence of Magnet Energy
Product (1800VA, 4-Pole,
50Hz)

An improved method for analysing the steady-state synchronous operation, in which the terminal voltage and current are included as subsidiary conditions in a finite element formulation, has been developed and validated against tests. In addition a design synthesis strategy has been established, and the influence of magnet material studied. Fig. 9 indicates how the volume of magnet and the total active volume of a motor, having a specified VA rating and full-load power factor, reduce with increasing magnet energy product.

GENERATORS (10,11)

Figs. 10 and 11 illustrate two different applications of permanent magnet generators. Fig. 10 is a safety lamp unit in which the generator is driven by an air-turbine. The original unit was rated at 55W, 12V when equipped with a cast alnico rotor. Some years ago the unit was uprated to 250W, 24V within the same lamination and stack length, by the use of a rotor equipped with polymer bonded SmCo magnets and by redesigning the inlet nozzles to the turbine. The generator has recently been redesigned around sintered NdFeB, to give exactly the same torque-speed characteristic, and at the same time to improve mechanical integrity.

Fig. 10 compares the new rotor with both the current and original models.

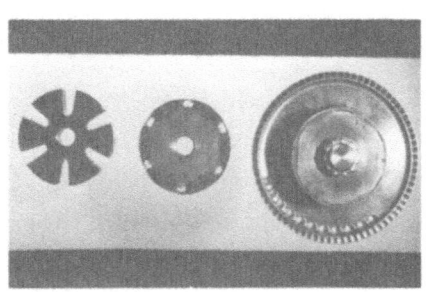

Fig. 10 - Air-Turbine Driven
         Generator

Fig. 11 illustrates the use of a pm generator as a rheostatic braking system for high-speed trains, the braking requirements of which are currently met by friction brakes. The rheostatic braking system is inherently anti-skid and would be virtually maintenance free. By closing the resistor switches at pre-selected speeds, leading to a short-circuit of the stator terminals at the lowest speed, a near constant braking effort is obtained down to low speed, when a simple friction brake would be blended in to maintain the braking effort. The intensity of braking can be varied by altering the circuit inductance. The 8-pole radial-field design shown in Fig. 11 requires 39kg of sintered NdFeB in order to develop a maximum torque of 4.8kNm without the temperature rise of the winding exceeding 155°C following three full service stops from 250km/hr separated by 10 mins intervals.

WORKHOLDING ASSEMBLIES (12)

Magnetic workholding devices are used extensively for gripping and lifting in a wide range of applications. To date, most designs have been

704

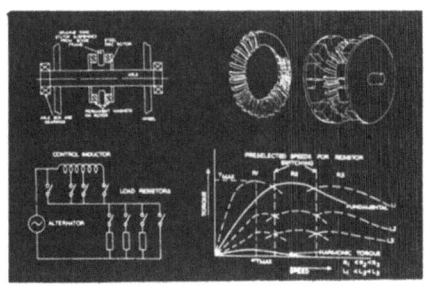 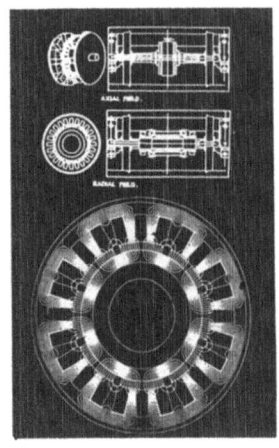

Fig. 11 - Rheostatic Braking of Railway Vehicle

based around alnico or ferrite. The usual requirement is for a high
attractive force when the assembly is in intimate contact with a
workpiece. The prime design goal, therefore, is a high flux density at
the contacting surfaces. However, there may be other important
considerations. The ability to exert a high attractive force against very
thin ferrous materials, for example, for which the pole-pitch is critical,
or in the case of switchable devices the ease of switching and the absence
of significant external field in the 'off' position.

A new range of such assemblies has been developed, and is shown in
Fig. 12. Sub-branded 'NEO-HOLD' the range comprises 'pot' magnets,
switchable 'holdfasts' and magnetic 'bases', which all employ bonded NdFeB
magnets, and a fine-pole 'chuck' which uses sintered NdFeB. Table 4 shows
that the magnet cost per unit of attractive force compares most favourably
with that for existing devices.

Product		Cost of attractive force ($/kg)	
		Alnico	NdFeB
Pot Magnets		0.28 - 0.06	0.04
Holdfasts		0.04 - 0.07	0.03
Magnetic base	Small	0.07	0.03
	Large	0.03	

Fig. 12 - Workholding Assemblies
Based on NdFeB Magnets   Table 4

FINITE ELEMENT FIELD ANALYSIS

Reliable design and analysis holds the key to the successful exploitation of NdFeB. Indeed finite element analysis has featured strongly in all the above studies. A 2-D finite element analysis package has been developed for implementation on an IBM PC and is now being made available to designers of voice-coil actuators and the like. It can handle planar or axisymmetric Poissonian field problems, formulated in terms of either the scalar or vector magnetic potential. It features highly interactive graphics, and is operated via menu selected commands. Comprehensive post-processing permits the calculation of global parameters such as force, inductance, flux-linkages etc.

Fig. 13 shows some typical examples of VDU images, related to a line-start synchronous motor.

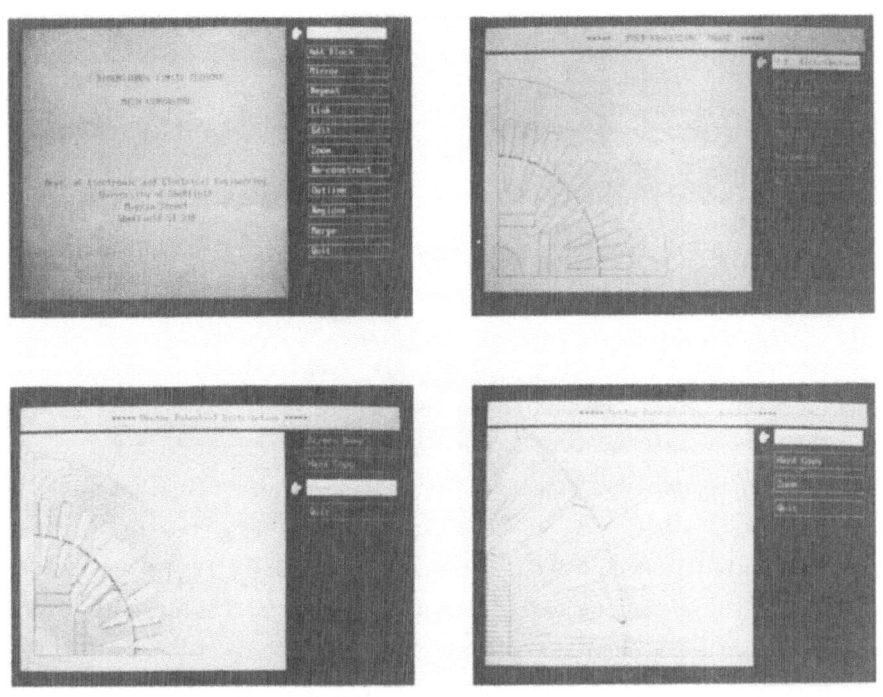

Fig. 13 - Typical Graphical Displays from 2-D Finite Element Package

TEMPERATURE CONTROLLED PERMEAMETER

Since current grades of NdFeB magnets exhibit such high temperature coefficients of remanence and coercivity which cause the normal second quadrant characteristic to become non-linear at elevated temperatures, and because the magnet working point must be limited at all times to the linear portion of the characteristic in order to avoid irreversible demagnetisation, it is essential that designers base their calculations on accurate data. To facilitate the measurement of magnet characteristics within the temperature range -20°C to +200°C a temperature controlled fixture has been constructed and incorporated in a computer controlled permeameter, as shown in Fig. 14. The fixture comprises pole-pieces attached via a thermal barrier to the poles of the electromagnet, thermostatically controlled cartridge heaters, and gas cooling ducts. B and H are measured by sensing coils and integrating fluxmeters.

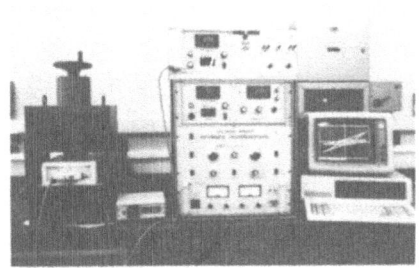 

Fig. 14 - Permeameter with Temperature Controlled Fixture

CONCLUSIONS (13)

Although still lacking large-scale application, especially in technologically more demanding areas where the strong temperature dependence of properties is a potentially serious drawback, NdFeB magnets are beginning to have an impact on the design of electrical machines and other permanent magnet excited devices. The work reported in this paper has demonstrated some of the potential advantages to be gained from employing NdFeB in various categories of machine. It has also highlighted the need for advanced design and analysis tools.

ACKNOWLEDGEMENTS

The authors acknowledge the support of the CEC within the CEAM project of the Stimulation Programme, as well as the contribution made by GEC Electromotors Ltd, Dowty Electrics Ltd, Wolf Safety Lamp Co Ltd, British Rail Technical Centre, Mullards Southport, Neill Tools Plc, and colleagues at the University of Sheffield.

REFERENCES

1. Williams, I.J., Howe, D., Staton, D. and Birch, T.S.,:'The design of a range of permanent magnet dc motors', IEE Conf. Pub. 254, Elec Machines & Design & App. 1985, 270-274.

2. Williams, I.J., Birch, T.S., Howe, D., and Staton, D.: 'Computer-aided design procedures for permanent magnet dc motors', Proc. Controls/ Motors/Drives 85 Conf. 1985,29-35.

3. Staton, D.A., Birch, T.S., and Howe, D.: 'Design optimisation of permanent magnet dc motors', Proc. INCEMADS 86, Int. Conf. on Electrical Machines and Drive Systems, 1986.

4. Mitchell,J.L., Birch, T.S., and Howe, D.:'Design and analysis of brushless dc motors', Proc. INCEMADS 86, Int. Conf. on Electrical Machines & Drive Systems, 1986.

5. Jenkins, M.K., Birch, T.S., and Howe, D.:'The influence of magnet mmf on static torque production in hybrid stepper motors', ibid, 1986.

6. Jenkins, M.K., Birch, T.S. and Howe, D.:'Static torque production in hybrid stepper motors: the influence of saturation and magnet mmf', 3rd Int. Conf. on 'Electrical Machines and Drives', IEE, London, Oct. 1987.

7. Chalmers, B.J., Devgan, S.K., Howe, D., and Low, W.F.:'Synchronous performance predictions for high-field permanent magnet synchronous motors', Proc. ICEM '86, 1986, 1067-1070.

8. Howe, D., and Low, W.F.: 'Methods for predicting steady-state operation of pm synchronous motors', Proceedings UPEC, April 1987.

9. Howe, D., and Low, W.F.: 'The finite element method for the direct simulation of the steady-state performance of a permanent magnet line-start synchronous motor', Proc. IEEE Intermag Conference, Tokyo, April 1987 and IEEE Transactions on Magnetics, Sept. 1987.

10. Tan, G., Howe, D., and Birch, T.S.:'Design and performance prediction of a permanent magnet air-turbo generator', Proc. 2nd Int. Conf. on Small & Special Elec. Machines, I.E.E. Conf. Publication, 202, 125-128.

11. Howe, D., Matthews, D.M.H., Birch, T.S., Jablonski, A.P. and Rash, N.M.: 'An electrodynamic braking system for railway vehicles based on an axle mounted permanent magnet alternator', Paper W2.3 presented at the 9th Int. Workshop on Rare-Earth Magnets, Bad Soden, Aug. 1987, 85-92.

12. Taylor, D., Manley, D.J., and Howe, D.:'Workholding assemblies based on NdFeB permanent magnets', 3rd Int. Conf. on 'The impact of NdFeB materials on permanent magnet users, producers and raw material suppliers', San Diego, California, Oct. 1987.

13. Howe, D., Birch, T.S., and Gray, P.: 'The potential for NdFeB in electrical machines', invited paper W2.1 presented at the 9th Int. Workshop on Rare-Earth Magnets, Bad Soden, Aug. 1987, 65-84.

PERMANENT-MAGNET A.C. AND D.C. MACHINES

B.J. Chalmers, E. Spooner, A.M. Sitzia, K.M. Richardson

Department of Electrical Engineering and Electronics

University of Manchester Institute of Science and Technology

P.O. Box 88, Manchester M60 1QD, UK

ABSTRACT

Design analysis procedures have been developed and validated for permanent-magnet synchronous motors. The potential for exploitation of the properties of Nd-Fe-B is demonstrated by design, construction and evaluation of alternative SmCo$_5$ and Nd-Fe-B rotors in a 7.5 kW machine, showing the performance advantages of the latter material.

The high-field properties of Nd-Fe-B raises the feasibility of slotless-stator brushless d.c. machines. Design approaches to the effective utilisation of magnets in such machines are developed and verified by construction and testing of an experimental machine. This is shown to yield an advantageous reduction of torque pulsation.

Problems of large permanent-magnet machines have been approached by study of new methods of in-situ magnetisation. Results obtained from an experimental magnetiser are reported, with a view to their application in production of a prototype 200kW motor.

PERMANENT MAGNET SYNCHRONOUS MOTORS

The increasing availability of high-field, permanent-magnet materials has created opportunities for the design of electrical machines with improved performance. The benefit may accrue in the form of either reduced size for a given output or improved performance for a given size and power. Designs have been produced for synchronous motors of interior type, in which magnets are located in internal slots within the rotor lamination. A cage winding is fitted for induction starting and damping. The magnet slots are usually bridged to form a single-piece lamination of adequate mechanical strength to withstand rotational forces. The majority of designs have rotors which are symmetrical about the polar axis, giving the same performance for both directions of rotation. This paper presents the results of a practical assessment of a design of this type.

Experimental Machines

Two different rotors have been tested within the same stator, with the following design details.

Stator   This is a standard 7.5 kW, 3-phase, 415 V, star-connected, 4-pole induction motor stator in a 132 S1 frame with 125 mm corelength and 165 mm stator bore.

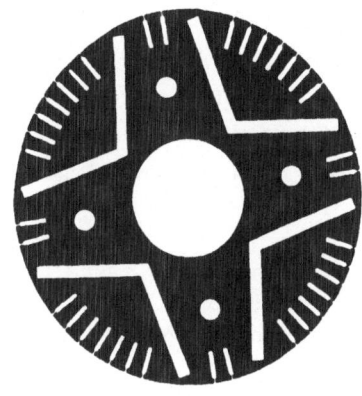

Figure 1.   Rotor lamination design

Rotors   The two rotors have the same lamination design, shown in Figure 1.   Each rotor is skewed by one slot pitch.   The magnets are accommodated in V-shaped internal slots, with alternate orientation to produce a four-pole field.   The effective magnet span is 120° (elec.) at the rotor surface.   The cage winding slots are pitched as for 60 per circle, with 6 slots omitted per pole as shown in Figure 1.   Magnet slot depth is 6 mm and lamination bridge depth 1.5 mm.

Rotor 1 has samarium cobalt ($SmCo_5$) magnets 5 mm thick and 45 mm wide, assembled in 12 mm lengths.   Magnet properties are B_r = 0.85 T and Hc = 600 kA/m, with 1.75 kg total magnet weight.

Rotor 2 has neodymium iron boron (Nd-Fe-B) magnets 5.4 mm thick and otherwise as for Rotor 1, with a weight of 1.73 kg.   Nominal magnet properties are B_r = 1.05 T and H_c = 765 kA/m.

Results

Rotor 1

The value of excitation emf E for rotor 1, measured on open circuit at 1500 rev/min, was 147.5 V per phase, or 255.5 V line.   Compared with the design objective of near equality with the nominal line voltage of 415 V, the reduced value of E was mainly attributed to the effect of the air space in the magnet slots.   To achieve operation at satisfactory values of power factor, the line voltage was correspondingly restricted and motoring tests were conducted at line voltages of 210, 300 and 400 V.

The procedure used for evaluation of the conventional d-q axis parameters from tests measurements was based on the assumption of constant values of E and stator resistance R. As for previous examples of this type of rotor, X_d was found to be substantially constant at 17.8 Ω, while X_q exhibited some saturation at high values of I_q. The unsaturated value of X_q, applicable to values of I_q in the normal load range, was 66.2 Ω.

Figure 2 shows the measured input current loci at the three supply voltages, illustrating varying degrees of over- and under-excitation. Also plotted are computed curves based on the above-mentioned constant values of machine parameters. Good correlation is evident at each supply voltage.

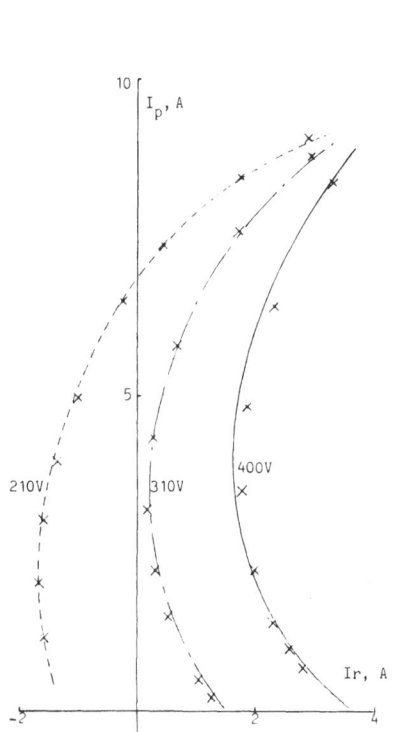 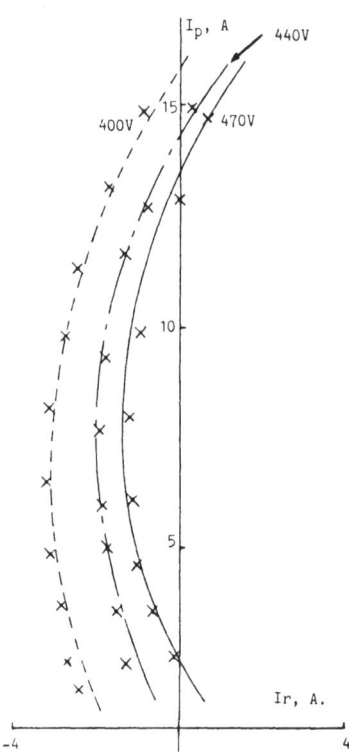

Fig.2. Current loci of SmCo5 rotor. Measured points and computed curves.

Fig.3. Current loci of Nd-Fe-B rotor. Measured points and computed curves.

Rotor 2

The Nd-Fe-B rotor, with its stronger and marginally thicker magnets and reduced clearance in the magnet slots, gave a significantly higher E of 263 V per phase (456 V line), enabling good performance to be achieved at near unity power factor with higher supply voltages than rotor 1. Tests were performed with supply voltages of 400, 440 and 470 V. From tests at 470 V, X_d was evaluated at 15.3 Ω with X_q unchanged at 66.2 Ω. Figure 3 gives the current loci for the three supply voltages. Once again, there is excellent agreement between measured results and those computed using constant parameters. The reduced value of X_d in this case is associated with increased magnet flux.

Overall Performance

The initial design objective was to achieve the same 7.5 kW output as the original induction motor in the same frame, corresponding to 47.8 Nm at 1500 rev/min, with improved performance. Rotor 1 failed to achieved this target, its maximum output being 7.4 kW, 46.9 Nm, with 400 V supply. Rotor 2 achieved the objective at each of the tested voltages, with excellent performance as tabulated below.

TABLE 1   Performance of Rotor 2

V, line	Measured Performance at 7.5 kW, 47.8 Nm				
	Current A	Input kW	PF	η%	T_{max}, Nm
470	10.0	8.09	0.994	92.7	64.4
440	10.7	8.08	0.991	92.8	60.9
400	12	8.23	0.990	91.1	54.1

Rotor 2 met the design objective, with higher power factor and efficiency relative to a standard induction motor. The latter has 13.5 A current, 86.3 % efficiency and 0.89 power factor with 415 V supply.

BRUSHLESS D.C. MOTOR WITH SLOTLESS STATOR

D.c. machine designs with permanent magnet excitation on the rotor offer the opportunity of eliminating mechanical commutation and rotor copper losses. The armature winding is arranged as several (usually three) stator phases, excited in sequence by bipolar switching of a d.c. supply. Slotted-stator versions of such a machine have been developed and applied to a variety of drive systems. The newer, high remanence and high coercivity neodymium–iron based materials can produce reasonably high flux densities even into large air–gaps. This suggests the feasibility of slotless–stator brushless d.c. machines.

Eliminating stator slotting removes the permeance component of torque ripple and so the need to skew the armature. This component of torque ripple alone has been estimated to be 10% peak–to–peak of mean torque in an unskewed slotted machine.

Experimental Machine Design

This study began with the design of an experimental machine for laboratory evaluation, based on Nd–Fe–B magnets.

The initial design procedure establishes the main parameters of the machine geometry. Having chosen an outside diameter (165 mm), a core length (50 mm), numbers of phases (3) and poles (4) and a magnet pole arc (120° electrical), the magnet and winding thicknesses are determined to give a desired torque from a chosen electric loading or current density.

The pole arc (120°) is the minimum possible without severe torque reduction and is chosen to minimise interpolar leakage and magnet volume.

The envisaged mode of operation of the machine is sequential bipolar switching of two stator phases out of three. The rotor follows the stator field synchronously and the angle between their respective mmf axes is defined as the torque angle (δ). The rotor moves through 60° (electrical) in any commutation cycle, from $\delta = 120°$ to $\delta = 60°$. Maximum torque is developed when $\delta = 90°$.

The magnets are Nd–Fe–B (27H) with a remanent flux density of 1.05 T and a coercivity of −765 kA/m.

This design procedure, based on a simple Lorentz force calculation at the mean winding radius, yields a winding thickness of 10.5 mm and a magnet thickness of 6 mm.

The stator core depth is 11.5 mm for an estimated maximum flux density of 1.3 T. There are 170 coils per phase per pole pair. The machine cross–section is shown in Figure 4. Mild steel pole–caps were used to allow the use of available magnet blocks without destroying the rectangular distribution of radial air–gap flux density. With a relatively thick air–gap winding, the pole caps do not lead to a high winding inductance as would be the case with a conventional slotted stator and a small air–gap.

Numerical Analysis of Experimental Machine

Torques and flux densities are computed for various rotor positions by finite–element analysis, using about 4100 elements. Two methods of torque calculation

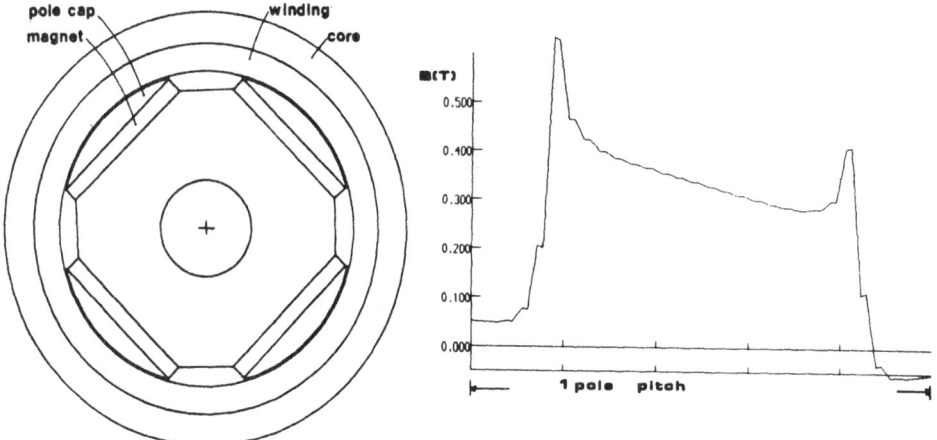

Fig.4. Cross-section of slotless stator machine.

Fig.5. Variation of radial flux density in air-gap at $\delta = 90^\circ$.

are employed at rotor positions corresponding to the beginning, middle and end of a commutation cycle, i.e. at $\delta = 120^\circ$, 90° and 60°. The first method is the integration of Maxwell stress expressions around a circular contour in the air-gap just outside the rotor. The other is the integration of the Lorentz force formula ($\underline{J}$ x $\underline{B}$) in each conductor region carrying current.

The torque on the rotor has its reaction directly on the stator winding since all the flux crossing the air-gap must necessarily pass through the winding. The computed torques are given in Table 2.

TABLE 2 Computed torques

	Torque (Nm)			
	$\delta=120^\circ$	$\delta=90^\circ$	$\delta=60^\circ$	mean
Maxwell Stress	4.2	5.0	4.1	4.6
Lorenz Force	4.1	4.9	4.0	4.5

Agreement between the two methods is good (within 3%).

The phase current is assumed to be constant throughout the commutation cycle at a value of 4.25 A which corresponds to an mmf of 720 A per phase per pole or a modest current density of 2A/mm^2.

The mean radial flux density at an average radius through the winding is about 0.23 T while that at the rotor pole face is about 0.38 T. The variation in air-gap flux density is shown in Figure 5 for δ = 90°.

The averages of computed torque values at δ = 60° and δ = 90° are shown as points A and B on the curves of Figure 6 for comparison with the measured torques.

Static Tests on Experimental Machine

The stator winding was wound on a former and glued to the inner bore of a plain cylindrical laminated core.

Static torque is applied to the rotor, with various values of stator current, and the rotor displacement observed. Two phases out of three are conducting to maintain the correspondence with normal dynamic operation.

Figure 6 shows the variation of torque with rotor position at various current levels. The smoothness of the curves confirms the absence of torque fluctuations which would be present in a machine with an unskewed slotted stator.

The agreement between computed and measured torques at 4.25 A and δ = 60° and 90° is good (within 10%). Differences between the computational model and reality include end effects, variations in geometry and material properties, and numerical error.

The variation of torque with phase current is approximately linear, reflecting the large effective air-gap and the relatively low armature reaction.

Analytical Design Optimisation

The main scope for design improvement lies in increasing the rotor pole arc to increase torque and reduce torque ripple, while ensuring that interpolar flux leakage is not excessive.

A revised design for 16.5 Nm peak torque, with unchanged overall dimensions, is based on an interpolar leakage factor (defined as the ratio of total gap length to interpolar arc length) of 1/1.5. This compares with 1/2.9 for the experimental design, indicating a large margin for increasing the pole arc.

The current density is increased to a more realistic value of 6 A/mm^2., while saturation in the core can still be avoided at working currents.

Various designs are feasible under these criteria, so one is selected which has reasonably small magnet and winding thicknesses. The chosen design has a winding thickness of 10 mm and magnet thickness of 5.6 mm. The rotor pole arc is now 150° (electrical). The stator core depth is 12.5 mm.

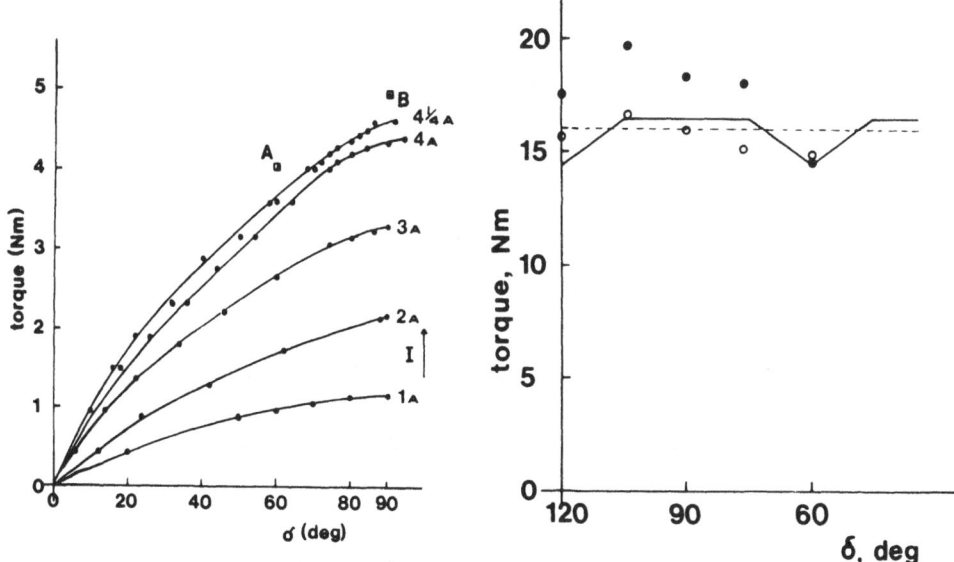

Fig.6. Variation of static torque with rotor position and current.

Fig.7. Variation of static torque with rotor position.

The fault current necessary to demagnetise the magnets is estimated to be six times rated current (corresponding to 36 A/mm^2), demonstrating substantial immunity to demagnetisation by overload current.

Performance Prediction of Optimised Machine

Finite–element field analysis was again employed to obtain torque and flux density distributions in the improved machine design at five different rotor positions.

The theoretical torque variation with rotor position, deduced from simple analytical application of the Lorentz force formula, is trapezoidal, shown by the solid line in Figure 7. Computed torques, by Maxwell stress and Lorentz force integrations, are also shown in Figure 7, for the five rotor positions.

The differences between the two sets of points are caused by numerical error, probably due to the rapidly–varying nature of the field in regions of the air–gap where the Maxwell stress integration takes place.

The mean computed torque, by the Lorentz method, which is considered to be the more accurate in this case, is about 15.7 Nm and the torque ripple is about 12% peak–to–peak of the mean torque. The previous design was subject to a torque ripple of 20% peak–to–peak over a commutation cycle. This demonstrates the benefit of a larger pole arc, even with reduced magnet and winding thicknesses.

MAGNETISATION PROCEDURES FOR LARGE MACHINES

The introduction of Nd–Fe–B based permanent magnet materials creates the opportunity to design and build electrical machines of any size excited either completely or partly by permanent magnets without compromising the armature design. Previously available materials could generate airgap flux densities of up to around 0.7 T but for large machines the optimum flux density for an efficient and compact armature is closer to 1.0 T. Currently available grades of Nd–Fe–B with remanent flux density exceeding 1.2 T should be capable of providing the required airgap flux density if the overall magnetic circuit is suitably designed. Future developments of the material toward a remanence of 1.5 T will make it a straightforward matter to achieve an airgap flux density of 1.0 T using a compact and economical field system.

Although these materials offer new opportunities to design large machines, the potential problems of assembly must be addressed first. Two severe problems are expected to be encountered in the assembly of large permanent magnet machines: rotor threading, a procedure which will become increasingly problematical as the sizes of magnetised components to be handled increases, and the assembly of large numbers of adjacent magnets to build the poles. These problems will be greatly eased if the magnets can be energised after the machine has been assembled.

The small machines which presently use permanent magnets can be designed to withstand faults without demagnetisation. In the large permanent magnet machines which may be contemplated since the advent of Nd–Fe–B, the greater pole pitch and the generally higher fault levels to which larger machines may be subjected make occasional demagnetisation likely. A simple means of remagnetising without dismantling and preferably without removing the machine from site may be regarded as a prerequisite to building large permanent magnet machines.

Two concepts have been examined in connection with a project which is aimed at converting a 200kW, 500 rpm d.c. traction motor to full or partial permanent magnet excitation. First, a set of measurements has been made of the field required to magnetise several grades of Nd–Fe–B at elevated temperature with a view to finding an optimum temperature for carrying out in–situ magnetisation. Secondly, an iron–cored magnetiser has been constructed and operated at flux densities exceeding 3 T in the working region using a process of dynamic flux confinement which will be employed in magnetising the large machine. Although the magnetisation methods under study are intended initially for the large d.c. machine, they should be readily applicable to other types.

The tests at elevated temperature have shown that, although raising the temperature does reduce the field for which some magnetisation is obtained, full magnetisation actually requires a higher field. It is usual to magnetise individual small magnets in air using a flux density of about 3.5 T, although the additional magnetisation gained by exceeding 2 T is rather small. A magnetiser capable of delivering between 2.0 and 2.3 T within an electrical machine should be entirely adequate.

The experimental magnetiser constructed for carrying out the measurement of magnetising force requirement is illustrated in Figure 8. In addition to the details shown it incorporates a heating and temperature measuring system housed within the screens. The experience gained in the design of the electromagnetic parts of the equipment is proving of value in developing a means of magnetising the magnets installed within the large d.c. machine.

Fig.8. View of experimental magnetiser test rig.

Additional tests with the experimental magnetiser have simulated the flux shunting and eddy current screening effects expected to occur within the solid iron sections of a machine magnetic circuit. Figure 9 shows the dummy steel shell used for these tests. Satisfactory magnetisation was found to be possible despite the flux shunting and screening effects.

Fig.9. Experimental arrangement of magnet surrounded by steel plate.

Several options exist for magnetisation within a machine. The flux plot shown in Figure 10 represents the situation where the entire set of magnets in the machine is being energised simultaneously. In order to create the necessary field at the magnets a current of 150000 Amp/pole is required. Alternatively, bands of magnets located at different positions on the poles (see Figure 11) may be magnetised separately. The pulse energy requirement should be much less in this case.

Currently, installation of magnets is proceeding. The construction of a major back-to-back rig for testing of the 200kW machine is nearing completion.

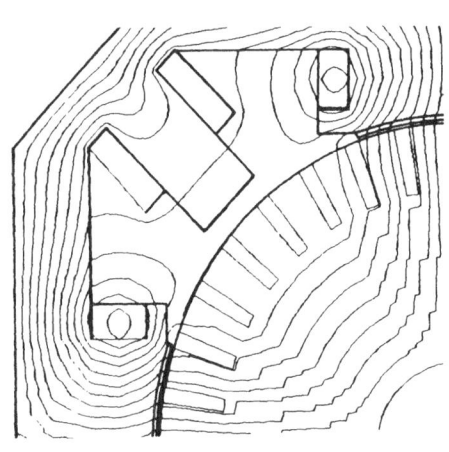

Fig.10. Finite element flux plots of 200kW motor during magnetisation.

Fig.11. Magnets on pole face of 200kW motor.

REFERENCES

Brown, J.E. and Chalmers, B.J. 1988 Locus diagrams and performance characteristics of various types of synchronous motor. Intern. Conf. on Electrical Machines, Pisa.

Chalmers, B.J., Devgan, S.K., Howe, D. and Low, W.F. 1986. Synchronous performance prediction for high−field permanent−magnet synchronous motors. Intern. Conf. on Electrical Machines, Munich.

Chalmers, B.J. and Devgan, S.K. 1988. Assessment of high−field permanent−magnet rotors by measurement of flux distribution in air. Intern. Conf. on Electrical Machines, Pisa.

Chalmers, B.J. and Devgan, S.K. 1988. Comparative performances of 7.5 kW permanent−magnet synchronous motors with SmCo$_5$ and Nd−Fe−B magnets. Intern. Conf. on Electrical Machines, Pisa.

Chalmers, B.J., Hamed, S.A. and Baines, G.D. 1985. Parameters and performance of a high−field permanent−magnet synchronous motor for variable−frequency operation. I.E.E Proc. 132, Pt.B, No.3, 117−124.

Hanitsch, R., Sitzia, A.M., Hemead, E. and Chalmers, B.J. 1988. Improved brushless motors with cup−type winding and Nd−Fe−B magnets. Intern. Conf. on Electrical Machines, Pisa.

Howe, D., Rash, N.M. and Spooner, E. 1985. A permanent−magnet alternator for use as an electro−dynamic brake. I.E.E Conf. on Electrical Machine Design and Application, London.

Richardson, K.M. and Spooner, E. 1987. Magnetisation procedures for Nd−Fe−B magnets in large electrical machines. I.E.E. Intern. Conf. on Electrical Machines and Drives, London.

Sitzia, A.M. and Chalmers, B.J. 1987. Brushless d.c. motors with slotless stator. Electrical Drive Symposium, Cagliari.

Sitzia, A.M. and Chalmers, B.J. 1987. Electromagnetic design of brushless d.c. motor with slotless stator. I.E.E. Intern. Conf. on Electrical Machines and Drives, London.

Spooner, E. and Chalmers, B.J. 1988. Toroidally−wound, axial−flux, permanent−magnet brushless d.c. motors. Intern. Conf. on Electrical Machines, Pisa.

Spooner, E. and Richardson, K.M. 1987. The properties of permanent−magnet materials for the excitation of large electrical machines, Electrical Drive Symposium, Cagliari.

COMPUTER AIDED DESIGN OF ND-FE-B PERMANENT MAGNET MOTORS.

R.Belmans   D.Verdyck   W.Geysen
Laboratory for Electrical Machines and Drives
K.U.Leuven
Kard.Mercierlaan 94
B-3030 LEUVEN-HEVERLEE
Belgium

ABSTRACT:

This paper deals with the various aspects encountered in the design of permanent magnet motors using Nd-Fe-B.

First the magnetic behaviour is analysed, not only accounting for the torque characteristic but also for the Fourier spectrum of the flux density distribution, which is extremely important for the prediction of the audible noise of the machine.

Second, the temperature distribution is calculated. If the motor is symmetrical with respect to the axial direction, a two dimensional approach is sufficient. However if a firm asymmetrical behaviour is found, e.g. due to the installation of a heat sink at one of the machine ends, a full three dimensional analysis has to be carried out.

Third, the mechanical analysis is very important with respect to the knowledge of the natural frequencies, which together with the electromagnetic forces obtained from the flux density distribution, form the basis of the audible noise prediction.

Therefore we can conclude that the combined Finite Element - CAD approach of the phenomena (thermal, magnetical and mechanical) in permanent magnet machines offers the possibility to fully describe and analyse the behaviour of these motors and to design them in such a way that full advantage is taken from the excellent characteristics of the Nd-Fe-B permanent magnets.

INTRODUCTION

When using a new and very promising material as Neodymium-Iron-Boron, one has to be very carefull in order to fully exploit the possibilities of the material, accounting for the problems due to its properties.

The design of electrical machines using Nd-Fe-B is not straightforward due to the interaction between the three major technological problems appearing in this problem, being magnetical, thermal and mechanical.

The mathematical technique, which is best fitted for the analysis of all these problems, is the finite element method, which has to be coupled with Computer Aided Design techniques in order to supply to the computer the input data (motor geometry, material properties, grid generation, etc.) and to allow an easy, graphical representation of the results (Belmans et al.).

The magnetical design of the machine has to take into account the calculation of the flux density distribution in the machine. However, the quantities derived from the field as e.g. inductances, torque and iron losses are far much more important. One of the very special problems is the audible noise, which is generated by the magnetic forces in the machine airgap. These forces can also be calculated from the analysis of the field density distribution in the machine airgap.

The mechanical design is related to the forces acting on the rotor and on the magnets. However these are not critical in small machines as discussed within CEAM. In order to predict the audible noise behaviour, the natural frequencies of the stator assembly have to be calculated.

Due to the temperature behaviour of the magnetic properties of Nd-Fe-B, the characteristics of the motor will depend on the temperature of the magnets. As the variation of these properties is different from classical ferrite magnets, the stability behaviour of the motors has to be considered carefully as already indicated in literature (J.Koch et.al., 1986). It is found that the stability of Nd-Fe-B based machines is critical in the high temperature regio. However in classical ferrite based designs, problems will occur at low temperature. This is e.g. a very important conclusion when dealing with motors for the automobile industry.

However the aim of the numerical analysis is not only to analyse the overall behaviour of the machine due to temperature variation, but also to actually calculate the temperature distribution in various machine parts. This is very important for machines using Nd-Fe-B as due to the higher power density in the machine, the temperature of the insulation material may become critical.

In order to illustrate this approach, we will discuss some examples which were carried out in the CEAM programme.

MAGNETICAL ANALYSIS

The magnetical analysis and optimisation of permanent magnet machines using classical ferrite or Sm-Co magnets was already performed for multipole assemblies with burried magnets. These machines are very promising for inverter fed applications (A.Hameed et al., 1984). This analysis was performed in close collaboration between the University of Liverpool (Prof.K.J.BINNS) and our laboratory.

These calculations were repeated using Nd-Fe-B permanent magnets. From this analysis some important conclusions may be drawn with respect to the material.

On Figure 1 the magnetisation curves of Sm-Co and Nd-Fe-B are given for two temperatures (20°C and 120°C). These curves are used to calculated the field distribution in the machine. On Figure 2 we see the field at no-load using Nd-Fe-B at 20°C.

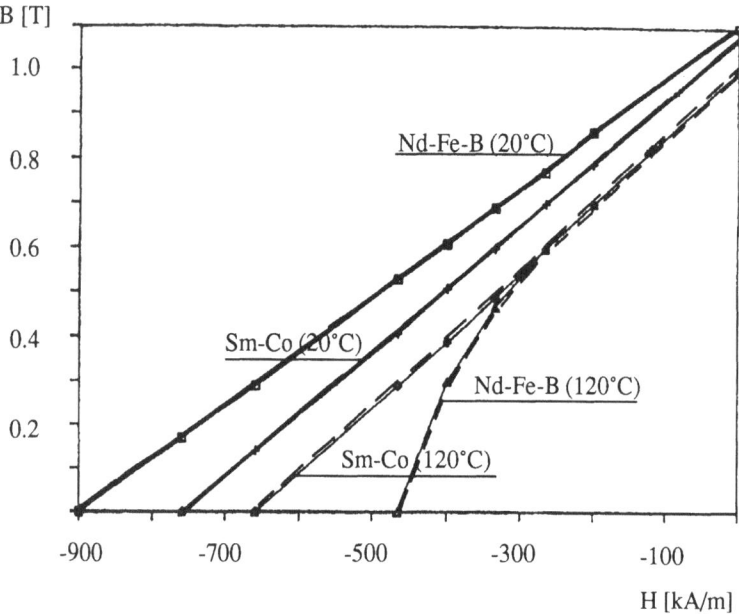

Figure 1. Magnetisation curve of Sm-Co and Nd-Fe-B at 20 °C and 120C.

From the field, the torque may be calculated. On Figure 3, the torque is shown for the various problems. We can see that Nd-Fe-B offers distinct advantages at room temperature. However, Sm-Co matches its performances at 80°C.

This example clearly indicates that not only the room temperature characteristics of the material have to be accounted for, but even more its behaviour with respect to temperature which depends on the load of the machine.

In order to see whether or not a high audible noise level has to be expected, one has to calculate the Fourier spectrum of the field density distribution and compare it with the harmonic content of an equivalent induction motor (Belmans et al., 1987)

The harmonic components are due to three phenomena appearing simultaneously in the

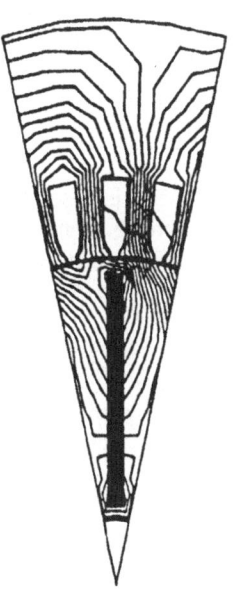

Figure 2. Field at no-load using Nd-Fe-B at 20 ℃.

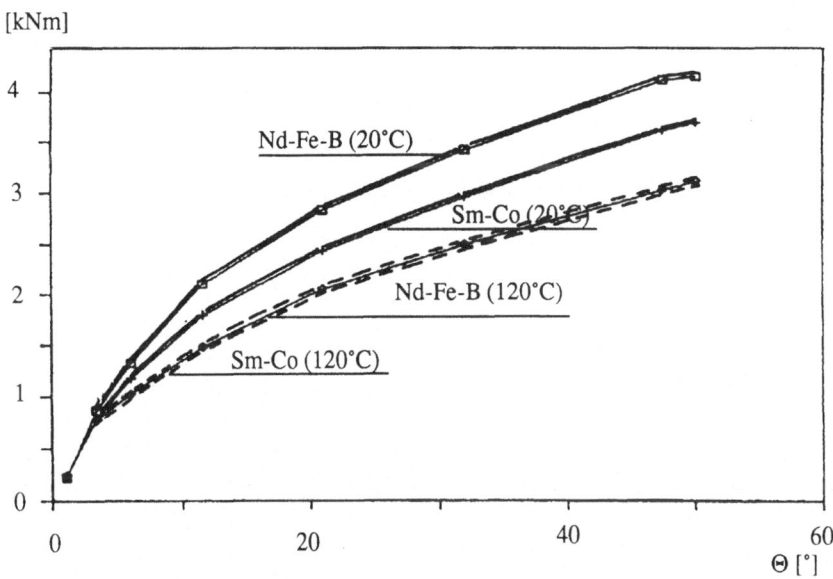

Figure 3. Torque for Sm-Co and Nd-Fe-B.

machine:

    - the slotted nature of the stator (a);

    - saturation of the iron (b);

    - permanent magnet field (c).

    On Figure 4 we see the split up of the Fourier spectrum. Only the two first components (due to slotting and due to saturation) would be present in an induction motor having an equivalent flux density distribution. Therefore it is obvious that a higher audible noise level has to be expected when using permanent magnet machines in stead of induction machines.

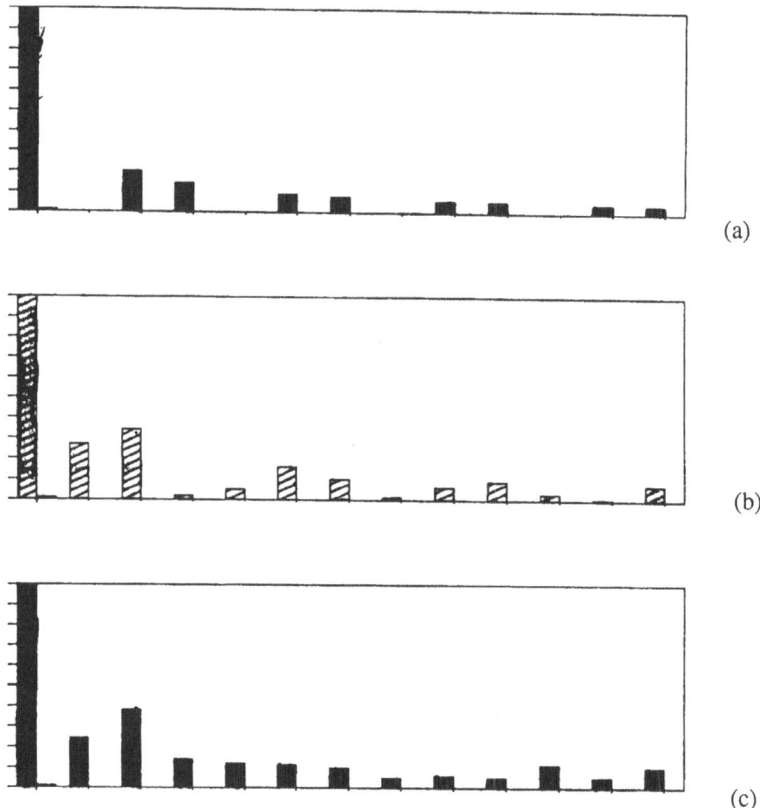

Figure 4. Analysis of the Fourier spectrum of the flux density distribution.

It may be noticed that this kind of analysis does not account for magnetostrictive effects. This is due to the fact that the flux density distribution values in rotating electrical machines are too low to generate these phenomena.

THERMAL ANALYSIS

The calculation of the temperature distribution in an electric machine is difficult and the results firmly depend on the exact knowledge of the material parameters and of the loss distribution (Vandenput et.al., 1987)

If the machine is air cooled and has no internal fan, it is mostly sufficient to analyse a two dimensional cut supposing that there is no axial heat transport.

However, if the machine possesses some extra cooling system in order to increase the output per unit weight, which is extremely important for aircraft applications, a two dimensional analysis is not sufficient.

The motor designed and presented by dr.D.HOWE of the University of Sheffield is clearly an axially non symmetrical assembly. This motor is equipped with an heat sink at one side. This heat sink is kept at 40°C and is directly coupled to the aluminium encasing of the motor. The stator lamination are in direct contact with the encasing, and the stator slots are filled with insulation material and copper windings.

In the rotor practically no losses are generated. Furthermore the rotor is supposed to be thermally isolated from the stator.

On Figure 5 we can see the three dimensional grid of the stator assembly.

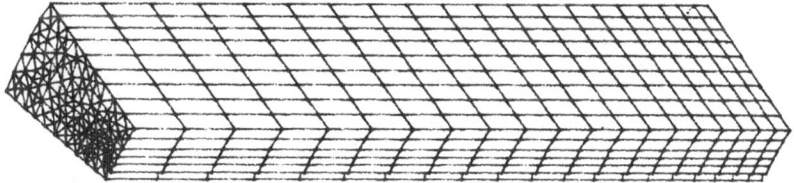

Figure 5. Three dimensional grid of the stator assembly.

Due to the anisotropic thermal behaviour of the lamination, two different thermal conductivity coefficients have to be taken into account:

$k_{radial} = 30.0 \ W/K/m^2$

$k_{axial} = 3.5 \ W/K/m^2$

In order to account for the cooling of the machine due to convection to the surrounding air, convective boundary conditions with

$h = 10 \ W/m^2$

are taken into account, representing a natural air cooling.

The highest temperature equals 132°C and is found at the top of the winding at the opposite end of the heat sink (Figure 6).

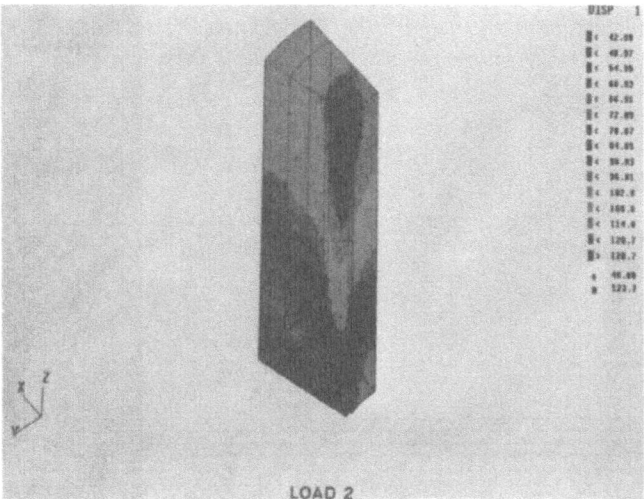

Figure 6. Three dimensional temperature distribution.

Due to the anisotropy, practically no heat flows in the axial direction in the laminations. All heat is carried radially from the copper winding to the aluminium encasing and from there on to the heat sink.

The cooling effect of the surrounding air is negligable.

MECHANICAL ANALYSIS

The knowledge of the natural frequencies is very important as high audible noise levels have to be expected if one of the natural frequencies coincides with one of the frequencies of the electromagnetic forces.

From the motor analysed by Prof.K.J.BINNS of the University of Liverpool, we calculated the natural frequencies of the stator assembly.

The grid used for these calculations is shown in Figure 7.

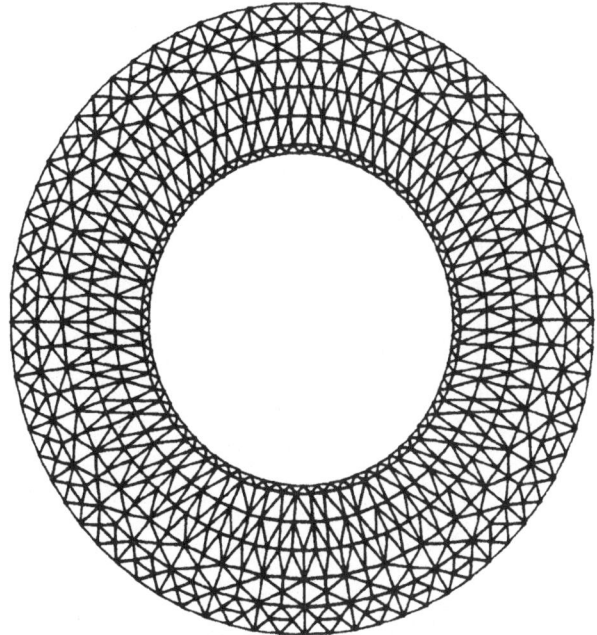

Figure 7. Grid of the mesh used.

As shown in Table 1, the natural frequencies have not equal distances. In Figure 8 the relative amplitudes of the various natural frequencies are shown.

Number	Frequency (Hz)
1	1181
2	2870
3	5302
4	7152
5	8769
6	9133
7	11729
8	13393
9	14918
10	17490
11	17994

Table 1. Natural frequencies of the stator assembly.

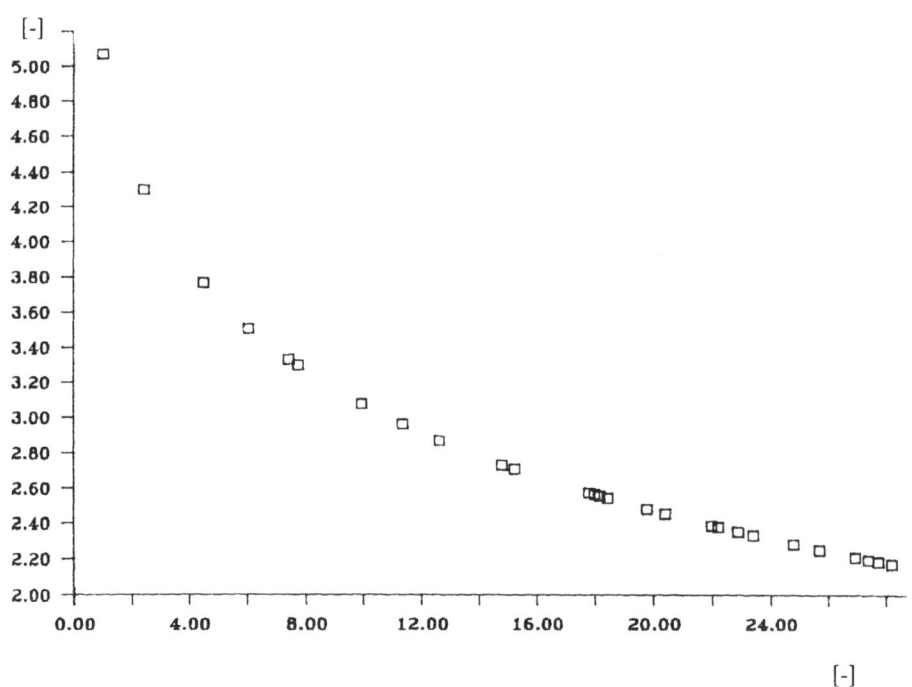

Figure 8. Relative amplitudes of the various natural frequencies (/1181 Hz).

Due to the mounting of the machine, boundary conditions are imposed on the structure. Together with the exact representation of stator slots and other geometrical problems, the possibility of accounting for the machine mounting is probably the most important advantage of the finite element calculation technique in comparison with classical analytical methods (Belmans et al., 1986 and 1987). We supposed that the lower part of the stator (45° both sides of the central line) was not able to move, as the machine was foot mounted.

On Figure 9 the first eigenmode is shown. Figure 10 represents the third eigenmode.

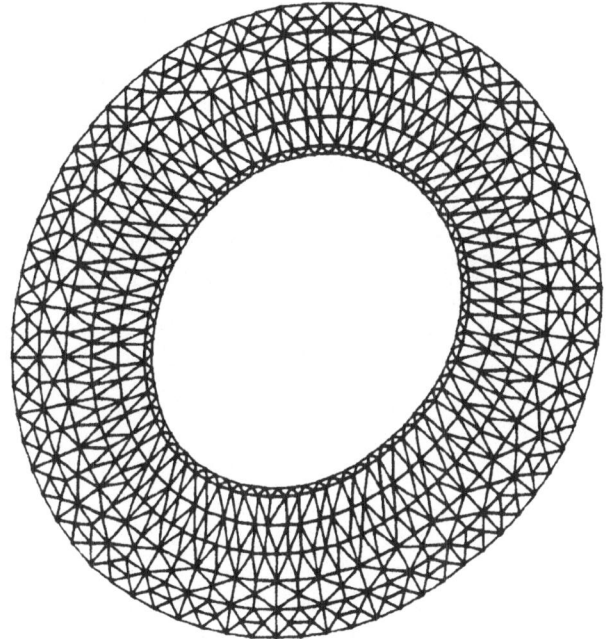

Figure 9. First eigenmode.

We can clearly see that the eigenmodes are influenced by the foot mounting of the machine.

CONCLUSIONS

The combined Finite Element - CAD approach of the phenomena (thermal, magnetical and mechanical) in permanent magnet machines offers the possibility to fully describe and analyse the behaviour of these motors and to design them in such a way that full advantage is taken from the excellent characteristics of the Nd-Fe-B permanent magnets.

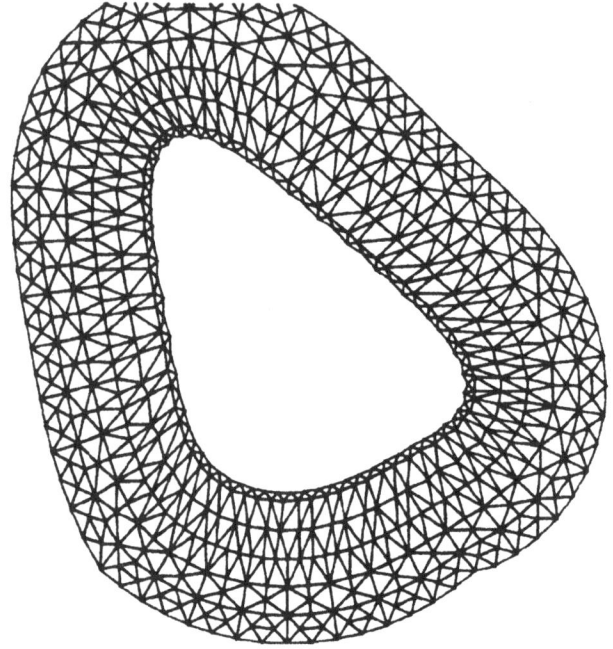

Figure 10. Third eigenmode.

ACKNOWLEDGMENTS

Apart from the Commission of the European Communities to which the authors are mostly gratefull for the support in CEAM, they wish to thank the Council of the Belgian National Science Foundation for granting a project dealing with CAD of electrical machines.

REFERENCES

Belmans R., Cornelissen P., Vandenput A., Geysen W., 1986, "CAD-Finite element combination for calculating natural frequencies of machine stators," *Proceedings of the International Conference on Evolution and Modern Aspects of Induction Motors,* July 8-11, p.219-223., Turino, Italy.

Belmans R., De Witte J., Vandenput A., Geysen W., 1987, "The comparison of CAD-Finite element calculations and modal analysis measurements for noise prediction of induction motors," *Proceedings BICEM, Beijing (China),* August 10-14, p.889-892.

Belmans R., Paternoster R., Vandenput A., Geysen W., Binns K.J., 1987, "Influence of the use of Nd-Fe-B in permanent magnet motors," *Proceedings of the Symposium of Electrical Drive,*

Cagliari, Italy, September 15-17, p.71-75.

Belmans R., Vandenput A., Verdyck D., Geysen W., 1988, "CAD of electrical machines," *Proceedings 1988 ASME International Computers in Engineering Conference*, New-York City, July 31-August 3, 1988.

Hameed A., 1984, "The finite element modelling and design optimisation of a multipole permanent magnet synchronous machine," *Ph.D.Thesis, K.U.Leuven*, Belgium.

Hameed A., Binns K.J., Vandenput A., Geysen W., 1984, "The finite element computation of the field in a permanent magnet machine", *Proceedings IMACS*, May 16-18, Liège, Belgium, p.4A.1.1-4A.1.4.

Hameed A., Vandenput A., Geysen W., Binns K.J., 1984, "Analysis of the magnetic field in a multipole permanent magnet machine by a finite element method," *Proceedings ICEM*, September 18-21, Lausanne, Switzerland, p.138-140.

Koch J., Plauman H.J., Ruschmeyer K., 1986, "Permanent magnetisch erregte Gleich-strommotoren," Valvo Unternehmensbereich Bauelemente der Philips GmbH, *Dr.A.Hütig Verlag GmbH, Heidelberg*.

Vandenput A., Belmans R., De Backker K., 1987, "Het algemeen eindige elementenpakket SYSTUS toegepast bij het ontwerp van elektrische motoren, *Cad/Cam*, Oktober, p.27-31.

ADVANTAGES AND DISADVANTAGES OF USING

NEODYMIUM IRON BORON IN ELECTRIC MOTORS

M. Bradford

ERA Technology Ltd, Cleeve Road,
Leatherhead, Surrey, KT22 7SA, England.

ABSTRACT

The potential use of neodymium iron boron in electric motors has been examined by studying material properties and by undertaking comparative design studies. Data has been collated on the new alloys and competing materials from manufacturers' catalogues, and the similarities and differences between the materials have been stated. In addition to the well-known temperature limitations and corrosion problems of the new alloys, this work identified the unusual thermal expansion properties of neodymium iron boron. A number of designs have been made using both outline design and detailed design procedures, the latter being based on finite element techniques. The designs have been for both brushed and brushless motors, for the power range from 300W to 2000W. The benefits of reduced size, weight and inertia of using neodymium iron boron compared with other materials for the same output have been calculated, and the extent to which these benefits diminish with increased temperature have been stated. The possibility of increasing the output from a given frame size has been demonstrated by designing, building and testing a motor with neodymium iron boron magnets.

INTRODUCTION

ERA Technology Ltd is the largest independent contract R & D organisation in the UK serving the electrical industry. The company has a particular capability in electrical machine research, design and development, and in the investigation of modern materials to enhance machine performance, including insulation materials, soft magnetic materials for electrical steels, and new permanent magnet materials.

In January 1986, ERA started a major project concerning permanent magnet machines. The main objective of the work was to develop design methods for a range of PM machine types and sizes with alternative PM materials. The work is continuing for a consortium of electric motor manufacturers from the UK and involves the collaboration of the University of Sheffield and the University of Manchester Institute of Science and Technology in the UK.

The work includes derivation and verification of performance calculation methods for a given design of machine. Design procedures are being derived and verified so as to be able to generate a machine design to give a specified performance.

The study has involved alternative permanent magnet materials including ferrites, samarium cobalt and neodymium iron boron. Design procedures have been produced for brushed dc PM motors, brushless dc PM motors and line start synchronous motors covering a range of outputs up to about 5kW.

When CEAM commenced, ERA proposed to undertake additional work leading to the development of a permanent magnet machine using neodymium iron boron high energy permanent magnets. The performance target was to be a significant improvement over that obtainable from samarium cobalt machines. A prototype motor has been built to this design, and has been tested to verify the design procedure. The performance obtained from this motor, compared with a commercially available motor and a general outline of the design of the advanced motor is described in this final report.

The work undertaken in the CEAM programme has consisted of three main tasks, namely:

- data collection
- machine design
- machine build and test.

These activities are described in detail in the following sections of this report.

DATA COLLECTION

A search of commercial brochures and literature has been undertaken regarding the availability and properties of new high-energy magnets and conventional magnets. The intention was to obtain representative data rather than a complete set of data for all magnets world-wide. Data has been collated on a wide range of Nd-Fe-B, SmCo and ferrite materials in order to provide suitable inputs to design procedures. This data includes thermal, mechanical and electrical parameters as well as the magnetic values, and allows important comparisons of the various properties to be made. An example of the data obtained is shown in Table 1.

Table 1

Major physical parameters of PM materials

	Cast/sintered				Bonded		
	Alnico	Ferrite	SmCo	Nd-Fe-B	Ferrite	SmCo	Nd-Fe-B
Resistivity, $\mu\Omega$m	0.55	10^{10}	0.7	1.5	10^{11}	1.4	180
Operating temperature °C	500	350	120-350	100-175	100	100-150	100-125
Density, kg m^{-3}	7150	4850	8300	7350	3750	6750	6500
Bend strength, MNm^{-2}	55-300	50-170	110	260	100	80	50
Compressive strength, MNm^{-2}	-	1400	300-900	740-1500	-	440	-
Tensile strength, MNm^{-2}	170	40	45	80	-	-	-
Hardness	650	350-800	550	650	60$_D$ Shore	190	100
Expansion Coefficient, 10^{-6}/°C	10-12.5	14// 10$\perp$	4-10// 10-12$\perp$	+3 // -5 $\perp$	5-175	10//	3.8
Thermal conductivity, Wm^{-1}K^{-1}	25	1.25	11	8.5	-	6	-
Specific heat, J kg^{-1}K^{-1}	450	725	360	420	-	-	-

An extensive set of data was sent to all Group 2 and 3 CEAM members in January 1987. Comments were received from some Group 2 members.

The data indicated the following relationships:

o   Nd-Fe-B and Sm Co are very similar in respect of density, resistivity, magnetising field, thermal conductivity, specific heat and hardness.

o   As is well known, Nd-Fe-B has a lower operating temperature, and is more susceptible to corrosion than SmCo.

o   Nd-Fe-B has a particularly high temperature coefficient for Hc.

o   Nd-Fe-B has a negative coefficient of thermal expansion normal to the direction of magnetisation whereas all other magnetic and constructional materials have positive coefficients. This could result in assembly problems.

There was a general lack of data on bonded permanent magnets.

DESIGN PROCEDURES

The design procedures employed include:

a) initial designs using conventional design procedures for permanent magnet slotted armature machines in order to determine leading dimensions, and to examine choices of pole number and the effect on magnet radial length, tooth and back iron thickness, etc.

b) a detailed design procedure using finite element techniques, in order to determine precise dimensions and electrical and mechanical parameters.

The outline designs were based on an iterative procedure firstly to calculate armature diameter from a power equation with an assumed initial airgap flux density; then magnet length from the demagnetisation conditions; and finally airgap flux density from the magnet length. This procedure enables alternative features such as pole number, magnet radial length etc., to be studied. This program has been used to examine the magnet dimensions for a variety of dc permanent magnet motors. It has shown the possibility of problems in Nd-Fe-B machine design and fabrication, in particular:

a) the need to fabricate thin arcs of magnet in order to minimise magnet weight and take full advantage of the magnetic properties of Nd-Fe-B. Such arcs can most effectively be fabricated from multiple segments in both the axial and circumferential directions.

b) the need to employ either high flux densities in the back iron or to adopt a large radial depth of back iron to reduce those flux densities. The former can result in high iron loss and the latter in increased overall dimensions.

OUTLINE DESIGNS

Various outline designs were undertaken for ferrite, SmCo and Nd-Fe-B materials. As an example it is appropriate for these materials to compare the leading dimensions of a 4-pole, 2500 rev/min, 2000 W, brushed motor, with a current density of $4A/mm^2$, a maximum armature reaction mmf of 10 times the working value, and a D/L ratio of 1. The main data are set out in Table 2. (at 100°C).

TABLE 2	Comparison of dimensions for 2000 W brushed motor.		
	Ferrite	Samarium cobalt	Neodymium iron boron
	FXD 460	Recoma 20	Crumax 301
Magnet length (mm)	14.2	2.9	2.7
Armature length (mm)	117	93	90
Gap flux density (T)	0.20	0.50	0.56
Magnet mass (kg)	2.17	0.46	0.36
Motor mass (kg)	15.0	8.5	8.0

The relatively minor difference between the SmCo and Nd-Fe-B designs is due to the knee field strength at 100°C (the assumed operating temperature) being only about 2.5% greater for Crumax 301 than for Recoma 20, which is a 2:17 SmCo. The potential advantage of using the Nd-Fe-B material is therefore the expected benefits in reduced cost in the future.

A design was undertaken for a brushless motor using Crumax 301 with slightly different design constraints and with a magnet retaining ring equal in radial depth to the airgap (0.7mm). The magnet length was increased to 5.55mm, but the armature length (and diameter) were reduced to 78mm.

The design method has been checked by comparing data for existing motors with those designs run at ERA. Generally good agreement has been obtained for the leading dimensions.

Designs have also been developed for two brushed dc motors rated at 300 W, one with ferrite and one with Nd-Fe-B magnets. Details of the motor parameters, magnet materials, motor dimensions, and performance are given in Tables 3a-c. The mesh build up used in the FE analysis is shown in Fig 1 along with a particular solution of the vector potential field at a low current density in Fig 2. A drawing of the Nd-Fe-B design is shown in Fig.3.

Table 3 (a) Main parameters of Brushed dc Motors

Rating	300 W continuous
Speed	2600 rev/min
Current density	5 A/mm^2
Voltage	24 V
Number of poles	2
Demagnetisation factor	5
Diameter/length ratio	0.62

Table 3 (b) Motor Details

	Ferrite Design	Nd-Fe-B Design
Magnet length (mm)	7.9	1.4
Armature diameter (mm)	54.8	44.6
Armature length (mm)	88.3	72.9
Slot depth (mm)	12.2	9.9
Gap length (mm)	0.8	0.8
Yoke depth (mm)	6.0	11.1
Overall diameter (mm)	84.2	71.2
Magnet volume (cm^3)	104.3	10.8
Gap flux density (T)	0.21	0.49
Flux/pole (mWb)	1.64	2.47
Current (A)	15.5	14.8
Number of turns	68	48
Losses (W)	73	54
Efficiency (%)	80.4	84.7
Volume (p.u)	1	0.59
Inertia (p.u)	1	0.36
Number of armature slots	12	12

Table 3 (c)　　　　　　　　Magnet Material Properties

	Ferrite	Nd-Fe-B
Type	FXD 330	NeIGT 27 H
Remanence, B_r (T)	0.32 at 85°	0.99 at 85°C
Flux density at knee, B_k (T)	0.34 at 20°C	0.90 at 85°C
Magnetising force knee, H_k (kA/m)	240 at 20°C	845 at 85°c
μ_{rec}	1.1	1.1

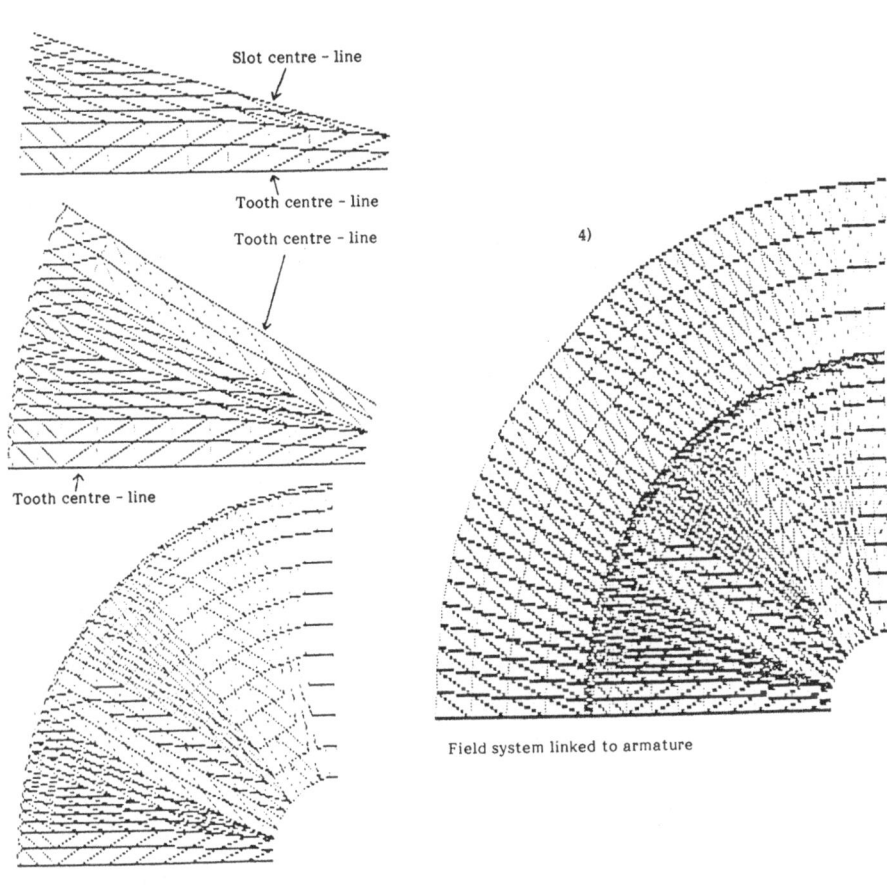

1)

Slot centre – line

Tooth centre – line

2)

Tooth centre – line

Tooth centre – line

4)

Tooth centre – line

3)

Field system linked to armature

Quarter of armature

Fig.1 Mesh build-up for F E analysis

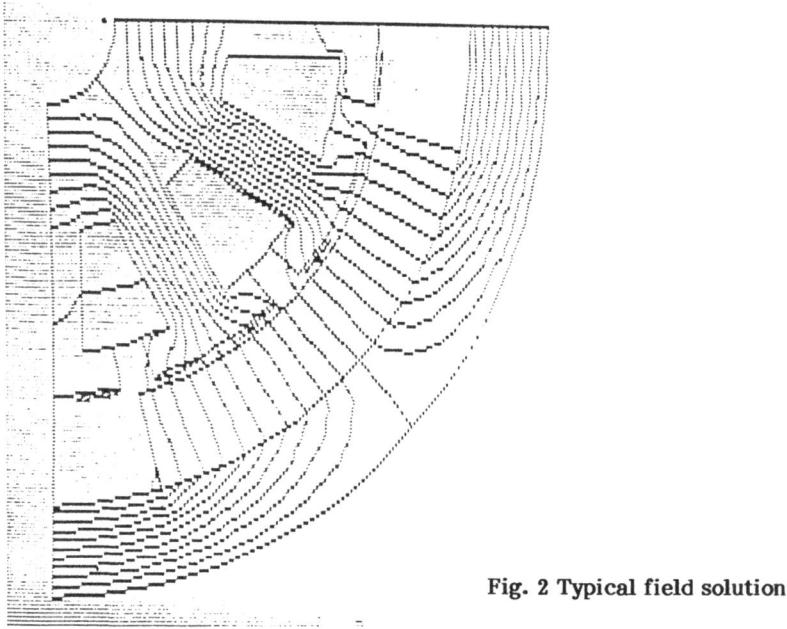

Fig. 2 Typical field solution

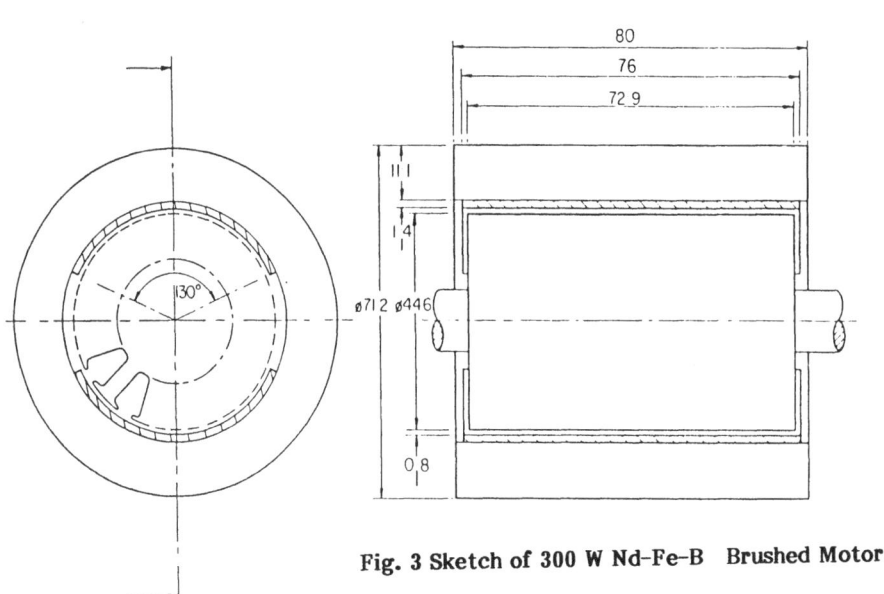

Fig. 3 Sketch of 300 W Nd-Fe-B Brushed Motor

MOTOR DESIGN, BUILD AND TEST

An advanced permanent magnet high torque dc brushed motor has been designed, built and tested.   A brushed motor has been selected in order to focus attention specifically on the electromagnetic design.   The objective has been to design an Nd-Fe-B motor within the same overall dimensions as an existing ferrite design which has the following performance characteristics:

Continuous rating	300W
Short time rating	750W (for 20 seconds)

but with greatly increased output performance. The magnet dimensions are 3.2mm radial length, and 90mm axial length, with an armature diameter of 56mm. The magnets have been assembled within the yoke from narrow premagnetised strips. Fig 4 shows a drawing of the magnet and barrel, and Fig 5 is a photograph of the magnet system.

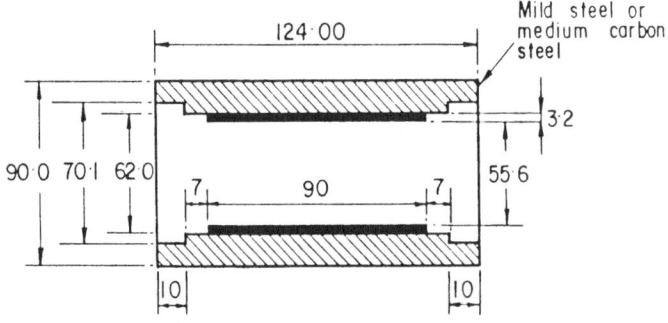

Fig. 4 Sketch of pm Shell for demonstration Motor

Fig. 5 Photograph of PM shell of demonstration motor

The design, build and test of the advanced PM motor has been undertaken in close collaboration with a leading UK manufacturer of small high-performance dc motors and a UK supplier of neodymium iron boron magnets. The design and calculated performance data for the Nd-Fe-B demonstration motor are set out in Table 4.

TABLE 4 Data for Nd-Fe-B prototype brushed motor.

Magnet properties (at 120°C)

	B_k (at knee)	0.85	Tesla
	H_k (at knee)	684x10^3	A/m
	B_r	0.95	Tesla
	μ_{rec}	1.1	

Dimensions

Armature diameter	54.0mm
Armature length	87.5mm
D/L ratio	0.62
Magnet length	3.2mm
Slot depth/radius	0.444
Gap length	0.8mm
Magnet arc/pole pitch	0.717
Magnet overhang factor	1.03
Number of poles	2

Performance data

Speed	2600 rev/min
Output	750 W
Current density	5 A/mm^2
Demagnetisation ratio	5
Voltage	24V
Current	34.2A
Gap flux density	0.56 Tesla average
Flux/pole	4.16 mWb
Efficiency	91.2
Current loading	12187 A/m

The test data for the original ferrite design and the new Nd-Fe-B design are set out in Fig. 6 and Table 5.

ADDITIONAL RELATED ACTIVITIES

In parallel with the CEAM study a number of related activities have been undertaken.

a) A prototype low-speed low-power permanent magnet generator has been designed, built and tested.

Fig. 6 Test data for ferrite and NdFe B motors

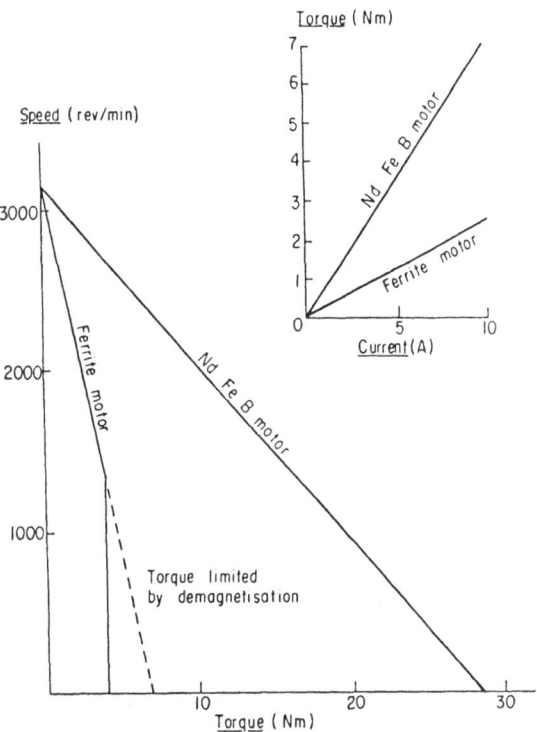

Table 5. Test data for original ferrite and new NdFe B motors

	Ferrite	NdFe B
Rated torque, Nm	1.1	2.8
Rated speed, rev/min	2625	1500
Rated output, W	300	440
Total loss, W	90	80
Torque/amp, Nm/A	0.24	0.70
Peak torque, Nm	4.0	14.5
No load loss, W (at 3000 rev/min)	15	63

b) A seminar on magnetic materials, including Nd-Fe-B was held by ERA in London on 4/5 November 1987. This was attended by 110 delegates, including representatives from 11 European nations.

c) A report on the possible future use of Nd-Fe-B in motors, generators and transducers in automobile equipment has been completed and sent to the European Commission on completion of a contract. This report is now available to the European Community via the CEC in Brussels.

The automotive report concluded that sufficient stocks of neodymium were available to satisfy any reasonable forecast of demand in the automobile industry, but the major problem to be overcome would be that of material price, where £10/kg may be needed for widespread use. The number of applications for small PM motors in cars is in excess of 30, and up to 70 motors and actuators could be in executive cars of the 1990s. Improvement in motor efficiency and power/weight ratios would be particularly beneficial in increasing fuel economy.

CONCLUSIONS

Neodymium iron boron can provide major improvements in motor performance if the disadvantages of temperature limitation and corrosion can be overcome. The former problem can be reduced in severity by careful attention to the thermal balance of the machine, adequate cooling and heat paths, and even by the use of heat pipes to directly cool the magnets; this is the subject of a separate study at ERA. The second problem of corrosion can be generally removed by the use of suitable coatings. With those difficulties on one side, various performance improvements can be obtained with neodymium iron boron, such as an efficiency increase of 18%, power/weight ratios of up to 800 W/kg, losses reduced by 26%, inertia reduced to 36%, motor volume reduced to 59% and motor mass reduced to 53%.

THREE PHASE PULSE WIDTH MODULATION WAVEFORM GENERATOR FOR USE

WITH PERMANENT MAGNET MOTORS

P.R. Kamdar

GEC Research laboratories, Hirst Research Centre

East Lane, Wembley, Middlesex, U.K.

ABSTRACT

A brief review is given on the development of a semi-custom chip to perform Pulse Width Modulation (PWM) waveform generation. This device can be used up to ultrasonic frequencies (>16 KHz) for variable speed electric drives using permanent magnet motors.

Work was also undertaken to develop a high speed hybrid switch for use at 415V a.c.

INTRODUCTION

We are concerned mainly with exploring system aspects relevant to Nd-Fe-B magnet motors.

The lower cost and improved performance of power devices now generally available has led to a rapid growth in the a.c permanent magnet (PM) motor drives market. The PWM waveform generator developed (the MA818) is the first fully digital stand-alone IC suitable for use in such drive

The MA818 is a versatile microprocessor controlled IC for the generation of three phase a.c waveforms. Fully digital derivation of PWM (pulse-width-modulation) waveforms gives very good operational accuracies and temperature stability. The MA818 has a standard MOTEL$^{\cent}$ interface suitable for connection to both Intel and Motorola microprocessors.

The upper power limit for use at ultrasonic frequencies is set by the available power switching devices. MOSFETs are suitable for 240V a.c as are some special bipolar transistors. However, at present no suitably fast devices (except for the recently launched 1KV 20A device by Toshiba) are commercially available for 415V a.c. In order to meet higher power requirements, we have also developed a hybrid switch to be used at 415V a.c 30A.

OPERATING FEATURES OF THE MA818 PWM WAVEFORM GENERATOR

The MA818 has been designed as a general purpose waveform generator to control variable speed a.c machines. It is an entirely digital device which interfaces to a digital controller of some sort (typically a microprocessor), a memory for waveform store (ROM - read only memory) and to the isolated drivers for a three-phase bridge inverter. With relation to the speed control of inverter driven a.c PM motor, the MA818 is situated in the control system as shown in figure 1.

A bridge inverter drives the three-phase a.c machine. The inverter consists of six switches that can be turned 'on' and 'off' electrically, and are typically power electronic devices such as MOSFETs, bipolar transistors or GTO's. The arrangement of these six switches is shown in figure 2. This configuration is such that by suitable control of the six switches a three-phase a.c output can be constructed from a d.c input. The frequency, voltage and waveshape of the a.c outputs can be varied as desired by suitable control. This process may not be readily apparent from an inverter bridge which can only produce an output which is either high (+ve d.c) or low (-ve d.c). However, a modulation process called pulse-width-modulation enables a switching signal consisting of high and low states to contain a low frequency a.c component of any desired waveform, frequency and amplitude.

The MA818 PWM strategy

Pulse-width-modulation (PWM) is a highly desirable method used to drive PM motors, as it offers great flexibility of the output waveform. Driving an PM motor requires that we maintain a low harmonic waveform, and as the speed is varied, we must also vary the a.c voltage to the motor so as to prevent saturation of the field. The PWM strategy in the MA818 provides both of these requirements at low cost and high reliability.

The PWM strategy in the MA818 is built around double-edged regular sampling as shown in figure 3. The sampling is uniform and asynchronous.

On the MA818, each of the three phase outputs duplicates this process with three in-phase triangle waveforms and three 120° out-of-phase waveforms defined by data stored in the external waveform store. The frequency of the carrier waveform, and the frequency and amplitude of the required waveforms are all controlled by the MA818 using an external reference clock input (usually 10 MHz). Figure 4 shows the internal structure of the MA818.

Waveform definition and storage

The required waveform used to construct the PWM sequence by the MA818 is held in a memory external to the device. Generally a ROM (read only memory) or EPROM (erasable/programmable ROM) are used. The size of memory required is dependant on the digital discretisation or coarseness of the digital representation of the required waveform. The maximum size of memory addressable is 1536 X 4 bits and so normally a 2K X 8 bit EPROM is adequate.

The waveform-store (see figure 5) contains one half-cycle of the required waveform (ie. 0° to 180°) and this data is assumed to be symmetrical around 90° and positive in amplitude throughout the 180° span. The MA818 samples the waveform-store and internally generates the three values of the instantaneous amplitudes for each of the three phases.

The MA818 directly connects to an EPROM or ROM and is totally self-contained. No external signals are required. For the MOSFET inverter drives, the external store contains half-sinewave plus 1/6th third harmonic component. This ensures maximum output sinewave voltages to the PM machine between lines for a given d.c inverter supply voltage. Any other waveforms can be programmed as required. It can be seen that the MA818 allows the optimisation of output waveform for individual motor or other load requirements. This represents a significant advance from other systems which have fixed output waveforms.

Input and output control features of the MA818

Description of the complete operation of the MA818 is not possible in this text, however some of the salient operating features of the device are :

(a) The device generates three-phase PWM sequences digitally from a set clock frequency and a digital control.

(b) Outputs can be interfaced directly to power electronic switches in a variable speed drive.

(c) The waveform contained in the PWM sequences is defined by an external ROM - not fixed by the MA818.

(d) The PWM sequence has a constant carrier frequency (asynchronous PWM) which is user selectable from 600 Hz up to 19.5 KHz for silent inverter operation.

(e) The power frequency of the PWM waveform is range selectable from 0-50 Hz to 0-3 KHz.

(f) The minimum pulse width in the PWM sequence is user defined from 0.1 μSec to over 200 μSec(the MA818 drops pulses if too short), see figure 6.

(g) The delay time between switching of the complementary outputs in each phase is user defined from 0.1 μSec to over 100 μSec, see figure 6.

(h) The device is capable of controlling inverters using all types of fully controlled power electronic switch such as MOSFETs, bipolar transistors and GTO thyristors.

Interfacing the MA818

The MA818 has a standard MOTEL$^\phi$ interface which allows the direct connection of all commonly used motorola or Intel microprocessors or microcontrollers. One 20-bit register is used to initialise the MA818, and a 23-bit register to control it during actual running of the motor. The register contents are shown in figure 7. The 8-bit registers R_0 - R_2 are used for data entry, and R_3 & R_4 transfer this data from the temporary storage latch to the initialisation and control latches respectively.

The MA818 requires CPU time only during initialisation or when a change of frequency is desired in the run mode. Hence other tasks, such as sensing of feedback from the motor current sensors can be accomodated to regulate motor torque demands and so improve the overall efficiency of the motor drive systems using PM motors.

A typical programming flowchart for the MA818 is illustrated in figure 8.

HIGH VOLTAGE, HIGH SPEED POWER SWITCH

The hybrid switch combines MOSFETs and bipolars and is capable of PWM at frequencies up to 16 KHz. It can switch from a 700V d.c supply at a design current of 30A, and can tolerate 100% overload for short periods depending on the heat sink size. Turn-on and turn-off times are about 1 µSec each. The hybrid switch is considerably cheaper than paralleled MOSFETs and uses a MOSFET gate drive arrangement as shown in figure 9.

This switch has obvious application to silent synchronous drives using Nd-Fe-B magnets.

CONCLUSION

A three phase PWM pattern generator (MA818) has been developed to control six power electronic switches in an inverter variable speed drive. The versatile MA818 is fully programmable and is controlled via an 8-bit address/data bus which directly connects via its MOTEL$^{\cent}$ interface to any standard Motorola or Intel microprocessors. The versatility of the MA818 makes it useful for applications such as PM drives, a.c induction machine drives, uninturruptible power supplies (UPS), d.c to d.c converters, etc.

In conjunction with the above, a high voltage, high speed hybrid power switch was developed to operate at 700V d.c 30A for use in 415V a.c inverter drive system.

REFERENCES

MA818 three phase PWM waveform generator data sheet, Marconi Electronic Devices Ltd, Lincoln, U.K.

MA818 three phase PWM waveform generator application notes, 1988, Marconi Electronic Devices Ltd, Lincoln, U.K.

¢ MOTEL is a registered trade mark of Motorola Corp. and Intel Corp.

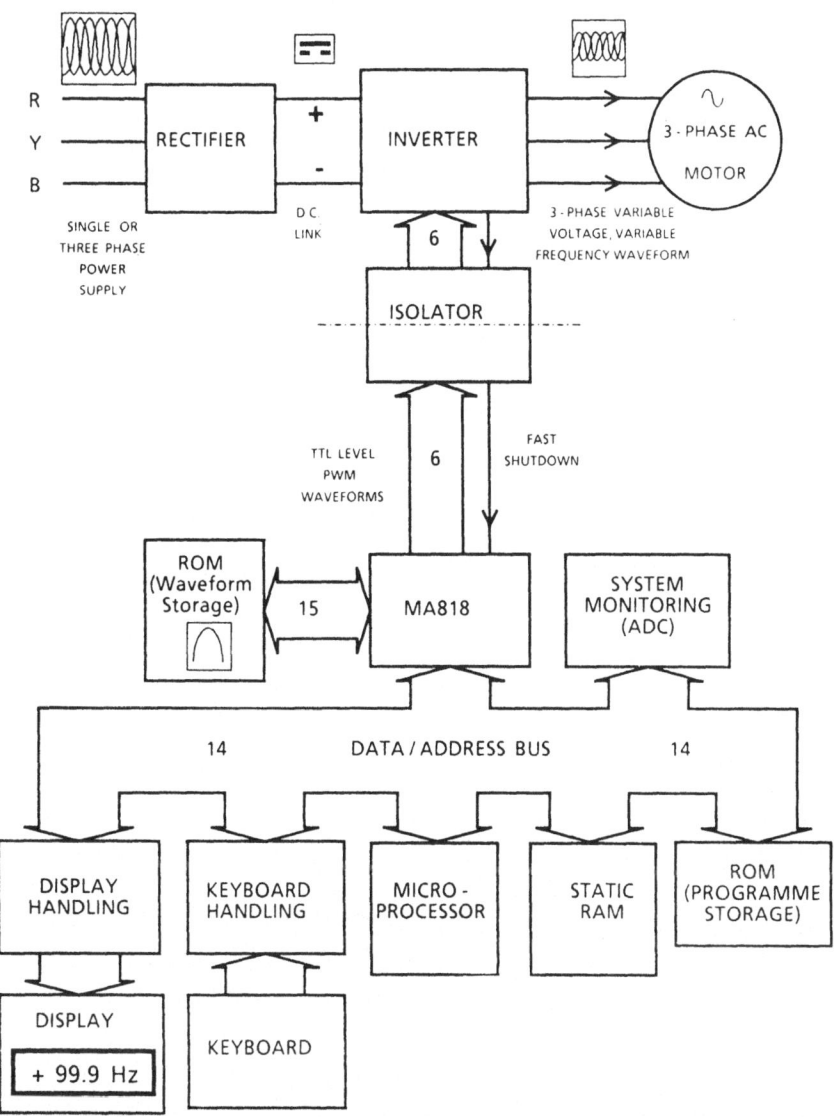

FIGURE1: *Block Diagram for an AC Permanent Magnet Motor Control Unit Using the MA818*

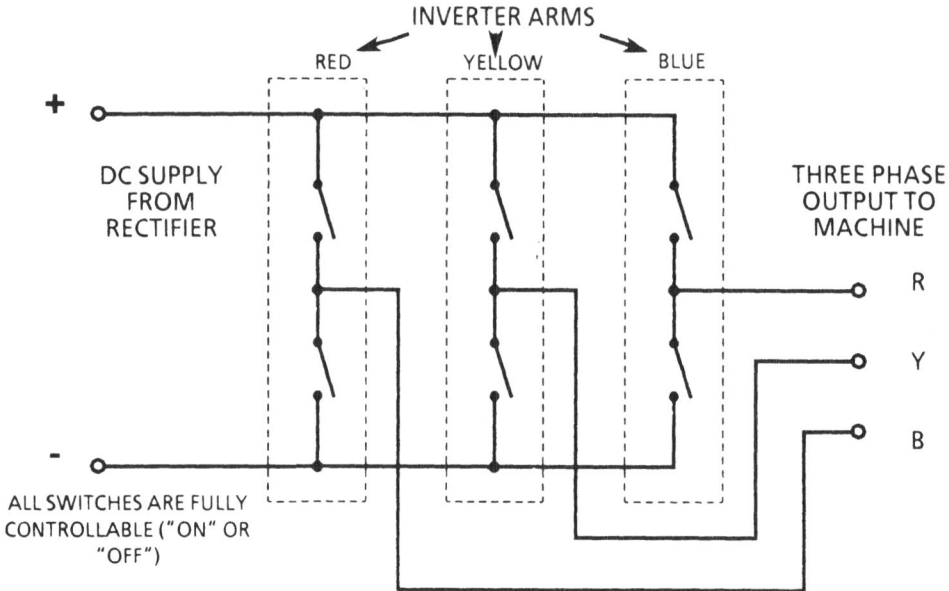

NOTE: IN THIS DIAGRAM THE POWER DEVICES ARE REPRESENTED BY SWITCHES

FIGURE 2: *Switch Configuration for a Three Phase Invertor*

FIGURE 3 : *Asynchronous PWM Generation with Uniform or Double-Edged Regular Sampling, as used on the MA818*

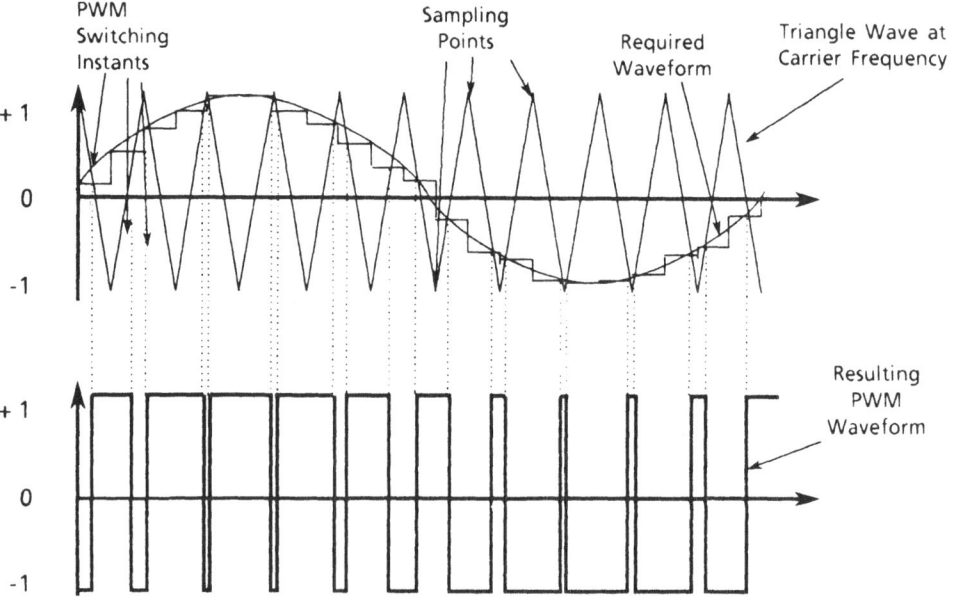

FIGURE 4: *MA818 Internal Block Diagram*

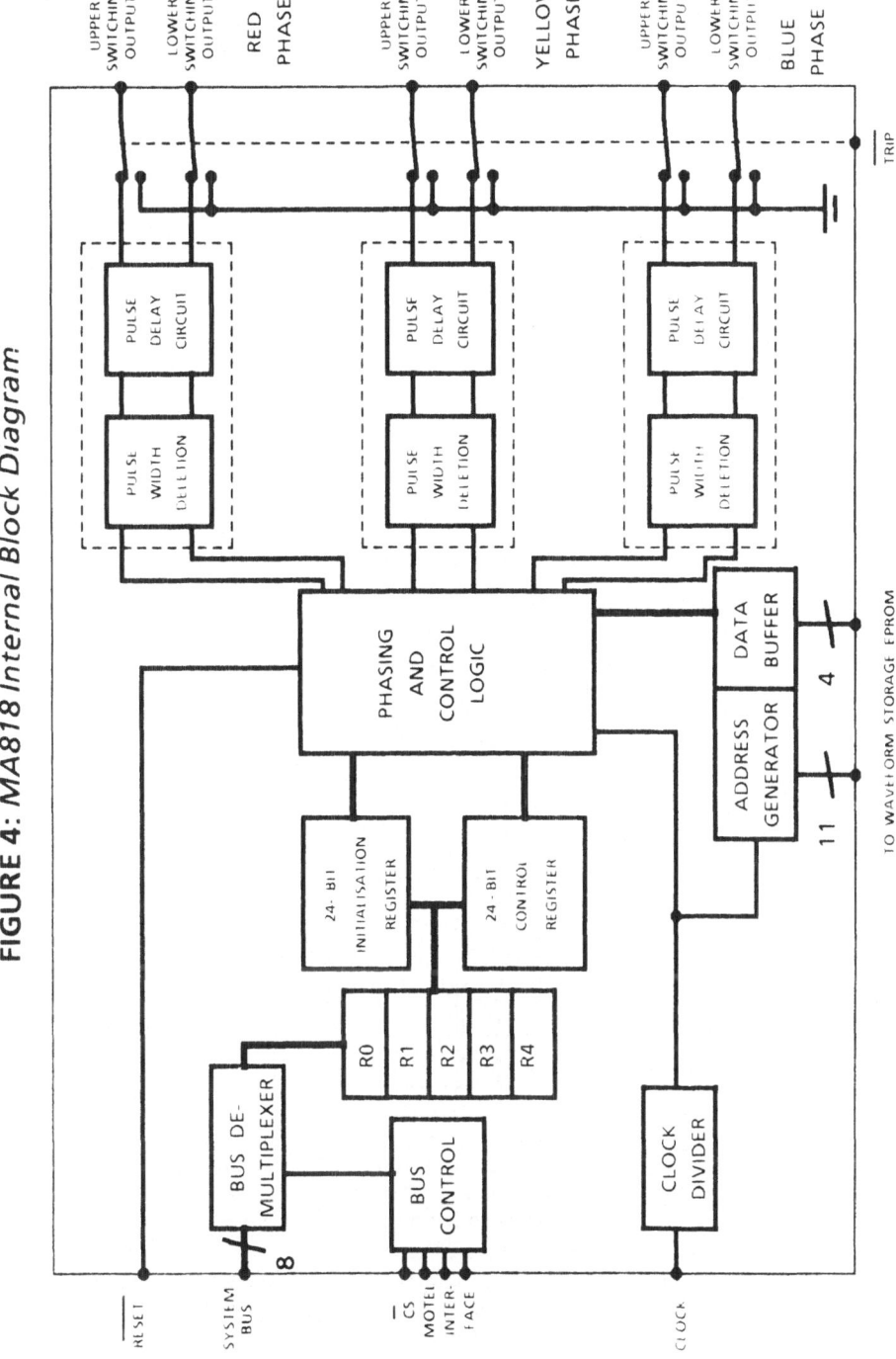

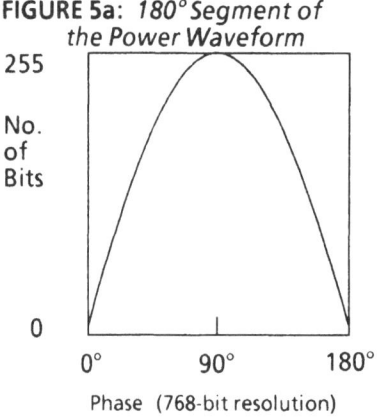

FIGURE 5a: *180° Segment of the Power Waveform*

Phase (768-bit resolution)

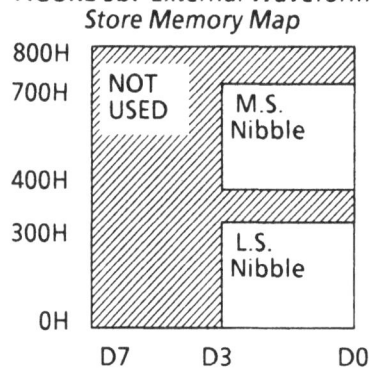

FIGURE 5b: *External Waveform Store Memory Map*

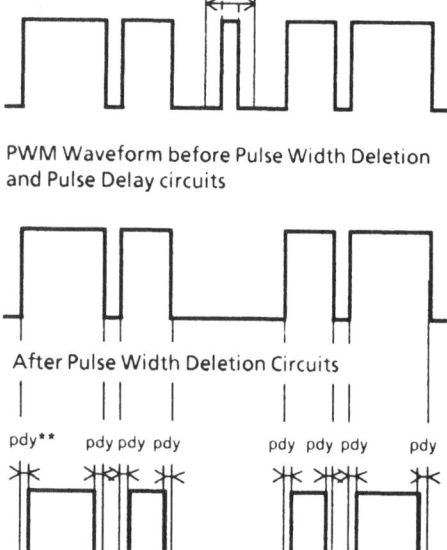

PWM Waveform before Pulse Width Deletion and Pulse Delay circuits

After Pulse Width Deletion Circuits

pdy**   pdy pdy pdy   pdy pdy pdy   pdy

After Pulse Delay Circuits : Note that some pulses may actually be smaller in width than the value of pdt set. This should be taken account of when setting the value of pdt and pdy.

* pdt = Pulse Width Deletion Time
** pdy = Pulse Delay Time

Figure 6: *Diagram to Illustrate the Digital Processing of the PWM Outputs.*

755

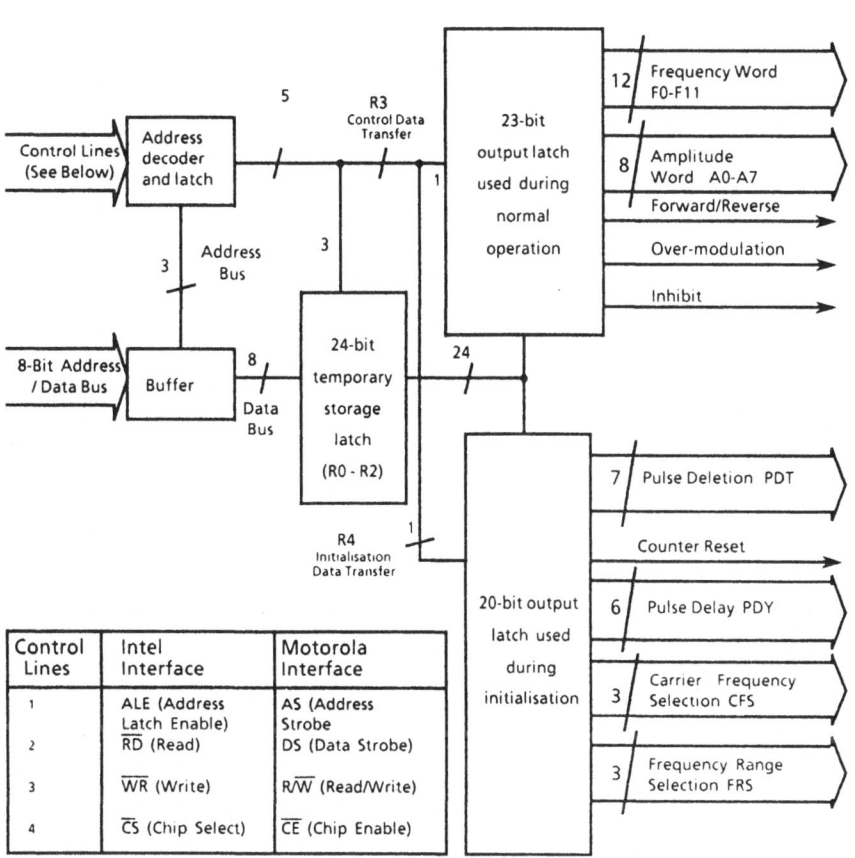

FIGURE 7 : *Internal Register Structure of the MA818*

TYPICAL PROGRAMMING ROUTINE FOR MA818

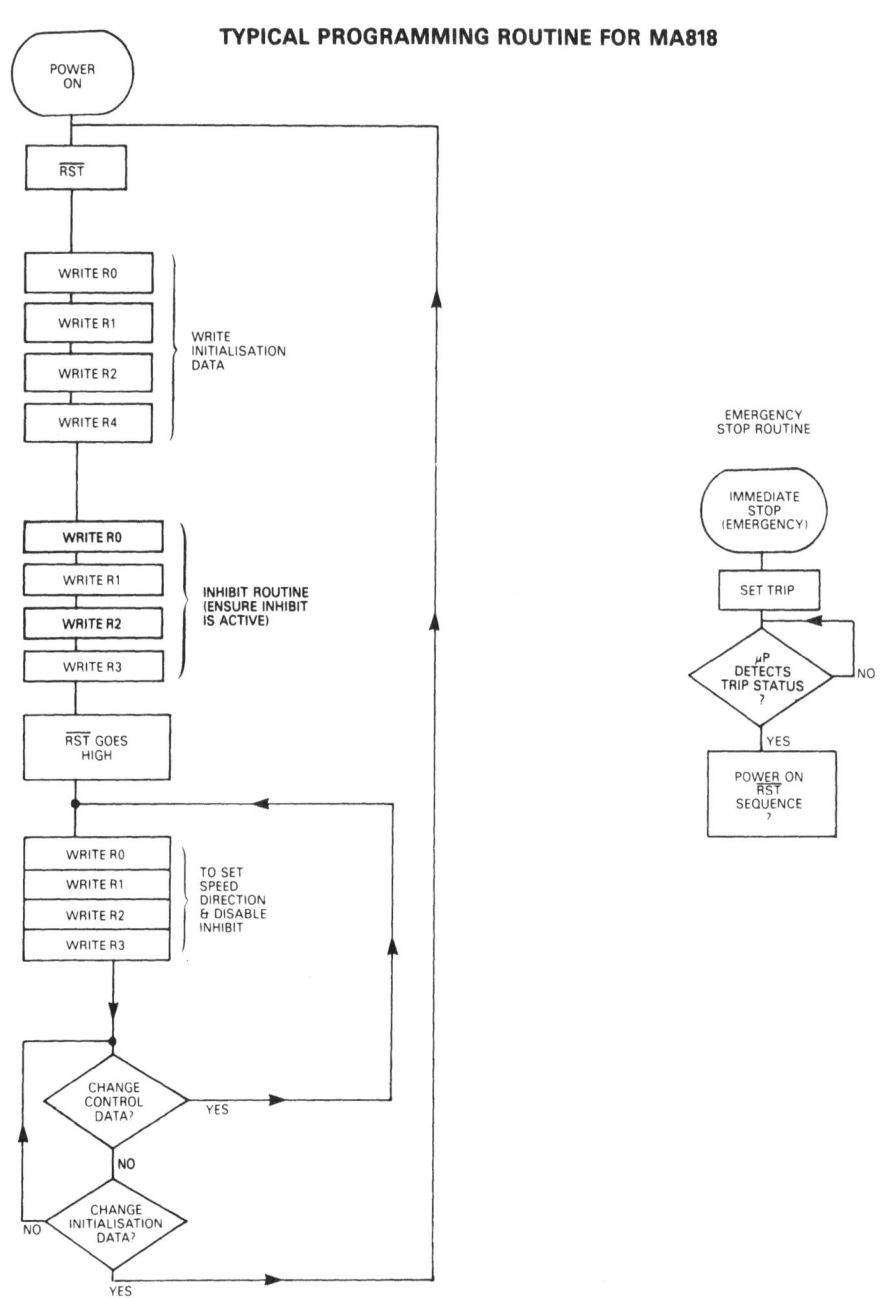

Figure 8

757

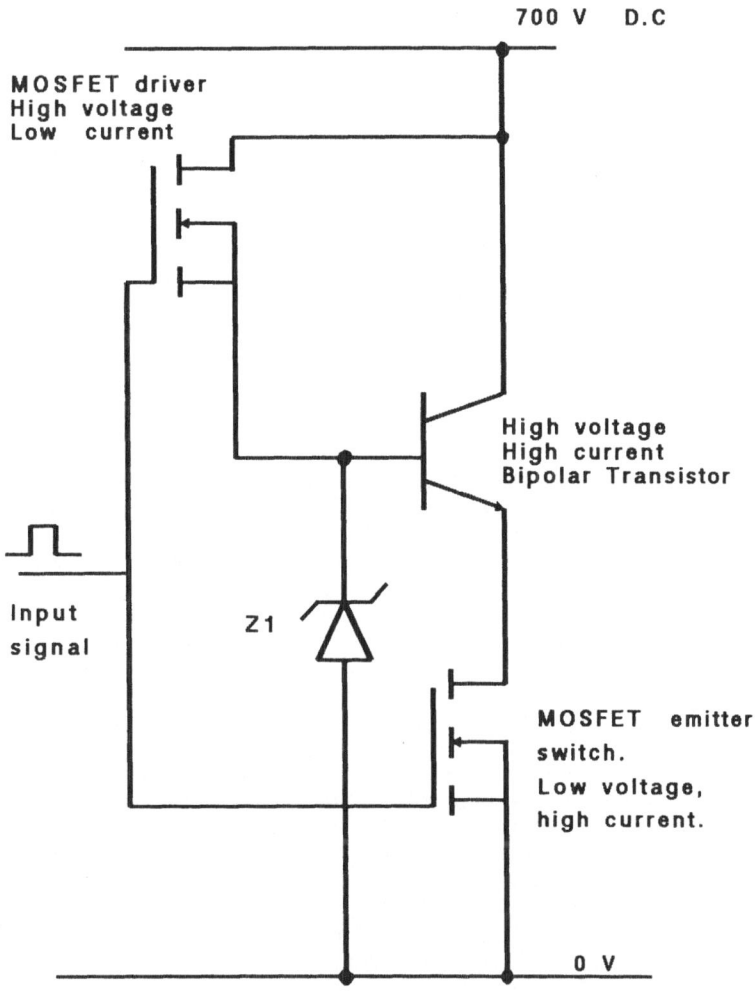

700 V   D.C

MOSFET driver
High voltage
Low current

High voltage
High current
Bipolar Transistor

Input
signal

Z1

MOSFET emitter
switch.
Low voltage,
high current.

0 V

Figure 9   High Performance Hybrid switch

ASSEMBLY OF A DISC-TYPE MOTOR WITH HIGH ENERGY
PERMANENT MAGNETS

T. Kindervater, A. Thoma, KVS

Berlin, West Germany

ABSTRACT

The new Nd-Fe-B magnets are extremely
difficult to handle. In order to use these magnets
in some electromechanical applications special
tools of nonmagnetic material and special assembly
procedures had to be developed. The following
article describes the way how to handle this high
energy magnets.

INTRODUCTION

KVS joined the CEAM - Group in March 1987.
Within the CEAM Contract period, we worked on the
following fields.
a) we undertook a series of outline designs to
   obtain a special device for the assembly of
   high energy permanent magnets.
b) we set up a test facility to figure out the
   strength of attraction or repulsion of the
   Nd-Fe-B magnets.
c) we built the assembly fixture and assembled 48
   magnets on two rotor-discs.
d) we assembled the whole machine and made first
   test measurements including balance of
   rotation.

METHODS AND TECHNIQUE

In order to assemble all the Nd-Fe-B magnets
on a rotor-disc of a new disc-type motor, we had to
develop a special method. The material of the
assembling device is brass.

Special attention was paid to the outline of the
assembling device. On Figure 1 there is the correct
outline in millimeters. The angle of the two legs
(see Figure 1) had to be figured out exactly, to
ensure the capability to assemble the magnets on
the rotor-disc. The distance between the magnets on
the edges is less than 1 mm (see Figure 2).
Also a special assembling procedure was developed
to ensure a safe assembling operation. Figure 3 to
5 give an idea how the assembling procedure works.

Phase 1:

> The magnet will be inserted in a pocked
> milled place and will be fixed with a
> clamping screw. The rotor-disc has to be
> centered with a guide pin.

Phase 2:

> The tool with the magnet inside drives
> toward the rotor-disc.

Phase 3:

> When the magnet is finally in the right
> position (check with the guide pin), the
> clamping screw has to be opened and the
> magnet itself has to be fixed with an
> antimagnetic screw on the surface of the
> rotor-disc (see also Figure 6 and 7).

In order to calculate the size of the clamping screw, the strength of attraction or repulsion of the magnets had to be figured out. With a small setup (ironplate and springbalance), we find out that the strength is about a 100 N for one magnet. We computed the size of the front screw and decided to use M6.

After having finished the assembling procedure of the magnets, we assembled the whole motor. Figure 8 gives an exploded view to each single part of the motor. The rotor-discs with 24 magnets each are shown. In the foreground the dummy-disc is seen, which was also manufactured in our company. This gives the chance to investigate a single or double sided magnetsystem. The size and the dimensions of the motor were completely computed and designed by TU Berlin.

ACHIEVMENTS

We developed a method to assemble the high energy magnets without losses. We also improved the ability to handle the Nd-Fe-B magnets with different materials and methods in a special field of application.

CEAM - COLLABORATION

Collaboration has been maintained with the TU Berlin.

PATENTS AND PUBLICATIONS

No patents property have been applied for. One joint paper was prepared together with TU Berlin.

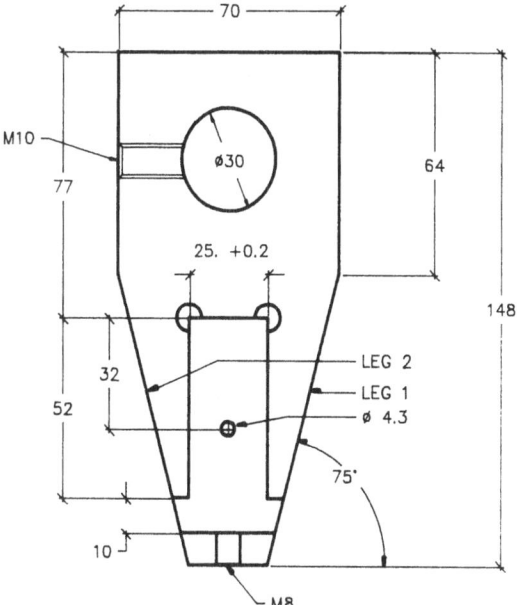

Figure 1 : Outline of the assembling device

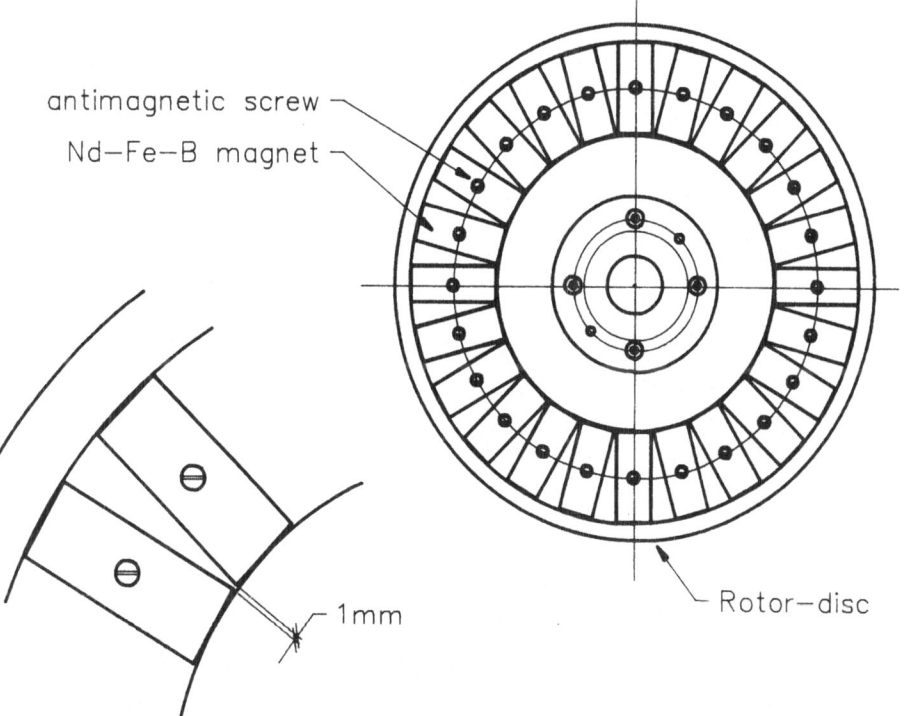

Figure 2 : Rotor−disc with the assembled magnets

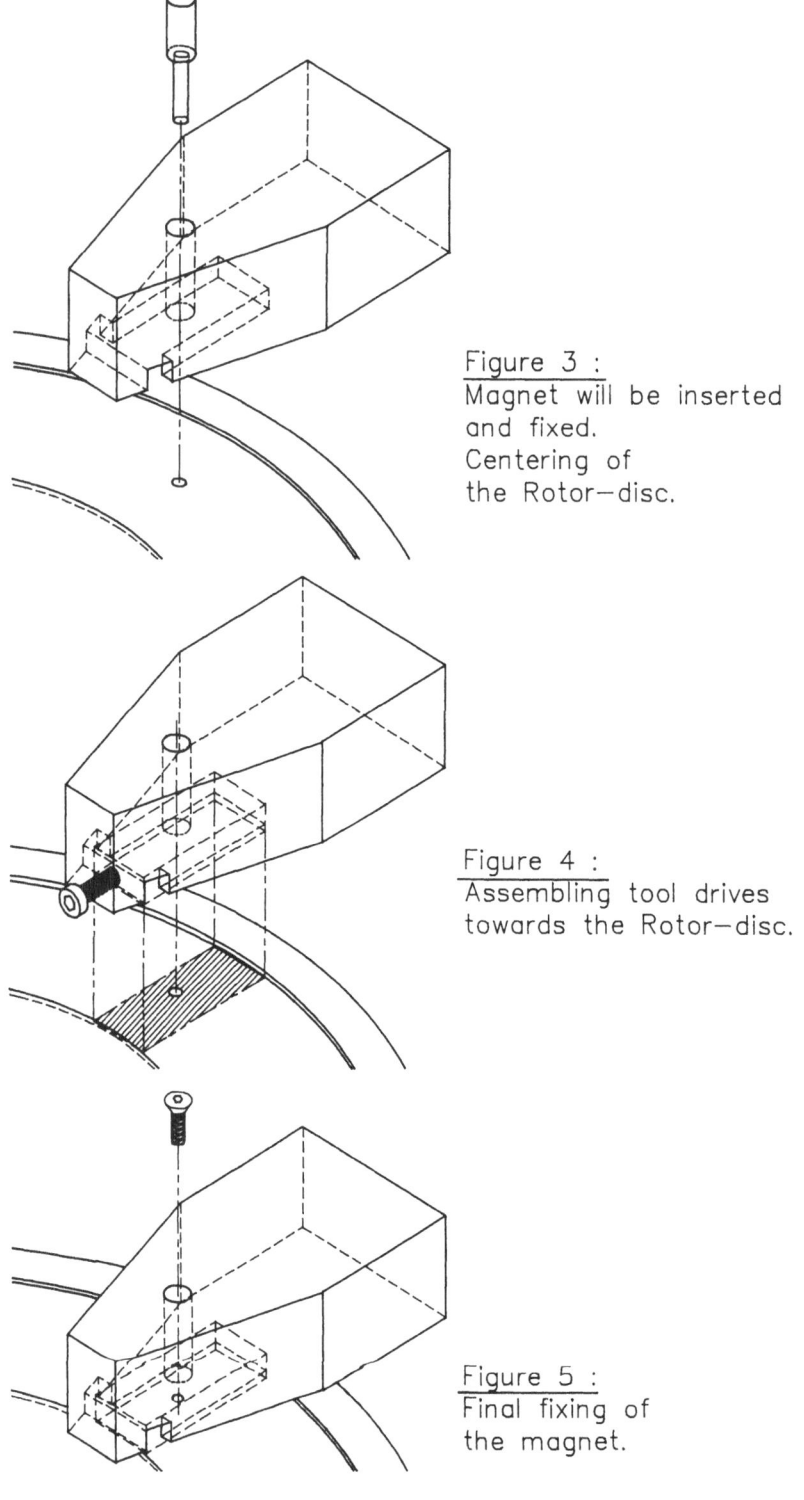

Figure 3 :
Magnet will be inserted
and fixed.
Centering of
the Rotor-disc.

Figure 4 :
Assembling tool drives
towards the Rotor-disc.

Figure 5 :
Final fixing of
the magnet.

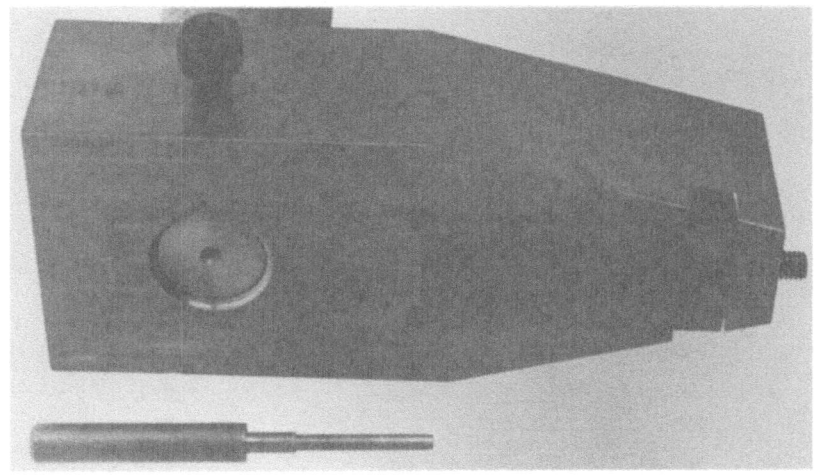

Figure 6: Assembling device with the inserted magnet

Figure 7: Part of the assembling procedure

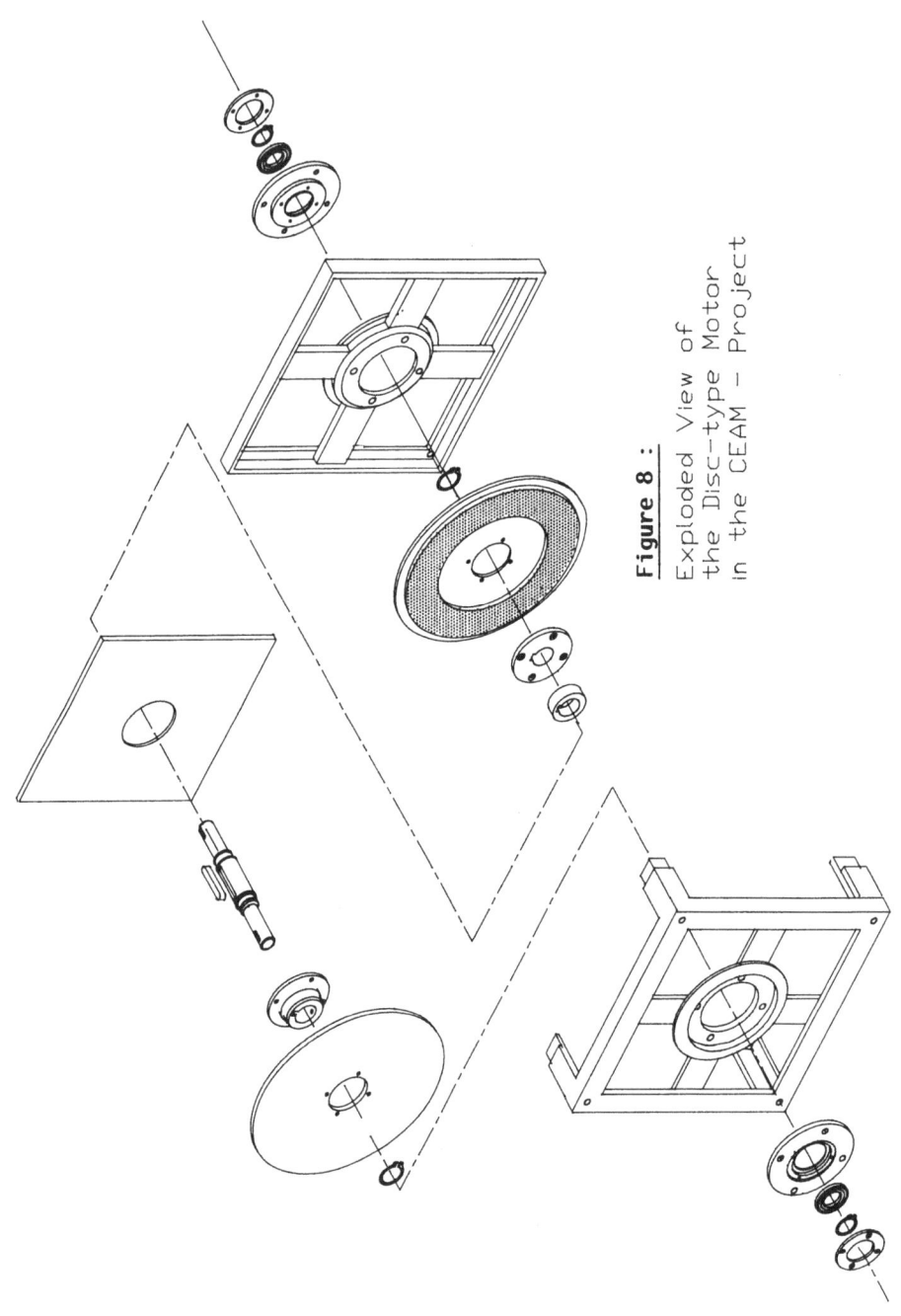

Figure 8 :

Exploded View of
the Disc-type Motor
in the CEAM – Project

MAGNETISING MULTI-POLE ND-FE-B MAGNETS

E.M.H.Kamerbeek**, A.J.C.van der Borst* and J.Koorneef*

Philips Research Laboratories
** 5600 JA Eindhoven, The Netherlands / D-5100 Aachen, West Germany
* 5600 JA Eindhoven, The Netherlands

ABSTRACT

In small axial-field motors, such as brushless DC motors and stepping motors, disc-type rotors of Nd-Fe-B have been applied successfully. Two methods for magnetising multi-pole rotor discs are investigated. The thickness of the discs dealt with is 0.7 to 1 mm, and the diameter 25 to 35 mm. It appears that numbers of poles up to 50 can satisfactorily be magnetised by using air-cored coils, which are excited with short pulses of high current. For higher pole numbers, up to 100, the use of an iron-cored magnetiser seems to be more adequate. However, preliminary results show that full magnetisation of a 100-pole rotor was not obtained.

INTRODUCTION

Multi-pole magnets are applied in electrical motors on a large scale (Howe et al., 1986). In small motors the magnets usually consist of only one piece that has been magnetised in a multi-pole field. Basic shapes for the one-piece magnets are rings in radial-field motors, discs in axial-field motors and strips in linear motors.

In this paper we will discuss the conditions necessary to magnetise multi-pole patterns in rotor discs made of the novel magnet material Nd-Fe-B. This high-quality material is very suitable for disc type rotors as applied in axial-field brushless DC motors and in stepping motors (Kamerbeek, 1987), because the direction of magnetisation coincides with the anisotropy axis, so that an optimal value for the remanence is achieved. Moreover, the disc can be made very thin, which yields a low-inertia rotor.

Photo 1 gives two examples of novel motor models that were designed and built using Nd-Fe-B disc rotors

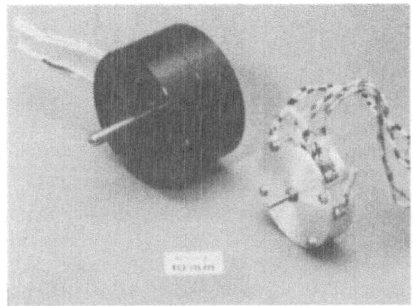

Photo 1   Novel motor models with Nd-Fe-B disc rotors. On the right: a 6-pole brushless DC motor; stator core fabricated from amorphous metal alloy, three-phase winding consists of etched multi-layer coils, electronic commutation with the help of rotational-voltage sensing; phase peak torque: 1.5 mNm at 0.3 A, phase resistance 2 Ω, phase inductance < 0.1 mH. On the left: a two-phase stepping motor; stator consists of toothed amorphous metal alloy, step angle : 3.6 °;holding torque (one phase excited): > 150 mNm.

The right one is a brushless DC motor. The rotor disc used has 6 poles, its diameter is 25 mm and its thickness 1 mm. The motor on the left is a stepping motor with a 3.6 ° step angle. The rotor contains 50 poles, its diameter is 35 mm and its thickness 0.70 mm. For further technical data of both motors see the photo caption.

In the following we first deal with some general technical aspects of magnetising Nd-Fe-B material, and also with the capacitor-discharge circuits used for generating short high-current pulses. In the main part of the paper we will describe and discuss the air-cored and iron-cored magnetising tools that we have developed for magnetising disc-shaped rotors having high pole numbers. Finally, we will make some concluding remarks with respect to the advantages and disadvantages of the magnetising methods dealt with.

MAGNETISING ND-FE-B MATERIAL

Initial magnetisation curve and demagnetisation curve

For Nd-Fe-B material very high magnetising fields are required.Normally, the flux density in the material must be increased beyond 3 T and the corresponding field strength beyond 1.5 MA/m. Fig.1 shows the initial magnetisation curve and the demagnetisation curve of Nd-Fe-B (NEOMAX 30H).

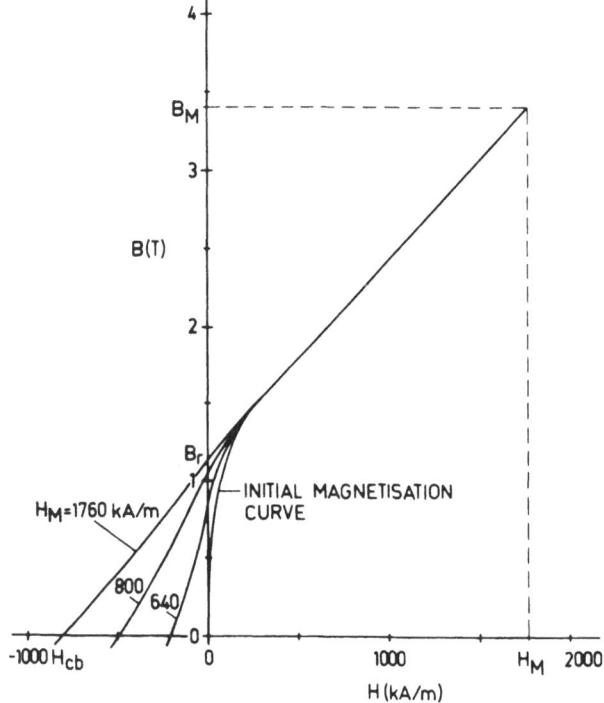

Fig.1   Initial magnetisation curve and demagnetisation curves of Nd-Fe-B material.

The remanence B_r and the coercivity H_{cb} resulting after magnetising depend on the value of B_M reached during magnetising, as shown in Fig.1. We found that the magnet material NEOMAX 30H is fully magnetised if B is increased to $B_M = 3.4$ T ($H_M = 1.8$ MA/m).

Calculation of required external field

If a spherical sample is to be magnetised in a uniform field produced by an air-cored coil, the required external field H_e can easily be found in a graphical way from the initial curve, provided that the demagnetising factor N of the sample is known. Denoting the working point at B_M with P_1 (see Fig.2), we have the relations

$$B_M = \mu_0 H_M + \mu_0 M \,(H = H_M), \quad H_M = H_e + H_d, \quad \text{and} \quad H_d = -NM \,(H = H_M) \tag{1}$$

where M is the magnetisation and H_d the demagnetising field of the magnet.

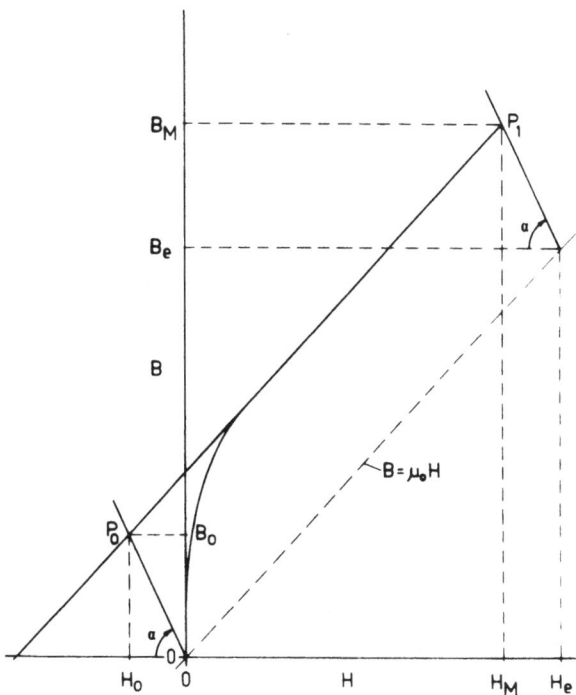

Fig.2 Graphical determination of the external field H_e.

From these equations it follows that

$$\frac{B_M - \mu_0 H_e}{\mu_0 H_e - \mu_0 H_M} = \frac{1-N}{N} = k. \tag{2}$$

After the magnetising process it holds for working point P_0 that

$$\frac{B_0}{-\mu_0 H_0} \, k. \tag{3}$$

where $k = \tan \alpha$, α being the angle of the load line. Of course, in magnetising multi-pole discs we do not use spherical samples and uniform external fields. However, we can still use the above construction. Then the point P_0 becomes a function of the space coordinates, and B and H correspond to the field components in the direction of the easy axis of the magnet. From the field H_e we can calculate the value of the current that must be supplied to the magnetising coil. In the following section a simple example will be given.

When the magnet is magnetised in an iron-cored coil, a task may be to calculate the needed ampere-turns ni. Schematically, Fig.3 shows an iron-cored magnetiser.

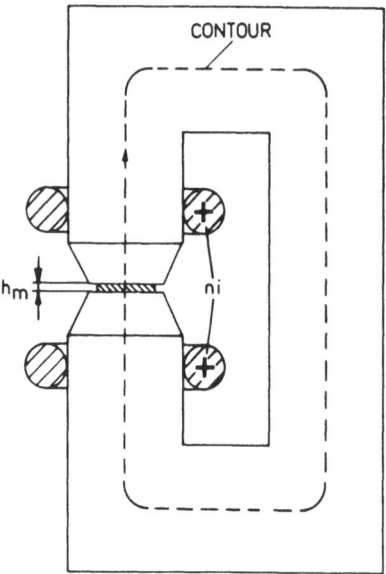

Fig.3   Schematic representation of an iron-cored magnetiser.

The disc to be magnetised is placed in the airgap between the pole pieces. For the working point P_1, where $B = B_M$ and $H = H_M$, we have

$$ni = H_M h_m + \int_{iron} H_c dc,$$ (4)

where h_m is the thickness of the magnet, and H_c the component of H along the contour. In magnetising Nd-Fe-B magnets the integral yields a notable contribution to the ampere-turns, because saturation of the pole pieces cannot be avoided in the environment of the magnet. By using a high-saturation soft-magnetic material, such as Co-Fe, the ni-value can be minimised. The flux passing through the airgap can be estimated from the relation

$$\phi_{gap} = A_m B_M + (A_p - A_m) \mu_0 H_M,$$ (5)

where A_m is the cross section of the magnet and A_p is the surface area of the pole piece. It is clear that an accurate calculation of the needed ampere-turns can only be performed by using numerical field calculations (Nakata et al., 1984)

Eddy-current effects

If short current pulses have to be used to prevent excessive heating of the coil, eddy currents can cause considerable problems, for they hinder a fast penetration of the field into the electrically con-ducting material. So there is always a minimum pulse time t_{pmin}, whose value depends on the material properties, the magnetiser geometry and the pulse shape. The minimum pulse time can be estimated by using the formula for the skin depth for sinusoidal steady-state diffusion into a linear material

$$\delta = \sqrt{\frac{2}{\mu_0 \mu_r \sigma \omega}} \,,$$ (6)

where μ_r is the relative permeability and σ is the electric conductivity. In our case we use a single pulse which we approximate by a sinusoidal shape. The radian frequency ω from eq.(6) is now related to the pulse time t_p by $\omega = \pi / t_p$. Of course, magnetising is a transient phenomenon, but then skin depth δ can also be used as a parameter characterising the depth of penetration.

If during magnetisation the field must penetrate into the material over a diffusion length L_d, then in any case it must hold that $\delta > L_d$ or

$$t_p > \frac{\mu_0 \mu_r \sigma \pi}{2} \, L_d^2 .$$ (7)

Introducing for Nd-Fe-B the average values $\mu_r = 2.2$, $\sigma = 0.7 \times 10^6$ $(\Omega m)^{-1}$ we obtain the final condition

$$t_{pmin} = 3.0 \, L_d^2 .$$ (8)

For instance, if $L_d = 1$ cm, then t_p must be more than 0.3 ms. However, if L_d is only 1 mm, then the pulse time can be further reduced by a factor of 100 ($t_p > 3$ μs). It is clear that in practice the skin depth can also be reduced by increasing the level of the penetrating field, because the permeability μ_r decreases at higher field levels (see Fig.1).(It must be noted, however,that the transient phenomenon is also in-fluenced by accompanying thermal effects.)

To investigate the eddy-current effects we performed some experiments and made further cal-culations for small disc magnets ($D_m = 6$ mm, $h_m = 1$ mm) which were magnetised in an air-cored annular coil; see Fig.4a.

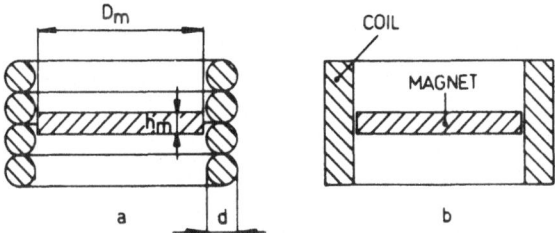

Fig.4   Coil used for magnetising a small disc-shaped Nd-Fe-B sample. a: Real configuration. b: Configuration used for the calculations.

According to the above, the flux density in the centre of the magnet must be increased to $B_M = 3.4$ T for full magnetisation. The field produced by the uniformly magnetised disc alone will be at that point

$$B_d = \frac{\mu_0 M h_m}{\sqrt{D_m^2 + h_m^2}} \,, \tag{9}$$

where D_m is the magnet diameter, h_m the magnet thickness and $M = M(H_M)$. From eq.(9) it follows that $B_d \simeq 0.2$ T; hence, the field excited by the coil must be $B_i = B_M - B_d = 3.2$ T. By using an elementary formula we then calculate that the current in the coil must be 5.3 kA. Further we calculate from eq.(8) that the pulse time must be greater then 30 μs ($L_d = 3$ mm). From our measurements we could conclude that full magnetisation still occurs with pulses that have a pulse time of about 70 μs. Unfortunately, our capacitor discharge circuit did not allow a further reduction of the pulse time.

In addition we did some numerical field calculations of the eddy-current phenomena for a simplified model of the above air-cored magnetiser, shown in Fig.4b. The coil shape was simplified to a rectangular copper section, and the skin effect in the coil was neglected. As hysteresis effects could not be dealt with in the computer program (PE2D), we replaced the hard-magnetic disc by an isotropic soft-magnetic one, whose B-H curve is represented by the initial magnetisation curve of the original magnet. For the sinusoidal pulse we chose a peak value i_p of 21.6 kA. Under static conditions this current will produce a flux density B_i of about 3.4 T in the centre of the magnet. Fig.5 shows how the field B_M penetrates into the 'magnet' for two different pulse times t_p.

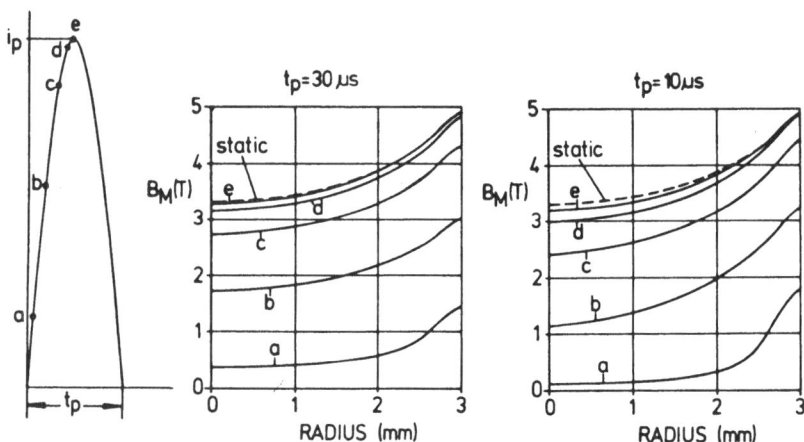

Fig.5 Penetration of the field B_M into a small disc. $i_p = 21.6$ kA. Profiles are plotted for given current levels as denoted in the figure. a: $t_p = 30$ μs, b:$t_p = 10$ μs.

We observe that in case a ($t_p = 30$ μs) the eddy-current effect is present, but the static field value is still reached when $i = i_p$. In case b ($t_p = 10$ μs), however, the eddy-current effect is more pronounced, and the static field value is not reached. We conclude that the estimation of t_{pmin} using eq.(8), may be somewhat too pessimistic for thin discs. It should be noted that, because of the above simplifications, the numerical calculations can give only a first approximation of the eddy-current phenomena during a realistic magnetisation procedure.

Maximum pulse time due to temperature rise of the coil

During magnetisation the coil is heated up by the current pulse, and too high a temperature rise ΔT will destroy the coil. In the following we make a rough calculation of the temperature rise due to a single sinusoidal current pulse. Assuming that the heating is adiabatic, and neglecting the skin effect in the wire, we obtain

$$\Delta T = \frac{8\rho_e i_p^2 t_p}{\pi^2 c_h d^4} ,$$

(10)

where ρ_e = average coil resistivity, c_h = average specific heat capacity, d = wire diameter, i_p = peak current. In Fig.6 i_p is plotted as a function of t_p for ΔT = 250 K and d=0.25, 0.5 and 1 mm. In this figure we used the values $\rho_e = 1.72 \times 10^{-8}$ Ωm and $c_h = 3.47 \times 10^6$ J/(m³K) (values at room temperature). Air-cored magnetisers for small pole pitches need a small wire size. From the above it becomes clear that these magnetisers can only operate at the required current level if very short current pulses can be generated.

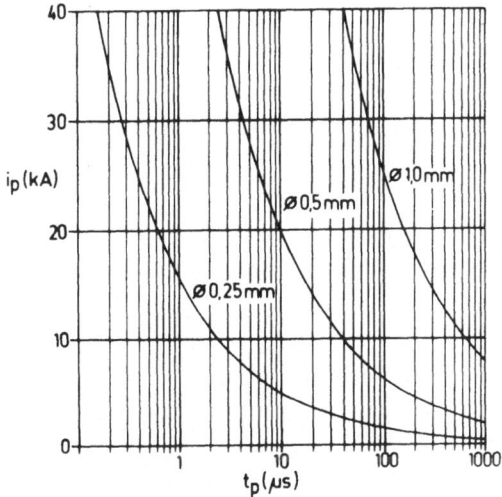

Fig.6 Relation between i_p and t_p for ΔT = 250 K and d=0.25, 0.5, 1 mm.

Capacitor discharge circuit

The principle of the circuit used is shown in Fig.7. As a result of the discharge diode D1 and the rectifying diode D2, and the fulfilment of the condition $R \ll 2\sqrt{L/C}$, the coil current will be a weakly damped sinusoidal pulse, whose peak value and pulse time can be approximated by

$$i_p = U_{c0}\sqrt{L/C} , \quad t_p = \pi\sqrt{LC} ,$$

(11)

where $U_{c0} = U_c$ (t=0).
Provided that the discharge resistance fulfils the condition R1 C $\gg t_p$, the capacitor voltage will be reversed almost to the value $-U_{c0}$ at t = t_p Then the capacitor is discharged via the diode D1 and resist-

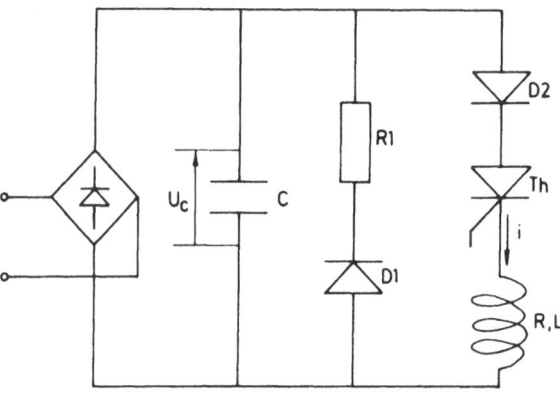

Fig.7   Principle of the capacitor-discharge circuit used for the generation of short high current-pulses.

ance R1. At our request an experimental capacitor discharger based on the above principle has been developed by Steingroever GmbH of Cologne. With the help of this device we were able to produce current pulses up to 28 kA having a pulse time of about 70 μs.

MAGNETISERS FOR MULTI-POLE DISCS

Air-cored magnetisers

The magnetiser coil is wound from copper wire, and consists of two halves in between which the disc is clamped, as shown in Fig.8. The direction and distribution of the currents must agree as well

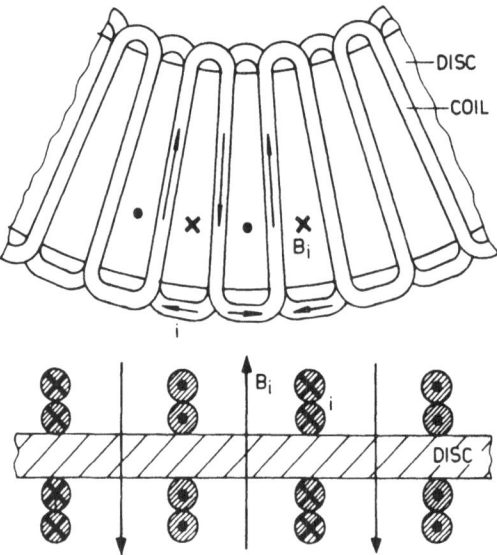

Fig.8   Basic construction of an air-cored magnetiser for multi-pole disc rotors.

as possible with the magnetisation-current distribution ($\nabla \times$ **M**) resulting after magnetising. In order to achieve high shape accuracy, high mechanical strength (in view of the high Lorentz forces), and good electric insulation, the wires are imbedded in a structure of a synthetic material.

Using a computer program we calculate the field B_i produced by a given current in the magnetising coil. In this way we find the dependence of B_i on the coil geometry, the wire diameter d and the number of poles p. Fig.9 gives an example of the calculated field profiles for a 8-pole and a 50-pole magnetising coil both with d = 0.5 mm.

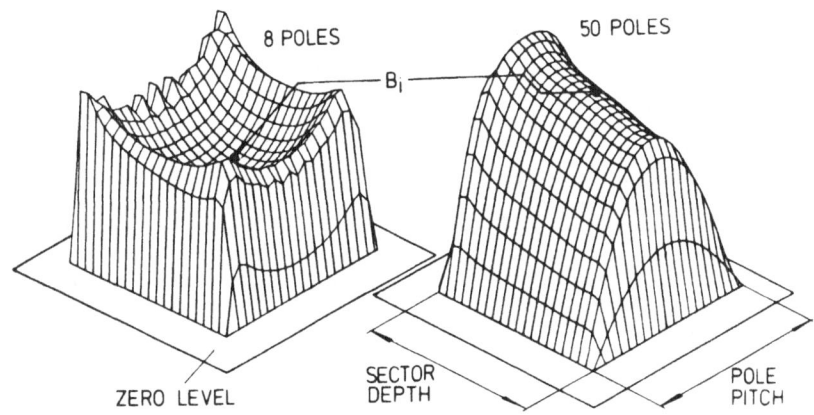

Fig.9   Calculated B_i-field profiles for a 8-pole and a 50-pole magnetiser

From these plots we obtain the coil current i_{p0} needed for achieving a flux density $B_i = 3.4$ T in the centre of the pole; see Fig.10. According to Fig.2, at that point the flux density B_M will then be some-what higher than 3.4 T. Next we determine from Fig.6 the maximum pulse time t_{pmax} for $i_p = i_{p0}$ and the wire diameter chosen. If now the actual pulse time t_p can be made shorter than t_{pmax}, the design is usable, provided that $t_p > t_{pmin}$, the latter being calculated from eq.(8).

In practice it is often required to have a narrow transition region between the poles. Then the relative steepness of the magnetising field B_i must be maximised. As a first approximation this steepness is proportional to the ratio $i_{pmax}\tau/(i_{p0}d)$, where τ is the average pole pitch and i_{pmax} follows from t_p and d. Fig.11 shows the influence of the wire diameter on the field profile at the pole centre radius, i_p being equal to i_{p0} (p = 20, disc size: $\phi35 \times \phi21 \times 0.7$mm).

Finally, we give a short description of the magnetisers that have been used for magnetising the rotors of the motors shown in photo 1. The 6-pole magnetiser, shown in photo 2, is characterised by d = 1 mm, $i_{p0} = 12.9$ kA, $t_{pmax} = 380$ μs and $t_{pmin} = 90$ μs.

As $t_{pmin} > 70$ μs, we did not use the short pulse magnetiser but another type delivering a critical damped pulse shape (peak current is 11.6 kA at t = 90 μs). Photo 3 gives an impression of the magnetisation pattern induced in the disc.

From a comparison of calculated with measured data for the normal component of the flux density along the disc surface, it can be concluded that in the centre of the disc full magnetisation is achieved ($M = M_{sat}$). We also did a test with a sinusoidal pulse characterized by $i_p = 13$ kA and $t_p = 70$ μs. Then

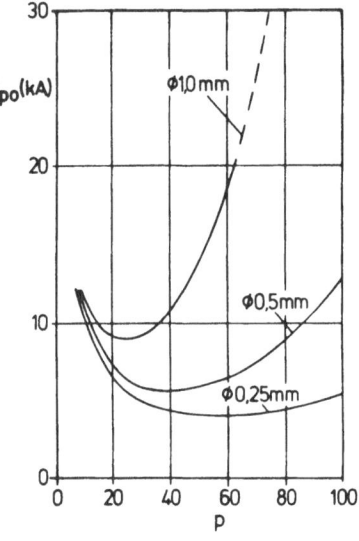

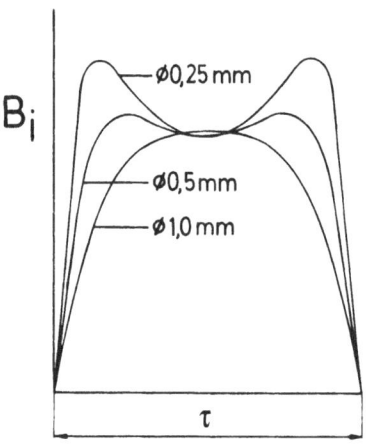

Fig.10 Current i_{p0} as a function of pole number and different wire diameters ($B_i = 3.4$ T and disc size $= \phi\, 35 \times \phi\, 21 \times 0.7$ mm)

Fig.11 Profile of B_i-field at centre radius for $d = 0.25, 0.5, 1$ mm ($i_p = i_{p0}$).

we observe that the disc is not fully magnetised in the centre because of the eddy-current phenomenon. This is in agreement with the fact that the condition $t_p > t_{pmin}$ is not satisfied.

The second magnetiser, built for the 50-pole rotor, is shown in photo 2 on the right. This device is characterised by the quantities $d = 0.5$ mm, $i_{p0} = 6$ kA, $t_{pmax} \approx 110\ \mu s$, $t_{pmin} = 2.3\ \mu s$. The pulse time used is $t_p = 70\ \mu s$, yielding $i_{pmax} = 7.4$ kA. The magnetisation pattern is shown in photo 4. Here too we can prove that in the pole center $M = M_{sat}$.

Photo 2   Air-cored magnetisers. On the left a 6-pole magnetiser, on the right a 50-pole magnetiser.

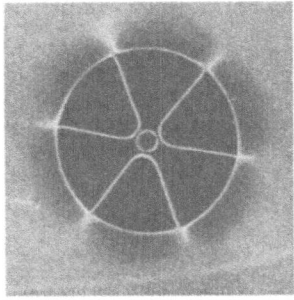

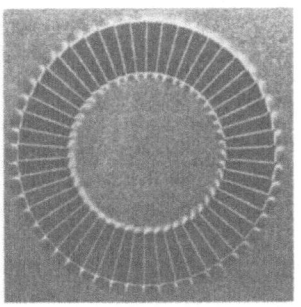

Photo 3   6-pole pattern magnetised with
an air-cored magnetiser.

Photo 4   50-pole pattern magnetised
with an air-cored magnetiser.

Increasing the pole number in the above disc to 100 would cause severe problems as to temperature rise and required pulse time. In that case the above estimations, based on $\Delta T = 250$ K and full magnetisation,   yield   for $d = 0.25$ mm : $i_{p0} = 5.4$ kA, $t_{pmax} = 8\,\mu s$, $t_{pmin} = 0.6\,\mu s$,   and   for $d = 0.50$ mm : $i_{p0} = 12.8$ kA, $t_{pmax} = 24\,\mu s$, $t_{pmin} = 0.6\,\mu s$. Thus in principle pulse times shorter than $24\,\mu s$ are required. At the moment we do not have equipment that can generate such short high-current pulses.

Iron-cored magnetiser

To overcome the technical problems arising in magnetising short pole pitches by means of air-cored magnetisers, we also investigated the use of a special iron-cored magnetiser. Here, we will discuss only the principle and the first results of this method, because the study has not yet been completed.

In this procedure the disc is first fully magnetised in one direction, and then locally remagnetised in the opposite direction by using two toothed cylindrical pieces of iron, see photo 5, which are clamped with the disc in between. This assembly is placed in the airgap of a normal iron-cored magnetiser. It is obvious that with this method no short high-current pulses are required, so that problems as to eddy-current effects and temperature rise are avoided.

We shall explain the remagnetising procedure in more detail using a simplified model. A fully magnetised strip is placed between two toothed pieces of iron. Next, a high magnetic potential is applied over the airgap, producing an external field that counteracts the original magnetisation of the disc; see Fig.12. The intention now is that in the tooth region the magnetisation M_{sat} is reversed to $-M_{sat}$, and that in the slot region the magnetisation M is not affected. In fig.13 the procedure is elucidated in a graphical way.

The B-H curve of the magnet material is plotted for two quadrants, and the working points before remagnetising in the slot region and the tooth region, are denoted by P_{s0} and P_{t0} respectively. Because of the slotting we have $\alpha_s < \pi/2$, while $\alpha_t = \pi/2$ .When the magnetic potential is applied, the intended result is obtained if working point P_t shifts beyond P_{t2}, and working point P_s does not pass point P_{s1}. Using the construction dealt with in Fig.2, we find that P_{s1} corresponds to an external field $H_{1e} < H_1$. The conditions to be met during remagnetising therefore are: $H_M > H_2$ and $H_{se} < H_{1e}$, see Fig.13. From

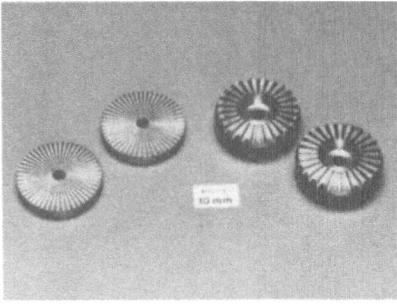

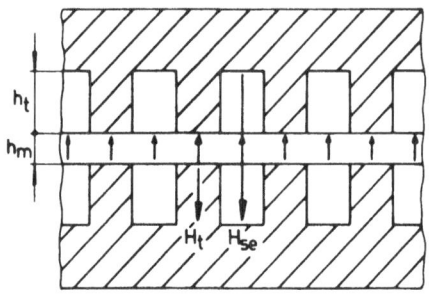

Photo 5 Pole pieces of an iron-cored magnetiser. On the left: pole pieces of a 100-pole magnetiser; on the right: pole pieces of a 50-pole magnetiser.

Fig.12 Schematic two-dimensional model of an iron-cored magnetiser for multi-pole magnets.

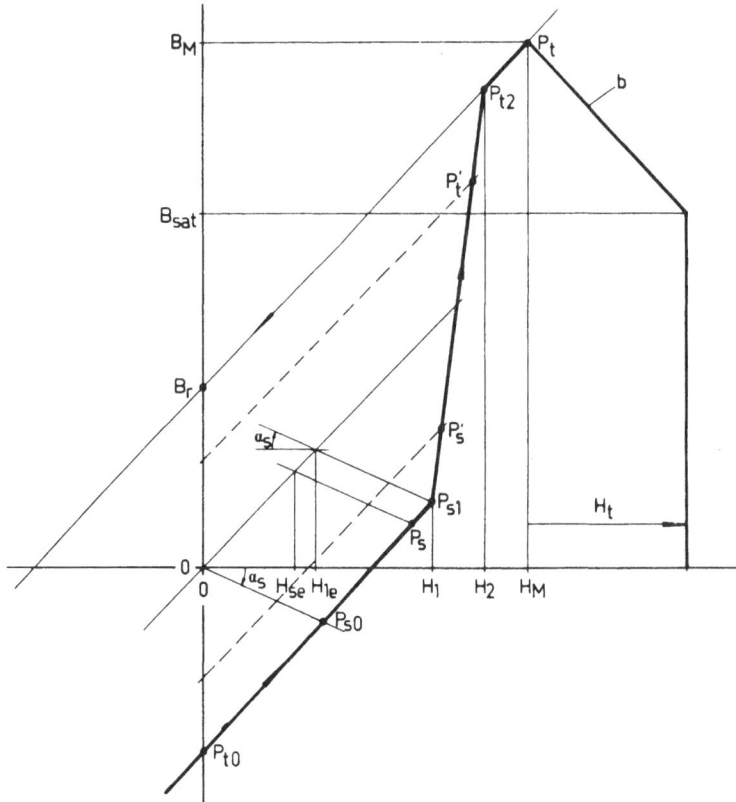

Fig.13 Principle of an iron-cored multi-pole magnetiser explained in a graphical way.

an elementary calculation it follows that $H_{1e} = H_1 - B_r/(\mu_0(1 + \tan \alpha_s))$. Further we use the approximation:

$$H_{se}(h_m + 2h_t) \simeq h_m H_M + 2h_t H_t, \tag{12}$$

where H_t is the field strength in the tooth and h_t the height of the tooth. The field H_t is easily found from B_M and the B-H curve b of the tooth material, see Fig. 13. As $B_M > B_{sat}$, we have $H_t = (B_M - B_{sat})/\mu_0$. Introducing the parameter $\beta = h_m/(h_m + 2h_t)$, we obtain the conditions to be met for fully remagnetising:

$$H_M > H_2 \text{ and } \beta H_M + (1 - \beta) H_t < H_1 - B_r/(\mu_0(1 + \tan \alpha_s)), \tag{13}$$

It is obvious that this condition can be satisfied only if $\beta < 1$, and can be more easily satisfied if $\Delta H = H_2 - H_1$ and H_t are smaller, and H_M lies closer to H_2.

When eq.(13) cannot be fulfilled, then a multi-pole pattern with a lower magnetisation level than M_{sat} is still achievable. However, we must then ensure that the working points $P_s{}'$ and $P_t{}'$ have an equal distance to the line $B_e = \mu_0 H_e$. so that these points lie on recoil lines with equal but opposite values of residual magnetisation M; see Fig. 13.

In the above way we magnetised a 100-pole disc (size: ϕ 35 x 0.7 mm), shown in photo 6.

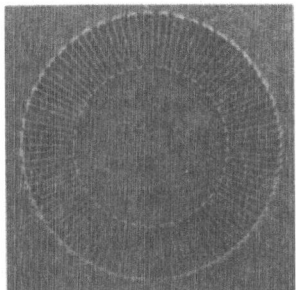

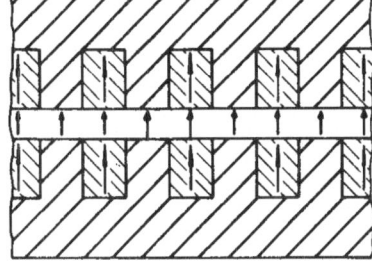

Photo 6   100-pole pattern magnetised with an iron-cored magnetiser.

Fig. 14   Improvement of a multi-pole magnetiser by using slots filled with Nd-Fe-B magnets.

In agreement with the above, we obtained the best result by using a material with a high anisotropy (ΔH small). However, the saturation level M_{sat} was not reached. Further investigations are going on, using numerical field calculations in order to find the effects of saturation and tooth geometry.

In principle, the above method can be improved by filling up the slots in the pole pieces with uniformly magnetised Nd-Fe-B, see Fig. 14 . Then $\alpha_s \to \pi/2$, whereas β is not affected. The fulfilment of the conditions (13) must then be easier, which can also be concluded from the construction shown in Fig. 13.

CONCLUDING REMARKS

Pole numbers up to 50 can be fully magnetised with air-cored magnetisers, provided that short pulses of high current are available. Sinusoidal pulses with a peak current of 13 kA and a pulse time of 70 μs are necessary. An advantage is that the final magnets can easily be separated from the ironless magnetiser. A pole number of 100 could be magnetised by using an iron-cored magnetiser. An advantage then is that the generation of the DC coil current is not a technical problem. However, full magnetisation does not seem to be achievable. A disadvantage of the method is that the final magnets are attracted to the magnetiser, so that special measures are required to handle them.

REFERENCES

Howe, D., Birch,T.S., and Williams, I.J., 1986. New design opportunities for permanent magnet excited machines. Proc. ERA Seminar : 'Permanent Magnets are good for your Wealth' (16 September 1986), London.
Kamerbeek, E.M.H., 1987. The significance of magnet properties for applications in electromechanical devices. Proc. ERA Seminar : 'Magnetic Materials Attract Business' (4-5 November 1987), London.
Nakata, T., Takahashi, N. and Fujiwara, K., 1984. Numerical design methods for magnetisers. Journal of Magnetism and Magnetic Materials, 41, 418-420.

NEW DESIGN OF A HYBRID STEPPING MOTOR
USING Nd-Fe-B PERMANENT MAGNETS

G. Ioppolo

Motori ed Apparecchiature Elettriche,
Offanengo, Italy

EXTENDED ABSTRACT

The outstanding remanence and coercivity of the new Nd-Fe-B permanent magnets are very advantageous for stepping motors and, within the present programme, a hybrid stepping motor incorporating such magnets, has been developed, built and tested.
A ring magnet with axial magnetization of 2mm thickness was used. The internal diameter of the ring magnet was 7,5mm and the outside diameter was 20,4mm. The specifications of this motor are given in the Table 1.

Parameter		Value
step angle		1,8 degree(+/- 3%)
rated phase current(A)		1,4
holding torque*	(Ncm)	40
detent torque	(Ncm)	3
rotor inertia	(Ncm2)	75
length	(mm)	38,5
outside diameter	(mm)	51

Table 1. **Motor Parameters (two phases energized at rated current)**

Due to the temperature characteristics and magnetic properties of Nd-Fe-B, the behaviour of the hybrid stepping motor depends on the working temperature of the magnets.
Figure 1. indicates the holding torque versus current. The holding torque was measured at 20 degree C and at an temperature of 100 degree C. A torque reduction of about 14% was measured. It is anticipated that when improved Nd-Fe-B magnets become available, their use should reduce the irreversible flux losses and consequently the torque decrease will be acceptable. Suitable bonded magnets were not available at the time the prototype design was initiated.

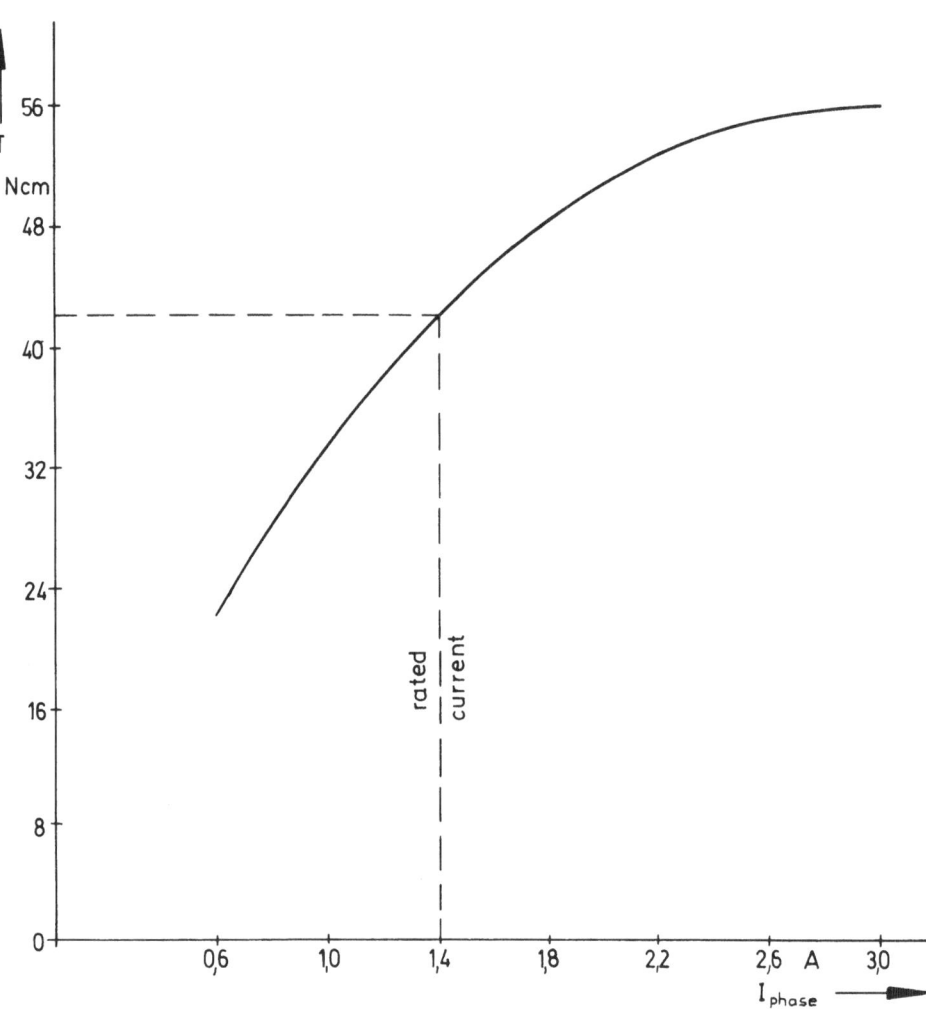

Fig. 1   Holding torque v.s. current

FeNdB for Slotless, Brushless D.C. Machines

I. Kitzmann, K.R. Schulz

Dornier GmbH

P.O.Box 1420

D-7990 Friedrichshafen (FRG)

Once a decision has been made, more or less intuitively, one can be happy if the intuition is verified in the course of time. When starting in the CEAM-project, we were looking for a possible innovation to be realized by using FeNdB-magnets. The basic idea was not to start with known d.c. motors designed optimally for e.g. hard-ferrite magnets, and adapting them for FeNdB substituting the hard-ferrite, but looking for less familiar designs, not useful for "classical" permanent magnets, but potentially for FeNdB.

How to get the proper design? Not with the help of CAD, at least not at the beginning. So other means have to be considered. First, it is necessary to have at least a good feeling of the trends in material properties to be sure, hopefully, that the chosen design allows for development with the improving materials. Concerning permanent magnets, Zijlstra has shown /1/ that the energy product of permanent magnets steadily increases, through newly developed materials. Starting with 2 kJ/m^3 one hundred years ago, having reached about 350 kJ/m^3 today, magnets of the year 2000 should have about 800 kJ/m^3, if extrapolation of the state-of-the-art will be realized. Considering the acceptance problems FeNdB had when it was young, who can imagine how to use these future magnets at the best of their properties?

But better permanent magnets alone do not satisfy the challenge of new designs, as the magnetic circuit is strongly influenced by the soft magnetic materials. The improvement of silicon steel, for instance, with regard to saturation, induction, and core losses is low, almost stagnant. Since decades, the saturation is in the region of 1.5 T up to 2 T, depending on other disadvantages the designer will accept at higher saturation; the gain by using CoFe-alloys up to 2.5 T is questionable, especially when design-to-cost is mandatory.

Trying to extrapolate the magnetic properties of electrical steel for the next decade, we come out where we are, unless new and applicable soft magnetic materials are found. This is the new challenge!

So design elements, leading to severe saturation of the electrical steel, must be omitted. One way is at least getting rid of the teeth and slots normally used. The teeth serve a.o. as concentrators for the magnetic flux, and because of the small air-gap the magnetic circuit is subject to armature-reaction and other non-linearities. Teeth are responsible for magnetostriction-induced noise, heat, and unbalanced forces on the rotor especially at small air-gaps. The solution is a toothless motor with a homogeneous air-gap filled with copper to such an extent that the remaining mechanical clearance allows safe operation of the motor. Such a cylindrical-type motor is described in /3, 4/, the latter collaborative work being stimulated through CEAM. A new design by W. Volkrodt is presented in /5/: a disc-type machine with a toroidal stator. One speciality is the lack of bulky end-winding, with the benefit of reduced ohmic losses and gain of space. An intensive study of this toroidal machine is being performed at UMIST, Manchester; first results are presented in /6/.

Beyond the opportunities of slotless d.c.-motors as outlined in /7/ there is a high resistance against demagnetization caused by over-current. Over-current factors in the order of 5...10 are permissible with respect to the nominal current, determined by heat balance in continuous service. A high peak-to-average ratio is required for servo-applications; the reduced winding-inductance in slotless motors allows the effective use of switch-mode current-control, thereby lowering the time-constant.

The highest benefit of modern magnets will be for electrical drives. It was presented by Zijlstra /2/ that there is still plenty of room for the improvement of electrical drives. He comes out with the simple recipe: 1/3 magnet, 1/3 iron, 1/3 copper, to get a higher power density. This relation is far from being fulfilled in conventional drives.
We suggest a fourth ingredient to be added to the recipe: electronics. There is a strong trend for both signal processing and power control, towards higher integration, higher power capability and switching speed and lower losses. Future drives demand interdisciplinary, not simply

additive combination of electromechanics, sensors and electronics in one machine, in one body.

Doing this, there is a need for a design with a high power-conversion efficiency, otherwise the designer will run into problems with cooling of the motor, especially with respect to the temperature-sensitive electronics. The reward for the successful designer can be an electric drive running at a 2-wire d.c. power bus, and being controlled by a fiber-optics-link, with an unequalled simplicity and comfort for the user, and free from electromagnetic interference problems. We are convinced that the "slotless and brushless" approach is one way in the right direction.

From our point of view, CEAM was a successful stimulator for future-oriented applications of FeNdB. The fruitful cooperation between UMIST and us was catalyzed by CEAM.

References:

/1/ H. Zijlstra
"Introduction to Permanent Magnets"
Proc. Workshop Meeting "Nd-Fe Permanent Magnets" Brussels, Oct. 25, 1985, Ed. by I.V. Mitchell, Commission of the European Communities, pp. 5-11

/2/ H. Zijlstra
"Application of Permanent Magnets in Electromechanical Devices", lecture given at the "Seminar über moderne Magnetwerkstoffe" Ruhr-Universität Bochum, Aug. 15, 1985 (unpublished)

/3/ W. Jaffe
"More 'teeth' for toothless motors"
Machine Design, Aug. 8, 1985, pp. 18-20

/4/ B.J. Chalmers and I. Kitzmann
"D.C. Motors now both slotless and brushless"
CEAM News Letter, No. 8, Sep. 1987, pp. 1-2

/5/ W. Volkrodt
"Neue Wege im Elektromaschinenbau"
Elektro-Jahr 1985, pp. 20-38, Vogel-Verlag, Würzburg

/6/ E. Spooner and B.J. Chalmers
"Toroidally-wound, axial-flux, permanent-magnet brushless dc motors"
Intern. Conf. on Electrical Machines, ICEM '88, Pisa, Sept. 1988,
Vol. III, pp. 81-86

7   B.J. Chalmers et al.
"Permanent-Magnet A.C. and D.C. Machines"
CEAM Final Report, 1989, pp. 613-624

THE ROLE OF NEODYMIUM-IRON-BORON IN RELATION TO OTHER MATERIALS FOR PERMANENT MAGNET MACHINES WITH EMBEDDED ROTOR MAGNETS

K. J. BINNS

UNIVERSITY OF LIVERPOOL, UNITED KINGDOM

ABSTRACT

The performance of permanent magnet machines excited by neodymium magnets is assessed by comparison with samarium cobalt and ferrite magnets. Both 4 pole and 6 pole machines have been constructed using the 'high field' configuration developed and patented by the Liverpool group. The characteristics of a drive system involving an ultrasonically modulated inverter and a permanent magnet machine are demonstrated.

INTRODUCTION

Interest in permanent magnet machines for drive systems was almost non-existent fifteen to twenty years ago. Alnico magnets were used, for example in exciters, and still are. Ferrite magnets were found in automobile equipment like windscreen washer motors. Research was being carried out and patents applied for in the area of drive systems but only a small handful of people were involved (Binns, 1976), (Binns and Jabbar, 1978), (Binns, 1980).

It was only when samarium cobalt magnets, with their high stored energy, became available that Industry recognised that permanent magnet machines were not simply low power special devices of minimal commercial importance.

Contracts were placed to develop prototype machines for a variety of applications. It was demonstrably clear that the magnet could be used to up-rate conventional systems (4), including those using induction and reluctance motors, but the cost of the magnets was generally found to be too high for widespread use. Many designs were proposed in which the expensive material was under utilised.

In 1983 two firms, one in the USA and the other in Japan, announced the development of neodymium-iron-boron magnets and in the short interval since then many other magnet producers have developed and produced these magnets. The cost is currently comparable with samarium-cobalt but it is hoped that the cost will fall ultimately by a factor of about four.

The emergence of rare earth permanent magnet materials such as neodymium-iron-boron has meant that electrical drives involving permanent magnet machines are becoming increasingly competitive. Since the rotor is virtually loss free, they can be designed for higher specific outputs and efficiencies than their induction motor equivalents. There is also a control benefit in that they can be operated in an open-loop if desired, thereby avoiding the need for feedback.

COMPARISON OF NEODYMIUM-IRON-BORON MAGNETS WITH SAMARIUM COBALT AND FERRITE MAGNETS

There are many designs of rotating machine involving permanent magnets, some with magnets fixed to the rotor surface and others with radial magnets but the most efficient use of magnet material is achieved with non-radial magnets making use of the configuration shown in Figure 2. This form of combination can be used over the full range of magnetic energy density varying from the ferrites to the rare earth (Binns, 1984), (Binns and Wong, 1984), (Binns, Riley and Wong, 1985), (Binns and Riley, 1986).

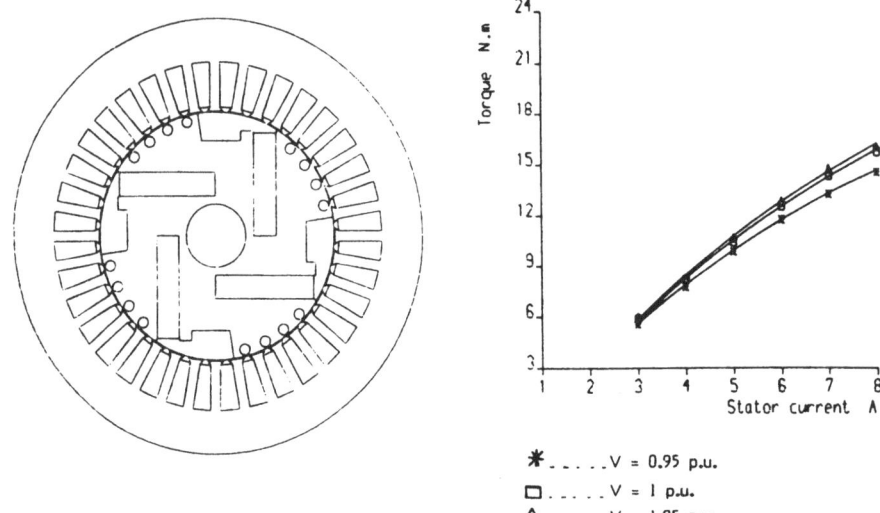

Fig. 1: Configuration of a high-field permanent magnet synchronous machine.

Fig. 2: Variation of torque as a function of the stator current for the Sm - Co₅ machine.

This particular design, when appropriately applied, results in an air-gap density exceeding that in the magnet itself. This results in a high torque whilst keeping magnet volume to a minimum. The output per unit magnet volume and the output per unit rotor volume are both relatively high. Susceptibility to demagnetisation is minimal because of the saturation of the iron flux path in the rotor.

In the light of the characteristics of three materials, ferrite, samarium-cobalt and neodymium-iron-boron, a comparison of achievable performance is made. A geometry with high performance potential (5) is used for the comparison since it makes possible a high air-gap field and hence a high torque to machine volume. Fig. 1 shows a cross section of the machine which can operate as line start synchronous motor or as a so-called 'brushless d.c. machine' if position sensing is incorporated.

The performance capability of this machine has already (6) been discussed but in this paper the design has been optimised for each of the three materials.

For a comparison to be worthwhile the dimensions have to be chosen to maximise performance for a given magnet.

Table 1 shows some of the important characteristics of the 3 materials chosen. Note the progressive increase in remanence and coercivity from the relative low values for the anisotropic ferrite to the much higher values for samarium cobalt and neodymium. However, the ferrites are, of course, relatively very cheap. The maximum operating temperature is most important for electrical machines and here neodymium magnets have a significant limitation. The temperature coefficient indicating a reversible loss of energy with temperature rise up to the maximum operating point shows ferrite as the most temperature dependent.

Table 1. Magnetic and Physical Properties of the Permanent Magnets Used

Magnetic Material	Residual Induction B_r, Tesla	Coercive Force $-H_c$, KA/m	Peak Energy Density $B.H_{max}$, KJ/m^3	Max. Operating temperature $^\circ$C	Reversible temp.coeff % per $^\circ$C
Ferrite	0.4	247	29	350	-0.20
SmCO$_5$	0.82	597	122	250	-0.04
Nd Fe B	0.97	493	178	140	-0.12
	1.03	772	205	100	-0.12

Using finite element software (7) the rotor is designed for a given stator. The stator is that of a standard induction motor having a four-pole winding. The three rotors have been constructed and tested for synchronous performance in the same stator, the winding of which was rated at 8A and was identical for the three rotors.

The measured torque for three voltage levels is shown in Fig. 2 as a function of load current for a 50 Hz mains supply. The unit of voltage is that induced in the winding when the machine is run as a generator. With this type of high field machine and $SmCO_5$ magnets there is little to be gained in torque by using a voltage higher than the generated emf. NeFeB magnets on the other hand have a higher flux carrying capacity. The torque is higher for a given current and operation at higher voltage gives a benefit, see Fig. 3.

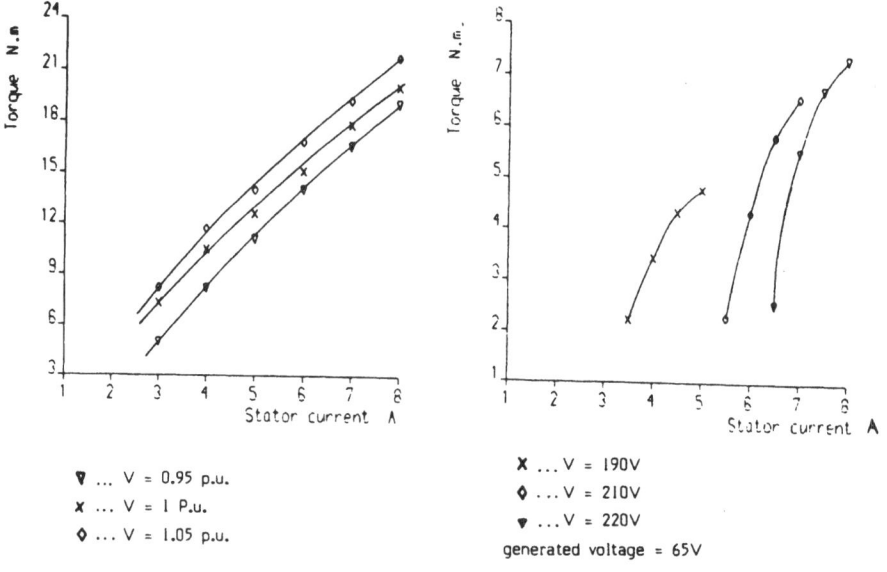

▼ ... V = 0.95 p.u.
x ... V = 1 P.u.
◊ ... V = 1.05 p.u.

x ...V = 190V
◊ ...V = 210V
▼ ...V = 220V
generated voltage = 65V

Fig. 3: Variation of torque as a function of the stator current for the Nd-Fe-B machine

Fig. 4: Variation of torque as a function of the stator current for the ferrite machine

The ferrite magnet rotor has a totally different characteristic and has to be operated well beyond the generated voltage in a regime where reluctance action is significant. The test rig had quite a high minimal torque and pull out prescribed the upper limit, see Fig. 4.

A torque comparison for the three materials using rated voltages as defined, gives a clear comparison of output, see Fig. 5.

The efficiency is an important parameter and a comparison for the three materials at the rated voltages is presented in Fig. 6.

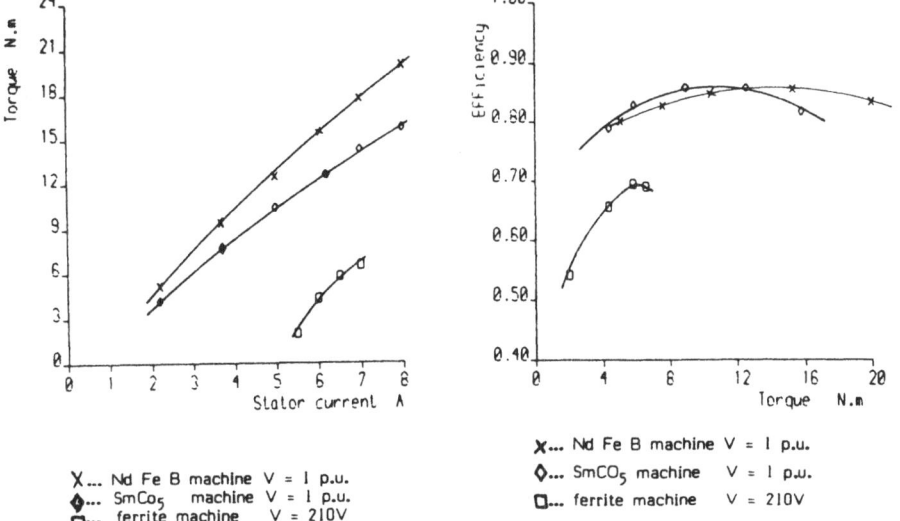

X... Nd Fe B machine V = 1 p.u.
◆... SmCo₅ machine V = 1 p.u.
□... ferrite machine V = 210V

Fig. 5: Comparison of the torque characteristics for the three machines used.

X... Nd Fe B machine V = 1 p.u.
◆... SmCO₅ machine V = 1 p.u.
□... ferrite machine V = 210V

Fig. 6: Efficiency torque characteristics for the three machines used.

The ferrite rotor efficiency is relatively low whilst the other two magnet materials are comparable in efficiency. The neodymium magnets have the advantage at higher outputs. The power factor is important particularly for inverter fed operation. Both rare earth rotors retain a power factor close to unity. The SmCO$_5$ rotor has a power factor which changes from leading to lagging as the output increases. For the ferrite rotor it improves with load but is much lower in comparison.

Fig. 8 shows a computed field distribution at 6A load showing the flux lines for a repeatable section of the machine. It is clear that the flux density in the air gap over the region of one pole is much higher than that in the magnet, a feature of this configuration. The flux emerging from each pole of the rotor is also greater than the flux from each magnet because some of the air gap flux from the rotor passes through two magnets. The magnet dimensions are chosen so as to produce a high air gap flux without undue saturation.

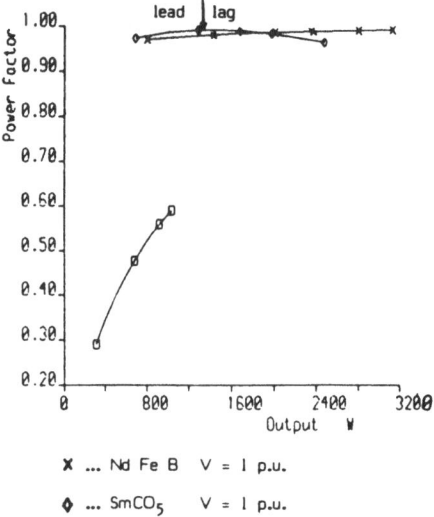

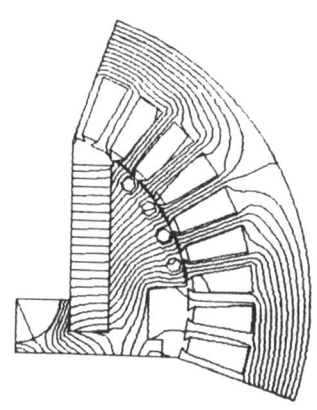

Fig. 7: Power factor - output
characteristics for the three
machines used.

Fig. 8: Flux distribution for the
Nd Fe B machine at full load

As explained earlier the different geometries for the 3 rotors
were obtained by using computer software which gives torque, flux
densities etc. as well as field maps, the latter serving mainly as an
aid to understanding and presentation.

The comparison of power capability and performance characteristics
such as efficiency and power factor is clearly brought out in the
results. One must, however, bear in mind the cost comparison for the
magnets. In a total drive system the cost of ferrite magnets is less
than one percent. But for the rare earth magnets the cost is 50 per
cent or more of the total motor cost and is a significant proportion of
the total cost of a drive system.

On the other hand the use of rare earth magnets can lead to a
drive systemwith a higher torque/weight ration than any other practical
alternative. The high efficiency and near unity power factor are
exactly what is required for the ideal inverter fed system. There is
no doubt that the neodymium magnets give the best performance but the
working temperature is limited.

Tests have been carried out which confirm that the ferrite magnets can be magnetised easily without a special rig, simply making use of the stator winding. These tests will be extended to cover the magnetisation of rare earth magnets. This technique has enormous advanatages in that the machine assembly is made much simpler by using unmagnetised material in the construction. The use of elevated temperatures to ease magnetisation of temperature dependent rare earths has an important role to play.

In conclusion, the characteristics of the motor for the three important types of magnet have been demonstrated. The cost effectiveness of the use of relatively expensive magnets can be assessed.

ULTRASONICALLY MODULATED DRIVE SYSTEM FOR PERMANENT MAGNET MACHINES

The drive system described here can form the basis of a general purpose variable speed drive which is expandable up to ratings of tens of kilowatts. It marries together the recent developments in permanent magnetic machine technology and inverter development involving ultrasonic modulation frequencies.

The machine is operated in a synchronous as opposed to a so-called brushless d.c. mode thereby avoiding the expense and inconvenience of a position sensor.

High efficiency permanent magnet machines have a low effective synchronous reactance and are, therefore, not well suited to direct connection to pulse width modulated (PWM) voltage source inverters which are switched at frequencies of a few kHz, since they will draw excessive harmonic currents. The inverter used in the drive discussed here is, therefore, modulated at a much higher ultrasonic frequency so eliminating the need for any supply filtering. As a result the supply-induced harmonic loss and acoustic noise in the machine is minimal.

The three phase inverter supply is designed using a modular approach to allow for simple progressive expansion of the drive rating. The system utilises a number of identical modules that can be paralleled up to the required volt-amp switching capability for each phase. The power electronic design of each module is such that good parallel load sharing, both in the transients of the switching and in the steady state, is achieved. Each unit has its own electrically isolated drive circuit so that paralleling can be simply achieved by connecting the outputs together and using common power rails and TTL - control signals. This approach has distinct advantages in terms of manufacturing and development costs because only one basic design is needed for an entire range of drives.

A block diagram of the permanent magnet drive is shown in Figure 1. The overall drive control and supervision is performed using an eight bit microprocessor system, which sets the voltage and frequency required from the inverter via a digital pulse width modulation circuit.

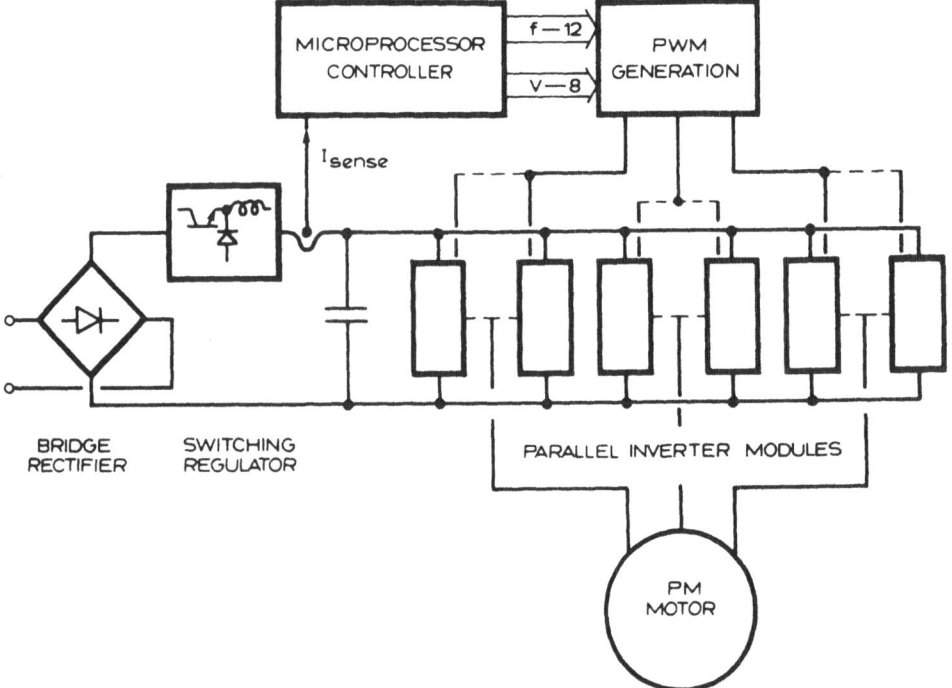

Fig. 9: Drive Schematic

A switching regulator is included to maintain a constant d.c. rail voltage to the inverter and, therefore, ensuring consistency in the voltage supply to the permanent magnet motor, this being important for the stability of the drive.

The permanent magnet machine is of the 'high field' type and a range of designs to correspond with the inverter ratings is available. The magnets used would for preference be neodymium-iron-boron but designs are available which make use of samarium cobalt and anisotropic ferrite magnets.

The basic power circuit of the inverter output module is shown in Figure 10. Because the rating requirement of the individual modules is only a fraction of the total inverter rating low power MOSFETs can be economically employed as the main switching devices. The major advantages of these devices are their low switching losses and their thermal stability when paralleled. In addition a simple drive circuit can be used because of the high impedance between the gate and source.

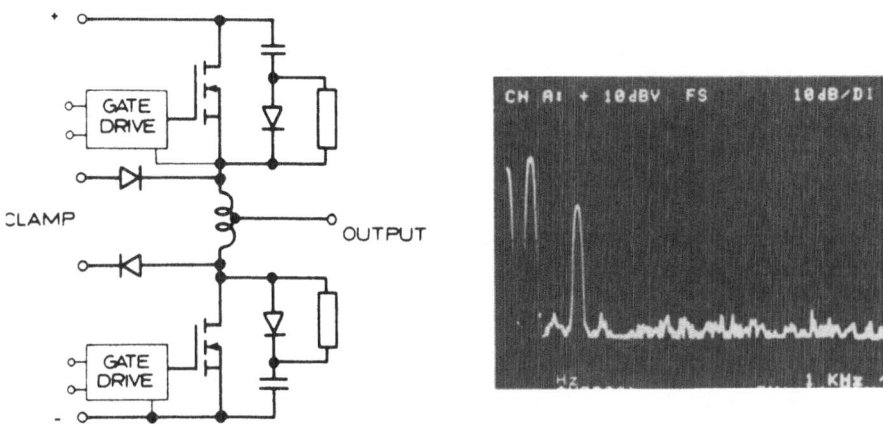

Fig. 10: Basic Module Power Circuit   Fig. 11: Pole Voltage Waveform
                                                    Spectrum

Switching-aid components are included to ensure good load sharing between the paralleled units during the period of the switching transients without enforcing any restrictions on the geometric layout of the power devices. Two external voltage clamping circuits are used to limit the peak voltages across the MOSFETs as they turn-off and transfer the stored energy in the central inductance. The clamp circuits are common to all the modules of the inverter. The MOSFETs internal drain-substrate diodes are utilised and it is, therefore, necessary to restrict the rate of the voltage across the devices to prevent secondary breakdown within the internal parasitic transistor. A small turn-off switching-aid network is used for this purpose which restricts the dV/dt across the devices to 1V/nS.

This protection has the additional advantage that at this value of dV/dt the displacement currents through the heatsink mounting insulation are negligible and grounded heatsinks can be used. The inclusion of these networks, however, only results in a small amount of additonal loss.

The switching losses with the MOSFETs themselves however are considerably reduced and the devices can be loaded to their full current rating, resulting in a typical module rating equivalent to a sinusoidal output of up to 1KVA. The isolated gate frive circuits are based upon the pulse transformer technique reported by P. Wood (Wood, 1985).

An entirely digital method of generating the pulse width modulation control signals is used, which has the advantages of low cost, reliability and can be directly interfaced to the microprocessor system controller. Prototypes of two altenative circuits were constructed, one based on a large dedicated digital circuit and the other on a single chip microprocessor in conjunction with a smaller digital circuit. The eventual aim is to include the digital circuits into single logic arrays. Of the two, the single chip microprocessor system is found to be the lower cost option since the circuitry external to the microprocessor will fit into a cheap programmable logic array (PLA). The single dedicated circuit will require a larger and thus more expensive ULA.

The specification of the PWM generation circuit is that it should be able to accept independent frequency and voltage demands over a frequency range of 0-400 Hz to a resolution of 0.1 Hz and with a voltage resolution of 0.5%. Additional features of the specification include a programmable pulse drop facility and rate limiting on the frequency and voltage inputs. The modulation frequency remains fixed at approximately 20 KHz.

The synthesised output voltage is that of a sine wave with an added one sixth of third harmonic, to maximise the available output voltage without introducing additional harmonic distortion. An example of the frequency spectrum of the pole voltage waveform generated by the control circuit is shown in Figure 11, where the added third harmonic and the absence of any other harmonics can be clearly seen.

The results that are presented here are for a relatively low power prototype that demonstrates the basic principles of the drive. The inverter modules use low cost IRF640 MOSFETs and have a sinusoidal output capability of 350VA. Two modules are paralleled per phase to give a total inverter rating of 2.1 kVA. The permanent magnet machine has a rotor containing samarium-cabalt magnets. The stator has a four-pole winding within a D60 frame and has a continuous torque rating of 3 Nm.

The permanent magnet machine is designed to operate at its most efficient when it is excited at close to unity power factor. This condition was maintained by adjusting the voltage fed to the machine as load varied at any given frequency. The efficiencies of the permanent magnet machines and the entire drive system are presented in Figures 12 and 13 for various frequencies when operated at this optimum condition.

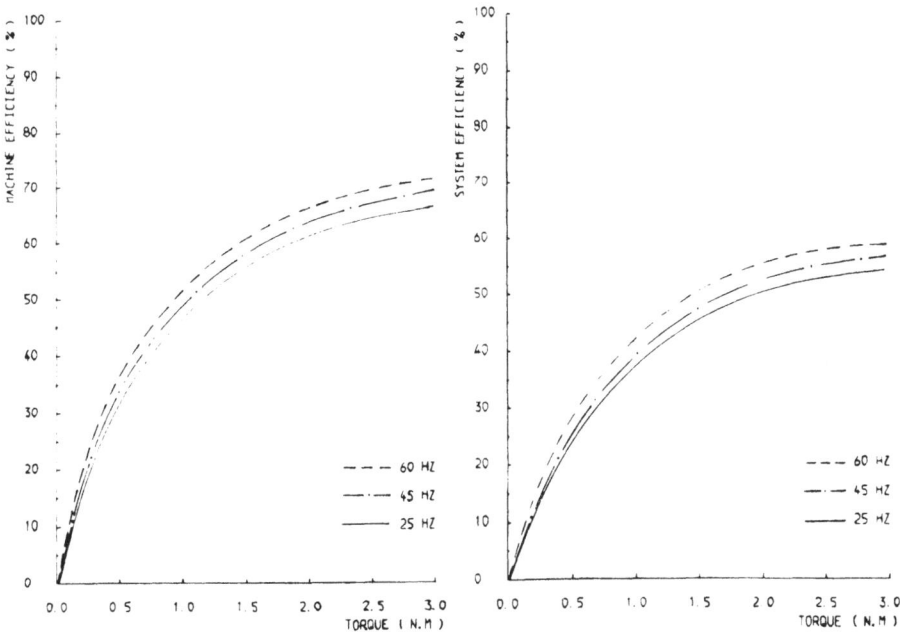

Fig. 12: Efficiency Torque
Characteristics of Permanent Magnet
Machine

Fig. 13: Efficiency Torque
Characteristics of Drive System

The presence of any additional harmonic losses induced in the
machine by the switching action of the inverter was investigated by
comparing the performance of the machine on the inverter supply with
its operation on a sinusoidal supply derived from an alternator.

The losses in the machine for the two supplies are shown in Figure
14 which demonstrates that the ultrasonically modulated inverter
introduces minimal additional loss and consequently there is no need to
derate the machine.

The combination of a rare earth permanent magnet machine and an
ultrasonically modulated inverter produces a drive system which is both
efficient, economical and flexible. The machine can be designed with a
high power to weight ratio and does not need to be derated when
connected to the inverter supply. Audio acoustic noise generated by
the inverter is eliminated and the system is, therefore, suited to many
applications involving quiet drives, in particular fan drives and
ventilation systems.

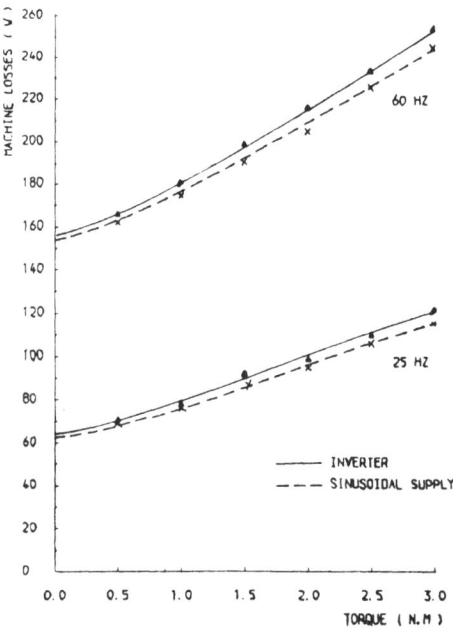

Fig. 14: Comparison of Machine Losses when Excited
from a Sinusoidal Supply and Inverter

Other applications involve systems where a precisely controlled speed is
required. Because of the use of a modular approach in the inverter
construction the drive rating can be easily expanded. From a single pole
module design an entire range of drive ratings can be produced and
manufactured cheaply in quantity. The use of digital control for the system
leads to greater reliability whilst reducing the controller costs.

EQUIPMENTS CONSTRUCTED

We have constructed two rotors for synchronous machines; one is 4 pole (see Fig. 15) and the other 6 pole (see Fig. 16). These machines are designed using our CAD packages and both incorporate neodymium-iron-boron magnets.

Fig. 15, 4 pole rotor with neodymium magnets

Fig. 16, 6 pole rotor with neodymium magnets

We have also constructed a high speed test facility making use of fluid film bearings (see Fig. 17).

Fig. 17, high speed spindle incorporating permanent magnet drive

Our research has involved the replacement of the conventional position sensor by an implicit sensing arrangement incorporated within the machine. The research equipment is shown in Fig. 18.

Fig. 18, equipment for studying the replacement of the conventional
position sensor

Research has also been carried out on permanent magnet generators. We have designed a permanent magnet rotor involving neodymium magnets to function as an automobile generator (see Fig. 19).

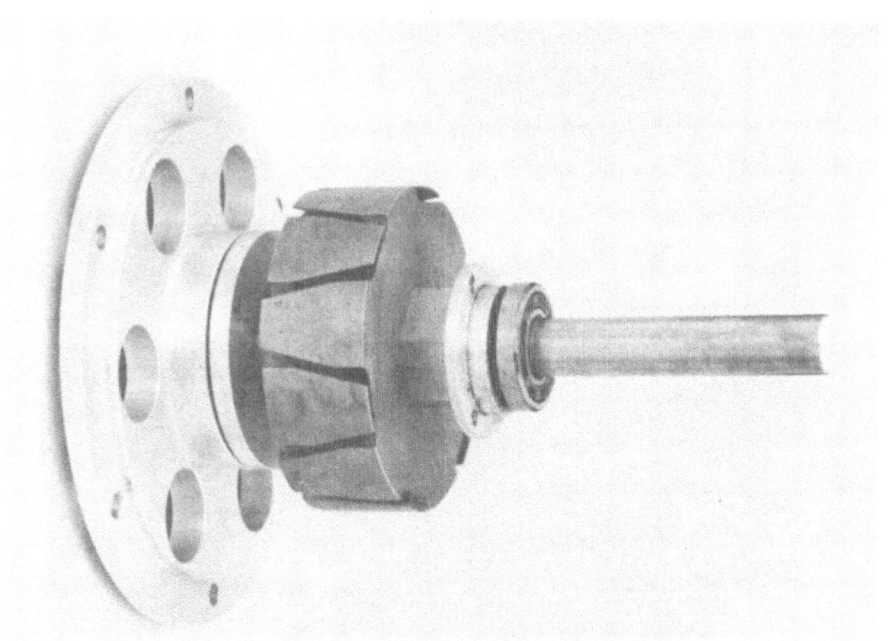

Fig. 19, embricated permanent magnet motor for automobile applications

REFERENCES

1. Binns, K. J. 1976. Alternator current generators or motors.   U.K. Patent 1437348.

2. Binns, K. J. and Jabbar, M. A. 1978. Some recent developments in permanent-magnet machines.   Proc. I.C.E.M., Brussels.

3. Binns, K. J. 1980. Alternating current rotating electrical machine. U.K. Patent No. 1560971, U.S.A. Patent No. 4 188 554,

4. Binns, K. J. and Jabbar, M. A. 1981. Comparison of performance characteristics of a class of high-field permanent-magnet machines for different magnet materials.   I.E.E. Conference (S.S.E.M.), London.

5. Binns, K. J. 1984. Permanent magnet motors for inverter-fed drives. Conference on Drives, Motors, Controls, Brighton, p.p. 101-105.

6. Binns, K. J. and Wong, T. M. 1984. Analysis and performance of a high-field permanent-magnet synchronous machine.   Proc. I.E.E., 131, Pt. B, No. 6, p.p. 252-258.

7. Binns, K. J., Riley, C. P. and Wong, T. M. 1985. Some design aspects of high-output permanent magnet synchronous machines with non-radial magnets.   IEE Conference, Publication No. 254, E.M.D.A., London.

8. Binns, K. J. and Riley, C. P. 1986. The scope for development of permanent magnet machines in the light of new materials. Proc. ICEM, Part 3, p.p 1060-1063, Munich.

9. Binns, K. J. and Jabbar, M. A. 1981. High-field self-starting permanent-magnet synchronous.   Proc. I.E.E., 128, Pt. B, No. 3, p.p 157-160.

10. Binns, K. J. and Wong, T. M. 1984. Development of a high performance permanent magnet machine.   I.C.E.M. Conference, Lausanne, p.565.

11. Binns, K. J. and Wong, T. M. 1984. Analysis and performance of a high-field permanent-magnet synchronous machine.   Proc. I.E.E., 131, Pt. B, No. 6, p.p. 252-258.

12. Wood, P. 1985. Transformer isolated HEXFET driver provides very large duty and cycle ratios.   Application note 950A, International Rectifier.

SUMMARY OF DISCUSSIONS :

ROTATING MACHINES

E.M.H. Kamerbeek and R. Hanitsch

Integrated design using Nd-Fe-B magnets are cost effective for fractional and subfractional horse power motors. It was reported (9.1, 9.6, 9.8, 9.11) that the trend towards high speed and high flux density drive systems will continue, because these drives are energy efficient.

New line start synchronous motors were developed (9.11) using Neodymium magnets. This type of design with buried magnets of 4 pole and 6 pole lay-out have been extensively tested and their performance found to be superior to machines using samarium cobalt in the same frame size. Because of the system aspects some of the group members also worked on invertors. Most of the new machines have been run on ultrasonically modulated invertors to give a quiet and efficient variable speed drive.

The high remanence and high coercivity at room temperature makes Nd-Fe-B magnet excitation particularly attractive in industrial drives (9.3, 9.5, 9.7). However, handling and assembly of magnetised parts proved to be difficult. Therefore special assembly strategies were developed by some of the group members (9.7). Examples of small brushless-dc and stepper motors have been constructed and tested (9.9). They showed the benefit of using the new material.

The sensitivity of the coercivity of Nd-Fe-B to high temperatures calls for increased attention to the thermal aspects of a new design. Improved CAD techniques have been reported (9.2, 9.4). Many of the participants also studied this problem and details of improved magnet specifications and properties provided by some of the Group (e.g. 9.5) were taken into account to refine the characteristics of the prototypes.

Beside the work on brushless-dc motors detailed studies were made on brushed-dc motors. Nd-Fe-B was considered for both radial- and axial-field designs of slotted and slott-less configurations. Therefore disc-type and cup-type machines were built (9.1, 9.7) and different designs of windings were studied. Among the advantages found for machines equipped with Nd-Fe-B are improved performance factors such as: higher torque per frame size; improved efficiency; better dynamic response; reduced volume; reduced weight.

Group members also designed and tested (9.2) a linear stepper motor and an air driven generator.

SECTION III

− APPLICATIONS

CHAPTER 10

STATIC PERMANENT MAGNET DEVICES

NdFeB MAGNETS : A NOVEL MAGNETIC DESIGN FOR THE

ELECTRON CYCLOTRON RESONANCE ION SOURCES (E.C.R.I.S.)

J. Chavanne, J. Laforest, R. Pauthenet[†]

Centre National de la Recherche Scientifique
Laboratoire Louis Néel 166X, 38042 Grenoble Cedex, France

ABSTRACT

The outstanding remanence and coercivity of the new NdFeB permanent magnets are very interesting in static applications. New concepts of calculations with these magnets can be applied : as in SmCo$_5$ magnets, the magnetization can be considered as rigid, the density of the surface Amperian current is constant, the relative permeability is approximately 1 and the induction calculations are linear. The high remanence of the NdFeB magnets and a good coercivity permit to use them in the production of the longitudinal and radial magnetic fields in a multiply charged ion source (E.C.R.I.S.). Such a source has been constructed and the magnetic configuration is presented.

INTRODUCTION

The use of permanent magnets in specific systems has many advantages when the value of the necessary induction is not too high. They do not need any power supply or any cooling fluid to operate. Moreover their induction is very stable. But their highest advantage is the possibility to produce an induction which does not depend on the size of the permanent magnets. This ability leads to the reduction of the dimensions of the devices. The rare earth permanent magnets which can be put together are good examples of such advantages.

In the first part we present the properties of the rare-earth permanent magnets and the methods of calculation used for such materials. In the second part a novel magnetic design for an E.C.R.I.S. is presented : all the magnetic inductions are produced by NdFeB magnets.

THE RARE EARTH PERMANENT MAGNETS

For more than fifteen years very powerful permanent magnets have been developed. They are based on magnetically anisotropic alloys of rare earth and transition metals. The first one discovered has the basic composition SmCo$_5$ and the most recent one is Nd$_2$Fe$_{14}$B. Their characteristics are shown in table 1 which also contains data on the hard ferrites, for comparison. The rare-earth materials are not only interesting for their oustanding values of remanent induction, B_R, and energy product,

Table 1 : Typical values at room temperature of the magnetic properties of different permanent magnet materials.

Materials	B_R	H_{cJ}	H_{cB}	$(BH)_{max}$
Hard ferrites	0.4 T	160 kA/m	150 kA/m	26 kJ/m^3
	4000 G	1950 Oe	1850 Oe	3.3 MG.Oe
SmCo$_5$-type	0.95 T	1600 kA/m	680 kA/m	180 kJ/m^3
	9500 G	2000 Oe	8500 Oe	22 MG.Oe
Sm$_2$Co$_{17}$-type	1.05 T	880 kA/m	720 kA/m	210 kJ/m^3
	10500 G	11000 Oe	9000 Oe	26 MG.Oe
NdFeB-type	1.2 T	1400 kA/m	880 kA/m	280 kJ/m^3
	12000 G	17500 Oe	11000 Oe	35 MG.Oe

$(BH)_{max}$, but also for the large values of their intrinsic coercivity H_{cJ} which is related to their very high magnetocrystalline anisotropy (around 10^4 kA/m). As a consequence the polarization, $\vec{J}$ is strongly fixed to the easy direction of the magnetization of the magnet. If a magnetic field of 800kA/m is applied perpendicular to this direction, $\vec{J}$ theoretically rotates only by 2 or 3 degrees. This allows us to use such two magnets at right angles in order to increase the value of the induction in devices. We can approximate the value of J as remaining constant ($= J_R$) for internal fields $|\vec{H}_i| < |\vec{H}_{cJ}|$. This important fact leads us to consider the magnetization vector $\vec{M}$ ($= \vec{J}/\mu_0$) as rigid. For magnets with $\mu_0 H_{cJ} > J_R$, the (B,H) curve can be considered as linear, $H_{cB} = B_R/\mu_0$; the relative permeabilities $\mu_R//$ and $\mu_R\perp$ are equal to 1. In figure 1 we have illustrated the validity of these hypotheses by comparing the idealized curves with the real properties of a NdFeB permanent magnet : VACODYM 370 from Vacuumschmelze. The difference between the approximation and the true values is less than 4%. Within these hypotheses the superimposition of the

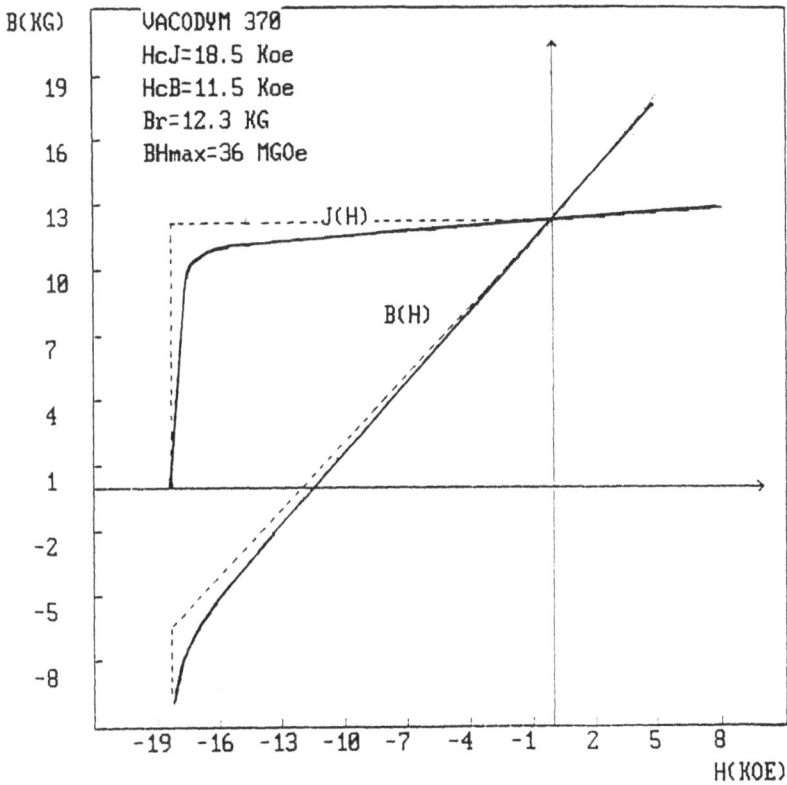

Fig. 1 : The demagnetization curves for a NdFeB permanent magnet at room temperature. The dotted lines correspond to the hypothesis $J = J_R = $ constant.

induction of magnets is a linear superimposition of the induction produced by each of them. This rigidity of the magnetization and the linearity property of the induction greatly simplify the computation of the magnetic induction in devices with rare earth permanent magnets.

THE METHODS OF CALCULATION OF THE INDUCTION

Most of the devices are made of prismatic elementary magnets. Inside each of them, the magnetization is assumed to be uniform and equal to the remanent magnetization M_R. The field configuration created by the magnet is the same as the configuration of a distribution of surface amperian currents with density $\vec{i}_s = \vec{M}_R \wedge \vec{n}$, where $\vec{n}$ is a unit vector perpendicular to the surface. On the face of a rectangular parallelepiped of NdFeB magnets polarized parallel to the long side of the face, the surface Amperian current density is equal to 10000 A/cm. This method allows us to calculate the induction $\vec{B}$, inside and outside the permanent magnet. The field configuration is also equivalent to that produced by a distribution of magnetic charges with a surface density $\sigma = \vec{M}_R \cdot \vec{n}$. No magnetic charges exist inside the magnet because we postulate that the magnetization is uniform. Using the methods of electrostatics the calculation gives directly the values of the field $\vec{H}$ inside and outside the magnet.

<u>Parallelepiped magnet</u> (Pauthenet 1984)

It is equivalent to two sheets one with $\sigma = M_R$ and the other with $\sigma = -M_R$. The components of the induction B_x, B_y, B_z corresponding to a charged rectangular surface M_R as shown on figure 2 can be expressed as :

$$B_x = \frac{\mu_o M_R}{4\pi} \; \text{Log} \; [\frac{(y+b)+ \sqrt{(x-a)^2+(y+b)^2+z^2}}{(y-b)+ \sqrt{(x-a)^2+(y-b)^2+z^2}} \cdot \frac{(y-b)+ \sqrt{(x+a)^2+(y-b)^2+z^2}}{(y+b)+ \sqrt{(x+a)^2+(y+b)^2+z^2}}]$$

$$B_y = \frac{\mu_o M_R}{4\pi} \; \text{Log} \; [\frac{(x+a)+ \sqrt{(x+a)^2+(y-b)^2+z^2}}{(x-a)+ \sqrt{(x-a)^2+(y-b)^2+z^2}} \cdot \frac{(x-a)+ \sqrt{(x-a)^2+(y+b)^2+z^2}}{(x+a)+ \sqrt{(x+a)^2+(y+b)^2+z^2}}]$$

$$B_z = \frac{\mu_o M_R}{4\pi} \; [\text{Arc tg} \; \frac{(x+a)(y+b)}{z\sqrt{(x+a)^2+(y+b)^2+z^2}} + \text{Arc tg} \; \frac{(x-a)(y-b)}{z\sqrt{(x-a)^2+(y-b)^2+z^2}}$$

$$- \text{Arc tg} \; \frac{(x-a)(y+b)}{z\sqrt{(x-a)^2+(y+b)^2+z^2}} - \text{Arc tg} \; \frac{(x+a)(y-b)}{z\sqrt{(x+a)^2+(y-b)^2+z^2}}]$$

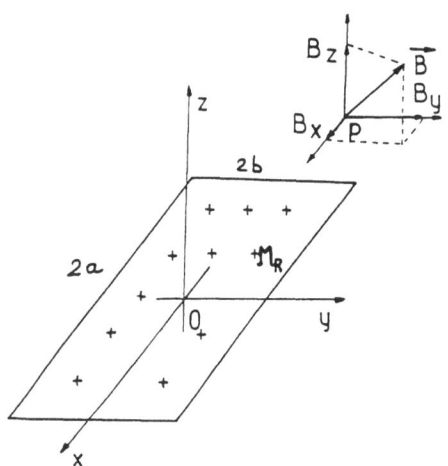

Fig. 2 : The components B_x, B_y, B_z of the induction B corresponding to a charged rectangular surface with magnetic charges density M_R.

It is easy to make field calculations using these expressions. The magnetic induction produced in a static device made with prismatic magnets is the linear resultant of the contributions of the different sheets of the device. Such a method has been used by use for the multipoles.

Annular magnet

In the surface Amperian currents method an annular magnet is equivalent to two co-axial coils : one with a radius a with $\vec{i}_s = -\vec{M}_R \wedge \vec{n}$ and the other with a radius b with $\vec{i}_s = \vec{M}_R \wedge \vec{n}$.

The magnetic induction produced by an equivalent coil of a radius a and a length L has been calculated (figure 3). The radial and axial

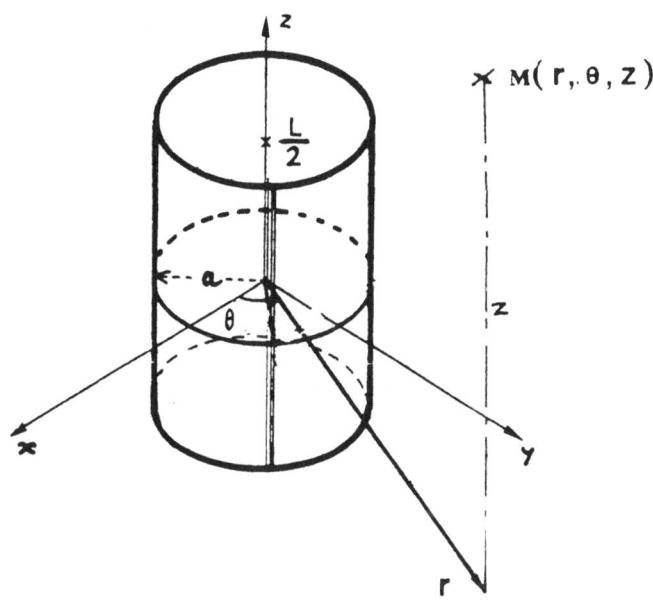

Fig. 3 : Equivalent coil for the calculation of the induction B.

components can be expressed analytically with the Elliptic Integrals K(k), E(k) and Π(k) respectively of the 1st, 2nd, and 3rd type. Their expressions in cylindrical coordinates (r, θ z) are given by :

$$B_r = \frac{\mu_o M R}{\pi} \sqrt{\frac{a}{r}} \left[\frac{(1-k_1^2/2) K(k_1) - E(k_1)}{k_1} - \frac{(1-k_2^2/2) K(k_2) - E(k_2)}{k_2} \right]$$

$$B_z = \frac{\mu_o M R}{2\pi} \left[- \frac{z - \frac{L}{2}}{\sqrt{(a+r)^2 + (z-\frac{L}{2})^2}} \left[K(k_1) + \frac{a-r}{a+r} \Pi(n, k_1) \right] \right.$$

$$\left. + \frac{z - \frac{L}{2}}{\sqrt{(a+r)^2 + (z+\frac{L}{2})^2}} \left[K(k_2) + \frac{a-r}{a+r} \Pi(n, k_2) \right] \right]$$

$$B_\theta = 0$$

$$\text{with } n = \frac{4ar}{(a+r)^2} \quad ; \quad k_1^2 = \frac{4ar}{(a+r)^2 + (z-\frac{L}{2})^2} \quad ; \quad k_2^2 = \frac{4ar}{(a+r)^2 + (z+\frac{L}{2})^2}$$

Such a method has been used to calculate by computation the longitudinal field of the "magnetic bottle" in a novel ion source.

Application to a novel NdFeB magnetic structure design for a multicharged ion source.

The multiply charged ions are used for Atomic and Nuclear Physics in particle accelerators as external injectors. In the Electron Cyclotron Resonance Ion Sources (ECRIS) developed at the Centre d' Etudes Nucléaires in Grenoble (Geller, 1972), a cold plasma of electrons and weakly charged ions is confined by a magnetic configuration called "magnetic bottle". This plasma is then heated with electrons at a frequency $\geq$ 10 GHz giving rise to multicharged ions. The "magnetic bottle" consists of the super-imposition of two magnetic fields. The longitudinal one shows a maximum at each end of the confinement box (magnetic lenses) and the radial one presents a very large gradient. Formerly, these two fields were produced by electric coils using a power supply of 3 MWatts. In 1980, R. Pauthenet proposed a radial magnetic structure with six $SmCo_5$ permanent magnets as main poles (Geller et al 1980). As a result, the size of the ion source was decreased and the electric power was reduced to 100kW. A new sketch has been designed using a magnetic flux concentrator which consists of prismatic magnets adjacent to the 6 main poles (figure 4).

(Pauthenet et al. 1982). This system allows a more efficient confinement permitting more ionization by an increase of the heating microwave power. The use of permanent magnets to produce the radial field was possible only with materials having a large anisotropy and coercivity to withstand the longitudinal applied field ($\leq$ 1 Tesla).

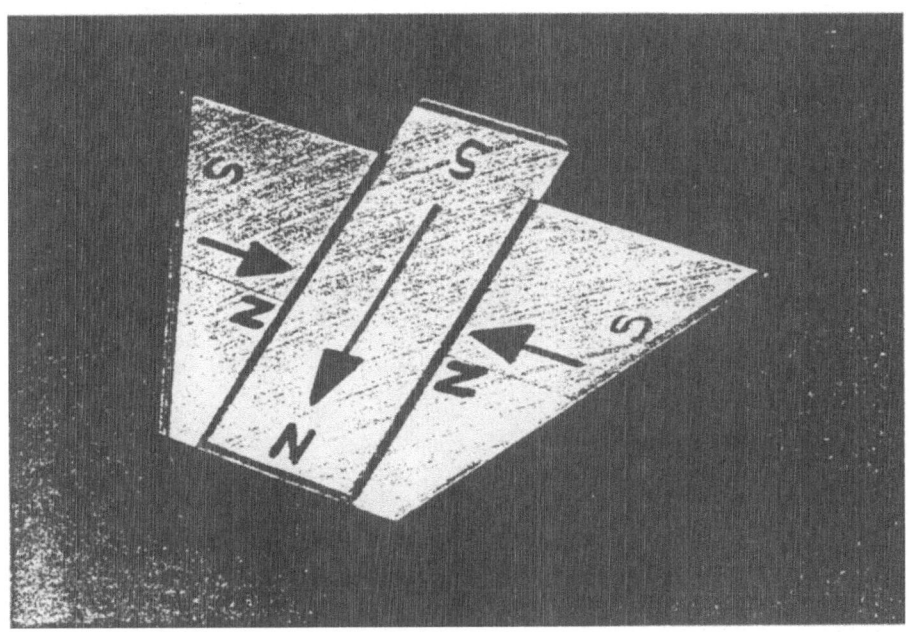

Fig. 4 : Basic element of the hexapole using adjacent permanent magnets as flux concentrator.

The discovery of NdFeB permanent magnets brings an actual breakthrough of the E.C.R. Ion Sources. They have larger remanent induction than $SmCo_5$ ($B_r \geq 1.15$ T) with a quite good anisotropy and coercivity ($H_{cJ} = 18$kOe). The replacement of the $SmCo_5$ magnets by the NdFeB ones allows an increase of the microwave frequencies up to 25 GHz. The flux distribution and variation of the hexapolar induction are plotted in figure 5. Maximum values of 0.8 Tesla with $SmCo_5$ and of 1.2 Tesla with NdFeB magnets (Chavanne et al 1987) have been obtained .

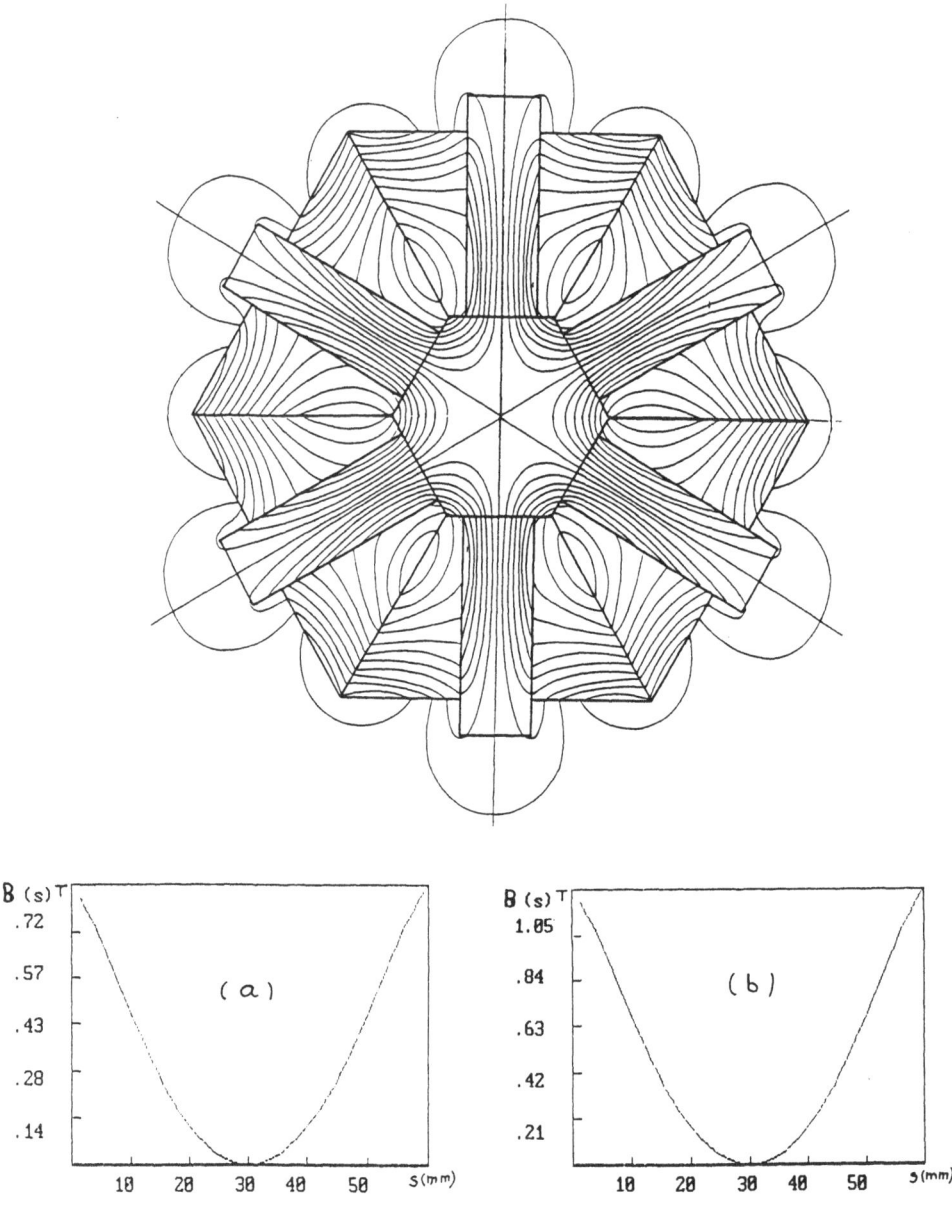

Fig. 5 : Flux distribution in rare-earth permanent magnet hexapole and variation of the modulus of the induction inside the hexapole (a) SmCo$_5$ magnets ; (b) NdFeB magnets.

Moreover, the most spectacular result is the possibility to make the longitudinal field using NdFeB magnets in a 10 GHz ECRIS ; this was not possible with any $SmCo_5$ magnets. A source with magnetic fields entirely produced by NdFeB magnets has been calculated and constructed in collaboration with the Departement de Physique Atomique et Développement des Sources Ioniques (R. Geller) and the Grand Accélérateur National d'Ions Lourds in Caen (P. Sortais). No power supply is needed and there is no limit in the high voltage operating system. The source is schematized in figure 6. The longitudinal field of 0.4 Tesla is produced by two perma-

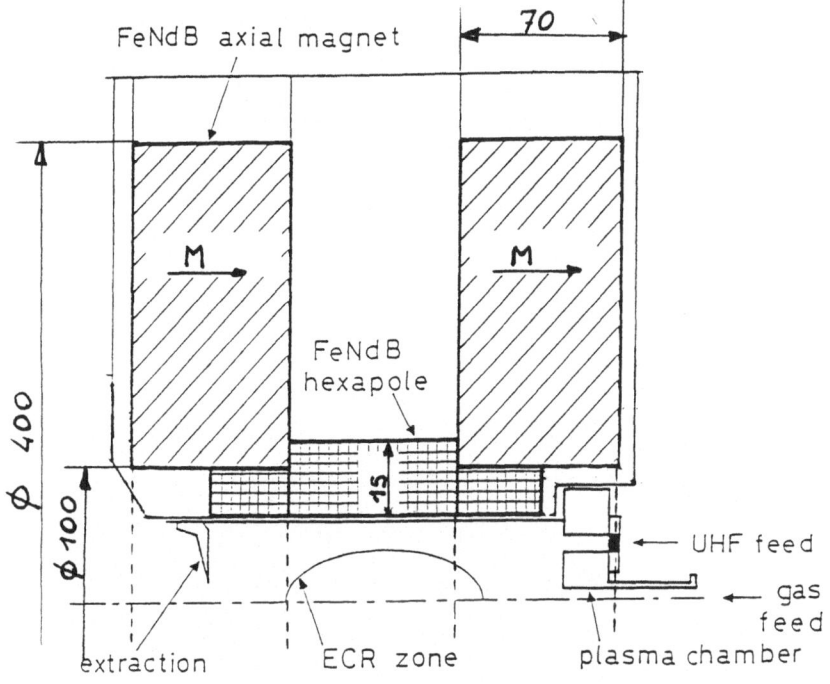

Fig. 6 : Schematic view of the novel ECR Ion Sources.

nent magnet rings. The internal diameter is 100 mm, the external one 400 mm and the length 70mm. Figure 7 shows one of the NdFeB rings during the assembling of the 6 elementary blocks composed with smaller magnets. These 6 blocks have been magnetized separately in a Bitter coil of 280 mm diameter in a 5 Tesla field at the Service National des Champs Intenses (SNCI) in Grenoble. The height of the hexapole has been reduced : 15 mm instead of 70 mm with $SmCo_5$ magnets. The special system to assemble the hexapole is shown in figure 8. The magnetic forces on each block during the

Fig. 7 : Longitudinal induction : one of the two rings during the assembling.

Fig. 8 : Radial induction : the hexapole during the assembling.

fitting are about 800 kg. After the mounting the longitudinal attractive force between the two rings is roughly 3 metric tons ! The performances of this source have been presented at the International Conference on Ion Sources, Michigan State University (P. Sortais et al. 1987).

A minimal size, no electric consumption and no problem to use it at a very high voltage make this type of multicharged ion sources very attractive.

CONCLUSION

New permanent magnet applications are now possible with the rare earth magnets. New concepts have been applied to calculate static systems. As the magnetization $\vec{M}$, can be considered as rigid, permanent magnets with different orientations can be assembled. This, associated with the high remanent induction of NdFeB magnets has permitted the reduction of weight of the static devices ans has made new applications possible.

ACKNOWLEDGEMENTS

The authors thank R. Geller and P. Sortais for helpful discussions and acknowledge the support of the Commission of the European Communities within the C.E.A.M. project of the Stimulation Programme. One of the authors (J.C.) is partly supported by a fellowship from Aimants Ugimag S.A. Saint Pierre d'Allevard, France.

REFERENCES

Chavanne J., Laforest J. and Pauthenet R. 1987.Static devices with new permanent magnet Proc. Materials Research Society V96 (Ed. S.G. Sankar, J.F. Herbst, N.C. Koon) p.307.
Geller R., 1972, I.E.E.E. Trans. Nucl. Sci., 19 (2), p.200
Geller R., Jacquot B. and Pauthenet R. 1980. Rev. Phys. Appli. 15 p.955
Geller R., Jacquot B., Lamy M. and Debernardi J. 1982. Proc. 4th Int. Workshop on E.C.R. Ion Sources (C.E.N.Grenoble)
Pauthenet R. 1984.New permanent magnet devices. J. de Physique C1-285.
Sortais P., Debernardi J., Geller R., Ludwig P., Pauthenet R. 1987 Study of a magnetic structure for an ECRIS built exclusively with permanent magnets. International Conference on Ion Sources (Michigan State University)

PERMANENT MAGNET SYSTEM FOR MAGNETIC RESONANCE IMAGING

W. Baran*, P.R. Locher**, W. Süsse*** and H. Zijlstra****

*Krupp Widia GmbH
Münchener Strasse 90
D-4300 Essen 1, Germany
**Philips Research Laboratories
P.O. Box 80.000
5600 JA Eindhoven, The Netherlands
***Krupp Industrietechnik GmbH
Franz-Schubert-Str. 1-3
D-4100 Duisburg 14, Germany
****CSC Scientific Consultancy
Eindhoven, The Netherlands

ABSTRACT

The design of a Permanent Magnet System for Magnetic Resonance Imaging is presented. The materials used are Neodymium-Iron-Boron and Iron. The magnet is of the so-called prismatic type, which refers to the design of the xy cross section. Along the z direction, the magnet has a full length of 150 cm and it is symmetrically composed of 4 parts, separated by 3 adjustable narrow air gaps. Access is along this direction and the access height is 80 cm. The field strength is 0.20 T and the field is expected to be homogeneous up to 100 ppm in an ellipsoidal region with a smallest diameter of 30 cm, and up to 10 ppm in a similar region with a smallest diameter of 18 cm.

A description is given of how the pieces of permanent magnet material can be magnetized and assembled with each other and with the iron yoke. A constructional drawing of the magnet is given, showing facilities to adjust the narrow gaps. The magnet has a weight of 7 tonnes.

INTRODUCTION

During the last decade, Magnetic Resonance Imaging (MRI) has grown to a useful and already established technique of Medical Imaging. It has found a place supplementary to older techniques such as X-ray Computer Tomography and Ultrasound.

The most critical and expensive part of MRI equipment is the magnet. Its magnetic field has to be very homogeneous over the region of a human being that has to be imaged. The field has to be very stable as well. Typically, the inhomogeneities should be less than $1 : 10^4$, or even $1 : 10^5$, and the instabilities during 1 second should be less than $1 : 10^7$ or even $1 : 10^8$.

Up to now, the vast majority of the MRI installations is equipped with a Superconducting Magnet, operating at the very low temperature of liquid helium. The most serious alternative is a Permanent Magnet. A disadvantage of a Permanent Magnet is its relatively low magnetic field strength of about 0.2 Tesla. This makes certain applications not possible (spectroscopy), but the normal images can be of acceptable quality, or better than that.

A MRI installation equipped with a Permanent Magnet System may be a useful instrument in the common hospitals for which a Superconducting Magnet is perhaps too expensive or difficult for other reasons.

We present here a design of a Permanent Magnet System using Neodymium-Iron-Boron. The basic idea was given by Zijlstra (1985). This has been worked out by the other authors as to the precise geometry (Locher) and as to the construction of the magnet (Baran and Süsse).

GEOMETRY

The main steps in designing the geometry of the magnet, for an excellent homogeneity, are the following.

1) Starting point is the concept by Zijlstra (1985) for a magnet of infinite length in the z direction.
   Fig. 1 shows such a magnet. The ratio (p):(1), indicated in the figure can be chosen quite arbitrarily. The properties of NdFeB and the desired value of about 0.2 T for the field strength brought us to the choice (p):(1)=(4):(1). Another ratio that can be chosen freely is (q):(p). We chose (q):(p)=(1.6):(4). If the blocks of NdFeB have an ideally homogeneous magnetization, fixed in both magnitude and direction, no matter what the counteracting magnetic field inside the material is, then the magnetic field inside the big air space is purely homogeneous, provided the surrounding iron (low carbon steel) is ideal too. The iron is said to be ideal if the permeability is infinite in all directions. Fig. 2 shows the magnetic field lines (B field) of this ideal magnet.
2) The NdFeB material is no longer idealized, but the magnet is still infinitely long in the z direction. If the configuration of step 1 is left unchanged, the homogeneity of the field will now be poor, because the counteracting field (H) is different in the NdFeB blocks on the side (left and right) as compared to the central blocks (top and bottom). This is corrected by a slight modification of the magnetization in the side blocks as compared to the central blocks. As a result of this correction the homogeneity is again excellent.

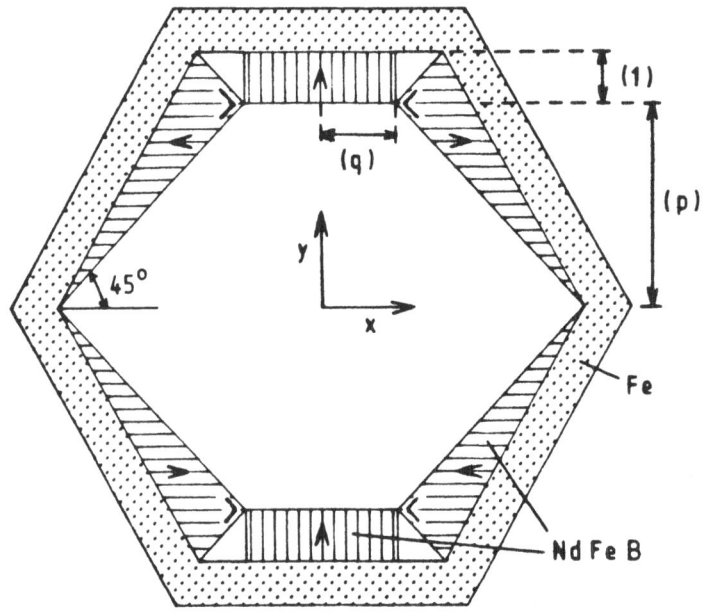

Fig. 1 The prismatic magnet. The arrows in the NdFeB blocks indicate the direction of the magnetization.

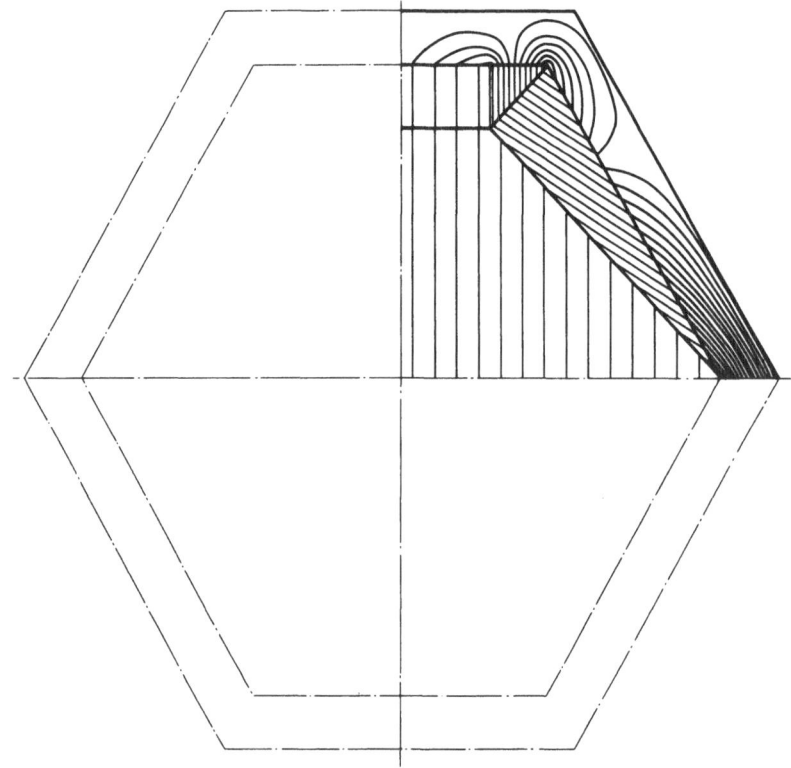

Fig. 2 Flux line pattern showing the homogeneity of the magnetic field.

3) The magnet has a finite length in the z direction. In comparison to the infinitely long magnet, the two parts stretching to minus infinity and to plus infinity have been omitted. This results in a severe inhomogeneity.

This inhomogeneity is now counteracted by the introduction of three small air gaps, one in the centre and the other two symmetrically at a distance of 21 cm of the centre, in the z direction.

Computer calculations of the magnetic field strenghts and their potential have been done
1) with a model of magnetic surface charges, and
2) with a computer package called "PADDY".

In the model of the charges, the field is calculated just as the Coulomb field of electric charges. This model assumes a knowledge of the magnetizations. These are not all precisely known, especially not in the iron. The model, therefore, does not give accurate results, but it yields a first approximation very quickly (short computation time).

More accurate calculations have been done using PADDY. In PADDY, the three-dimensional space is divided into meshes; the magnetic potentials inside the meshes are solved in an iterative way. The package has been made by a mathematical software group of Philips in Eindhoven, The Netherlands.

Results

In the drawing of the xy- and the zy-cross-section (Fig. 3a and b), a central region is indicated in which the field strength deviates less than $1 : 10^4$ (100 ppm). It has an elliptical shape and its diameters are 54 cm in the x direction, 41 cm in the y direction, and 30 cm in the z direction.

In a somewhat smaller region, the homogeneity is even $1 : 10^5$ (10 ppm). The diameters of that region are 18 cm in the x direction, 20 cm in the y direction, and 22 cm in the z direction.

These results have been obtained by carefully considering what happens if one varies the following three parameters:
1) the position of the outer gap,
2) the width of the outer gap, and
3) the width of the inner gap.

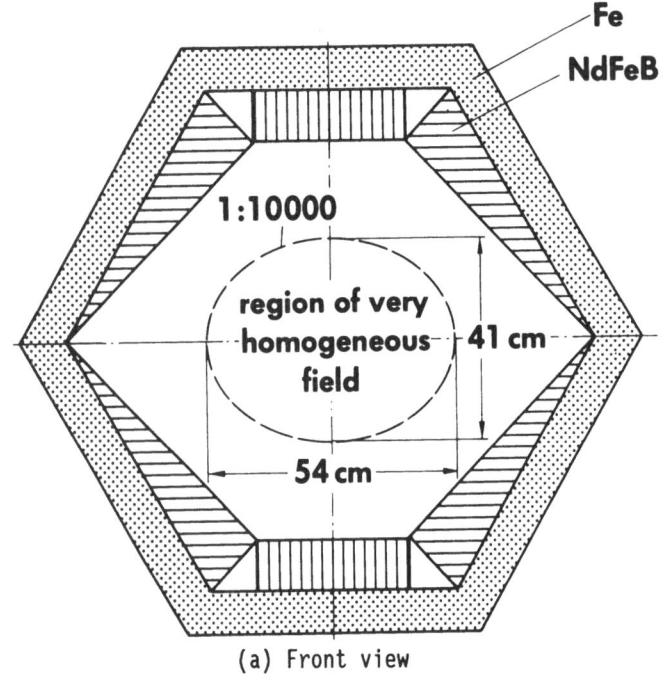

(a) Front view

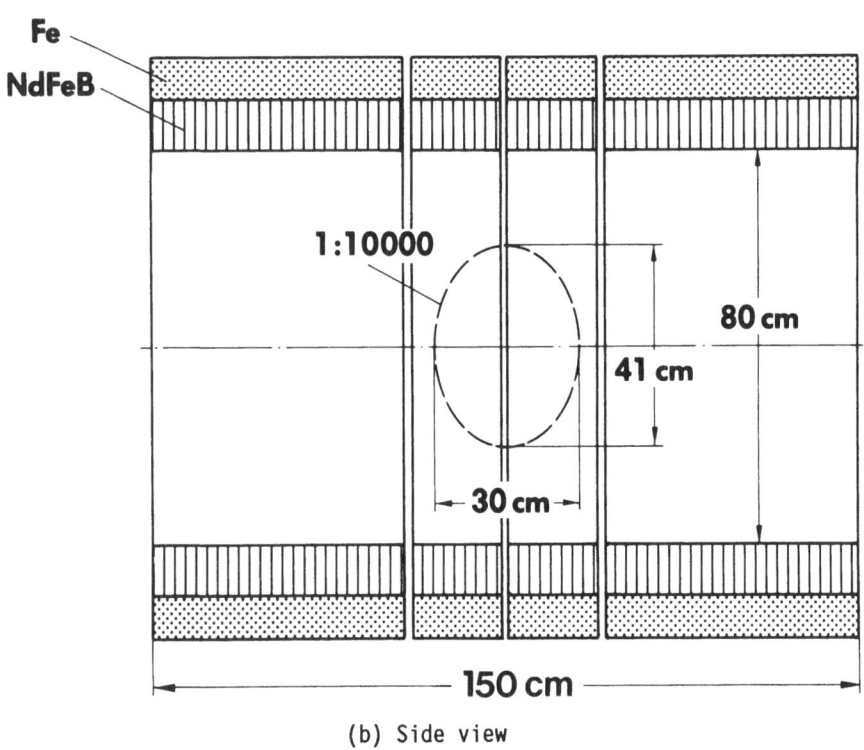

(b) Side view

Fig. 3   Permanent magnet system for Magnetic Resonance Imaging (schematic).

Details about the geometrical design

At <u>step 1</u>, we said that the iron had to be ideal. This is not critical, however. Taking realistic iron, the results of the computer calculations differ only slightly from the results obtained in the ideal case.

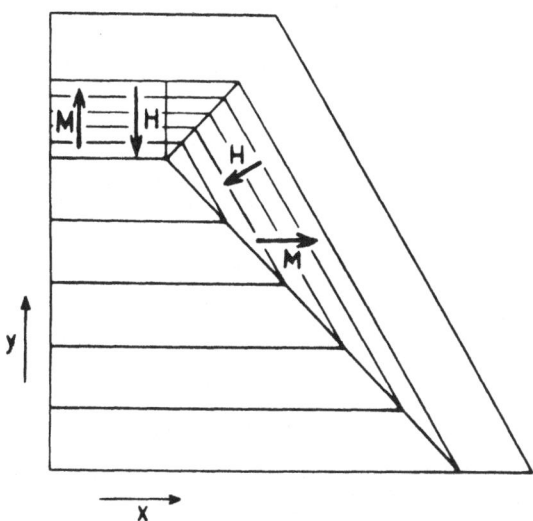

Fig. 4 Equipotential lines. The potential is here the scalar potential of the magnetic field H.

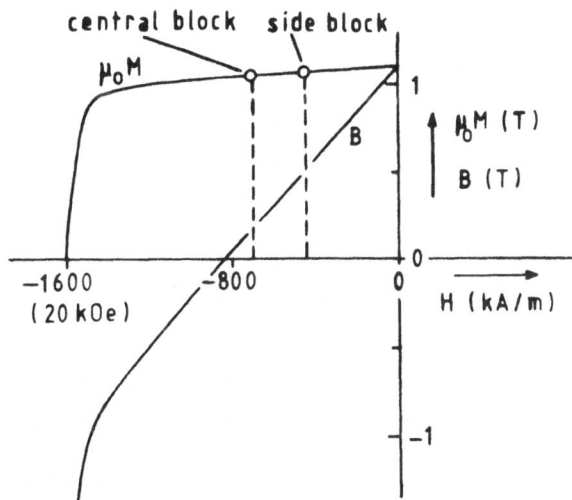

Fig. 5 The magnetization M of NdFeB as a function of the field H. The B field inside NdFeB is plotted as well. The values of H in the NdFeB blocks of our magnet are indicated by the dotted lines.

The statements of <u>step 2</u> can be made clear with the help of Fig. 4, in which lines are drawn of constant potential of the field H. From these lines one can deduce how strong the field H is at the several regions: H is strong if the potential lines are close to each other. One can see that the two blocks of NdFeB feel a different H.

Fig. 5 shows that different values of H lead to slightly different values of the magnetization. From Fig. 4 one can also see that the side block feels a H that makes an angle with M. This results in a slight rotation of M. The amount of rotation is determined by the perpendicular relative permeability, which is 1.1512.

We have compensated these effects by a small rotation and a small reduction of the magnetization of the side blocks. As to the rotation: My/Mx=0.04321 for H=0. The reduction is by a factor 0.9789.

<u>Step 3</u> finally considers the 3-dimensional magnet. The truncation of the magnet at z=75 cm and at z=-75 cm results in a strong inhomogeneity if no gaps are present. Without gaps, the inhomogeneity is so strong that the field deviates 100 ppm already at z=2 cm, and 1000 ppm at z=7 cm. The homogeneity is improved considerably by the introduction of the gaps. The procedure of finding the position and the magnitude of the gaps is as follows:
We try to make the field B constant along the z axis, thereby hoping that it will then become constant in a region outside this axis as well. The field B is expanded into a power series:

$$B = B_0 + B_2 \cdot (z/z_{ref})^2 + B_4 \cdot (z/z_{ref})^4 + \ldots,$$

where z_{ref} is an arbitrary reference value of z, e.g. 15 cm.
We now try to minimize the coefficients B_2, B_4, and B_6 by varying the position and the magnitude of the gaps. This is a problem of 3 unknowns to be solved from 3 non-linear equations.

We were able to make B_2 and B_4 practically zero and to reduce B_6, but we did not yet succeed in making B_6 zero, although this is perhaps possible. Nevertheless, we have obtained a reasonably good homogeneity along the z axis as shown in Fig. 6.

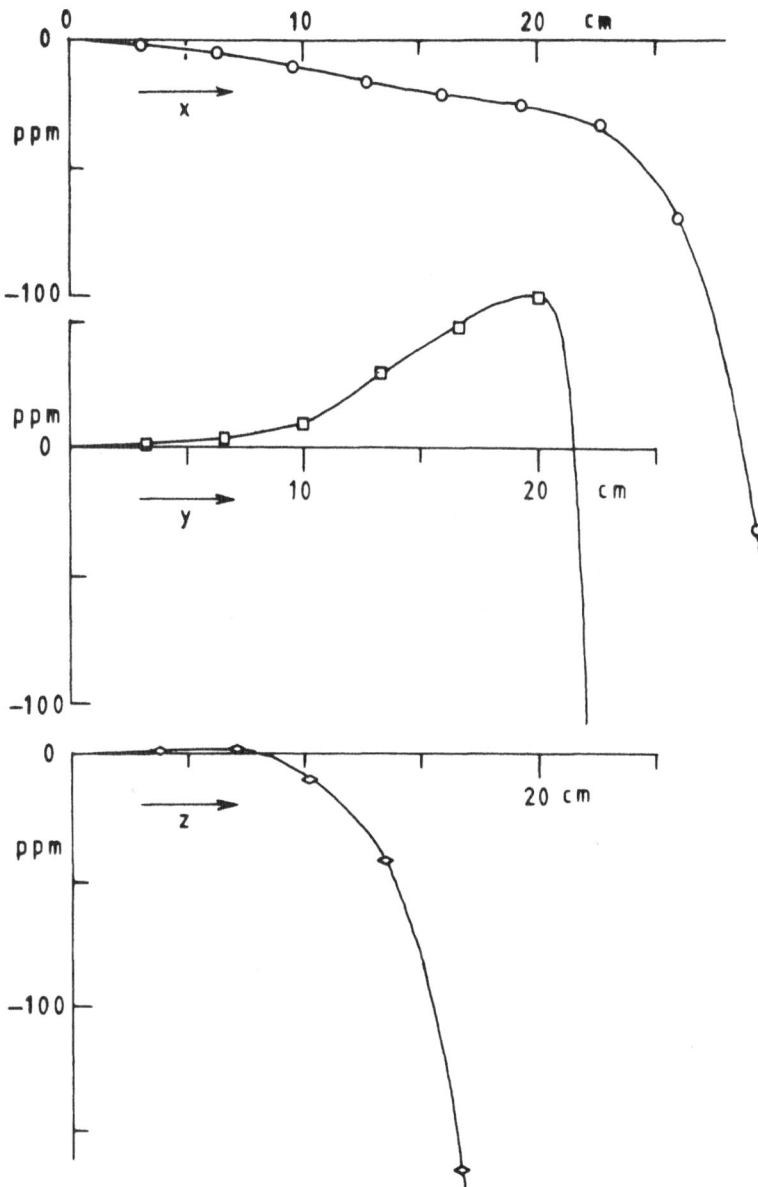

Fig. 6 The magnetic field strength on the three main axes, x, y, and z. The ppm values give the difference of the field with the value at the magnet centre.

For the 3 unknowns, we obtained the following values. The centres of the outer gaps are at 20.8 and -20.8 cm. The width of these gaps is 1.820

cm. The width of the central gap is 1.333 cm. The accuracy as to which one should be able to adjust the gaps is 0.1 mm for the inner gap and 0.2 or 0.3 mm for the outer gap. If this would turn out to be too difficult in practice, one should of course take other measures, namely apply a shimming procedure. Shimming is likely to be necessary anyhow, to compensate for geometrical tolerances and for inhomogeneities in the materials. It can be done with pieces of iron and/or permanent magnet material.

Fig. 6 also shows how the field behaves along the x axis and along the y axis. The introduction of the gaps appears to have considerably improved the homogeneity along these two directions as well, which is just what we hoped for.

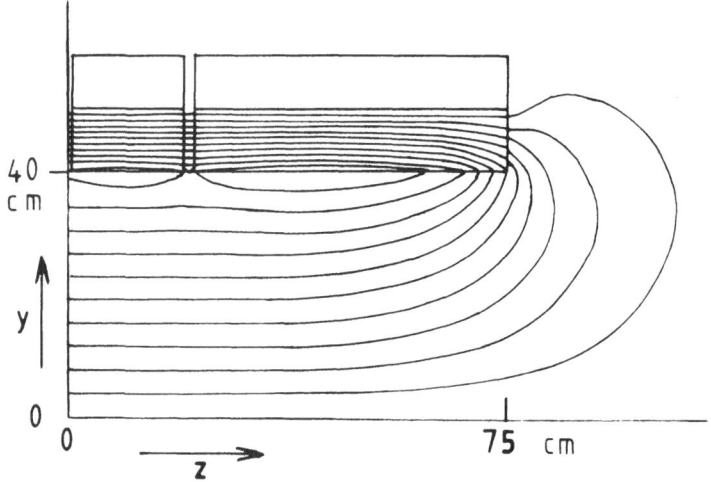

Fig. 7   Plane x=0 with lines of constant potential of the field H.

Fig. 7 gives lines of constant potential of the field H in the zy plane through the centre of the magnet (the plane x=0). We present it to show how the field behaves outside the central region. One may compare it with Fig. 4, which also presents equi-potential lines, but for the xy plane.

MATERIALS

NdFeB, currently the best permanent magnet material, is particularly suitable for engineering a compact magnet system for MRI. Fig. 8 shows its demagnetization curve (marked 4) in comparison with those of other permanent magnet materials. Besides a very high remanence value of approx. 1,200 mT - higher than that of all other permanent magnets - and reversible permeability approaching unity, NdFeB possesses adequately high coercivity H_{cJ}

which prevents demagnetization by the internal magnetic field even if the operating point of the magnet system is very low.

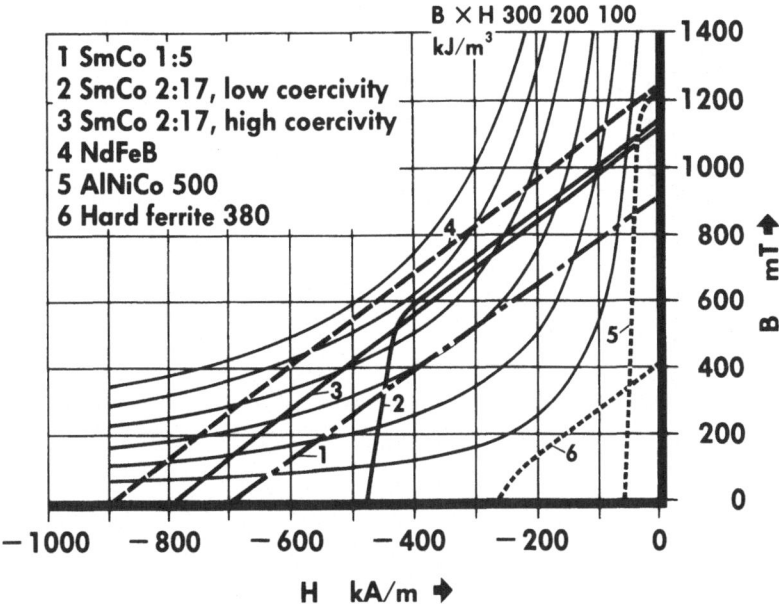

Fig. 8 Demagnetization curves.

NdFeB's low Curie temperature of approx. 310°C, which is disadvantageous in many applications, and the associated marked temperature dependence of saturation polarization J_S (TK(J_S)=-0.13%/K) and hence also of the air-space field strength have hardly any detrimental effect as the temperature changes to be expected are only small and the magnet system follows such changes only very slowly on account of its large mass. The temperature coefficient of coercivity H_{cJ}, which at -0.8%/K is particularly high and can lead to an irreversible decrease in air-space field strength, is likewise of no consequence in magnet systems for MRI which are housed in closed, possibly even temperature-controlled rooms.

According to the virgin curve of magnetization marked 4 in Fig. 9, magnetic fields of approx. 1,200 kA/m are required to completely magnetize NdFeB. At 1.5 $\mu\Omega$m, its electrical resistance is greater by a factor of 2 to 3 than that of the other metallic permanent magnets. This is significant for magnetization with pulsed fields as the depth of penetration - the depth down to which a pulsed magnetic field of a given duration can penetrate the material practically unweakened by opposing fields generated by induced eddy currents - is proportional to electrical resistance.

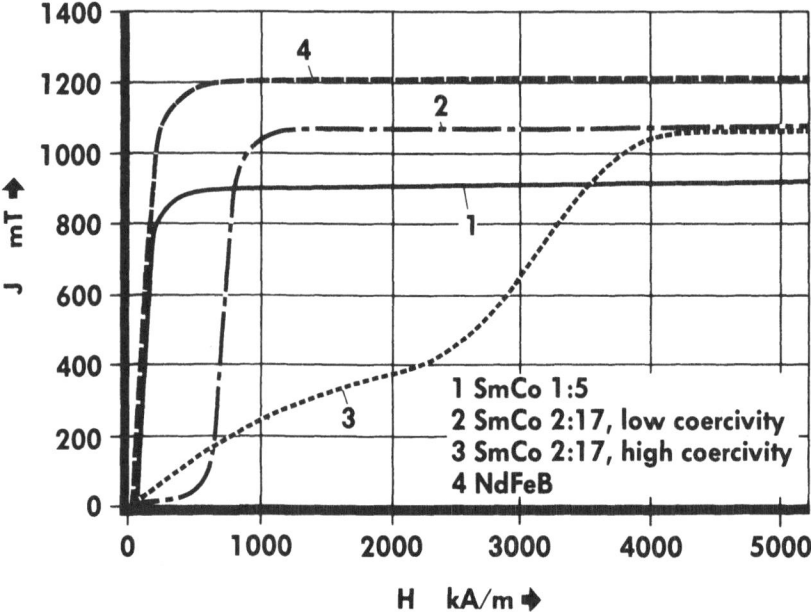

Fig. 9   Virgin curves of some permanent magnet materials.

Neither especially high permeability nor extremely high saturization polarization is required of the magnetically soft material used for the return path. A normal low-carbon steel (engineering steel) is thus complete-ly adequate.

ENGINEERING OF THE MAGNET SYSTEM

The magnet system designed in accordance with the field calculations is shown in Fig. 10a and b. It consists of four rings each containing two cuboidal and four prismatic permanent magnets in a hexagonal iron return path. The two outer rings have an appreciably larger axial length than the two inside ones.

The individual operations which are necessary in the manufacture of such a magnet system are outlined in the following. They need to be well thought out as the NdFeB magnets which can be produced and magnetized using the techniques now available are relatively small while the forces acting between fully magnetized magnets are very large.

The starting materials are bricks of NdFeB measuring 50x50x20 mm ground to close tolerances. These are magnetized, graded according to flux values and subsequently thermally demagnetized. Magnets whose flux values deviate by more than 1% from the mean are rejected.

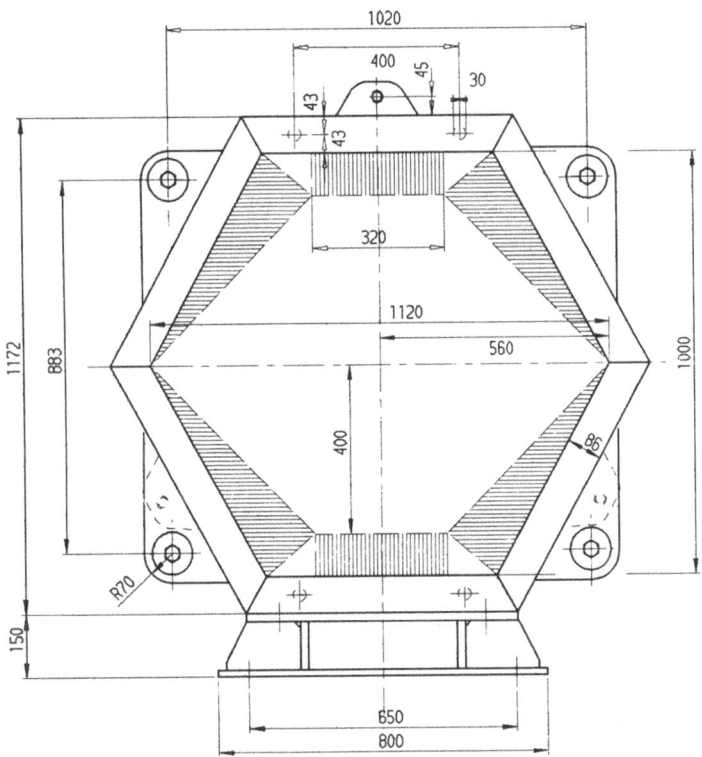

Fig. 10a Permanent magnet MRI apparatus. Front view.

The next operation is to cement these bricks into larger units using a two-part adhesive. Eight rectangular blocks and sixteen stepped shapes are the result.

Once the adhesive has set, these parts are subsequently ground to their final cuboidal or prismatic shape. Great care has to be taken to ensure that the preferred direction of magnetization imparted to the magnetic bricks during their manufacture coincides with the desired direction of polarization in the finished magnet system. In the case of the cuboidal magnets this is relatively straightforward as polarization and hence the preferred direction of magnetization must be exactly perpendicular to the end faces. The stepped shapes, on the other hand, have to be aligned with great care on the grinding bench in order to guarantee the desired angle of polarization relative to the surfaces of the prisms (see Fig. 4). A grinding error that puts the preferred direction of magnetization out of position in the magnet cannot be remedied during magnetization. The magnetizing field influences the degree but not the direction of polarization, it being determined by the preferred direction of magnetization imparted to the magnet bricks.

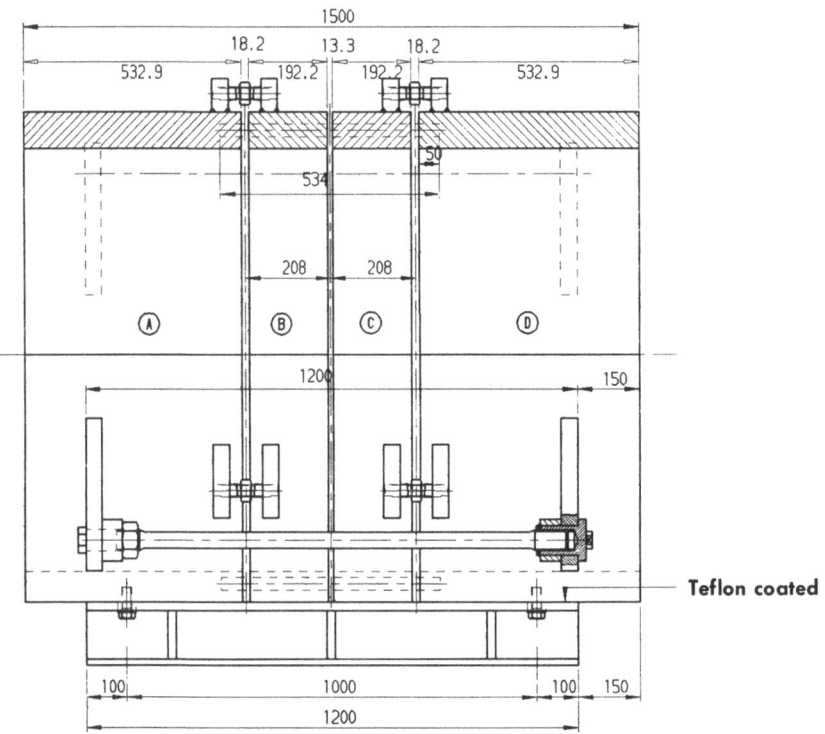

Fig. 10b Permanent magnet MRI apparatus. Side view.

The magnetization of the eight cuboidal and sixteen prismatic magnets is best effected in a normal-conducting air-core coil manufactured from aluminium strip. Such a coil with 400 windings, a length of 800 mm and a clearance of 600x700 mm yields a field strength of 2,000 kA/m at a current of 4,000 A. Of this field strength, approx. 800 kA/m is required to compensate the demagnetizing field. The remaining 1,200 kA/m is sufficient, according to Fig. 9, to completely magnetize the NdFeB material. The current is best taken from a set of 1200 12-V batteries. An energization period of 10 s is sufficient to set up the magnetic field and completely penentrate the magnets. Owing to the size of the forces occurring during magnetization the magnets have to be fixed firmly in position in the coil.

A hydraulic device is used to fit the magnetized magnets to the hexagonal iron rings. It must be able not only to take the weight of the magnets but also to withstand the forces of magnetic attraction and repulsion, which can amount to as much as 1 tonne. Using this device, the magnets, whose bonding surface is covered with two-part adhesive, can be pressed into the final positions. Only when the adhesive has hardened may the next magnet be pushed in.

Once all four rings have been fitted with magnets, the AB and CD pairs each consisting of a wide and a narrow ring are pressed together using three screws to a spacing of 18.2 mm, sliding on a well oiled iron plate coated with Teflon (see Fig. 10a and b). The two halves, likewise sliding on the iron plate, are subsequently spaced 13.3 mm apart using four screws. Once set, these spacings are fixed by non-magnetic discs.

Should field measurements show that the spacings have to be adjusted in order to improve field homogeneity, this is relatively straightforward to accomplish. The individual rings can also be slightly inclined and then screwed onto the base plate in their final positions.

REFERENCES

Zijlstra, H. 1985. Permanent Magnet Systems for NMR Tomography. Philips Journal of Research 40, 259-288 (1985).

SUMMARY OF DISCUSSIONS :

STATIC PERMANENT MAGNET DEVICES

P.R. Locher

A new range of workholding tools has been developed and reported on (9.2). The evidence is that with careful design procedures, Nd-Fe-B has a magnificent future in the clamping\holding-devices industry.

Using Nd-Fe-B, a new hexapole device was constructed (10.1) for use in an electric cyclotron resonance ion source. With the new magnet material, stronger hexapole fields can be obtained. This allows the construction of a more powerful ion source for nuclear physics investigations.

Another big achievement of the application group (10.2) was the design of a permanent magnet configuration for whole-body magnetic resonance imaging (MRI) using Nd-Fe-B material. From the study it became clear that about 5 tonnes of magnet were required to replace approximately 100 tonnes ferrite magnets. The device is designed to produce in its centre volume a highly homogeneous magnetic field with a flux density of 0.2 Tesla. This novel permanent magnet-based MRI is an attractive alternative for medical diagnostic examinations.

APPENDIX .i.

LIST OF CODEST MEMBERS

LIST OF CODEST MEMBERS

Monsieur Aigrain
S.A. Thomson
Bd. Haussman, 173
PARIS Cédex 08 F-75739
France

Tel 33 1 45 61 96 00
Tlx 042 204 780 TCSF

Professor Thor Bak
Jette Sogren Nielsen
Danish Research Administration
Holmens Kanal 7
1060 COPENHAGEN K
Denmark

Tel 45 1 11 43 00
Tlx 15 089 fd dk
Fax 45 1 15 02 05

Professor N. Cabibbo
Presidente INFN
Piazzale A. Moro 2
ROMA I-00185
Italy

Tel 39 6 49911
Tlx 614125 INFNUP I
Fax 39 665 47 924

Professor U. Colombo
Presidente E.N.E.A.
Viale Regina Margherita 125
ROMA I-00198
Italy

Tel 39 6 85 10 07
Tlx 043 610 167 ENEA I
Fax 39 6 852 858 04

Dr. De Waard
Rijksuniversiteit Groningen
Laboratorium voor Algemene
Natuurkunde
Westersingel 34
GRONINGEN NL 9718
The Netherlands

Tel 31 50 63 47 37
Tlx 044 77 391 RUISO

Dr. Ferry-Borges
President
Laboratorio Nacional de
Engenharia
Avenue do Brasil 101
LISBOA P-1799
Portugal

Tel 88 21 31
Tlx 0404 16 760

Monsieur C. Fréjacquees
Président
Centre National de la Recherche
Scientifique
Quai Anatole France, 15
PARIS F-75700
France

Tel 33 1 47 53 15 15
Tlx 042 260 034 CNRS F

Dr. M.W. Geerlings
Akzo N.V.
Verperweg, 76
AARNHEM SB 6800
The Netherlands

Tel 31 85 66 29 30
Tlx 45 438

Monsieur Gros François
Directeur Honoraire de
l'Institut Pasteur
Rue du Dr Roux, 25
PARIS F-75724
France

Tel 33 1 45 68 84 73
Tlx 042 250 609
Fax 33 1 430 698 35

Professor Dr. B. Hess
Vice President, M.P.G.
M.P.I. für Ernährungsphysiologie
Rheinlanddam 201
DORTMUND D-4600
W. Germany

Tel 49 231 1206 385
Tlx 041 822 7147
Fax 49 231 1206 464

Professor F.C. Kafatos
Director - Institute of Molecular
Biology and Technology
Research Center of Crete
P.O. Box 527
HERAKLIO 7110 - Crete
Greece

Tel 30 81 23 46 53
Tlx 0601 262 728 MPUC

Professor F.C. Kafatos
Biological Laboratories
Harvard University
16 Divinity Ave
CAMBRIDGE MA 021138
U.S.A.

Professor Pedro Pascual
Secretaria de Estado
Universidades e Investigacion
Serrano, 150
28006 MADRID
Spain

Tel 34 1 26 15 400

Sir David Phillips
SERC
Polaris House
North Star Avenue
SWINDON SN2 1ET
United Kingdom

Tel 44 793 411 036
Tlx 44 94 66 GTN 1434
Fax 44 793 411 400

Professor I. Prigogine
Campus de la Plaine - U.L.B.
Code Postale 231
Boulevard du Triomphe
BRUXELLES B-1050
Belgium

Tel 02 640 00 15 x5540
Tlx 23 069

Professor Luigi Rossi-Bernardi
Presidente del C.N.R.
Piazzale A. Moro, 7
ROMA
Italy

Tel 39 649 06 78
Tlx 043 61 00 72
Fax 39 6 490 134

Dr. Pierce Ryan
An Foras Taluntais
Sandymount Avenue 19
DUBLIN 4
Ireland

Tel 353 1 68 81 88
Tlx 0500 30 459

Mr. M.P. Seck
Centre Universitaire de Luxembourg
Départment des Sciences
162a Avenue de la Faïencerie
LUXEMBOURG L-1511
G.D. Luxembourg

Tel 352 21 595

Professor Dr. E. Seibold
Geologisches Institut
Der Universität van Freibourg
Albertstrabe, 23
D-78 FREIBOURG
W. Germany

Tel 761 203 23 56

Sir Peter Swinnerton-Dyer
University Grants Committee
Park Crescent 14
LONDON WIN 4DH
England

Tel 44 1 636 77 99

Fax 01 631 4227

Professor Dr. Max Syrbe
Präsident der Fraunhofer
Gesellschaft
MÜNCHEN 19 D-8000
W. Germany

Tel 49 89 12 05 201
Tlx 521 5382
Fax 89 12 05 317

Professor van Overstraeten
Interuniversitair
Micro-Elektronica Centrum vzw
Kapeldreef, 75
LEUVEN B-3030
Belgium

Tel 016 281 372
Tlx 26 152
Fax 016 22 94 00

APPENDIX .ii.

NAMES AND ADDRESSES OF
CEAM PARTICIPANTS

MATERIALS

1 - 01	**D. GIVORD**	Laboratoire Louis Néel C.N.R.S. 166 X 38042 GRENOBLE CEDEX France
1 - 02	**J.M. MOREAU**	Laboratoire de Structure de la Matière 9, rue de l'Arc en Ciel B.P. 908 74019 ANNECY-LE-VIEUX CEDEX France
1 - 03	**V.G. RIVLIN** **R.I. SAUNDERSON**	Fulmer Research Limited Hollybush Hill, Stoke Poges Slough BERKSHIRE, SL2 4QD U.K.
1 - 04	**W. RODEWALD**	Vacuumschmelze Grüner Weg 37 6450 HANAU 1 F.R.G.
1 - 05	**R. CHAMBEROD**	D.R.F./S.P.H./Métallurgie Physique C.E.N.G. 85 X 38041 GRENOBLE CEDEX France
1 - 06	**J. ETOURNEAU**	Laboratoire de Chimie du Solide C.N.R.S. 351, cours de la Libération 33405 TALENCE CEDEX France
1 - 07	**C. ALLIBERT**	L.T.P.C.M. E.N.S.E.E.G. B.P. 75 38402 SAINT MARTIN D'HERES CEDEX France
1 - 08	**Y. BERTHIER** **F. HARTMANN-BOUTRON**	Labo. de Spectrométrie Physique B.P. 87 38402 SAINT MARTIN D'HERES CEDEX
1 - 09	**D. FRUCHART**	Laboratoire de Cristallographie C.N.R.S. 166 X 38042 GRENOBLE CEDEX France

1 - 10	**J.P. SANCHEZ**	Centre de Recherches Nucléaires 23, rue du Loess 67037 STRASBOURG France
1 - 11	**M. BOGE**	D.R.F. - S.P.H. - M.D.I.H. C.E.N.G. 85 X 38041 GRENOBLE CEDEX France
1 - 12	**R. FRUCHART**	Labo. des Matériaux et de Génie Physique E.N.S.P.G. B.P. 46 38402 SAINT MARTIN D' HERES CEDEX France
1 - 13	**J. BARTOLOME**	I.C.M.A. Faculdad de Ciencias Ciudad Universitaria 50009 ZARAGOZA Spain
1 - 14	**R. GILLET**	D.M.G. - S.E.M. - L.E.M.M. C.E.N.G. 85 X 38041 GRENOBLE CEDEX France
1 - 21	**J.M.D. COEY**	Physics Department Trinity College 2 DUBLIN Ireland
1 - 22	**J.J.M. FRANSE**	Natuurkundig Laboratorium UvA Valckenierstraat 65 1018 XE AMSTERDAM The Netherlands
1 - 23	**G. CZJZEK**	Kerforschungszentrum Karlsruhe I.N.F.P. Postfach 3640 7500 KARLSRUHE F.R.G.
1 - 24	**G. ASTI** **L. PARETI**	IST Maspec del C.N.R. Via Chiavari, 18 A 43100 PARMA Italy

1 - 25	M. ROSENBERG	Institüt f. Experimentalphysik VI Ruhr-Universität Bochum 4630 BOCHUM F.R.G.
1 - 26	K.H.J. BUSCHOW	Philips Research Labs. P.O. Box 80000 5600 JA EINDHOVEN The Netherlands
1 - 27	W.D. CORNER	Physics Department University of Durham South Road DURHAM DH1 3LE U.K.
1 - 28	A. KOSTIKAS D. NIARCHOS	Institute of Materials Science Attikis P.O.B. 60228 15310 AGHIA PARASKEVI Greece
1 - 29	H. KRONMÜLLER E.T. HENIG	Max-Planck Inst. f. Metallforsch. Institüt für Physik Heisenbergstr. 1 7000 STUTTGART 80 F.R.G.
1-30	P.J. GRUNDY	Department of Physics University of Salford SALFORD M5 4WT U.K.
1 - 31	S. ABELL R. HARRIS	Metallurgy and Materials University of Birmingham Elms road BIRMINGHAM B15 2TT U.K.
1 - 33	D.G. PETTIFOR	Department of Mathematics Imperial College 180 Queen's Gate LONDON SW7 2BZ U.K.
1 - 34	G.C. HADJIPANAYIS	Department of Physics University of Heraklion Crete Greece
1 - 35	H.R. KIRCHMAYR	Institüt f. experimental Physik T.U. Wien Wiedner Hauptstr. 8 1040 WIEN Austria

| 1 - 36 | C. HEIDEN | Institüt f. Angew. Physik
Univ. Giessen
Heinrich-Buff-Ring 16
6300 GIESSEN
F.R.G. |

MAGNET PROCESSING

| 2 - 01 | R. HARRIS | Metallurgy and Materials
University of Birmingham
Elms Road
BIRMINGHAM B15 2TT
U.K. |

| 2 - 02 | H.A. DAVIES | School of Materials
University of Sheffield
Mappin Street
SHEFFIELD S1 3JD
U.K. |

| 2 - 03 | A.J. WARD
S.M. BOND | S.G. Magnets, Tesla House
85, Ferry Lane
RAINHAM, ESSEX RM13 9YH
U.K. |

| 2 - 04 | D.W.A. MURPHY
D. KENNEDY | Rare Earth Products
Waterloo Road
Widnes
CHESHIRE WA8 0QH
U.K. |

| 2 - 05 | R. HÄHN
S. SATTELBERGER | Gesellschaft f. Elektrometallurgie
Höfener Str. 45
Postfach 2844
8500 NURNBERG 80
F.R.G. |

| 2 - 06 | A. CARTOCETI | I.O.S.
Viale Dell' Industria, 23
21023 MALCESSO (VA)
Italy |

| 2 - 07 | J. ORMEROD | Mullard Southport
Balmoral Drive
Southport
MERSEYSIDE PR9 8PZ
U.K. |

2 - 08	**A.G. CLEGG**	Magnet Centre, Physics Division Sunderland Polytechnic Chester Road SUNDERLAND, SR2 7EE U.K.
2 - 09	**M. WARD**	Lucas Eng. & Systems Ltd Lucas Res. Center Shirley, Solihull WEST MIDLANDS B90 4JJ U.K.
2 - 10	**A.J. WALKDEN J. VINCENT**	G.E.C. Research Hirst Research Center East Lane MIDDX WEMBLEY U.K.
2 - 11	**J.M.D. COEY**	Physics Department Trinity College 2 DUBLIN Ireland
2 - 13	**H. NAGEL**	Thyssen Edelstahlwerke AG Magnetfabrik Dortmund Ostkirchstr. 177 4600 DORTMUND 41 F.R.G.
2 - 14	**S. TORI**	Labo. Materiali e Processi Ing. c. Olivetti c., S.P.A. Via G. Jervis, 77 10015 IVREA (TO) Italy

APPLICATIONS

3 - 01	**R. HANITSCH**	T.U. Berlin FB 19 Einsteinhufer 11 1000 BERLIN 10 F.R.G.
3 - 02	**D. HOWE**	Dept. of Electronic & Elec. Eng. University of Sheffield Mappin Street SHEFFIELD, S1 3JD U.K.

3 - 03	B.J. CHALMERS E. SPOONER	Dept. of Electr. Eng. & Electron. University of Manchester Inst. of Science and Technology MANCHESTER, M60 1QD U.K.
3 - 04	W. GEYSEN R. BELMANS	K.U. Leuven Kard. Mercierlaan, 94 3030 HEVERLEE Belgium
3 - 05	M. BRADFORD	ERA Technology Ltd Cleeve Road LEATHERHEAD, SURREY, KT22 7SA U.K.
3 - 06	A.J. WALKDEN KAMDAR	G.E.C. Research Hirst Res. Center East Lane MIDDX WEMBLEY U.K.
3 - 09	A. THOMA T. KINDERVATER	Kindervater & Sohn KG Maybachufer 48-51 1000 BERLIN 44 F.R.G.
3 - 10	I. KITZMANN	Dornier System GmbH Department MERP P.O. Box 1360 7990 FRIEDRICHSHAFEN F.R.G.
3 - 11	E.M.H. KAMERBEEK	Philips Research Laboratory Post Box 1980 D100 AACHEN F.R.G.
3 - 12	R. PAUTHENET[†] J. LAFOREST	Laboratoire Louis Néel C.N.R.S. 166 X 38042 GRENOBLE CEDEX France
3 - 13	W. BARAN	Krupp Widia GmbH Munchener Str. 90 4300 ESSEN 1 F.R.G.
3 - 14	G. IOPPOLO	M.A.E. Via Circonvollazione Sud 5 26010 OFFANENGO Italy

3 - 16   **K.J. BINNS**   Dept. of Electr. Engin. & Electro.
                                     University of Liverpool
                                     P.O. Box 147
                                     LIVERPOOL, L69 3BX
                                     U.K.

3 - 17   **R.P. VAN STAPELE**   Philips Research Laboratories
                                     P.O. Box 80000
                                     5600 JA EINDHOVEN
                                     The Netherlands

APPENDIX .iii.

LIST OF PUBLICATIONS OF CEAM PARTICIPANTS (BY SUBCONTRACT)

SUB-CONTRACT N° 1.01 - D. GIVORD

1 - Crystal chemistry and magnetic properties of the $R_2Fe_{14}B$ family of compounds
D. Givord, H.S. Li
Proc. European Workshop "Nd-Fe permanent magnets. Their present and future applications", 131 (1984).

2 - Magnetic properties and crystal structure of $Nd_2Fe_{14}B$
D. Givord, H.S. Li, J.M. Moreau
Sol. Stat. Commun., 50, 497 (1984).

3 - Magnetic properties of $Y_2Fe_{14}B$ and $Nd_2Fe_{14}B$ single crystals
D. Givord, H.S. Li, R. Perrier de la Bathie
Sol. Stat. Commun., 51, 857 (1984).

4 - Structural and magnetic properties in $R_2Fe_{14}B$ compounds
D. Givord, H.S. Li, J.M. Moreau, R. Perrier de la Bâthie,
E. Du Trémolet de Lacheisserie
Physica, 130B, 323 (1985).

5 - Polarized neutron study of the compounds $Y_2Fe_{14}B$ and $Nd_2Fe_{14}B$
D. Givord, H.S. Li et F. Tasset
J. Appl. Phys., 57, 4100 (1985).

6 - 3d-4f magnetic interactions and crystalline electric field in the $R_2Fe_{14}B$ compounds : magnetization measurements and Mössbauer study of $Gd_2Fe_{14}B$
M. Bogé, J.M.D. Coey, G. Czjzek, D. Givord, C. Jeandey, H.S. Li,
J.L. Oddou
Sol. Stat. Commun. 55, 295 (1985).

7 - $Nd_5Fe_{18}B_{18}(Nd_{1.11}Fe_4B_4)$. A new Nowotny-like phase. Structural and magnetic properties
D. Givord, J.M. Moreau, P. Tenaud
Sol. Stat. Commun., 55, 303 (1985).

8 - Rare earth-transition metal permanent magnets
D. Givord, J. Laforest, H.S. Li, A Liénard, R. Perrier de la Bâthie,
P. Tenaud
J. de Physique, C6, 213 (1985).

9 - Determination of the degree of crystallites orientation in permanent magnets by X-ray scattering and magnetic measurements
D. Givord, A. Liénard, R. Perrier de la Bâthie, P. Tenaud, T. Viadieu
J de Physique, C6, 313 (1985).

10 - Exchange interactions and hyperfine field at the rare-earth nuclei in $R_2Fe_{14}B$ compounds
Y. Berthier, M. Bogé, G Czjzek, <u>D. Givord</u>, C. Jeandey, <u>H.S. Li</u>,
J.L. Oddou
J. Magn. Magn. Mat. 54-57 (1986) 589

11 - A ^{155}Gd Mössbauer study of a $Gd_2Fe_{14}B$ single crystal
M. Bogé, G. Czjzek, <u>D. Givord</u>, C. Jeandey, <u>H.S. Li</u>, J.L. Oddou
J. Phys. F 16 (1986) L67.

12 - Modulation of the Fe tetrahedra driven by interactions with R atoms in $R_{1+\epsilon}Fe_4B_4$ compounds
<u>D. Givord</u>, J.M. Moreau, <u>P. Tenaud</u>
J. Less Comm. metals 115 (1986) L7.

13 - Analysis of hysterisis loops in Nd-Fe-B sintered magnets
<u>D. Givord</u>, <u>P. Tenaud</u>, <u>T. Viadieu</u>
J. Appl. Phys. 60 (1986) 3263.

14 - A note on exchange and crystal field interactions in $R_2Fe_{14}B$ compounds : $Yb_2Fe_{14}B$
P. Burlet, J.M.D. Coey, J.R. Gavigan, <u>D. Givord</u>, C. Meyer
Sol. St. Comm. 60 (1986) 723.

15 - Refinement of the crystal structure of $R_{1+\epsilon}Fe_4B_4$ compounds (R - Nd, Gd)
<u>D. Givord</u>, J.M. Moreau, <u>P. Tenaud</u>
J. Less Comm. Metals 123 (1986) 109.

16 - Magnetic properties of the R-Fe-B ternary compounds
<u>D. Givord</u>, <u>H.S. Li</u>, J.M. Moreau, <u>P. Tenaud</u>
J. Magn. Magn. Mat. 54-57 (1986) 445.

17 - Texture in Nd-Fe-B magnets analysed on the basis of the determination of $Nd_2Fe_{14}B$ single crystals easy growth axis
<u>P. Tenaud</u>, A Chamberod, F. Vanoni
Sol. St. Comm., 63(4), (1987), 303.

18 - Exchange and CEF interactions in $R_2Fe_{14}B$ compounds
J.M. Cadogan, J.M.D. Coey, J.P. Gavigan, <u>D. Givord</u>, <u>H.S. Li</u>
J. Appl. Phys. 61(8) (1987) 3974.

19 - 3d magnetism in $R_2Fe_{14}B$ compounds
F. Bolzoni, J.P. Gavigan, <u>D. Givord</u>, <u>H.S. Li</u>, D. Moze
J. Magn. Magn. Mat., 66 (1987) 158.

20 - ^{155}Gd and ^{161}Dy Mössbauer study of $R_{1+\epsilon}Fe_4B_4$ alloys (R - Gd, Dy)
H.R. Rechenberg, M. Bogé, C. Jeandey, J.L. Oddou, J.P. Sanchez,
P. Tenaud
Solid State Comm., 64 (1987) 277.

21 - Magnetic properties of $Pr_2(Fe_{1-x}Co_x)_{14}B$ compounds
F. Bolzoni, J.M.D. Coey, J. Gavigan, D. Givord, O. Moze, L. Pareti
T. Viadieu
J. Magn. Magn. Mat., 66 (1987) 123.

22 - Magnetic viscosity in Nd-Fe-B sintered magnets
D. Givord, A Lienard, P. Tenaud, T. Viadieu
J. Magn. Magn. Mat., 67(3) (1987) L281.

23 - Magnetic viscosity in different Nd-Fe-B magnets
D. Givord, P. Tenaud, T. Viadieu, G. Hadjipanayis
J. Appl. Phys., 61(8) (1987) 3454.

24 - High field magnetization measurements on $Tb_2Fe_{14}B$ and
$Er_2Fe_{14}B$ single crystals
J.P. Gavigan, D. Givord, H.S. Li, O. Yamada, H. Maruyama, M. Sagawa,
S. Hirosawa
J. Magn. Magn. Mat., 70 (1987) 416.

25 - Dependence of domain width on crystal thickness in $Nd_2Fe_{14}B$ single
crystals
R. Szymczack, D. Givord, H. S. Li
Acta Phys. Pol. A., A72(1) (1987) 113.

26 - Evidence in R-M intermetallics for a systematic dependence of R-M
exchange interactions on the nature of the R atom
E. Belorizky, M.A. Fremy, J.P. Gavigan, D. Givord, H.S. Li,
J. Appl. Phys., 61 (1987) 3971.

27 - Magnetic properties of R-Fe-B magnets
J.P. Gavigan, D. Givord, H.S. Li, P. Tenaud, T. Viadieu
Proceeding of the 3rd International Conference on Physics of Magnetic
Materials World Scientific Publishing Co (Singapore) (1987), 201.

28 - A new approach to the analysis of magnetization measurements in rare
earth transition metal compounds. Application to $Nd_2Fe_{14}B$
J.M. Cadogan, J.P. Gavigan, D. Givord, H.S. Li,
J. Phys. F (Met. Phys.) (1988) (à paraître).

29 - Magnetic transitions and anomalous behaviour of Pr in $Pr_2(Fe_{1-x}Co_x)_{14}B$
J.P. Gavigan, H.S. Li, J.M.D. Coey, T. Viadieu, L. Pareti, F. Bolzoni,
O. Moze
To be presented at ICM'88 Paris.

30 - Coercivity mechanisms in ferrite and rare earth transition metal
sintered magnets ($SmCo_5$, Nd-Fe-B)
D. Givord, P. Tenaud, T. Viadieu
I.E.E.E. Trans. Mag., (1988) (to appear).

31 - Dependence of the coercive field and magnetic viscosity coefficient
in NdFeB magnets on the magnetic history of the sample
D. Givord, C. Heiden, A. Hoehler, P. Tenaud, T. Viadieu, K. Zeibig
I.E.E.E. Trans. Mag., (1988) (to appear).

32 - 3d magnetism in R-M and $R_2M_{14}B$ compounds (M = Fe, Co ; R = Rare Earth)
J.P. Gavigan, D. Givord, H.S. Li, J. Voiron
J. Magn. Magn. Mat., (to appear).

33 - A study of exchange and cristalline electric field interactions in
$Nd_2Co_{14}B$: comparison with $Nd_2Fe_{14}B$
H.S. Li, J.P. Gavigan, J.M. Cadogan, D. Givord, J.M.D. Coey
J. Magn. Magn. Mat., (1988) (to appear).

34 - Analysis of high field magnetization measurements on $R_2Fe_{14}B$ single
crystals (R = Tb, Dy, Ho, Er and Tm)
D. Givord, H.S. Li, J.M. Gadogan, J.P. Gavigan, O. Yamada, H. Maruyama,
M. Sagawa, S. Hirosawa
J. App. Phys., (1988) (to appear).

35 - Preparation and ^{57}Fe Mössbauer study of $PrCo_3FeB$, $NdCo_3Fe_B$, $SmCo_3FeB$
and $SmCo_2Fe_2B$
Y. Gros, M.A. Fremy, F. Hartmann-Boutron, C. Meyer, P. Tenaud
Submitted to J. Magn. Magn. Mat.

36 - Mössbauer study of compounds $RCo_{4-x}Fe_xB$ and RFe_4B
Y. Gros, F. Hartmann-Boutron, C. Meyer, M.A. Fremy, P. Tenaud
Communication presented at ICM'88 Paris (July 1988).

37 - Magnetostriction in high pulsed magnetic field on a single crystal of
$Nd_2Fe_{14}B$
C. Marquina, M.R. Ibarra, P.A. Algarabel, A. del Moral, D. Givord and
S. Zemirli
Accepted for presentation at the International Symposium on High Field
Magnetism (K.U. Leuven 1988).

SUB-CONTRACT N° 1.02 - J.M. MOREAU

1 - Magnetic properties and crystal structure of $Nd_2Fe_{14}B$
    D. Givord, H.S. Li, <u>J.M. Moreau</u>
    Sol. Stat. Commun., 50 (1984) 497.

2 - Structural and magnetic properties in $R_2Fe_{14}B$ compounds
    D. Givord, H.S. Li, <u>J.M. Moreau</u>, R. Perrier de la Bathie,
    E. du Trémolet de Lacheisserie
    Physica, 130B (1985) 323.

3 - $Nd_5Fe_{18}B_{18}(Nd_{1.11}Fe_4B_4)$. A new Nowotny-like phase. Structural and
    magnetic properties
    D. Givord, <u>J.M. Moreau</u>, P. Tenaud
    Sol. Stat. Commm., 55 (1985) 303.

4 - Refinement of the crystal structure of $R_{1+\epsilon}Fe_4B_4$ compounds (R = Nd, Gd)
    D. Givord, <u>J.M. Moreau</u>, P. Tenaud
    J. Less Comm. Metals, 123 (1986) 109.

5 - Magnetic properties of the R-Fe-B ternary compounds
    D. Givord, H.S. Li, <u>J.M. Moreau</u>, P. Tenaud
    J. Magn. Magn. Mat., 54-57 (1986) 445.

6 - Modulation of the Fe tetrahedra driven by interactions with R atoms in
    $R_{1+\epsilon}Fe_4B_4$ compounds
    D. Givord, <u>J.M. Moreau</u>, P. Tenaud
    J. Less Comm. Metals, 115 (1986) L7.

7 - Evidence in R-M intermetallics for a systematic dependence of R-M
    exchange ineractions on the nature of the R atom
    E. Belorizky, <u>M.A. Fremy</u>, J.P. Gavigan, D. Givord, H.S. Li
    J. Appl. Phys., <u>61</u> (1987) 3971.

8 - Crystallographic and Magnetic properties of a new series $RFe_{10}SiCo_5$
    (R = Ce, Pr, Nd, Sm)
    <u>J. Leroy</u>, <u>J.M. Moreau</u>, <u>C. Bertrand</u>, <u>M.A. Fremy</u>
    J. Less Comm. Metals, 136 (1987) 19.

9 - A new quaternary phase stabilized by Carbon : $DyFe_2SiC$
    <u>L. Paccard</u>, <u>D. Paccard</u>, <u>C. Bertrand</u>
    J. Less Comm. Metals, 135 (1987) L5.

10- $Dy_2Fe_2Si_2$: A new structure derived from Ge_2Os and stabilized by Carbon
    <u>L. Paccard</u>, <u>D. Paccard</u>
    J. Less. Comm. Metals, 136 (1988) 297.

11 - Preparation and ^{57}Fe Mössbauer study of $PrCo_3FeB$, $NdCo_3FeB$, $SmCo_3FeB$
and $SmCo_2Fe_2B$
Y. Gros, M.A. Fremy, F. Hartmann-Boutron, C. Meyer, P. Tenaud
Submitted to J. Magn. Magn. Mat.

12 - Mössbauer study of compounds $RCo_{4-x}Fe_xB$ and RFe_4B
Y. Gros, F. Hartmann-Boutron, C. Meyer, M.A. Fremy, P. Tenaud
Communication presented at ICM'88 Paris (July 1988).

SUB-CONTRACT N° 1.04 - RODEWALD

1 - Magnetic properties of $Nd_{15-x}Dy_xFe_{77}B_8$ alloys
W. Rodewald
Proc. 8th Intern. Workshop on RE magnets, Dayton, U.S.A. (1985) 737.

2 - Magnetization and aging of sintered Nd-Fe-B magnets
W. Rodewald
J. Less Comm. Metals, 111 (1985) 77.

3 - Measurement of angle deviation of easy axis in anisotropic permanent
magnets
E. Adler, W Sattler, J. Spencer
Int. Conf. Insertion devices for synchrotron sources, Oct. 85,
Stanford, U.S.A.

4 - A contribution to the understanding of coercivity and its temperature
dependence in sintered $SmCo_5$ and $Nd_2Fe_{14}B$ magnets
E. Adler, P. Hamann
Proc. 8th Intern. Workshop on RE Magnets, Dayton, U.S.A. (1985) 747.

5 - On the theory of nucleation fields in uniaxal ferromagnets
G. Herzer, W. Fernengel, E. Adler
J. Magn. Magn. Mat., 58 (1986) 48.

6 - The magnetizability and the internal hysteresis loops of $SmCo_5$ and
$Nd_2Fe_{14}B$ permanent magnets
E. Adler, W. Fernengel
Digest Intermag (1986) G.C-07.

7 - TEM studies of sintered Fe-Nd-B magnets
P. Schrey
I.E.E.E. Trans.. Mag. MAG-22 (1986) 913.

8 - How to match powder metallurgy and physics of magnetic materials
E. Adler, G.W. Reppel, W. Rodewald, H. Warlimont
in Pulvermetallurgie in Wissenschaft und Praxis (in press).

9 - Magnetic properties of sintered Nd-Fe-Al-B magnets
W. Rodewald
Proc. 9th Int. Workshop on R.E. Magnets and their Appl Bad Soden
31 August -2 Sept. 1987. 609.

10 - The effect of surface oxidation on the demagnetization curve of
sintered Nd-Fe-B permanent magnets
R. Blanck, E. Adler
Proc. 9th. Int. Workshop on RE-magnets and Their Appl., Bad Soden,
31st August - 2nd Sept. 1987, p. 537.

11 - The magnetizability and the internal hysteresis loops of SmCo$_5$ and
Nd-Fe-B permanent magnets
W. Fernengel, E. Adler
Proc. 5th Int. Symposium on Magnetic Anisotropy and Coercivity, Bad
Soden, 31st August - 2nd Sept. 1987, p. 247.

12 - The coercivity of the virgin magnetization curve of sintered Nd-Fe-B
permanent magnets
W. Fernengel
Proc. 5th Int. Symposium on Magnetic Anisotropy and Coercivity, Bad
Soden, 31st August - 2nd Sept. 1987, p. 247.

13 - Properties of sintered Nd-Fe-TM-B magnets
W. Rodewald, W. Fernengel
I.E.E.E. Trans. Magnetics (in press).

14 - On the precipitation of intermetallics phases on Nb-containing
Nd-Fe-B magnets
P. Schrey
J. Magn. Magn. Mat., (in press).

15 - Erste Ergebnisse der europäischen Zusammenarbeit bei der Erforschung
moderner Dauermagnete
E. Adler, W. Rodewald, W. Fernengel
52. Physikertagung, Karlsruhe 1988, Hauptwortrag BAI.

16 - Magnetische Nachwirkung auf inneren und äußeren Hystereseschleifen
von gesinterten Nd-Fe-B Dauermagneten
W. Fernengel, E. Adler
52 Physikertagung, Karlsruhe, Vortrag AM-4.5

17 - Wirkung von Mo auf das Gefüge und die magnetischen Eigenschaften
gesinterter Nd-Fe-Co-B Magnete
W. Rodewald, P. Schrey
52. Physikertagung, Karlsruhe, Bortrag AM-4.6

SUB-CONTRACT N° 1.05 - R. CHAMBEROD

1 - Texture in Nd-Fe-B magnets analysed on the basis of the determination
of Nd2Fe14B single crystals easy growth axis
P. Tenaud, A. Chamberod, F. Vanoni
Solid State Commun. (USA), 63(4) (1987) 303.

SUB-CONTRACT N° 1.06 - J. ETOURNEAU

1 - De nouveaux matériaux ferromagnétiques : étude des composés
Nd $Co_{9-x}Fe_xSi_2$ ($0 < x \leq 5$).
B. Chevalier, G. Gurov, L. Fournes et J. Etourneau
2ème Congrés International de la Soc. Franç. de Chimie SFC, Paris,
8-12 Septembre 1986.

2 - New ferromagnetic compounds : the ternary silicides $NdCo_{9-x}Fe_xSi_2$
($0 \leq x \leq 5$)
M. Chevalier, G. Gurov, L. Fournes, J. Etourneau
J. Chem. Res., 5 (1987) 138.

3 - Iron-rich pseudobinary alloys with the $ThMn_{12}$ structure obtained by
melt spinning : $Gd(Fe_nAl_{12-n})$ n = 6, 8, 10
Wang Xiang Zhong, B. Chevalier, J. Etourneau, T. Berlureau,
J.M.D. Coey, J.M. Cadogan
J. Less Comm. Metals, 138 (1988) 235.

4 - Magnetic properties of Nd $(Co_{1-x}Fe_x)_9 Si_2$ ($0 < x \leq 0.55$) from magnetization,
NMR and Mössbauer studies.
Y. Berthier, B. Chevalier, J. Etourneau, R. Rechenberg and J.P. Sanchez
J. Magn. Magn. Mat., (to be published).

5 - Hydrogen absorption and desorption in Nd_2Fe_{17} and Sm_2Fe_{17}
Wang Xiang Zhong, K. Donnelly, J.M.D. Coey, B. Chevallier, J. Etourneau,
T. Berlureau
J. Mat. Sci., 23 (1988) 329.

SUB-CONTRACT N° 1.08 - Y. BERTHIER - F. HARTMANN

1 - A note on exchange and crystal field interaction in $R_2Fe_{14}B$ compounds : $Yb_2Fe_{14}B$
P. Burlet, J.M.D. Coey, J.R. Gavigan, D. Givord, C. Meyer
Sol. St. Comm., 60 (1986) 723.

2 - Exchange interactions and hyperfine field at the rare-earth nuclei in $R_2Fe_{14}B$ compounds
Y. Berthier, M. Bogé, G. Czjzek, D. Givord, C. Jeandey, H.S. Li, J.L. Oddou
J. Magn. Magn. Mat., 54-57 (1986) 589.

3 - Evidence in R-M intermetallics for a systematic dependence of R-M exchange interactions on the nature of the R atom
E. Belorizky, M.A. Fremy, J.P. Gavigan, D. Givord, H.S. Li
J. Appl. Phys., 61 (1987) 3971.

4 - Preparation and ^{57}Fe Mössbauer study of $PrCo_3FeB$, $NdCo_3FeB$, $SmCo_3FeB$ and $SmCo_2Fe_2B$
Y. Gros, M.A. Fremy, F. Hartmann-Boutron, C. Meyer, P. Tenaud
Submitted to J. Magn. Magn. Mat.

5 - Mössbauer study of compounds $RCo_{4-x}Fe_xB$ and RFe_4B
Y. Gros, F. Hartmann-Boutron, C. Meyer, M.A. Fremy, P. Tenaud
Communication presented at ICM'88 Paris (July 1988).

6 - A study of crystal field effects in $Yb_2Fe_{14}B$ by ^{174}Yb Mössbauer spectroscopy
C. Meyer, J.P. Gavigan, G. Czjzek, H.J. Bornemann
Communication presented at I.C.M. Paris (July 1988).

7 - Magnetic properties of $Nd(Co_{1-x}Fe_x)_9Si_2$ ($0<x\leq0.55$) from magnetization, NMR and Mössbauer studies.
Y. Berthier, B. Chevalier, J. Etourneau, R. Rechenberg and J.P. Sanchez
J. Magn. Magn. Mat., (to be published).

SUB-CONTRACT N° 1.09 - D. FRUCHART

1 - Structural and magnetic properties of $RE_2Fe_{14}B$ hydrides
D. Fruchart, P. Wolfers, P. Vulliet, A. Yaouanc, R. Fruchart, P. L'Héritier
Proc. European Workshop "Nd-Fe permanent magnets their present and future applications" (1984) p.173.

2 - Properties of hydrided $RE_2Fe_{14}B$ compounds
P. Dalmas de Réotier, D. Fruchart, P. Wolfers, P. Vulliet, A. Yaouanc,
R. Fruchart, P. L'Héritier
J. de Physique, C6 (1985) 249.

3 - Crystallographic and magnetic studies of $Nd_2Fe_{14}B$ and $Y_2Co_{14}B$
D. Le Roux, H. Vincent, P. L'Héritier, R. Fruchart
J. de Phys., 46 C6 (1985) 243.

4 - Magnetization, ^{57}Fe and ^{161}Dy Mössbauer study of $Dy_2Fe_{14}BH_x$ with
$0 \leq x \leq 4.7$
L.P. Ferreira, R. Guillen, P. Vulliet, A. Yaouanc, D. Fruchart,
P. Wolfers, P. L'Héritier, R. Fruchart
J. Magn. Magn. Mat., 53 (1985) 145.

5 - ^{57}Fe Mössbauer study of $RE_2Fe_{14}B$
P. Dalmas de Réotier, D. Fruchart, P. Wolfers, R. Guillen, P. Vulliet,
A. Yaouanc, R. Fruchart, P. L'Héritier
J. de Physique, C6 (1985) 323.

6 - Hydrogen dependence of the intrinsic magnetic properties of $R_2Fe_{14}B$
J.M.D. Coey, A. Yaouanc, D. Fruchart, R. Fruchart, P. L'Héritier
Intern. Symp. on the properties and Applications of Metal Hydrides V,
Maubuisson, France, (May 25-30 1986).

7 - Effect of hydrogen on the magnetic properties of $Y_2Fe_{14}B$
J.M.D. Coey, A. Yaouanc, D. Fruchart
Sol. Stat. Comm., 58 (1986) 413.

8 - Mössbauer spectroscopy of $R_2Fe_{14}B$
R. Fruchart, P. L'Héritier, P. Dalmas de Reotier, D. Fruchart, P. Wolfers,
J.M.D. Coey, L.P. Ferreira, R. Guillen, P. Vulliet and A. Yaouanc
J. Phys. F. : Met. Phys., 17 (1987) 483.

9 - Positive Muon Spectroscopy of $Nd_2Fe_{14}B$ and $Pr_2Fe_{14}B$
A. Yaouanc, J. Budnick, E. Albert, M. Hamma, A. Weidinger, R. Fruchart,
P. L'Héritier, D. Fruchart, P. Wolfers
Phys. Rev. Lett., 67 (1987) L286.

10- Structural and magnetic properties of $R_2Fe_{14}BH_x$
D. Fruchart, L. Pontonnier, F. Vaillant, J. Bartolome, J.M. Fernandez,
J.A. Puertolas, C. Rillo, J.M. Regnard, A. Yaouanc, R. Fruchart,
P. L'Héritier
EMMA 87, to be published in I.E.E.E. Trans. Mag.

11 - Hydrogen induced changes of valency and hybridization in Ce
intermetallic compounds
D. Fruchart, F. Vaillant, A. Yaouanc, J.M.D. Coey, R. Fruchart,
P. L'Héritier, T. Riester, J. Osterwalder, L. Schlapbach
J. Less Comm. Metals, 130 (1987) 97.

12 - Hydrogen dependence of the intrinsic magnetic propeties of $Re_{12}Fe_{14}B$
hydrides
J.M.D. Coey, P. Vulliet, A. Yaouanc, D. Fruchart, P. Wolfers,
P. L'Héritier, R. Fruchart
J. Less Comm. Metals, 131 (1987) 419.

13 - Structural and magnetic properties of $RE_2Fe_{14}BH$ $(D)_x$, RE = Y, Ce, Er
D. Fruchart, L. Pontonnier, F. Vaillant, P. Wolfers, A. Yaouanc,
J.M.D. Coey, R. Fruchart, P. L'héritier, P. Dalmas de Réotier
J. Less Comm. Met., 129 (1987) 133.

14 - Spin rotation in $R_2Fe_{14}BH_x$
J.P. Regnard, A Yaouanc, D. Fruchart, D. Le Roux, P. L'héritier,
R. Fruchart, J.M.D. Coey, J.P. Gavigan
J. Appl. Phys., 61 (1987) 3565.

15 - Dynamical susceptibility of $Ho_2Fe_{14}B$ single crystal : spin rotation
and domain wall motions
C. Rillo, J. Chaboy, R. Navarro, J. Bartolomé, D. Fruchart, A. Yaouanc,
B. Chenevier, M. Sagawa, S. Hirosawa
Accepted in $4h^{th}$ Joint MMM-Intermag Conference 1988. To be published
in J. Appl. Phys.

16 - Effects of hydrogen absorption on the 3d and 4f anisotropy in
$RE_2Fe_{14}B$ (RE = Y, Nd, Ho, Tm)
L. Pareti, O. Moze, M. Solzi, D. Fruchart, P. L'Héritier and A. Yaouanc
Accepted for publication in the Journal of Less-Common Met. (1988).

SUB-CONTRACT N° 1.10 - J.P. SANCHEZ

1 - Mössbauer spectroscopy in the $RE_2Fe_{14}BH_x$ materials
J.M. Friedt, J.P. Sanchez, P. L'Héritier, R. Fruchart
Proc. Workshop on Nd-Fe permanent magnets (Ed. I.V. Mitchell)
CEC. Brussels, 1984, p. 179

2 - Spin reorientation phenomena in $RE_2Fe_{14}B$ (RE = Ce, Dy, Er) alloys
from ^{57}Fe and ^{161}Dy Mössbauer spectroscopies
A. Vasquez, J.M. Friedt, J.P. Sanchez, P. L'Héritier, R. Fruchart
Sol. Stat. Comm., 55 N°9 (1985) 783.

3 - Mössbauer spectroscopy on $RE_2Fe_{14}BH$ (RE : Y, Ce, Dy, Er) alloys
J.M. Friedt, A. Vasquez, J.P. Sanchez, P. L'Héritier, R. Fruchart
Hyperfine interactions, 28 (1986) 611.

4 - ^{166}Er Mössbauer spectroscopy in the $Er_2Fe_2BH_x$ alloys
J.P. Sanchez, J.M. Friedt, A. Vasquez, P. L'Héritier, R. Fruchart
Sol. Stat. Commun., 57 (1986) 309.

5 - Magnetism and crystal field properties of the $RE_2Fe_{14}BH_x$ alloys
(RE - Y, Ce, Dy, Er) from Mössbauer spectroscopy
J.M. Friedt, A. Vasquez, J.P. Sanchez, P. L'Héritier, R. Fruchart
J. Phys. F., 16 (1986) 651.

6 - Spin reorientation phenomena in $(Er_{11-x}Gd_x)_2Fe_{14}B$ alloys
A. Vasquez, J.P. Sanchez
J. Less Comm. Metals, 127 (1987) 71.

7 - Effective magnetic fields on Fe in $R_{1+\epsilon}Fe_4B_4$ alloys
H.R. Rechenberg and J.P. Sanchez
Stat. Commun., 62 (1987) 461.

8 - New permanent magnets investigated by Mössbauer spectroscopy
J.P. Sanchez
Proc. XII School on Physics, 1987 part 2 : Condensed matter studied by
nuclear methods. (Ed. Qk; krolas and K. Tomala) (Institute of Nuclear
Physics and Jagellomian University, Krakow) p. 156.

9 - Crystal-field interactions and spin reorientation in $(Er_{1-x}Dy_x)_2Fe_{14}B$
H.R. Rechenberg, J.P. Sanchez, P. L'Héritier, R. Fruchart
Phys. Rev., B36, 4 (1987) 1865.

10- ^{155}Gd and ^{161}Dy Mössbauer study of $R_{1+\epsilon}Fe_4B_4$ alloys (R - Gd, Dy)
H.R. Rechenberg, M. Bogé, C. Jeandey, J.L. Oddou, J.P. Sanchez, P. Tenaud
Solid State Comm., 64 (1987) 277.

11- Magnetic properties of $Nd(Co_{1-x}Fe_x)_9Si_2$ alloys from magnetization, NMR
and Mössbauer studies
Y. Berthier, B. Chevalier, J. Etourneau, H.R. Rechenberg, J.P. Sanchez
J. Magn. Magn. Mat., to be published.

SUB-CONTRACT N° 1.11 - M. BOGE

1 - Structural and magnetic properties of $RE_2Fe_{14}B$ hydrides
D. Fruchart, P. Wolfers, P. Vulliet, A. Yaouanc, R. Fruchart,
P. L'Héritier
Proc. European Workshop "Nd-Fe permanent magnets their present and
future applications" (1984) p.173.

2 - 3d-4f magnetic interactions and crystalline electric field in the
$R_2Fe_{14}B$ compounds : magnetization measurements and Mössbauer study of
$Gd_2Fe_{14}B$
M. Bogé, J.M.D. Coey, G. Czjzek, D. Givord, C. Jeandey, H.S. Li,
J.L. Oddou
Sol. Stat. Commun., 55 (1985) 295.

3 - ^{57}Fe Mössbauer study of $RE_2Fe_{14}B$
P. Dalmas de Réotier, D. Fruchart, P. Wolfers, R. Guillen, P. Vulliet,
A. Yaouanc, R. Fruchart, P. L'Héritier
J. de Physique, C6 (1985) 323.

4 - Properties of hydrided $RE_2Fe_{14}B$ compounds
P. Dalmas de Réotier, D. Fruchart, P. Wolfers, P. Vulliet, A. Yaouanc,
R. Fruchart, P. L'Héritier
J. de Physique, C6 (1985) 249.

5 - Magnetization, ^{57}Fe and ^{161}Dy Mössbauer study of $Dy_2Fe_{14}BH_x$ with
$0 \leq x \leq 4.7$
L.P. Ferreira, R. Guillen, P. Vulliet, A. Yaouanc, D. Fruchart, P.
Wolfers, P. L'Héritier, R. Fruchart
J. Magn. Magn. Mat., 53 (1985) 145.

6 - Exchange interactions and hyperfine field at the rare-earth nuclei in
$R_2Fe_{14}B$ compounds
Y. Berthier, M. Bogé, G. Czjzek, D. Givord, C. Jeandey, H.S. Li,
J.L. Oddou
J. Magn. Magn. Mat., 54-57 (1986) 589.

7 - A ^{155}Gd Mössbauer study of a $Gd_2Fe_{14}B$ single crystal
M. Bogé, G. Czjzek, D. Givord, C. Jeandey, H.S. Li, J.L. Oddou
J. Phys. F, 16, L67 (1986).

8 - Hydrogen dependence of the intrinsic magnetic properties of $R_2Fe_{14}B$
J.M.D. Coey, A. Yaouanc, D. Fruchart, R. Fruchart, P. L'Héritier
Inter. Symp. on the properties and Applications of Metal Hydrides V,
Maubuisson, France, (May 25-30 1986).

9 - ^{155}Gd and ^{161}Dy Mössbauer study of $R_{1+\epsilon}Fe_4B_4$ alloys (R = Gd, Dy)

H.R. Rechenberg, M. Bogé, C. Jeandey, J.L. Oddou, J.P. Sanchez,

P. Tenaud

Solid State Comm., 64 (1987) 277.

10 - Mössbauer spectroscopy of $R_2Fe_{14}B$

R. Fruchart, P. L'Héritier, P. Dalmas de Réotier, D. Fruchart, P.

Wolfers, J.M.D. Coey, L.P. Ferreira, R. Guillen, P. Vulliet, A. Yaouanc

J. Phys. F. : Met. Phys., 17 (1987) 483.

11 - Hydrogen dependence of the intrinsic magnetic properties of $Re_{12}Fe_{14}B$ hydrides

J.M.D. Coey, P. Vulliet, A. Yaouanc, D. Fruchart, P. Wolfers,

P. L'Héritier, R. Fruchart

J. Less Comm. Metals, 131 (1987) 419.

12 - Dynamical susceptibility of $Ho_2Fe_{14}B$ single crystal : spin rotation

and domain wall motions

C. Rillo, J. Chaboy, R. Navorro, J. Bartolomé, D. Fruchart, A. Yaouanc,

B. Chenevier, M. Sagawa, S. Hirosawa

Accepted in 4th Joint MMM-Intermag Conference 1988. To be published

in J. Appl. Phys.

SUB-CONTRACT N° 1.12 - R. FRUCHART

1 - Structural and magnetic properties of $RE_2Fe_{14}B$ hydrides

D. Fruchart, P. Wolfers, P. Vulliet, A. Yaouanc, R. Fruchart,

P. L'Héritier

Proc. European Workshop "Nd-Fe permanent magnets their present and

future applications" (1984) p.173.

2 - Mössbauer spectroscopy in the $RE_2Fe_{14}BH_x$ materials

J.M. Friedt, J.P. Sanchez, P. L'Héritier, R. Fruchart

Proc. European Workshop "Nd-Fe permanent magnets their present and

future applications" (1984) p.179.

3 - The structure of a New Magnetic Phase Related to the Sigma Phase :

Iron Neodymium Boride $Nd_2Fe_{14}B$

C.B. Shoemaker, D.P. Shoemaker, R. Fruchart

Acta Cryst. C40 (1984) 1665.

4 - Une nouvelle série d'hydrures métalliques ferromagnétiques de type
$Nd_2Fe_{14}BH_x$ ($0 \leq x \leq 5$)
P. L'Héritier, P. Chaudouët, R. Madar, A. Rouault, J.P. Senateur,
R. Fruchart
C.R. Acad. Sc. Paris, 299-II (1984) 849.

5 - L'analogie des modes d'insertion et d'empilement dans les phases
$Nd_2Fe_{14}B$, Mn_5 SiC et Fe_5SiB_2
P. L'Héritier, P. Chaudouët, R. Fruchart, C.B. Shoemaker, D.P.
Shoemaker,
J. Solid State Chem., 59 (1985) 54.

6 - Properties of hydrided $RE_2Fe_{14}B$ compounds
P. Dalmas de Reotier, D. Fruchart, P. Wolfers, P. Vulliet, A. Yaouanc,
R. Fruchart and P. L'Héritier
Journal de Physique, (1985) C6-249.

7 - Crystallographic and magnetic study of solid solutions
$Gd_2(Fe_{14-x}Co_x)B$, $Gd_2(Co_{14-x}Mn_x)B$, $Dy_2Fe_{14}BH_x$, $Y_2Fe_{14}BH_x$ and $Lu_2Fe_{14}BH_x$
P. L'Héritier and R. Fruchart
Journal de Physique, 46 (1985) C6-319.

8 - Spin reorientation phenomena in $RE_2Fe_{14}B$ (RE = Ce, Dy, Er) alloys
from [57]Fe and [161]Dy Mössbauer spectroscopies
A. Vasquez, J.M. Friedt, J.P. Sanchez, P. L'Héritier, R. Fruchart
Sol. Stat. Commun., 55 n° 9 (1985) 783.

9 - Magnetization, [57]Fe and [161]Dy Mössbauer study of $Dy_2Fe_{14}BH_x$ with
$0 \leq x \leq 4.7$
L.P. Ferreira, R. Guillen, P. Vulliet, A. Yaouanc, D. Fruchart,
P. Wolfers, P. L'Héritier, R. Fruchart
J. Magn. Magn. Mat., 53 (1985) 145.

10 - [57]Fe Mössbauer study of $RE_2Fe_{14}B$
P. Dalmas de Reotier, D. Fruchart, P. Wolfers, R. Guillen, P. Vulliet,
A. Yaouanc, R. Fruchart, P. L'Héritier
J. de Phys., C6, (1985) 323.

11 - Crystallographic and magnetic studies of $Nd_2Fe_{14}B$ and $Y_2Co_{14}B$
D. Le Roux, H. Vincent, P. L'Héritier, R. Fruchart
J. de Phys. 46, C6 (1985) 243.

12 - [166]Er Mössbauer spectroscopy in the $Er_2Fe_{14}BH_x$ alloys
J.P. Sanchez, J.M. Friedt, A. Vasquez, P. L'Héritier, R. Fruchart
Sol. Stat. Commun., 57 (1986) 309.

13 - Mössbauer spectroscopy on $RE_2Fe_{14}BH_x$ (RE = Y, Ce, Dy, Er) alloys
J.M. Friedt, A. Vasquez, J.P. Sanchez, P. L'Héritier, R. Fruchart
Hyperfine Interactions, 28 (1986) 611.

14 - Magnetism and crystal field properties of the $RE_2Fe_{14}BH_x$ alloys
(RE = Y, Ce, Dy, Er) from Mössbauer spectroscopy
J.M. Friedt, A. Vasquez, J.P. Sanchez, P. L'Héritier, R. Fruchart
J. Phys. F : Met. Phys., 16 (1986) 651.

15 - Hydrogen dependence of the intrinsic magnetic properties of $RE_2Fe_{14}B$
J.M.D. Coey, A. Yaouanc, D. Fruchart, R. Fruchart, P. L'Héritier
Intern. Symp. on the properties and Applications of Metal Hydrides V,
Maubuisson, France, (May 25-30 1986).

16 - Spin rotation in $R_2Fe_{14}BH_x$
J.P. Regnard, A. Yaouanc, D. Fruchart, D. Le Roux, P. L'Héritier,
R. Fruchart, J.M.D. Coey, J.P. Gavigan
J. Appl. Phys., 61 (1987) 3565.

17 - Structural and magnetic properties of $R_2Fe_{14}BH_x$
D. Fruchart, L. Pontonnier, F. Vaillant, J. Bartolomé, J.M. Fernandez,
J.A. Puertolas, C. Rillo, J.M. Regnard, A. Yaouanc, R. Fruchart,
P. L'Héritier
EMMA 87, to be published in I.E.E.E. Trans. Mag.

18 - Mössbauer spectroscopy of $R_2Fe_{14}B$
R. Fruchart, P. L'Héritier, P. Dalmas de Réotier, D. Fruchart,
P. Wolfers, J.M.D. Coey, L.P. Ferreira, R. Guillen, P. Vulliet and
A. Yaouanc
J. Phys. F. : Met. Phys., 17 (1987) 483.

19 - Hydrogen dependence of the intrinsic magnetic properties of $Re_2Fe_{14}B$
hydrides
J.M.D. Coey, P. Vulliet, A. Yaouanc, D. Fruchart, P. Wolfers,
P. L'Héritier, R. Fruchart
J. Less. Comm. Metals, 131 (1987) 419.

20 - Positive muon spectroscopy of $Nd_2Fe_{14}B$ and $Pr_2Fe_{14}B$
A. Yaouanc, J. Budnick, E. Albert, M. Hamma, A. Weidinger,
R. Fruchart, P. L'Héritier, D. Fruchart, P. Wolfers
Phys. Rev. Lett., 67 (1987) L286.

21 - Crystal field interactions and spin reorientation in $(Er_{1-x}Dy_x)_2Fe_{14}B$
H.R. Rechenberg, J.P. Sanchez, P. L'Héritier, R. Fruchart
Phys. Rev. B, 36 (1987) 1865.

22 - Structural and magnetic properties of $RE_2Fe_{14}BH (D)_x$, RE = Y, Ce, Er
D. Fruchart, L. Pontonnier, F. Vaillant, P. Wolfers, A. Yaouanc,
J.M.D. Coey, R. Fruchart, P. L'Héritier, P. Dalmas de Réotier
J. Less Comm. Met., 129 (1987) 133.

23 - Hydrogen induced changes of valency and hybridization in Ce
intermetallic compounds
D. Fruchart, F. Vaillant, A. Yaouanc, J.M.D. Coey, R. Fruchart,
P. L'Héritier, T. Riesterer, J. Osterwalder, L. Schlapbach
J. Less Comm. Met., 130 (1987) 97.

24 - Structural and magnetic properties of $R_2Fe_{14}BH_x$
D. Fruchart, L. Pontonier, F. Vaillant, J. Bartolomé, J.M. Fernandez,
J.A. Puertolas, C. Rillo, J.M. Reganrd, A. Yaouanc, R. Fruchart,
P. L'Héritier
EMMA 87, to be published in I.E.E.E. Trans. Mag.

25 - Effects of hydrogen absorption on the 3d and 4f anisotropy in $RE_2Fe_{14}B$
(RE = Y, Nd, Ho, Tm)
L. Paretti, O. Moze, M. Solzi, D. Fruchart, P. L'Héritier and A. Yaouanc
Accepted for publications in the Journal of Less-Common Met. (1988).

PATENTS

1 - Nouveaux hydrures de terre rare / fer / bore et terre rare / cobalt /
bore magnetiques, leur procédé de fabrication et de fabrication des
produits deshydridrurés pulvérulents correspondants, leurs applications
R. Fruchart, R. Madar, A. Rouault, P. L'Héritier, P. Taunier,
D. Boursier, D. Fruchart, P. Chaudouët
Institut National de la Propriété Industrielle (Paris), déposé le 29
juin 1984, N° 84 10387
European Patent Office (The Hague, Netherlands), 20 June, 1985,
N° 85 401230.9 (Applicant : CNRS.)
Magnetic Rare Earth / Iron / Boron and Rare Earth / Cobalt / Boron
Hydrides, the process for their manufacture of the corresponding
pulverulent deshydrogenated products, and their applications
U.S. Patent and Trademark Office, Washington D.C., Filed June 19, 1985,
serial N° 746, 360 (in course).

SUB-CONTRACT N° 1.13 - J. BARTOLOME

1 - Magnetostriction and thermal expansion of $RE_2Fe_{14}B$
   M.R. Ibarra, P.A. Algarabel, A Alberdi, J. Bartolomé and A. del Moral
   J. Appl. Phys. 61 (1987), 3451.

2 - Structural and magnetic properties of $R_2Fe_{14}BH_x$,
   D. Fruchart, L. Pontonnier, F. Vaillant, J. Bartolomé, J.M. Fernandez,
   J.A. Puertolas, C. Rillo, J.M. Regnard, A. Yaouanc, R. Fruchart,
   P. L'Héritier
   EMMA 87, to be published in I.E.E.E. Trans. Mag.

3 - Dynamical suceptibility of $Ho_2Fe_{14}B$ single crystal : spin rotation and
   domain wall motions
   C. Rillo, J. Chaboy, R. Navarro, J. Bartolomé, D. Fruchart, A. Yaouanc,
   B. Chenevier, M. Sagawa, S. Hirosawa
   Accepted in 4hth Joint MMM-Intermag Conference 1988. To be published in
   J. Appl. Phys.

4 - A.C. Initial magnetic susceptibility and spin reorientation transition
   in $(Er_xR_{1-x})_2Fe_{14}B$ magnets (R = Nd and Dy)
   M.R. Ibarra, C. Marquina, P.A. Algarel, J.L. Arnaudas, A. del Moral
   Submitted to Solid State Commun.

5 - A study of the magnetic anisotropy in $(Er_xR_{1-x})_2Fe_{14}B$ intermetallic
   compounds
   M.R. Ibarra, C. Marquina, J.L. Arnaudas, A. del Moral
   Accepted for presentation at the ICM-88 Conference

6 - Magnetostriction in high pulsed magnetic field on a single crystal of
   $Nd_2Fe_{14}B$
   C. Marquina, M.R. Ibarra, P.A. Algarabel, A. del Moral, D. Givord and
   S. Zemirli
   Accepted for presentation at the International Symposium on High Field
   Magnetism (K.U. Leuven 1988)

7 - Spin reorientation processes and Crystal Electric Field interaction in
   hard magnetic pseudoternaries $(Er_xNd_{1-x})_2Fe_{14}B$
   M.R. Ibarra, C. Marquina, J.L. Arnaudas, P.A. Algarabel and A. del Moral
   Accepted for presentation at the 4th Joint MMM - Intermag Conference
   (Vancouver 1988)

8 - High pulsed magnetic field measurements of the magnetic anisotropy in $(Er_xDy_{1-x})_2Fe_{14}B$ compounds
P.A. Algarabel, M.R. Ibarra, C. Marquina, G. Masuri, D. Moze,
L. Pareti, M. Solzi, J.L. Arnaudas and A. del Moral
Accepted for presentation at the Int. Symp. on High Field Magnetism
(K.U. Leuven 1988).

SUB-CONTRACT N° 1.14 - R. GILLET

1 - The $Nd_{15}Fe_{77}B_8$ microstructure : some effects of oxygen for different solidification rates
D. Cochet-Muchy and S. Paidassi
Presented at the International Conference on Magnetism, I.C.M. Paris
(July 1988).

SUB-CONTRACT N° 1.21 - J.M.D. COEY

1 - Aspects of the intrinsic magnetic properties of $R_2Fe_{14}B$ alloys
J.M.D. Coey, J.M. Cadogan, D.H. Ryan
Workshop on Nd-Fe permanent magnets. their present and future
applications, I.V. Mitchell (Ed), C.E.E. (1984) 143.

2 - Crystal fields in $Nd_2Fe_{14}B$
J.M. Cadogan, J.M.D. Coey
Phys. Rev. B, 30 (1984) 7326.

3 - 3d-4f magnetic interactions and crystalline electric field in the
$R_2Fe_{14}B$ compounds : magnetization measurements and Mössbauer study of
$Gd_2Fe_{14}B$
M. Bogé, J.M.D. Coey, G. Czjzek, D. Givord, C. Jeandey, H.S. Li,
J.L. Oddou
Sol. Stat. Commun., 55 (1985) 295.

4 - Hydrogen dependence of the intrinsic magnetic properties of $R_2Fe_{14}B$
J.M.D. Coey, A. Yaouanc, D. Fruchart, R. Fruchart, P. L'Héritier
Intern. Symp. on the properties and Applications of Metal Hydrides V,
Maubuisson, France, May 25-30 (1986).

5 - Mössbauer spectra of $Nd_2Fe_{14}B$ and related compounds
Q. Ling, J.M. Cadogan, J.M.D. Coey
Hyperfine interactions, 28 (1986) 665.

6 - Hydrogen absorption and desorption in $Nd_2Fe_{14}B$
J.M. Cadogan, J.M.D. Coey
Appl. Phys. Lett., 48 (1986) 442.

7 - A note on exchange and crystal field interactions in $R_2Fe_{14}B$
compounds : YbFe14B
P. Burlet, J.M.D. Coey, J.P. Gavigan, D. Givord, C. Meyer
Sol. St. Comm., 60 (1986) 723.

8 - Intrinsic magnetic properties of compounds with the $Nd_2Fe_{14}B$ structure
J.M.D. Coey
J. Less. Comm. Metals, 126 (1986) 21.

9 - Effect of hydrogen on the magnetic properties of $Y_2Fe_{14}B$
J.M.D. Coey, A. Yaouanc, D. Fruchart
Sol. Stat. Comm., 58 (1986) 413.

10 - New Permanent Magnet Materials
J.M.D. Coey
Physica Scripta, T19 (1987) 426.

11 - 3d magnetism in $R_2Fe_{14}B$ compounds
F. Bolzoni, J.P. Gavigan, D. Givord, H.S. Li, D. Moze
J. Magn. Magn. Mat., 66 (1987) 158.

12 - High field magnetization measurements on $Tb_2Fe_{14}B$ and $Er_2Fe_{14}B$ single
crystals
J.P. Gavigan, D. Givord, H.S. Li, O. Yamada, H. Maruyama, M. Sagawa,
S. Hirosawa
J. Magn. Magn. Mat., 70 (1987), 416.

13 - Evidence in R-M intermetallics for a systematic dependence of R-M
exchange interactions on the nature of the R atom
E. Belorizky, M.A. Fremy, J.P. Gavigan, D. Givord, H.S. Li
J. Appl. Phys., 61 (1987) 3971.

14 - Magnetic properties of R-Fe-B magnets
J.P. Gavigan, D. Givord, H.S. Li, P. Tenaud, T. Viadieu
Proceeding of the 3rd International Conference on Physics of Magnetic
Materials World Scientific Publishing Co (Singapore) (1987), 201.

15 - Structural and magnetic properties of $RE_2Fe_{14}BH$ (D)$_x$, RE = Y, Ce, Er
D. Fruchart, L. Pontonnier, F. Vaillant, P. Wolfers, A. Yaouanc,
J.M.D. Coey, R. Fruchart, P. L'Héritier, P. Dalmas de Réotier
J. Less Comm. Met., 129 (1987) 133.

16 - Hydrogen induced changes of valency and hybridization in Ce
intermetallic compounds
D. Fruchart, F. Vaillant, A. Yaouanc, J.M.D. Coey, R. Fruchart,
P. L'Héritier, T. Riester, J. Osterwalder, L. Schlapbach
J. Less Comm. Metals, 130 (1987) 97.

17 - Hydrogen dependence of the intrinsic magnetic properties of $Re_{12}Fe_{14}B$
Hydrides
J.M.D. Coey, P. Vulliet, A. Yaouanc, D. Fruchart, P. Wolfers,
P. L'Héritier, R. Fruchart
J. Less Comm. Metals, 131 (1987) 419.

874

18 - Spin rotation in $R_2Fe_{14}BH_x$
J.P. Regnard, A. Yaouanc, D. Fruchart, D. Le Roux, P. L'Héritier,
R. Fruchart, J.M.D. Coey, J.P. Gavigan
J. Appl. Phys., 61 (1987) 3565.

19 - Exchange and CEF interactions in $R_2Fe_{14}B$ compounds
J.M. Cadogan, J.M.D. Coey, J.P. Gavigan, D. Givord, H.S. Li
J. Appl. Phys., 61(8) (1987) 3974.

20 - Magnetic properties of $Pr_2(Fe_{1-x}Co_x)_{14}B$ compounds
F. Bolzoni, J.M.D. Coey, J.P. Gavigan, D. Givord, O. Moze, L. Pareti,
T. Viadieu
J. Magn. Magn. Mat., 66 (1987) 123.

21 - Mössbauer spectroscopy of $R_2Fe_{14}B$
R. Fruchart, P. L'Héritier, P. Dalmas de Reotier, D. Fruchart,
P. Wolfers, J.M.D. Coey, L.P. Ferreira, R. Guillen, P. Vulliet,
A. Yaouanc
J. Phys. F : Metal Phys., 17 (1987) 483.

22 - Canted spin structures in $R_2Fe_{14}B$ (R = Tm, Er)
J.M. Cadogan
J. Less Common. Metal, 135 (1987) 269.

23 - Iron-Rich pseudodinary alloys with the $ThMn_{12}$ structure obtained by
melt spinnning : $Gd(Fe_nAl_{12-n})$ n - 6, 8, 10
Wang Xiang Zhong, B. Chevalier, J. Etourneau, T. Berlureau,
J.M.D. Coey, J.M. Cadogan
J. Less Comm. Metals, 138 (1988) 235.

24 - Hydrogen absorption and desorption in Nd_2Fe_{17} and Sm_2Fe_{17}
Wang Xiang Zhong, K. Donnelly, J.M.D. Coey, B. Chevalier, J. Etourneau,
T. Berlureau
J. Mat. Sci., 23 (1988) 329.

25 - A new approach to the analysis of magnetization measurements in rare
earth transition metal compounds. Application to $Nd_2Fe_{14}B$
J.M. Cadogan, J.P. Gavigan, D. Givord, H.S. Li
J. Phys. F. (Met. Phys.) (1988) (à paraître).

26 - 3d magnetism in R-M and $R_2M_{14}B$ compounds (M = Fe, Co ; R : Rare Earth)
J.P. Gavigan, D. Givord, H.S. Li, J. Voiron
J. Magn. Magn. Mat. (to appear).

27 - Analysis of high field magnetization measurements on $R_2Fe_{14}B$ single
crystals (R = Tb, Dy, Ho, Er and Tm)
D. Givord, H.S. Li, J.M. Cadogan, J.P. Gavigan, O. Yamada, H. Maruyama,
M. Sagawa, S. Hirosawa
J. App. Phys., (1988) (to appear).

28 - Spin reorientation transitions in $Dy(Fe_{11}Ti)$
H.S. Li, Hu Bo-Ping, J.M.D. Coey
Sol. Stat. Comm., 66 (1988) 133.

29 - A study of exchange and Crystalline Electric Field Interactions in
$Nd_2Co_{14}B$: Comparison with $Nd_2Fe_{14}B$
H.S. Li, J.P. Gavigan, J.M. Cadogan, D. Givord, J.M.D. Coey
J. Magn. Magn. Mat., (1988) to appear.

30 - Effect of hydrogen on the Curie temperature of $Nd_2(Fe_{15}M_2)$,
M = Al, Si, Co.
Hu Bo-Ping, J.M.D. Coey
J. Less. Comm. Metals, (1988) to appear.

31 - Intrinsic Magnetic Poperties of Compounds of the iron-rich $ThMn_{12}$
structure alloys $R(Fe_{11}Ti)$; R= Y, Nd, Sm, Gd, Tb, Dy, Mo, Er, Tm, Lu
Hu Bo-ping, H.S. Li, J.P. Gavigan, J.M.D. Coey
to be submitted to J. Phys. F.

32 - A study of crystal field effects in $Yb_2Fe_{14}B$ by [174]Yb Mössbauer
spectroscopy
C. Meyer, J.P. Gavigan, G. Czjzek, H.J. Bornemann
Communication presented at ICM Paris (July 88).

33 - Magnetic transitions and anomalous behaviour of Pr in $Pr_2(Fe_{1-x}Co_x)_{14}B$
J.P. Gavigan, H.S. Li, J.M.D. Coey, T. Viadieu, L. Pareti, F. Bolzoni,
O. Moze
To be presented at ICM Paris (July 88).

34 - First order magnetization process in $Sm(Fe_{11}Ti)$
H.S. Li, B.P. Hu, J.P. Gavigan, J.M.D. Coey, L. Pareti, O. Moze
To be presented at ICM Paris (July 88).

SUB CONTRACT N° 1.22 - J.J.M. FRANSE

1 - Magnetic properties of ternary rare-earth compounds of the type $R_2Fe_{14}B$
S. Sinnema, R.J. Radwanski, J.J.M. Franse, D.B. Mooij, K.H.J. Buschow
J. Magn. Magn. Mat., 44 (1984) 333.

2 - Magnetic measurements on $R_2Fe_{14}B$ and $R_2Co_{14}B$ compounds in high fields
S. Sinnema, J.J.M. Franse, R.J. Radwanski, K.H.J. Buschow, D.B. Mooij
J. de Phys., 46 (9) C6 (1985) 301.

3 - Magnetic and crystallographic properties of ternary rare earth compounds
of the type $R_2Co_{14}B$
K.H.J. Buschow, D.B. Mooij, S. Sinnema, R.J. Radwanski, J.J.M. Franse
J. Magn. Magn. Mat., 51 (1985) 211.

4 - Magnetic properties of $(Nd, Tb)_{6.7}Fe_{7.5}B_{7.8}$ compounds
Yang Fu-Ming, En Ke, Zhao Xi-Chao, F.R. de Boer and S. Sinnema
J. Less. Comm. Met., 124 (1986) 269.

5 - High-field magnetization measurements on R-T compounds
S. Sinnema, R. Verhoef, J.J.M. Franse and F.R. de Boer
Proc. 5th Int. Symp. on Magn. Anisotropy in R.E. Trans. metals alloys
Bad Soden (3 Sept. 1987). 69.

6 - Temperature dependence of the magnetocrystalline anisotropy energy in
$Nd_2Fe_{14}B$
R.J. Radwanski, K. Krop, J.J.M. Franse and R. Verhoef
Proc. 5th Int. Symp. on Magn. Anisotropy in R.E. Trans. Metals alloys
Bad Soden (3 Sept. 1987) 95.

7 - Ordering phenomena in $Er_2Fe_{14-x}Al_xB$ compounds
P.H. Quang, T.H. Anh, N.H. Luong, L.T. Tai, T. Hein and J.J.M. Franse
Proc. 5th Int. Symp. on Magn. anisotropy in R.E. Trans. Metals alloys
Bad Soden (3 Sept. 1987). 206.

8 - Crystal Growth and Characterization of RE_2T_{17} and $RE_2Fe_{14}B$ Intermetallics
S. Sinnema, R. Verhoef, J.J.M. Franse and A.A. Menovsky
Proc. 7th Am. Conf. Cryst. Growth, Monterey California (1987) to appear.

9 - Comparison of the magnetic properties of MM-Fe-B compounds and Nd-Fe-B
compounds
T.D. Hein, L.T. Tai, R. Grössinger, R. Krewenka, F.R. de Boer and
F.F. Bekker
J. Less-Common Met., 127 (1987) 111.

10- Magnetic properties of a series of novel ternary intermetallics
$(RFe_{10}V_2)$
F.R. de Boer, Huang Ying-kai, D.B. de Mooij and K.H.J. Buschow
J. Less Comm. Met., 135 (1987) 199.

11 - Magnetic anisotropy in $Pr_2(Fe_{1-x}Co_x)_{14}B$ compounds
R. Grössinger, R. Krewenka, H.R. Kirchmayr, S. Sinnema, Yang Fu-Ming,
Huang Ying-Kai, F.R. de Boer and K.H.J. Buschow
J. Less-Comm. Met., 132-2 (1987) 265.

12 - Effective anisotropy constants in rare earth-3d intermetallics
R.J. Radwanski, J.J.M. Franse and S. Sinnema
Proc. Int. Symp. Magn. Intermet. Comp., Kyoto (1987),
J. Magn. Magn. Mat., 70 (1987) 313.

13 - Moment reorientations in RE_2TM_{17} and $RE_2TM_{14}B$ compounds
R. Verhoef, S. Sinnema, J.J.M. Franse and R.J. Radwanski
Published in : Wissenschaftliche Zeitschrift der Hochshule für
Verkehrswesen "Friedlich List" Dresden, Sondernheft 31, Dresden, 1987,
Supplement p.9.

14 - Magnetocrystalline anisotropy study of some $R_2Fe_{14}B$ based quasiternary
compounds
Yang Fu-Ming, Zhao Ru-Wen, Li Xin-Wen, Huang Ying-Kai, F.R. de Boer and
R.J. Radwanski
European Magnetic Materials And Applications Conference, Salford
(England), 1987

15 - High-Field Magnetization Process in $R_2Fe_{14}B$ compounds
Yang Fu-Ming, Huang Ying-Kai, K.H.J. Buschow, H. Heerooms,
S. Sinnema and F.R. de Boer
Proc. ISPMM, Sendai (1987) to appear.

16 - Rare Earth contribution to the magnetocristalline anisotropy energy in
$R_2Fe_{14}B$
R.J. Radwanski and J.J.M. Franse
Phys. Rev. B... (1987).

17 - Magnetic and crystallographic properties of ternary rare-earth
compounds of the type $R_2Fe_{14}C$,
F.R. de Boer, Huang Ying-Kai, Zhang Zhi-Dong, D.B. de Mooij and
K.H.J. Buschow
to be published.

18 - Effective anisotropy constants in rare earth-3d intermetallics
R.J. Radwanski, J.J.M. Franse and S. Sinnema
Proc. Int. Symp. Magn. Intermet. Comp., Kyoto (1987),
J. Magn. Magn. Mat., 70 (1987) 313.

19 - Moment reorientations in RE_2TM_{17} and $RE_2TM_{14}B$ compounds
   R. Verhoef, S. Sinnema, J.J.M. Franse and R.J. Radwanski
   Published in : Wissenschaftliche Zeitschrift der Hochshule für
   Verkehrswesen "Friedlich List" Dresden, Sondernheft 31, Dresden, 1987,
   Supplement p.9.

SUB-CONTRACT N° 1.23 - G. CZJZEK

1 - 3d-4f magnetic interactions and crystalline electric field in the $R_2Fe_{14}B$
   compounds : magnetization measurements and Mössbauer study of $Gd_2Fe_{14}B$
   M. Bogé, J.M.D. Coey, G. Czjzek, D. Givord, C. Jeandey, H.S. Li,
   J.L. Oddou
   Sol. Stat. Commun., 55 (1985) 295.

2 - Exchange interactions and hyperfine field at the rare-earth nuclei in
   $R_2Fe_{14}B$ compounds
   Y. Berthier, M. Bogé, G. Czjzek, D. Givord, C. Jeandey, H.S. Li,
   J.L. Oddou
   J. Magn. Magn. Mat., 54-57 (1986) 589.

3 - A ^{155}Gd Mössbauer study of a $Gd_2Fe_{14}B$ single crystal
   M. Bogé, G. Czjzek, D. Givord, C. Jeandey, H.S. Li, J.L. Oddou
   J. Phys. F, 16 (1986) L67.

4 - A study of crystal field effects in $YB_2Fe_{14}B$ by ^{174}Yb Mössbauer
   spectroscopy
   C. Meyer, J.P. Gavigan, G. Czjzek, H.J. Bornemann
   Communication presented at ICM Paris (July 1988).

SUB-CONTRACT N° 1.24 - G. ASTI

1 - Anisotropy measurements of Nd-Fe-B by the singular point detection
   technique
   G. Asti, F. Bolzoni, F. Leccabue, L. Pareti, R. Panizzieri
   Proc. Workshop on Nd-Fe permanent magnets present and future
   applications. Brussels, 1984, Ed I.V. Mitchell. 161.

2 - Singular point detection of discontinuous magnetization processes
   G. Asti, F. Bolzoni
   J. Appl. Phys., 58 (1985) 1924.

3 - Curie temperatures of some substituted $Nd_2Fe_{14}B$ tetragonal alloys
   F. Leccabue, J.L. Sanchez, L. Pareti, F. Bolzoni, R. Panizzieri
   Phys. Stat. Sol. a, 91 (1985) K63.

4 - Direct observations of first-order magnetization processes in single-crystal $Nd_2Fe_{14}B$
L. Pareti, F.Bolzoni, O. Moze
Phys. Rev. B, 32 (1985) 7604.

5 - High field magnetization processes in $Pr_2Fe_{14}B$
L. Pareti, H. Szymczack, H.K. Lachowicz
Phys. Stat. Sol. a, 92 (1985) K65.

6 - Elastic constants of the $(RE)_2Fe_{14}B$ tetragonal compound (RE = Y, Nd and Pr)
G. Turilli, H. Szymczak, H.K. Lachowicz
Phys. Stat. Sol. a, 90 (1985) K143.

7 - Magnetic anisotropy of carbon doped $Nd_2Fe_{14}B$
F. Bolzoni, F. Leccabue, L. Pareti, J.L. Sanchez
J. de Physique, (1985) C6-305.

8 - Magnetic and Mössbauer study of $Y_{2-x}Nd_xFe_{14}B$
F. Bolzoni, A. Deriu, F. Leccabue, L. Pareti, J.L. Sanchez
J. Magn. Magn. Mat., 54-57 (1986) 595.

9 - 3d magnetism in $R_2Fe_{14}B$ compounds
F. Bolzoni, J.P. Gavigan, D. Givord, H.S. Li, O. Moze,
J. Magn. Magn. Mat., 66 (1987) 158.

10 - Magnetic properties of $Pr_2(Fe_{1-x}Co_x)_{14}B$ compounds
F Bolzoni, J.M.D. Coey, J. Gavigan, D. Givord, O. Moze, L. Pareti,
T. Viadieu
J. Magn. Magn. Mat., 65 (1987) 123.

11 - 3d magnetism in $Y_2Fe_{14-x}Me_xB$ with Me = Co, Ni, Mn, Cr
L. Pareti, M. Solzi, F. Bolzoni, O. Moze, R. Panizzieri
Sol. Stat. Commun., 61, 12 (1987) 761.

12 - Magnetocrystalline anisotropy in $Nd_{2-x}Tb_xFe_{14}B$
L. Pareti, F. Bolzoni, M. Solzi, K.H.J. Buschow
J. Less Comm. Metals, 132 (1987) L5.

13 - Magnetocrystalline anisotropy of Ni and Mn substituted $Nd_2Fe_{14}B$ compounds
F. Bolzoni, F. Leccabue, O. Moze, L. Pareti, M. Solzi
J. Magn. Magn. Mat., 67 (1987) 373.

14 - 3d and 4f magnetism in $Nd_2Fe_{14-x}Co_xB$ and $Y_2Fe_{14-x}Co_xB$ compounds
F. Bolzoni, F. Leccabue, O. Moze, L. Pareti, M. Solzi
J. Appl. Phys., 61, 12 (1987) 5369.

15 - First-order field-induced magnetization transitions in single-crystal $Nd_2Fe_{14}B$
F. Bolzoni, O. Moze, L. Pareti
J. Appl. Phys., 62, 2 (1987) 615.

16 - Magnetic anisotropy of RE-magnets
G. Asti, F. Bolzoni, L. Pareti
I.E.E.E. Trans. Mag., 23 n°5 (1987) 2521.

17 - Magnetic anisotropy and magnetization process in rare-earth metal compounds
G. Asti
V Int. Symp. on magnetic anisotropy and coercivity in RE-TM alloys. Bad Soden, FRG Sept.3, 1987.

18 - Magnetic structure and preferential site occupation in Mn and Cr substituted $Y_2Fe_{14}B$ compounds
O. Moze, L. Pareti, M. Solzi, F. Bolzoni, W.I.P. David, W.T.A. Harrison and A. Hewat
J. less Comm. Metals., 136 (1988) 375.

19 - Neutron diffraction and magnetic anisotropy study of Y-Fe-Ti Intermetallic compounds
O. Moze, L. Pareti, M. Solzi, W.I.F. David
Sol. Stat. Comm., 66, 5 (1988) 465.

20 - Magneto-crystalline anisotropy in $Pr_{1-x}Y_xCo_5$
L. Pareti, O. Moze, M. Solzi, F. Bolzoni
J. Appl. Phys., 63 (1988) 172.

21 - High pulsed magnetic field measurements of the magnetic anisotropy in $(Er_{1-x}Dy)Fe_{14}B$ compounds
P.A. Algarabel, M.R. Ibarra, C. Marquina, G. Marusi, O. Moze, L. Pareti, M. Solsi, J.I. Arnaudas, A. del Moral
To be presented at the 2nd International Symposium On High Field Magnetism, 1988, Leuven.

22 - Magnetic transitions and anomalous behaviour of Pr in $Pr_2(Fe_{1-x}Co_x)_{14}B$
J.P. Gavigan, H.S. Li, J.M.D. Coey, T. Viadieu, L.Pareti, F. Bolzoni, O. Moze
To be presented at ICM'88 Paris.

23 - First order magnetization Process in $Sm(Fe_{11}Ti)$
H.S. Li, B.P. Hu, J. Gavigan, J.M.D. Coey, L. Pareti, O. Moze
To be presented at ICM'88. Paris.

24 - 4f and 4g Site anisotropy in $Nd_2Fe_{14}B$ and $Pr_2Fe_{14}B$ compounds
O. Moze, G. Marusi, L. Pareti, M. Solzi, W.I.F. David
To be presented at the International Conference On Neutron Scattering,
1988, Grenoble.

25 - Neutron diffraction determination of the crystal structure of the
intermetallic compounds $Y-Fe_{10}V_2$
O. Moze, L. Pareti, M. Solzi, W.I.F. David
To be presented at the International Conference On Neutron Scattering,
1988, Grenoble.

26 - Effects of hydrogen absorption on the 3d and 4f anisotropy in
$RE_2Fe_{14}B$ (RE - Y, Nd, Ho, Tm)
L. Pareti, O. Moze, M. Solzi, D. Fruchart, P. L'Héritier and A. Yaouanc
Accepted for publication in the Journal of Less-Common Met..(1988).

27 - Rare-Earth and transition metal magnetic anisotropy in some
intermetallic compounds of the tetragonal $ThMn_{12}$ structure
M. Solzi, L. Pareti, O. Moze
To be presented at ICM'88. Paris.

SUB-CONTRACT N° 1.25 M. ROSENBERG

1 - A NMR and Mössbauer study of $Nd_2Fe_{14}B$
M. Rosenberg, P. Deppe, M. Wojcik, H. Stadelmaier
J. Appl. Phys., 57 (8) (1985) 4124.

2 - Magnetic properties and Mössbauer effect in $R_5Fe_{18}B_{18}(R_{1-x}Fe_4B_4)$
H.R. Rechenberg, A Paduan - Filho, J.P. Missel, P. Deppe, M. Rosenberg
Sol. State Comm., 59 (1986) 541.

3 - A Mössbauer spectroscopy study of $R_2Fe_{14}B$ alloys
M. Rosenberg, P. Deppe, H. Stadelmaier
Hyperfine Interactions, 28 (1986) 503.

4 - NMR and Mössbauer study of $R_2TM_{14}B$ with R - Y, La, Th, and TM = Fe, Co
K. Erdmann, P. Deppe, M. Rosenberg, K.H.J. Buschow
J. Appl. Phys., 61, 8 (1987) 4340.

5 - A ^{57}Fe Mössbauer study of $Nd_2(Fe_{1-x}Co_x)_{14}B$
P. Deppe, M. Rosenberg, S. Hirosawa, M. Sagawa
J. Appl. Phys., 61, 8 (1987) 4337.

6 - ^{59}Co spin echo NMR in the $Co_{3-x}Fe_xB$ system
M. Wojcik, E. Jedryka, K. Nesteruk, M. Rosenberg, J.D. Livingston
J. Appl. Phys., 61, 8 (1987) 3650.

7 - A Mössbauer spectroscopy study of $Nd_5Fe_2B_6$
P. Deppe, M. Rosenberg, K.H.J. Buschow
Sol. Stat. Comm., 64 (N°9) (1987) 1247.

8 - Magnetic properties of $YCo_{12}B_6$ and $GdCo_{12}B_6$ intermetallics
M. Rosenberg, M. Mittag, K.H.J. Buschow
presented at the 32nd M.M.M. Chicago 1987
to appear in J. Appl Phys. (April 88).

9 - ANMR study of $R_2TM_{14}B$ compounds with R = Sm, Gd, Lu, and TM = Fe and Co
K. Erdmann, M. Rosenberg and K.H.J. Buschow
presented at the 32nd M.M.M. Chicago 1987
to appear in J. Appl. Phys. (April 88).

10- ANMR Study of $YCo_{12}B_6$ and $GdCo_{12}B_6$ intermetallics
K. Erdman, M. Rosenberg, K.H.J. Buschow
presented at the 32nd M.M.M. Chicago 1987
to appear in J. Appl Phys. (April 88).

SUB-CONTRACT N° 1.26 - K.H.J. BUSCHOW

1 - Magnetic properties of ternary rare-earth compounds of the type $R_2Fe_{14}B$
S. Sinnema, R.J. Radwanski, J.J.M. Franse, D.B. de Mooij, K.H.J. Buschow
J. Magn. Magn. Mat., 44 (1984) 333.

2 - Note on the crystal field induced magnetic anisotropy in several
permanent magnet materials
K.H.J. Buschow, J.W.C. De Vries, R.C. Thiel
Physica, 132B (1985) 13.

3 - ^{161}Dy Mössbauer effect in $Dy_2Fe_{14}B$
P.C.M. Gubbens, A.M. van der Kraan, K.H.J. Buschow
Phys. Stat. Sol. (b), 130 (1985) 575.

4 - The Fe-rich isothermal section of Nd-Fe-B at 900°^{C}
K.H.J. Buschow, D.B. de Mooij, H.M. van Noort
Philips J. Res., 40 (1985) 227.

5 - ^{57}Fe Mössbauer spectroscopy study of the magnetic properties of $R_2Fe_{14}B$ compounds (R - Ce, Nd, Cd, Y)
H.M. van Noort, D.B. de Mooij, K.H.J. Buschow
J. Appl. Phys., 57 (1985) 5414.

6 - On the site preference of 3d atoms in compounds of the $R_2(Co_{1-x}Fe_x)_{14}B$ type
H.M. Van Noort, K.H.J. Buschow
J. Less Comm. Metals, L9 (1985) 113.

7 - The temperature dependence of the anisotropy field in $R_2Fe_{14}B$ compounds (R - Y, La, Ce, Pr, Nd, Gd, Ho, Lu)
R. Grössinger, X.K. Sun, R. Eibler, K.H.J. Buschow, H.R. Kirchmayr
J. de Phys., 46 (1985) 221.

8 - Magnetic and structural properties of $Nd_2Fe_{14}B$, $Th_2Fe_{14}B$ and related materials
K.H.J. Buschow, H.M. van Noort, D.B. de Mooij
J. Less Comm. Metals, 109 (1) (1985) 79.

9 - Magnetic properties and ^{57}Fe Mössbauer effect of $ErFe_4B$, $TmFe_4B$ and $LuFe_4B$
H.M. Van Noort, D.B. de Mooij, K.H.J. Buschow
J. Less Comm. Metals, 111 (1985) 87.

10 - Magnetic and crystallographic properties of ternary rare earth compounds of the type $R_2Co_{14}B$
K.H.J. Buschow, D.B. de Mooij, S. Sinnema, R.J. Radwanski, J.J.M. Franse
J. Magn. Magn. Mat., 51 (1985) 211.

11 - Magnetic measurement on $R_2Fe_{14}B$ and $R_2Co_{14}B$ compounds in high fields
R. Sinnema, J.J.M. Franse, R.J. Radwanski, K.H.J. Buschow, D.B. de Mooij
J. de Physique, 46 (9) (1985) C6-301.

12 - Temperature dependence of anisotropy fields and initial susceptibilities in $R_2Fe_{14}B$ compounds
R. Grössinger, X.K. Sun, R. Eibler, K.H.J. Buschow, H.R. Kirchmayr
J. Magn. Magn. Mat., 58 (1986) 55.

13 - Phase relationships, magnetic and crystallographic properties of Nd-Fe-B alloys
K.H.J. Buschow, D.B. de Mooij, J.L.C. Daams, H.M. Van Noort
J. Less Comm. Metals, 115 (1986) 357.

14 - Note on the hard magnetic properties of sintered Nd-Fe-B permanent magnets
R. Grössinger, R. Krewernka, R. Eibler, H.R. Kirchmayr, J. Ormerod, K.H.J. Buschow
J. Less Comm. Metals, 118 (1986) 167.

15 - New permanent magnet materials
K.H.J. Buschow
Materials Science Report, 1 (1986) 1-63.

16 - Magnetic phase transition and magnetic anisotropy in $Nd_2Fe_{14-x}Co_xB$ compounds
R. Grössinger, R. Krewenka, X.K. Sun, K.H.J. Buschow, R. Eibler, H.R. Kirchmayr
J. Less Comm. Met., 124 (1986) 165-172.

17 - A novel ternary Nd-Fe-B compound
D.B. de Mooij, K.H.J. Buschow
Philips J. Res., 41 (1986) 400.

18 - ^{57}Fe Mössbauer investigation of ternary compounds on the $R_2Fe_{14}B$ type
H.M. Van Noort, D.B. de Mooij, K.H.J. Buschow
J. Less comm. Metals, 115 (1986) 155.

19 - Invar effect in $R_2Fe_{14}B$ compounds (R - La, Gd, Nd, Sm, Gd, Er)
K.H.J. Buschow
J. Less Comm. Metals, 118 (1986) 349.

20 - Properties of metastable ternary compounds and amorphous alloys in the Nd-Fe-B system
K.H.J. Buschow, D.B. de Mooij, H.M. van Noort
J. Less Comm. Metals, 125 (1986) 135.

21 - Magnetocrystalline anisotropy in $Nd_{2-x}Tb_xFe_{14}B$
L. Pareti, F. Bolzoni, M. Solzi, K.H.J. Buschow
J. Less Comm. Metals, 132 (1987) L5.

22 - Spontaneous volume magnetostriction in $R_2Fe_{14}B$ compounds
K.H.J. Buschow, R. Grössinger
J. Less Commm. Metals, 135 (1) (1987) 39.

23 - A new class of ferromagnetic materials : $RFe_{10}V_2$
B. de Mooij, K.H.J. Buschow
Philips J. Res., 42 (1987) 246.

24 - A Metastable compound in the Y-Fe-B system
   D.B. de Mooij, J.L.C. Daams, K.H.J. Buschow
   Philips J. Res., 42 (1987) 339.

25 - Magnetic properties of a series of novel ternary intermetallics
   ($RFe_{10}V_2$)
   F.R. de Boer, Huang Ying-Kai, D.B. de Mooij, K.H.J. Buschow
   J. Less Comm. Metals, 135 (1987) 199.

26 - Physical properties of ternary RE-Fe Based alloys and their relationship
   to permanent magnet applications
   K.H.J. Buschow
   Proc. of the Mat. Res. Soc. Sympos. Proc. Series, V96 (1987) 1.

27 - N.M.R. and Mössbauer study of $R_2TM_{14}B$ with R = Y, La, Th and TM = Fe, Co
   K. Erdmann, P. Deppe, M. Rosenberg, K.H.J. Buschow
   J. Appl. Phys., 61, 8 (1987) 4340.

28 - High-field magnetization process in $R_2Fe_{14}B$ compounds
   Yan Fu-Ming, Huang Ying-Kai, K.H.J. Buschow, H. Heerooms, S. Sinnema,
   and F.R. de Boer
   Proc. ISPMM, Sendai (1987) to appear.

29 - Kerr effect in $R_2Fe_{14}B$ and $R_2Co_{14}B$ compounds
   P.P.J. van Engelen, K.H.J. Buschow
   J. Magn. Magn. Mat., 66 (1987) 291.

30 - Magnetic interaction and crystal field in $RFe_{10}V_2$ and related compounds
   P.C.M. Gubbens, A.M. van der Kraan, K.H.J. Buschow
   Proc. 5th Int. Symp. on Magn. anisotropy in R.E. Trans. Metal alloys
   Bad Soden (3 Sept. 1987), 117.

31 - Crystal field and magnetic anisotropy in ternary R-3d-B compounds
   K.H.J. Buschow, H.H.A. Smit and R.C. Thiel
   Proc. 5th Int. Symp. on Magn. anisotropy in R.E. trans. Metal alloys
   Bad Soden (3 Sept. 1987), 39.

32 - Physico-chemical properties of ternary rare-earth alloys and their
   relation to permanent magnet applications
   K.H.J. Buschow
   9th Int. Workshop on R.E. Magnets and their Appl. Bad Soden
   (31 Aug. -2 Sept. 1987), 453.

33 - ^{169}Tm Mössbauer effect in $Tm_2Fe_{14}B$
   P.C.M. Gubbens, A.M. van der Kraan, R.P. van Stapele, K.H.J. Buschow
   J. Magn. Magn. Mat., 68 (2) (1987) 238-42.

34 - A Mössbauer spectroscopy study of $Nd_5Fe_2B_6$
P. Deppe, M. Rosenberg, K.H.J. Buschow
Sol. Stat. Comm., 64 (N°9) (1987) 1247.

35 - Magnetic anisotropy in $Pr_2(Fe_{1-x}Co_x)_{14}B$ compounds
R. Grössinger, R. Krewenka, H.R. Kirchmayr, S. Sinnema, Yang Fu-Ming, Huang Ying-Kai, F.R. de Boer, K.H.J. Buschow
J. Less Comm. Metals, 132 (1987) 265.

36 - Note on the coercivity in Nd-Fe-B magnets
R. Grössinger, R. Krewenka, H.R. Kirchmayr, P. Naastepad,
K.H.J. Buschow
J. Less Comm. Met., 134 (1987) L17-L21.

37 - A NMR Study of $R_2TM_{14}B$ compounds with R= Sm, Gd, Lu and TM = Fe and Co
K. Erdmann, M. Rosenberg, K.H.J. Buschow
Presented at the 32nd M.M.M. Chicago 1987.
To appear in J. Appl. Phys. (April 1988).

38 - A NMR Study of $YCo_{12}B_6$ and $GdCo_{12}B_6$ intermetallics
K. Erdmann, M. Rosenberg, K.H.J. Buschow
Presented at the 32nd M.M.M. Chicago 1987.
To appear in J. Appl. Phys. April 1988.

39 - Magnetic Properties of $YCo_{12}B_6$ and $GdCo_{12}B_6$ intermetallics
M. Rosenberg, M. Mittag, K.H.J. Buschow
To appear in J. Appl Phys. (April 1988).

40 - Some novel ternary $ThMn_{12}$-type compounds
D.B. de Mooij, and K.H.J. Buschow
J. Less Comm. Metals, 136 (1988) 207.

41 - Magnetic anisotropy in the system $La_2Fe_{14-x}Co_xB$ and its relation to the system $Nd_2Fe_{14-x}Co_xB$
R. Grössinger, H. Kirchmayr and K.H.J. Buschow
J. Less Comm. Metals, 136 (1988) 367.

42 - Note on the crystallographic and magnetic structure of $YFe_{10}V_2$
R.B. Helmholdt, J.M. Vleggar and K.H.J. Buschow
J. Less Comm. Metals, 138 (1988) L11.

43 - $RFe_{10}V_2$ compounds studied by ^{57}Fe, ^{161}Dy, ^{166}Er and ^{169}Tm Mössbauer spectroscopy
P.M.C. Gubbens, A.M. Van der Kraan and K.H.J. Buschow
Hyperfine Interactions, 40 (1988) 389.

44 - Magnetic and crystallographic properties of ternary rare-earth
compounds of the type $R_2Fe_{14}B$
F.R. de Boer, Huang Ying-Kai, Zhang Zhi-Dong, <u>D.B. de Mooij</u> and
<u>K.H.J. Buschow</u>
To be published.

PATENTS

1 - Magnetic material based on Fe, B and rare earth elements (rare earth
doped Fe_3B).

2 - Magnetic material based on Fe, Co, B and rare earth elements ($R_2Co_{14}B$
and $R_2Co_{14-x}Fe_xB$).

3 - Magnetic material based on Fe(Co), rare earth elements and a nonmagnetic
element T (Ti, V, Cr, Si, Wo, M) having the approximate composition
$R(Fe_{1-x}T_x)_{12}$.

4 - Intermetallic compounds based on rare earths, iron and carbon suitable
for permanent magnet applications.

Applications made but not yet published.

SUB-CONTRACT N° 1.27 - W.D. CORNER

1 - Magnetic Domains and Domain Wall Energies in Rare Earth-Iron-Boron
Intermetallics
<u>W.D. Corner</u> and <u>J.M. Hawton</u>
J. Magn. Magn. Mat., <u>72</u> (1988) 59.

2 - Magnetocrystalline anisotropy of $R_2Fe_{14}B$, R = Dy, Ho, Gd, measured
using high field torque magnetometry
<u>M.J. Ma Ion</u> and <u>W.D. Corner</u>
J. Magn. Magn. Mat., <u>72</u> (1988) 52.

3 - An automated torque magnetometer for use in a 13 T superconducting
solenoid
<u>M.J. Hawton</u>, <u>W.D. Corner</u>
J. Phys. E : Sci. Instrum, 20 (1987) 406.

4 - Magnetic domains and domain wall energies in Rare-Iron-Boron
intermetallics
<u>W.D. Corner</u> and <u>M.J. Hawton</u>
J. Magn. Magn. Mat., 72 (1988) 59.

5 - Magnetocrystalline anisotropy of $R_2Fe_{14}B$, R = Dy, Ho, Gd, measured using high field torque magnetometry
M.J. Hawton, W.D. Corner
J. Magn. Magn. Mat., 72 (1988) 52.

SUB-CONTRACT N° 1.28 - A. KOSTIKAS

1 - Magnetic properties of the $R_{1+e}Fe_4B_4$ compounds (R = rare earth) from magnetization and Mössbauer measurements
D. Niarchos, G. Zouganelis, A. Kostikas, A. Simopoulos
Sol. Stat. Commun. 59 (1986) 389.

2 - Temperature variation of the spin reorientation in $Er_{1-x}Dy_xFe_{14}B$ alloys
D. Niarchos, A. Simopoulos
Sol. Stat. Commun. 59 (1986) 669.

3 - Spin reorientation in $(RE)_2(TM)_{14}B$ alloys (RE = Rare Earth, TM = Transition Metal)
A. Simopoulos, D. Niarchos
Hyperfine Interactions 40 (1988) 429.

SUB-CONTRACT N° 1.29 - H. KRONMÜLLER & E.T. HENIG

1 - Determination of intrinsic magnetic parameters of $Nd_2Fe_{14}B$ from magnetic measurements of sintered $Nd_{15}Fe_{77}B_8$ magnets
K.D. Durst, H. Kronmüller
J. Magn. Magn. Mat. 59 (1986) 86.

2 - Phase relations in the system Fe-Nd-B
G. Schneider, E.Th. Henig, G. Petzow, and H.H. Stadelmaier
Z. Metallkde. 77 (1986) 755.

3 - Reaction Sintering of the Magnetically hard phase $Fe_{14}Nd_2B$
G. Schneider, E. Th. Henig, H.H. Stadelmaier, G. Petzow
Proc. of the 1986 Int. Powder Metallurgy Conf. P/M 86, Düsseldorf, FRG (1986) 1197.

4 - A $CaCu_5$-type Iron-Neodymium Phase stabilized by rapid solidification
H.H. Stadelmaier, G. Schneider, M. Ellner,
J. Less-Common Metals 115 (1986) L11.

5 - Konstitutionelle Untersuchungen zu den Homogenitätsbereichen von $(Nd_xSE_{1-x})_2Fe_{14}B$
B. Grieb
Diploma Thesis, Universität Stuttgart (1986).

6 - Phasengleichgewichte im Hartmagnetsystem Fe-Nd-B
G. Schneider, E.Th. Henig
DGM Hauptversammlung, Göttingen, May (1986).

7 - Hartmagnete au Fe-Nd-B-Basis : Konstitution, Herstellung und Eigenschaften
E.Th. Henig, G. Schneider
Pulverkolloquium Hochleistungswerkstoffe, Stuttgart, 4.6.1987.

8 - The coercive field of sintered and melt-spun Nd-Fe-B magnets
K.D. Durst, H. Kronmüller
J. Magn. Magn. Mat. 68 (1987) 63.

9 - Magnetic hardening mechanisms in Fe-Nd-B type permanent magnets
K.D. Durst, H. Kronmüller, G. Schneider
Proc. 5th Intern. Symp. on Magn. Anisotropy in R.E. Transition Metal Alloys ; Bad Soden (3 Sept. 1987), 209.

10 - Intrinsic magnetic properties of $Fe_{14}Nd_2B$ single crystals
S. Hock and H. Kronmüller
Proc. 5th Intern. Symp. on Magn. Anisotropy in R.E. Transition Metal Alloys ; Bad Soden (3 Sept. 1987), 275.

11 - The phase diagram of Fe-Nd-B and the Optimization of the Microstructure of Sintered Magnets
G. Schneider, E.Th. Henig, H.H. Stadelmaier, G. Petzow
Proceeding of the 5th Int. Symp. on Magn. anisotropy and coercivity in Rare Earth-Transition metal alloys, Bad Soden, FRG (1987) 347-362.

12 - Magnetic Hardering Mechanisms in Fe-Nd-B Type Permanent Magnets
K.D. Durst, H. Kronmüller, G. Schneider
Proceeding of the 5th Int. Symp. on Magn. Anisotropy and Coercivity in rare earth - Transition Metal Alloys, Bad Soden, FRG (1987) 209.

13 - Homogeneity ranges and phase equilibria around $(Nd, Re)_2Fe_{14}B$ with RE - Tb, Dy
B. Grieb, E.Th. Henig, G. Schneider, G. Petzow
Proceeding of the 5th Int. Symp. on magn anisotropy and coercivity in Rare-Earth - Transition metal alloys, Bad Soden, FRG (1987) 395.

14 - Angular dependence of the coercive field in sintered $Fe_{77}Nd_{15}B_8$ magnets
     H. Kronmüller, K.D. Durst, G. Martinek
     J. Magn. Magn. Mat. 69 (1987) 149.

15 - On the relation between the microstructure and soft and hard magnetic
     materials
     H. Kronmüller
     Proc. of the Int. Symposium on Physics of Magnetic Materials, Sendai,
     Japan, 1987, p.17.

16 - High intrinsic coercivity in Iron-Rare Earth-Carbon-Boron alloys
     through the Carbide or Boro-Carbide $Fe_{14}R_2X$ (X = B_xC_{1-x})
     N.C. Liu, H.H. Stadelmaier, G. Schneider
     Mat. Letters 4 (1987) 377.

17 - The Binary system Iron-Neodymium
     G. Schneider, E.Th. Henig, G. Petzow, H.H. Stadelmaier
     Z. Metallkde, 78 (1987) 694.

18 - Metastable solidification of Fe-Rich Iron-Neodymium-Boron alloys
     E.Th. Henig, G. Schneider, H.H. Stadelmaier
     Z. Metallkde, 78 (1987) 818.

19 - High Intrinsic coercivities in Iron-Rare Earth-Carbon-Boron Alloys
     through the carbide or Boro-Carbide $Fe_{14}R_2X$ (X = B_xC_{1-x})
     N.C. Liu, H.H. Stadelmaier, G. Schneider
     J. Appl. Phys. 61 (1987) 3574.

20 - Phase equilibria in Fe-Nd-B and related systems
     H.H. Stadelmaier, G. Schneider, E.Th. Henig
     Proceeding of the ASM Materials Week' 87, Cincinnati, Ohio (1987),
     publi. by ASM International 8710-001, P.1

21 - Züchtung und magnetische Eigenschaften von $Fe_{14}Nd_2B$-Einkristallen
     S. Hock, H. Kronmüller
     Verhandlungen der Deutschen Physikalischen Gesellschaft, 52. Physiker-
     tagung Karlsruhe, A.G. magnetismus, 15.3. 1988 10.

22 - Winkelabhängigkeit des Koerzitivfeldes bei Fe-Nd-B Permanentmagneten
     G. Martinek, K.D. Durst, H. Kronmüller
     Verhandlungen der Deutschen Physikalischen Gesellschaft, 52.
     Physikertagung Karlsruhe, A.G. Magnetismus, 15.3. 1988, 10.

23 - Analysis of the magnetic hardering mechanism in RE-Fe-B permanent
     magnets
     H. Kronmüller, K.D. Durst, M. Sagawa
     J. Magn. Magn. Mat., to be published.

24 - Konstitution und Sinterverhalten von Hartmagnetwerkstoffen auf Fe-Nd-B-Basis
G. Schneider
Universität Stuttgart (1988).

PATENT :

1 - Optimierung der Gefügestruktur des Fe-Nd-B-Basis Sintermagneten
E.Th. Henig, G. Petzow, G. Schneider, K.D. Durst, H. Kronmüller,
W. Draxler, F.J. Esper, A. Meller
Deutsche Patentanmeldung, eingereicht am 2.9.1987

2 - Sintermagnet auf Basis Fe-Nd-B
E.Th. Henig, G. Petzow, G. Schneider, A. Büchel, K.D. Durst,
H. Kronmüller, W. Draxler, F.J. Esper
Deutsche Patentanmeldung, eingereicht am 26.11.1987

SUB-CONTRACT N° 1.30 - P.J. GRUNDY

1 - Electron Microscope Study of Precipitation in a Niobium-containing
(Nd, Dy)FeB Sintered Magnet
S.F.M. Parker, P.J. Grundy and J. Fidler
J. Magn. Magn. Mat., 66 (1987) 74.

2 - Precipitation in NdFeB type Magnet Materials
S.F.H. Parker, R.J. Pollard and P.J. Grundy
I.E.E.E. Trans. Magn., MAG-23 (1987) 2103.

3 - Effect of Zirconium Additions on the Microstructural and magnetic
Properties of NdFeB Based magnets
R.J. Pollard, P.J. Grundy, S.F.H. Parker and D.G. Lord
Paper AA05, EMMA Conference, Salford, Sep. 1987, to be published in
I.E.E.E. Trans. Magn., March 1988.

SUB-CONTRACT N° 1.33 - D.G. PETTIFOR

1 - New alloys from the quantum engineer
D.G. Pettifor
New Scientist, 110, n°1510, 48 (1986).

2 - Structure maps for pseudo-binary and ternary phases
D.G. Pettifor
Mat. Sc. and Tech. (1988).

3 - Structure maps in magnetic alloy design
D.G. Pettifor
Physica, B149 (1988).

4 - The calculated electronic and structural properties of the transition metal monoborides
P. Mohn and D.G. Pettifor
J. Phys. C (in press).

5 - The calculated electronic and magnetic properties of tetragonal transition metal semi borides
P. Mohn
J. Phys. C (in press).

SUB-CONTRACT N° 1.34 - G.C. HADJIPANAYIS

1 - New-iron-rare-earth based permanent magnet materials
G.C. Hadjipanayis, R.C. Hazelton, K.R. Lawless
Appl. Phys. Lett., 43, 797 (1983).

2 - Magnetic hysteresis in rapidly quenched rare-earth alloys
G.C. Hadjipanayis, R.C. Hazelton, K.R. Lawless, D.J. Sellmyer
J. Magn. Magn. Mat., 40, 278 (1984).

3 - Cobalt-free-permanent magnet materials based on iron-rare-earth alloys
G.C. Hadjipanayis, R.C. Hazelton, K.R. Lawless
J. Appl. Phys., 55, 2073 (1984).

4 - Magnetic hardering in rapidly quenched Fe-Pr alloys
D.J. Sellmyer, A.U. Ahmed, G.C. Hadjipanayis
J. Appl. Phys., 55, 2088 (1984).

5 - Thermomagnetic data in iron-rare-earth boron alloys
G.C. Hadjipanayis, Y.F. Tao, K.R. Lawless, D.J. Sellmyer
Mat. Lett., 3, 270 (1985).

6 - Magnetic properties of melt-spun iron-rare-earth boron alloys
G.C. Hadjipanayis, C.P. Wong, Y.F. Tao
Rapidly quenched metals, S Steep, H Warlimont, Eds. Elsevier Science Publishers B.V., (1985).

7 - Magnetic and structural studies of rapidly quenched iron-rare-earth-metalloid alloys
Y.F. Tao, G.C. Hadjipanayis
J. Appl. Phys., 57, 4103 (1985).

8 - Formation of $Fe_{14}La_2B$ phase in as-cast and melt-spun samples
G.C. Hadjipanayis, Y.F. Tao, G. Gudimetta
Appl. Phys. Lett., 47, 757 (1985).

9 - Microstructure and magnetic properties of iron-rare-earth magnets
G.C. Hadjipanayis, Y.F. Tao, K.R. Lawless
8th Inter. Workshop on Rare-earth magnets and their applications, Dayton, Ohio (May 1985), p. 657.

10 - The microstructure and magnetic properties of melt-spun $Fe_{76}Nd_{16}B_8$ magnetic materials
G.C. Hadjipanayis, R.C. Dickenson, K.R. Lawless
J. Magn. Magn. Mat. (in press.).

11 - Magnetic properties of melt-spun and sintered Fe-Nd-B magnets at elevated temperatures
G.C. Hadjipanayis, Y.F. Tao
J. de Phys., C6-237, (1985).

12 - Low field magnetic properties of Fe-Nd based permanent magnets
K.V. Rao, R. Malmhall, G.C. Hadjipanayis
J. de Phys., C6-229 (1985).

13 - The microstructure and magnetic properties of melt spun $Fe_{76}Nd_{16}B_8$ magnetic materials
G.C. Hadjipanayis, R.C. Dickenson, K.R. Lawless
J. Magn. Magn. Mat. 54-57 (1986) 557.

14 - Magnetic propeties of amorphous iron-rare earth-boron alloys
S.H. Aly, G.N. Nicolaides, Y.F. Tao, G.C. Hadjipanayis
J. of Phys. F 16 (1986) L121.

15 - Magnetic properties of Fe-R-B powders
K. Gudimetta, C.N. Christodoulou, G.C Hadjipanayis
Appl. Phys. Lett. 48 (1986) 670.

16 - Microstructure and magnetic properties of iron based rare earth magnets
G.C. Hadjipanayis, Y.F. Tao, K.R. Lawless
I.E.E.E. Trans. Mag. Mag 22 (1986) 1845.

17 - Magnetic viscosity in different Nd-Fe-B magnets
    D. Givord, P. Tenaud, T. Viadieu, G.C. Hadjipanayis
    J. Appl. Phys. 61 (1987) 3454.

SUB-CONTRACT N° 1.35 - H. KIRCHMAYR

1 - The anisotropy of Nd-Fe-B magnets
    R. Grössinger, P. Obitsch, X.K. Sun, R. Eibler, H. Kirchmayr,
    F. Rothwarf, H. Sassik
    Mat. Lett., 2, 539 (1984).

2 - Investigation of the magnetic properties of Nd-Fe-B based hard magnetic
    materials
    R. Grössinger, G. Hilscher, H. Kirchmayr, H. Sassik, R. Strnat,
    G. Wiesinger
    Physica, 130B, 307 (1985).

3 - Magnetic investigations of $(Nd_{1-x}R_x)Fe_{77}B_8$ and $Nd_{15}Fe_{85-x}B_x$
    X.K. Sun, R. Grössinger, R. Eibler, H.R. Kirchmayr
    Physica, 130B, 300 (1985).

4 - Reversal of magnetization in Nd-Fe-B magnets
    A. Handstein, J. Schneider, D. Stephan, W. Fischer, U. Heinecke,
    R. Grössinger, H. Sassik, H.R. Kirchmayr
    Mat. Lett., 3, 200 (1985).

5 - The temperature dependence of the hystersis loop and the magnetic
    anisotropy of some selected Nd-Fe-B magnets
    R. Grössinger, R. Krewenka, K.S.V.L. Narasimhan, M. Sagawa
    J. Magn. Magn. Mat., 51, 160 (1985).

6 - Magnetic anisotropy of (Nd-RE)-Fe-B alloys
    R. Grössinger, X.K. Sun, R. Eibler, H.R. Kirchmayr
    Proc. of the 8th Intern. Workshop on Rare-earth Magnets and
    Applications, Dayton, Ed. K. Strnat, p. 553 (1985).

7 - High temperature behaviour of technical Nd-Fe-B permanent magnets
    R. Grössinger, H.R. Kirchmayr, R. Krewenka, K.S.V.L. Narasimhan,
    M. Sagawa
    Proc. of the 8th Intern. Workshop on Rare-Earth Magnets and
    Applications, Dayton, Ed. K. Strnat, p. 565 (1985).

8 - Temperature dependence of magnetic properties of Nd-Fe-B magnets
    H. Schneider, R. Grössinger, R. Krewenka, V. Heinecke, H. Sassik
    Mat. Lett., 3, 401 (1985).

9 - The temperature dependence of the anisotropy field in $R_2Fe_{14}B$ compounds
(R = Y, La, Ce, Pr, Nd, Gd, Ho, Lu)
R. Grössinger, X.K. Sun, R. Eibler, K.H.J. Buschow, H.R. Kirchmayr
J. de Phys., 46, C6-221 (1985).

10 - Temperature dependence of anisotropy field and initial susceptibility
in $R_2Fe_{14}B$ compounds
R. Grössinger, X.K. Sun, R. Eibler, K.H.J. Buschow, H.R. Kirchmayr
J. Magn. Magn. Mat., 58 (1986) 55.

11 - Magnetic phase transition and magnetic anisotropy in $Nd_2Fe_{14-x}Co_xB$
compounds
R. Grössinger, R. Krewenka, X.K. Sun, K.H.J. Buschow, R. Eibler,
H.R. Kirchmayr
J. Less Comm. Met. 124 (1986), 165-172.

12 - Investigation of the hysteresis loop and the magnetic anisotropy of
Nd-Fe-B based permanent magnets.
R. Grössinger, R. Krewenka, K.S.V.L. Narashimhan, H.R. Kirchmayr
I.E.E.E. Trans. Mag. MAG.22 (1986) 760.

13 - Magnetic and anisotropy studies on Nd-Fe-B based permanent magnets
G. Hilscher, R. Grössinger, S. Heisz, H. Sassik, G. Wiesinger
J. Magn. Magn. Mat., 54 (1986) 577.

14 - The hard magnetic properties of sintered Nd-Fe-B permanent magnets
R. Grössinger, R. Krewernka, H.R. Kirchmayr, J. Ormerod, K.H.J. Buschow
J. Less Comm. Metals 118 (1986) 167.

15 - Rapidly solidified hard magnetic materials
R. Grössinger, G. Hilscher
Proc. of Summer school on "Amorphous metals" ed. Hm. Matyja,
P.G. Zielinski (1986) p.285.

16 - Rare earth permanent magnets
R. Grössinger, G. Hilscher, H.R. Kirchmayr
Radex Rundschau 2/3 (1986) 120.

17 - Effects of hydrogen absorption on the magnetic properties of
$Nd_{15}Fe_{77}B_8$
G. Wiesinger, G. Hilscher, R. Grössinger
J. Less Comm. Metals 131 (1987) 400.

18 - Comparison of the magnetic properties of MM-Fe-B and Nd-Fe-B compounds
T.H. Hien, L.T. Tai, R. Grössinger, R. Krewenka, F.R. De Boer,
F.F. Bekker
J. Less Comm. Met. 127 (1987) 111-116.

19 - Does a Co substitution really improve the temperature dependence of
Nd-Fe-B based permanent magnets
R. Grössinger
Proc. 9th Int Workshop on rare earth magnets (3 Sept. 1987), II, p.15.

20 - Temperature dependence of the coercivity and the anisotropy of
Nd-Fe-B based magnets
R. Grössinger, R. Krewenka, F. Haslinger, M. Sagawa, H.R. Kirchmayr
I.E.E.E. Trans. Mag. MAG-23 (1987) 2114.

21 - Anisotropy and hysteresis studies of highly substituted Nd-Fe-B based
permanent magnets
R. Grössinger, H. Harada, A. Keresztes, H.R. Kirchmayr, M. Tokunaga
I.E.E.E. Trans. Mag. MAG-23 (1987) 2117.

22 - The effect of the spin reorientation upon the demagnetization curves
of Nd-Fe-B magnets
S. Heisz, G. Hilscher, H.R. Kirchmayr, H. Harada, M. Tokunaga
I.E.E.E. Trans. Mag. MAG-23 (1987) 3110.

23 - The effect of substitution of Al on the magnetic properties of
$Nd_{16}Fe_{78}B_8$ permanent magnets
R. Grössinger, F. Haslinger, Zhang Shougong, R. Eibler, Liu Yinglie,
J. Schneider, A. Handstein, H.R. Kirchmayr
EMMA 87, to be published in I.E.E.E. Trans. Mag.

24 - Magnetic anisotropy in $Pr_2(Fe_{2-x}Co_x)_{14}B$ compounds
R. Grössinger, R. Krewenka, H.R. Kirchmayr, S. Sinnema, Yang Fu-Ming,
Huang Ying-kai, F.R. de Boer, K.H.J. Buschow
J. Less Comm. Metals 132 (1987) 265-272.

25 - The initial magnetization process of melt spun Nd-Fe-B material
S. Heisz, G. Hilscher, G. Wiesinger, H. Sassik
Proc. of the 9th Intern. Workshop on Rare-Earths Magnets and their
Applications, Bad Soden, F.R.G., (Aug. 31 - Sept. 2, 1987) 267.

26 - The origin of graduated magnetization curves of Nd-Fe-B magnets
S. Heisz, G. Hilscher
J.M.M.M. 67 (1987) 20.

27 - Spontaneous volume magnetostriction in $R_2Fe_{14}B$ compounds
K.H.J. Buschow, R. Grössinger
J. Less Comm. Metals, 135 (1) (1987) 39.

28 - Note on the coercivity in Nd-Fe-B magnets
R. Grössinger, R. Krewenka, H.R. Kirchmayr, P. Naastepad,
K.H.J. Buschow
J. Less Comm. Met. 134 (1987) L17-L21.

29 - Magnetic anisotropy and magnetization process in RE-Fe-B magnets
J. Schneider, A. Handstein, R. Grössinger, S. Heisz
Proc. of 3rd Int. Conf. on "Physics of Magnetic Materials" (1987)
p.225.

30 - The coercivity and anisotropy of (Nd-X)-(Fe,Y)-B (X = Dy, Y = Co, Al)
based permanent magnets
R. Grössinger, A. Keresztes, H. Harada, Z. Shougong
Proc. 9th Int. Workshop on rare earth magnets (Sept. 1987) I, p. 523.

31 - Nd-Fe-B type permanent magnets. Recent developments in basic and
applied research
G. Wiesinger, G. Hilscher
Intern. Conf. on the Application of the Mössbauer effect, Melbourne
(Aug. 1987) Hyperfine Interactions 40 (1988) 235-48.

32 - Magnetic anisotropy in the system $La_2Fe_{14-x}Co_xB$ and its relation to
the system $Nd_2Fe_{14-x}Co_xB$
R. Grössinger, H.R. Kirchmayr, and K.H.J. Buschow
J. Less Comm. Metals, 136 (1988) 367.

SUB-CONTRACT N° 1.36 - C . HEIDEN

1 - Investigation of magnetization creep of Co_5Sm samples
S. Stieler, C. Heiden, K. Kuntze, D. Kohake
I.E.E.E. trans. Magn., MAG-20, 1581 (1984).

2 - Magnetization creep and thermal aftereffect in rare earth-transition
metal-permanent magnets and P.M. systems
S. Stieler, K. Kuntze, D. Kohake
Proc. of the 8th Intern. workshop on Rare-Earth Magnets and
Applications, Dayton, Ed. K. Strnat, p. 179 (1985).

3 - Temperature dependence of magnetization reversal of sintered Nd-Fe-B
    magnets
    K. Kuntze, D. Kohake, R. Beranek, S. Steiler, C. Heiden
    J. de Phys., C6-253, 9 (1985).

4 - Dependence of the coercive field and magnetic viscosity coefficient in
    NdFeB magnets on the magnetic history of the sample
    D. Givord, C. Heiden, A. Höhler, P. Tenaud, T. Viadieu and K. Zeibig
    Proc. EMMA-Conf., Salford, (1987).

SUB-CONTRAT N° 2.01 - R. HARRIS

1 - The effect of ageing at 600°C on the microhardness and Hc_i values of a
Nd-Fe-B alloys
I.R. Harris, T. Bailey
Proc. European Workshop "Nd-Fe permanent magnets. Their present and
future applications", 143 (1984).

2 - The hydrogen decrepitation of a $Nd_{15}Fe_{77}B_8$ magnetic alloy
I.R. Harris, C. Noble, T. Bailey
J. Less Comm. metals, 106, L1 (1985).

3 - Microhardness studies of a Nd-Fe-B permanent magnet alloy
T. Bailey, I.R. Harris
J. Mat. Sc. Lett., 4, 151 (1985).

4 - The ageing behaviour of the microhardness and intrinsic coercivity of
a Nd-Fe-B alloy
T. Bailey, I.R. Harris
J. Mat. Sc. Lett., 4, 645 (1985).

5 - The production of a Nd-Fe-B permanent magnet by a hydrogen
decrepitation/attritor milling route
P.J. McGuiness, I.R. Harris, E. Rozendaal, J. Ormerod, M. Ward
J. Mat. Sci., 21, pp 4107-4110 (1986).

6 - Hydrogen absorption/desorption studies (HADS) on $Nd_{16}Fe_{76}B_8$ and $Nd_2Fe_{14}B$
I.R. Harris, P.J. McGuiness, D.G.R. Jones, J.S. Abell
Physica Scripta, Vol. T19, pp 435-440 (1987).

7 - The potential of hydrogen in permanent magnet production
I.R. Harris
J. Less Common Metals, 131, pp 245-262 (1987).

8 - The production of Nd-Fe-B magnets by the hydrogen decrepitation/jet
milling (HD/JM) process
E. Rozendaal, J. Ormerod, P.J. McGuiness, I.R. Harris
9th International Workshop on Rare Earth Magnets and their Applications.
Bad Soden. pp 275-285 (1987).

9 - The potential of the HD-process in permanent magnet production
I.R. Harris
9th International Workshop on Rare Earth Magnets and their Applications.
Bad Soden. pp 249-265 (1987).

10 - The use of hydrogen in the production and characterisation of NdFeB
     magnets.
     P.J. McGuiness, I.R. Harris
     Intermag-MMM 1988.

11 - Hydrogen absorption and desorption in NdFeB alloys
     P.J. McGuiness, I.R. Harris, U.D. Scholz, H. Nagel
     Presented at the "International Symposium on Metal-Hydrogen Systems,
     Fundamentals and Application". Max Plank Institute (1988).

12 - Modification of the microstructure of Nd15Fe77B8 alloy by controled
     solidification
     J.S. Abell, I.R. Harris
     IEEE Transactions on Magnets, Vol. 24, N° 2 pp 1620-1622, March (1988).

13 - A study of Nd-Fe-B magnets produced using a combination of hydrogen
     decrepitation and jet milling
     P.J. McGuiness, I.R. Harris, E. Rozendaal, J. Ormerod
     To be published in J. Mat. Sci.

PATENT :

1 - Magnets
    European Patent - ER : 88302927 - 4
    21 March 1988
    I.R. Harris, H. Safi

SUB-CONTRACT N° 2.02 - H.A. DAVIES

1 - Crystallite Size Determinations for Melt Spun Fe-Nd-B Permanent Magnet
    Alloys
    G.E. Carr, H.A. Davies and R.A. Buckley
    Presented at 6th International Conference on Rapidly Quenched Metals,
    Montréal, Canada, Aug. 1987. To be published March 1988 in Materials
    Science and Engineering.

SUB-CONTRACT N° 2.07 - J. ORMEROD

1 - Note on the hard magnetic properties of sintered Nd-Fe-B permanent
    magnets
    R. Grössinger, J. Krewernka, R. Eibler, H.R. Kirchmayr, J. Ormerod,
    K.H.J. Buschow
    J. Less Comm. Metals 118 (1986) 167.

2 - The production of a Nd-Fe-B permanent magnet by a hydrogen
decrepitation/attritor milling route
P.J. McGuiness, I.R. Harris, E. Rozendaal, J. Ormerod, M. Ward
J. Mat. Sci., 21 (1986) 4107-4110.

3 - The production of Nd-Fe-B magnets by the hydrogen decrepitation/jet
milling (HD/JM) process
E. Rozendaal, J. Ormerod, P.J. McGuiness, I.R. Harris
9th International Workshop on Rare Earth Magnets and their Applications
Bad Soden (1987) 275-285.

4 - A study of Nd-Fe-B magnets produced using a combination of hydrogen
decrepitation and jet milling
P.J. McGuiness, I.R. Harris, E. Rozendaal, J. Ormerod

SUB-CONTRACT N° 2.08 - A.G. CLEGG

1 - A technical and commercial assement of Nd-Fe-B permanent magnets for
users, producers and raw materials suppliers
H.D. Olmstead, A.G. Clegg, R.J. Parker, P. Wheeler
Gorham Int. Inc. Maine, U.S.A. (1986).

2 - Measurements of Magnet Parameters and Stability of Magnets for Motors
A.G. Clegg, I. Coulson, G. Hilton
22nd Universities power Engineering Conference (UPEC) paper 9-15 (1987).

3 - Permanent Magnets in Theory and Practice
M. Mc Caig[†], A.G. Clegg
Whiley and Pentech Press, p. 415 (1987).

SUB-CONTRACT N° 2.09 - M. WARD

1 - The production of a Nd-Fe-B permanent magnet by a hydrogen decrepitation/
attritor milling route
P.J. McGuiness, I.R. Harris, E. Rozendaal, J. Ormerod, M. Ward
J. Mat. Sci., 21, pp 4107-4110 (1986).

SUB-CONTRAT N° 2.11 - J.M.D. COEY

1 - Influence of Quench rate and hydrogen absorption on the magnetic
    properties of melt-spun $Nd_{15}Fe_{77}B8$
    J.M. Cadogan, D.H. Ryan, J.M.D. Coey
    Presented at the 6th Int. Conf. on Rapid Quench. Metal, Montreal (1987)
    to be published in Materials Science and Engineering 89 (1988).

SUB-CONTRACT N° 2.13 - H. NAGEL

1 - The influence of Nd-Fe-B alloy hydrogen absorption on permanent magnet
    production
    U.D. Scholz, W.E. Krönert, H. Nagel
    Proceedings of the 9th Int. Workshop on R-E Magnets and their
    applications.

2 - Hydrogen absorption and desorption in NdFeB alloys
    P.J. McGuiness, I.R. Harris, U.D. Scholz, H. Nagel
    Presented at the "International Symposium on Metal-Hydrogen Systems,
    Fundamentals and Application". Max Plank Institute (1988).

SUB-CONTRACT N° 3.01 - R. HANITSCH

1 - A cup armature brushless DC motor
   R. Hanitsch, J.C. Chang
   Proc. 21th Universities Power Engineering Conf.
   London, April 1986, p. 371.

2 - Brushless DC motors
   R. Hanitsch
   ERA-Technology Seminar, London, Sept. 1986, 3.2.1, 3.2.5.

3 - Permanent magnet motor with neodymium-iron-boron magnets
   R. Hanitsch, L. Neubauer
   Beijing Intern. Conf. on Electrical Machines, Aug. 1987, Beijing.

4 - Disc-type motor with high energy permanent magnet
   R. Hanitsch
   3rd Intern. Conf. on Electrical machines and Drives
   I.E.E., London Nov. 1987.

5 - Beitrag zu Entwicklung von Keinmotoren mit Hochenergie-Permanent-Magneten
   R. Hanitsch
   Intl. wissenschaftlich-technische Tagung 7./9. Juni 1988, Dresden

6 - Improved brushless Motor with cup-type winding and Nd-Fe-B magnets
   R. Hanitsch, A. Sitzia, E Hemead, B.J. Chalmers
   Proc.   Int. Conf. on Electrical Machines, ICEM' 88
   12-14 Sept. 1988 Pisa.

SUB-CONTRACT N° 3.02 - D. HOWE

1 - The computer-aided design of permanent magnet dc motors
   D. Howe, D.A. Staton, T.S. Birch, I.J. Williams
   I.E.E. Conference, E.M.D.A., London P. 213 (1985).

2 - Computer-aided design procedures for permanent magnet dc motors
   I.J. Williams, T.S. Birch, D. Howe, D.A. Staton
   Proc. Conference on Drives, Motors, Controls, Brighton, P. 29 (1985).

3 - A permanent magnet alternator for use as an electrodynamic brake
   D. Howe, N.M. Rash, E. Spooner
   I.E.E. Conference, E.M.D.A., p. 270 (1985).

4 - Synchronous performance prediction for high-field permanent magnet
    synchronous motors
    B.J. Chalmers, S.K. Devgan, D. Howe, W.F. Low
    Presented at I.C.E.M. '86 Conference, Munich (1986).

5 - The influence of magnet MMF on static torque production in hybrid
    stepper motors
    M.K. Jenkins, T.S. Birch, D. Howe
    "Proc. of Intern. Conf. on Electrical Machines and Drives Systems",
    Romania (16-17 Sept. 1986).

6 - Design and analysis of brushless D.C. motors
    J.K. Mitchell, T.S. Birch, D. Howe
    "Proc. of the intern. Conf. on Electrical Machines and Drive
    Systems", Romania (16-17 Sept. 1986).

7 - Design optimisation of permanent magnet D.C. motors
    D.A. Staton, T.S. Birch, D. Howe
    "Proc. of the Intern. Conf. on Electrical Machines and Drive
    systems", Romania (16 -17 Sept.1986).

8 - New design opportunities for permanent magnet excited machines
    D. Howe, T.S. Birch, I.J. Williams
    Presented at the seminar "Permanent Magnets are good for your
    wealth", London (16 Sept. 1986).

9 - The finite element method for the direct simulation of the steady-state
    performance of a permanent magnet line-start synchronous motors
    W.F. Low, D. Howe
    I.E.E.E. Intermag 1987, Tokyo, Japan.

10 - Methods for predicting the steady-state operation of permanent magnet
    synchronous motors
    W.F. Low, D. Howe
    Presented at UPEC'87 Conference (Sunderland U.K.).

11 - The potential for Nd-Fe-B in electrical machines
    D. Howe, T.S. Birch, P. Gray
    9th Intern, Workshop on Rare-Earth magnets and their Applications,
    Bad Soden, F.R.G. (31 August-2 Sept. 1987), 65.

12 - An electrodynamic braking system for railway vehicles based on an
    axle-mounted permanent magnet alternator
    D. Howe, D.M. Matthew, T.S. Birsch, A.P. Jablonski, N.M. Rash
    9th Intern. Workshop on Rare-Earth Magnets and their Applications,
    Bad Soden, F.R.G. (31 August-2 Sept. 1987), 85.

13 - Static torque production in hydrid stepping motors : the influence of
     saturation and magnet mmf
     M.K. Jenkins, T.S. Birch, D. Howe
     I.E.E. Intern. Conf. on electrical Machines and Drives, London,
     (November 1987).

14 - Workholding assemblies based on NdFeB permanent magnets
     D. Taylor, D.J. Manley, D. Howe
     3th Intern, GORHAM Conf., San Diego, (Oct. 1987).

15 - Permanent magnet motors and drive systems
     D. Howe
     Proc. of Conference entitled Magnetism : Commercial Applications of
     current research, Salford (Sept. 1987), 65.

16 - Permanent magnet excited motors and drive systems for industry and
     transport
     D. Howe
     Electric Vehicle developments 7 n°1 (1988) 10.

17 - Aspects of magnetic circuit design for induction machines with
     permanent magnet excitation
     D. Howe
     Proc. Electric Energy Conf. - Electrical machines and devices Adelaide
     (Oct. 1987).

18 - Electrical drives systems
     D. Howe
     Presented at actuator 88 Conference, Bremen, (9-10 June 1988).

19 - The impact of high permanent magnets on Electrical drive systems
     D. Howe
     Presented at I.E.E.E. Conf. on Magnetic Materials and applications,
     Purdue University, U.S.A. (May 1988).

SUB-CONTRACT N° 3.03 - B.J. CHALMERS

1 - Parameters and performance of a high-field permanent magnet synchronous
    motor for variable-frequency operation
    B.J. Chalmers, S.A. Hamed, G.D. Baines
    I.E.E. Proc., 132, Pt B n°3 (1985) 117.

2 - A permanent magnet alternator for use as an electro-dynamic brake
    D. Howe, N.M. Rash, E. Spooner
    I.E.E. Conference, E.M.D.A., p. 270 (1985).

3 - Synchronous performance prediction for high-field permanent-magnet
    synchronous motors
    B.J. Chalmers, S.K. Devgan, D. Howe, W.F. Low
    Intern. Conf. on Electrical Machines, Munich, (Sept. 1986).

4 - Brushless d.c. motors with slotless stator
    A.M. Sitzia, B.J. Chalmers
    Electrical Drive Symposium, Cagliari, Italy, (Sept. 1987).

5 - The properties of permanent-magnet materials for the excitation of
    large electrical machines
    E. Spooner, K.M. Richardson
    Electrical Drive Symposium, Cagliari, Italy, (Sept. 1987).

6 - Electromagnetic design of brushless d.c. motor with slotless stator
    A.M. Sitzia, B.J. Chalmers
    I.E.E. Intern. Conf. on Electrical Machines and Drives, London,
    (November 1987).

7 - Magnetization procedures for Nd-Fe-B magnets in large electrical
    machines
    E. Spooner, K.M. Richardson
    I.E.E. Intern. Conf. on electrical Machines and Drives, London,
    (November 1987).

8 - Assessment of high-field permanent magnet rotors by measurement of
    flux distribution in air
    B.J. Chalmers and S.K. Devgan
    Intern. Conf. on Electrical Machines, Pisa, (Sept. 1988).

9 - Comparative performances of 7.5kW permanent-magnet synchronous motors
    with Sm Co$_5$ and Nd-Fe-B magnets
    B.J. Chalmers and S.K. Devgan
    Intern. Conf. on Electrical Machines, Pisa, (Sept. 1988).

10 - Improved brushless motors with cup-type winding and Nd-Fe-B magnets
     R. Hanitsch, A.M. Sitzia, E. Hemead and B.J. Chalmers
     Intern. Conf. on Electrical Machines, Pisa, (Sept. 1988).

11 - Toroidally-wound, axial-flux, permanent-brushless dc motors
     E. Spooner and B.J. Chalmers
     Intern. Conf. on Electrical Machines, Pisa, (Sept. 1988).

12 - Locus diagrams and performance characteristics of various types of
     synchronous motors
     J.E. Brown and B.J. Chalmers
     Intern. Conf. on Electricales Machines, Pisa, (Sept. 1988).

SUB-CONTRACT N° 3.04 - W. GEYSEN

1 - The finite element computation of the field in a permanent magnet
machine
A. Hameed, K.J. Binns, A. Vandenput, W. Geysen
Proc. I.M.A.C.S. TC-1, Liège (1984). p.4-1-1.

2 - Influence of the use of Nd-Fe-B in permanent magnet on the behaviour
of permanent magnet machines
R. Belmans, R. Paternoster, A. Vandenput, W. Geysen, K.J. Binns
Proc. International Drive Symposium, Cagliary, Italy (1987), 71.

3 - Calculation of the natural frequencies of an electrical machine stator
using a CAD-Finite element approach
R. Belmans, J. de White, D. Vandenput, W. Geysen
BICEM, 1987, Being (China), Aug. 1987.

4 - CAD of electrical machines
R. Belmans, A. Vandenput, D. Verdyck, W. Geysen
1988 ASME Computer in Engineering Conference, New-York City, 07/31/88-
08/03/88.

SUB-CONTRACT N° 3.05 - M. BRADFORD

1 - The present and future influence of neodymuim-iron-boron in automotive
products
J.G.W. West, M. Bradford
"Permanent Magnets are good for your wealth", Seminar Proceeding, ERA
Report 86-0185.

2 - Introduction of new permanent magnet materials into electric motors
and generators for automobiles
M. Bradford, J. Cameron, A.S. Mills
Final Report of Contract N° MSM2-0216-UK (H), November 1987.

3 - An appraisal of technical and commercial parameters for neodymium-iron-
boron magnets
M. Bradford
ERA Report 87-0017.

SUB-CONTRACT N° 3.11 - E.M.H. KAMERBEEK

1 - The signifiance of magnet properties for applications in electro-
mechanical devices
E.M.H. Kamerbeek
Proc. ERA's Seminar "Permanent Magnets are good for your Wealth",
Nov. 1987 (rep. 86-185).

SUB-CONTRACT N° 3.12 - R. PAUTHENET

1 - Static devices with new permanent magnets
J. Chavanne, J. Laforest and R. Pauthenet
Proc. of the Mat. Res. Soc. Symposium Proc. Series, V96 (1987) 307.

2 - Nd-Fe-B magnets for hexapolar field configuration
J. Chavanne, J. Laforest and R. Pauthenet
Presented at EMMA87, 14-16 Sept. 1987. Salford, UK.
I.E.E.E. Trans. Mag. 24 n° 2 (1988) 1617.

SUB-CONTRACT N° 3-13 - W. BARAN

1 - Overview of present and potential applications of permanent magnets :
other than motors
W. Baran
Proc. European Workshop "Nd-Fe permanent magnets.
Their present and future applications", (1984) 189.

2 - Overview of present and potential applications of permanent magnets :
other than motors
W. Baran
Proc. of the Intern. Workshop on "Rare-earth magnets and their
applications", Dayton, Ohio, (May 1985) 365.

3 - Impact of improved materials on permanent magnet applications
W. Baran
Metal Powder Report, V. 42, n° 6, (June 1987) 426.

4 - Design of a permanent magnet system for an ECR-Type ion source
W. Baran
Proc. of the Intern. Workshop on "Rare-earth magnets and their
applications", Bad Soden, FRG (Aug.-Sept. 1987) 55.

SUB-CONTRACT N° 3.16 - K.J. BINNS

1 - The finite element computation of the field in a permanent magnet machine
   A. Hameed, K.J. Binns, A. Vandenput, W. Geysen
   Proc. I.M.A.C.S. TC-1, Liège (1984).

2 - Development of a high performance permanent magnet machine
   K.J. Binns, T.M. Wong
   I.C.E.M. Conference, Lausanne, p. 565 (1984).

3 - Permanent magnet motors for inverter-fed drives
   K.J. Binns
   Conference on Drives, Motors, Controls, Brighton, p. 101 (1984).

4 - Analysis and performance of high-field permanent-magnet synchronous machine
   K.J. Binns, T.M. Wong
   Proc. I.E.E., 131, 252 (1984).

5 - The efficient evaluation of torque and field gradient in permanent magnet machines with small air-gap
   K.J. Binns, C.P. Riley, T.M. Wong
   I.E.E. Trans. Magn. MAG-21, 2435 (1985).

6 - Some design aspects of high-output permanent magnet synchronous machines with non-radial magnets
   K.J. Binns, C.P. Riley, T.M. Wong
   I.E.E. Conference, E.M.D.A., London (1985).

7 - Multistacked imbricated rotors with permanent magnet excitation design for new magnet materials
   T.S. Low, K.J. Binns
   Proc. I.E.E., Part B, p. 205, V133, N° 4 (1986).

8 - The scope for development of permanent magnet machines in the light of new materials
   K.J. Binns, C.P. Riley
   Proc. ICEM, Part 3, p. 1060, Munich, (Sept. 1986).

9 - Handbook of electric machines (ed. by S.A. Nasar), permanent Magnet Machines, Chapt. 9
   K.J. Binns
   McGraw-Hill, (March 1987).

10 - The role of neodymium-iron-boron in relation to other materials for
permanent magnet machines.
K.J. Binns, F. Chaaban, P.J.G. Lisboa, P.H. Mellor
3rd Intern. Conf. on Electrical Machines and Drives
I.E.E., London, (Nov. 1987).

11 - A Nd-Fe-B excited permanent magnet motor-design and performance
T.S. Low, M.F. Rahman, L.B. Wee, K.J. Binns
3rd Intern. Conf. on Electrical Machines and Drives
I.E.E., London. (Nov 1987).

12 - Influence of the use of Nd-Fe-B in permanent magnet motors
R. Belmans, R. Paternoster, A. Vandenput, W. Geysen, K.J. Binns
Proc. Symposium on Electrical Drives. Cagliari (Italie) (1987) 71.

SUB-CONTRACT N° 3.17 - H. ZIJLSTRA & R.P. Van STAPELE

1 - Introduction to permanent magnets
H. Zijlstra
Nd-Fe Permanent Magnets. Their Present and Future Applications,
Ed. I.V. Mitchell, Elsevier, (1985) 5.

2 - Permanent magnet systems for magnetic resonance imaging
H. Zijlstra
9th Intern. Conf. on Magnet Technology (Zürich). (1985).

3 - Application of permanent magnets in electro-mechanical power
converters ; the impact of Nd-Fe-B magnets
H. Zijlstra
J. de Physique, 46 (1985) C6-3.

4 - Permanent magnet systems for NMR tomography
H. Zijlstra
Philips J. Res., 40, (1985) 259.

5 - The simulation of Anisotropic and Hysteretic Materials in Magnetic
Field Computations
H. Zijlstra
Proc. Ninth Intern. Workshop on Rare-Earth Magnets (Bad Soden, 1987)
p. 17.

APPENDIX .iv .

SUBJECT INDEX

$YFe_{11}Ti$	2.3
$Y_2Fe_{14-x}Me_xB$	2.3

APPENDIX .v.

LIST OF AUTHORS

AUTHOR LIST		Sect.III Chapter No.	Page No.		
J.S. Abell	see Harris	6.3	489		
P.A. Algarabel		2.7	240		
J. Allemand		1.5	98		
C.H. Allibert		4.2	358		
J.I. Arnaudas	see Algarabel	2.7	240		
M.M. Ashraf	see Rowlinson	8.3	670		
G. Asti	see Pareti	2.3	188		
W. Baran		10.2	818		
J. Bartolomé	see Algarabel	2.7	240		
G. Bava	see Tori	7.4	628		
R. Belmans		9.4	720		
R. Beranek	see Stieler	5.7	449		
Y. Berthier		3.1	265		
C. Bertrand	see Allemand	1.5	98		
K.J. Binns		9.11	785		
T.S. Birch	see Howe	9.2	695		
M. Bogé		3.5	310		
F. Bolzoni	see Pareti	2.3	188		
H-J. Bornemann	see Czjzek	3.6	323		
M. Bradford		9.5	732		
R.A. Buckley	see Davies	6.7	543		
K.H.J. Buschow		1.3	2.4	63	203
J.M. Cadogan	see Coey	1.4	76		
G.E. Carr	see Davies	6.7	543		
A. Cartocetti		7.1	581		
J. Chaboy	see Algarabel	2.7	240		
B.J. Chalmers		9.3	708		
A. Chamberod		5.6	436		
J. Chavanne		10.1	807		
B. Chevalier		1.9	134		
C. Christides		1.6	109		
A.G. Clegg		7.5	640		
D. Cochet-Muchy		1.8	4.3	130	369
R. Coehoorn		6.8	558		
J.M.D. Coey			17		
J.M.D. Coey		1.4	76		
J.M.D. Coey	see Chevalier	1.9	134		
P. Collins	see Pettifor	1.1	39		
W.D. Corner		5.5	424		
I.M. Coulson	see Clegg	7.5	640		
G. Czjzek		3.6	323		
F.R. de Boer	see Franse	2.2	174		
P. Dalmas de Reotier	see D. Fruchart	2.5	214		
H.A. Davies		6.7	543		
D.B. de Mooij	see Buschow	1.3	63		
A. Del Moral	see Algarabel	2.7	240		
P. Deppe	see Rosenberg	3.2	276		
E.J. Devlin	see Harris	6.3	489		
J.P.W. Duchateau	see Coehoorn	6.8	558		
D-D. Durst	see Kronmüller	5.2	392		

AUTHOR LIST		Sect.III Chapter No.	Page No.	
J. Koorneef	see Kamerbeek	9.8	765	
A. Kostikas	see Christides	1.6	109	
R. Krewenka	see Buschow	2.4	203	
H. Kronmüller		5.2	392	
W. Kurtz	see Stieler	5.7	449	
J. LaForest	see Chavanne	10.1	807	
D. Le Roux	see D. Fruchart	2.5	214	
J. Le Roy	see Allemand	1.5	98	
R. Leeb	see Buschow	2.4	203	
H.S. Li	see Coey	1.4	76	
H.S. Li	see Gavigan	2.1	163	
P.R. Locher	see Baran	10.2	818	
D.G. Lord	see Grundy	5.3	405	
P. l'Héritier	see D. Fruchart	2.5	214	
P. l'Héritier	see R. Fruchart	2.6	230	
A. Manaf	see Davies	6.7	543	
C. Marquina	see Algarabel	2.7	240	
G. Marusi	see Pareti	2.3	188	
K.J.A. Mawella	see Davies	6.7	543	
O. Mayerhofer	see Buschow	2.4	203	
P.J. McGuinness	see Harris	6.3	489	
P.J. McGuinness	see Rozendaal	6.4	510	
A. Menovsky	see Franse	2.2	174	
C. Meyer	see Gros	3.3	288	
A.P. Miodownik	see Rivlin	1.2	53	
S. Miraglia	see D. Fruchart	2.5	214	
I.V. Mitchell			1	
K.J. Mitchell	see Howe	9.2	695	
P. Mohn	see Pettifor	1.1	39	
J.M. Moreau	see Allemand	1.5	98	
O. Moze	see Pareti	2.3	188	
D.W.A. Murphy	see Kennedy	6.1	467	
H. Nagel	see Scholz	6.5	521	
R. Navarro	see Algarabel	2.7	240	
D. Niarchos	see Christides	1.6	109	
J.P. Nozieres		8.2	659	
J.L. Oddou	see Bogé	3.5	310	
J. Ormerod	see Rozendaal	6.4	510	
D. Paccard	see Allemand	1.5	98	
L. Paccard	see Allemand	1.5	98	
S. Païdassi	see Cochet-Muchy	1.8 4.3	130 369	
L. Pareti		2.3	188	
S.F.H. Parker	see Grundy	5.3	405	
R. Pauthenet	see Chavanne	10.1	807	
R. Perrier de la Bathie	see Nozieres	8.2	659	
D.G. Pettifor		1.1	39	
R.J. Pollard	see Grundy	5.3	405	
L. Pontonnier	see D. Fruchart	2.5	214	
R.J. Radwanski	see Franse	2.2	174	
K.M. Richardson	see Chalmers	9.3	708	